Periodic Table of Elements

Representative elements

Halogens Noble gases

Transition elements

Period number	Group 1A 1	Group 2A 2	3 3B	4 4B	5 5B	6 6B	7 7B	8 8B	9 8B	10 8B	11 1B	12 2B	13 Group 3A	14 Group 4A	15 Group 5A	16 Group 6A	17 Group 7A	18 Group 8A
1	1 H 1.008																	2 He 4.003
2	3 Li 6.941	4 Be 9.012											5 B 10.81	6 C 12.01	7 N 14.01	8 O 16.00	9 F 19.00	10 Ne 20.18
3	11 Na 22.99	12 Mg 24.31											13 Al 26.98	14 Si 28.09	15 P 30.97	16 S 32.07	17 Cl 35.45	18 Ar 39.95
4	19 K 39.10	20 Ca 40.08	21 Sc 44.96	22 Ti 47.87	23 V 50.94	24 Cr 52.00	25 Mn 54.94	26 Fe 55.85	27 Co 58.93	28 Ni 58.69	29 Cu 63.55	30 Zn 65.41	31 Ga 69.72	32 Ge 72.64	33 As 74.92	34 Se 78.96	35 Br 79.90	36 Kr 83.80
5	37 Rb 85.47	38 Sr 87.62	39 Y 88.91	40 Zr 91.22	41 Nb 92.91	42 Mo 95.94	43 Tc (99)	44 Ru 101.1	45 Rh 102.9	46 Pd 106.4	47 Ag 107.9	48 Cd 112.4	49 In 114.8	50 Sn 118.7	51 Sb 121.8	52 Te 127.6	53 I 126.9	54 Xe 131.3
6	55 Cs 132.9	56 Ba 137.3	57* La 138.9	72 Hf 178.5	73 Ta 180.9	74 W 183.8	75 Re 186.2	76 Os 190.2	77 Ir 192.2	78 Pt 195.1	79 Au 197.0	80 Hg 200.6	81 Tl 204.4	82 Pb 207.2	83 Bi 209.0	84 Po (209)	85 At (210)	86 Rn (222)
7	87 Fr (223)	88 Ra (226)	89† Ac (227)	104 Rf (261)	105 Db (262)	106 Sg (266)	107 Bh (264)	108 Hs (265)	109 Mt (268)	110 Ds (271)	111 Rg (272)	112 Cn (285)	113 Nh (286)	114 Fl (289)	115 Mc (289)	116 Lv (293)	117 Ts (294)	118 Og (294)

*Lanthanides

58 Ce 140.1	59 Pr 140.9	60 Nd 144.2	61 Pm (145)	62 Sm 150.4	63 Eu 152.0	64 Gd 157.3	65 Tb 158.9	66 Dy 162.5	67 Ho 164.9	68 Er 167.3	69 Tm 168.9	70 Yb 173.0	71 Lu 175.0

†Actinides

90 Th 232.0	91 Pa 231.0	92 U 238.0	93 Np (237)	94 Pu (244)	95 Am (243)	96 Cm (247)	97 Bk (247)	98 Cf (251)	99 Es (252)	100 Fm (257)	101 Md (258)	102 No (259)	103 Lr (262)

Alkali metals

Alkaline earth metals

Metals   Metalloids   Nonmetals

Atomic Masses of the Elements

Name	Symbol	Atomic Number	Atomic Mass[a]	Name	Symbol	Atomic Number	Atomic Mass[a]
Actinium	Ac	89	(227)[b]	Mendelevium	Md	101	(258)
Aluminum	Al	13	26.98	Mercury	Hg	80	200.6
Americium	Am	95	(243)	Molybdenum	Mo	42	95.94
Antimony	Sb	51	121.8	Moscovium	Mc	115	(289)
Argon	Ar	18	39.95	Neodymium	Nd	60	144.2
Arsenic	As	33	74.92	Neon	Ne	10	20.18
Astatine	At	85	(210)	Neptunium	Np	93	(237)
Barium	Ba	56	137.3	Nickel	Ni	28	58.69
Berkelium	Bk	97	(247)	Nihonium	Nh	113	(286)
Beryllium	Be	4	9.012	Niobium	Nb	41	92.91
Bismuth	Bi	83	209.0	Nitrogen	N	7	14.01
Bohrium	Bh	107	(264)	Nobelium	No	102	(259)
Boron	B	5	10.81	Oganesson	Og	118	(294)
Bromine	Br	35	79.90	Osmium	Os	76	190.2
Cadmium	Cd	48	112.4	Oxygen	O	8	16.00
Calcium	Ca	20	40.08	Palladium	Pd	46	106.4
Californium	Cf	98	(251)	Phosphorus	P	15	30.97
Carbon	C	6	12.01	Platinum	Pt	78	195.1
Cerium	Ce	58	140.1	Plutonium	Pu	94	(244)
Cesium	Cs	55	132.9	Polonium	Po	84	(209)
Chlorine	Cl	17	35.45	Potassium	K	19	39.10
Chromium	Cr	24	52.00	Praseodymium	Pr	59	140.9
Cobalt	Co	27	58.93	Promethium	Pm	61	(145)
Copernicium	Cn	112	(285)	Protactinium	Pa	91	231.0
Copper	Cu	29	63.55	Radium	Ra	88	(226)
Curium	Cm	96	(247)	Radon	Rn	86	(222)
Darmstadtium	Ds	110	(271)	Rhenium	Re	75	186.2
Dubnium	Db	105	(262)	Rhodium	Rh	45	102.9
Dysprosium	Dy	66	162.5	Roentgenium	Rg	111	(272)
Einsteinium	Es	99	(252)	Rubidium	Rb	37	85.47
Erbium	Er	68	167.3	Ruthenium	Ru	44	101.1
Europium	Eu	63	152.0	Rutherfordium	Rf	104	(261)
Fermium	Fm	100	(257)	Samarium	Sm	62	150.4
Flerovium	Fl	114	(289)	Scandium	Sc	21	44.96
Fluorine	F	9	19.00	Seaborgium	Sg	106	(266)
Francium	Fr	87	(223)	Selenium	Se	34	78.96
Gadolinium	Gd	64	157.3	Silicon	Si	14	28.09
Gallium	Ga	31	69.72	Silver	Ag	47	107.9
Germanium	Ge	32	72.64	Sodium	Na	11	22.99
Gold	Au	79	197.0	Strontium	Sr	38	87.62
Hafnium	Hf	72	178.5	Sulfur	S	16	32.07
Hassium	Hs	108	(265)	Tantalum	Ta	73	180.9
Helium	He	2	4.003	Technetium	Tc	43	(99)
Holmium	Ho	67	164.9	Tellurium	Te	52	127.6
Hydrogen	H	1	1.008	Tennessine	Ts	117	(294)
Indium	In	49	114.8	Terbium	Tb	65	158.9
Iodine	I	53	126.9	Thallium	Tl	81	204.4
Iridium	Ir	77	192.2	Thorium	Th	90	232.0
Iron	Fe	26	55.85	Thulium	Tm	69	168.9
Krypton	Kr	36	83.80	Tin	Sn	50	118.7
Lanthanum	La	57	138.9	Titanium	Ti	22	47.87
Lawrencium	Lr	103	(262)	Tungsten	W	74	183.8
Lead	Pb	82	207.2	Uranium	U	92	238.0
Lithium	Li	3	6.941	Vanadium	V	23	50.94
Livermorium	Lv	116	(293)	Xenon	Xe	54	131.3
Lutetium	Lu	71	175.0	Ytterbium	Yb	70	173.0
Magnesium	Mg	12	24.31	Yttrium	Y	39	88.91
Manganese	Mn	25	54.94	Zinc	Zn	30	65.41
Meitnerium	Mt	109	(268)	Zirconium	Zr	40	91.22

[a]Values for atomic masses are given to four significant figures.
[b]Values in parentheses are the mass number of an important radioactive isotope.

BASIC
CHEMISTRY

BASIC CHEMISTRY

Sixth Edition

Karen Timberlake
William Timberlake

 Pearson

Courseware Portfolio Manager: Jessica Moro
Director of Portfolio Management: Jeanne Zalesky
Content Producer: Melanie Field
Managing Producer: Kristen Flathman
Courseware Director, Content Development: Barbara Yien
Courseware Portfolio Management Analyst: Coleen Morrison
Courseware Editorial Assistant: Harry Misthos
Rich Media Content Producer: Ziki Dekel
Full-Service Vendor: Pearson CSC
Full-Service Project Manager: Rose Kernan
Copyeditor: Karen Slaght
Courseware Portfolio Analyst, Content Development, Art: Jay McElroy
Design Manager: Mark Ong

Cover and Interior Designer: Tamara Newnam
Photo and Illustration Project Manager: Stephanie Marquez
  and Alicia Elliott/Imagineering Art
Rights & Permissions Project Manager: Matt Perry
Rights & Permissions Management: Ben Ferrini
Photo Researcher: Namrata Aggarwal
Manufacturing Buyer: Stacey Weinberger
Director of Field Marketing: Tim Galligan
Director of Product Marketing: Allison Rona
Field Marketing Manager: Christopher Barker
Product Marketing Manager: Elizabeth Ellsworth
Cover Photo Credit: Getty Images/Westend61

Library of Congress Cataloging-in-Publication Data

Names: Timberlake, Karen, author. | Timberlake, William, author.
  Title: Basic chemistry / Karen Timberlake, William Timberlake.
  Description: Sixth edition. | New York, NY : Pearson, [2019] | Includes index.
  Identifiers: LCCN 2018048212 | ISBN 9780134878119
  Subjects: LCSH: Chemistry--Textbooks.
  Classification: LCC QD31.3 .T54 2019 | DDC 540--dc23
  LC record available at https://lccn.loc.gov/2018048212

ISBN 10:   0-134-87811-6;
ISBN 13: 978-0-134-87811-9 (Student edition)

ISBN 10:   0-134-98699-7;
ISBN 13: 978-0-13498699-9 (Looseleaf Edition)

ISBN 10:   0-135-24461-7;
ISBN 13: 978-0-13524461-6 (NASTA)

P Pearson

www.pearson.com

4 2019

Brief Contents

Contents

4
Atoms and Elements 100

5
Electronic Structure of Atoms and Periodic Trends 125

6
Ionic and Molecular Compounds 156

7
Chemical Quantities 183

8
Chemical Reactions 213

9
Chemical Quantities in Reactions 239

10
Bonding and Properties of Solids and Liquids 269

14
Acids and Bases 431

15
Oxidation and Reduction 476

16
Nuclear Chemistry 508

17
Organic Chemistry 540

18
Biochemistry 592

Applications and Activities

KEY MATH SKILLS

CORE CHEMISTRY SKILLS

Interactive Videos

About the Authors

KAREN TIMBERLAKE is Professor Emerita of Chemistry at Los Angeles Valley College, where she taught chemistry for allied health and preparatory chemistry for 36 years. She received her bachelor's degree in chemistry from the University of Washington and her master's degree in biochemistry from the University of California at Los Angeles.

Professor Timberlake has been writing chemistry textbooks for more than 40 years. During that time, her name **has become associated with the strategic use of pedagogical tools that promote student success in chemistry and the application of chemistry to real-life situations**. More than one million students have learned chemistry using texts, laboratory manuals, and study guides written by Karen Timberlake. In addition to *Basic Chemistry*, sixth edition, she is also the author of *General, Organic, and Biological Chemistry: Structures of Life*, sixth edition, with the accompanying *Study Guide and Selected Solutions Manual*, and *Chemistry: An Introduction to General, Organic, and Biological Chemistry*, thirteenth edition, with the accompanying *Study Guide and Selected Solutions Manual*, *Laboratory Manual*, and *Essential Laboratory Manual*.

Professor Timberlake belongs to numerous scientific and educational organizations including the American Chemical Society (ACS) and the National Science Teachers Association (NSTA). She has been the Western Regional Winner of Excellence in College Chemistry Teaching Award given by the Chemical Manufacturers Association. She received the McGuffey Award in Physical Sciences from the Textbook Authors Association for her textbook *Chemistry: An Introduction to General, Organic, and Biological Chemistry*, eighth edition. She received the "Texty" Textbook Excellence Award from the Textbook Authors Association for the first edition of *Basic Chemistry*. She has participated in education grants for science teaching including the Los Angeles Collaborative for Teaching Excellence (LACTE) and a Title III grant at her college. She speaks at conferences and educational meetings on the use of student-centered teaching methods in chemistry to promote the learning success of students.

Her husband, William Timberlake, who is the coauthor of this text, is Professor Emeritus of Chemistry at Los Angeles Harbor College, where he taught preparatory and organic chemistry for 36 years. He received his bachelor's degree in chemistry from Carnegie Mellon University and his master's degree in organic chemistry from the University of California at Los Angeles.

When the Professors Timberlake are not writing textbooks, they relax by playing tennis, ballroom dancing, hiking, traveling, trying new restaurants, cooking, and enjoying care of their grandchildren, Daniel and Emily.

DEDICATION

- Our son, John, daughter-in-law, Cindy, grandson, Daniel, and granddaughter, Emily, for the precious things in life

- The wonderful students over many years whose hard work and commitment always motivated us and put purpose in our writing

Preface

Welcome to the sixth edition of *Basic Chemistry*. This chemistry text was written and designed to prepare you for science-related professions, such as engineering, nursing, medicine, environmental or agricultural science, or for careers such as laboratory technology. This text assumes no prior knowledge of chemistry. Our main objective in writing this text is to make the study of chemistry an engaging and positive experience for you by relating the structure and behavior of matter to real life. This new edition introduces more problem-solving strategies, Analyze the Problem with Connect features, Try It First and Engage features, and conceptual and challenge problems.

It is our goal to help you become a critical thinker by understanding scientific concepts that will form a basis for making important decisions about issues concerning health and the environment. Thus, we have utilized materials that

- help you to learn and enjoy chemistry
- relate chemistry to careers that interest you
- develop problem-solving skills
- promote learning and success in chemistry

Active Reading Features for Successful Learning

In this sixth edition of Basic Chemistry, we have increased our emphasis on providing features that promote student interaction with the textual content. We continue to develop features based on new research on learning theory and extend them throughout the text as a part of our study plan Strategies and Practices for Active Reading in Chemistry (SPARC).

With the success of students involved in active learning in the classroom, we see the opportunity to develop a parallel plan of reading and learning experiences using our textbook. As chemistry textbook authors we are interested in connecting cognitive science and learning research to improve student reading and learning. SPARC is the combination of utilizing reading strategies that increase learning and success in chemistry.

Strategies for Learning New Information

1. **Combine graphics with words**
   Students improve learning by receiving information in different ways. In this text, we combine text and questions with illustrations using *macro-to-micro art*, tables, graphs, diagrams, videos, photos, and concept maps.

2. **Connect abstract concepts with concrete representations**
   New concepts are illustrated and explained with *real-life examples, career stories and updates, Chemical Links to Health and the Environment,* and *Applications.* Prompts of *Key Math Skills* and *Core Chemistry Skills* alert students to the fundamental ideas in each chapter.

Strategies for Connecting New Information

3. **Ask inquiring questions**
   Engage questions throughout each chapter ask students "why," "how," and "what if," requiring them to link new information with prior knowledge.

4. **Alternate problems containing solutions with similar problems that students must solve**
   Many *Sample Problems* with *Try It First* reminders throughout each chapter contain step-by-step solutions that guide students through the process of problem-solving. An abundance of *Practice Problems, Understanding the Concepts, Additional Problems,* and *Challenge Problems* provide students with similar problem-solving experience. *Answers* are provided for immediate feedback.

Strategies for Recalling and Retrieving Information

5. **Provide opportunities to practice recall and retrieval**
   Throughout each chapter, students are encouraged to practice the recall and retrieval of material by repeating practice every few days, weeks, and even months using *Self Tests, Practice Problems, Understanding the Concepts, Additional Problems,* and *Challenge Problems. Review* prompts remind students of key ideas in previous chapters.

6. **Combine different but related topics and skills**
   Better learning can be achieved by alternating different topics and types of content. In this text, problem sets including *Understanding the Concepts, Additional Practice Problems, Challenge Problems,* and *Combining Ideas* provide students practice of different topics or skills, rather than focusing on one topic or skill.

7. **Assess to maintain and improve retention**
   Every chapter provides many types of assessment such as *Self Tests, Practice Problems, Understanding the Concepts, Additional Practice Problems, Challenge Problems,* and *Combining Ideas.* By practicing information retrieval, checking progress, and reviewing, students improve their success on exams.

New and Updated for the Sixth Edition

New and updated features have been added throughout this sixth edition, including the following:

- **NEW! Chapter Openers** provide timely examples and engaging, topical examples of the chemistry that is part of contemporary professions.
- **NEW! Chapter Openers** include references to new **Update** features at the end of the chapter that continue the story.
- **NEW! Review** heads are now listed at the beginning of each section to emphasize the Key Math Skills and Core Chemistry Skills from previous chapters required for learning new chemistry principles.
- **NEW! Pictorial Representations** using photos and graphs are added to increase the understanding of new topics.
- **NEW! Sample Problems** show **Steps** to guide the student through problem solving.
- **NEW! Self Test** icons in Sample Problems encourage students to use problem-solving strategies immediately as they review the content in a section.
- **NEW!** Expanded **Self Test** questions in **Sample Problems** provide students with additional self-testing practice.
- **NEW!** Margin icons **Practice Problems** encourage students to work related Practice Problems and self-assess as they study.
- **NEW!** Expanded **Engage Questions** and **Answers** are included in the chapter.
- **NEW!** Additional **Practice Problems** and **Challenge Problems** are added to help student practice testing and increase understanding of the concepts in the chapter.
- **NEW! Three-Dimensional Representations**, including ball-and-stick and space-filling models, are included to illustrate the shapes of molecules and polyatomic ions.
- **NEW!** Foreground colors in color palette are now ADA accessible.
- **NEW!** Multiple art pieces contain separate captions in boxes for each art.
- **NEW! Concept Maps** are tinted with color screens for emphasis of topics.
- **NEW! Interactive Videos** are added to illustrate more step-by-step problem-solving strategies.

Chapter Organization of the Sixth Edition

In each textbook we write, we consider it essential to relate every chemical concept to real-life issues. Because a chemistry course may be taught in different time frames, it may be difficult to cover all the chapters in this text. However, each chapter is a complete package, which allows some chapters to be skipped or the order of presentation to be changed.

Chapter 1, Chemistry in Our Lives, discusses the Scientific Method in everyday terms and guides students in developing a study plan for learning chemistry, with a section of Key Math Skills that reviews the basic math, including scientific notation, needed in chemistry calculations.

- The Chapter Opener tells the story of a murder and features the work and career of forensic scientists.
- The Update feature describes the forensic evidence that helps to solve the murder and includes Applications.
- An updated Section 1.3, Studying and Learning Chemistry, expands the discussion of study strategies that improve learning and understanding of content.
- A new Decimal Place Value Chart is added in Section 1.4, Key Math Skills for Chemistry, to clarify decimal place values.
- In Section 1.4, Interpreting Graphs, the format for the x and y axes is standardized.
- Key Math Skills are: Identifying Place Values, Using Positive and Negative Numbers in Calculations, Calculating Percentages, Solving Equations, Interpreting Graphs, and Writing Numbers in Scientific Notation.

Chapter 2, Chemistry and Measurements, looks at measurement and emphasizes the need to understand numerical relationships of the metric system. Significant figures are discussed in the determination of final answers. Prefixes from the metric system are used to write equalities and conversion factors for problem-solving strategies. Density is discussed and used as a conversion factor.

- The Chapter Opener tells the story of a patient with high blood pressure and features the work and career of a registered nurse.
- The Update describes the patient's status and follow-up visit with his doctor.
- In Section 2.5, conversion factors with multiple units are added in the Practice Problems.
- In Section 2.6, Steps as guides to problem solving are added to Sample Problems 2.3, 2.4, and 2.5.
- In Section 2.6, a new type of Sample Problem and new Practice Problems for the conversion of units in a fraction are added.
- Sample Problems relate problem solving to health-related topics such as the measurements of blood volume, omega-3 fatty acids, radiological imaging, body fat, cholesterol, and medication orders.
- Applications feature questions about measurements, daily values for minerals and vitamins, and equalities and conversion factors for medications.
- Key Math Skill is: Rounding Off.
- Core Chemistry Skills are: Counting Significant Figures, Using Significant Figures in Calculations, Using Prefixes, Writing Conversion Factors from Equalities, Using Conversion Factors, and Using Density as a Conversion Factor.

Chapter 3, Matter and Energy, classifies matter and states of matter, describes temperature measurement, and discusses energy, specific heat, and energy in nutrition. Physical and chemical properties and physical and chemical changes are discussed.

- The Chapter Opener describes diet and exercise for an overweight adolescent at risk for type 2 diabetes and features the work and career of a dietitian.
- The Update describes the diet prepared with a dietitian for weight loss.
- Figures of Decomposition of Salt, and Separation of Mixtures by Filtration are moved to Section 3.2 for clarity of content.
- In Section 3.5, Specific Heat, a new Sample Problem using heat exchange data is added along with new Practice Problems.
- Practice Problems and Sample Problems include high temperatures used in cancer treatment, the energy produced by a high-energy shock output of a defibrillator, body temperature lowering using a cooling cap, ice bag therapy for muscle injury, dental implants, and energy values for food.
- Core Chemistry Skills are: Identifying Physical and Chemical Changes, Converting Between Temperature Scales, Using Energy Units, Calculating Specific Heat, and Using the Heat Equation.
- The interchapter problem set, Combining Ideas from Chapters 1 to 3, completes the chapter.

Chapter 4, Atoms and Elements, introduces elements and atoms and the periodic table. The names and symbols for the newest elements 113, Nihonium, Nh, 115, Moscovium, Mc, 117, Tennessine, Ts, and 118, Oganesson, Og, are included on the periodic table. Atomic numbers and mass numbers are determined for isotopes. Atomic mass is calculated using the masses of the naturally occurring isotopes and their abundances.

- The Chapter Opener and Update discuss the improvement in crop production and feature the work and career of a farmer.
- Atomic number and mass number are used to calculate the number of protons and neutrons in an atom.
- The number of protons and neutrons are used to calculate the mass number and to write the atomic symbol for an isotope.
- Figure 4.3 now includes Francium (Fr) in Group 1A (1), and Figure 4.4 now includes Tennessine (Ts).
- Core Chemistry Skills are: Counting Protons and Neutrons, Writing Atomic Symbols for Isotopes, and Calculating Atomic Mass.

Chapter 5, Electronic Structure of Atoms and Periodic Trends, uses the electromagnetic spectrum to explain atomic spectra and develop the concept of energy levels and sublevels. Electrons in sublevels and orbitals are represented using orbital diagrams and electron configurations. Periodic properties of elements, including atomic size, ionization energy and metallic character, are related to their valence electrons. Small periodic tables illustrate the trends of periodic properties.

- The Chapter Opener and Update discuss the development of new products of metals, plastics, and semiconductors, and career of a materials engineer.

- The electromagnetic spectrum is described with everyday examples and a diagram.
- The three-dimensional representations of the s, p, and d orbitals are drawn.
- The trends in periodic properties are described for valence electrons, atomic size, ionization energy, and metallic character.
- A photo of infrared radiation used to keep food warm, and a photo of gamma knife radiation used to kill cancer cells are added.
- Table 5.2 for electron capacity in sublevels is reordered with Energy Level $n = 1$ at the top and Energy Level $n = 4$ at the bottom.
- Core Chemistry Skills are: Writing Electron Configurations, Using the Periodic Table to Write Electron Configurations, and Identifying Trends in Periodic Properties.

Chapter 6, Ionic and Molecular Compounds, describes the formation of ionic and covalent bonds. Chemical formulas are written, and ionic compounds—including those with polyatomic ions—and molecular compounds are named.

- The Chapter Opener describes the chemistry of aspirin and features the work and career of a pharmacist.
- The Update describes several types of compounds at a pharmacy and includes Applications.
- New art is added or updated to provide everyday examples of the content.
- New material on polyatomic ions compares the names of *ate* ions and *ite* ions, the charge of sulfate and sulfite, phosphate and phosphite, carbonate and hydrogen carbonate, and the formulas and charges of halogen polyatomic ions with oxygen.
- Core Chemistry Skills are: Writing Positive and Negative Ions, Writing Ionic Formulas, Naming Ionic Compounds, and Writing the Names and Formulas for Molecular Compounds.

Chapter 7, Chemical Quantities, discusses Avogadro's number, the mole, and molar masses of compounds, which are used in calculations to determine the mass or number of particles in a quantity of a substance. The mass percent composition of a compound is calculated and used to determine its empirical and molecular formula.

- The Chapter Opener and Update describe the diagnosis and treatment of a pet and the work and career of a veterinarian.
- Core Chemistry Skills are: Converting Particles to Moles, Calculating Molar Mass, Using Molar Mass as a Conversion Factor, Calculating Mass Percent Composition, Calculating an Empirical Formula, and Calculating a Molecular Formula.
- The interchapter problem set, Combining Ideas from Chapters 4 to 7, completes the chapter.

Chapter 8, Chemical Reactions, shows students how to balance chemical equations, and discusses how to classify chemical reactions into types: combination, decomposition, single replacement, double replacement, combustion, and oxidation–reduction.

- The Chapter Opener and Update discuss tests and treatment for emphysema and the work and career of an exercise physiologist.
- Core Chemistry Skills are: Balancing a Chemical Equation, Classifying Types of Chemical Reactions, and Identifying Oxidized and Reduced Substances.

Chapter 9, Chemical Quantities in Reactions, describes the mole and mass relationships among the reactants and products and provides calculations of limiting reactants and percent yields. The chapter concludes with a discussion of energy in reactions.

- The Chapter Opener describes insecticides and pharmaceuticals used on a ranch and discusses the career of an environmental scientist.
- The Update describes the collection of soil and water samples for testing for insecticides.
- Mole and mass relationships among the reactants and products are examined along with calculations of percent yield and limiting reactants.
- Material including new problems with three equations and calculations using Hess's Law is rewritten for clarity.
- Core Chemistry Skills are: Using Mole–Mole Factors, Converting Grams to Grams, Calculating Quantity of Product from a Limiting Reactant, Calculating Percent Yield, and Using the Heat of Reaction.

Chapter 10, Bonding and Properties of Solids and Liquids, introduces Lewis structures for molecules and ions with single and multiple bonds as well as resonance structures. Electronegativity leads to a discussion of the polarity of bonds and molecules. Lewis structures and VSEPR theory illustrate covalent bonding and the three-dimensional shapes of molecules and ions. The intermolecular forces between particles and their impact on states of matter and changes of state are described. The energy involved with changes of state is calculated.

- The Chapter Opener and Update describe the processing of a tissue sample and the work and career of a histologist.
- New three-dimensional representations of ball-and-stick models and space-filling models are added to illustrate shapes of molecules and polyatomic ions.
- Lewis structures are drawn for molecules and ions with single, double, and triple bonds. Resonance structures are drawn if two or more Lewis structures are possible.
- Shapes and polarity of bonds and molecules are predicted using VSEPR theory.
- Intermolecular forces in compounds are discussed including ionic bonds, hydrogen bonds, dipole–dipole attractions, and dispersion forces.
- Core Chemistry Skills are: Drawing Lewis Symbols, Drawing Lewis Structures, Drawing Resonance Structures, Predicting Shape, Using Electronegativity, Identifying Polarity of Molecules, Identifying Intermolecular Forces, and Calculating Heat for Change of State.
- The interchapter problem set, Combining Ideas from Chapters 8 to 10, completes the chapter.

Chapter 11, Gases, discusses the properties of gases and calculates changes in gases using the gas laws: Boyle's, Charles's, Gay-Lussac's, Avogadro's, Dalton's, and the Ideal Gas Law. Problem-solving strategies enhance the discussion and calculations with gas laws including chemical reactions using the ideal gas law.

- The Chapter Opener and Update feature the work and career of a respiratory therapist, who uses oxygen to treat a child with asthma.
- Applications include calculations of mass or pressure of oxygen in uses of hyperbaric chambers.
- Core Chemistry Skills are: Using the Gas Laws, Using the Ideal Gas Law, Calculating Mass or Volume of a Gas in a Chemical Reaction, and Calculating Partial Pressure.

Chapter 12, Solutions, describes solutions, electrolytes, saturation and solubility, insoluble salts, concentrations, and osmosis. The concentrations of solutions are used to determine volume or mass of solute. The volumes and molarities of solutions are used in calculations for dilutions and titration. Properties of solutions, freezing and boiling points, osmosis, and dialysis are discussed.

- The Chapter Opener describes a patient with kidney failure and dialysis treatment and features the work and career of a dialysis nurse.
- The Update discusses dialysis treatment and electrolyte levels in the dialysate fluid.
- A new example of suspensions used to purify water in treatment plants is added.
- New art illustrates the freezing point decrease and boiling point increase for aqueous solutions with increasing number of moles of solute in one kilogram of water.
- Core Chemistry Skills are: Using Solubility Rules, Calculating Concentration, Using Concentration as a Conversion Factor, Calculating the Quantity of a Reactant or Product for a Chemical Reaction in Solution, and Calculating the Freezing Point/Boiling Point of a Solution.

Chapter 13, Reaction Rates and Chemical Equilibrium, looks at the rates of reactions and the equilibrium condition when forward and reverse rates for a reaction become equal. Equilibrium expressions for reactions are written and equilibrium constants are calculated. The equilibrium constant is used to calculate the concentration of a reactant or product at equilibrium. Le Châtelier's principle is used to evaluate the impact on concentrations when stress is placed on a system at equilibrium. The concentrations of solutes in a solution is used to calculate the solubility product constant (K_{sp}).

- The Chapter Opener and Update discuss the equilibrium of CO_2 in the ocean and feature the work and career of a chemical oceanographer.
- Core Chemistry Skills are: Writing the Equilibrium Expression, Calculating an Equilibrium Constant, Calculating Equilibrium Concentrations, Using Le Châtelier's Principle, Writing the Solubility Product Expression, Calculating a Solubility Product Constant, and Calculating the Molar Solubility.

Chapter 14, Acids and Bases, discusses acids and bases and their strengths, and conjugate acid–base pairs. The dissociation of strong and weak acids and bases is related to their strengths as acids or bases. The dissociation of water leads to the water dissociation expression, K_w, the pH scale, and the calculation of pH. Chemical equations for acids in reactions are balanced and titration of an acid is illustrated. Buffers are discussed along with their role in the blood. The pH of a buffer is calculated.

- The Chapter Opener describes a blood sample for an emergency room patient sent to the clinical laboratory for analysis of blood pH and CO_2 gas and describes the work and career of a medical laboratory technologist.
- The Update describes the symptoms and treatment for acid reflux disease (GERD).
- Key Math Skills are: Calculating pH from $[H_3O^+]$, and Calculating $[H_3O^+]$ from pH.
- Core Chemistry Skills are: Identifying Conjugate Acid–Base Pairs, Calculating $[H_3O^+]$ and $[OH^-]$ in Solutions, Writing Equations for Reactions of Acids and Bases, Calculating Molarity or Volume of an Acid or Base in a Titration, and Calculating the pH of a Buffer.
- The interchapter problem set, Combining Ideas from Chapters 11 to 14, completes the chapter.

Chapter 15, Oxidation and Reduction, looks at the characteristics of oxidation and reduction reactions. Oxidation numbers are assigned to the atoms in elements, molecules, and ions to determine the components that lose electrons during oxidation and gain electrons during reduction. The half-reaction method is utilized to balance oxidation–reduction reactions. The production of electrical energy in voltaic cells and the requirement of electrical energy in electrolytic cells are diagrammed using half-cells. The activity series is used to determine the spotaneous direction of an oxidation–reduction reaction.

- The Chapter Opener and Update discuss the reactions involved in teeth whitening and the work and career of a dentist.
- New material and art on lithium-ion batteries is added.
- Core Chemistry Skills are: Assigning Oxidation Numbers, Using Oxidation Numbers, Identifying Oxidizing and Reducing Agents, Using Half-Reactions to Balance Redox Equations, and Identifying Spontaneous Reactions.

Chapter 16, Nuclear Chemistry, looks at the types of radiation emitted from the nuclei of radioactive atoms. Nuclear equations are written and balanced for both naturally occurring radioactivity and artificially produced radioactivity. The half-lives of radioisotopes are discussed, and the amount of time for a sample to decay is calculated. Radioisotopes important in the field of nuclear medicine are described. Fission and fusion and their role in energy production are discussed.

- The Chapter Opener and Update describe a stress test using a radioactive isotope and feature the work and career of a radiation technologist.

- Core Chemistry Skills are: Writing Nuclear Equations, and Using Half-Lives.
- The interchapter problem set, Combining Ideas from Chapters 15 and 16, completes the chapter.

Chapter 17, Organic Chemistry, compares inorganic and organic compounds, and describes the condensed structural and line-angle formulas of alkanes, alkenes, alcohols, ethers, aldehydes, ketones, carboxylic acids, esters, amines, and amides.

- The Chapter Opener and Update describe emergency treatment for burns and feature the work and career of a firefighter/emergency medical technician.
- The properties of organic and inorganic compounds are compared in Table 17.1.
- Line-angle formulas are added to Table 17.2 IUPAC Names and Formulas of the First 10 Alkanes.
- More line-angle structures are included in text examples, sample problems, questions, and problems.
- The two-dimensional and three-dimensional representations of methane and ethane are illustrated using condensed structural formulas, expanded structural formulas, ball-and-stick models, space-filling models, and wedge–dash models.
- Core Chemistry Skills are: Naming and Drawing Alkanes, Writing Equations for Hydrogenation and Polymerization, Naming Aldehydes and Ketones, Naming Carboxylic Acids, Forming Esters, and Forming Amides.

Chapter 18, Biochemistry, looks at the chemical structures and reactions of chemicals that occur in living systems. We focus on four types of biomolecules—carbohydrates, lipids, proteins, and nucleic acids—as well as their biochemical reactions.

- The Chapter Opener and Update describe diagnosis and treatment of diabetes and feature the work and career of a diabetes nurse.
- Monosaccharides are classified as aldo or keto pentoses or hexoses.
- Haworth structures are drawn for monosaccharides, disaccharides, and polysaccharides.
- The shapes of proteins are related to the activity and regulation of enzyme activity.
- The genetic code is described and utilized in the process of protein synthesis.
- Core Chemistry Skills are: Drawing Haworth Structures, Identifying Fatty Acids, Drawing Structures for Triacylglycerols, Drawing the Structure for an Amino Acid at Physiological pH, Identifying the Primary, Secondary, Tertiary, and Quaternary Structures of Proteins, Writing the Complementary DNA Strand, Writing the mRNA Segment for a DNA Template, and Writing the Amino Acid for an mRNA Codon.
- The interchapter problem set, Combining Ideas from Chapters 17 and 18, completes the chapter.

Acknowledgments

The preparation of a new text is a continuous effort of many people. We are thankful for the support, encouragement, and dedication of many people who put in hours of tireless effort to produce a high-quality book that provides an outstanding learning package. The editorial team at Pearson has done an exceptional job. We want to thank Jeanne Zalesky, Director, Courseware Portfolio Management, and Editor Jessica Moro, who supported our vision of this sixth edition and the development of strategies based on learning theory research.

We appreciate all the wonderful work of Melanie Field, Content Producer, who skillfully brought together files, art, web site materials, and all the things it takes to prepare a book for production. We appreciate the work of Rose Kernan at SPi Global, who brilliantly coordinated all phases of the manuscript to the final pages of a beautiful book. Thanks to Mark Quirie, manuscript and accuracy reviewer, and Karen Slaght, who analyzed and edited the manuscripts and pages to make sure the words and problems were correct to help students learn chemistry. Their keen eyes and thoughtful comments were extremely helpful in the development of this text.

We appreciate the contributions from Dr. John Timberlake that connected recent learning theory research with our effort to encourage students to incorporate active reading in their study plan.

Thanks to Kristen Flathman, Managing Producer, Coleen Morrison, Courseware Analyst, and Barbara Yien, Courseware Director, for their excellent review of pages and helpful suggestions.

We am especially proud of the art program in this text, which lends beauty and understanding to chemistry. We would like to thank Jay McElroy, Art Courseware Analyst, and Stephanie Marquez and Alicia Elliott, Photo and Illustration Project Managers, Mark Ong, Design Manager, and Tamara Newnam, Cover and Interior Designer, whose creative ideas provided the outstanding design for the cover and pages of the book. We appreciate the tireless efforts of Namrata Aggarwal, Photo Researcher, and Matt Perry, Rights and Permissions Project Manager, in researching and selecting vivid photos for the text so that students can see the beauty of chemistry. Thanks also to *Bio-Rad Laboratories* for their courtesy and use of *KnowItAll ChemWindows*, drawing software that helped us produce chemical structures for the manuscript. The macro-to-micro illustrations designed by Jay McElroy and Imagineering Art give students visual impressions of the atomic and molecular organization of everyday things and are a fantastic learning tool. We also appreciate all the hard work in the field put in by the marketing team and Allison Rona, Marketing Manager.

We are extremely grateful to an incredible group of peers for their careful assessment of all the new ideas for the text; for their suggested additions, corrections, changes, and deletions; and for providing an incredible amount of feedback about improvements for the book. We admire and appreciate every one of you.

If you would like to share your experience with chemistry, or have questions and comments about this text, We would appreciate hearing from you.

Karen Timberlake
William Timberlake
Email: khemist@aol.com

Reviewers

Sixth Edition Reviewers

Jennifer Bell
Daytona State College, Main

Karen Endebrock
Daytona State College, Deland

Paul Forster
University of Nevada, Las Vegas

Ann Mills
Minnesota West Community and Technical College, Worthington

Phil Nubel
Waubonsee Community College

Amy Whiting
North Central Texas College

Accuracy Reviewer

Mark Quirie
Algonquin College

BASIC
CHEMISTRY

Help students master the math and problem solving they will use in their future careers

Basic Chemistry introduces students to the essential scientific and mathematical concepts of chemistry while providing the scaffolded support they need. With accessible language and a moderate pace, the text is easy-to-follow for first-time chemistry students. The **6th Edition** incorporates sound pedagogy and the best principles from learning design theory to create an updated learning program designed for today's students. The applied focus helps students connect chemistry with their interests and potential careers. Enhanced digital tools and additional practice problems in **Mastering Chemistry** ensure students master the basic quantitative and science skills needed to succeed in this course and beyond.

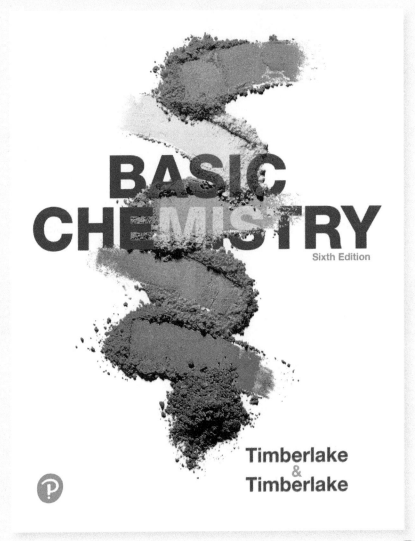

BASIC CHEMISTRY

Sixth Edition

Timberlake
&
Timberlake

Art and Videos that are more understandable than ever before

Sodium chloride

Chemical change

Sodium metal   and   Chlorine gas

FIGURE 3.3 The decomposition of salt, NaCl, produces the elements sodium and chlorine.

P. 74

Updated!
Art program incorporates sound pedagogy and the best learning design principles based on the way today's students learn.

In-art captions replace long legends, and the flow and size of the art is updated to help increase student understanding.

Water molecule

A water molecule, H₂O, consists of two atoms of hydrogen (white) for one atom of oxygen (red).

Hydroxide peroxide molecule

A hydroxide peroxide molecule, H₂O₂, consists of two atoms of hydrogen (white) for every two atoms of oxygen (red).

P. 69

Engage students in learning chemistry

Chemistry Link to Health

Toxicology and Risk–Benefit Assessment

Each day, we make choices about what we do or what we eat, often without thinking about the risks associated with these choices. We are aware of the risks of cancer from smoking or the risks of lead poisoning, and we know there is a greater risk of having an accident if we cross a street where there is no light or crosswalk.

The LD_{50} of caffeine is 192 mg/kg.

A basic concept of toxicology is the statement of Paracelsus that the dose is the difference between a poison and a cure. To evaluate the level of danger from various substances, natural or synthetic, a risk assessment is made by exposing laboratory animals to the substances and monitoring the health effects. Often, doses very much greater than humans might ordinarily encounter are given to the test animals.

Many hazardous chemicals or substances have been identified by these tests. One measure of toxicity is the LD_{50}, or lethal dose, which is the concentration of the substance that causes death in 50% of the test animals. A dosage is typically measured in milligrams per kilogram (mg/kg) of body mass or micrograms per kilogram (mcg/kg) of body mass.

Other evaluations need to be made, but it is easy to compare LD_{50} values. Parathion, a pesticide, with an LD_{50} of 3 mg/kg, would be highly toxic. This means that 3 mg of parathion per kg of body mass would be fatal to half the test animals. Table salt (sodium chloride) with an LD_{50} of 3300 mg/kg would have a much lower toxicity. You would need to ingest a huge amount of salt before any toxic effect would be observed. Although the risk to animals can be evaluated in the laboratory, it is more difficult to determine the impact in the environment since there is also a difference between continued exposure and a single, large dose of the substance.

TABLE 2.9 lists some LD_{50} values and compares substances in order of increasing toxicity.

TABLE 2.9 Some LD_{50} Values for Substances Tested in Rats

Substance	LD_{50} (mg/kg)
Table sugar	29 700
Boric acid	5140
Baking soda	4220
Table salt	3300
Ethanol	2080
Aspirin	1100
Codeine	800
Oxycodone	480
Caffeine	192
DDT	113
Cocaine (injected)	95
Dichlorvos (pesticide strips)	56
Ricin	30
Sodium cyanide	6
Parathion	3

P. 52

Updated! Chemistry Links to Health and Chemistry Links to the Environment

appear throughout the text and relate chemistry concepts to real-life topics in health, the environment, and medicine. High-interest topics include weight loss and weight gain, hyperglycemia and hypoglycemia, antacids, gout and kidney stones, sweeteners, and essential amino acids. Follow-up questions also appear throughout the text.

Updated! Interactive Videos

give students an opportunity to reinforce what they just learned by showing how chemistry works in real life and introducing a bit of humor into chemical problem solving and demonstrations. Topics include Using Conversion Factors, Mass Calculations for Reactions, Concentration of Solutions, Balancing Nuclear Equations, and Chemical v. Physical Change.

Tools to help students succeed

Features in This Text That Help You Study and Learn Chemistry

This text has been designed with study features to support your learning. On the inside of the front cover is a periodic table of the elements. On the inside of the back cover are tables that summarize useful information needed throughout your study of chemistry. Each chapter begins with *Looking Ahead*, which outlines the topics in the chapter. At the beginning of each section, a *Learning Goal* describes the topics to learn. *Review* icons in the margins refer to Key Math Skills or Core Chemistry Skills from previous chapters that relate to new material in the chapter. *Key Terms* are bolded when they first appear in the text and are summarized at the end of each chapter. They are also listed and defined in the comprehensive *Glossary and Index*, which appears at the end of the text. *Key Math Skills* and *Core Chemistry Skills* that are critical to learning chemistry are indicated by icons in the margin and summarized at the end of each chapter.

REVIEW

KEY MATH SKILL

CORE CHEMISTRY SKILL

ENGAGE

P. 7

Marginal notes and end-of-chapter problems deepen the connection between key math skills, core chemistry skills, textual content, practice problems, and why they are so important to success in the course.

Chemistry Primer

Balancing a Chemical Equation

Constants | Periodic Table

According to the Law of Conservation of Mass, matter cannot be created nor destroyed in a chemical reaction.

Therefore, a chemical equation must show the same number of each kind of atom in the reactants as it does in the products.

As shown in the the figure, the following balanced equation has 6 atoms of hydrogen and 2 atoms of nitrogen on each side of the arrow:

$$3H_2 + N_2 \rightarrow 2NH_3$$

In General Chemistry, you will be presented with unbalanced chemical equations for which you must supply the coefficients. This is called balancing the chemical equation.

▼ Part A

What are the resulting coefficients when you balance the chemical equation for the combustion of ethane, C_2H_6?

In this reaction, ethane is burned in the presence of oxygen (O_2) to form carbon dioxide (CO_2) and water (H_2O).

$$\_\_\_C_2H_6(g) + \_\_\_O_2(g) \rightarrow \_\_\_CO_2(g) + \_\_\_H_2O(g)$$

Recall that the coefficients of the final balanced equation should be whole numbers. Thus, you might need to multiply through the equation by a factor of two to obtain whole numbers in your last step.

If you have trouble balancing the equation below, use the first hint to view a video of a similar equation being balanced. Then, use the rest of the hints to help you balance the equation, step-by-step.

Express the coefficients as integers separated by commas.

▶ View Available Hint(s)

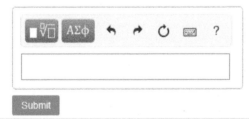

Submit

The Chemistry Primer in Mastering Chemistry helps students remediate their chemistry math skills and prepare for their first college chemistry course. Scaled to students' needs, remediation is only suggested to students that perform poorly on an initial assessment. Remediation includes tutorials, wrong-answer specific feedback, video instruction, and stepwise scaffolding to build students' abilities.

Build students' problem-solving skills

ENGAGE 14.8

If you know the $[H_3O^+]$ of a solution, how do you use the K_w to calculate the $[OH^-]$?

SAMPLE PROBLEM 14.6   Calculating the $[H_3O^+]$ of a Solution

TRY IT FIRST

A vinegar solution has a $[OH^-] = 5.0 \times 10^{-12}$ M at 25 °C. What is the $[H_3O^+]$ of the vinegar solution? Is the solution acidic, basic, or neutral?

SOLUTION

STEP 1 State the given and needed quantities.

ANALYZE THE PROBLEM	Given	Need	Connect
	$[OH^-] = 5.0 \times 10^{-12}$ M	$[H_3O^+]$	$K_w = [H_3O^+][OH^-]$

STEP 2 Write the K_w for water and solve for the unknown $[H_3O^+]$.

$$K_w = [H_3O^+][OH^-] = 1.0 \times 10^{-14}$$

Solve for $[H_3O^+]$ by dividing both sides by $[OH^-]$.

$$\frac{K_w}{[OH^-]} = \frac{[H_3O^+][OH^-]}{[OH^-]}$$

$$[H_3O^+] = \frac{1.0 \times 10^{-14}}{[OH^-]}$$

STEP 3 Substitute the known $[OH^-]$ into the equation and calculate.

$$[H_3O^+] = \frac{1.0 \times 10^{-14}}{[5.0 \times 10^{-12}]} = 2.0 \times 10^{-3} \text{ M}$$

Because the $[H_3O^+]$ of 2.0×10^{-3} M is larger than the $[OH^-]$ of 5.0×10^{-12} M, the solution is acidic.

ENGAGE 14.9

Why does the $[H_3O^+]$ of an aqueous solution increase if the $[OH^-]$ decreases?

SELF TEST 14.6

a. What is the $[H_3O^+]$ of an ammonia cleaning solution with $[OH^-] = 4.0 \times 10^{-4}$ M? Is the solution acidic, basic, or neutral?

b. The $[H_3O^+]$ of tomato juice is 6.3×10^{-5} M. What is the $[OH^-]$ of the juice? Is the tomato juice acidic, basic, or neutral?

ANSWER

a. $[H_3O^+] = 2.5 \times 10^{-11}$ M, basic   **b.** $[OH^-] = 1.6 \times 10^{-10}$ M, acidic

---

PRACTICE PROBLEMS

Try Practice Problems 14.37 to 14.42

NEW! Pedagogical features in worked **Sample Problems** throughout the text help students build stronger problem-solving skills, setting them up for success in this and future courses.

TRY IT FIRST! Feature encourages students to solve the problem before looking at the solution.

UPDATED! Connect feature added to **Analyze the Problem boxes** specify information that relates the **Given and Need** boxes to help students identify and connect the components within a word problem and show **STEPs** as guides that set up a solution strategy.

NEW! Engage questions remind students to pause and answer a question related to the material.

UPDATED! Self Tests provide students with immediate problem solving and feedback with answers.

NEW! Practice Problems suggest problems to work as students study the section.

Core Chemistry Skills and Key Math Skills Tutorials in Mastering Chemistry

provide assignable practice problems related to the in-text feature boxes, ensuring that students master the basic quantitative and science skills they need to succeed in the course.

Give students anytime, anywhere access with Pearson eText

Pearson eText is a simple-to-use, mobile-optimized, personalized reading experience available within Mastering. It allows students to easily highlight, take notes, and review key vocabulary all in one place—even when offline. Seamlessly integrated videos, rich media, and interactive self-assessment questions engage students and give them access to the help they need, when they need it. Pearson eText is available within Mastering when packaged with a new book; students can also purchase Mastering with Pearson eText online.

For instructors not using Mastering, Pearson eText can also be adopted on its own as the main course material.

Figure 3.1

The decomposition of salt, NaCl, produces the elements sodium and chlorine.
Question How do elements and compounds differ?

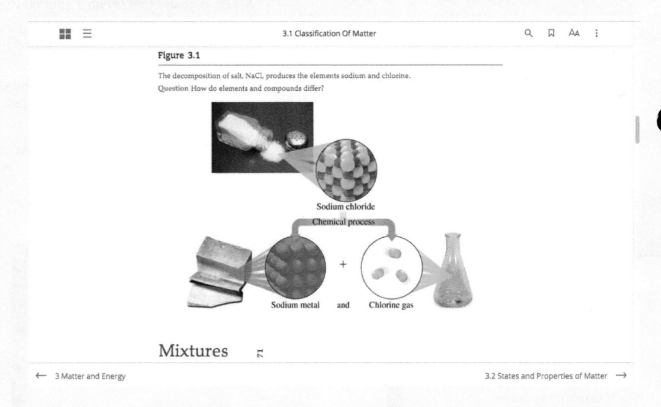

Mixtures

Improve learning with Dynamic Study Modules

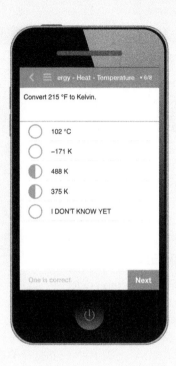

Dynamic Study Modules in Mastering Chemistry help students study effectively—and at their own pace—by keeping them motivated and engaged. The assignable modules rely on the latest research in cognitive science, using methods—such as adaptivity, gamification, and intermittent rewards—to stimulate learning and improve retention.

Each module poses a series of questions about a course topic. These question sets adapt to each student's performance and offer personalized, targeted feedback to help them master key concepts. With **Dynamic Study Modules**, students build the confidence they need to deepen their understanding, participate meaningfully, and perform better—in and out of class.

Instructor support you can rely on

Instructor Resources

Instructor Resources

Resources by Chapter
Links to downloadable resources for each chapter.

TestBank

Solutions Manual
The Solutions Manual for this product is available ONLY in digital format. Please contact your Pearson rep to request the files.

Basic Chemistry includes a full suite of instructor support materials in the Instructor Resources area in Mastering Chemistry.

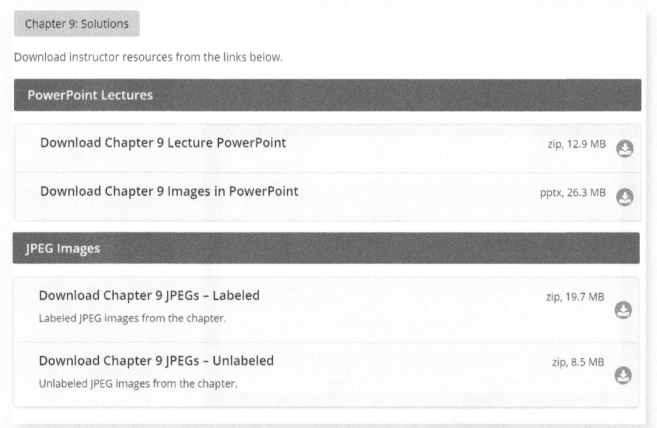

Chapter 9: Solutions

Download instructor resources from the links below.

PowerPoint Lectures

Download Chapter 9 Lecture PowerPoint — zip, 12.9 MB

Download Chapter 9 Images in PowerPoint — pptx, 26.3 MB

JPEG Images

Download Chapter 9 JPEGs – Labeled — zip, 19.7 MB
Labeled JPEG images from the chapter.

Download Chapter 9 JPEGs – Unlabeled — zip, 8.5 MB
Unlabeled JPEG images from the chapter.

Resources include customizable PowerPoint lecture and image presentations; all images and worked examples from the text; and a test bank.

Chemistry in Our Lives

A call came in to 911 from a man who arrived home from work to find his wife, Gloria, lying on their living room floor. When the police arrive, they pronounce the woman dead. There is no blood at the scene, but the police do find a glass on the side table that contains a small amount of liquid. In an adjacent laundry room, the police find a half-empty bottle of antifreeze, which contains the toxic compound ethylene glycol. The bottle, glass, and liquid are bagged and sent to the forensic laboratory. At the morgue, Gloria's height is measured as 1.673 m, and her mass is 60.5 kg.

Sarah, a forensic scientist, uses scientific procedures and chemical tests to examine the evidence from law enforcement agencies. She analyzes blood, stomach contents, and the unknown liquid from Gloria's home, as well as the fingerprints on the glass. She also looks for the presence of drugs, poisons, and alcohol.

CAREER

Forensic Scientist

Most forensic scientists work in crime laboratories that are part of city or county legal systems. They analyze bodily fluids and tissue samples collected by crime scene investigators. In analyzing these samples, forensic scientists identify the presence or absence of specific chemicals within the body to help solve criminal cases. Some of the chemicals they look for include alcohol, illegal or prescription drugs, poisons, arson debris, metals, and various gases such as carbon monoxide. To identify these substances, they use a variety of instruments and highly specific methodologies. Forensic scientists analyze samples from criminal suspects, athletes, and potential employees. They also work on cases involving environmental contamination and animal samples for wildlife crimes. Forensic scientists usually have a bachelor's degree that includes courses in math, chemistry, and biology.

UPDATE   Forensic Evidence Helps Solve the Crime

In the forensic laboratory, Sarah analyzes Gloria's stomach contents and blood for toxic compounds. You can view the results of the tests on the forensic evidence in the **UPDATE Forensic Evidence Helps Solve the Crime**, page 21, and determine if Gloria ingested a toxic level of ethylene glycol (antifreeze).

Molecules of NO_2

The chemical reaction of NO with oxygen in the air forms NO_2, which produces the reddish brown color of smog.

Antacid tablets undergo a chemical reaction when dropped into water.

ENGAGE 1.1

Why is water a chemical?

Toothpaste is a combination of many chemicals.

PRACTICE PROBLEMS

Try Practice Problems 1.1 to 1.6

1.1 Chemistry and Chemicals

LEARNING GOAL Define the term chemistry, and identify chemicals.

Now that you are in a chemistry class, you may be wondering what you will be learning. What questions in science have you been curious about? Perhaps you are interested in what smog is or how aspirin relieves a headache. Just like you, chemists are curious about the world we live in.

How does car exhaust produce the smog that hangs over our cities? One component of car exhaust is nitrogen oxide (NO), which forms in car engines where high temperatures convert nitrogen gas (N_2) and oxygen gas (O_2) to NO. In the atmosphere, the $NO(g)$ reacts with $O_2(g)$ to form $NO_2(g)$, which has a reddish brown color of smog. In chemistry, reactions are written in the form of equations:

$$N_2(g) + O_2(g) \longrightarrow 2NO(g)$$
$$2NO(g) + O_2(g) \longrightarrow 2NO_2(g)$$
Smog

Why does aspirin relieve a headache? When a part of the body is injured, substances called prostaglandins are produced, which cause inflammation and pain. Aspirin acts to block the production of prostaglandins, reducing inflammation and pain. Chemists in the medical field develop new treatments for diabetes, genetic defects, cancer, AIDS, and other diseases. For the forensic scientist, the nurse, the dietitian, the chemical engineer, or the agricultural scientist, chemistry plays a central role in understanding problems and assessing possible solutions.

Chemistry

Chemistry is the study of the composition, structure, properties, and reactions of matter. *Matter* is another word for all the substances that make up our world. Perhaps you imagine that chemistry takes place only in a laboratory where a chemist is working in a white coat and goggles. Actually, chemistry happens all around you every day and has an impact on everything you use and do. You are doing chemistry when you cook food, add bleach to your laundry, or start your car. A chemical reaction has taken place when silver tarnishes or an antacid tablet fizzes when dropped into water. Plants grow because chemical reactions convert carbon dioxide, water, and energy to carbohydrates. Chemical reactions take place when you digest food and break it down into substances that you need for energy and health.

Chemicals

A **chemical** is a substance that always has the same composition and properties wherever it is found. All the things you see around you are composed of one or more chemicals. Often the terms *chemical* and *substance* are used interchangeably to describe a specific type of matter.

Every day, you use products containing substances that were developed and prepared by chemists. Soaps and shampoos contain chemicals that remove oils on your skin and scalp. In cosmetics and lotions, chemicals are used to moisturize, prevent deterioration of the product, fight bacteria, and thicken the product. Perhaps you wear a ring or watch made of gold, silver, or platinum. Your breakfast cereal is probably fortified with iron, calcium, and phosphorus, whereas the milk you drink is enriched with vitamins A and D. When you brush your teeth, the substances in toothpaste clean your teeth, prevent plaque formation, and stop tooth decay. Some of the chemicals used to make toothpaste are listed in **TABLE 1.1**.

TABLE 1.1 Chemicals Commonly Used in Toothpaste	
Chemical	**Function**
Calcium carbonate	Used as an abrasive to remove plaque
Sorbitol	Prevents loss of water and hardening of toothpaste
Sodium lauryl sulfate	Used to loosen plaque
Titanium dioxide	Makes toothpaste white and opaque
Sodium fluorophosphate	Prevents formation of cavities by strengthening tooth enamel
Methyl salicylate	Gives toothpaste a pleasant wintergreen flavor

Branches of Chemistry

The field of chemistry is divided into several branches. *General chemistry* is the study of the composition, properties, and reactions of matter. *Organic chemistry* is the study of substances that contain the element carbon. *Biological chemistry* is the study of the chemical reactions that take place in biological systems. Today chemistry is often combined with other sciences, such as geology and physics, to form cross-disciplines such as geochemistry and physical chemistry. *Geochemistry* is the study of the chemical composition of ores, soils, and minerals of the surface of the Earth and other planets. *Physical chemistry* is the study of the physical nature of chemical systems, including energy changes.

A geochemist collects newly erupted lava samples from Kilauea Volcano, Hawaii.

PRACTICE PROBLEMS

1.1 Chemistry and Chemicals

In every chapter, odd-numbered exercises in the *Practice Problems* are paired with even-numbered exercises. The answers for the orange-shaded, odd-numbered *Practice Problems* are given at the end of each chapter. The complete solutions to the odd-numbered *Practice Problems* are in the *Study Guide and Student Solutions Manual*.

1.1 Write a one-sentence definition for each of the following:
   a. chemistry   **b.** chemical

1.2 Ask two of your friends (not in this class) to define the terms in problem 1.1. Do their answers agree with the definitions you provided?

Applications

1.3 Obtain a bottle of multivitamins, and read the list of ingredients. What are four chemicals from the list?

1.4 Obtain a box of breakfast cereal, and read the list of ingredients. What are four chemicals from the list?

1.5 Read the labels on some items found in a drugstore. What are the names of some chemicals contained in those items?

1.6 Read the labels on products used to wash your dishes. What are the names of some chemicals contained in those products?

1.2 Scientific Method: Thinking Like a Scientist

LEARNING GOAL Describe the scientific method.

Linus Pauling won the Nobel Prize in Chemistry in 1954.

When you were very young, you explored the things around you by touching and tasting. As you grew, you asked questions about the world in which you live. What is lightning? Where does a rainbow come from? Why is the sky blue? As an adult, you may have wondered how antibiotics work or why vitamins are important to your health. Every day, you ask questions and seek answers to organize and make sense of the world around you.

When the late Nobel Laureate Linus Pauling (1901–1994) described his student life in Oregon, he recalled that he read many books on chemistry, mineralogy, and physics. "I mulled over the properties of materials: why are some substances colored and others not, why are some minerals or inorganic compounds hard and others soft?" He said, "I was building up this tremendous background of empirical knowledge and at the same time asking a great number of questions." Linus Pauling won two Nobel Prizes: the first, in 1954, was in chemistry for his work on the nature of chemical bonds and the determination of the structures of complex substances; the second, in 1962, was the Peace Prize, for his opposition to the spread of nuclear weapons.

The Scientific Method

The process of trying to understand nature is unique to each scientist. However, the **scientific method** is a process that scientists use to make observations in nature, gather data, and explain natural phenomena (see the figure on the next page).

1. **Make Observations** The first step in the scientific method is to make **observations** about nature and ask questions about what you observe. When an observation always seems to be true, it may be stated as a *law* that predicts that behavior and is often

Scientific Method

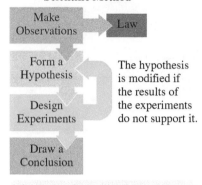

The scientific method develops a conclusion about nature using observations, hypotheses, and experiments.

measurable. However, a law does not explain that observation. For example, we can use the *Law of Gravity* to predict that if we drop our chemistry book, it would fall on the floor, but this law does not explain why our book falls.

2. **Form a Hypothesis** A scientist forms a **hypothesis**, which gives a possible explanation of an observation or a law. The hypothesis must be stated in such a way that it can be tested by experiments.

3. **Design Experiments** To determine if a hypothesis is valid, **experiments** are done to find a relationship between the hypothesis and the observations. The results of the experiments may confirm the validity of the hypothesis. However, experiments may also show that the hypothesis is invalid, which means that it is modified or discarded. Then new experiments will be designed to test the new hypothesis.

4. **Draw a Conclusion** When many experiments give consistent results, we may draw the **conclusion** that the hypothesis is valid. Even then, more experiments are done to test the hypothesis. If new experimental results indicate the hypothesis is not valid, it is modified or replaced.

Chemistry Link to Health

Early Chemist: Paracelsus

For many centuries, chemistry has been the study of changes in matter. From the time of the ancient Greeks to the sixteenth century, alchemists described matter in terms of four components of nature: earth, air, fire, and water. By the eighth century, alchemists believed that they could change metals such as copper and lead into gold and silver. Although these efforts failed, the alchemists provided information on the chemical reactions involved in the extraction of metals from ores. The alchemists also designed some of the first laboratory equipments and developed early laboratory procedures. These early efforts were some of the first observations and experiments using the scientific method.

Paracelsus (1493–1541) was a physician and an alchemist who thought that alchemy should be about preparing new medicines. Using observation and experimentation, he proposed that a healthy body was regulated by a series of chemical processes that could be unbalanced by certain chemical compounds and rebalanced by using minerals and medicines. For example, he determined that inhaled

dust caused lung disease in miners. He also thought that goiter was a problem caused by contaminated water, and he treated syphilis with compounds of mercury. His opinion of medicines was that the right dose makes the difference between a poison and a cure. Paracelsus changed alchemy in ways that helped establish modern medicine and chemistry.

Swiss physician and alchemist Paracelsus (1493–1541) believed that chemicals and minerals could be used as medicines.

Through observation you may think that you are allergic to cats.

ENGAGE 1.2

Why would the following statement, "Today I placed two tomato seedlings in the garden, and two more in a closet. I will give all the plants the same amount of water and fertilizer," be considered an experiment?

Using the Scientific Method in Everyday Life

You may be surprised to realize that you use the scientific method in your everyday life. Suppose you visit a friend in her home. Soon after you arrive, your eyes start to itch and you begin to sneeze. Then you observe that your friend has a new cat. Perhaps you form the hypothesis that you are allergic to cats. To test your hypothesis, you leave your friend's home. If the sneezing stops, perhaps your hypothesis is correct. You test your hypothesis further by visiting another friend who also has a cat. If you start to sneeze again, your experimental results support your hypothesis, and you come to the conclusion that you are allergic to cats. However, if you continue sneezing after you leave your friend's home, your hypothesis is not supported. Now you need to form a new hypothesis, which could be that you have a cold.

SAMPLE PROBLEM 1.1   Scientific Method

TRY IT FIRST

Identify each of the following as an observation, a hypothesis, an experiment, or a conclusion:

a. During an assessment in the emergency room, a nurse writes that the patient has a resting pulse of 30 beats/min.

b. Repeated studies show that lowering sodium in the diet leads to a decrease in blood pressure.

c. A nurse thinks that an incision from a recent surgery that is red and swollen is infected.

SOLUTION

a. observation   **b.** conclusion   **c.** hypothesis

Nurses make observations in the hospital.

SELF TEST 1.1

Identify each of the following as an observation, a hypothesis, an experiment, or a conclusion:

a. Drinking coffee at night keeps me awake.
b. I will try drinking coffee only in the morning.
c. If I stop drinking coffee in the afternoon, I will be able to sleep at night.
d. When I drink decaffeinated coffee, I sleep better at night.
e. I am going to drink only decaffeinated coffee.
f. I sleep better at night because I stopped drinking caffeinated drinks.

ANSWER

a. observation   **b.** experiment   **c.** hypothesis
d. observation   **e.** experiment   **f.** conclusion

PRACTICE PROBLEMS

Try Practice Problems 1.7 to 1.10

PRACTICE PROBLEMS

1.2 Scientific Method: Thinking Like a Scientist

1.7 Identify each activity, **a** to **f**, as an observation, a hypothesis, an experiment, or a conclusion.
At a popular restaurant, where Chang is the head chef, the following occurred:
a. Chang determined that sales of the house salad had dropped.
b. Chang decided that the house salad needed a new dressing.
c. In a taste test, Chang prepared four bowls of sliced cucumber, each with a new dressing: sesame seed, olive oil and balsamic vinegar, creamy Italian, and blue cheese.
d. Tasters rated the sesame seed salad dressing as the favorite.
e. After two weeks, Chang noted that the orders for the house salad with the new sesame seed dressing had doubled.
f. Chang decided that the sesame seed dressing improved the sales of the house salad because the sesame seed dressing enhanced the taste.

Customers rated the sesame seed dressing as the best.

1.8 Identify each activity, **a** to **f**, as an observation, a hypothesis, an experiment, or a conclusion.
Lucia wants to develop a process for dyeing shirts so that the color will not fade when the shirt is washed. She proceeds with the following activities:
a. Lucia notices that the dye in a design fades when the shirt is washed.
b. Lucia decides that the dye needs something to help it combine with the fabric.

c. She places a spot of dye on each of four shirts and then places each one separately in water, salt water, vinegar, and baking soda and water.
d. After one hour, all the shirts are removed and washed with a detergent.
e. Lucia notices that the dye has faded on the shirts in water, salt water, and baking soda, whereas the dye did not fade on the shirt soaked in vinegar.
f. Lucia thinks that the vinegar binds with the dye so it does not fade when the shirt is washed.

Applications

1.9 Identify each of the following as an observation, a hypothesis, an experiment, or a conclusion:
a. One hour after drinking a glass of regular milk, Jim experienced stomach cramps.
b. Jim thinks he may be lactose intolerant.
c. Jim drinks a glass of lactose-free milk and does not have any stomach cramps.
d. Jim drinks a glass of regular milk to which he has added lactase, an enzyme that breaks down lactose, and has no stomach cramps.

1.10 Identify each of the following as an observation, a hypothesis, an experiment, or a conclusion:
a. Sally thinks she may be allergic to shrimp.
b. Yesterday, one hour after Sally ate a shrimp salad, she broke out in hives.
c. Today, Sally had some soup that contained shrimp, but she did not break out in hives.
d. Sally realizes that she does not have an allergy to shrimp.

1.3 Studying and Learning Chemistry

LEARNING GOAL Identify strategies that are effective for learning. Develop a study plan for learning chemistry.

Here you are taking chemistry, perhaps for the first time. Whatever your reasons for choosing to study chemistry, you can look forward to learning many new and exciting ideas.

Strategies to Improve Learning and Understanding

Success in chemistry utilizes good study habits, connecting new information with your knowledge base, rechecking what you have learned and what you have forgotten, and retrieving what you have learned for an exam. Let's take a look at ways that can help you study and learn chemistry. Suppose you were asked to indicate if you think each of the following common study habits is helpful or not helpful:

	Helpful	Not helpful
Highlighting		
Underlining		
Reading the chapter many times		
Memorizing the key words		
Testing practice		
Cramming		
Studying different ideas at the same time		
Retesting a few days later		

Learning chemistry requires us to place new information in our long-term memory, which allows us to remember those ideas for an exam, a process called *retrieval*. Thus, our study habits need to help us to recall knowledge. The study habits that are not very helpful in retrieval include highlighting, underlining, reading the chapter many times, memorizing key words, and cramming. If we want to recall new information, we need to connect it with prior knowledge. By doing practice tests, you develop the skills needed to retrieve new information. We can determine how much we have learned by going back a few days later and retesting. Another useful learning strategy is to study different ideas at the same time, which allows us to connect those ideas and to differentiate between them. Although these study habits may take more time and seem more difficult, they help us find the gaps in our knowledge and connect new information with what we already know.

Tips for Using New Study Habits for Successful Learning

1. **Do not keep rereading text or notes.** Reading the same material over and over will make that material seem familiar but does not mean that you have learned it. You need to test yourself to find out what you do and do not know.
2. **Ask yourself questions as you read.** Asking yourself questions as you read requires you to interact continually with new material. The *Engage* features in the margin of this text are helpful in asking you to think about new material. By linking new material with long-term knowledge, you make pathways for retrieving new material.
3. **Self-test by giving yourself quizzes.** Using problems in the text or sample exams, practice taking tests frequently.
4. **Study at a regular pace rather than cramming.** Once you have tested yourself, go back in a few days and practice testing and retrieving information again. We do not recall all the information when we first read it. By frequent quizzing and retesting, we identify what we still need to learn. Sleep is also important for strengthening the associations between newly learned information. Lack of sleep may interfere with retrieval of information as well. So staying up all night to cram for your chemistry exam is not a good idea. Success in chemistry is a combined effort to learn new information and then to retrieve that information when you need it for an exam.

Students learn by continuously asking and answering questions as they study new material.

5. **Study different topics in a chapter, and relate the new concepts to concepts you know.** We learn material more efficiently by relating it to information we already know. By increasing connections between concepts, we can retrieve information when we need it.

Helpful	**Not helpful**
Testing practice	Highlighting
Studying different ideas at the same time	Underlining
	Reading the chapter many times
Retesting a few days later	Memorizing the key words
	Cramming

ENGAGE 1.3

Why is self-testing helpful for learning new concepts?

SAMPLE PROBLEM 1.2 Strategies for Learning Chemistry

TRY IT FIRST

Predict which student, **a**, **b**, or **c**, will be most successful on the exam.

a. Bill, who reads the chapter four times
b. Jennifer, who reads the chapter two times and works all the problems at the end of each section
c. Mark, who reads the chapter the night before the exam

SOLUTION

b. Jennifer, who reads the chapter two times and works all the problems at the end of each section, interacts with the content in the chapter using self-testing to make connections between concepts and practicing retrieving information learned previously.

SELF TEST 1.2

What are two more ways that Jennifer could improve her retrieval of information?

ANSWER

1. Jennifer could wait two or three days and practice working the problems in each section again to determine how much she has learned. Retesting strengthens connections between new and previously learned information for longer lasting memory and more efficient retrieval.
2. Jennifer could also ask questions as she reads and try to study at a regular pace to avoid cramming.

Features in This Text That Help You Study and Learn Chemistry

This text has been designed with study features to support your learning. On the inside of the front cover is a periodic table of the elements. On the inside of the back cover are tables that summarize useful information needed throughout your study of chemistry. Each chapter begins with *Looking Ahead*, which outlines the topics in the chapter. At the beginning of each section, a *Learning Goal* describes the topics to learn. *Review* icons in the margins refer to Key Math Skills or Core Chemistry Skills from previous chapters that relate to new material in the chapter. *Key Terms* are bolded when they first appear in the text and are summarized at the end of each chapter. They are also listed and defined in the comprehensive *Glossary and Index*, which appears at the end of the text. *Key Math Skills* and *Core Chemistry Skills* that are critical to learning chemistry are indicated by icons in the margin and summarized at the end of each chapter.

Before you begin reading, obtain an overview of a chapter by reviewing the topics in *Looking Ahead*. As you prepare to read a section of the chapter, look at the section title, and turn it into a question. Asking yourself questions about new topics builds new connections to material you have already learned. For example, for Section 1.1, "Chemistry and Chemicals," you could ask, "What is chemistry?" or "What are chemicals?" At the beginning of

REVIEW

KEY MATH SKILL

CORE CHEMISTRY SKILL

ENGAGE

ANALYZE THE PROBLEM	Given	Need	Connect

Illustrating the atoms of aluminum in aluminum foil is an example of macro-to-micro art.

each section, a *Learning Goal* states what you need to understand. As you read the text, you will see *Engage* questions in the margin, which remind you to pause your reading and test yourself with a question related to the material. The *Answers to Engage Questions* are at the end of each chapter.

Several *Sample Problems* are included in each chapter. The *Try It First* feature reminds you to work the problem before you look at the *Solution*. It is helpful to try to work a problem first because it helps you link what you know to what you need to learn. The *Analyze the Problem* feature includes *Given*, the information you have; *Need*, what you have to accomplish; and *Connect*, how you proceed. Sample Problems include a Solution that shows the steps you can use for problem solving. Work the associated *Self Test* problems, and compare your answer to the one provided.

At the end of each chapter section, you will find a set of *Practice Problems* that allows you to apply problem solving immediately to the new concepts. Throughout each section, *Practice Problem* icons remind you to try the indicated Practice Problems as you study. The *Applications* in the Practice Problems relate the content to health, careers, and real-life examples. The problems are paired, which means that each of the odd-numbered problems is matched to the following even-numbered problem. At the end of each chapter, the *Answers* to all the odd-numbered problems are provided. If the answers match yours, you most likely understand the topic; if not, you need to study the section again.

Throughout each chapter, boxes titled *Chemistry Link to Health* and *Chemistry Link to the Environment* help you relate the chemical concepts you are learning to real-life situations. Many of the figures and diagrams use macro-to-micro illustrations to depict the atomic level of organization of ordinary objects, such as the atoms in aluminum foil. These visual models illustrate the concepts described in the text and allow you to "see" the world in a microscopic way. *Interactive Video* suggestions illustrate content as well as problem solving.

At the end of each chapter, you will find several study aids that complete the chapter. *Chapter Reviews* provide a summary in easy-to-read bullet points, and *Concept Maps* visually show the connections between important topics. *Understanding the Concepts* are problems that use art and models to help you visualize concepts and connect them to your background knowledge. *Additional Practice Problems* and *Challenge Problems* provide additional exercises to test your understanding of the topics in the chapter.

After some chapters, problem sets called *Combining Ideas* test your ability to solve problems containing material from more than one chapter.

Many students find that studying with a group can be beneficial to learning. In a group, students motivate each other to study, fill in gaps, and correct misunderstandings by teaching and learning together. Studying alone does not allow the process of peer correction. In a group, you can cover the ideas more thoroughly as you discuss the reading and problem solve with other students.

Making a Study Plan

As you embark on your journey into the world of chemistry, think about your approach to studying and learning chemistry. You might consider some of the ideas in the following list. Check those ideas that will help you successfully learn chemistry. Commit to them now. *Your* success depends on *you*.

My study plan for learning chemistry will include the following:

_____ reading the chapter before class

_____ going to class

_____ reviewing the Learning Goals

_____ keeping a problem notebook

_____ reading the text

_____ working the Practice Problems as I read each section

_____ answering the Engage questions

_____ trying to work the Sample Problem before looking at the Solution

_____ studying different topics at the same time

_____ organizing a study group

_____ seeing the professor during office hours
_____ reviewing Key Math Skills and Core Chemistry Skills
_____ attending review sessions
_____ studying as often as I can

SAMPLE PROBLEM 1.3   A Study Plan for Learning Chemistry

TRY IT FIRST

Which of the following activities should you include in your study plan for learning chemistry successfully?

a. reading the chapter over and over until you think you understand it
b. going to the professor's office hours
c. doing the Self Test associated with each Sample Problem
d. waiting to study until the night before the exam
e. trying to work the Sample Problem before looking at the Solution
f. retesting on new information a few days later

SOLUTION

b, **c**, **e**, and **f**

SELF TEST 1.3

Which of the following will help you learn chemistry?

a. skipping review sessions
b. working Practice Problems as you read each section
c. staying up all night before an exam
d. reading the assignment before class
e. highlighting the key ideas in the text

ANSWER

b and **d**

PRACTICE PROBLEMS

Try Practice Problems 1.11 to 1.14

PRACTICE PROBLEMS

1.3 Studying and Learning Chemistry

1.11 What are four things you can do to help yourself to succeed in chemistry?

1.12 What are four things that would make it difficult for you to learn chemistry?

1.13 A student in your class asks you for advice on learning chemistry. Which of the following might you suggest?
  a. forming a study group
  b. skipping class
  c. asking yourself questions while reading the text
  d. waiting until the night before an exam to study
  e. answering the Engage questions

1.14 A student in your class asks you for advice on learning chemistry. Which of the following might you suggest?
  a. studying different topics at the same time
  b. not reading the text; it's never on the test
  c. attending review sessions
  d. working the problems again after a few days
  e. keeping a problem notebook

1.4 Key Math Skills for Chemistry

LEARNING GOAL Review math concepts used in chemistry: place values, positive and negative numbers, percentages, solving equations, and interpreting graphs.

During your study of chemistry, you will work many problems that involve numbers. You will need various math skills and operations. We will review some of the key math skills that are particularly important for chemistry. As we move through the chapters, we will also reference the key math skills as they apply.

Identifying Place Values

Decimal Place Value Chart

Thousands	Hundreds	Tens	Ones	Decimal point	Tenths	Hundredths	Thousandths
4	8	2	.		5	7	

In the number 8.034, how do you know the 0 is in the tenths place?

Identifying Place Values

The name of each place value depends on its position relative to the decimal point. For the number 482.57 in the Decimal Place Value Chart, the place values increase 10 times moving to the left and decrease 10 times by moving to the right. The digits on the left of the decimal point on the place value chart are 4 hundreds (400), 8 tens (80), and 2 ones (2), or four hundred eighty-two (482). The digits on the right of the decimal point are 5 tenths (.5), and 7 hundredths (.07), or fifty-seven hundredths (.57).

A premature baby has a mass of 2518 g. We can assign the place values for the number 2518 as follows:

Digit	Place Value
2	thousands
5	hundreds
1	tens
8	ones

A silver coin has a mass of 6.407 g. We can assign the place values for the number 6.407 as follows:

Digit	Place Value
6	ones
4	ten**ths**
0	hundred**ths**
7	thousand**ths**

Note that place values ending with the suffix *ths* refer to the decimal places to the right of the decimal point.

SAMPLE PROBLEM 1.4 Identifying Place Values

TRY IT FIRST

A bullet found at a crime scene has a mass of 15.24 g. What are the place values for each of the digits in the mass of the bullet?

SOLUTION

Digit	Place Value
1	tens
5	ones
2	tenths
4	hundredths

SELF TEST 1.4

Identify the place values for the digits in each of the following:

a. the victim's height of 1.673 m   **b.** the victim's mass of 60.5 kg

ANSWER

a.

Digit	Place Value
1	ones
6	tenths
7	hundredths
3	thousandths

b.

Digit	Place Value
6	tens
0	ones
5	tenths

Try Practice Problems 1.15 and 1.16

Using Positive and Negative Numbers in Calculations

A *positive number* is any number that is greater than zero and has a positive sign (+). Often the positive sign is understood and not written in front of the number. For example, the number +8 is usually written as 8. A *negative number* is any number that is less than zero and is written with a negative sign (−). For example, a negative eight is written as −8.

Multiplication and Division of Positive and Negative Numbers

When two positive numbers or two negative numbers are multiplied, the answer is positive (+).

$$2 \times 3 = 6 \quad \text{The + sign (+6) is understood.}$$
$$(-2) \times (-3) = 6$$

When a positive number and a negative number are multiplied, the answer is negative (−).

$$2 \times (-3) = -6$$
$$(-2) \times 3 = -6$$

The rules for the division of positive and negative numbers are the same as the rules for multiplication. When two positive numbers or two negative numbers are divided, the answer is positive (+).

$$\frac{6}{3} = 2 \quad \frac{-6}{-3} = 2$$

When a positive number and a negative number are divided, the answer is negative (−).

$$\frac{-6}{3} = -2 \quad \frac{6}{-3} = -2$$

Addition of Positive and Negative Numbers

When positive numbers are added, the sign of the answer is positive.

$$3 + 4 = 7 \quad \text{The + sign (+7) is understood.}$$

When negative numbers are added, the sign of the answer is negative.

$$-3 + (-4) = -7$$

When a positive number and a negative number are added, the smaller number is subtracted from the larger number, and the result has the same sign as the larger number.

$$12 + (-15) = -3$$

Subtraction of Positive and Negative Numbers

When two numbers are subtracted, change the sign of the number to be subtracted and follow the rules for addition shown above.

$$12 - (+5) = 12 + (-5) = 7$$
$$-12 - (-5) = -12 + (+5) = -7$$

Calculator Operations

On your calculator, there are four keys that are used for basic mathematical operations. The change sign +/− key may be used to change the sign of a number.

To practice these basic calculations on the calculator, work through the problem going from the left to the right, doing the operations in the order they occur. On some calculators, a negative number is entered by pressing the number and then pressing the change sign +/− key. However, this may not work on all calculators. Because each type of calculator is slightly different, make sure that you know how your particular calculator works. At the end, press the equals = key or ANS or ENTER.

Multiplication

Division

Subtraction

Equals

Change sign

Addition

Addition and Subtraction		Multiplication and Division	
Example 1:	$15 - 8 + 2 =$	Example 3:	$2 \times (-3) =$
Solution:	$15\boxed{-}8\boxed{+}2\boxed{=}9$	Solution:	$2\boxed{\times}3\boxed{+/-}\boxed{=}-6$
Example 2:	$4 + (-10) - 5 =$	Example 4:	$\dfrac{8 \times 3}{4} =$
Solution:	$4\boxed{+}10\boxed{+/-}\boxed{-}5\boxed{=}-11$	Solution:	$8\boxed{\times}3\boxed{\div}4\boxed{=}6$

Calculating Percentages

To determine a percentage, divide the parts by the total (whole) and multiply by 100%. For example, if an aspirin tablet contains 325 mg of aspirin (active ingredient) and the tablet has a mass of 545 mg, what is the percentage of aspirin in the tablet?

$$\frac{325 \text{ mg aspirin}}{545 \text{ mg tablet}} \times 100\% = 59.6\% \text{ aspirin}$$

When a value is described as a percent (%), it represents the number of parts of an item in 100 of those items. If the percent of red balls is 5, it means there are 5 red balls in every 100 balls. If the percent of green balls is 50, there are 50 green balls in every 100 balls.

$$5\% \text{ red balls} = \frac{5 \text{ red balls}}{100 \text{ balls}} \qquad 50\% \text{ green balls} = \frac{50 \text{ green balls}}{100 \text{ balls}}$$

A bullet casing at a crime scene is marked as evidence.

SAMPLE PROBLEM 1.5 Calculating a Percentage

TRY IT FIRST

A bullet found at a crime scene may be used as evidence in a trial if the percentage of metals is a match to the composition of metals in a bullet from the suspect's ammunition. A forensic scientist's analysis of the bullet shows that it contains 13.9 g of lead, 0.3 g of tin, and 0.9 g of antimony. What is the percentage of each metal in the bullet? Express your answers to the ones place.

SOLUTION

A percentage is calculated by dividing the mass of each element by the total mass of the metals in the bullet and multiplying by 100%.

Total mass $= 13.9 \text{ g} + 0.3 \text{ g} + 0.9 \text{ g} = 15.1 \text{ g}$

Percentage of lead

$$\frac{13.9 \text{ g}}{15.1 \text{ g}} \times 100\% = 92\% \text{ lead}$$

Percentage of tin

$$\frac{0.3 \text{ g}}{15.1 \text{ g}} \times 100\% = 2\% \text{ tin}$$

Percentage of antimony

$$\frac{0.9 \text{ g}}{15.1 \text{ g}} \times 100\% = 6\% \text{ antimony}$$

A bullet seized from the suspect's ammunition has a composition of lead 11.6 g, tin 0.5 g, and antimony 0.4 g.

a. What is the percentage of each metal in the bullet? Express your answers to the ones place.

b. Could the bullet removed from the suspect's ammunition be considered as evidence that the suspect was at the crime scene mentioned in Sample Problem 1.5?

ANSWER

a. The bullet from the suspect's ammunition is lead 93%, tin 4%, and antimony 3%.

b. The composition of this bullet does not match the bullet from the crime scene and cannot be used as supporting evidence.

PRACTICE PROBLEMS

Try Practice Problems 1.23 and 1.24

Solving Equations

In chemistry, we use equations that express the relationship between certain variables. Let's look at how we would solve for x in the following equation:

$$2x + 8 = 14$$

Our overall goal is to rearrange the items in the equation to obtain x on one side.

KEY MATH SKILL

Solving Equations

1. *Place all like terms on one side.* The numbers 8 and 14 are like terms. To remove the 8 from the left side of the equation, we subtract 8. To keep a balance, we need to subtract 8 from the 14 on the other side.

$$2x + 8 - 8 = 14 - 8$$
$$2x = 6$$

ENGAGE 1.7

Why is the number 8 subtracted from both sides of this equation?

2. *Isolate the variable you need to solve for.* In this problem, we obtain x by dividing both sides of the equation by 2. The value of x is the result when 6 is divided by 2.

$$\frac{2x}{2} = \frac{6}{2}$$
$$x = 3$$

3. *Check your answer.* Check your answer by substituting your value for x back into the original equation.

$$2(3) + 8 = 14$$
$$6 + 8 = 14$$
$$14 = 14 \quad \text{Your answer } x = 3 \text{ is correct.}$$

Summary: To solve an equation for a particular variable, be sure you perform the same mathematical operations on *both* sides of the equation.

If you eliminate a symbol or number by subtracting, you need to subtract that same symbol or number from both sides.

If you eliminate a symbol or number by adding, you need to add that same symbol or number to both sides.

If you cancel a symbol or number by dividing, you need to divide both sides by that same symbol or number.

If you cancel a symbol or number by multiplying, you need to multiply both sides by that same symbol or number.

When we work with temperature, we may need to convert between degrees Celsius and degrees Fahrenheit using the following equation:

$$T_F = 1.8(T_C) + 32$$

A plastic strip thermometer changes color to indicate body temperature.

To obtain the equation for converting degrees Fahrenheit to degrees Celsius, we subtract 32 from both sides.

$$T_F = 1.8(T_C) + 32$$
$$T_F - 32 = 1.8(T_C) + \cancel{32} - \cancel{32}$$
$$T_F - 32 = 1.8(T_C)$$

To obtain T_C by itself, we divide both sides by 1.8.

$$\frac{T_F - 32}{1.8} = \frac{\cancel{1.8}(T_C)}{\cancel{1.8}} = T_C$$

SAMPLE PROBLEM 1.6   Solving Equations

TRY IT FIRST

Solve the following equation for V_2:

$$P_1 V_1 = P_2 V_2$$

INTERACTIVE VIDEO

Pearson eText

Solving Equations

ENGAGE 1.8

Why is the numerator divided by P_2 on both sides of the equation?

PRACTICE PROBLEMS

Try Practice Problems 1.19 and 1.20

SOLUTION

$$P_1 V_1 = P_2 V_2$$

To solve for V_2, divide both sides by the symbol P_2.

$$\frac{P_1 V_1}{P_2} = \frac{\cancel{P_2} V_2}{\cancel{P_2}}$$
$$V_2 = \frac{P_1 V_1}{P_2}$$

SELF TEST 1.6

Solve each of the following equations for m:

a. $q = m \times \Delta T \times SH$ **b.** $D = \dfrac{m}{V}$

ANSWER

a. $m = \dfrac{q}{\Delta T \times SH}$ **b.** $m = D \times V$

KEY MATH SKILL

Interpreting Graphs

Interpreting Graphs

A graph is a diagram that represents the relationship between two variables. These quantities are plotted along two perpendicular axes, which are the x axis (horizontal) and y axis (vertical).

Example

In the graph Volume of a Balloon versus Temperature, the volume of a gas in a balloon is plotted against its temperature.

Title

Look at the title. What does it tell us about the graph? The title indicates that the volume of a balloon was measured at different temperatures.

Vertical Axis

Look at the label and the numbers on the vertical (y) axis. The label indicates that the volume of the balloon was measured in liters (L). The numbers, which are chosen to include the low and high measurements of the volume of the gas, are evenly spaced from 22.0 L to 32.0 L.

Horizontal Axis

The label on the horizontal (x) axis indicates that the temperatures of the balloon, measured in kelvins (K), are evenly spaced from 270 K to 390 K.

Volume of a Balloon versus Temperature

ENGAGE 1.9

Why are the numbers on the vertical and horizontal axes placed at regular intervals?

Points on the Graph

Each point on the graph represents a volume in liters that was measured at a specific temperature. When these points are connected, a line is obtained.

Interpreting the Graph

From the graph, we see that the volume of the gas increases as the temperature of the gas increases. This is called a *direct relationship*. Now we use the graph to determine the volume at various temperatures. For example, suppose we want to know the volume of the gas at 320 K. We would start by finding 320 K on the *x* axis and then drawing a line up to the plotted line. From there, we would draw a horizontal line that intersects the *y* axis and read the volume value where the line crosses the *y* axis as shown on the graph above.

SAMPLE PROBLEM 1.7 Interpreting a Graph

TRY IT FIRST

A nurse administers Tylenol to lower a child's fever. The graph shows the body temperature of the child plotted against time.

a. What is measured on the vertical axis?
b. What is the range of values on the vertical axis?
c. What is measured on the horizontal axis?
d. What is the range of values on the horizontal axis?

SOLUTION

a. body temperature, in degrees Celsius
b. 37.0 °C to 39.8 °C
c. time, in minutes, after Tylenol was given
d. 0 min to 30 min

Child's Body Temperature versus Time

SELF TEST 1.7

a. Using the graph in Sample Problem 1.7, what was the child's temperature 15 min after Tylenol was given?
b. How many minutes elapsed for the temperature to decrease from 39.4 °C to 38.0 °C?
c. What was the decrease, in degrees Celsius, between 5 min and 20 min?

ANSWER

a. 37.6 °C   **b.** 8 min   **c.** 0.9 °C

PRACTICE PROBLEMS

Try Practice Problems 1.21 and 1.22, 1.25 and 1.26

PRACTICE PROBLEMS

1.4 Key Math Skills for Chemistry

1.15 What is the place value for the bold digit?
    a. 7.3**2**88   **b.** 16.1**2**34   **c.** 0.2**8**0
    d. 4**6**75.99   **e.** 0.090**1**

1.16 What is the place value for the bold digit?
    a. 975.**6**89   **b.** 375.8**8**   **c.** 4**6**.1000
    d. 0.**8**128   **e.** 2.7**1**0

1.17 Evaluate each of the following:
    a. $15 - (-8) =$ _____   **b.** $-8 + (-22) =$ _____
    c. $4 \times (-2) + 6 =$ _____   **d.** $75 - (-44) - 65 =$ _____
    e. $\dfrac{-25 + (-11)}{-3 + 6} =$ _____

1.18 Evaluate each of the following:
    a. $-11 - (-9) =$ _____   **b.** $34 + (-55) =$ _____
    c. $\dfrac{-56}{8} =$ _____   **d.** $0.04 + (-0.08) =$ _____
    e. $\dfrac{-2 \times (-4)}{2} + 12 =$ _____

1.19 Solve each of the following for a:
    a. $4a + 4 = 40$ **b.** $\dfrac{a}{6} = 7$
    c. $\dfrac{55 - a + 15}{4} = a$ **d.** $\dfrac{6a - 27}{3} = -a$

1.20 Solve each of the following for b:
    a. $2b + 7 = b + 10$ **b.** $3b - 4 = 24 - b$
    c. $\dfrac{-16 - 2b + 8b}{2} = b$ **d.** $\dfrac{3 \times (-4b) + 32}{2} = -4b$

Use the following graph for problems 1.21 and 1.22:

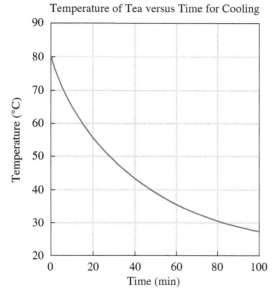

Temperature of Tea versus Time for Cooling

1.21 **a.** What does the title indicate about the graph?
    b. What is measured on the vertical axis?
    c. What is the range of values on the vertical axis?
    d. Does the temperature increase or decrease with an increase in time?

1.22 **a.** What is measured on the horizontal axis?
    b. What is the range of values on the horizontal axis?
    c. What is the temperature of the tea after 20 min?
    d. How many minutes were needed to reach a temperature of 45 °C?

Applications

1.23 **a.** A clinic had 25 patients on Friday morning. If 21 patients were given flu shots, what percentage of the patients received flu shots? Express your answer to the ones place.
    b. An alloy contains 56 g of pure silver and 22 g of pure copper. What is the percentage of silver in the alloy? Express your answer to the ones place.
    c. A collection of coins contains 11 nickels, 5 quarters, and 7 dimes. What is the percentage of dimes in the collection? Express your answer to the ones place.
    d. A saline solution has a mass of 44 g, of which 2.2 g is sodium chloride. What percent of the solution is sodium chloride? Express your answer to the ones place.

1.24 **a.** At a local hospital, 35 babies were born in May. If 22 were boys, what percentage of the newborns were boys? Express your answer to the ones place.
    b. An alloy contains 67 g of pure gold and 35 g of pure zinc. What is the percentage of zinc in the alloy? Express your answer to the ones place.
    c. A collection of coins contains 15 pennies, 14 dimes, and 6 quarters. What is the percentage of pennies in the collection? Express your answer to the ones place.
    d. In the hospital storeroom, 18 of 56 intravenous (IV) fluid bags contain dextrose solutions. The bags with dextrose are what percent of the total? Express your answer to the ones place.

Use the following graph for problems 1.25 and 1.26:

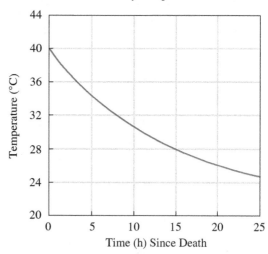

Post-Mortem Body Temperature versus Time

1.25 **a.** What does the title indicate about the graph?
    b. What is measured on the vertical axis?
    c. What is the range of values on the vertical axis?
    d. Does the temperature increase or decrease with an increase in time?

1.26 **a.** What is measured on the horizontal axis?
    b. What is the range of values on the horizontal axis?
    c. How many hours were needed to reach a temperature of 28 °C?
    d. The coroner measured Gloria's body temperature at 9 P.M. as 34 °C. What was the time of her death?

1.5 Writing Numbers in Scientific Notation

LEARNING GOAL Write a number in scientific notation.

In chemistry, we often work with numbers that are very large and very small. We might measure something as tiny as the width of a human hair, which is about 0.000 008 m. Or perhaps we want to count the number of hairs on the average human scalp, which is about 100 000 hairs. In this text, we add spaces between sets of three digits when it helps make the places easier to count. However, we will see that it is more convenient to write large and small numbers in *scientific notation*.

1×10^5 hairs

8×10^{-6} m

Humans have an average of 1×10^5 hairs on their scalps. Each hair is about 8×10^{-6} m wide.

Standard Number	Scientific Notation
0.000 008 m	8×10^{-6} m
100 000 hairs	1×10^5 hairs

A number written in **scientific notation** has two parts: a coefficient and a power of 10. For example, the number 2400 is written in scientific notation as 2.4×10^3. The coefficient, 2.4, is obtained by moving the decimal point to the left to give a number that is at least 1 but less than 10. Because we moved the decimal point three places to the left, the power of 10 is a positive 3, which is written as 10^3. When a number greater than 1 is converted to scientific notation, the power of 10 is positive.

Standard Number		Scientific Notation	
2400.	=	2.4	$\times$	10^3

$\leftarrow$ 3 places   Coefficient   Power of 10

In another example, 0.000 86 is written in scientific notation as 8.6×10^{-4}. The coefficient, 8.6, is obtained by moving the decimal point to the right. Because the decimal point is moved four places to the right, the power of 10 is a negative 4, written as 10^{-4}. When a number less than 1 is written in scientific notation, the power of 10 is negative.

Standard Number		Scientific Notation		
0.00086	=	8.6	$\times$	10^{-4}

4 places $\rightarrow$ Coefficient   Power of 10

TABLE 1.2 gives some examples of numbers written as positive and negative powers of 10. The powers of 10 are a way of keeping track of the decimal point in the number. **TABLE 1.3** gives several examples of writing measurements in scientific notation.

TABLE 1.2 Some Powers of 10

Standard Number	Multiples of 10	Scientific Notation	
10 000	$10 \times 10 \times 10 \times 10$	1×10^4	Some positive powers of 10
1000	$10 \times 10 \times 10$	1×10^3	
100	10×10	1×10^2	
10	10	1×10^1	
1	0	1×10^0	
0.1	$\dfrac{1}{10}$	1×10^{-1}	Some negative powers of 10
0.01	$\dfrac{1}{10} \times \dfrac{1}{10} = \dfrac{1}{100}$	1×10^{-2}	
0.001	$\dfrac{1}{10} \times \dfrac{1}{10} \times \dfrac{1}{10} = \dfrac{1}{1000}$	1×10^{-3}	
0.0001	$\dfrac{1}{10} \times \dfrac{1}{10} \times \dfrac{1}{10} \times \dfrac{1}{10} = \dfrac{1}{10\,000}$	1×10^{-4}	

A chickenpox virus has a diameter of 3×10^{-7} m.

TABLE 1.3 Some Measurements Written as Standard Numbers and in Scientific Notation

Measured Quantity	Standard Number	Scientific Notation
Volume of gasoline used in the United States each year	550 000 000 000 L	5.5×10^{11} L
Diameter of Earth	12 800 000 m	1.28×10^7 m
Average volume of blood pumped in 1 day	8500 L	8.5×10^3 L
Time for light to travel from the Sun to Earth	500 s	5×10^2 s
Mass of a typical human	68 kg	6.8×10^1 kg
Mass of stirrup bone in ear	0.003 g	3×10^{-3} g
Diameter of a chickenpox (*Varicella zoster*) virus	0.000 000 3 m	3×10^{-7} m
Mass of bacterium (mycoplasma)	0.000 000 000 000 000 000 1 kg	1×10^{-19} kg

---

SAMPLE PROBLEM 1.8   Writing a Number in Scientific Notation

TRY IT FIRST

Write each of the following in scientific notation:

a. 3500   **b.** 0.000 016

SOLUTION

ANALYZE THE PROBLEM	Given	Need	Connect
	standard number	scientific notation	coefficient is at least 1 but less than 10

a. 3500

STEP 1   **Move the decimal point to obtain a coefficient that is at least 1 but less than 10.** For a number greater than 1, the decimal point is moved to the left three places to give a coefficient of 3.5.

STEP 2   **Express the number of places moved as a power of 10.** Moving the decimal point three places to the left gives a power of 3, written as 10^3.

STEP 3   **Write the product of the coefficient multiplied by the power of 10.**
3.5×10^3

b. 0.000 016

STEP 1   **Move the decimal point to obtain a coefficient that is at least 1 but less than 10.** For a number less than 1, the decimal point is moved to the right five places to give a coefficient of 1.6.

STEP 2   **Express the number of places moved as a power of 10.** Moving the decimal point five places to the right gives a power of negative 5, written as 10^{-5}.

STEP 3   **Write the product of the coefficient multiplied by the power of 10.**
1.6×10^{-5}

SELF TEST 1.8

Write each of the following in scientific notation:

a. 425 000   **b.** 0.000 000 86   **c.** 0.007 30   **d.** 978×10^5

ANSWER

a. 4.25×10^5 **b.** 8.6×10^{-7} **c.** 7.30×10^{-3} **d.** 9.78×10^7

PRACTICE PROBLEMS

Try Practice Problems 1.27 and 1.28

Scientific Notation and Calculators

You can enter a number in scientific notation on many calculators using the [EE or EXP] key. After you enter the coefficient, press the [EE or EXP] key and enter the power 10. To enter a negative power of 10, press the [+/−] key or the [−] key, depending on your calculator.

Number to Enter	Procedure	Calculator Display
4×10^6	4 [EE or EXP] 6	4 06 or 4⁰⁶ or 4E06
2.5×10^{-4}	2.5 [EE or EXP] [+/−] 4	2.5−04 or 2.5⁻⁰⁴ or 2.5E−04

When a calculator answer appears in scientific notation, the coefficient is shown as a number that is at least 1 but less than 10, followed by a space or E and the power of 10. To express this display in scientific notation, write the coefficient value, write $\times$ 10, and use the power of 10 as an exponent.

Calculator Display	Expressed in Scientific Notation
7.52 04 or 7.52⁰⁴ or 7.52E04	7.52×10^4
5.8−02 or 5.8⁻⁰² or 5.8 E−02	5.8×10^{-2}

ENGAGE 1.12

Describe how you enter a number in scientific notation on your calculator.

On many calculators, a number is converted into scientific notation using the appropriate keys. For example, the number 0.000 52 is entered, followed by pressing the 2nd or 3rd function key (2nd F) and the SCI key. The scientific notation appears in the calculator display as a coefficient and the power of 10.

0.000 52 [2ndF][SCI]   =   5.2−04 or 5.2⁻⁰⁴ or 5.2E−04   = 5.2×10^{-4}
Calculator display

Converting Scientific Notation to a Standard Number

When a number written in scientific notation has a positive power of 10, the standard number is obtained by moving the decimal point to the right for the same number of places as the power of 10. Placeholder zeros are used, as needed, to give additional places.

Scientific Notation   **Standard Number**

4.3×10^2 =   4.30   =   430

For a number with a negative power of 10, the standard number is obtained by moving the decimal point to the left for the same number of places as the power of 10. Placeholder zeros are added in front of the coefficient as needed.

Scientific Notation   **Standard Number**

$$2.5 \times 10^{-5} = 000002.5 = 0.000\ 025$$

SAMPLE PROBLEM 1.9   Writing Scientific Notation as a Standard Number

TRY IT FIRST

Write each of the following as a standard number:

a. 7.2×10^{-3} **b.** 2.4×10^{5}

SOLUTION

a. To write the standard number for an exponential number with a negative power of 10, move the decimal point to the left the same number of places (three) as the power of 10. Add placeholder zeros before the coefficient as needed.

$$7.2 \times 10^{-3} = 0007.2\,m = 0.0072$$

b. To write the standard number for an exponential number with a positive power of 10, move the decimal point to the right the same number of places (five) as the power of 10. Add placeholder zeros following the coefficient as needed.

$$2.4 \times 10^{5} = 2.40000 = 240\,000$$

SELF TEST 1.9

Write each of the following as a standard number:

a. 7.25×10^{-4} **b.** 3.05×10^{6}

ANSWER

a. 0.000 725   **b.** 3 050 000

PRACTICE PROBLEMS
Try Practice Problems 1.29 and 1.30

PRACTICE PROBLEMS

1.5 Writing Numbers in Scientific Notation

1.27 Write each of the following in scientific notation:
  a. 55 000
  b. 480
  c. 0.000 005
  d. 0.000 14
  e. 0.0072
  f. 670 000

1.28 Write each of the following in scientific notation:
  a. 180 000 000
  b. 0.000 06
  c. 750
  d. 0.15
  e. 0.024
  f. 1500

1.29 Write each of the following as a standard number:
  a. 1.2×10^{4}
  b. 8.25×10^{-2}
  c. 4×10^{6}
  d. 5.8×10^{-3}

1.30 Write each of the following as a standard number:
  a. 3.6×10^{-5}
  b. 8.75×10^{4}
  c. 3×10^{-2}
  d. 2.12×10^{5}

UPDATE   Forensic Evidence Helps Solve the Crime

Using a variety of laboratory tests, Sarah finds ethylene glycol in Gloria's blood. The quantitative tests indicate that Gloria had ingested 125 g of ethylene glycol. Sarah determines that the liquid in a glass found in Gloria's home was ethylene glycol that had been added to an alcoholic beverage. Ethylene glycol is a clear, sweet-tasting, thick liquid that is odorless and mixes with water. It is easy to obtain since it is used as antifreeze in automobiles and in brake fluid. Because the initial symptoms of ethylene glycol poisoning are similar to being intoxicated, the victim is often unaware of its presence.

If ingestion of ethylene glycol occurs, it can cause depression of the central nervous system, cardiovascular damage, and kidney failure. If discovered quickly, hemodialysis may be used to remove ethylene glycol from the blood. A toxic amount of ethylene glycol is 1.5 g of ethylene glycol/kg of body mass. Thus, 75 g could be fatal for a 50-kg (110-lb) person.

Sarah determines that fingerprints on the glass containing the ethylene glycol were those of Gloria's husband. This evidence along with the container of antifreeze found in the home led to the arrest and conviction of the husband for poisoning his wife.

Applications

1.31 Identify each of the following comments in the police report as an observation, a hypothesis, an experiment, or a conclusion:
  a. Gloria may have had a heart attack.
  b. Test results indicate that Gloria was poisoned.
  c. The liquid in the glass was analyzed.
  d. The antifreeze in the pantry was the same color as the liquid in the glass.

1.32 Identify each of the following comments in the police report as an observation, a hypothesis, an experiment, or a conclusion:
  a. Gloria may have committed suicide.
  b. Sarah ran blood tests to identify any toxic substances.
  c. The temperature of Gloria's body was 34 °C.
  d. The fingerprints found on the glass were determined to be her husband's.

1.33 A container was found in the home of the victim that contained 120 g of ethylene glycol in 450 g of liquid. What was the percentage of ethylene glycol? Express your answer to the ones place.

1.34 If the toxic quantity is 1.5 g of ethylene glycol per 1000 g of body mass, what percentage of ethylene glycol is fatal?

CONCEPT MAP

CHEMISTRY IN OUR LIVES

deals with	uses the	is learned by	uses key math skills
Substances	**Scientific Method**	**Reading the Text**	**Identifying Place Values**
called	*starting with*	**Practicing Problem Solving**	**Using Positive and Negative Numbers**
Chemicals	**Observations**	**Self-Testing**	**Calculating Percentages**
	that lead to	**Working with a Group**	
	Hypothesis		**Solving Equations**
	Experiments	**Answering the Engage Questions**	**Converting Between Standard Numbers and Scientific Notation**
	Conclusion	**Trying It First**	

CHAPTER REVIEW

1.1 Chemistry and Chemicals

LEARNING GOAL Define the term chemistry, and identify chemicals.

- Chemistry is the study of the composition, structure, properties, and reactions of matter.
- A chemical always has the same composition and properties wherever it is found.

1.2 Scientific Method: Thinking Like a Scientist
LEARNING GOAL Describe the scientific method.

- The scientific method is a process of explaining natural phenomena beginning with making observations, forming a hypothesis, and performing experiments.
- After repeated successful experiments, a conclusion may be drawn that the hypothesis is valid.

1.3 Studying and Learning Chemistry

LEARNING GOAL Identify strategies that are effective for learning. Develop a study plan for learning chemistry.

- A plan for learning chemistry utilizes the features in the text that help develop a successful approach to learning chemistry.

- By using the Learning Goals, Reviews, Engage questions, Analyze the Problems, Try It First, Sample Problems, and Self Tests in the chapter, and Practice Problems at the end of each section, you can successfully learn the concepts of chemistry.

1.4 Key Math Skills for Chemistry

LEARNING GOAL Review math concepts used in chemistry: place values, positive and negative numbers, percentages, solving equations, and interpreting graphs.
- Solving chemistry problems involves a number of math skills: identifying place values, using positive and negative numbers, calculating percentages, solving equations, and interpreting graphs.

1.5 Writing Numbers in Scientific Notation
LEARNING GOAL Write a number in scientific notation.

- A number written in scientific notation has two parts, a coefficient and a power of 10.
- When a number greater than 1 is converted to scientific notation, the power of 10 is positive.
- When a number less than 1 is written in scientific notation, the power of 10 is negative.

KEY TERMS

chemical A substance that has the same composition and properties wherever it is found.

chemistry The study of the composition, structure, properties, and reactions of matter.

conclusion An explanation of an observation that has been validated by repeated experiments that support a hypothesis.

experiment A procedure that tests the validity of a hypothesis.

hypothesis An unverified explanation of a natural phenomenon.

observation Information determined by noting and recording a natural phenomenon.

scientific method The process of making observations, proposing a hypothesis, and testing the hypothesis; after repeated experiments validate the hypothesis, a conclusion may be drawn as to its validity.

scientific notation A form of writing large and small numbers using a coefficient that is at least 1 but less than 10, followed by a power of 10.

KEY MATH SKILLS

The chapter section containing each Key Math Skill is shown in parentheses at the end of each heading.

Identifying Place Values (1.4)
- The place value identifies the numerical value of each digit in a number.

Example: Identify the place value for each of the digits in the number 456.78.

Answer:

Digit	Place Value
4	hundreds
5	tens
6	ones
7	tenths
8	hundredths

Using Positive and Negative Numbers in Calculations (1.4)

- A *positive number* is any number that is greater than zero and has a positive sign (+), which is often omitted. A *negative number* is any number that is less than zero and is written with a negative sign (−).
- When two positive numbers are added, multiplied, or divided, the answer is positive.
- When two negative numbers are multiplied or divided, the answer is positive. When two negative numbers are added, the answer is negative.
- When a positive and a negative number are multiplied or divided, the answer is negative.
- When a positive and a negative number are added, the smaller number is subtracted from the larger number, and the result has the same sign as the larger number.
- When two numbers are subtracted, change the sign of the number to be subtracted, then follow the rules for addition.

Example: Evaluate each of the following:
  a. $-8 - 14 =$ _____   **b.** $6 \times (-3) =$ _____
Answer: **a.** -22 **b.** -18

Calculating Percentages (1.4)

- A percentage is the part divided by the total (whole) multiplied by 100%.

Example: A drawer contains 6 white socks and 18 black socks. What is the percentage of white socks?

Answer: $\dfrac{6 \text{ white socks}}{24 \text{ total socks}} \times 100\% = 25\%$ white socks

Solving Equations (1.4)

An equation in chemistry often contains an unknown. To rearrange an equation to obtain the unknown factor by itself, you keep it balanced by performing matching mathematical operations on both sides of the equation.

- If you eliminate a number or symbol by subtracting, subtract that same number or symbol from both sides.
- If you eliminate a number or symbol by adding, add that same number or symbol to both sides.
- If you cancel a number or symbol by dividing, divide both sides by that same number or symbol.
- If you cancel a number or symbol by multiplying, multiply both sides by that same number or symbol.

Example: Solve the equation for a: $3a - 8 = 28$
Answer: *Add 8 to both sides* $3a - 8 + 8 = 28 + 8$
                                 $3a = 36$

         Divide both sides by 3 $\dfrac{3a}{3} = \dfrac{36}{3}$
                                 $a = 12$
Check: $3(12) - 8 = 28$
                           $36 - 8 = 28$
                               $28 = 28$

Your answer $a = 12$ is correct.

Interpreting Graphs (1.4)

- A graph represents the relationship between two variables.
- The quantities are plotted along two perpendicular axes, which are the x axis (horizontal) and y axis (vertical).
- The title indicates the components of the x and y axes.
- Numbers on the x and y axes show the range of values of the variables.
- The graph shows the relationship between the component on the y axis and that on the x axis.

Example: Solubility of Sugar in Water versus Temperature

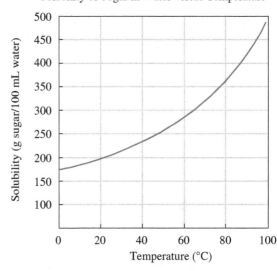

  a. Does the amount of sugar that dissolves in 100 mL of water increase or decrease when the temperature increases?
  b. How many grams of sugar dissolve in 100 mL of water at 70 °C?
  c. At what temperature (°C) will 275 g of sugar dissolve in 100 mL of water?

Answer: **a.** increase
  b. 320 g
  c. 55 °C

Writing Numbers in Scientific Notation (1.5)

- A number written in scientific notation consists of a coefficient and a power of 10.

A number is written in scientific notation by:

- Moving the decimal point to obtain a coefficient that is at least 1 but less than 10.
- Expressing the number of places moved as a power of 10. The power of 10 is positive if the decimal point is moved to the left, negative if the decimal point is moved to the right.

Example: Write the number 28 000 in scientific notation.
Answer: Moving the decimal point four places to the left gives a coefficient of 2.8 and a positive power of 10, 10^4. The number 28 000 written in scientific notation is 2.8×10^4.

UNDERSTANDING THE CONCEPTS

The chapter sections to review are shown in parentheses at the end of each problem.

1.35 A "chemical-free" shampoo includes the following ingredients: water, cocamide, glycerin, and citric acid. Is the shampoo truly "chemical-free"? (1.1)

1.36 A "chemical-free" sunscreen includes the following ingredients: titanium dioxide, vitamin E, and vitamin C. Is the sunscreen truly "chemical-free"? (1.1)

1.37 According to Sherlock Holmes, "One must follow the rules of scientific inquiry, gathering, observing, and testing data, then formulating, modifying, and rejecting hypotheses, until only one remains." Did Holmes use the scientific method? Why or why not? (1.2)

1.38 In *A Scandal in Bohemia*, Sherlock Holmes receives a mysterious note. He states, "I have no data yet. It is a capital mistake to theorize before one has data. Insensibly one begins to twist facts to suit theories, instead of theories to suit facts." What do you think Holmes meant? (1.2)

1.39 For each of the following, indicate if the answer has a positive or negative sign: (1.4)

   a. Two negative numbers are added.
   b. A positive and negative number are multiplied.

1.40 For each of the following, indicate if the answer has a positive or negative sign: (1.4)

   a. A negative number is subtracted from a positive number.
   b. Two negative numbers are divided.

Applications

1.41 Classify each of the following statements as an observation or a hypothesis: (1.2)

   a. A patient breaks out in hives after receiving penicillin.
   b. Dinosaurs became extinct when a large meteorite struck Earth and caused a huge dust cloud that severely decreased the amount of light reaching Earth.
   c. A patient's blood pressure was 110/75.

1.42 Classify each of the following statements as an observation or a hypothesis: (1.2)

   a. Analysis of 10 ceramic dishes showed that four dishes contained lead levels that exceeded federal safety standards.
   b. Marble statues undergo corrosion in acid rain.
   c. A child with a high fever and a rash may have chickenpox.

ADDITIONAL PRACTICE PROBLEMS

1.43 Select the correct phrase(s) to complete the following statement: If experimental results do not support your hypothesis, you should (1.2)

   a. pretend that the experimental results support your hypothesis
   b. modify your hypothesis
   c. do more experiments

1.44 Select the correct phrase(s) to complete the following statement: A hypothesis is supported when (1.2)

   a. one experiment proves the hypothesis
   b. many experiments validate the hypothesis
   c. you think your hypothesis is correct

1.45 Identify each of the following as an observation, a hypothesis, an experiment, or a conclusion: (1.2)

   a. During an assessment in the emergency room, a nurse writes that the patient has a resting pulse of 30 beats/min.
   b. A nurse thinks that an incision from a recent surgery that is red and swollen is infected.
   c. Repeated studies show that lowering sodium in the diet leads to a decrease in blood pressure.

1.46 Identify each of the following as an observation, a hypothesis, an experiment, or a conclusion: (1.2)

   a. Drinking coffee at night keeps me awake.
   b. If I stop drinking coffee in the afternoon, I will be able to sleep at night.
   c. I will try drinking coffee only in the morning.

1.47 Which of the following would be part of a successful study plan? (1.3)

   a. skipping class and just reading the text
   b. working the Sample Problems as you go through a chapter

   c. self-testing
   d. reading through the chapter, but working the problems later
   e. reading the assignment before class

1.48 Which of the following would be part of a successful study plan? (1.3)

   a. studying all night before the exam
   b. highlighting important ideas in the text
   c. working problems in a notebook for easy reference
   d. forming a study group and discussing the problems together
   e. reviewing Key Math Skills and Core Chemistry Skills

1.49 Evaluate each of the following: (1.4)

   a. $4 \times (-8) =$ _____   **b.** $-12 - 48 =$ _____
   c. $\dfrac{-168}{-4} =$ _____

1.50 Evaluate each of the following: (1.4)

   a. $-95 - (-11) =$ _____   **b.** $\dfrac{152}{-19} =$ _____
   c. $4 - 56 =$ _____

1.51 A bag of gumdrops contains 16 orange gumdrops, 8 yellow gumdrops, and 16 black gumdrops. (1.4)

   a. What is the percentage of yellow gumdrops? Express your answer to the ones place.
   b. What is the percentage of black gumdrops? Express your answer to the ones place.

1.52 On the first chemistry test, 12 students got As, 18 students got Bs, and 20 students got Cs. (1.4)

   a. What is the percentage of students who received Bs? Express your answer to the ones place.
   b. What is the percentage of students who received Cs? Express your answer to the ones place.

1.53 Write each of the following in scientific notation: (1.5)

    a. 120 000   **b.** 0.000 000 34

    c. 0.066   **d.** 2700

1.54 Write each of the following in scientific notation: (1.5)

    a. 0.0042   **b.** 310

    c. 890 000 000   **d.** 0.000 000 056

1.55 Write each of the following as a standard number: (1.5)

    a. 2.6×10^{-5} **b.** 6.5×10^{2}

    c. 3.7×10^{-1} **d.** 5.3×10^{5}

1.56 Write each of the following as a standard number: (1.5)

    a. 7.2×10^{-2} **b.** 1.44×10^{3}

    c. 4.8×10^{-4} **d.** 9.1×10^{6}

Applications

1.57 Identify each of the following as an observation, a hypothesis, an experiment, or a conclusion: (1.2)

    a. A patient has a high fever and a rash on her back.

    b. A nurse tells a patient that her baby who gets sick after drinking milk may be lactose intolerant.

    c. Numerous studies have shown that omega-3 fatty acids lower triglyceride levels.

1.58 Identify each of the following as an observation, a hypothesis, an experiment, or a conclusion: (1.2)

    a. Every spring, you have congestion and a runny nose.

    b. An overweight patient decides to exercise more to lose weight.

    c. Many research studies have linked obesity to heart disease.

CHALLENGE PROBLEMS

The following problems are related to the topics in this chapter. However, they do not all follow the chapter order, and they require you to combine concepts and skills from several sections. These problems will help you increase your critical thinking skills and prepare for your next exam.

1.59 Classify each of the following as an observation, a hypothesis, an experiment, or a conclusion: (1.2)

    a. The bicycle tire is flat.

    b. If I add air to the bicycle tire, it will expand to the proper size.

    c. When I added air to the bicycle tire, it was still flat.

    d. The bicycle tire has a leak in it.

1.60 Classify each of the following as an observation, a hypothesis, an experiment, or a conclusion: (1.2)

    a. A big log in the fire does not burn well.

    b. If I chop the log into smaller wood pieces, it will burn better.

    c. The small wood pieces burn brighter and make a hotter fire.

    d. The small wood pieces are used up faster than burning the big log.

1.61 Solve each of the following for x: (1.4)

    a. $2x + 5 = 41$ **b.** $\dfrac{5x}{3} = 40$

1.62 Solve each of the following for z: (1.4)

    a. $3z - (-6) = 12$ **b.** $\dfrac{4z}{-12} = -8$

Use the following graph for problems 1.63 and 1.64:

Solubility of Carbon Dioxide in Water versus Temperature

1.63 **a.** What does the title indicate about the graph? (1.4)

    b. What is measured on the vertical axis?

    c. What is the range of values on the vertical axis?

    d. Does the solubility of carbon dioxide increase or decrease with an increase in temperature?

1.64 **a.** What is measured on the horizontal axis? (1.4)

    b. What is the range of values on the horizontal axis?

    c. What is the solubility of carbon dioxide in water at 25 °C?

    d. At what temperature does carbon dioxide have a solubility of 0.20 g/100 g water?

ANSWERS TO ENGAGE QUESTIONS

1.1 A chemical has the same composition and properties wherever it is found.

1.2 An experiment is done to find a relationship between the hypothesis and the observations.

1.3 Self-testing practices the retrieval of new information and helps connect to prior knowledge.

1.4 The value 0 is located in the first place to the right of the decimal point, which is the tenths place.

1.5 Addition of a negative number and a positive number gives an answer that has the sign of the larger value. Addition of two negative numbers gives a larger number with a negative sign.

1.6 The value of 100% represents exactly 100 units of the whole.

1.7 Subtracting 8 from both sides of the equation places like terms on the same sides.

1.8 Dividing both sides of the equation solves the equation for the unknown V_2.

1.9 A graph represents the relationship between two variables plotted in equal intervals along the x axis (horizontal) and the y axis (vertical).

1.10 The decimal point is moved five places to the left to give a coefficient that is at least 1 but less than 10 and a power of 10 that is a positive 5.

1.11 The decimal point is moved five places to the right to give a coefficient that is at least 1 but less than 10, and a power of 10 that is a negative 5.

1.12 Enter the coefficient, press the EE or EXP key, and enter the power of 10. To enter a negative power of 10, press the +/− key or the − key.

ANSWERS TO SELECTED PROBLEMS

1.1   **a.** Chemistry is the study of the composition, structure, properties, and reactions of matter.
     b. A chemical is a substance that has the same composition and properties wherever it is found.

1.3   Many chemicals are listed on a multivitamin bottle, such as vitamin A, vitamin B_3, vitamin B_{12}, vitamin C, and folic acid.

1.5   Typical items found in a drugstore, and some of the chemicals they contain, are as follows:

    Antacid tablets: calcium carbonate, cellulose, starch, stearic acid, silicon dioxide

    Mouthwash: water, alcohol, thymol, glycerol, sodium benzoate, benzoic acid

    Cough suppressant: menthol, beta-carotene, sucrose, glucose

1.7   **a.** observation   **b.** hypothesis
     c. experiment   **d.** observation
     e. observation   **f.** conclusion

1.9   **a.** observation   **b.** hypothesis
     c. experiment   **d.** experiment

1.11 There are several things you can do that will help you successfully learn chemistry: forming a study group, retesting, working Sample Problems before reading the given answer, working the Self Test items, answering Engage questions as you study, working Practice Problems and checking Answers, reading the assignment ahead of class, and keeping a problem notebook.

1.13 **a, c,** and **e**

1.15 **a.** thousandths   **b.** ones   **c.** tenths
     d. hundreds   **e.** hundredths

1.17 **a.** 23   **b.** -30 **c.** -2
     d. 54   **e.** -12

1.19 **a.** The graph shows the relationship between the temperature of a cup of tea and time.
     b. temperature, in °C
     c. 20 °C to 90 °C
     d. decrease

1.21 **a.** 9   **b.** 42   **c.** 14   **d.** 3

1.23 **a.** 84%   **b.** 72%   **c.** 30%   **d.** 5%

1.25 **a.** This graph shows the relationship between body temperature and time since death.
     b. temperature, in °C
     c. 20 °C to 44 °C
     d. decrease

1.27 **a.** 5.5×10^4 **b.** 4.8×10^2 **c.** 5×10^{-6}
     d. 1.4×10^{-4} **e.** 7.2×10^{-3} **f.** 6.7×10^5

1.29 **a.** 12 000   **b.** 0.0825
     c. 4 000 000   **d.** 0.0058

1.31 **a.** hypothesis   **b.** conclusion
     c. experiment   **d.** observation

1.33 27% ethylene glycol

1.35 No. All of the ingredients are chemicals.

1.37 Yes. Sherlock's investigation includes making observations (gathering data), formulating a hypothesis, testing the hypothesis, and modifying it until one of the hypotheses is validated.

1.39 **a.** negative   **b.** negative

1.41 **a.** observation   **b.** hypothesis   **c.** observation

1.43 **b** and **c**

1.45 **a.** observation   **b.** hypothesis   **c.** conclusion

1.47 **b, c,** and **e**

1.49 **a.** -32 **b.** -60 **c.** 42

1.51 **a.** 20%   **b.** 40%

1.53 **a.** 1.2×10^5 **b.** 3.4×10^{-7}
     c. 6.6×10^{-2} **d.** 2.7×10^3

1.55 **a.** 0.000 026   **b.** 650
     c. 0.37   **d.** 530 000

1.57 **a.** observation   **b.** hypothesis   **c.** conclusion

1.59 **a.** observation   **b.** hypothesis
     c. experiment   **d.** conclusion

1.61 **a.** 18   **b.** 24

1.63 **a.** The graph shows the relationship between the solubility of carbon dioxide in water and temperature.
     b. solubility of carbon dioxide (g CO_2/100 g water)
     c. 0 to 0.35 g of CO_2/100 g of water
     d. decrease

Chemistry and Measurements

During the past few months, Greg has had an increased number of headaches, dizzy spells, and nausea. He goes to his doctor's office, where Sandra, the registered nurse, completes the initial part of his exam by recording several measurements: weight 164 lb, height 5 ft 8 in., temperature 37.2 °C, and blood pressure 155/95. Normal blood pressure is 120/80 or lower.

When Greg sees his doctor, he is diagnosed with high blood pressure, or hypertension. The doctor prescribes 80. mg of Inderal (propranolol), which is available in 40.-mg tablets. Inderal is a beta blocker, which relaxes the muscles of the heart. It is used to treat hypertension, angina (chest pain), arrhythmia, and migraine headaches.

Two weeks later, Greg visits his doctor again, who determines that Greg's blood pressure is now 152/90. The doctor increases the dosage of Inderal to 160. mg. Sandra informs Greg that he needs to increase his daily dosage from two tablets to four tablets.

CAREER

Registered Nurse

In addition to assisting physicians, registered nurses work to promote patient health and prevent and treat disease. They provide patient care and help patients cope with illness. They take measurements such as a patient's weight, height, temperature, and blood pressure; make conversions; and calculate drug dosage rates. Registered nurses also maintain detailed medical records of patient symptoms and prescribed medications.

UPDATE Greg's Visit with His Doctor

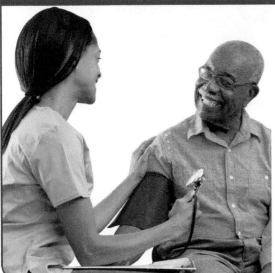

A few weeks later, Greg complains to his doctor that he is feeling tired. He has a blood test to determine if his iron level is low. You can see the results of Greg's blood serum iron level in the **UPDATE Greg's Visit with His Doctor**, page 59, and determine if Greg should be given an iron supplement.

2.1 Units of Measurement

LEARNING GOAL Write the names and abbreviations for the metric and SI units used in measurements of volume, length, mass, temperature, and time.

Think about your day. You probably took some measurements. Perhaps you checked your weight by stepping on a bathroom scale. If you made rice for dinner, you added two cups of water to one cup of rice. If you did not feel well, you may have taken your temperature. Whenever you take a measurement, you use a measuring device such as a scale, a measuring cup, or a thermometer.

Scientists and health professionals throughout the world use the **metric system** of measurement. It is also the common measuring system in all but a few countries in the world. The **International System of Units (SI)**, or Système International, is the official system of measurement throughout the world except for the United States. In chemistry, we use metric units and SI units for volume, length, mass, temperature, and time, as listed in TABLE 2.1.

Your weight on a bathroom scale is a measurement.

TABLE 2.1 Units of Measurement and Their Abbreviations

Measurement	Metric	SI
Volume	liter (L)	cubic meter (m^3)
Length	meter (m)	meter (m)
Mass	gram (g)	kilogram (kg)
Temperature	degree Celsius (°C)	kelvin (K)
Time	second (s)	second (s)

Suppose you walked 1.3 mi to campus today, carrying a backpack that weighs 26 lb. The temperature was 72 °F. Perhaps you weigh 128 lb and your height is 65 in. These measurements and units may seem familiar to you because they are stated in the U.S. system of measurement. However, in chemistry, we use the *metric system* in making our measurements. Using the metric system, you walked 2.1 km to campus, carrying a backpack that has a mass of 12 kg, when the temperature was 22 °C. You have a mass of 58.0 kg and a height of 1.7 m.

1.7 m (65 in.)

22 °C (72 °F)

58.0 kg (128 lb)

12 kg (26 lb)

2.1 km (1.3 mi)

There are many measurements in everyday life.

1000 mL = 1 L
946.4 mL = 1 qt

FIGURE 2.1 In the metric system, volume is based on the liter.

Volume

Volume (V) is the amount of space a substance occupies. The metric unit for volume is the **liter (L)**, which is slightly larger than a quart (qt). In a laboratory or a hospital, chemists work with metric units of volume that are smaller and more convenient, such as the **milliliter (mL)**. There are 1000 mL in 1 L (see FIGURE 2.1). Some relationships between units for volume are

$$1 \text{ L} = 1000 \text{ mL} \qquad 1 \text{ L} = 1.057 \text{ qt} \qquad 946.4 \text{ mL} = 1 \text{ qt}$$

FIGURE 2.2 Length in the metric system (SI) is based on the meter, which is slightly longer than a yard.

Length

The metric and SI unit of length is the **meter (m)**. The **centimeter (cm)**, a smaller unit of length, is commonly used in chemistry and is about equal to the width of your little finger (see **FIGURE 2.2**). Some relationships between units for length are

$$1\ m = 100\ cm \qquad 1\ m = 39.37\ in. \qquad 1\ m = 1.094\ yd \qquad 2.54\ cm = 1\ in.$$

FIGURE 2.3 On an electronic balance, the digital readout gives the mass of a nickel, which is 5.01 g.

Mass

The **mass** of an object is a measure of the quantity of material it contains. The SI unit of mass, the **kilogram (kg)**, is used for larger masses such as body mass. In the metric system, the unit for mass is the **gram (g)**, which is used for smaller masses. There are 1000 g in 1 kg. One pound (lb) is equal to 453.6 g. Some relationships between units for mass are

$$1\ kg = 1000\ g \qquad 1\ kg = 2.205\ lb \qquad 453.6\ g = 1\ lb$$

You may be more familiar with the term *weight* than with mass. Weight is a measure of the gravitational pull on an object. On Earth, an astronaut with a mass of 75.0 kg has a weight of 165 lb. On the Moon, where the gravitational pull is one-sixth that of Earth, the astronaut has a weight of 27.5 lb. However, the mass of the astronaut is the same as on Earth, 75.0 kg. Scientists measure mass rather than weight because mass does not depend on gravity. In a chemistry laboratory, an electronic balance is used to measure the mass in grams of a substance (see **FIGURE 2.3**).

FIGURE 2.4 A thermometer is used to determine temperature.

Temperature

Temperature tells us how hot something is, tells us how cold it is outside, or helps us determine if we have a fever (see **FIGURE 2.4**). In the metric system, temperature is measured using Celsius temperature. On the **Celsius (°C) temperature scale**, water freezes at 0 °C and boils at 100 °C, whereas on the Fahrenheit (°F) scale, water freezes at 32 °F and boils at 212 °F. In the SI system, temperature is measured using the **Kelvin (K) temperature scale** on which the lowest possible temperature is 0 K, known as absolute zero. A unit on the Kelvin scale is called a kelvin (K) and is not written with a degree sign.

Time

We typically measure time in units such as years (yr), days, hours (h), minutes (min), or seconds (s). Of these, the SI and metric unit of time is the **second (s)**. The standard now used to determine a second is an atomic clock. Some relationships between units for time are

$$1\ day = 24\ h \qquad 1\ h = 60\ min \qquad 1\ min = 60\ s$$

A stopwatch is used to measure the time of a race.

SAMPLE PROBLEM 2.1 Units of Measurement

TRY IT FIRST

On a typical day, a nurse encounters several situations involving measurement. State the name and type of measurement indicated by the units in each of the following:

a. A patient has a temperature of 38.5 °C.
b. A physician orders 1.5 g of cefuroxime for injection.
c. A physician orders 1 L of a sodium chloride solution to be given intravenously.
d. A medication is to be given to a patient every 4 h.

SOLUTION

a. A degree Celsius is a unit of temperature.
b. A gram is a unit of mass.
c. A liter is a unit of volume.
d. An hour is a unit of time.

SELF TEST 2.1

State the names and types of measurements for an infant that is 54.6 cm long with a mass of 5.2 kg.

ANSWER

A centimeter (cm) is a unit of length. A kilogram (kg) is a unit of mass.

PRACTICE PROBLEMS

Try Practice Problems 2.1 to 2.8

PRACTICE PROBLEMS

2.1 Units of Measurement

2.1 Write the abbreviation for each of the following:
  a. gram   **b.** degree Celsius
  c. liter   **d.** pound
  e. second

2.2 Write the abbreviation for each of the following:
  a. kilogram   **b.** kelvin
  c. quart   **d.** meter
  e. centimeter

2.3 State the type of measurement in each of the following statements:
  a. I put 12 L of gasoline in my gas tank.
  b. My friend is 170 cm tall.
  c. Earth is 385 000 km away from the Moon.
  d. The horse won the race by 1.2 s.

2.4 State the type of measurement in each of the following statements:
  a. I rode my bicycle 15 km today.
  b. My dog weighs 12 kg.
  c. It is hot today. It is 30 °C.
  d. I added 2 L of water to my fish tank.

2.5 State the name of the unit and the type of measurement indicated for each of the following quantities:
  a. 4.8 m   **b.** 325 g   **c.** 1.5 mL
  d. 4.8×10^2 s   **e.** 28 °C

2.6 State the name of the unit and the type of measurement indicated for each of the following quantities:
  a. 0.8 L   **b.** 3.6 cm   **c.** 4 kg
  d. 3.5 h   **e.** 373 K

Applications

2.7 On a typical day, medical personnel may encounter several situations involving measurement. State the name and type of measurement indicated by the units in each of the following:
  a. The clotting time for a blood sample is 12 s.
  b. A premature baby weighs 2.0 kg.
  c. An antacid tablet contains 1.0 g of calcium carbonate.
  d. An infant has a temperature of 39.2 °C.

2.8 On a typical day, medical personnel may encounter several situations involving measurement. State the name and type of measurement indicated by the units in each of the following:
  a. During open-heart surgery, the temperature of a patient is lowered to 29 °C.
  b. The circulation time of a red blood cell through the body is 20 s.
  c. A patient with a persistent cough is given 10. mL of cough syrup.
  d. The amount of iron in the red blood cells of the body is 2.5 g.

2.2 Measured Numbers and Significant Figures

REVIEW
Writing Numbers in Scientific
  Notation (1.5)

LEARNING GOAL Identify a number as measured or exact; determine the number of significant figures in a measured number.

When you make a measurement, you use some type of measuring device. For example, you may use a meterstick to measure your height, a scale to check your weight, or a thermometer to take your temperature.

Measured Numbers

Measured numbers are the numbers you obtain when you measure a quantity such as your height, weight, or temperature. Suppose you are going to measure the lengths of the objects in **FIGURE 2.5**. To report the length of the object, you observe the numerical values of the marked lines at the end of the object. Then you can *estimate* by visually dividing the space between the marked lines. This estimated value is the final digit in a measured number.

For example, in Figure 2.5a, the end of the object is between the marks of 4 cm and 5 cm, which means that the length is more than 4 cm but less than 5 cm. If you estimate that the end of the object is halfway between 4 cm and 5 cm, you would report its length as 4.5 cm. Another student might report the length of the same object as 4.4 cm because people do not estimate in the same way.

The metric ruler shown in Figure 2.5b is marked at every 0.1 cm. Now you can determine that the end of the object is between 4.5 cm and 4.6 cm. Perhaps you report its length as 4.55 cm, whereas another student reports its length as 4.56 cm. Both results are acceptable.

In Figure 2.5c, the end of the object appears to line up with the 3-cm mark. Because the end of the object is on the 3-cm mark, the estimated digit is 0, which means the measurement is reported as 3.0 cm.

(a) 4.5 cm

(b) 4.55 cm

(c) 3.0 cm

FIGURE 2.5 The lengths of the rectangular objects are measured in centimeters.

Significant Figures

In a measured number, the **significant figures (SFs)** are all the digits including the estimated digit. Nonzero numbers are always counted as significant figures. However, a zero may or may not be a significant figure depending on its position in a number. **TABLE 2.2** gives the rules and examples of counting significant figures.

CORE CHEMISTRY SKILL
Counting Significant Figures

TABLE 2.2 Significant Figures in Measured Numbers

Rule	Measured Number	Number of Significant Figures
1. A number is a *significant figure* if it is		
a. a nonzero digit	4.5 g	2
	122.35 m	5
b. a zero between nonzero digits	205 °C	3
	5.008 kg	4
c. a zero at the end of a decimal number	50. L	2
	16.00 mL	4
d. in the coefficient of a number written in scientific notation	4.8×10^5 m	2
	5.70×10^{-3} g	3
2. A zero is *not significant* if it is		
a. at the beginning of a decimal number	0.0004 s	1
	0.075 cm	2
b. used as a placeholder in a large number without a decimal point	850 000 m	2
	1 250 000 g	3

PRACTICE PROBLEMS
Try Practice Problems 2.9 to 2.14

Significant Zeros and Scientific Notation

In this text, we will place a decimal point after a significant zero at the end of a number. For example, if a measurement is written as 500. g, the decimal point after the second zero indicates that *both zeros* are significant. To show this more clearly, we can write it as 5.00×10^2 g. When the first zero in the measurement 300 m is a significant zero, but the second zero is not, the measurement is written as 3.0×10^2 m. We will assume that all zeros at the end of large standard numbers without a decimal point are not significant. Therefore, we write 400 000 g as 4×10^5 g, which has only one significant figure.

SAMPLE PROBLEM 2.2 Significant Figures

TRY IT FIRST

Write the number of significant figures in each of the following measured numbers:

a. 0.000 250 m **b.** 70.040 g **c.** 1.020×10^6 L

SOLUTION

ANALYZE THE PROBLEM	Given	Need	Connect
	measured number	number of significant figures	Table 2.2, rules for significant figures

a. Nonzero digits and zeros at the end of a decimal number are significant. 0.000 250 m has three SFs (Rules 1a, 1c).

b. Nonzero digits, zeros between nonzero digits, or at end of a decimal number are significant. 70.040 g has five SFs (Rules 1a, 1b, 1c).

c. All digits in the coefficient of a number written in scientific notation are significant. 1.020×10^6 L has four SFs (Rule 1d).

SELF TEST 2.2

Write the number of significant figures in each of the following measured numbers:

a. 0.040 08 m **b.** 6.00×10^3 g

ANSWER

a. four SFs (Rules 1a, 1b)
b. three SFs (Rule 1d)

Exact Numbers

Exact numbers *are those numbers obtained by counting items or using a definition that compares two units in the same measuring system.* Suppose a friend asks you how many classes you are taking. You would answer by counting the number of classes in your schedule. Suppose you want to state the number of seconds in one minute. Without using any measuring device, you would give the definition: There are 60 s in 1 min. *Exact numbers are not measured, do not have a limited number of significant figures, and do not affect the number of significant figures in a calculated answer.* For more examples of exact numbers, see **TABLE 2.3**.

The number of baseballs is counted, which means 2 is an exact number.

TABLE 2.3 Examples of Some Exact Numbers

Counted Numbers	Defined Equalities	
Items	Metric System	U.S. System
8 doughnuts	1 L = 1000 mL	1 ft = 12 in.
2 baseballs	1 m = 100 cm	1 qt = 4 cups
5 capsules	1 kg = 1000 g	1 lb = 16 oz

For example, a mass of 42.2 g and a length of 5.0×10^{-3} cm are measured numbers because they are obtained using measuring tools. There are three SFs in 42.2 g because all nonzero digits are always significant. There are two SFs in 5.0×10^{-3} cm because all the digits in the coefficient of a number written in scientific notation are significant. However, a quantity of three eggs is an exact number that is obtained by counting. In the equality 1 kg = 1000 g, the masses of 1 kg and 1000 g are both exact numbers because this is a definition in the metric system.

ENGAGE 2.2

Why is the 4 in 4 hats an exact number, whereas the 6.24 in 6.24 cm has three significant figures?

SAMPLE PROBLEM 2.3 Measured and Exact Numbers

TRY IT FIRST

Identify each of the following numbers as measured or exact, give the number of significant figures (SFs) in each of the measured numbers, and explain:

a. 0.170 L   **b.** 4 knives
c. 6.3×10^{-6} s   **d.** 1 m = 100 cm

SOLUTION

ANALYZE THE PROBLEM	Given	Need	Connect
	numbers	identify exact or measured, number of SFs	Table 2.2, rules for SFs in measured numbers

> **STEP 1** A measured number is the result of a measurement. An exact number is obtained by counting or by definition.

   a. measured   **b.** exact, counted
   c. measured   **d.** exact, definition

> **STEP 2** Significant figures are the nonzero digits and any zeros after nonzero digits in a decimal number. In numbers written in scientific notation, the numbers in the coefficient are all significant.

   a. three SFs
   c. two SFs

SELF TEST 2.3

Identify each of the following numbers as measured or exact, and give the number of significant figures (SFs) in each of the measured numbers:

a. 0.020 80 kg   **b.** 5.06×10^{4} h   **c.** 4 chemistry books

ANSWER

a. measured; four SFs   **b.** measured; three SFs   **c.** exact

PRACTICE PROBLEMS

Try Practice Problems 2.19 to 2.24

PRACTICE PROBLEMS

2.2 Measured Numbers and Significant Figures

2.9 Use the metric ruler to measure the length in each of the following, and write the measurement, the estimated digit, and the number of significant figures:

(a)

(b)

(c)

2.10 Determine the volume, in milliliters, of each liquid in the graduated cylinders by reading the lowest point of the curve of the liquid in the diagrams (a), (b), and (c) using the correct number of significant figures.

(a)   (b)   (c)

2.11 How many significant figures are in each of the following?
a. 11.005 g   b. 0.000 32 m
c. 36 000 000 km   d. 1.80×10^4 kg
e. 0.8250 L   f. 30.0 °C

2.12 How many significant figures are in each of the following?
a. 20.60 mL   b. 1036.48 kg
c. 4.00 m   d. 20 °C
e. 60 800 000 g   f. 5.0×10^{-3} L

2.13 In which of the following pairs do both numbers contain the same number of significant figures?
a. 11.0 m and 11.00 m
b. 0.0250 m and 0.205 m
c. 0.000 12 s and 12 000 s
d. 250.0 L and 2.5×10^{-2} L

2.14 In which of the following pairs do both numbers contain the same number of significant figures?
a. 0.005 75 g and 5.75×10^{-3} g
b. 405 K and 405.0 K
c. 150 000 s and 1.50×10^4 s
d. 3.8×10^{-2} L and 3.80×10^5 L

2.15 Indicate if the zeros are significant in each of the following measurements:
a. 0.0038 m   b. 5.04 cm
c. 800. L   d. 3.0×10^{-3} kg
e. 85 000 g

2.16 Indicate if the zeros are significant in each of the following measurements:
a. 20.05 °C   b. 5.00 m
c. 0.000 02 g   d. 120 000 yr
e. 8.05×10^2 L

2.17 Write each of the following in scientific notation with two significant figures:
a. 5000 L   b. 30 000 g
c. 100 000 m   d. 0.000 25 cm

2.18 Write each of the following in scientific notation with two significant figures:
a. 5 100 000 g   b. 26 000 s
c. 40 000 m   d. 0.000 820 kg

2.19 Identify the numbers in each of the following statements as measured or exact:
a. A patient has a mass of 67.5 kg.
b. A patient is given 2 tablets of medication.
c. In the metric system, 1 L is equal to 1000 mL.
d. The distance from Denver, Colorado, to Houston, Texas, is 1720 km.

2.20 Identify the numbers in each of the following statements as measured or exact:
a. There are 31 students in the laboratory.
b. The oldest known flower lived 1.20×10^8 yr ago.
c. The largest gem ever found, an aquamarine, has a mass of 104 kg.
d. A laboratory test shows a blood cholesterol level of 184 mg/dL.

2.21 Identify the measured number(s), if any, in each of the following pairs of numbers:
a. 3 hamburgers and 6 oz of hamburger
b. 1 table and 4 chairs
c. 0.75 lb of grapes and 350 g of butter
d. 60 s = 1 min

2.22 Identify the exact number(s), if any, in each of the following pairs of numbers:
a. 5 pizzas and 50.0 g of cheese
b. 6 nickels and 16 g of nickel
c. 3 onions and 3 lb of onions
d. 5 miles and 5 cars

Applications

2.23 Identify each of the following as measured or exact, and give the number of significant figures (SFs) in each measured number:
a. The mass of a neonate is 1.607 kg.
b. The Daily Value (DV) for iodine for an infant is 130 mcg.
c. There are 4.02×10^6 red blood cells in a blood sample.
d. In November, 23 babies were born in a hospital.

2.24 Identify each of the following as measured or exact, and give the number of significant figures (SFs) in each measured number:
a. An adult with the flu has a temperature of 103.5 °F.
b. A blister (push-through) pack of prednisone contains 21 tablets.
c. The time for a nerve impulse to travel from the feet to the brain is 0.46 s.
d. A brain contains 1.20×10^{10} neurons.

REVIEW

Identifying Place Values (1.4)

Using Positive and Negative
  Numbers in Calculations (1.4)

2.3 Significant Figures in Calculations

LEARNING GOAL Give the correct number of significant figures for a calculated answer.

In the sciences, we measure many things: the length of a bacterium, the volume of a gas sample, the temperature of a reaction mixture, or the mass of iron in a sample. The number of significant figures in measured numbers determines the number of significant figures in the calculated answer.

Using a calculator will help you perform calculations faster. However, calculators cannot think for you. It is up to you to enter the numbers correctly, press the correct function keys, and give the answer with the correct number of significant figures.

Rounding Off

Suppose you decide to buy carpeting for a room that has a length of 5.52 m and a width of 3.58 m. To determine how much carpeting you need, you would calculate the area of the room by multiplying 5.52 times 3.58 on your calculator. The calculator shows the number 19.7616 in its display. Because each of the original measurements has only three significant figures, the calculator display (19.7616) should be *rounded off* to three significant figures, 19.8.

$$5.52 \text{ m} \quad \times \quad 3.58 \text{ m} \quad = \quad 19.7616 \quad = \quad 19.8 \text{ m}^2$$

| Three SFs | Three SFs | Calculator display | Final answer, rounded off to three SFs |

Therefore, you can order carpeting that will cover an area of 19.8 m^2.

Each time you use a calculator, it is important to look at the original measurements and determine the number of significant figures that can be used for the answer. You can use the following rules to round off the numbers shown in a calculator display.

Rules for Rounding Off

1. If the first digit to be dropped is *4 or less*, it and all following digits are simply dropped from the number.
2. If the first digit to be dropped is *5 or greater*, the last retained digit of the number is increased by 1.

A technician uses a calculator in the laboratory.

TABLE 2.4 Rules for Rounding Off		
Number to Round Off	**Three Significant Figures**	**Two Significant Figures**
8.4234	8.42 (drop 34)	8.4 (drop 234)
14.780	14.8 (drop 80, increase the last retained digit by 1)	15 (drop 780, increase the last retained digit by 1)
3256	3260* (drop 6, increase the last retained digit by 1, add 0) (3.26 × 10³)	3300* (drop 56, increase the last retained digit by 1, add 00) (3.3 × 10³)
*The value of a large number is retained by using placeholder zeros to replace dropped digits.		

SAMPLE PROBLEM 2.4 Rounding Off

TRY IT FIRST

Round off each of the following numbers to three significant figures:

a. 35.7823 m   **b.** 0.002 621 7 L   **c.** 3.8268 × 10³ g

SOLUTION

ANALYZE THE PROBLEM	Given	Need	Connect
	measured numbers	three significant figures	Table 2.4, rules for rounding off

STEP 1 **Identify the digits to be dropped.**

a. drop 823   b. drop 17   c. drop 68

STEP 2 **Increase the last retained digit by 1 if the first digit dropped is 5 or greater. Keep the last retained digit if the first digit dropped is 4 or less.**

a. Increase the last retained digit by 1 to give 35.8 m.
b. Retain digits to give 0.002 62 L.
c. Increase the last retained digit by 1 to give 3.83×10^3 g.

SELF TEST 2.4

Round off each of the numbers in Sample Problem 2.4 to two significant figures.

ANSWER

a. 36 m   b. 0.0026 L   c. 3.8×10^3 g

PRACTICE PROBLEMS

Try Practice Problems 2.25 to 2.28

Multiplication and Division with Measured Numbers

CORE CHEMISTRY SKILL

Using Significant Figures in Calculations

In multiplication or division, the final answer is written so that it has the same number of significant figures (SFs) as the measurement with the fewest SFs.

Perform the following operations with measured numbers:

$$\frac{2.8 \times 67.40}{34.8} =$$

When the problem has multiple steps, the numbers in the numerator are multiplied and then divided by each of the numbers in the denominator.

2.8 $\boxed{\times}$ 67.40 $\boxed{\div}$ 34.8 $\boxed{=}$ 5.422988506 = 5.4
Two SFs   Four SFs   Three SFs   Calculator display   Answer, rounded off to two SFs

ENGAGE 2.4

Why is the answer for the multiplication of 0.3×52.6 written with one significant figure?

Because the calculator display has more digits than the significant figures in the measured numbers allow, we need to round it off. Using the measured number that has the smallest number (two) of significant figures, 2.8, we round off the calculator display to an answer with two SFs.

Adding Significant Zeros

Sometimes, a calculator display gives a small whole number. For example, suppose the calculator display is 4, but you used measurements that have three significant numbers. Then two significant zeros are *added* to give 4.00 as the correct answer.

Three SFs
$$\frac{8.00}{2.00} = \qquad 4. \quad = \quad 4.00$$
Three SFs   Calculator display   Final answer, two zeros added to give three SFs

A calculator is helpful in working problems and doing calculations faster.

SAMPLE PROBLEM 2.5 Significant Figures in Multiplication and Division

TRY IT FIRST

Perform the following calculations with measured numbers, and give the answers with the correct number of significant figures:

a. 56.8×0.37 b. $\dfrac{(2.075)(0.585)}{(8.42)(0.0245)}$ c. $\dfrac{25.0}{5.00}$

SOLUTION

ANALYZE THE PROBLEM	Given	Need	Connect
	multiplication and division	answer with the correct number of SFs	rules for multiplication/division

STEP 1 Determine the number of significant figures in each measured number.

a.
Three SFs Two SFs
a. 56.8 × 0.37

b.
Four SFs Three SFs
b. $\dfrac{(2.075)\,(0.585)}{(8.42)\,(0.0245)}$
Three SFs Three SFs

c.
Three SFs
c. $\dfrac{25.0}{5.00}$
Three SFs

STEP 2 Perform the indicated operations on the calculator.

a. 21.016
Calculator display

b. 5.884313345
Calculator display

c. 5.
Calculator display

STEP 3 Round off (or add zeros) to give the same number of significant figures as the measurement having the fewest significant figures.

a. 21

b. 5.88

c. 5.00

ENGAGE 2.5

Why does the calculator answer of 5 in part **c.** require two significant zeros to give an answer of 5.00?

SELF TEST 2.5

Perform the following calculations with measured numbers, and give the answers with the correct number of significant figures:

a. 45.26 × 0.010 88

b. 2.6 ÷ 324

c. $\dfrac{4.0 \times 8.00}{16}$

ANSWER

a. 0.4924

b. 0.0080 or 8.0×10^{-3}

c. 2.0

PRACTICE PROBLEMS

Try Practice Problems 2.29 and 2.30

Addition and Subtraction with Measured Numbers

In addition or subtraction, the final answer is written so that it has the same number of decimal places as the measurement having the fewest decimal places.

2.045	Thousandths place
+34.1	Tenths place (fewer decimal places)
36.145	Calculator display
36.1	Answer, rounded off to the tenths place

When numbers are added or subtracted to give an answer ending in zero, the zero does not appear after the decimal point in the calculator display. For example, $14.5\ \text{g} - 2.5\ \text{g} = 12.0\ \text{g}$. However, if you do the subtraction on your calculator, the display shows 12. To write the correct answer, a significant zero is written after the decimal point.

ENGAGE 2.6

Why is the answer 57.7 correct for the addition of 55.2 and 2.506?

SAMPLE PROBLEM 2.6 Decimal Places in Addition and Subtraction

TRY IT FIRST

Perform the following calculations with measured numbers, and give each answer with the correct number of decimal places:
a. 104 + 7.8 + 40   **b.** 153.247 − 14.82

SOLUTION

	Given	Need	Connect
ANALYZE THE PROBLEM	addition and subtraction	answer with the correct number of decimal places	rules for addition/ subtraction

STEP 1 Determine the number of decimal places in each measured number.

a.	104	Ones place	**b.**	153.247 Thousandths place
	7.8	Tenths place		−14.82 Hundredths place
	+40	Tens place (fewest decimal places)		(fewer decimal places)

STEP 2   Perform the indicated operations on the calculator.

a.   *151.8*
Calculator
display

b.   *138.427*
Calculator
display

STEP 3   Round off (or add zeros) to give the same number of decimal places as the measured number having the fewest decimal places.

a. 150 Rounded off to the tens place   b. 138.43 Rounded off to the hundredths place

SELF TEST 2.6

Perform the following operations with measured numbers, and give each answer with the correct number of decimal places:

a. 82.45 + 1.245 + 0.000 56   b. 4.259 − 3.8

ANSWER

a. 83.70   b. 0.5

PRACTICE PROBLEMS
Try Practice Problems 2.31 and 2.32

PRACTICE PROBLEMS

2.3 Significant Figures in Calculations

2.25 Round off each of the following calculator answers to three significant figures:
   a. 1.854 kg   b. 88.2038 L
   c. 0.004 738 265 cm   d. 8807 m
   e. 1.832×10^5 s

2.26 Round off each of the calculator answers in problem 2.25 to two significant figures.

2.27 Round off or add zeros to each of the following to give three significant figures:
   a. 56.855 m   b. 0.002 282 g
   c. 11 527 s   d. 8.1 L

2.28 Round off or add zeros to each of the following to give two significant figures:
   a. 3.2805 m
   b. 1.855×10^2 g
   c. 0.002 341 mL
   d. 2 L

2.29 Perform each of the following operations, and give each answer with the correct number of significant figures:
   a. 45.7×0.034 b. $0.002\ 78 \times 5$
   c. $\dfrac{34.56}{1.25}$ d. $\dfrac{(0.2465)\ (25)}{1.78}$
   e. $(2.8 \times 10^4)\ (5.05 \times 10^{-6})$ f. $\dfrac{(3.45 \times 10^{-2})\ (1.8 \times 10^5)}{(8 \times 10^3)}$

2.30 Perform each of the following operations, and give each answer with the correct number of significant figures:
   a. 400×185
   b. $\dfrac{2.40}{(4)\ (125)}$

   c. $0.825 \times 3.6 \times 5.1$
   d. $\dfrac{(3.5)\ (0.261)}{(8.24)\ (20.0)}$
   e. $\dfrac{(5 \times 10^{-5})\ (1.05 \times 10^4)}{(8.24 \times 10^{-8})}$
   f. $\dfrac{(4.25 \times 10^2)\ (2.56 \times 10^{-3})}{(2.245 \times 10^{-3})\ (56.5)}$

2.31 Perform each of the following operations, and give each answer with the correct number of decimal places:
   a. 45.48 + 8.057
   b. 23.45 + 104.1 + 0.025
   c. 145.675 − 24.2
   d. 1.08 − 0.585
   e. 2300 + 196.11
   f. 145.111 − 22.9 + 34.49

2.32 Perform each of the following operations, and give each answer with the correct number of decimal places:
   a. 5.08 + 25.1
   b. 85.66 + 104.10 + 0.025
   c. 24.568 − 14.25
   d. 0.2654 − 0.2585
   e. 66.77 + 17 − 0.33
   f. 460 − 33.77

2.4 Prefixes and Equalities

LEARNING GOAL Use the numerical values of prefixes to write a metric equality.

The special feature of the metric system is that a **prefix** can be placed in front of any unit to increase or decrease its size by some factor of 10. For example, the prefixes *milli* and *micro* are used to make the smaller units, milligram (mg) and microgram (μg).

The U.S. Food and Drug Administration (FDA) has determined the Daily Values (DV) for nutrients for adults and children age 4 or older. Examples of these recommended Daily Values, some of which use prefixes, are listed in **TABLE 2.5**.

The prefix *centi* is like cents in a dollar. One cent would be a "centidollar" or 0.01 of a dollar. That also means that one dollar is the same as 100 cents. The prefix *deci* is like dimes in a dollar. One dime would be a "decidollar" or 0.1 of a dollar. That also means that one dollar is the same as 10 dimes. **TABLE 2.6** lists some of the metric prefixes, their symbols, and their numerical values.

CORE CHEMISTRY SKILL
Using Prefixes

TABLE 2.5 Daily Values (DV) for Selected Nutrients

Nutrient	Amount Recommended
Calcium	1.0 g
Copper	2 mg
Iodine	150 μg (150 mcg)
Iron	18 mg
Magnesium	400 mg
Niacin	20 mg
Phosphorus	800 mg
Potassium	3.5 g
Selenium	70. μg (70. mcg)
Sodium	2.4 g
Zinc	15 mg

TABLE 2.6 Metric and SI Prefixes

Prefix	Symbol	Numerical Value	Scientific Notation	Equality
Prefixes That Increase the Size of the Unit				
peta	P	1 000 000 000 000 000	10^{15}	$1\,Pg = 1 \times 10^{15}\,g$ $1\,g = 1 \times 10^{-15}\,Pg$
tera	T	1 000 000 000 000	10^{12}	$1\,Ts = 1 \times 10^{12}\,s$ $1\,s = 1 \times 10^{-12}\,Ts$
giga	G	1 000 000 000	10^{9}	$1\,Gm = 1 \times 10^{9}\,m$ $1\,m = 1 \times 10^{-9}\,Gm$
mega	M	1 000 000	10^{6}	$1\,Mg = 1 \times 10^{6}\,g$ $1\,g = 1 \times 10^{-6}\,Mg$
kilo	k	1 000	10^{3}	$1\,km = 1 \times 10^{3}\,m$ $1\,m = 1 \times 10^{-3}\,km$
Prefixes That Decrease the Size of the Unit				
deci	d	0.1	10^{-1}	$1\,dL = 1 \times 10^{-1}\,L$ $1\,L = 10\,dL$
centi	c	0.01	10^{-2}	$1\,cm = 1 \times 10^{-2}\,m$ $1\,m = 100\,cm$
milli	m	0.001	10^{-3}	$1\,ms = 1 \times 10^{-3}\,s$ $1\,s = 1 \times 10^{3}\,ms$
micro	μ*	0.000 001	10^{-6}	$1\,\mu g = 1 \times 10^{-6}\,g$ $1\,g = 1 \times 10^{6}\,\mu g$
nano	n	0.000 000 001	10^{-9}	$1\,nm = 1 \times 10^{-9}\,m$ $1\,m = 1 \times 10^{9}\,nm$
pico	p	0.000 000 000 001	10^{-12}	$1\,ps = 1 \times 10^{-12}\,s$ $1\,s = 1 \times 10^{12}\,ps$
femto	f	0.000 000 000 000 001	10^{-15}	$1\,fs = 1 \times 10^{-15}\,s$ $1\,s = 1 \times 10^{15}\,fs$

*In medicine, the abbreviation *mc* is used for the prefix *micro* because the symbol μ may be misread, which could result in a medication error. Thus, 1 μg would be written as 1 mcg.

ENGAGE 2.7
Why is 60. mg of vitamin C the same as 0.060 g of vitamin C?

PRACTICE PROBLEMS
Try Practice Problems 2.33 to 2.40

The relationship of a prefix to a unit can be expressed by replacing the prefix with its numerical value. For example, when the prefix *kilo* in kilometer is replaced with its value of 1000, we find that a kilometer is equal to 1000 m. Other examples follow:

1 **kilo**meter (1 km) = **1000** meters (1000 m = 10^{3} m)

1 **kilo**liter (1 kL) = **1000** liters (1000 L = 10^{3} L)

1 **kilo**gram (1 kg) = **1000** grams (1000 g = 10^{3} g)

An endoscope has a video camera with a width of 1 mm attached to the end of a thin cable.

Equalities Between Prefixes

Equalities can also be written for two prefix units. Suppose you need an equality for milliliters and deciliters. From Table 2.6, using L as the base unit, we can write the equalities 1 L = 10 dL and 1 L = 1000 mL and then write an equality for dL and mL.

$$1 \text{ L} = 10 \text{ dL} = 1000 \text{ mL} \quad \text{or} \quad 1 \text{ dL} = 100 \text{ mL}$$

SAMPLE PROBLEM 2.7 Prefixes and Equalities

TRY IT FIRST

The camera in an endoscope has a width of 1 mm. Complete each of the following equalities involving millimeters:

a. 1 m = _____ mm   **b.** 1 cm = _____ mm

SOLUTION

a. 1 m = 1000 mm   **b.** 1 cm = 10 mm

SELF TEST 2.7

a. What is the relationship between liters and centiliters?
b. What is the relationship between millimeters and micrometers?

ANSWER

a. 1 L = 100 cL   **b.** 1 mm = 1000 μm (mcm)

First quantity   Second quantity

$$\underset{\underset{\text{Number + unit}}{\uparrow \qquad \uparrow}}{1 \qquad \text{m}} = \underset{\underset{\text{Number + unit}}{\uparrow \qquad \uparrow}}{100 \qquad \text{cm}}$$

This example of an equality shows the relationship between meters and centimeters.

ENGAGE 2.8

Why can the relationship of centimeters and meters be written as 1 m = 100 cm or 0.01 m = 1 cm?

Measuring Length

An ophthalmologist may measure the diameter of the retina of an eye in centimeters (cm), whereas a surgeon may need to know the length of a nerve in millimeters (mm). When the prefix *centi* is used with the unit meter, it becomes *centimeter*, a length that is one-hundredth of a meter (0.01 m). When the prefix *milli* is used with the unit meter, it becomes *millimeter*, a length that is one-thousandth of a meter (0.001 m). There are 100 cm and 1000 mm in a meter.

If we compare the lengths of a millimeter and a centimeter, we find that 1 mm is 0.1 cm; there are 10 mm in 1 cm. These comparisons are examples of **equalities**, which show the relationship between two units that measure the same quantity. Examples of equalities between different metric units of length follow:

$$1 \text{ m} = 100 \text{ cm} = 1 \times 10^2 \text{ cm}$$
$$1 \text{ m} = 1000 \text{ mm} = 1 \times 10^3 \text{ mm}$$
$$1 \text{ cm} = 10 \text{ mm} = 1 \times 10^1 \text{ mm}$$

Some metric units for length are compared in **FIGURE 2.6**.

FIGURE 2.6 The metric length of 1 m is the same length as 10 dm, 100 cm, or 1000 mm.

Measuring Volume

Volumes smaller than 1 L are common in the health sciences. When a liter is divided into 10 equal portions, each portion is a deciliter (dL). There are 10 dL in 1 L. Laboratory results for bloodwork are often reported in mass per deciliter. **TABLE 2.7** lists normal laboratory test values for some substances in the blood.

FIGURE 2.7 A plastic intravenous fluid container contains 1000 mL.

TABLE 2.7 Some Normal Laboratory Test Values	
Substance in Blood	**Normal Range**
Albumin	3.5–5.4 g/dL
Ammonia	20–70 µg/dL (mcg/dL)
Calcium	8.5–10.5 mg/dL
Cholesterol	105–250 mg/dL
Iron (male)	80–160 µg/dL (mcg/dL)
Protein (total)	6.0–8.5 g/dL

When a liter is divided into a thousand parts, each of the smaller volumes is a milliliter (mL). In a 1-L container of physiological saline, there are 1000 mL of solution (see **FIGURE 2.7**). Examples of equalities between different metric units of volume follow:

$$1 \text{ L} = 10 \text{ dL} \qquad = 1 \times 10^1 \text{ dL}$$
$$1 \text{ L} = 1000 \text{ mL} \qquad = 1 \times 10^3 \text{ mL}$$
$$1 \text{ dL} = 100 \text{ mL} \qquad = 1 \times 10^2 \text{ mL}$$
$$1 \text{ mL} = 1000 \text{ } \mu\text{L (mcL)} = 1 \times 10^3 \text{ } \mu\text{L (mcL)}$$

The **cubic centimeter** (abbreviated as **cm³** or **cc**) is the volume of a cube whose dimensions are 1 cm on each side. A cubic centimeter has the same volume as a milliliter, and the units are often used interchangeably.

$$1 \text{ cm}^3 = 1 \text{ cc} = 1 \text{ mL}$$

When you see *1 cm*, you are reading about length; when you see *1 cm³* or *1 cc* or *1 mL*, you are reading about volume. A comparison of units of volume is illustrated in **FIGURE 2.8**.

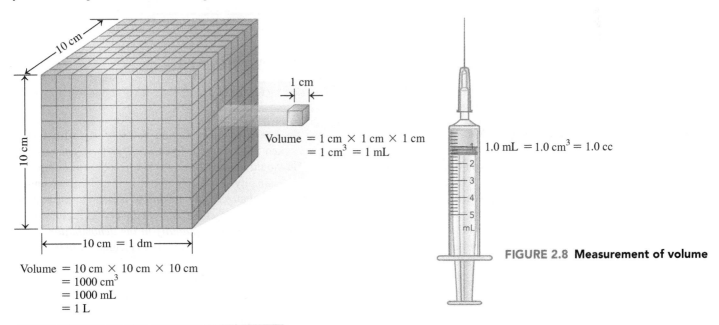

Volume = 1 cm × 1 cm × 1 cm
= 1 cm³ = 1 mL

1.0 mL = 1.0 cm³ = 1.0 cc

FIGURE 2.8 Measurement of volume

Volume = 10 cm × 10 cm × 10 cm
= 1000 cm³
= 1000 mL
= 1 L

A cube measuring 10 cm on each side has a volume of 1000 cm³ or 1 L; a cube measuring 1 cm on each side has a volume of 1 cm³ (cc) or 1 mL.

A syringe contains 1.0 mL, which is the same as 1.0 cm³ and 1.0 cc.

Measuring Mass

When you go to the doctor for a physical examination, your mass is recorded in kilograms, whereas the results of your laboratory tests are reported in grams, milligrams (mg), or micrograms (μg or mcg). A kilogram is equal to 1000 g. One gram represents the same mass as 1000 mg, and one mg equals 1000 μg (or 1000 mcg). Examples of equalities between different metric units of mass follow:

$$1 \text{ kg} = 1000 \text{ g} \qquad = 1 \times 10^3 \text{ g}$$
$$1 \text{ g} = 1000 \text{ mg} \qquad = 1 \times 10^3 \text{ mg}$$
$$1 \text{ mg} = 1000 \text{ } \mu\text{g (mcg)} = 1 \times 10^3 \text{ } \mu\text{g (mcg)}$$

PRACTICE PROBLEMS

Try Practice Problems 2.41 to 2.44

PRACTICE PROBLEMS

2.4 Prefixes and Equalities

2.33 Write the abbreviation for each of the following units:
a. milligram
b. deciliter
c. kilometer
d. picogram

2.34 Write the abbreviation for each of the following units:
a. gigagram
b. megameter
c. microliter
d. femtosecond

2.35 Write the complete name for each of the following units:
a. cL   b. kg   c. ms   d. Pm

2.36 Write the complete name for each of the following units:
a. dL   b. Ts   c. mcg   d. pm

2.37 Write the numerical value for each of the following prefixes:
a. centi   b. tera   c. milli   d. deci

2.38 Write the numerical value for each of the following prefixes:
a. giga   b. micro   c. mega   d. nano

2.39 Use a prefix to write the name for each of the following:
a. 0.1 g
b. 10^{-6} g
c. 1000 g
d. 0.01 g

2.40 Use a prefix to write the name for each of the following:
a. 10^9 m
b. 10^6 m
c. 0.001 m
d. 10^{-12} m

2.41 Complete each of the following metric relationships:
a. 1 m = _____ cm
b. 1 m = _____ nm
c. 1 mm = _____ m
d. 1 L = _____ mL

2.42 Complete each of the following metric relationships:
a. 1 Mg = _____ g
b. 1 mL = _____ μL
c. 1 g = _____ kg
d. 1 g = _____ mg

2.43 For each of the following pairs, which is the larger unit?
a. milligram or kilogram
b. milliliter or microliter
c. m or km
d. kL or dL
e. nanometer or picometer

2.44 For each of the following pairs, which is the smaller unit?
a. mg or g
b. centimeter or nanometer
c. millimeter or micrometer
d. mL or dL
e. centigram or megagram

REVIEW

Calculating Percentages (1.4)

CORE CHEMISTRY SKILL

Writing Conversion Factors from Equalities

2.5 Writing Conversion Factors

LEARNING GOAL Write a conversion factor for two units that describe the same quantity.

Many problems in chemistry and the health sciences require you to change from one unit to another unit. Suppose you worked 2.0 h on your homework, and someone asked you how many minutes that was. You would answer 120 min. You must have multiplied 2.0 h × 60 min/h because you knew the equality (1 h = 60 min) that related the two units. When you expressed 2.0 h as 120 min, you changed only the unit of measurement used to express the time. *Any equality can be written as fractions called* **conversion factors** *with one of the quantities in the numerator and the other quantity in the denominator.* Two conversion factors are always possible from any equality. Be sure to include the units when you write the conversion factors.

Two Conversion Factors for the Equality: 1 h = 60 min

$$\frac{\text{Numerator} \longrightarrow}{\text{Denominator} \longrightarrow} \qquad \frac{60 \text{ min}}{1 \text{ h}} \text{ and } \frac{1 \text{ h}}{60 \text{ min}}$$

PRACTICE PROBLEMS

Try Practice Problems 2.45 and 2.46

These factors are read as "60 minutes per 1 hour" and "1 hour per 60 minutes." The term *per* means "divide." Some common relationships are given in **TABLE 2.8**.

The numbers in any definition between two metric units or between two U.S. system units are exact. Because numbers in a definition are exact, they are not used to determine

TABLE 2.8 Some Common Equalities

Quantity	Metric (SI)	U.S.	Metric–U.S.
Length	1 km = 1000 m	1 ft = 12 in.	2.54 cm = 1 in. (exact)
	1 m = 1000 mm	1 yd = 3 ft	1 m = 39.37 in.
	1 cm = 10 mm	1 mi = 5280 ft	1 m = 1.094 yd
			1 km = 0.6214 mi
Volume	1 L = 1000 mL	1 qt = 4 cups	946.4 mL = 1 qt
	1 dL = 100 mL	1 qt = 2 pt	1 L = 1.057 qt
	1 mL = 1 cm^3	1 gal = 4 qt	473.2 mL = 1 pt
	1 mL = 1 cc*		3.785 L = 1 gal
			5 mL = 1 tsp*
			15 mL = 1 T (tbsp)*
Mass	1 kg = 1000 g	1 lb = 16 oz	1 kg = 2.205 lb
	1 g = 1000 mg		453.6 g = 1 lb
	1 mg = 1000 mcg*		28.35 g = 1 oz
Time	1 h = 60 min	1 h = 60 min	
	1 min = 60 s	1 min = 60 s	

*Used in medicine.

significant figures. For example, the equality of 1 g = 1000 mg is a definition, which means that both of the numbers 1 and 1000 are exact.

When an equality consists of a metric unit and a U.S. unit, one of the numbers in the equality is obtained by measurement and counts toward the significant figures in the answer. For example, the equality of 1 lb = 453.6 g is obtained by measuring the grams in exactly 1 lb. In this equality, the measured quantity 453.6 g has four significant figures, whereas the 1 is exact. An exception is the relationship of 1 in. = 2.54 cm, which has been defined as exact.

Metric Conversion Factors

We can write two metric conversion factors for any of the metric relationships. For example, from the equality for meters and centimeters, we can write the following factors:

Metric Equality	Conversion Factors
1 m = 100 cm	$\dfrac{100 \text{ cm}}{1 \text{ m}}$ and $\dfrac{1 \text{ m}}{100 \text{ cm}}$

Both are proper conversion factors for the relationship; one is just the inverse of the other. *The usefulness of conversion factors is enhanced by the fact that we can turn a conversion factor over and use its inverse.* The numbers 100 and 1 in this equality, and its conversion factors are both *exact* numbers.

Metric–U.S. System Conversion Factors

Suppose you need to convert from pounds, a unit in the U.S. system, to kilograms in the metric system. A relationship you could use is

1 kg = 2.205 lb

The corresponding conversion factors would be

$$\frac{2.205 \text{ lb}}{1 \text{ kg}} \quad \text{and} \quad \frac{1 \text{ kg}}{2.205 \text{ lb}}$$

FIGURE 2.9 illustrates the contents of some packaged foods in both U.S. and metric units.

ENGAGE 2.10

Why does the equality 1 day = 24 h have two conversion factors?

FIGURE 2.9 In the United States, the contents of many packaged foods are listed in both U.S. and metric units.

Equalities and Conversion Factors Stated Within a Problem

An equality may also be stated within a problem that applies only to that problem. For example, the speed of a car in kilometers per hour or the price of onions in dollars per pound would be specific relationships for that problem only. From each of the following statements, we can write an equality and two conversion factors, and identify each number as exact or give the number of significant figures.

The car was traveling at a speed of 85 km/h.

Equality	Conversion Factors	Significant Figures or Exact
1 h = 85 km	$\dfrac{85\ km}{1\ h}$ and $\dfrac{1\ h}{85\ km}$	The 85 km is measured: It has two significant figures. The 1 h is exact.

The price of onions is $1.24 per pound.

Equality	Conversion Factors	Significant Figures or Exact
1 lb = $1.24	$\dfrac{\$1.24}{1\ lb}$ and $\dfrac{1\ lb}{\$1.24}$	The $1.24 is measured: It has three significant figures. The 1 lb is exact.

Conversion Factors from Dosage Problems

Equalities stated within dosage problems for medications can also be written as conversion factors. Keflex (cephalexin), an antibiotic used for respiratory and ear infections, is available in 250-mg capsules. Vitamin C, an antioxidant, is available in 500-mg tablets. These dosage relationships can be used to write equalities from which two conversion factors can be derived.

One capsule contains 250 mg of Keflex.

Equality	Conversion Factors	Significant Figures or Exact
1 capsule = 250 mg of Keflex	$\dfrac{250\ mg\ Keflex}{1\ capsule}$ and $\dfrac{1\ capsule}{250\ mg\ Keflex}$	The 250 mg is measured: It has two significant figures. The 1 capsule is exact.

Keflex (cephalexin), used to treat respiratory infections, is available in 250-mg capsules.

One tablet contains 500 mg of vitamin C.

Equality	Conversion Factors	Significant Figures or Exact
1 tablet = 500 mg of vitamin C	$\dfrac{500\ mg\ vitamin\ C}{1\ tablet}$ and $\dfrac{1\ tablet}{500\ mg\ vitamin\ C}$	The 500 mg is measured: It has one significant figure. The 1 tablet is exact.

Conversion Factors from a Percent, ppm, and ppb

A percent (%) is written as a conversion factor by choosing a unit and expressing the numerical relationship of the parts of this unit to 100 parts of the whole. For example, a person might have 18% body fat by mass. The percent can be written as 18 mass units of body fat in every 100 mass units of body mass. Different mass units such as grams (g), kilograms (kg), or pounds (lb) can be used, but both units in the factor must be the same.

The thickness of the skin fold at the abdomen is used to determine the percentage of body fat.

Equality	Conversion Factors	Significant Figures or Exact
100 kg of body mass = 18 kg of body fat	$\dfrac{18\ kg\ body\ fat}{100\ kg\ body\ mass}$ and $\dfrac{100\ kg\ body\ mass}{18\ kg\ body\ fat}$	The 18 kg is measured: It has two significant figures. The 100 kg is exact.

ENGAGE 2.11

How is a percent used to write an equality and two conversion factors?

When scientists want to indicate very small ratios, they use numerical relationships called *parts per million* (ppm) or *parts per billion* (ppb). The ratio of parts per million is the same as the milligrams of a substance per kilogram (mg/kg). The ratio of parts per billion equals the micrograms per kilogram (μg/kg, mcg/kg).

Ratio	Units
parts per million (ppm)	milligrams per kilogram (mg/kg)
parts per billion (ppb)	micrograms per kilogram (μg/kg, mcg/kg)

For example, the maximum amount of lead that is allowed by the FDA in glazed pottery bowls is 2 ppm, which is 2 mg/kg.

Equality	Conversion Factors	Significant Figures or Exact
1 kg of glaze = 2 mg of lead	$\dfrac{2 \text{ mg lead}}{1 \text{ kg glaze}}$ and $\dfrac{1 \text{ kg glaze}}{2 \text{ mg lead}}$	The 2 mg is measured: It has one significant figure. The 1 kg is exact.

Conversion Factors with Powers

Sometimes we use a conversion factor that is squared or cubed. This is the case when we need to calculate an area or a volume.

Distance = length

Area   = length $\times$ length = length^2

Volume = length $\times$ length $\times$ length = length^3

Suppose you want to write the equality and the conversion factors for the relationship between an area in square centimeters and in square meters. To square the equality 1 m = 100 cm, we square both the number and the unit on each side.

Equality: 1 m = 100 cm

Area: $(1 \text{ m})^2 = (100 \text{ cm})^2$ or $1 \text{ m}^2 = (100)^2 \text{ cm}^2$

From the new equality, we can write two conversion factors as follows:

Conversion factors: $\dfrac{(100 \text{ cm})^2}{(1 \text{ m})^2}$ and $\dfrac{(1 \text{ m})^2}{(100 \text{ cm})^2}$

In the following example, we show that the equality 1 in. = 2.54 cm can be squared to give area or can be cubed to give volume. *Both the number and the unit must be squared or cubed.*

Measurement	Equality	Conversion Factors
Length	1 in. = 2.54 cm	$\dfrac{2.54 \text{ cm}}{1 \text{ in.}}$ and $\dfrac{1 \text{ in.}}{2.54 \text{ cm}}$
Area	$(1 \text{ in.})^2 = (2.54 \text{ cm})^2$ $(1 \text{ in.})^2 = (2.54)^2 \text{ cm}^2 = 6.45 \text{ cm}^2$	$\dfrac{(2.54 \text{ cm})^2}{(1 \text{ in.})^2}$ and $\dfrac{(1 \text{ in.})^2}{(2.54 \text{ cm})^2}$
Volume	$(1 \text{ in.})^3 = (2.54 \text{ cm})^3$ $(1 \text{ in.})^3 = (2.54)^3 \text{ cm}^3 = 16.4 \text{ cm}^3$	$\dfrac{(2.54 \text{ cm})^3}{(1 \text{ in.})^3}$ and $\dfrac{(1 \text{ in.})^3}{(2.54 \text{ cm})^3}$

SAMPLE PROBLEM 2.8   Equalities and Conversion Factors in a Problem

TRY IT FIRST

Write the equality and two conversion factors, and identify each number as exact or give the number of significant figures for each of the following:

a. The medication that Greg takes for his high blood pressure contains 40. mg of propranolol in 1 tablet.

b. Cold-water fish such as salmon contains 1.9% omega-3 fatty acids by mass.

Propranolol
Tablets BP 40

Propranolol is used to lower high blood pressure.

Salmon contains high levels of omega-3 fatty acids.

The maximum amount of mercury allowed in tuna is 0.5 ppm.

PRACTICE PROBLEMS
Try Practice Problems 2.47 to 2.56

c. The U.S. Environmental Protection Agency (EPA) has set the maximum level for mercury in tuna at 0.5 ppm.

d. The area of $(1 \text{ cm})^2$ is the same as $(10 \text{ mm})^2$.

SOLUTION

a.

Equality	Conversion Factors	Significant Figures or Exact
1 tablet = 40. mg of propranolol	$\dfrac{40. \text{ mg propranolol}}{1 \text{ tablet}}$ and $\dfrac{1 \text{ tablet}}{40. \text{ mg propranolol}}$	The 40. mg is measured: It has two significant figures. The 1 tablet is exact.

b.

Equality	Conversion Factors	Significant Figures or Exact
100 g of salmon = 1.9 g of omega-3 fatty acids	$\dfrac{1.9 \text{ g omega-3 fatty acids}}{100 \text{ g salmon}}$ and $\dfrac{100 \text{ g salmon}}{1.9 \text{ g omega-3 fatty acids}}$	The 1.9 g is measured: It has two significant figures. The 100 g is exact.

c.

Equality	Conversion Factors	Significant Figures or Exact
1 kg of tuna = 0.5 mg of mercury	$\dfrac{0.5 \text{ mg mercury}}{1 \text{ kg tuna}}$ and $\dfrac{1 \text{ kg tuna}}{0.5 \text{ mg mercury}}$	The 0.5 mg is measured: It has one significant figure. The 1 kg is exact.

d.

Equality	Conversion Factors	Significant Figures or Exact
$(1 \text{ cm})^2 = (10 \text{ mm})^2$	$\dfrac{(10 \text{ mm})^2}{(1 \text{ cm})^2}$ and $\dfrac{(1 \text{ cm})^2}{(10 \text{ mm})^2}$	Both 1 cm and 10 mm are exact, by definition.

SELF TEST 2.8

Write the equality and its corresponding conversion factors, and identify each number as exact or give the number of significant figures for each of the following:

a. Levsin (hyoscyamine), used to treat stomach and bladder problems, is available as drops with 0.125 mg of Levsin per 1 mL of solution.

b. The EPA has set the maximum level of cadmium in rice as 0.4 ppm.

ANSWER

a. 1 mL of solution = 0.125 mg of Levsin

$$\frac{0.125 \text{ mg Levsin}}{1 \text{ mL solution}} \text{ and } \frac{1 \text{ mL solution}}{0.125 \text{ mg Levsin}}$$

The 0.125 mg is measured: It has three significant figures. The 1 mL is exact.

b. 1 kg of rice = 0.4 mg of cadmium

$$\frac{0.4 \text{ mg cadmium}}{1 \text{ kg rice}} \text{ and } \frac{1 \text{ kg rice}}{0.4 \text{ mg cadmium}}$$

The 0.4 mg is measured: It has one significant figure. The 1 kg is exact.

2.5 Writing Conversion Factors

2.45 Why can two conversion factors be written for an equality such as 1 m = 100 cm?

2.46 How can you check that you have written the correct conversion factors for an equality?

2.47 Write the equality and two conversion factors for each of the following pairs of units:
  a. milligrams and kilograms
  b. nanograms and grams
  c. liters and kiloliters
  d. seconds and milliseconds
  e. millimeters and decimeters
  f. cubic meters and cubic centimeters

2.48 Write the equality and two conversion factors for each of the following pairs of units:
  a. centimeters and inches
  b. kilometers and miles
  c. pounds and grams
  d. liters and deciliters
  e. grams and picograms
  f. square centimeters and square inches

2.49 Write the equality and two conversion factors, and identify the numbers as exact or give the number of significant figures for each of the following:
  a. One yard is 3 ft.
  b. One kilogram is 2.205 lb.
  c. A car goes 27 mi on 1 gal of gas.
  d. Sterling silver is 93% silver by mass.
  e. One minute is 60 s.
  f. One cubic meter is 1×10^6 cm^3.

2.50 Write the equality and two conversion factors, and identify the numbers as exact or give the number of significant figures for each of the following:
  a. One liter is 1.057 qt.
  b. At the store, oranges are $1.29 per lb.
  c. There are 7 days in 1 week.
  d. One deciliter contains 100 mL.
  e. An 18-carat gold ring contains 75% gold by mass.
  f. One square foot is 144 in.^{2}

2.51 Write the equality and two conversion factors, and identify the numbers as exact or give the number of significant figures for each of the following:
  a. A bee flies at an average speed of 3.5 m per second.
  b. The Daily Value (DV) for potassium is 3.5 g.

  c. An automobile traveled 26.0 km on 1 L of gasoline.
  d. The pesticide level in plums was 29 ppb.
  e. Silicon makes up 28.2% by mass of Earth's crust.

2.52 Write the equality and two conversion factors, and identify the numbers as exact or give the number of significant figures for each of the following:
  a. The Daily Value (DV) for iodine is 150 mcg.
  b. The nitrate level in well water was 32 ppm.
  c. Gold jewelry contains 58% gold by mass.
  d. The price of a liter of milk is $1.65.
  e. A metric ton is 1000 kg.

Applications

2.53 Write the equality and two conversion factors, and identify the numbers as exact or give the number of significant figures for each of the following:
  a. A calcium supplement contains 630 mg of calcium per tablet.
  b. The Daily Value (DV) for vitamin C is 60 mg.
  c. The label on a bottle reads 50 mg of atenolol per tablet.
  d. A low-dose aspirin contains 81 mg of aspirin per tablet.

2.54 Write the equality and two conversion factors, and identify the numbers as exact or give the number of significant figures for each of the following:
  a. The label on a bottle reads 10 mg of furosemide per 1 mL.
  b. The Daily Value (DV) for selenium is 70. mcg.
  c. An IV of normal saline solution has a flow rate of 85 mL per hour.
  d. One capsule of fish oil contains 360 mg of omega-3 fatty acids.

2.55 Write an equality and two conversion factors for each of the following medications:
  a. 10 mg of Atarax per 5 mL of Atarax syrup
  b. 0.25 g of Lanoxin per 1 tablet of Lanoxin
  c. 300 mg of Motrin per 1 tablet of Motrin

2.56 Write an equality and two conversion factors for each of the following medications:
  a. 2.5 mg of Coumadin per 1 tablet of Coumadin
  b. 100 mg of Clozapine per 1 tablet of Clozapine
  c. 1.5 g of Cefuroxime per 1 mL of Cefuroxime

2.6 Problem Solving Using Unit Conversion

LEARNING GOAL Use conversion factors to change from one unit to another.

The process of problem solving in chemistry often requires one or more conversion factors to change a given unit to the needed unit. For the problem, the unit of the given and the unit of the needed are identified. From there, the problem is set up with one or more conversion factors used to convert the given unit to the needed unit as seen in Sample Problem 2.9.

Given unit × one or more conversion factors = needed unit

SAMPLE PROBLEM 2.9 Using Conversion Factors

TRY IT FIRST

Greg's doctor has ordered a PET scan of his heart. In radiological imaging, dosages of pharmaceuticals are based on body mass. If Greg weighs 164 lb, what is his body mass in kilograms?

SOLUTION

STEP 1 State the given and needed quantities.

ANALYZE THE PROBLEM	Given	Need	Connect
	164 lb	kilograms	U.S.–metric conversion factor

STEP 2 Write a plan to convert the given unit to the needed unit.

$$\text{pounds} \quad \boxed{\text{U.S.–metric factor}} \quad \text{kilograms}$$

STEP 3 State the equalities and conversion factors.

$$1 \text{ kg} = 2.205 \text{ lb}$$
$$\frac{2.205 \text{ lb}}{1 \text{ kg}} \quad \text{and} \quad \frac{1 \text{ kg}}{2.205 \text{ lb}}$$

STEP 4 **Set up the problem to cancel units and calculate the answer.** Write the given, 164 lb, and multiply by the conversion factor that has lb in the denominator (bottom number) to cancel lb in the given.

ENGAGE 2.12

Why would you use the conversion factor $\frac{1 \text{ lb}}{16 \text{ oz}}$, not $\frac{16 \text{ oz}}{1 \text{ lb}}$, to convert 6.4 oz to pounds?

$$\underbrace{164 \text{ lb}}_{\text{Given}} \quad \times \quad \underbrace{\frac{1 \text{ kg}}{2.205 \text{ lb}}}_{\substack{\text{Conversion factor} \\ \text{to cancel given unit}}} \quad = \quad \underbrace{74.4 \text{ kg}}_{\text{Answer}}$$

Unit for answer goes here

The given unit lb cancels out, and the needed unit kg is in the numerator. *The unit you want in the final answer is the one that remains after all the other units have canceled out.* This is a helpful way to check that you set up a problem properly.

$$\text{lb} \times \frac{\text{kg}}{\text{lb}} = \text{kg} \quad \text{Unit needed for answer}$$

The calculator display gives the numerical answer, which is adjusted to give a final answer with the proper number of significant figures (SFs). The value of 74.4 combined with the unit, kg, gives the final answer of 74.4 kg.

INTERACTIVE VIDEO

Conversion Factors

$$\underset{\text{Three SFs}}{164} \quad \boxed{\times} \quad \underset{\text{Four SFs}}{\overset{\text{Exact}}{\frac{1}{2.205}}} \quad \boxed{=} \quad 164 \quad \boxed{\div} \quad 2.205 \quad \boxed{=} \quad \underset{\substack{\text{Calculator} \\ \text{display}}}{74.37641723} \quad \boxed{=} \quad \underset{\substack{\text{Three SFs} \\ \text{(rounded off)}}}{74.4}$$

SELF TEST 2.9

a. A total of 2500 mL of a boric acid antiseptic solution is prepared from boric acid concentrate. How many quarts of boric acid have been prepared?
b. A vial contains 65 mg of phenobarbital per 1 mL of solution. How many milliliters are needed for an order of 40. mg of phenobarbital?

ANSWER
a. 2.6 qt   **b.** 0.62 mL

PRACTICE PROBLEMS
Try Practice Problems 2.57 to 2.62

Using Two or More Conversion Factors

In problem solving, two or more conversion factors are often needed to complete the change of units. In setting up these problems, one factor follows the other. Each factor is arranged to cancel the preceding unit until the needed unit is obtained. Once the problem is set up to cancel units properly, the calculations can be done without writing intermediate results. In this text, when two or more conversion factors are required, the final answer will be based on obtaining a final calculator display and rounding off (or adding zeros) to give the correct number of significant figures as shown in Sample Problem 2.10.

CORE CHEMISTRY SKILL
Using Conversion Factors

SAMPLE PROBLEM 2.10   Using Two Conversion Factors

TRY IT FIRST

Greg has been diagnosed with diminished thyroid function. His doctor prescribes a dosage of 0.150 mg of Synthroid to be taken once a day. If tablets in stock contain 75 mcg of Synthroid, how many tablets are required to provide the prescribed medication?

SOLUTION

STEP 1   State the given and needed quantities.

ANALYZE THE PROBLEM	Given	Need	Connect
	0.150 mg of Synthroid	number of tablets	metric conversion factor, clinical conversion factor

STEP 2   Write a plan to convert the given unit to the needed unit.

milligrams → Metric factor → micrograms → Clinical factor → number of tablets

STEP 3   State the equalities and conversion factors.

$$1 \text{ mg} = 1000 \text{ mcg}$$
$$\frac{1000 \text{ mcg}}{1 \text{ mg}} \text{ and } \frac{1 \text{ mg}}{1000 \text{ mcg}}$$

$$1 \text{ tablet} = 75 \text{ mcg of Synthroid}$$
$$\frac{75 \text{ mcg Synthroid}}{1 \text{ tablet}} \text{ and } \frac{1 \text{ tablet}}{75 \text{ mcg Synthroid}}$$

STEP 4   Set up the problem to cancel units and calculate the answer. The problem can be set up using the metric factor to cancel milligrams, and then the clinical factor to obtain the number of tablets as the final unit.

$$0.150 \text{ mg Synthroid} \times \frac{1000 \text{ mcg}}{1 \text{ mg}} \times \frac{1 \text{ tablet}}{75 \text{ mcg Synthroid}} = 2 \text{ tablets}$$

Three SFs, Exact, Exact, Two SFs

ENGAGE 2.13
How are two conversion factors used in a problem setup?

One teaspoon of cough syrup is measured for a patient.

SAMPLE PROBLEM 2.11   Using a Percent as a Conversion Factor

TRY IT FIRST

A person who exercises regularly has 16% body fat by mass. If this person weighs 155 lb, what is the mass, in kilograms, of body fat?

SOLUTION

STEP 1   State the given and needed quantities.

ANALYZE THE PROBLEM	Given	Need	Connect
	155 lb body weight	kilograms of body fat	U.S.–metric conversion factor, percent conversion factor

STEP 2   Write a plan to convert the given unit to the needed unit.

pounds of body weight   U.S.–metric factor   kilograms of body mass   Percent factor   kilograms of body fat

Exercising regularly helps reduce body fat.

STEP 3   State the equalities and conversion factors.

1 kg of body mass = 2.205 lb of body weight   100 kg of body mass = 16 kg of body fat

$$\frac{2.205 \text{ lb body weight}}{1 \text{ kg body mass}} \text{ and } \frac{1 \text{ kg body mass}}{2.205 \text{ lb body weight}} \qquad \frac{16 \text{ kg body fat}}{100 \text{ kg body mass}} \text{ and } \frac{100 \text{ kg body mass}}{16 \text{ kg body fat}}$$

STEP 4   Set up the problem to cancel units and calculate the answer.

$$155 \text{ lb body weight} \times \frac{1 \text{ kg body mass}}{2.205 \text{ lb body weight}} \times \frac{16 \text{ kg body fat}}{100 \text{ kg body mass}} = 11 \text{ kg of body fat}$$

Three SFs   Four SFs (Exact)   Exact (Two SFs)   Two SFs

For problems that use several conversion factors, other plans may be possible. For example, for this problem, the pounds of body weight could be converted to pounds of body fat using the percent factor, which can be converted to kilograms of body fat using the metric–U.S. factor for kilograms and pounds.

Conversion of Units in a Fraction

Sometime units in an equality are written as a fraction that relates the two units. For example, we measure fuel consumption in a car in units of miles per gallon. Perhaps, the mileage for a car is 39.0 mi per gal (39.0 mi/gal). Someone driving a car in Europe would express the use of fuel in kilometers per liter (km/L). If we want to express mi/gal in units of km/L, we one use conversion factor to change miles to kilometers, and another to change gallons to liters as shown in Sample Problem 2.12.

SAMPLE PROBLEM 2.12 Conversion of Units in a Fraction

TRY IT FIRST

Your new car has a fuel consumption of 39.0 mi/gal. Convert this fuel consumption to km/L.

SOLUTION

STEP 1 State the given and needed quantities.

	Given	Need	Connect
ANALYZE THE PROBLEM	$\dfrac{39.0 \text{ mi}}{1 \text{ gal}}$	$\dfrac{\text{km}}{\text{L}}$	U.S.–metric conversion factors, km/mi, L/gal

STEP 2 Write a plan to convert the given unit to the needed unit.

$$\dfrac{\text{miles}}{\text{gallon}} \quad \begin{array}{c}\text{U.S.–metric}\\\text{factor}\end{array} \quad \dfrac{\text{kilometers}}{\text{gallon}} \quad \begin{array}{c}\text{U.S.–metric}\\\text{factor}\end{array} \quad \dfrac{\text{kilometers}}{\text{liter}}$$

STEP 3 State the equalities and conversion factors.

$$1 \text{ km} = 0.6214 \text{ mi}$$

$$\dfrac{0.6214 \text{ mi}}{1 \text{ km}} = \dfrac{1 \text{ km}}{0.6214 \text{ mi}}$$

$$1 \text{ gal} = 3.785 \text{ L}$$

$$\dfrac{3.785 \text{ L}}{1 \text{ gal}} = \dfrac{1 \text{ gal}}{3.785 \text{ L}}$$

STEP 4 Set up the problem to cancel units and calculate the answer.

The problem can be set up using the U.S.–metric factor to convert miles to kilometers, and another U.S.–metric factor to convert gallons to liters to obtain the fraction km/L.

$$\underset{\text{Exact}}{\overset{\text{Three SFs}}{\dfrac{39.0 \text{ mi}}{1 \text{ gal}}}} \times \underset{\text{Four SFs}}{\overset{\text{Exact}}{\dfrac{1 \text{ km}}{0.6214 \text{ mi}}}} \times \underset{\text{Four SFs}}{\overset{\text{Exact}}{\dfrac{1 \text{ gal}}{3.785 \text{ L}}}} = \underset{\text{Exact}}{\overset{\text{Three SFs}}{\dfrac{16.6 \text{ km}}{1 \text{ L}}}}$$

SELF TEST 2.12

Convert each of the following fractions to the indicated units:

a. 2 mg/kg to oz/lb   **b.** 50. m/s to mi/h

ANSWER

a. 3×10^{-5} oz/lb   **b.** 110 mi/h

PRACTICE PROBLEMS

Try Practice Problems 2.63 to 2.68

Chemistry Link to Health

Toxicology and Risk–Benefit Assessment

Each day, we make choices about what we do or what we eat, often without thinking about the risks associated with these choices. We are aware of the risks of cancer from smoking or the risks of lead poisoning, and we know there is a greater risk of having an accident if we cross a street where there is no light or crosswalk.

The LD$_{50}$ of caffeine is 192 mg/kg.

A basic concept of toxicology is the statement of Paracelsus that the dose is the difference between a poison and a cure. To evaluate the level of danger from various substances, natural or synthetic, a risk assessment is made by exposing laboratory animals to the substances and monitoring the health effects. Often, doses very much greater than humans might ordinarily encounter are given to the test animals.

Many hazardous chemicals or substances have been identified by these tests. One measure of toxicity is the LD$_{50}$, or lethal dose, which is the concentration of the substance that causes death in 50% of the test animals. A dosage is typically measured in milligrams per kilogram (mg/kg) of body mass or micrograms per kilogram (mcg/kg) of body mass.

Other evaluations need to be made, but it is easy to compare LD$_{50}$ values. Parathion, a pesticide, with an LD$_{50}$ of 3 mg/kg, would be highly toxic. This means that 3 mg of parathion per kg of body mass would be fatal to half the test animals. Table salt (sodium chloride) with an LD$_{50}$ of 3300 mg/kg would have a much lower toxicity. You would need to ingest a huge amount of salt before any toxic effect would be observed. Although the risk to animals can be evaluated in the laboratory, it is more difficult to determine the impact in the environment since there is also a difference between continued exposure and a single, large dose of the substance.

TABLE 2.9 lists some LD$_{50}$ values and compares substances in order of increasing toxicity.

TABLE 2.9 Some LD$_{50}$ Values for Substances Tested in Rats

Substance	LD$_{50}$ (mg/kg)
Table sugar	29 700
Boric acid	5140
Baking soda	4220
Table salt	3300
Ethanol	2080
Aspirin	1100
Codeine	800
Oxycodone	480
Caffeine	192
DDT	113
Cocaine (injected)	95
Dichlorvos (pesticide strips)	56
Ricin	30
Sodium cyanide	6
Parathion	3

PRACTICE PROBLEMS

2.6 Problem Solving Using Unit Conversion

2.57 Perform each of the following conversions using metric conversion factors:
  a. 44.2 mL to liters
  b. 8.65 m to nanometers
  c. 5.2×10^8 g to megagrams
  d. 0.72 ks to milliseconds

2.58 Perform each of the following conversions using metric conversion factors:
  a. 4.82×10^{-5} L to picoliters
  b. 575.2 dm to kilometers
  c. 5×10^{-4} kg to micrograms
  d. 6.4×10^{10} ps to seconds

2.59 Perform each of the following conversions using metric and U.S. conversion factors:
  a. 3.428 lb to kilograms
  b. 1.6 m to inches
  c. 4.2 L to quarts
  d. 0.672 ft to millimeters

2.60 Perform each of the following conversions using metric and U.S. conversion factors:
  a. 0.21 lb to grams
  b. 11.6 in. to centimeters
  c. 0.15 qt to milliliters
  d. 35.41 kg to pounds

2.61 Use metric conversion factors to solve each of the following problems:
  a. If a student is 175 cm tall, how tall is the student in meters?
  b. A cooler has a volume of 5000 mL. What is the capacity of the cooler in liters?
  c. A hummingbird has a mass of 0.0055 kg. What is the mass, in grams, of the hummingbird?
  d. A balloon has a volume of 3500 cm^3. What is the volume in liters?

2.62 Use metric conversion factors to solve each of the following problems:
  a. The Daily Value (DV) for phosphorus is 800 mg. How many grams of phosphorus are recommended?
  b. A glass of orange juice contains 3.2 dL of juice. How many milliliters of orange juice are in the glass?
  c. A package of chocolate instant pudding contains 2840 mg of sodium. How many grams of sodium are in the pudding?
  d. A jar contains 0.29 kg of olives. How many grams of olives are in the jar?

2.63 Solve each of the following problems using one or more conversion factors:
   a. A container holds 0.500 qt of liquid. How many milliliters of lemonade will it hold?
   b. What is the mass, in kilograms, of a person who weighs 175 lb?
   c. An athlete has 15% body fat by mass. What is the weight of fat, in pounds, of a 74-kg athlete?
   d. A plant fertilizer contains 15% nitrogen (N) by mass. In a container of soluble plant food, there are 10.0 oz of fertilizer. How many grams of nitrogen are in the container?

Agricultural fertilizers applied to a field provide nitrogen for plant growth.

2.64 Solve each of the following problems using one or more conversion factors:
   a. Wine is 12% alcohol by volume. How many milliliters of alcohol are in a 0.750-L bottle of wine?
   b. Blueberry high-fiber muffins contain 51% dietary fiber by mass. If a package with a net weight of 12 oz contains six muffins, how many grams of fiber are in each muffin?
   c. A jar of crunchy peanut butter contains 1.43 kg of peanut butter. If you use 8.0% of the peanut butter for a sandwich, how many ounces of peanut butter did you take out of the container?
   d. In a candy factory, the nutty chocolate bars contain 22.0% pecans by mass. If 5.0 kg of pecans were used for candy last Tuesday, how many pounds of nutty chocolate bars were made?

Applications

2.65 Using conversion factors, solve each of the following problems:
   a. You have used 250 L of distilled water for a dialysis patient. How many gallons of water is that?
   b. A patient needs 0.024 g of a sulfa drug. There are 8-mg tablets in stock. How many tablets should be given?
   c. The daily dose of ampicillin for the treatment of an ear infection is 115 mg/kg of body mass. What is the daily dose for a 34-lb child?
   d. You need 4.0 oz of a steroid ointment. How many grams of ointment does the pharmacist need to prepare?

2.66 Using conversion factors, solve each of the following problems:
   a. The physician has ordered 1.0 g of tetracycline to be given every six hours to a patient. If your stock on hand is 500-mg tablets, how many will you need for one day's treatment?
   b. An intramuscular medication is given at 5.00 mg/kg of body mass. What is the dose for a 180-lb patient?
   c. A physician has ordered 0.50 mg of atropine, intramuscularly. If atropine were available as 0.10 mg/mL of solution, how many milliliters would you need to give?
   d. During surgery, a patient receives 5.0 pt of plasma. How many milliliters of plasma were given?

2.67 Using conversion factors, solve each of the following problems:
   a. A nurse practitioner prepares 500. mL of an IV of normal saline solution to be delivered at a rate of 80. mL/h. What is the infusion time, in hours, to deliver 500. mL?
   b. A nurse practitioner orders Medrol to be given 1.5 mg/kg of body mass.
      i. What is the dosage of Medrol in g/lb of body weight?
      ii. If a child weighs 72.6 lb and the available stock of Medrol is 20. mg/mL, how many milliliters does the nurse administer to the child?
   c. A dosage for a patient is 25.0 mg of a drug/kg of body mass.
      i. What is that dosage in oz of drug/lb of body weight?
      ii. What is the dose, in ounces, required for a 164-lb patient?

2.68 Using conversion factors, solve each of the following problems:
   a. A nurse practitioner prepares an injection of promethazine, an antihistamine used to treat allergic rhinitis. If the stock bottle is labeled 25 mg/mL and the order is a dose of 12.5 mg, how many milliliters will the nurse draw up in the syringe?
   b. An order for ampicillin is 25 mg/kg of body mass.
      i. What is the dosage of ampicillin in g/lb of body weight?
      ii. If stock on hand is 250 mg/capsule, how many capsules should be given to a 67-lb child?
   c. The dosage of an asthma drug is 5 mg/kg of body mass.
      i. What is that dosage in g/lb of body weight?
      ii. What is the dose, in milligrams, required for a 48-lb child?

2.7 Density

LEARNING GOAL Calculate the density of a substance; use the density to calculate the mass or volume of a substance.

The mass and volume of any object can be measured. If we compare the mass of the object to its volume, we obtain a relationship called **density**.

$$\text{Density} = \frac{\text{mass of substance}}{\text{volume of substance}}$$

Every substance has a unique density, which distinguishes it from other substances. For example, lead has a density of 11.3 g/mL, whereas cork has a density of 0.26 g/mL. From these densities, we can predict if these substances will sink or float in water. *If an object is less dense than a liquid, the object floats when placed in the liquid.* If a substance, such as cork, is less dense than water, it will float. However, a lead object sinks because its density is greater than that of water (see **FIGURE 2.10**).

ENGAGE 2.14

If a piece of iron sinks in water, how does its density compare to that of water?

FIGURE 2.10 Objects that sink in water are more dense than water; objects that float are less dense.

Cork (D = 0.26 g/mL)

Ice (D = 0.92 g/mL)

Water (D = 1.000 g/mL)

Aluminum (D = 2.70 g/mL)

Lead (D = 11.3 g/mL)

Density is used in chemistry in many ways. If we calculate the density of a pure metal as 10.5 g/mL, then we could identify it as silver, but not gold or aluminum. Metals such as gold and silver have higher densities, whereas gases have low densities. In the metric system, the densities of solids and liquids are usually expressed as grams per cubic centimeter (g/cm^3) or grams per milliliter (g/mL). The densities of gases are usually stated as grams per liter (g/L). **TABLE 2.10** gives the densities of some common substances.

TABLE 2.10 Densities of Some Common Substances

Solids (at 25 °C)	Density (g/cm^3)	Liquids (at 25 °C)	Density (g/mL)	Gases (at 0 °C)	Density (g/L)
Cork	0.26	Gasoline	0.74	Hydrogen	0.090
Body fat	0.909	Ethanol	0.79	Helium	0.179
Ice (at 0 °C)	0.92	Olive oil	0.92	Methane	0.714
Muscle	1.06	Water (at 4 °C)	1.000	Neon	0.902
Sugar	1.59	Urine	1.003–1.030	Nitrogen	1.25
Bone	1.80	Plasma (blood)	1.03	Air (dry)	1.29
Salt (NaCl)	2.16	Milk	1.04	Oxygen	1.43
Aluminum	2.70	Blood	1.06	Carbon dioxide	1.96
Iron	7.86	Mercury	13.6		
Copper	8.92				
Silver	10.5				
Lead	11.3				
Gold	19.3				

Calculating Density

We can calculate the density of a substance from its mass and volume as shown in Sample Problem 2.13.

SAMPLE PROBLEM 2.13 Calculating Density

TRY IT FIRST

High-density lipoprotein (HDL) is a type of cholesterol, sometimes called "good cholesterol," that is measured in a routine blood test. If a 0.258-g sample of HDL has a volume of 0.215 mL, what is the density, in grams per milliliter, of the HDL sample?

SOLUTION

STEP 1 State the given and needed quantities.

ANALYZE THE PROBLEM	Given	Need	Connect
	0.258 g of HDL, 0.215 mL	density (g/mL) of HDL	density expression

STEP 2 Write the density expression.

$$\text{Density} = \frac{\text{mass of substance}}{\text{volume of substance}}$$

STEP 3 Express mass in grams and volume in milliliters.

Mass of HDL sample = 0.258 g

Volume of HDL sample = 0.215 mL

STEP 4 Substitute mass and volume into the density expression and calculate the density.

Three SFs

$$\text{Density} = \frac{0.258 \text{ g}}{0.215 \text{ mL}} = \frac{1.20 \text{ g}}{1 \text{ mL}} = 1.20 \text{ g/mL}$$

Three SFs   Three SFs

SELF TEST 2.13

a. Low-density lipoprotein (LDL), sometimes called "bad cholesterol," is also measured in a routine blood test. If a 0.380-g sample of LDL has a volume of 0.362 mL, what is the density, in grams per milliliter, of the LDL sample?

b. Osteoporosis is a condition in which bone deteriorates to cause a decreased bone mass. If a bone sample has a mass of 2.15 g and a volume of 1.40 cm^3, what is its density in grams per cubic centimeter?

ANSWER

a. 1.05 g/mL   **b.** 1.54 g/cm^3

PRACTICE PROBLEMS

Try Practice Problems 2.69 to 2.72

Density of Solids Using Volume Displacement

The volume of a solid can be determined by volume displacement. When a solid is completely submerged in water, it displaces a volume that is equal to the volume of the solid. In **FIGURE 2.11**, the water level rises from 35.5 mL to 45.0 mL after the zinc object is added. This means that 9.5 mL of water is displaced and that the volume of the object is 9.5 mL.

The density of the zinc is calculated using volume displacement as follows:

Four SFs

$$\text{Density} = \frac{68.60 \text{ g Zn}}{9.5 \text{ mL}} = 7.2 \text{ g/mL}$$

Two SFs   Two SFs

Problem Solving Using Density

Density can be used as a conversion factor. For example, if the volume and the density of a sample are known, the mass in grams of the sample can be calculated as shown in Sample Problem 2.14.

CORE CHEMISTRY SKILL

Using Density as a Conversion Factor

The mass of a zinc object is determined.

The volume of the zinc object is determined by volume displacement.

FIGURE 2.11 Calculation of the density of a solid

Chemistry Link to Health

Bone Density

Our bones' density is a measure of their health and strength. Our bones are constantly gaining and losing calcium, magnesium, and phosphate. In childhood, bones form at a faster rate than they break down. As we age, bone breakdown occurs more rapidly than new bone forms. As bone loss increases, bones begin to thin, causing a decrease in mass and density. Thinner bones lack strength, which increases the risk of fracture. Hormonal changes, disease, and certain medications can also contribute to the bone thinning. Eventually, a condition of severe bone thinning known as *osteoporosis*, may occur. *Scanning electron micrographs* (SEMs) show (**a**) normal bone and (**b**) bone with osteoporosis due to loss of bone minerals.

Bone density is often determined by passing low-dose X-rays through the narrow part at the top of the femur (hip) and the spine (**c**). These locations are where fractures are more likely to occur, especially as we age. Bones with high density will block more of the X-rays compared to bones that are less dense. The results of a bone density test are compared to a healthy young adult as well as to other people of the same age.

Recommendations to improve bone strength include calcium and vitamin D supplements. Weight-bearing exercise such as walking and lifting weights can also improve muscle strength, which in turn increases bone strength.

(**a**) Normal bone

(**b**) Bone with osteoporosis

(**c**) Viewing a low-dose X-ray of the spine

SAMPLE PROBLEM 2.14 Problem Solving Using Density

TRY IT FIRST

Greg has a blood volume of 5.9 qt. If the density of blood is 1.06 g/mL, what is the mass, in grams, of Greg's blood?

SOLUTION

STEP 1 State the given and needed quantities.

ANALYZE THE PROBLEM	Given	Need	Connect
	5.9 qt of blood	grams of blood	U.S.–metric conversion factor, density conversion factor

STEP 2 Write a plan to calculate the needed quantity.

quarts → U.S.–metric factor → milliliters → Density factor → grams

STEP 3 Write the equalities and their conversion factors including density.

1 qt = 946.4 mL   1 mL of blood = 1.06 g of blood

$$\frac{946.4 \text{ mL}}{1 \text{ qt}} \text{ and } \frac{1 \text{ qt}}{946.4 \text{ mL}} \qquad \frac{1.06 \text{ g blood}}{1 \text{ mL blood}} \text{ and } \frac{1 \text{ mL blood}}{1.06 \text{ g blood}}$$

STEP 4 Set up the problem to calculate the needed quantity.

$$5.9 \text{ qt blood} \times \underset{\text{Exact}}{\overset{\text{Four SFs}}{\frac{946.4 \text{ mL}}{1 \text{ qt}}}} \times \underset{\text{Exact}}{\overset{\text{Three SFs}}{\frac{1.06 \text{ g blood}}{1 \text{ mL blood}}}} = \underset{\text{Two SFs}}{5900 \text{ g of blood}}$$

Two SFs

1 pt of blood contains 473.2 mL.

SELF TEST 2.14

a. During surgery, a patient receives 3.0 pt of blood. How many kilograms of blood (density = 1.06 g/mL) were needed for the transfusion?
b. A woman receives 1280 g of type A blood. If the blood has a density of 1.06 g/mL, how many liters of blood did she receive?

ANSWER
a. 1.5 kg   b. 1.21 L

PRACTICE PROBLEMS
Try Practice Problems 2.73 to 2.78

Specific Gravity

Specific gravity (sp gr) is a relationship between the density of a substance and the density of water. Specific gravity is calculated by dividing the density of a sample by the density of water, which is 1.000 g/mL at 4 °C. A substance with a specific gravity of 1.000 has the same density as water (1.000 g/mL).

$$\text{Specific gravity} = \frac{\text{density of sample}}{\text{density of water}}$$

PRACTICE PROBLEMS
Try Practice Problems 2.79 and 2.80

Specific gravity is one of the few unitless values you will encounter in chemistry. The specific gravity of urine helps evaluate the water balance in the body and the substances in the urine. In **FIGURE 2.12**, a hydrometer is used to measure the specific gravity of urine. The normal range of specific gravity for urine is 1.003 to 1.030. The specific gravity can decrease with *type 2 diabetes* and kidney disease. Increased specific gravity may occur with dehydration, kidney infection, and liver disease. In a clinic or hospital, a dipstick containing chemical pads is used to evaluate specific gravity.

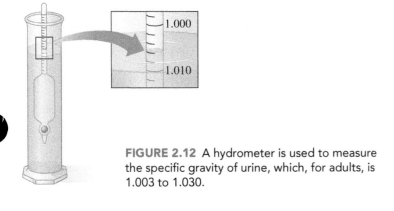

FIGURE 2.12 A hydrometer is used to measure the specific gravity of urine, which, for adults, is 1.003 to 1.030.

A dipstick is used to measure the specific gravity of a urine sample.

PRACTICE PROBLEMS

2.7 Density

2.69 Determine the density (g/mL) for each of the following:
   a. A 20.0-mL sample of a salt solution has a mass of 24.0 g.
   b. A cube of butter weighs 0.250 lb and has a volume of 130.3 mL.
   c. A gem has a mass of 4.50 g. When the gem is placed in a graduated cylinder containing 12.00 mL of water, the water level rises to 13.45 mL.
   d. A 3.00-mL sample of a medication has a mass of 3.85 g.

2.70 Determine the density (g/mL) for each of the following:
   a. The fluid in a car battery has a volume of 125 mL and a mass of 155 g.
   b. A plastic material weighs 2.68 lb and has a volume of 3.5 L.
   c. A 4.000-mL urine sample from a person suffering from diabetes mellitus has a mass of 4.004 g.
   d. A solid object has a mass of 1.65 lb and a volume of 170 mL.

2.71 What is the density (g/mL) of each of the following samples?
   a. A lightweight head on a golf club is made of titanium. The volume of a sample of titanium is 114 cm^3, and the mass is 514.1 g.

Lightweight heads on golf clubs are made of titanium.

   b. A syrup is added to an empty container with a mass of 115.25 g. When 0.100 pt of syrup is added, the total mass of the container and syrup is 182.48 g.

   115.25 g   182.48 g

   c. A block of aluminum metal has a volume of 3.15 L and a mass of 8.51 kg.

2.72 What is the density (g/mL) of each of the following samples?
   a. An ebony carving has a mass of 275 g and a volume of 207 cm^3.
   b. A 14.3-cm^3 sample of tin has a mass of 0.104 kg.
   c. A bottle of acetone (fingernail polish remover) contains 55.0 mL of acetone with a mass of 43.5 g.

2.73 Use the density values in Table 2.10 to solve each of the following problems:
   a. How many liters of ethanol contain 1.50 kg of ethanol?
   b. How many grams of mercury are present in a barometer that holds 6.5 mL of mercury?
   c. A sculptor has prepared a mold for casting a silver figure. The figure has a volume of 225 cm^3. How many ounces of silver are needed in the preparation of the silver figure?

2.74 Use the density values in Table 2.10 to solve each of the following problems:
   a. A graduated cylinder contains 18.0 mL of water. What is the new water level, in milliliters, after 35.6 g of silver metal is submerged in the water?
   b. A thermometer containing 8.3 g of mercury has broken. What volume, in milliliters, of mercury spilled?
   c. A fish tank holds 35 gal of water. How many kilograms of water are in the fish tank?

2.75 Use the density values in Table 2.10 to solve each of the following problems:
   a. What is the mass, in grams, of a cube of copper that has a volume of 74.1 cm^3?
   b. How many kilograms of gasoline fill a 12.0-gal gas tank?
   c. What is the volume, in cubic centimeters, of an ice cube that has a mass of 27 g?

2.76 Use the density values in Table 2.10 to solve each of the following problems:
   a. If a bottle of olive oil contains 1.2 kg of olive oil, what is the volume, in milliliters, of the olive oil?
   b. A cannonball made of iron has a volume of 115 cm^3. What is the mass, in kilograms, of the cannonball?
   c. A balloon filled with helium has a volume of 7.3 L. What is the mass, in grams, of helium in the balloon?

2.77 In an old trunk, you find a piece of metal that you think may be aluminum, silver, or lead. You take it to a lab, where you find it has a mass of 217 g and a volume of 19.2 cm^3. Using Table 2.10, what is the metal you found?

2.78 Suppose you have two 100-mL graduated cylinders. In each cylinder, there is 40.0 mL of water. You also have two cubes: one is lead, and the other is aluminum. Each cube measures 2.0 cm on each side. After you carefully lower each cube into the water of its own cylinder, what will the new water level be in each of the cylinders? Use Table 2.10 for density values.

Applications

2.79 Solve each of the following problems:
   a. A urine sample has a density of 1.030 g/mL. What is the specific gravity of the sample?
   b. A 20.0-mL sample of a glucose IV solution has a mass of 20.6 g. What is the density of the glucose solution?
   c. The specific gravity of a vegetable oil is 0.92. What is the mass, in grams, of 750 mL of vegetable oil?
   d. A bottle containing 325 g of cleaning solution is used to clean hospital equipment. If the cleaning solution has a specific gravity of 0.850, what volume, in milliliters, of solution was used?

2.80 Solve each of the following problems:
   a. A glucose solution has a density of 1.02 g/mL. What is its specific gravity?
   b. A 0.200-mL sample of very-low-density lipoprotein (VLDL) has a mass of 190 mg. What is the density of the VLDL?
   c. Butter has a specific gravity of 0.86. What is the mass, in grams, of 2.15 L of butter?
   d. A 5.000-mL urine sample has a mass of 5.025 g. If the normal range for the specific gravity of urine is 1.003 to 1.030, would the specific gravity of this urine sample indicate that the patient could have type 2 diabetes?

UPDATE Greg's Visit with His Doctor

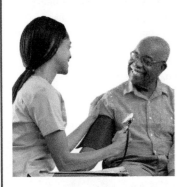

On Greg's visit to his doctor, he complains of feeling tired. Sandra, the registered nurse, withdraws 8.0 mL of blood, which is sent to the lab and tested for iron. When the iron level is low, a person may have fatigue and decreased immunity.

The normal range for serum iron in men is 80 to 160 mcg/dL. Greg's iron test shows a blood serum iron level of 42 mcg/dL, which indicates that Greg has iron-deficiency anemia. His doctor orders an iron supplement. One tablet of the iron supplement contains 50 mg of iron.

Applications

2.81 **a.** Write an equality and two conversion factors for Greg's serum iron level.
   b. How many micrograms of iron were in the 8.0-mL sample of Greg's blood?

2.82 **a.** Write an equality and two conversion factors for one tablet of the iron supplement.
   b. How many grams of iron will Greg consume in one week, if he takes two tablets each day?

Each tablet contains 50 mg of iron, which is given for iron supplementation.

CONCEPT MAP

CHEMISTRY AND MEASUREMENTS

Measurements

in chemistry involve

Metric Units | **Measured Numbers** | **Prefixes**

for measuring / have / that change the size of

Length (m) | **Significant Figures** | **Metric Units**

that require / to give

Mass (g) → **Density** | **Rounding Off Answers** | **Equalities**

or / used for

Volume (L) → **Density** and → **Specific Gravity** | **Adding Zeros** | **Conversion Factors**

Temperature (°C, K)

Time (s)

to change units in

Problem Solving

CHAPTER REVIEW

2.1 Units of Measurement

LEARNING GOAL Write the names and abbreviations for the metric and SI units used in measurements of volume, length, mass, temperature, and time.

- In science, physical quantities are described in units of the metric or International System of Units (SI).
- Some important units are liter (L) for volume, meter (m) for length, gram (g) and kilogram (kg) for mass, degree Celsius (°C) and kelvin (K) for temperature, and second (s) for time.

2.2 Measured Numbers and Significant Figures

LEARNING GOAL Identify a number as measured or exact; determine the number of significant figures in a measured number.

- A measured number is any number obtained by using a measuring device.
- An exact number is obtained by counting items or from a definition; no measuring device is needed.
- Significant figures are the numbers reported in a measurement including the estimated digit.
- Zeros in front of a decimal number or at the end of a nondecimal number are not significant.

2.3 Significant Figures in Calculations

LEARNING GOAL Give the correct number of significant figures for a calculated answer.

- In multiplication and division, the final answer is written so that it has the same number of significant figures as the measurement with the fewest significant figures.
- In addition and subtraction, the final answer is written so that it has the same number of decimal places as the measurement with the fewest decimal places.

2.4 Prefixes and Equalities

LEARNING GOAL Use the numerical values of prefixes to write a metric equality.

- A prefix placed in front of a metric or SI unit changes the size of the unit by factors of 10.

- Prefixes such as *centi*, *milli*, and *micro* provide smaller units; prefixes such as *kilo*, *mega*, and *tera* provide larger units.
- An equality shows the relationship between two units that measure the same quantity of volume, length, mass, or time.
- Examples of metric equalities are
  1 L = 1000 mL, 1 m = 100 cm, 1 kg = 1000 g, and 1 min = 60 s.

2.5 Writing Conversion Factors

LEARNING GOAL Write a conversion factor for two units that describe the same quantity.

- Conversion factors are used to express a relationship in the form of a fraction.
- Two conversion factors can be written for any relationship in the metric or U.S. system.
- A percentage is written as a conversion factor by expressing matching units as the parts in 100 parts of the whole.

2.6 Problem Solving Using Unit Conversion

LEARNING GOAL Use conversion factors to change from one unit to another.

- Conversion factors are useful when changing a quantity expressed in one unit to a quantity expressed in another unit.
- In the problem-solving process, a given unit is multiplied by one or more conversion factors that cancel units until the needed answer is obtained.

2.7 Density

LEARNING GOAL Calculate the density of a substance; use the density to calculate the mass or volume of a substance.

- The density of a substance is a ratio of its mass to its volume, usually g/mL or g/cm^3.
- The units of density can be used to write conversion factors that convert between the mass and volume of a substance.
- Specific gravity (sp gr) compares the density of a substance to the density of water, 1.000 g/mL.

KEY TERMS

Celsius (°C) temperature scale A temperature scale on which water has a freezing point of 0 °C and a boiling point of 100 °C.

centimeter (cm) A unit of length in the metric system; there are 2.54 cm in 1 in.

conversion factor A ratio in which the numerator and denominator are quantities from an equality or given relationship. For example, the two conversion factors for the equality 1 kg = 2.205 lb are written as

$$\frac{2.205 \text{ lb}}{1 \text{ kg}} \quad \text{and} \quad \frac{1 \text{ kg}}{2.205 \text{ lb}}$$

cubic centimeter (cm³, cc) The volume of a cube that has 1-cm sides; 1 cm³ is equal to 1 mL.

density The relationship of the mass of an object to its volume expressed as grams per cubic centimeter (g/cm³), grams per milliliter (g/mL), or grams per liter (g/L).

equality A relationship between two units that measure the same quantity.

exact number A number obtained by counting or by definition.

gram (g) The metric unit used in measurements of mass.

International System of Units (SI) The official system of measurement throughout the world, except for the United States, that modifies the metric system.

Kelvin (K) temperature scale A temperature scale on which the lowest possible temperature is 0 K.

kilogram (kg) A metric mass of 1000 g, equal to 2.205 lb. The kilogram is the SI standard unit of mass.

liter (L) The metric unit for volume that is slightly larger than a quart.

mass A measure of the quantity of material in an object.

measured number A number obtained when a quantity is determined by using a measuring device.

meter (m) The metric unit for length that is slightly longer than a yard. The meter is the SI standard unit of length.

metric system A system of measurement used by scientists and in most countries of the world.

milliliter (mL) A metric unit of volume equal to one-thousandth of a liter (0.001 L).

prefix The part of the name of a metric unit that precedes the base unit and specifies the size of the measurement. All prefixes are related on a decimal scale.

second (s) A unit of time used in both the SI and metric systems.

SI See International System of Units (SI).

significant figures (SFs) The numbers recorded in a measurement.

specific gravity (sp gr) A relationship between the density of a substance and the density of water:

$$\text{sp gr} = \frac{\text{density of sample}}{\text{density of water}}$$

temperature An indicator of the hotness or coldness of an object.

volume (V) The amount of space occupied by a substance.

KEY MATH SKILLS

The chapter section containing each Key Math Skill is shown in parentheses at the end of each heading.

Rounding Off (2.3)

Calculator displays are rounded off to give the correct number of significant figures.

- If the first digit to be dropped is *4 or less*, then it and all following digits are simply dropped from the number.
- If the first digit to be dropped is *5 or greater*, then the last retained digit of the number is increased by 1.

One or more significant zeros are added when the calculator display has fewer digits than the needed number of significant figures.

Example: Round off each of the following to three significant figures:
- **a.** 3.608 92 L
- **b.** 0.003 870 298 m
- **c.** 6 g

Answer: **a.** 3.61 L   **b.** 0.003 87 m   **c.** 6.00 g

CORE CHEMISTRY SKILLS

The chapter section containing each Core Chemistry Skill is shown in parentheses at the end of each heading.

Counting Significant Figures (2.2)

The significant figures (SFs) are all the measured numbers including the last, estimated digit.

- All nonzero digits
- Zeros between nonzero digits
- Zeros after a nonzero digit in a decimal number
- All digits in a coefficient of a number written in scientific notation

An *exact* number is obtained from counting or a definition and has no effect on the number of significant figures in the final answer.

Example: State the number of significant figures in each of the following:
- **a.** 0.003 045 mm
- **b.** 15 000 m
- **c.** 45.067 kg
- **d.** 5.30 × 10³ g
- **e.** 2 cans of soda

Answer: **a.** four SFs
b. two SFs
c. five SFs
d. three SFs
e. exact

Using Significant Figures in Calculations (2.3)

- In multiplication or division, the final answer is written so that it has the same number of significant figures as the measurement with the fewest SFs.

- In addition or subtraction, the final answer is written so that it has the same number of decimal places as the measurement having the fewest decimal places.

Example: Perform the following calculations using measured numbers, and give answers with the correct number of SFs or decimal places:

    a. 4.05 m × 0.6078 m   **b.** $\dfrac{4.50\ g}{3.27\ mL}$

    c. 0.758 g + 3.10 g   **d.** 13.538 km − 8.6 km

Answer: **a.** 2.46 m^2 **b.** 1.38 g/mL
         c. 3.86 g   **d.** 4.9 km

Using Prefixes (2.4)

In the metric and SI systems of units, a prefix attached to any unit increases or decreases its size by some factor of 10.

- When the prefix *centi* is used with the unit meter, it becomes centimeter, a length that is one-hundredth of a meter (0.01 m).
- When the prefix *milli* is used with the unit meter, it becomes millimeter, a length that is one-thousandth of a meter (0.001 m).

Example: Complete each of the following metric relationships:

    a. 1000 m = 1 __ m   **b.** 0.01 g = 1 __ g

Answer: **a.** 1000 m = 1 km   **b.** 0.01 g = 1 cg

Writing Conversion Factors from Equalities (2.5)

- A conversion factor allows you to change from one unit to another.
- Two conversion factors can be written for any equality in the metric, U.S., or metric–U.S. systems of measurement.
- Two conversion factors can be written for a relationship stated within a problem.

Example: Write two conversion factors for the equality:
    1 L = 1000 mL

Answer: $\dfrac{1000\ mL}{1\ L}$ and $\dfrac{1\ L}{1000\ mL}$

Using Conversion Factors (2.6)

In problem solving, conversion factors are used to cancel the given unit and to provide the needed unit for the answer.

- State the given and needed quantities.
- Write a plan to convert the given unit to the needed unit.
- State the equalities and conversion factors.
- Set up the problem to cancel units and calculate the answer.

Example: A computer chip has a width of 0.75 in. What is the width in millimeters?

Answer: $0.75\ \cancel{in.} \times \dfrac{2.54\ \cancel{cm}}{1\ \cancel{in.}} \times \dfrac{10\ mm}{1\ \cancel{cm}} = 19\ mm$

Using Density as a Conversion Factor (2.7)

Density is an equality of mass and volume for a substance, which is written as the *density expression.*

$$\text{Density} = \frac{\text{mass of substance}}{\text{volume of substance}}$$

Density is useful as a conversion factor to convert between mass and volume.

Example: The element tungsten used in light bulb filaments has a density of 19.3 g/cm^3. What is the volume, in cubic centimeters, of 250 g of tungsten?

Answer: $250\ \cancel{g} \times \dfrac{1\ cm^3}{19.3\ \cancel{g}} = 13\ cm^3$

UNDERSTANDING THE CONCEPTS

The chapter sections to review are shown in parentheses at the end of each problem.

2.83 In which of the following pairs do both numbers contain the same number of significant figures? (2.2)
    a. 2.0500 m and 0.0205 m
    b. 600.0 K and 60 K
    c. 0.000 75 s and 75 000 s
    d. 6.240 L and 6.240 × 10^{-2} L

2.84 In which of the following pairs do both numbers contain the same number of significant figures? (2.2)
    a. 3.44 × 10^{-3} g and 0.0344 g
    b. 0.0098 s and 9.8 × 10^4 s
    c. 6.8 × 10^3 m and 68 000 m
    d. 258.000 g and 2.58 × 10^{-2} g

2.85 Indicate if each of the following is an exact number or a measured number: (2.2)

    a. number of legs   **b.** height of table
    c. chairs at the table

2.86 Use the figure in problem 2.85 to indicate if each of the following is an exact number or a measured number: (2.2)
    a. area of tabletop   **b.** plates on the table
    c. length of tabletop

2.87 State the temperature on the Celsius thermometer to the correct number of significant figures: (2.3)

2.88 State the temperature on the Celsius thermometer to the correct number of significant figures: (2.3)

2.89 The length of this rug is 38.4 in. and the width is 24.2 in. (2.3, 2.6)

   a. What is the length of this rug, in centimeters?
   b. What is the width of this rug, in centimeters?
   c. How many significant figures are in the length measurement?
   d. Calculate the area of the rug, in square centimeters, to the correct number of significant figures. (Area = Length × Width)

2.90 A shipping box has a length of 7.00 in., a width of 6.00 in., and a height of 4.00 in. (2.3, 2.6)

   a. What is the length of the box, in centimeters?
   b. What is the width of the box, in centimeters?
   c. How many significant figures are in the width measurement?
   d. Calculate the volume of the box, in cubic centimeters, to the correct number of significant figures. (Volume = Length × Width × Height)

2.91 Each of the following diagrams represents a container of water and a cube. Some cubes float while others sink. Match diagrams **1**, **2**, **3**, or **4** with one of the following descriptions and explain your choices: (2.7)

Solid   Water

   1   **2**   **3**   **4**

   a. The cube has a greater density than water.
   b. The cube has a density that is 0.80 g/mL.
   c. The cube has a density that is one-half the density of water.
   d. The cube has the same density as water.

2.92 What is the density of the solid object that is weighed and submerged in water? (2.7)

2.93 Consider the following solids. The solids **A**, **B**, and **C** represent aluminum (D = 2.70 g/cm³), gold (D = 19.3 g/cm³), and silver (D = 10.5 g/cm³). If each has a mass of 10.0 g, what is the identity of each solid? (2.7)

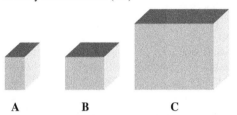

   A   **B**   **C**

2.94 A graduated cylinder contains three liquids **A**, **B**, and **C**, which have different densities and do not mix: mercury (D = 13.6 g/mL), vegetable oil (D = 0.92 g/mL), and water (D = 1.00 g/mL). Identify the liquids **A**, **B**, and **C** in the cylinder. (2.7)

2.95 The gray cube has a density of 4.5 g/cm³. Is the density of the green cube the same, lower than, or higher than that of the gray cube? (2.7)

2.96 The gray cube has a density of 4.5 g/cm³. Is the density of the green cube the same, lower than, or higher than that of the gray cube? (2.7)

ADDITIONAL PRACTICE PROBLEMS

2.97 Identify each of the following numbers as measured or exact, and write the number of significant figures (SFs) in each of the measured numbers: (2.2, 2.3)
a. 5.82 g
b. 1 L = 1000 mL
c. 0.009 050 m
d. 1.08×10^3 kg

2.98 Identify each of the following numbers as measured or exact, and write the number of significant figures (SFs) in each of the measured numbers: (2.2, 2.3)
a. 2.3055 cm
b. 3.40×10^{-4} km
c. 82 mg
d. 2.54 cm = 1 in.

2.99 Round off or add zeros to the following calculated answers to give a final answer with three significant figures: (2.2)
a. 0.000 012 58 L
b. 3.528×10^2 kg
c. 125 111 m
d. 34.9673 s

2.100 Round off or add zeros to the following calculated answers to give a final answer with three significant figures: (2.2)
a. 58.703 mL
b. 3×10^{-3} s
c. 0.010 826 g
d. 1.7484×10^3 ms

2.101 Perform each of the following operations, and write answers with the correct number of significant figures or decimal places: (2.2, 2.3)
a. 3.285 + 14.8 + 7.06
b. $(1.20 \times 10^{-4}) \times (2.234 \times 10^6)$
c. $46.7 \times \dfrac{23.22}{90.10}$
d. 85.2 + 6.081 − 5.608

2.102 Perform each of the following operations, and write answers with the correct numbers of significant figures or decimal places: (2.2, 2.3)
a. 3.7 + 4.888
b. $(4.09 \times 10^2) \times (6.33 \times 10^4)$
c. $3.376 \times 23 \times 0.123$
d. 0.0467 + 23.32 − 22.8

2.103 A dessert contains 137.25 g of vanilla ice cream, 84 g of fudge sauce, and 43.7 g of nuts. (2.3, 2.6)
a. What is the total mass, in grams, of the dessert?
a. What is the total weight, in pounds, of the dessert?

2.104 A fish company delivers 22 kg of salmon, 5.5 kg of crab, and 3.48 kg of oysters to your seafood restaurant. (2.3, 2.6)
a. What is the total mass, in kilograms, of the seafood?
a. What is the total number of pounds?

2.105 Perform each of the following conversions: (2.5)
a. 3.47 mm to m
b. 37 g to mg
c. 873 L to kL
d. 2.05×10^5 ns to s

2.106 Perform each of the following conversions: (2.5)
a. 4.8 L to mL
b. 77.8 g to kg
c. 0.008 67 mg to mcg
d. 4.6×10^9 mm to km

2.107 In France, grapes are 1.95 euros per kilogram. What is the cost of grapes, in dollars per pound, if the exchange rate is 1.14 dollars/euro? (2.6)

2.108 In Mexico, avocados are 48 pesos per kilogram. What is the cost, in cents, of an avocado that weighs 0.45 lb if the exchange rate is 18 pesos to the dollar? (2.6)

2.109 Bill's recipe for onion soup calls for 4.0 lb of thinly sliced onions. If an onion has an average mass of 115 g, how many onions does Bill need? (2.6)

2.110 The price of 1 lb of potatoes is $1.75. If all the potatoes sold today at the store bring in $1420, how many kilograms of potatoes did grocery shoppers buy? (2.6)

2.111 During a workout at the gym, you set the treadmill at a pace of 55.0 m/min. How many minutes will you walk if you cover a distance of 7500 ft? (2.6)

2.112 The distance between two cities is 1700 km. How long will it take, in hours, to drive from one city to the other if your average speed is 63 mi/h? (2.6)

2.113 Perform each of the following conversions: (2.5)
a. 40.5 in.^{2} to ft^2
b. 8.2 km^2 to m^2
c. 3.24 yd^2 to cm^2
d. 2.2×10^6 cm^3 to m^3

2.114 Perform each of the following conversions: (2.5)
a. 0.254 ft^2 to cm^2
b. 2.26 in.^{3} to cm^3
c. 0.22 cm^2 to mm^2
d. 7.8 m^2 to ft^2

2.115 Perform each of the following conversions: (2.6)
a. 65 mi/h to ft/s
b. $0.75/lb to cents/kg
c. 1.63 g/mL to lb/gal
d. 11 m/s to mi/h

2.116 Perform each of the following conversions: (2.6)
a. 76 km/L to mi/gal
b. 35 lb/in.^{2} to kg/cm^2
c. 55 mi/h to m/s
d. 2.5 mg/kg to oz/lb

2.117 The water level in a graduated cylinder initially at 215 mL rises to 285 mL after a piece of lead is submerged. What is the mass, in grams, of the lead (see Table 2.10)? (2.7)

2.118 A graduated cylinder contains 155 mL of water. A 15.0-g piece of iron and a 20.0-g piece of lead are added. What is the new water level, in milliliters, in the cylinder (see Table 2.10)? (2.7)

2.119 How many milliliters of gasoline have a mass of 1.2 kg (see Table 2.10)? (2.7)

2.120 What is the volume, in quarts, of 3.40 kg of ethanol (see Table 2.10)? (2.7)

Applications

2.121 The following nutrition information is listed on a box of crackers: (2.6)

Serving size 0.50 oz (6 crackers)

Fat 4 g per serving; Sodium 140 mg per serving
a. If the box has a net weight (contents only) of 8.0 oz, about how many crackers are in the box?
b. If you ate 10 crackers, how many ounces of fat did you consume?
c. How many servings of crackers in part **a** would it take to obtain the Daily Value (DV) for sodium, which is 2.4 g?

2.122 A dialysis unit requires 75 000 mL of distilled water. How many gallons of water are needed? (2.6)

2.123 To treat a bacterial infection, a doctor orders 4 tablets of amoxicillin per day for 10 days. If each tablet contains 250 mg of amoxicillin, how many ounces of the medication are given in 10 days? (2.6)

2.124 Celeste's diet restricts her intake of protein to 24 g per day. If she eats 1.2 oz of protein, has she exceeded her protein limit for the day? (2.6)

2.125 A doctor orders 5.0 mL of phenobarbital elixir. If the phenobarbital elixir is available as 30. mg per 7.5 mL, how many milligrams is given to the patient? (2.6)

2.126 A doctor orders 2.0 mg of morphine. The vial of morphine on hand is 10. mg/mL. How many milliliters of morphine should you administer to the patient? (2.6) ing—

CHALLENGE PROBLEMS

The following problems are related to the topics in this chapter. However, they do not all follow the chapter order, and may require you to combine concepts and skills from several sections. These problems will help you increase your critical thinking skills and prepare for your next exam.

2.127 A balance measures mass to 0.001 g. If you determine the mass of an object that weighs about 31 g, would you record the mass as 31 g, 31.1 g, 31.08 g, 31.075 g, or 31.0750? Explain your choice by writing two to three complete sentences that describe your thinking. (2.3)

2.128 When three students use the same meterstick to measure the length of a paper clip, they obtain results of 5.8 cm, 5.75 cm, and 5.76 cm. If the meterstick has millimeter markings, what are some reasons for the different values? (2.3)

2.129 A car travels at 55 mi/h and gets 11 km/L of gasoline. How many gallons of gasoline are needed for a 3.0-h trip? (2.6)

2.130 A sunscreen preparation contains 2.50% benzyl salicylate by mass. If a tube contains 4.0 oz of sunscreen, how many kilograms of benzyl salicylate are needed to manufacture 325 tubes of sunscreen? (2.6)

2.131 How many milliliters of olive oil have the same mass as 1.50 L of gasoline (see Table 2.10)? (2.7)

2.132 A 50.0-g silver object and a 50.0-g gold object are both added to 75.5 mL of water contained in a graduated cylinder. What is the new water level, in milliliters, in the cylinder (see Table 2.10)? (2.7)

Applications

2.133 **a.** An athlete with a body mass of 65 kg has 3.0% body fat. How many pounds of body fat does that person have? (2.6)
b. In liposuction, a doctor removes fat deposits from a person's body. If body fat has a density of 0.909 g/cm^3 and 3.0 L of fat is removed, how many pounds of fat were removed from the patient?

2.134 A mouthwash is 21.6% ethanol by mass. If each bottle contains 1.06 pt of mouthwash with a density of 0.876 g/mL, how many kilograms of ethanol are in 180 bottles of the mouthwash? (2.6, 2.7)

ANSWERS TO ENGAGE QUESTIONS

2.1 All the digits in the coefficient of a number written in scientific notation are significant.

2.2 The quantity 4 hats is obtained from counting. It is an exact number, whereas the value of 6.24 cm is obtained by measurement, which has three significant figures.

2.3 Because the first of the digits to be dropped (7) is 5 or greater, the last retained digit is increased by 1 to give 10.1.

2.4 The answer is rounded off to the same number of SFs as the measurement with the fewer SFs, which is one significant figure.

2.5 When the multiplication and division of the numbers with three significant figures give the whole number 5, two significant zeros are added to give the correct number of significant figures in the answer.

2.6 The answer from addition and subtraction is adjusted to the decimal place of the measurement with the fewest decimal places.

2.7 Because the numbers in the equality 1 g = 1000 mg are exact, 60. mg with two significant figures is the same as 0.060 g. 60. m̶g̶ × 1 g/1000 m̶g̶ = 0.060 g (2 SFs)

2.8 When we divide both sides of the equality 1 m = 100 cm by 100, we get 0.01 m = 1 cm.

2.9 1 mL is the same volume as a cube with sides of 1 cm × 1 cm × 1 cm = 1 cm^3.

2.10 The equality 1 day = 24 h can be written as two conversion factors, 24 h/1 day or 1 day/24 h.

2.11 The equality of a percent states the parts of a relationship to 100 parts of the whole, parts/100 parts × 100%. Two conversion factors can be written from this equality.

2.12 6.4 oz is multiplied by the conversion factor 1 lb/16 oz to cancel oz and give an answer in lb.

2.13 Two or more conversion factors can be used in a problem, as long as the units cancel to give the desired unit for the final answer.

2.14 Iron is more dense than water.

ANSWERS TO SELECTED PROBLEMS

2.1 **a.** g **b.** °C **c.** L **d.** lb **e.** s

2.3 **a.** volume **b.** length **c.** length **d.** time

2.5 **a.** meter, length **b.** gram, mass
c. milliliter, volume **d.** second, time
e. degree Celsius, temperature

2.7 **a.** second, time **b.** kilogram, mass
c. gram, mass **d.** degree Celsius, temperature

2.9 **a.** 2.2 cm, two SFs **b.** 3.90 cm, three SFs
c. 4.81 cm, three SFs

2.11 **a.** five SFs **b.** two SFs **c.** two SFs
d. three SFs **e.** four SFs **f.** three SFs

2.13 b and c

2.15 **a.** not significant **b.** significant **c.** significant
d. significant **e.** not significant

2.17 a. 5.0×10^3 L **b.** 3.0×10^4 g
c. 1.0×10^5 m **d.** 2.5×10^{-4} cm

2.19 a. measured **b.** exact
c. exact **d.** measured

2.21 a. 6 oz **b.** none
c. 0.75 lb, 350 g **d.** none (definitions are exact)

2.23 a. measured, four SFs **b.** measured, two SFs
c. measured, three SFs **d.** exact

2.25 a. 1.85 kg **b.** 88.2 L
c. 0.004 74 cm **d.** 8810 m
e. 1.83×10^5 s

2.27 a. 56.9 m **b.** 0.002 28 g
c. 11 500 s (1.15×10^4 s) **d.** 8.10 L

2.29 a. 1.6 **b.** 0.01
c. 27.6 **d.** 3.5
e. 0.14 (1.4×10^{-1}) **f.** 0.8 (8×10^{-1})

2.31 a. 53.54 **b.** 127.6
c. 121.5 **d.** 0.50
e. 2500 **f.** 156.7

2.33 a. mg **b.** dL
c. km **d.** pg

2.35 a. centiliter **b.** kilogram
c. millisecond **d.** petameter

2.37 a. 0.01 **b.** 1×10^{12}
c. 0.001 **d.** 0.1

2.39 a. decigram **b.** microgram
c. kilogram **d.** centigram

2.41 a. 100 cm **b.** 1×10^9 nm
c. 0.001 m **d.** 1000 mL

2.43 a. kilogram **b.** milliliter
c. km **d.** kL
e. nanometer

2.45 A conversion factor can be inverted to give a second conversion factor.

2.47 a. $1 \text{ kg} = 1 \times 10^6 \text{ mg}$; $\dfrac{1 \times 10^6 \text{ mg}}{1 \text{ kg}}$ and $\dfrac{1 \text{ kg}}{1 \times 10^6 \text{ mg}}$

b. $1 \text{ g} = 1 \times 10^9 \text{ ng}$; $\dfrac{1 \times 10^9 \text{ ng}}{1 \text{ g}}$ and $\dfrac{1 \text{ g}}{1 \times 10^9 \text{ ng}}$

c. $1 \text{ kL} = 1000 \text{ L}$; $\dfrac{1000 \text{ L}}{1 \text{ kL}}$ and $\dfrac{1 \text{ kL}}{1000 \text{ L}}$

d. $1 \text{ s} = 1000 \text{ ms}$; $\dfrac{1000 \text{ ms}}{1 \text{ s}}$ and $\dfrac{1 \text{ s}}{1000 \text{ ms}}$

e. $1 \text{ dm} = 100 \text{ mm}$; $\dfrac{100 \text{ mm}}{1 \text{ dm}}$ and $\dfrac{1 \text{ dm}}{100 \text{ mm}}$

f. $(1 \text{ m})^3 = (100 \text{ cm})^3 = 10^6 \text{ cm}^3$; $\dfrac{(100 \text{ cm})^3}{(1 \text{ m})^3}$ and $\dfrac{(1 \text{ m})^3}{(100 \text{ cm})^3}$

2.49 a. $1 \text{ yd} = 3 \text{ ft}$; $\dfrac{3 \text{ ft}}{1 \text{ yd}}$ and $\dfrac{1 \text{ yd}}{3 \text{ ft}}$

The 1 yd and 3 ft are both exact.

b. $1 \text{ kg} = 2.205 \text{ lb}$; $\dfrac{2.205 \text{ lb}}{1 \text{ kg}}$ and $\dfrac{1 \text{ kg}}{2.205 \text{ lb}}$

The 2.205 lb is measured: It has four SFs. The 1 kg is exact.

c. $1 \text{ gal} = 27 \text{ mi}$; $\dfrac{27 \text{ mi}}{1 \text{ gal}}$ and $\dfrac{1 \text{ gal}}{27 \text{ mi}}$

The 27 mi is measured: It has two SFs. The 1 gal is exact.

d. 100 g of sterling = 93 g of silver;

$\dfrac{93 \text{ g silver}}{100 \text{ g sterling}}$ and $\dfrac{100 \text{ g sterling}}{93 \text{ g silver}}$

The 93 g is measured: It has two SFs. The 100 g is exact.

e. $1 \text{ min} = 60 \text{ s}$; $\dfrac{60 \text{ s}}{1 \text{ min}}$ and $\dfrac{1 \text{ min}}{60 \text{ s}}$

The 1 min and 60 s are both exact.

f. $1 \text{ m}^3 = 1 \times 10^6 \text{ cm}^3$; $\dfrac{1 \times 10^6 \text{ cm}^3}{1 \text{ m}^3}$ and $\dfrac{1 \text{ m}^3}{1 \times 10^6 \text{ cm}^3}$

The 1 m and $1 \times 10^6 \text{ cm}^3$ are both exact.

2.51 a. $1 \text{ s} = 3.5 \text{ m}$; $\dfrac{3.5 \text{ m}}{1 \text{ s}}$ and $\dfrac{1 \text{ s}}{3.5 \text{ m}}$

The 3.5 m is measured: It has two SFs. The 1 s is exact.

b. 1 day = 3.5 g of potassium;

$\dfrac{3.5 \text{ g potassium}}{1 \text{ day}}$ and $\dfrac{1 \text{ day}}{3.5 \text{ g potassium}}$

The 3.5 g is measured: It has two SFs. The 1 day is exact.

c. $1 \text{ L} = 26.0 \text{ km}$; $\dfrac{26.0 \text{ km}}{1 \text{ L}}$ and $\dfrac{1 \text{ L}}{26.0 \text{ km}}$

The 26.0 km is measured: It has three SFs. The 1 L is exact.

d. 1 kg of plums = 29 mcg of pesticide;

$\dfrac{29 \text{ mcg pesticide}}{1 \text{ kg plums}}$ and $\dfrac{1 \text{ kg plums}}{29 \text{ mcg pesticide}}$

The 29 mcg is measured: It has two SFs. The 1 kg is exact.

e. 100 g of crust = 28.2 g of silicon;

$\dfrac{28.2 \text{ g silicon}}{100 \text{ g crust}}$ and $\dfrac{100 \text{ g crust}}{28.2 \text{ g silicon}}$

The 28.2 g is measured: It has three SFs. The 100 g is exact.

2.53 a. 1 tablet = 630 mg of calcium;

$\dfrac{630 \text{ mg calcium}}{1 \text{ tablet}}$ and $\dfrac{1 \text{ tablet}}{630 \text{ mg calcium}}$

The 630 mg is measured: It has two SFs. The 1 tablet is exact.

b. 1 day = 60 mg of vitamin C;

$\dfrac{60 \text{ mg vitamin C}}{1 \text{ day}}$ and $\dfrac{1 \text{ day}}{60 \text{ mg vitamin C}}$

The 60 mg is measured: It has one SF. The 1 day is exact.

c. 1 tablet = 50 mg of atenolol;

$\dfrac{50 \text{ mg atenolol}}{1 \text{ tablet}}$ and $\dfrac{1 \text{ tablet}}{50 \text{ mg atenolol}}$

The 50 mg is measured: It has one SF. The 1 tablet is exact.

d. 1 tablet = 81 mg of aspirin;

$\dfrac{81 \text{ mg aspirin}}{1 \text{ tablet}}$ and $\dfrac{1 \text{ tablet}}{81 \text{ mg aspirin}}$

The 81 mg is measured: It has two SFs. The 1 tablet is exact.

2.55 **a.** 5 mL of syrup $=$ 10 mg of Atarax;

$$\dfrac{10\ \text{mg Atarax}}{5\ \text{mL syrup}} \quad\text{and}\quad \dfrac{5\ \text{mL syrup}}{10\ \text{mg Atarax}}$$

b. 1 tablet $=$ 0.25 g of Lanoxin;

$$\dfrac{0.25\ \text{g Lanoxin}}{1\ \text{tablet}} \quad\text{and}\quad \dfrac{1\ \text{tablet}}{0.25\ \text{g Lanoxin}}$$

c. 1 tablet $=$ 300 mg of Motrin;

$$\dfrac{300\ \text{mg Motrin}}{1\ \text{tablet}} \quad\text{and}\quad \dfrac{1\ \text{tablet}}{300\ \text{mg Motrin}}$$

2.57 **a.** 0.0442 L   **b.** 8.65×10^9 nm
   c. 5.2×10^2 Mg   **d.** 7.2×10^5 ms

2.59 **a.** 1.555 kg   **b.** 63 in.
   c. 4.4 qt   **d.** 205 mm

2.61 **a.** 1.75 m   **b.** 5 L
   c. 5.5 g   **d.** 3.5 L

2.63 **a.** 473 mL   **b.** 79.4 kg
   c. 24 lb   **d.** 43 g

2.65 **a.** 66 gal   **b.** 3 tablets
   c. 1800 mg   **d.** 110 g

2.67 **a.** 6.3 h
   b. **i.** 6.8×10^{-4} g/lb   **ii.** 2.5 mL
   c. **i.** 4.00×10^{-4} oz/lb   **ii.** 0.0656 oz

2.69 **a.** 1.20 g/mL   **b.** 0.870 g/mL
   c. 3.10 g/mL   **d.** 1.28 g/mL

2.71 **a.** 4.51 g/mL   **b.** 1.42 g/mL   **c.** 2.70 g/mL

2.73 **a.** 1.9 L of ethanol   **b.** 88 g of mercury
   c. 83.3 oz of silver

2.75 **a.** 661 g   **b.** 34 kg   **c.** 29 cm^3

2.77 Because we calculate the density to be 11.3 g/cm^3, we identify the metal as lead.

2.79 **a.** 1.030   **b.** 1.03 g/mL
   c. 690 g   **d.** 382 mL

2.81 **a.** 1 dL of blood $=$ 42 mcg of iron;

$$\dfrac{42\ \text{mcg iron}}{1\ \text{dL blood}} \quad\text{and}\quad \dfrac{1\ \text{dL blood}}{42\ \text{mcg iron}}$$

   b. 3.4 mcg of iron

2.83 **c** and **d**

2.85 **a.** exact
   b. measured
   c. exact

2.87 61.5 °C

2.89 **a.** 97.5 cm   **b.** 61.5 cm
   c. three SFs   **d.** 6.00×10^3 cm^2

2.91 **a.** Diagram 3; a cube that has a greater density than the water will sink to the bottom.
   b. Diagram 4; a cube with a density of 0.80 g/mL will be about four-fifths submerged in the water.
   c. Diagram 1; a cube with a density that is one-half the density of water will be one-half submerged in the water.
   d. Diagram 2; a cube with the same density as water will float just at the surface of the water.

2.93 **A** would be gold; it has the highest density (19.3 g/cm^3) and the smallest volume.
B would be silver; its density is intermediate (10.5 g/cm^3) and the volume is intermediate.
C would be aluminum; it has the lowest density (2.70 g/cm^3) and the largest volume.

2.95 The green cube has the same volume as the gray cube. However, the green cube has a larger mass on the scale, which means that its mass/volume ratio is larger. Thus, the density of the green cube is higher than the density of the gray cube.

2.97 **a.** measured, three SFs   **b.** exact
   c. measured, four SFs   **d.** measured, three SFs

2.99 **a.** 0.000 012 6 L (1.26×10^{-5} L)
   b. 353 kg (3.53×10^2 kg)
   c. 125 000 m (1.25×10^5 m)
   d. 35.0 s

2.101 **a.** 25.1   **b.** 2.68×10^2 **c.** 12.0   **d.** 85.7

2.103 **a.** 265 g   **b.** 0.584 lb

2.105 **a.** 0.003 47 m   **b.** 37 000 mg
   c. 0.873 kL   **d.** 2.05×10^{-4} s

2.107 $1.01 per lb

2.109 16 onions

2.111 42 min

2.113 **a.** 0.281 ft^2 **b.** 8.2×10^6 m^2
   c. 2.71×10^4 cm^2 **d.** 2.2 m^3

2.115 **a.** 95 ft/s   **b.** 170 cents/kg
   c. 13.6 lb/gal   **d.** 25 mi/h

2.117 790 g

2.119 1600 mL (1.6×10^3 mL)

2.121 **a.** 96 crackers   **b.** 0.2 oz of fat   **c.** 17 servings

2.123 0.35 oz

2.125 20. mg

2.127 You would record the mass as 31.075 g. Because the balance will weigh to the nearest 0.001 g, the mass value would be reported to 0.001 g.

2.129 6.4 gal

2.131 1200 mL

2.133 **a.** 4.3 lb of body fat   **b.** 6.0 lb

Matter and Energy

Charles is 13 years old and overweight. His doctor is worried that Charles is at risk for type 2 diabetes and advises his mother to make an appointment with a dietitian. Daniel, a dietitian, explains to them that choosing the appropriate foods is important to living a healthy lifestyle, losing weight, and preventing or managing diabetes.

Daniel also explains that food contains potential or stored energy, and different foods contain different amounts of potential energy. For instance, carbohydrates contain 4 kcal/g (17 kJ/g), whereas fats contain 9 kcal/g (38 kJ/g). He then explains that diets high in fat require more exercise to burn the fats, as they contain more energy. When Daniel looks at Charles's typical daily diet, he calculates that Charles obtains 2500 kcal in one day, which exceeds the recommendation of the American Heart Association of 1800 kcal for boys 9 to 13 years of age. Daniel encourages Charles and his mother to include whole grains, fruits, and vegetables in their diet instead of foods high in fat. They also discuss food labels and the fact that smaller serving sizes of healthy foods are necessary to lose weight. Daniel also recommends that Charles exercise at least 60 min every day. Before leaving, Charles and his mother make an appointment for the following week to look at a weight loss plan.

CAREER

Dietitian

Dietitians specialize in helping individuals learn about good nutrition and the need for a balanced diet. This requires them to understand biochemical processes, the importance of vitamins and food labels, as well as the differences between carbohydrates, fats, and proteins in terms of their energy value and how they are metabolized. Dietitians work in a variety of environments, including hospitals, nursing homes, school cafeterias, and public health clinics. In these roles, they create specialized diets for individuals diagnosed with a specific disease or create meal plans for those in a nursing home.

UPDATE A Diet and Exercise Program

When Daniel sees Charles and his mother, they discuss a menu for weight loss. Charles is going to record his food intake and return to discuss his diet with Daniel. You can view the results in the **UPDATE A Diet and Exercise Program**, page 90, and calculate both the kilocalories that Charles consumes in one day and the weight that Charles has lost.

3.1 Classification of Matter

LEARNING GOAL Classify examples of matter as pure substances or mixtures.

Matter is anything that has mass and occupies space. Matter is everywhere around us: the orange juice we had for breakfast, the water we put in the coffee maker, the plastic bag we put our sandwich in, our toothbrush and toothpaste, the oxygen we inhale, and the carbon dioxide we exhale. All of this material is matter. The different types of matter are classified by their composition.

Pure Substances: Elements and Compounds

All matter is made of extremely small particles called *atoms*. Much of matter is made of atoms bonded together in definite arrangements called *molecules*. A **pure substance** is matter that consists of just one type of atom or one type of molecule. An **element**, the simplest type of a pure substance, is composed of only one type of atom such as silver, iron, or aluminum. Silver is composed of silver atoms, iron of iron atoms, and aluminum of aluminum atoms. A full list of the elements is found on the inside front cover of this text.

A **compound** is also a pure substance, but it consists of atoms of two or more elements always chemically combined in the same proportion. For example, in the compound water, there are two hydrogen atoms for every one oxygen atom, which is represented by the formula H_2O. This means that water always has the same composition of H_2O. Another compound that consists of a chemical combination of hydrogen and oxygen is hydrogen peroxide. It has two hydrogen atoms for every two oxygen atoms and is represented by the formula H_2O_2. Thus, water (H_2O) and hydrogen peroxide (H_2O_2) are different compounds even though they contain the same elements, hydrogen and oxygen.

Aluminum
atom

An aluminum can consists
of many atoms of aluminum.

ENGAGE 3.1

Why are elements and compounds both pure substances?

A water molecule, H_2O, consists of two atoms of hydrogen (white) for one atom of oxygen (red).

Hydroxide peroxide molecule

A hydroxide peroxide molecule, H_2O_2, consists of two atoms of hydrogen (white) for every two atoms of oxygen (red).

PRACTICE PROBLEMS

Try Practice Problems 3.1 and 3.2

Mixtures

In a **mixture**, two or more different substances are physically mixed. Much of the matter in our everyday lives consists of mixtures. The air we breathe is a mixture of mostly oxygen and nitrogen gases. The steel in buildings and railroad tracks is a mixture of iron, nickel, carbon, and chromium. The brass in doorknobs and musical instruments is a mixture of

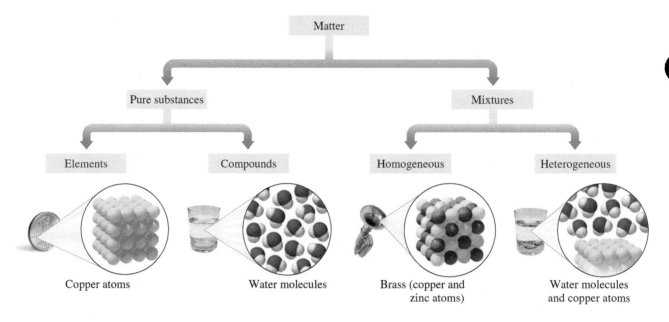

FIGURE 3.1 Matter is organized by its components: elements, compounds, and mixtures.

A cookie with raisins is a heterogeneous mixture.

ENGAGE 3.2

Why is a pizza heterogeneous, whereas vinegar is a homogeneous mixture?

copper and zinc (see **FIGURE 3.1**). Tea, coffee, and ocean water are mixtures too. The proportions of substances in a mixture are not consistent but can vary. For example, two sugar–water mixtures may look the same, but the one with the higher ratio of sugar to water would taste sweeter.

Types of Mixtures

Mixtures are classified further as homogeneous or heterogeneous. In a *homogeneous mixture*, also called a *solution*, the composition is uniform throughout the sample. We cannot see the individual components, which appear as one state. Familiar examples of homogeneous mixtures are air, which contains oxygen and nitrogen gases, and seawater, a solution of salt and water.

In a *heterogeneous mixture*, the components do not have a uniform composition throughout the sample. The components appear as two separate regions. A mixture of oil and water is heterogeneous because the oil floats on the surface of the water. Other examples of heterogeneous mixtures are a cookie with raisins and orange juice with pulp.

Chemistry Link to Health

Breathing Mixtures

The air we breathe is composed mostly of the gases oxygen (21%) and nitrogen (79%). The homogeneous breathing mixtures used by scuba divers differ from the air we breathe depending on the depth of the dive. Nitrox is a mixture of oxygen and nitrogen, but with more oxygen gas (up to 32%) and less nitrogen gas (68%) than air. A breathing mixture with less nitrogen gas decreases the risk of *nitrogen narcosis* associated with breathing regular air while diving. Heliox contains oxygen and helium, which is typically used for diving to more than 200 ft. By replacing nitrogen with helium, nitrogen narcosis does not occur. However, at dive depths over 300 ft, helium is associated with severe shaking and a drop in body temperature.

A breathing mixture used for dives over 400 ft is trimix, which contains oxygen, helium, and some nitrogen. The addition of some nitrogen lessens the problem of shaking that comes with breathing high levels of helium. Heliox and trimix are used only by professional, military, or other highly trained divers.

In hospitals, heliox may be used as a treatment for respiratory disorders and lung constriction in adults and premature infants. Heliox is less dense than air, which reduces the effort of breathing and helps distribute the oxygen gas to the tissues.

A nitrox mixture is used to fill scuba tanks.

SAMPLE PROBLEM 3.1 Classifying Mixtures

TRY IT FIRST

Classify each of the following as a pure substance (element or compound) or a mixture (homogeneous or heterogeneous):

a. copper in wire
b. chocolate-chip ice cream
c. nitrox, a combination of oxygen and nitrogen used to fill scuba tanks

SOLUTION

a. Copper, which is only one type of matter, is a pure substance. It is an element.
b. Chocolate-chip ice cream is two or more substances mixed together; it is a mixture. Because the distribution of substances in the ice cream is not uniform, it is heterogeneous.
c. Nitrox, consisting of nitrogen and oxygen, is a mixture. It has a uniform composition; it is homogeneous.

SELF TEST 3.1

a. A salad dressing is prepared with oil, vinegar, and chunks of blue cheese. Is this a homogeneous or heterogeneous mixture?
b. A mouthwash used to reduce plaque and clean gums and teeth contains several ingredients, such as menthol, alcohol, hydrogen peroxide, and a flavoring. Is this a homogeneous or heterogeneous mixture?

ANSWER

a. heterogeneous mixture
b. homogeneous mixture

PRACTICE PROBLEMS
Try Practice Problems 3.3 to 3.6

PRACTICE PROBLEMS

3.1 Classification of Matter

3.1 Classify each of the following pure substances as an element or a compound:
   a. a silicon (Si) chip
   b. hydrogen peroxide (H_2O_2)
   c. oxygen gas (O_2)
   d. rust (Fe_2O_3)
   e. methane (CH_4) in natural gas

3.2 Classify each of the following pure substances as an element or a compound:
   a. helium gas (He)
   b. sulfur (S)
   c. sugar ($C_{12}H_{22}O_{11}$)
   d. mercury (Hg) in a thermometer
   e. lye (NaOH)

3.3 Classify each of the following as a pure substance or a mixture:
   a. baking soda ($NaHCO_3$)
   b. a blueberry muffin
   c. ice (H_2O)
   d. zinc (Zn)
   e. trimix (oxygen, nitrogen, and helium) in a scuba tank

3.4 Classify each of the following as a pure substance or a mixture:
   a. a soft drink
   b. propane (C_3H_8)

   c. a cheese sandwich
   d. an iron (Fe) nail
   e. salt substitute (KCl)

Applications

3.5 A dietitian includes one of the following mixtures in the lunch menu. Classify each of the following as homogeneous or heterogeneous:
   a. vegetable soup
   b. tea
   c. fruit salad
   d. tea with ice and lemon slices

3.6 A dietitian includes one of the following mixtures in the lunch menu. Classify each of the following as homogeneous or heterogeneous:
   a. nonfat milk
   b. spaghetti and meatballs
   c. peanut butter sandwich
   d. cranberry juice

3.2 States and Properties of Matter

LEARNING GOAL Identify the states and the physical and chemical properties of matter.

On Earth, matter exists in one of three *physical forms* called the **states of matter**: *solids*, *liquids*, and *gases*. Water is a familiar substance that we routinely observe in all three states. In the solid state, water can be an ice cube or a snowflake. It is a liquid when it comes out of a faucet or fills a pool. Water forms a gas, or vapor, when it evaporates from wet clothes or boils in a pan. A **solid**, such as a pebble or a baseball, has a definite shape and volume. You can probably recognize several solids within your reach right now such as books, pencils, or a computer mouse. In a solid, strong attractive forces hold the particles close together. The particles in a solid are arranged in such a rigid pattern, their only movement is to vibrate in fixed positions. For many solids, this rigid structure produces a crystal such as that seen in amethyst. **TABLE 3.1** compares the three states of matter.

Amethyst, a solid, is a form of quartz that contains atoms of Si and O.

TABLE 3.1 A Comparison of Solids, Liquids, and Gases			
Characteristic	**Solid**	**Liquid**	**Gas**
Shape	Has a definite shape	Takes the shape of the container	Takes the shape of the container
Volume	Has a definite volume	Has a definite volume	Fills the volume of the container
Arrangement of Particles	Fixed, very close	Random, close	Random, far apart
Interaction Between Particles	Very strong	Strong	Essentially none
Movement of Particles	Vibrate in fixed positions	Move slowly around each other	Move rapidly, spread out
Examples	Ice, salt, iron	Water, oil, vinegar	Water vapor, helium, air

A **liquid** has a definite volume, but not a definite shape. In a liquid, the particles move in random directions but are sufficiently attracted to each other to maintain a definite volume, although not a rigid structure. Thus, when water, oil, or vinegar is poured from one container to another, the liquid maintains its own volume but takes the shape of the new container.

A **gas** does not have a definite shape or volume. In a gas, the particles are far apart, have little attraction to each other, and move rapidly to fill the shape and volume of their container. When you inflate a bicycle tire, the air, which is a gas, fills the entire volume of the tire. The propane gas in a tank fills the entire volume of the tank.

ENGAGE 3.3

Why does a gas take both the shape and volume of its container?

PRACTICE PROBLEMS

Try Practice Problems 3.7 and 3.8

Water as a liquid takes the shape of its container.

A gas takes the shape and volume of its container.

Physical Properties and Physical Changes

One way to describe matter is to observe its properties. For example, if you were asked to describe yourself, you might list characteristics such as your height and weight, the color of your eyes and skin, or the length, color, and texture of your hair.

Physical properties are those characteristics that can be observed or measured without affecting the identity of a substance. In chemistry, typical physical properties include the shape, color, melting point, boiling point, and physical state of a substance. For example, some of the physical properties of a penny include its round shape, orange-red color (from copper), solid state, and shiny luster. TABLE 3.2 gives more examples of physical properties of copper, which is found in pennies, electrical wiring, and copper pans.

Water is a substance that is commonly found in all three states: solid, liquid, and gas. When matter undergoes a **physical change**, its state, size, or appearance will change, but its composition does not change. The solid state of water, snow or ice, has a different appearance than its liquid or gaseous state, but all three states are water. In a physical change, no new substances are produced.

Physical change is used to separate mixtures because there are no chemical interactions between the components. For example, different coins, such as nickels, dimes, and quarters, can be separated by size; iron particles mixed with sand can be picked up with a magnet; and water is separated from cooked spaghetti by using a strainer (see FIGURE 3.2).

In the chemistry laboratory, mixtures are separated by various methods. Solids are separated from liquids by *filtration*, which involves pouring a mixture through a filter paper, set in a funnel. The solid (residue) remains in the filter paper, and the filtered liquid (filtrate) moves through. In *chromatography*, different components of a liquid mixture separate as they move at different rates up the surface of a piece of chromatography paper.

Copper, used in cookware, is a good conductor of heat.

TABLE 3.2	Some Physical Properties of Copper
State at 25 °C	Solid
Color	Orange-red
Odor	Odorless
Melting Point	1083 °C
Boiling Point	2567 °C
Luster	Shiny
Conduction of Electricity	Excellent
Conduction of Heat	Excellent

A mixture of a liquid and a solid is separated by filtration.

Different substances in a mixture of inks are separated as they travel at different rates up the surface of chromatography paper.

FIGURE 3.2 A mixture of spaghetti and water is separated using a strainer.

Chemical Properties and Chemical Changes

Chemical properties describe the ability of a substance to change into a new substance. For example, the rusting or corrosion of a metal, such as iron, is a chemical property. When a **chemical change** takes place, the original substances, iron (Fe) and oxygen (O_2), are converted into a new substance, rust (Fe_2O_3), which has different physical and chemical properties. Another example of a chemical change occurs when table salt, sodium chloride, is separated into sodium metal and chlorine gas, as seen in FIGURE 3.3. TABLE 3.3 summarizes physical and chemical properties and changes. TABLE 3.4 gives examples of some physical and chemical changes.

Sodium chloride

Chemical change

Sodium metal   and   Chlorine gas

FIGURE 3.3 The decomposition of salt, NaCl, produces the elements sodium and chlorine.

ENGAGE 3.4

Why is the decomposition of salt a chemical change?

CORE CHEMISTRY SKILL

Identifying Physical and Chemical Changes

A chemical change occurs when sugar is heated, forming a caramelized topping for flan.

ENGAGE 3.5

Why is the melting point of iron, 1538 °C, a physical property, whereas the heating of iron with oxygen to form rust, Fe_2O_3, is a chemical property?

A gold ingot is hammered to form gold leaf.

TABLE 3.3 Summary of Physical and Chemical Properties and Changes

	Physical	Chemical
Property	A characteristic of a substance: color, shape, odor, luster, size, melting point, or density.	A characteristic that indicates the ability of a substance to form another substance: paper can burn, iron can rust, silver can tarnish.
Change	A change in a physical property that retains the identity of the substance: a change of state, a change in size, or a change in shape.	A change in which the original substance is converted to one or more new substances: paper burns, iron rusts, silver tarnishes.

TABLE 3.4 Examples of Some Physical and Chemical Changes

Physical Changes	Chemical Changes
Water boils to form water vapor.	Water and cesium combine explosively.
Paper is cut into tiny pieces of confetti.	Paper burns with a bright flame and produces heat, ashes, carbon dioxide, and water vapor.
Sugar dissolves in water to form a sugar solution.	Heating sugar forms a smooth, caramel-colored substance.
Iron has a melting point of 1538 °C.	Iron, which is gray and shiny, combines with oxygen to form orange-red iron oxide (rust).

SAMPLE PROBLEM 3.2   Physical and Chemical Changes

TRY IT FIRST

Classify each of the following as a physical or chemical change:

a. A gold ingot is hammered to form gold leaf.
b. Gasoline burns in air.
c. Garlic is chopped into small pieces.
d. Milk left in a warm room turns sour.
e. A mixture of oil and water is separated.

SOLUTION

a. physical change   **b.** chemical change   **c.** physical change
d. chemical change   **e.** physical change

SELF TEST 3.2

Classify each of the following as a physical or chemical change:

a. Water freezes on a pond.
b. Gas bubbles form when baking powder is placed in vinegar.
c. A log is cut for firewood.
d. Butter melts in a warm room.

ANSWER

a. physical change   **b.** chemical change
c. physical change   **d.** physical change

INTERACTIVE VIDEO

 Pearson eText   Chemical vs. Physical Changes

PRACTICE PROBLEMS

Try Practice Problems 3.9 to 3.14

PRACTICE PROBLEMS

3.2 States and Properties of Matter

3.7 Indicate whether each of the following describes a gas, a liquid, or a solid:
   a. The breathing mixture in a scuba tank has no definite volume or shape.
   b. The neon atoms in a lighting display do not interact with each other.
   c. The particles in an ice cube are held in a rigid structure.

3.8 Indicate whether each of the following describes a gas, a liquid, or a solid:
   a. Lemonade has a definite volume but takes the shape of its container.
   b. The particles in a tank of oxygen are very far apart.
   c. Helium occupies the entire volume of a balloon.

3.9 Describe each of the following as a physical or chemical property:
   a. Chromium is a steel-gray solid.
   b. Hydrogen reacts readily with oxygen.
   c. A patient has a temperature of 40.2 °C.
   d. Ammonia will corrode iron.
   e. Butane gas in an igniter burns in oxygen.

3.10 Describe each of the following as a physical or chemical property:
   a. Neon is a colorless gas at room temperature.
   b. Apple slices turn brown when they are exposed to air.
   c. Phosphorus will ignite when exposed to air.
   d. At room temperature, mercury is a liquid.
   e. Propane gas is compressed to a liquid for placement in a small cylinder.

3.11 What type of change, physical or chemical, takes place in each of the following?
   a. Water vapor condenses to form rain.
   b. Cesium metal reacts explosively with water.
   c. Gold melts at 1064 °C.
   d. A puzzle is cut into 1000 pieces.
   e. Cheese is grated.

3.12 What type of change, physical or chemical, takes place in each of the following?
   a. Pie dough is rolled into thin pieces for a crust.
   b. A silver pin tarnishes in the air.
   c. A tree is cut into boards at a saw mill.
   d. Food is digested.
   e. A chocolate bar melts.

3.13 Describe each of the following properties for the element fluorine as physical or chemical:
   a. is highly reactive
   b. is a gas at room temperature
   c. has a pale, yellow color
   d. will explode in the presence of hydrogen
   e. has a melting point of −220 °C

3.14 Describe each of the following properties for the element zirconium as physical or chemical:
   a. melts at 1852 °C
   b. is resistant to corrosion
   c. has a grayish white color
   d. ignites spontaneously in air when finely divided
   e. is a shiny metal

3.3 Temperature

LEARNING GOAL Given a temperature, calculate the corresponding temperature on another scale.

Temperatures in science are measured and reported in units of degrees Celsius (°C). On the Celsius scale, the reference points are the freezing point of water, defined as 0 °C, and the boiling point, defined as 100 °C. In the United States, everyday temperatures are commonly reported in units of degrees Fahrenheit (°F). On the Fahrenheit scale, water freezes at 32 °F

REVIEW

Using Positive and Negative Numbers in Calculations (1.4)

Solving Equations (1.4)

Counting Significant Figures (2.2)

A digital ear thermometer is used to measure body temperature.

ENGAGE 3.6

Why is a degree Celsius a larger unit of temperature than a degree Fahrenheit?

and boils at 212 °F. A typical room temperature of 22 °C is the same as 72 °F. Normal human body temperature is 37.0 °C, which is the same temperature as 98.6 °F.

On the Celsius and Fahrenheit temperature scales, the temperature difference between freezing and boiling is divided into smaller units called *degrees*. On the Celsius scale, there are 100 degrees Celsius between the freezing and boiling points of water, whereas the Fahrenheit scale has 180 degrees Fahrenheit between the freezing and boiling points of water. That makes a degree Celsius almost twice the size of a degree Fahrenheit: 1 °C = 1.8 °F (see **FIGURE 3.4**).

$$180 \text{ degrees Fahrenheit} = 100 \text{ degrees Celsius}$$

$$\frac{180 \text{ degrees Fahrenheit}}{100 \text{ degrees Celsius}} = \frac{1.8 \text{ °F}}{1 \text{ °C}}$$

We can write a temperature equation that relates a Fahrenheit temperature and its corresponding Celsius temperature.

$$T_F = 1.8(T_C) + 32 \qquad \text{Temperature equation to obtain degrees Fahrenheit}$$

Changes °C to °F   Adjusts freezing point

FIGURE 3.4 A comparison of the Fahrenheit, Celsius, and Kelvin temperature scales between the freezing and boiling points of water.

In the equation, the Celsius temperature is multiplied by 1.8 to change °C to °F; then 32 is added to adjust the freezing point from 0 °C to the Fahrenheit freezing point, 32 °F. The values, 1.8 and 32, used in the temperature equation are exact numbers and are not used to determine significant figures in the answer.

To convert from degrees Fahrenheit to degrees Celsius, the temperature equation is rearranged to solve for T_C. First, we subtract 32 from both sides because we must apply the same operation to both sides of the equation.

$$T_F - 32 = 1.8(T_C) + \cancel{32} - \cancel{32}$$

$$T_F - 32 = 1.8(T_C)$$

CORE CHEMISTRY SKILL

Converting Between Temperature Scales

Second, we solve the equation for T_C by dividing both sides by 1.8.

$$\frac{T_F - 32}{1.8} = \frac{1.8(T_C)}{1.8}$$

$$T_C = \frac{T_F - 32}{1.8} \quad \text{Temperature equation to obtain degrees Celsius}$$

Scientists have learned that the coldest temperature possible is −273 °C (more precisely, −273.15 °C). On the *Kelvin* scale, this temperature, called *absolute zero*, has the value of 0 K. Units on the Kelvin scale are called kelvins (K); *no degree symbol is used*. Because there are no lower temperatures than absolute zero, the Kelvin scale has no negative temperature values. Between the freezing point of water, 273 K, and the boiling point, 373 K, there are 100 kelvins, which makes a kelvin equal in size to a degree Celsius.

$$1\text{ K} = 1\text{ °C}$$

We can write an equation that relates a Celsius temperature to its corresponding Kelvin temperature by adding 273 to the Celsius temperature. **TABLE 3.5** gives a comparison of some temperatures on the three scales.

$$T_K = T_C + 273 \quad \text{Temperature equation to obtain kelvins}$$

An antifreeze mixture in a car radiator will not freeze until the temperature drops to −37 °C. We can calculate the temperature of the antifreeze mixture in kelvins by adding 273 to the temperature in degrees Celsius.

$$T_K = -37\text{ °C} + 273 = 236\text{ K}$$

TABLE 3.5 A Comparison of Temperatures

Example	Fahrenheit (°F)	Celsius (°C)	Kelvin (K)
Sun	9937	5503	5776
A hot oven	450	232	505
Water boils	212	100	373
A high fever	104	40	313
Normal body temperature	98.6	37.0	310
Room temperature	70	21	294
Water freezes	32	0	273
A northern winter	−66	−54	219
Nitrogen liquefies	−346	−210	63
Absolute zero	−459	−273	0

ENGAGE 3.7

Show that −40. °C is the same temperature as −40. °F.

SAMPLE PROBLEM 3.3 Calculating Temperature

TRY IT FIRST

A dermatologist uses cryogenic nitrogen at −196 °C to remove skin lesions and some skin cancers. What is the temperature, in degrees Fahrenheit, of the nitrogen?

SOLUTION

STEP 1 State the given and needed quantities.

ANALYZE THE PROBLEM	Given	Need	Connect
	−196 °C	T in degrees Fahrenheit	temperature equation

STEP 2 Write a temperature equation.

$$T_F = 1.8(T_C) + 32$$

The low temperature of cryogenic nitrogen is used to destroy skin lesions.

STEP 3 **Substitute in the known values and calculate the new temperature.**

$$T_F = 1.8(-196) + 32 \quad \text{1.8 is exact; 32 is exact}$$

$$= -353 + 32 = -321\ {}^\circ\text{F} \quad \text{Answer to the ones place}$$

SELF TEST 3.3

a. In the process of making ice cream, rock salt is added to crushed ice to chill the ice cream mixture to $-11\ {}^\circ$C. What is that temperature in degrees Fahrenheit?
b. A hot tub reaches a temperature of 40.6 $^\circ$C. What is that temperature in degrees Fahrenheit?

ANSWER

a. 12 $^\circ$F **b.** 105 $^\circ$F

PRACTICE PROBLEMS
Try Practice Problems 3.15 to 3.18

SAMPLE PROBLEM 3.4 Calculating Degrees Celsius and Kelvins

TRY IT FIRST

In a type of cancer treatment called *thermotherapy*, temperatures as high as 113 $^\circ$F are used to destroy cancer cells. What is that temperature in degrees Celsius? In kelvins?

SOLUTION

STEP 1 **State the given and needed quantities.**

ANALYZE THE PROBLEM	Given	Need	Connect
	113 $^\circ$F	T in degrees Celsius, kelvins	temperature equations

STEP 2 **Write a temperature equation.**

$$T_C = \frac{T_F - 32}{1.8} \qquad T_K = T_C + 273$$

STEP 3 **Substitute in the known values and calculate the new temperature.**

$$T_C = \frac{(113 - 32)}{1.8} \quad \text{32 is exact; 1.8 is exact}$$

$$= \frac{81}{1.8} = 45\ {}^\circ\text{C}$$
Two SFs; Exact; Two SFs

Using the equation that converts degrees Celsius to kelvins, we substitute in degrees Celsius.

$$T_K = 45 + 273 = 318\ \text{K}$$
Ones place, Ones place, Ones place

SELF TEST 3.4

a. A child has a temperature of 103.6 $^\circ$F. What is this temperature on a Celsius thermometer?
b. A child who fell through the ice on a lake has hypothermia with a core temperature of 90. $^\circ$F. What is this temperature in degrees Celsius?

PRACTICE PROBLEMS
Try Practice Problems 3.19 and 3.20

ANSWER

a. 39.8 $^\circ$C **b.** 32 $^\circ$C

PRACTICE PROBLEMS

3.3 Temperature

3.15 Your friend who is visiting from Canada just took her temperature. When she reads 99.8 °F, she becomes concerned that she is quite ill. How would you explain this temperature to your friend?

3.16 You have a friend who is using a recipe for flan from a Mexican cookbook. You notice that he set your oven temperature at 175 °F. What would you advise him to do?

3.17 Calculate the unknown temperature in each of the following:
    a. 37.0 °C = _____ °F
    b. 65.3 °F = _____ °C
    c. −27 °C = _____ K
    d. 62 °C = _____ K
    e. 114 °F = _____ °C

3.18 Calculate the unknown temperature in each of the following:
    a. 25 °C = _____ °F
    b. 155 °C = _____ °F
    c. −25 °F = _____ °C
    d. 224 K = _____ °C
    e. 145 °C = _____ K

Applications

3.19 a. A patient with hyperthermia has a temperature of 106 °F. What does this read on a Celsius thermometer?
    b. Because high fevers can cause convulsions in children, the doctor needs to be called if the child's temperature goes over 40.0 °C. Should the doctor be called if a child has a temperature of 103 °F?

3.20 a. Water is heated to 145 °F. What is the temperature of the hot water in degrees Celsius?
    b. During extreme hypothermia, a child's temperature dropped to 20.6 °C. What was his temperature in degrees Fahrenheit?

Chemistry Link to Health

Variation in Body Temperature

Normal body temperature is considered to be 37.0 °C, although it varies throughout the day and from person to person. Oral temperatures of 36.1 °C are common in the morning and climb to a high of 37.2 °C between 6 P.M. and 10 P.M. Individuals who are involved in prolonged exercise may also experience elevated temperatures. Body temperatures of marathon runners can range from 39 °C to 41 °C as heat production during exercise exceeds the body's ability to lose heat. Temperatures above 37.2 °C for a person at rest are usually an indication of illness.

Hyperthermia occurs at body temperatures above 41 °C. Sweat production stops, and the skin becomes hot and dry. The pulse rate is elevated, and respiration becomes weak and rapid. The person can become lethargic and lapse into a coma. High body temperatures can lead to convulsions, particularly in children, which may cause permanent brain damage. Damage to internal organs is a major concern, and treatment, which must be immediate, may include immersing the person in an ice-water bath.

At the low temperature extreme of *hypothermia*, body temperature can drop as low as 28.5 °C. The person may appear cold and pale and have an irregular heartbeat. Unconsciousness can occur if the body temperature drops below 26.7 °C. Respiration becomes slow and shallow, and oxygenation of the tissues decreases. Treatment involves providing oxygen and increasing blood volume with glucose and saline fluids. Injecting warm fluids (37.0 °C) into the peritoneal cavity may restore the internal temperature.

3.4 Energy

LEARNING GOAL Identify energy as potential or kinetic; convert between units of energy.

Almost everything you do involves energy. When you are running, walking, dancing, or thinking, you are using energy to do *work*. In fact, **energy** is defined as the ability to do work. Suppose you are climbing a steep hill and you become too tired to go on. At that moment, you do not have the energy to do any more work. Now suppose you sit down and have lunch. In a while, you will have obtained energy from the food, and you will be able to do more work and complete the climb.

REVIEW

Rounding Off (2.3)

Using Significant Figures in Calculations (2.3)

Writing Conversion Factors from Equalities (2.5)

Using Conversion Factors (2.6)

ENGAGE 3.8

Why does a book have more potential energy when it is on the top of a high table than when it is on the floor?

PRACTICE PROBLEMS

Try Practice Problems 3.21 to 3.24

Water at the top of the dam has potential energy. When the water flows over the dam, potential energy is converted to kinetic energy.

CORE CHEMISTRY SKILL

Using Energy Units

Kinetic and Potential Energy

Energy can be classified as either kinetic energy or potential energy. **Kinetic energy** is the energy of motion. Any object that is moving has kinetic energy. **Potential energy** is determined by the position or composition of a substance. A boulder resting on top of a mountain has potential energy because of its location. If the boulder rolls down the mountain, the potential energy becomes kinetic energy. Water stored in a reservoir has potential energy. When the water goes over the dam and falls to the stream below, its potential energy is converted to kinetic energy. Foods and fossil fuels have potential energy. When you digest food or burn gasoline in your car, potential energy is converted to kinetic energy to do work.

Heat and Energy

Heat is the energy associated with the motion of particles. An ice cube feels cold because heat flows from your hand into the ice cube. The faster the particles move, the greater the heat or thermal energy of the substance. In the ice cube, the particles are moving very slowly. As heat is added, the motion of the particles in the ice cube increases. Eventually, the particles have enough energy to make the ice cube melt as it changes from a solid to a liquid.

Units of Energy

The SI unit of energy and work is the **joule (J)** (pronounced "jewel"). The joule is a small amount of energy, so scientists often use the kilojoule (kJ), 1000 joules. To heat water for one cup of tea, you need about 75 000 J or 75 kJ of heat. **TABLE 3.6** shows a comparison of energy in joules for several energy sources or uses.

You may be more familiar with the unit **calorie (cal)**, from the Latin *caloric*, meaning "heat." The calorie was originally defined as the amount of energy (heat) needed to raise the temperature of 1 g of water by 1 °C. Now, one calorie is defined as exactly 4.184 J. This equality can be written as two conversion factors:

$$1 \text{ cal} = 4.184 \text{ J (exact)} \qquad \frac{4.184 \text{ J}}{1 \text{ cal}} \quad \text{and} \quad \frac{1 \text{ cal}}{4.184 \text{ J}}$$

One *kilocalorie* (kcal) is equal to 1000 calories, and one *kilojoule* (kJ) is equal to 1000 joules. The equalities and conversion factors follow:

$$1 \text{ kcal} = 1000 \text{ cal} \qquad \frac{1000 \text{ cal}}{1 \text{ kcal}} \quad \text{and} \quad \frac{1 \text{ kcal}}{1000 \text{ cal}}$$

$$1 \text{ kJ} = 1000 \text{ J} \qquad \frac{1000 \text{ J}}{1 \text{ kJ}} \quad \text{and} \quad \frac{1 \text{ kJ}}{1000 \text{ J}}$$

TABLE 3.6 A Comparison of Energy for Various Resources and Uses

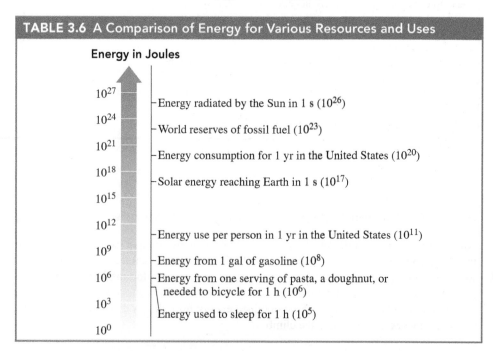

Energy in Joules

- 10^{27}
- 10^{24} — Energy radiated by the Sun in 1 s (10^{26})
- 10^{21} — World reserves of fossil fuel (10^{23})
- 10^{18} — Energy consumption for 1 yr in the United States (10^{20})
- 10^{15} — Solar energy reaching Earth in 1 s (10^{17})
- 10^{12}
- 10^{9} — Energy use per person in 1 yr in the United States (10^{11})
- 10^{6} — Energy from 1 gal of gasoline (10^{8})
- 10^{3} — Energy from one serving of pasta, a doughnut, or needed to bicycle for 1 h (10^{6})
- 10^{0} — Energy used to sleep for 1 h (10^{5})

SAMPLE PROBLEM 3.5   Energy Units

TRY IT FIRST

A defibrillator gives a high-energy shock of 360 J. What is this quantity of energy in calories?

A defibrillator provides electrical energy to heart muscle to re-establish normal rhythm.

SOLUTION

STEP 1 State the given and needed quantities.

ANALYZE THE PROBLEM	Given	Need	Connect
	360 J	calories	energy factor

STEP 2 Write a plan to convert the given unit to the needed unit.

$$\text{joules} \quad \boxed{\begin{array}{c}\text{Energy}\\\text{factor}\end{array}} \quad \text{calories}$$

STEP 3 State the equalities and conversion factors.

$$1 \text{ cal} = 4.184 \text{ J}$$
$$\frac{4.184 \text{ J}}{1 \text{ cal}} \quad \text{and} \quad \frac{1 \text{ cal}}{4.184 \text{ J}}$$

STEP 4 Set up the problem to calculate the needed quantity.

$$360 \text{ J} \quad \times \quad \overset{\text{Exact}}{\frac{1 \text{ cal}}{4.184 \text{ J}}} \quad = \quad 86 \text{ cal}$$

Two SFs   Exact   Two SFs

SELF TEST 3.5

a. When 1.0 g of glucose is metabolized in the body, it produces 3900 cal. How many joules are produced?

b. A swimmer expends 855 kcal during practice. How many kilojoules did the swimmer expend?

ANSWER

a. 16 000 J   **b.** 3580 kJ or 3.58×10^3 kJ

PRACTICE PROBLEMS

Try Practice Problems 3.25 to 3.28

PRACTICE PROBLEMS

3.4 Energy

3.21 Discuss the changes in the potential and kinetic energy of a roller-coaster ride as the roller-coaster car descends from the top of the ramp.

3.22 Discuss the changes in the potential and kinetic energy of a ski jumper taking the elevator to the top of the jump and going down the ramp.

3.23 Indicate whether each of the following statements describes potential or kinetic energy:
a. water at the top of a waterfall
b. kicking a ball

c. the energy in a lump of coal
d. a skier at the top of a hill

3.24 Indicate whether each of the following statements describes potential or kinetic energy:
a. the energy in your food
b. a tightly wound spring
c. a car speeding down the freeway
d. an earthquake

3.25 Convert each of the following energy units:
  a. 3500 cal to kcal
  b. 415 J to cal
  c. 28 cal to J
  d. 4.5 kJ to cal

3.26 Convert each of the following energy units:
  a. 8.1 kcal to cal
  b. 325 J to kJ
  c. 2550 cal to kJ
  d. 2.50 kcal to J

Applications

3.27 The energy needed to keep a 75-watt light bulb burning for 1.0 h is 270 kJ. Calculate the energy required to keep the light bulb burning for 3.0 h in each of the following energy units:
  a. joules   **b.** kilocalories

3.28 A person uses 750 kcal on a long walk. Calculate the energy used for the walk in each of the following energy units:
  a. joules   **b.** kilojoules

3.5 Specific Heat

LEARNING GOAL Calculate the specific heat for a substance. Use specific heat to calculate heat loss or gain.

Every substance has its own characteristic ability to absorb heat. When you bake a potato, you place it in a hot oven. If you are cooking pasta, you add the pasta to boiling water. You already know that adding heat to water increases its temperature until it boils. Certain substances must absorb more heat than others to reach a certain temperature.

The energy requirements for different substances are described in terms of a physical property called *specific heat*. The **specific heat (SH)** for a substance is defined as the amount of heat (q) in calories (or joules) needed to change the temperature of exactly 1 g of a substance by exactly 1 °C. To calculate the specific heat for a substance, we measure the heat in calories or joules, the mass in grams, and the temperature change written as ΔT.

Symbols Used in Specific Heat Equation		
Symbol	Meaning	Unit
SH	specific heat	$\dfrac{\text{cal (or J)}}{\text{g °C}}$
q	heat	calories (cal), joules (J)
m	mass	grams (g)
ΔT	temperature change	degrees Celsius (°C)

$$\text{Specific heat } (SH) = \frac{q}{\text{mass} \times \Delta T} = \frac{\text{cal (or J)}}{\text{g} \times \text{°C}}$$

The specific heat for water is written using our definition of the calorie and joule.

$$SH \text{ for H}_2\text{O}(l) = \frac{1.00 \text{ cal}}{\text{g °C}} = \frac{4.184 \text{ J}}{\text{g °C}}$$

ENGAGE 3.9

Why will 1 g of silver have a greater increase of temperature than 1 g of iron when each absorbs 400 J of heat?

If we look at **TABLE 3.7**, we see that 1 g of water requires 1.00 cal or 4.184 J to increase its temperature by 1 °C. The specific heat of water is about five times the specific heat of aluminum. Aluminum has a specific heat that is about twice that of copper. However, adding the same amount of heat (1.00 cal or 4.184 J) will raise the temperature of 1 g of aluminum

The high specific heat of water keeps temperatures more moderate in summer and winter.

TABLE 3.7 Specific Heats for Some Substances

Substance	cal/g °C	J/g °C
Elements		
Aluminum, Al(s)	0.214	0.897
Copper, Cu(s)	0.0920	0.385
Gold, Au(s)	0.0308	0.129
Iron, Fe(s)	0.108	0.452
Silver, Ag(s)	0.0562	0.235
Titanium, Ti(s)	0.125	0.523
Compounds		
Ammonia, NH₃(g)	0.488	2.04
Ethanol, C₂H₆O(l)	0.588	2.46
Sodium chloride, NaCl(s)	0.207	0.864
Water, H₂O(l)	1.00	4.184
Water, H₂O(s)	0.485	2.03

by about 5 °C and 1 g of copper by about 10 °C. The low specific heats of aluminum and copper mean they transfer heat efficiently, which makes them useful in cookware.

The high specific heat of water has a major impact on the temperatures in a coastal city compared to an inland city. A large mass of water near a coastal city can absorb or release five times the energy absorbed or released by the same mass of rock near an inland city. This means that in the summer a body of water absorbs large quantities of heat, which cools a coastal city, and then in the winter that same body of water releases large quantities of heat, which provides warmer temperatures. A similar effect happens with our bodies, which contain 70% water by mass. Water in the body absorbs or releases large quantities of heat to maintain body temperature.

SAMPLE PROBLEM 3.6   Calculating Specific Heat

TRY IT FIRST

What is the specific heat, in J/g °C, of lead if 57.0 J raises the temperature of 35.6 g of lead by 12.5 °C?

SOLUTION

STEP 1   State the given and needed quantities.

	Given	Need	Connect
ANALYZE THE PROBLEM	q = 57.0 J, m = 35.6 g of lead, ΔT = 12.5 °C	specific heat of lead (J/g °C)	specific heat equation

STEP 2   Write the relationship for specific heat. The specific heat (SH) is calculated by dividing the heat (q) by the mass (m) and by the temperature change (ΔT).

$$SH = \frac{\text{heat}}{\text{mass} \quad \Delta T} = \frac{q}{m \quad \Delta T}$$

STEP 3   Set up the problem to calculate the specific heat.

Three SFs

$$SH = \frac{57.0 \text{ J}}{35.6 \text{ g} \quad 12.5 \text{ °C}} = \frac{0.128 \text{ J}}{\text{g °C}}$$

Three SFs Three SFs   Three SFs

SELF TEST 3.6

a. What is the specific heat, in J/g °C, of sodium if 92.3 J is needed to raise the temperature of 3.00 g of sodium by 25.0 °C?
b. What is the specific heat, in J/g °C, of magnesium if 58.8 J is needed to raise the temperature of 4.65 g of sodium by 12.4 °C?

ANSWER
a. 1.23 J/g °C   **b.** 1.02 J/g °C

CORE CHEMISTRY SKILL
Calculating Specific Heat

PRACTICE PROBLEMS
Try Practice Problems 3.29 to 3.32

CORE CHEMISTRY SKILL
Using the Heat Equation

Heat Equation

When we know the specific heat of a substance, we can calculate the heat lost or gained by measuring the mass of the substance and the initial and final temperatures. We can substitute these measurements into the specific heat equation that is rearranged to solve for heat, which we call the **heat equation**.

$$SH = \frac{q}{m \times \Delta T}$$

$$m \times \Delta T \times SH = \frac{q}{\cancel{m} \times \cancel{\Delta T}} \times \cancel{m} \times \cancel{\Delta T}$$

We can write the heat equation as

$$q = m \times \Delta T \times SH$$

The heat lost or gained, in calories or joules, is obtained when the units of grams and °C in the numerator cancel grams and °C in the denominator of specific heat in the heat equation.

$$\text{cal} = \cancel{g} \times \cancel{°C} \times \frac{\text{cal}}{\cancel{g}\,\cancel{°C}}$$

$$\text{J} = \cancel{g} \times \cancel{°C} \times \frac{\text{J}}{\cancel{g}\,\cancel{°C}}$$

A cooling cap lowers the body temperature to reduce the oxygen required by the tissues.

SAMPLE PROBLEM 3.7 Calculating Heat Loss

TRY IT FIRST

During surgery or when a patient has suffered a cardiac arrest or stroke, lowering the body temperature will reduce the amount of oxygen needed by the body. Some methods used to lower body temperature include cooled saline solution, cool water blankets, or cooling caps worn on the head. How many kilojoules are lost when the body temperature of a surgery patient with a blood volume of 5500 mL is cooled from 38.5 °C to 33.2 °C? (Assume that the specific heat and density of blood is the same as for water.)

SOLUTION

STEP 1 State the given and needed quantities.

	Given	Need	Connect
ANALYZE THE PROBLEM	5500 mL of blood = 5500 g of blood, cooled from 38.5 °C to 33.2 °C	kilojoules removed	heat equation, specific heat of water

STEP 2 Calculate the temperature change (ΔT).

$$\Delta T = 38.5\,°C - 33.2\,°C = 5.3\,°C$$

STEP 3 Write the heat equation and needed conversion factors.

$$q = m \times \Delta T \times SH$$

$$SH_{\text{water}} = \frac{4.184\ \text{J}}{\text{g}\ °C}$$

$$\frac{4.184\ \text{J}}{\text{g}\ °C} \quad \text{and} \quad \frac{\text{g}\ °C}{4.184\ \text{J}}$$

$$1\ \text{kJ} = 1000\ \text{J}$$

$$\frac{1000\ \text{J}}{1\ \text{kJ}} \quad \text{and} \quad \frac{1\ \text{kJ}}{1000\ \text{J}}$$

STEP 4 Substitute in the given values and calculate the heat, making sure units cancel.

$$q = 5500\ \cancel{g} \times 5.3\ \cancel{°C} \times \overset{\text{Exact}}{\frac{4.184\ \cancel{J}}{\cancel{g}\ \cancel{°C}}} \times \overset{\text{Exact}}{\frac{1\ \text{kJ}}{1000\ \cancel{J}}} = 120\ \text{kJ}$$

Two SFs   Two SFs   Exact   Exact   Two SFs

SELF TEST 3.7

a. Dental implants use titanium because it is strong, is nontoxic, and can fuse to bone. How many calories are needed to raise the temperature of 16 g of titanium from 36.0 °C to 41.3 °C (see Table 3.7)?

b. Some cooking pans have a layer of copper on the bottom. How many kilojoules are needed to raise the temperature of 125 g of copper from 22 °C to 325 °C (see Table 3.7)?

ANSWER

a. 11 cal   **b.** 14.6 kJ

A dental implant contains the metal titanium that is attached to a crown.

Another use of the heat equation is to calculate the mass, in grams, of a substance by rearranging the heat equation to solve for mass (*m*) as shown in Sample Problem 3.8.

Rearrange the heat equation to solve for mass (*m*).

SAMPLE PROBLEM 3.8 Using the Heat Equation

TRY IT FIRST

When 655 J is added to a sample of ethanol, its temperature rises from 18.2 °C to 32.8 °C. What is the mass, in grams, of the ethanol sample (see Table 3.7)?

SOLUTION

STEP 1 State the given and needed quantities.

	Given	Need	Connect
ANALYZE THE PROBLEM	q = 655 J, heated from 18.2 °C to 32.8 °C	grams of ethanol	heat equation

STEP 2 Calculate the temperature change (ΔT).

$$\Delta T = 32.8\,°C - 18.2\,°C = 14.6\,°C$$

STEP 3 Write the heat equation.

$$q = m \times \Delta T \times SH$$

When the heat equation is rearranged for mass (*m*), the heat is divided by the temperature change and the specific heat.

$$m = \frac{q}{\Delta T \; SH}$$

STEP 4 Substitute in the given values and solve, making sure units cancel.

Three SFs

$$m = \frac{655\,\text{J}}{14.6\,°\text{C} \; \dfrac{2.46\,\text{J}}{\text{g}\,°\text{C}}} = 18.2\,\text{g}$$

Three SFs   Three SFs

Three SFs

SELF TEST 3.8

a. When 8.81 kJ is absorbed by a piece of iron, its temperature rises from 15 °C to 122 °C. What is the mass, in grams, of the piece of iron (see Table 3.7)?
b. A silver bar with a mass of 284 g at an initial temperature of 25.3 °C absorbs 1.13 kJ of heat. What is the final temperature of the silver bar (see Table 3.7)?

ANSWER

a. 182 g   b. 42.2 °C

PRACTICE PROBLEMS

Try Practice Problems 3.33 to 3.40

Heat Exchange: Heat Gain Equals Heat Loss

If a piece of metal is dropped into a container of cold water, the metal cools and the water warms until they are both at the same temperature. We assume that the heat lost by the metal is equal to the heat gained by the water. The heat equation allows us to calculate the heat gained by the water. Because the heat loss and heat gain are equal, we can use the heat equation again to calculate the specific heat of the metal.

SAMPLE PROBLEM 3.9   Heat Exchange

TRY IT FIRST

A 44-g piece of chromium metal at 58.3 °C is dropped into a beaker containing 20. g of water at 24.6 °C. All the heat lost by the metal heats the water. The final temperature of the water and metal is 31.0 °C. Calculate the specific heat, in J/g °C, of chromium.

SOLUTION

STEP 1   State the given and needed quantities.

	Given	Need	Connect
ANALYZE THE PROBLEM	44 g of metal, $\Delta T = 58.3\,°C - 31.0\,°C =$ 27.3 °C, 20. g of water, $\Delta T = 31.0\,°C - 24.6\,°C = 6.4\,°C$	specific heat of metal	heat gain equals heat loss, specific heat of water

STEP 2   Write the heat equation.

$$q \quad = \quad m \quad \times \quad \Delta T \quad \times \quad SH$$

For water,

$$q \quad = \quad 20.\,\cancel{g} \quad \times \quad 6.4\,\cancel{°C} \quad \times \quad \frac{4.184\ \text{J}}{\cancel{g}\,\cancel{°C}} \quad = 540\ \text{J}$$

Because the heat gained by water is equal to the heat lost by the metal, q for the metal is equal to 540 J.

STEP 3   Write the relationship for specific heat and calculate the specific heat.
The specific heat (SH) is calculated by dividing the heat (q) by the mass (m) and by the temperature change (ΔT).

Solving the heat equation for the specific heat of the metal:

$$SH \ = \ \frac{q}{m\ \ \Delta T} \ = \ \frac{540\ \text{J}}{44\ \text{g} \ \times \ 27.3\,°C} \ = \ \frac{0.45\ \text{J}}{\text{g}\ °C}$$

SELF TEST 3.9

A 50.-g piece of zinc metal at 62.0 °C is dropped into 24 g of water at 29.3 °C. All the heat lost by the metal is used to heat the water, and the final temperature of the water and metal is 34.6 °C. Calculate the specific heat, in J/g °C, of zinc.

ANSWER

$$\frac{0.39\ \text{J}}{\text{g}\ °C}$$

PRACTICE PROBLEMS

3.5 Specific Heat

3.29 If the same amount of heat is supplied to samples of 10.0 g each of aluminum, iron, and copper, all at 15.0 °C, which sample would reach the highest temperature (see Table 3.7)?

3.30 Substances A and B are the same mass and at the same initial temperature. When the same amount of heat is added to each, the final temperature of A is 75 °C and B is 35 °C. What does this tell you about the specific heats of A and B?

3.31 Calculate the specific heat (J/g °C) for each of the following:
   a. a 13.5-g sample of zinc heated from 24.2 °C to 83.6 °C that absorbs 312 J of heat
   b. a 48.2-g sample of a metal that absorbs 345 J with a temperature increase from 35.0 °C to 57.9 °C

3.32 Calculate the specific heat (J/g °C) for each of the following:
   a. an 18.5-g sample of tin that absorbs 183 J of heat when its temperature increases from 35.0 °C to 78.6 °C
   b. a 22.5-g sample of a metal that absorbs 645 J when its temperature increases from 36.2 °C to 92.0 °C

3.33 Use the heat equation to calculate the energy, in joules, for each of the following (see Table 3.7):
   a. required to heat 25.0 g of water from 12.5 °C to 25.7 °C
   b. required to heat 38.0 g of copper from 122 °C to 246 °C

3.34 Use the heat equation to calculate the energy, in joules, for each of the following (see Table 3.7):
   a. required to heat 5.25 g of water from 5.5 °C to 64.8 °C
   b. required to heat 10.0 g of silver from 112 °C to 275 °C

3.35 Calculate the mass, in grams, for each of the following using Table 3.7:
   a. a sample of gold that absorbs 225 J to increase its temperature from 15.0 °C to 47.0 °C
   b. a sample of iron that loses 8.40 kJ when its temperature decreases from 168.0 °C to 82.0 °C

3.36 Calculate the mass, in grams, for each of the following using Table 3.7:
   a. a sample of water, that absorbs 8250 J when its temperature increases from 18.4 °C to 92.6 °C
   b. a sample of silver that loses 3.22 kJ when its temperature decreases from 145 °C to 24 °C

3.37 Calculate the change in temperature (ΔT) for each of the following using Table 3.7:
   a. 20.0 g of iron that absorbs 1580 J
   b. 150.0 g of water that absorbs 7.10 kJ

3.38 Calculate the change in temperature (ΔT) for each of the following using Table 3.7:
   a. 115 g of copper that loses 2.45 kJ
   b. 22.0 g of silver that loses 625 J

3.39 **a.** A 13.0-g piece of glass at 64.8 °C is dropped into 38 g of water at 22.9 °C. All the heat lost by the glass is used to heat the water, and the final temperature of the water and glass is 25.6 °C. Calculate the specific heat, in J/g °C, of the glass.
   b. A 19.6-g piece of metal at 88.7 °C is dropped into 42 g of water at 24.2 °C. All the heat lost by the metal is used to heat the water, and the final temperature of the water and metal is 27.6 °C. Calculate the specific heat, in J/g °C, of the metal.

3.40 **a.** A 22.8-g piece of metal at 92.6 °C is dropped into 46 g of water at 24.7 °C. All the heat lost by the metal is used to heat the water, and the final temperature of the water and metal is 28.2 °C. Calculate the specific heat, in J/g °C, of the metal.
   b. A 23.4-g piece of metal at 96.9 °C is dropped into 52 g of water at 23.3 °C. All the heat lost by the metal is used to heat the water, and the final temperature of the water and metal is 24.4 °C. Calculate the specific heat, in J/g °C, of the metal.

3.6 Energy and Nutrition

LEARNING GOAL Use the energy values to calculate the kilocalories (kcal) or kilojoules (kJ) for a food.

The food we eat provides energy to do work in the body, which includes the growth and repair of cells. Carbohydrates are the primary fuel for the body, but if the carbohydrate reserves are exhausted, fats and then proteins are used for energy.

For many years in the field of nutrition, the energy from food was measured as Calories or kilocalories. The nutritional unit *Calorie, Cal* (with an uppercase C), is the same as 1000 cal, or 1 kcal. The international unit, kilojoule (kJ), is becoming more prevalent. For example, a baked potato has an energy content of 100 Calories, which is 100 kcal or 440 kJ. A typical diet that provides 2100 Cal (kcal) is the same as an 8800 kJ diet.

$$1 \text{ Cal} = 1 \text{ kcal} = 1000 \text{ cal}$$

$$1 \text{ Cal} = 4.184 \text{ kJ} = 4184 \text{ J}$$

In the nutrition laboratory, foods are burned in a steel container called a *calorimeter* to determine their *energy value* (kcal/g or kJ/g) (see **FIGURE 3.5**). A measured amount of water is added to fill the area surrounding the combustion chamber. The food sample

PRACTICE PROBLEMS

Try Practice Problems 3.41 and 3.42

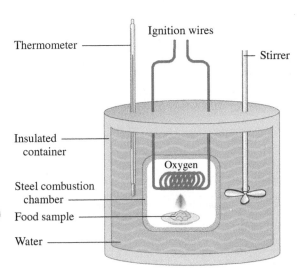

Thermometer
Ignition wires
Stirrer
Insulated container
Oxygen
Steel combustion chamber
Food sample
Water

FIGURE 3.5 Heat released from burning a food sample in a calorimeter is used to determine the energy value for the food.

is burned, releasing heat that increases the temperature of the water. From the known mass of the food and water, as well as the measured temperature increase, the energy value for the food is calculated. We assume that the energy absorbed by the calorimeter is negligible.

Energy Values for Foods

The **energy values** for food, listed in **TABLE 3.8**, are the kilocalories or kilojoules obtained from burning 1 g of carbohydrate, fat, or protein. Using these energy values, we can calculate the total energy for a food if the mass of each food type is known.

$$\text{kilocalories} = \cancel{g} \times \frac{\text{kcal}}{\cancel{g}} \qquad \text{kilojoules} = \cancel{g} \times \frac{\text{kJ}}{\cancel{g}}$$

On packaged food, the energy content is listed on the Nutrition Facts label, usually in terms of the number of Calories or kilojoules for one serving. The composition and energy content for some foods are given in **TABLE 3.9**. The total energy in kilocalories for each food type was calculated using energy values in kilocalories. Total energy in kilojoules was calculated using energy values in kilojoules. The energy for each food type was rounded off to the tens place.

TABLE 3.8 Typical Energy Values for the Three Food Types

Food Type	kcal/g	kJ/g
Carbohydrate	4	17
Fat	9	38
Protein	4	17

ENGAGE 3.11

What type of food provides the most energy per gram?

TABLE 3.9 Composition and Energy Content for Some Foods

Food	Carbohydrate (g)	Fat (g)	Protein (g)	Energy (kcal, kJ)
Apple, 1 medium	15	0	0	60 kcal (260 kJ)
Banana, 1 medium	26	0	1	110 kcal (460 kJ)
Beef, ground, 3 oz	0	14	22	220 kcal (900 kJ)
Broccoli, 3 oz	4	0	3	30 kcal (120 kJ)
Carrots, 1 cup	11	0	2	50 kcal (220 kJ)
Chicken, no skin, 3 oz	0	3	20	110 kcal (450 kJ)
Egg, 1 large	0	6	6	70 kcal (330 kJ)
Milk, nonfat, 1 cup	12	0	9	90 kcal (350 kJ)
Potato, baked	23	0	3	100 kcal (440 kJ)
Salmon, 3 oz	0	5	16	110 kcal (460 kJ)
Steak, 3 oz	0	27	19	320 kcal (1350 kJ)

Snack Crackers

Nutrition Facts
Serving Size 14 crackers (31g)
Servings Per Container About 7

Amount Per Serving

Calories 130   Calories from Fat 40

	% Daily Value*
Total Fat 4g	6%
Saturated Fat 0.5g	3%
Trans Fat 0g	
Polyunsaturated Fat 0.5%	
Monounsaturated Fat 1.5g	
Cholesterol 0mg	0%
Sodium 310mg	13%
Total Carbohydrate 19g	6%
Dietary Fiber Less than 1g	4%
Sugars 2g	
Proteins 2g	

The Nutrition Facts include the total Calories and the grams of carbohydrate, fat, and protein per serving.

SAMPLE PROBLEM 3.10   Calculating the Energy from a Food

TRY IT FIRST

While working on his diet log, Charles observed that the Nutrition Facts label for Snack Crackers states that one serving contains 19 g of carbohydrate, 4 g of fat, and 2 g of protein. If Charles eats one serving of Snack Crackers, what is the energy, in kilocalories, from each food type and the total kilocalories? (Round off the kilocalories for each food type to the tens place.)

SOLUTION

STEP 1   State the given and needed quantities.

ANALYZE THE PROBLEM	Given	Need	Connect
	19 g of carbohydrate, 4 g of fat, 2 g of protein	total number of kilocalories	energy values

STEP 2 Use the energy value for each food type to calculate the kilocalories, rounded off to the tens place.

Food Type	Mass	Energy Value	Energy
Carbohydrate	19 g	× $\dfrac{4 \text{ kcal}}{1 \text{ g}}$ =	80 kcal
Fat	4 g	× $\dfrac{9 \text{ kcal}}{1 \text{ g}}$ =	40 kcal
Protein	2 g	× $\dfrac{4 \text{ kcal}}{1 \text{ g}}$ =	10 kcal

STEP 3 Add the energy for each food type to give the total energy from the food.

Total energy = 80 kcal + 40 kcal + 10 kcal = 130 kcal

SELF TEST 3.10

a. Using the mass of carbohydrate listed in the Nutrition Facts, calculate the energy, in kilojoules, from the carbohydrate in one serving of Snack Crackers. (Round off the kilojoules to the tens place.)
b. Using the energy calculations from Sample Problem 3.10, calculate the percentage of energy obtained from fat in one serving of Snack Crackers.

ANSWER

a. 320 kJ
b. 31%

PRACTICE PROBLEMS

Try Practice Problems 3.43 to 3.50

Chemistry Link to Health

Losing and Gaining Weight

The number of kilocalories or kilojoules needed in the daily diet of an adult depends on gender, age, and level of physical activity. Some typical levels of energy needs are given in **TABLE 3.10**.

A person gains weight when food intake exceeds energy output. The amount of food a person eats is regulated by the hunger center in the hypothalamus, which is located in the brain. Food intake is normally proportional to the nutrient stores in the body. If these nutrient stores are low, you feel hungry; if they are high, you do not feel like eating.

A person loses weight when food intake is less than energy output. Many diet products contain cellulose, which has no nutritive value but provides bulk and makes you feel full. Some diet drugs depress the hunger center and must be used with caution because they excite the nervous system and can elevate blood pressure. Because muscular exercise is an important way to expend energy, an increase in daily exercise aids weight loss. **TABLE 3.11** lists some activities and the amount of energy they require.

TABLE 3.10 Typical Energy Requirements for Adults

Gender	Age	Moderately Active kcal (kJ)	Highly Active kcal (kJ)
Female	19–30	2100 (8800)	2400 (10 000)
	31–50	2000 (8400)	2200 (9200)
Male	19–30	2700 (11 300)	3000 (12 600)
	31–50	2500 (10 500)	2900 (12 100)

TABLE 3.11 Energy Expended by a 70.0-kg (154-lb) Adult

Activity	Energy (kcal/h)	Energy (kJ/h)
Sleeping	60	250
Sitting	100	420
Walking	200	840
Swimming	500	2100
Running	750	3100

One hour of swimming uses 2100 kJ of energy.

PRACTICE PROBLEMS

3.6 Energy and Nutrition

3.41 Calculate the kilocalories for each of the following:
  a. one stalk of celery that produces 125 kJ when burned in a calorimeter
  b. a waffle that produces 870 kJ when burned in a calorimeter

3.42 Calculate the kilocalories for each of the following:
  a. one cup of popcorn that produces 131 kJ when burned in a calorimeter
  b. a sample of butter that produces 23.4 kJ when burned in a calorimeter

3.43 Using the energy values for foods (see Table 3.8), determine each of the following (round off the answer for each food type to the tens place):
  a. the total kilojoules for one cup of orange juice that contains 26 g of carbohydrate, no fat, and 2 g of protein
  b. the grams of carbohydrate in one apple if the apple has no fat and no protein and provides 72 kcal of energy
  c. the kilocalories in one tablespoon of vegetable oil, which contains 14 g of fat and no carbohydrate or protein
  d. the grams of fat in one avocado that has 410 kcal, 13 g of carbohydrate, and 5 g of protein

3.44 Using the energy values for foods (see Table 3.8), determine each of the following (round off the answer for each food type to the tens place):
  a. the total kilojoules in two tablespoons of crunchy peanut butter that contains 6 g of carbohydrate, 16 g of fat, and 7 g of protein
  b. the grams of protein in one cup of soup that has 110 kcal with 9 g of carbohydrate and 6 g of fat
  c. the kilocalories in one can of cola if it has 40. g of carbohydrate and no fat or protein
  d. the total kilocalories for one cup of vanilla ice cream that contains 26 g of carbohydrate, 14 g of fat, and 4 g of protein

Applications

3.45 For dinner, Charles had one cup of clam chowder, which contains 16 g of carbohydrate, 12 g of fat, and 9 g of protein (see Table 3.8). How much energy, in kilocalories and kilojoules, is in the clam chowder? (Round off the answer for each food type to the tens place.)

3.46 For lunch, Charles consumed 3 oz of skinless chicken, 3 oz of broccoli, 1 medium apple, and 1 cup of nonfat milk (see Table 3.9). How many kilocalories did Charles obtain from the lunch?

3.47 A patient receives 3.2 L of intravenous (IV) glucose solution. If 100. mL of the solution contains 5.0 g of glucose (carbohydrate), how many kilocalories did the patient obtain from the glucose solution (see Table 3.8)?

3.48 A high-protein diet contains 70.0 g of carbohydrate, 5.0 g of fat, and 150 g of protein (see Table 3.8). How much energy, in kilocalories and kilojoules, does this diet provide? (Round off the answer for each food type to the tens place.)

3.49 When a 1.50-g sample of walnuts is burned in a calorimeter, the heat released increases the temperature of 450. g of water from 21.8 °C to 44.1 °C. What is the energy value (kJ/g) for the walnuts?

3.50 When a 1.2-g sample of chocolate-chip cookie is burned in a calorimeter, the heat released increases the temperature of 360 g of water from 23 °C to 40. °C. What is the energy value (kJ/g) for the chocolate-chip cookie?

UPDATE   A Diet and Exercise Program

After Charles has been on his diet and exercise plan for a month, he and his mother meet again with the dietitian. Daniel looks at the diary of food intake and exercise, and weighs Charles to evaluate how well the diet is working. The following is what Charles ate in one day:

Breakfast

1 banana, 1 cup of nonfat milk, 1 egg

Lunch

1 cup of carrots, 3 oz of ground beef, 1 apple, 1 cup of nonfat milk

Dinner

6 oz of skinless chicken, 1 baked potato, 3 oz of broccoli, 1 cup of nonfat milk

Applications

3.51 Using energy contents from Table 3.9, determine each of the following:
  a. the total kilocalories for each meal
  b. the total kilocalories for one day
  c. If Charles consumes 1800 kcal per day, he will maintain his weight. Would he lose weight on his new diet?
  d. If expending 3500 kcal is equal to a loss of 1.0 lb, how many days will it take Charles to lose 5.0 lb?

3.52 a. During one week, Charles swam for a total of 2.5 h and walked for a total of 8.0 h. If Charles expends 340 kcal/h swimming and 160 kcal/h walking, how many total kilocalories did he expend for one week (see Table 3.11)?
  b. For the amount of exercise that Charles did for one week in part **a**, if expending 3500 kcal is equal to a loss of 1.0 lb, how many pounds did he lose?
  c. How many hours would Charles have to walk to lose 1.0 lb?
  d. How many hours would Charles have to swim to lose 1.0 lb?

CONCEPT MAP

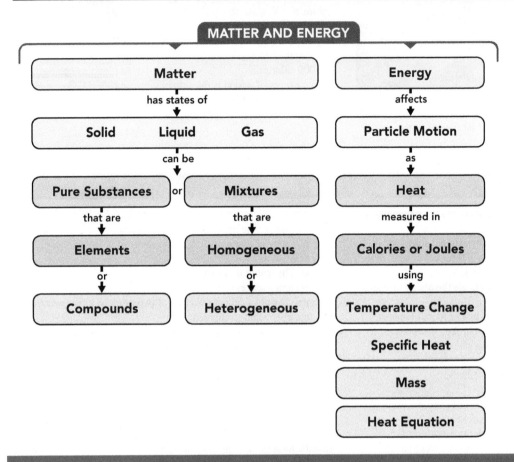

MATTER AND ENERGY

Matter — has states of → **Solid   Liquid   Gas** — can be → **Pure Substances** or **Mixtures**

Pure Substances — that are → **Elements** — or → **Compounds**

Mixtures — that are → **Homogeneous** — or → **Heterogeneous**

Energy — affects → **Particle Motion** — as → **Heat** — measured in → **Calories or Joules** — using → **Temperature Change**

Specific Heat

Mass

Heat Equation

CHAPTER REVIEW

3.1 Classification of Matter

LEARNING GOAL Classify examples of matter as pure substances or mixtures.

Aluminum atom

- Matter is anything that has mass and occupies space.
- Matter is classified as pure substances or mixtures.
- Pure substances, which are elements or compounds, have fixed compositions, and mixtures have variable compositions.
- The substances in mixtures can be separated using physical methods.

3.2 States and Properties of Matter

LEARNING GOAL Identify the states and the physical and chemical properties of matter.

- The three states of matter are solid, liquid, and gas.
- A physical property is a characteristic of a substance that can be observed or measured without affecting the identity of the substance.
- A physical change occurs when physical properties change, but not the composition of the substance.
- A chemical property indicates the ability of a substance to change into another substance.
- A chemical change occurs when one or more substances react to form a new substance with new physical and chemical properties.

3.3 Temperature

LEARNING GOAL Given a temperature, calculate the corresponding temperature on another scale.

- In science, temperature is measured in degrees Celsius (°C) or kelvins (K).
- On the Celsius scale, there are 100 units between the freezing point (0 °C) and the boiling point (100 °C) of water.
- On the Fahrenheit scale, there are 180 units between the freezing point (32 °F) and the boiling point (212 °F) of water. A Fahrenheit temperature is related to its Celsius temperature by the equation $T_F = 1.8(T_C) + 32$.
- The SI unit, kelvin, is related to the Celsius temperature by the equation $T_K = T_C + 273$.

3.4 Energy

LEARNING GOAL Identify energy as potential or kinetic; convert between units of energy.

- Energy is the ability to do work.
- Potential energy is determined by the position or by the composition of a substance; kinetic energy is the energy of motion.
- Common units of energy are the calorie (cal), kilocalorie (kcal), joule (J), and kilojoule (kJ).
- One calorie is equal to 4.184 J.

3.5 Specific Heat

LEARNING GOAL Calculate the specific heat for a substance. Use specific heat to calculate heat loss or gain.

- Specific heat is the amount of energy required to raise the temperature of exactly 1 g of a substance by exactly 1 °C.
- The heat lost or gained by a substance is determined by multiplying its mass, the temperature change, and its specific heat.

3.6 Energy and Nutrition

LEARNING GOAL Use the energy values to calculate the kilocalories (kcal) or kilojoules (kJ) for a food.

- The nutritional Calorie is the same amount of energy as 1 kcal or 1000 calories.
- The energy of a food is the sum of kilocalories or kilojoules from carbohydrate, fat, and protein.

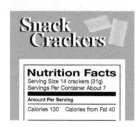

KEY TERMS

calorie (cal) The amount of heat energy that raises the temperature of exactly 1 g of water by exactly 1 °C.

chemical change A change during which the original substance is converted into a new substance that has a different composition and new physical and chemical properties.

chemical properties The properties that indicate the ability of a substance to change into a new substance.

compound A pure substance consisting of two or more elements, with a definite composition, that can be broken down into simpler substances only by chemical methods.

element A pure substance containing only one type of matter, which cannot be broken down by chemical methods.

energy The ability to do work.

energy value The kilocalories (or kilojoules) obtained per gram of the food types: carbohydrate, fat, and protein.

gas A state of matter that does not have a definite shape or volume.

heat The energy associated with the motion of particles in a substance.

heat equation A relationship that calculates heat (q) given the mass, specific heat, and temperature change for a substance.

joule (J) The SI unit of heat energy; 4.184 J = 1 cal.

kinetic energy The energy of moving particles.

liquid A state of matter that takes the shape of its container but has a definite volume.

matter The material that makes up a substance and has mass and occupies space.

mixture The physical combination of two or more substances that are not chemically combined.

physical change A change in which the physical properties of a substance change but its identity stays the same.

physical properties The properties that can be observed or measured without affecting the identity of a substance.

potential energy A type of energy related to position or composition of a substance.

pure substance A type of matter that has a definite composition.

solid A state of matter that has its own shape and volume.

specific heat (*SH*) A quantity of heat that changes the temperature of exactly 1 g of a substance by exactly 1 °C.

states of matter Three forms of matter: solid, liquid, and gas.

CORE CHEMISTRY SKILLS

The chapter section containing each Core Chemistry Skill is shown in parentheses at the end of each heading.

Identifying Physical and Chemical Changes (3.2)

- Physical properties can be observed or measured without changing the identity of a substance.
- Chemical properties describe the ability of a substance to change into a new substance.
- When matter undergoes a physical change, its state or its appearance changes, but its composition remains the same.
- When a chemical change takes place, the original substance is converted into a new substance, which has different physical and chemical properties.

Example: Classify each of the following as a physical or chemical property:

    a. Helium in a balloon is a gas.
    b. Methane, in natural gas, burns.
    c. Hydrogen sulfide smells like rotten eggs.

Answer: **a.** A gas is a state of matter, which makes it a physical property.
    b. When methane burns, it changes to different substances with new properties, which is a chemical property.
    c. The odor of hydrogen sulfide is a physical property.

Converting Between Temperature Scales (3.3)

- The temperature equation $T_F = 1.8(T_C) + 32$ is used to convert from Celsius to Fahrenheit and can be rearranged to convert from Fahrenheit to Celsius.
- The temperature equation $T_K = T_C + 273$ is used to convert from Celsius to Kelvin and can be rearranged to convert from Kelvin to Celsius.

Example: Convert 75.0 °C to degrees Fahrenheit.

Answer: $T_F = 1.8(T_C) + 32$

$$T_F = 1.8(75.0) + 32 = 135 + 32$$

$$= 167 \,°F$$

Example: Convert 355 K to degrees Celsius.

Answer: $T_K = T_C + 273$

To solve the equation for T_C, subtract 273 from both sides.

$$T_K - 273 = T_C + \cancel{273} - \cancel{273}$$

$$T_C = T_K - 273$$

$$T_C = 355 - 273$$

$$= 82 \,°C$$

Using Energy Units (3.4)

- Equalities for energy units include 1 cal = 4.184 J, 1 kcal = 1000 cal, and 1 kJ = 1000 J.
- Each equality for energy units can be written as two conversion factors:

$$\frac{4.184 \text{ J}}{1 \text{ cal}} \text{ and } \frac{1 \text{ cal}}{4.184 \text{ J}} \quad \frac{1000 \text{ cal}}{1 \text{ kcal}} \text{ and } \frac{1 \text{ kcal}}{1000 \text{ cal}} \quad \frac{1000 \text{ J}}{1 \text{ kJ}} \text{ and } \frac{1 \text{ kJ}}{1000 \text{ J}}$$

- The energy unit conversion factors are used to cancel given units of energy and to obtain the needed unit of energy.

Example: Convert 45 000 J to kilocalories.

Answer: Using the conversion factors above, we start with the given 45 000 J and convert it to kilocalories.

$$45\,000 \text{ J} \times \frac{1 \text{ cal}}{4.184 \text{ J}} \times \frac{1 \text{ kcal}}{1000 \text{ cal}} = 11 \text{ kcal}$$

Calculating Specific Heat (3.5)

- Specific heat (*SH*) is the amount of heat (*q*) that raises the temperature of 1 g of a substance by 1 °C.

$$SH = \frac{q}{m \times \Delta T} = \frac{\text{cal (or J)}}{\text{g} \,°\text{C}}$$

- To calculate specific heat, the heat lost or gained is divided by the mass of the substance and the change in temperature (Δ*T*).

Example: Calculate the specific heat, in J/g °C, for a 4.0-g sample of tin that absorbs 63 J when heated from 125 °C to 197 °C.

Answer: $q = 63$ J, $m = 4.0$ g, $\Delta T = 197\,°\text{C} - 125\,°\text{C} = 72\,°\text{C}$

$$SH = \frac{q}{m \times \Delta T} = \frac{63 \text{ J}}{4.0 \text{ g} \times 72\,°\text{C}} = \frac{0.22 \text{ J}}{\text{g} \,°\text{C}}$$

Using the Heat Equation (3.5)

- The quantity of heat absorbed or lost by a substance is calculated using the heat equation.

$$q = m \times \Delta T \times SH$$

- Heat, in calories, is obtained when the specific heat of a substance in cal/g °C is used.
- Heat, in joules, is obtained when the specific heat of a substance in J/g °C is used.
- To cancel, the unit grams is used for mass, and the unit °C is used for temperature change.

Example: How many joules are required to heat 5.25 g of titanium from 85.5 °C to 132.5 °C?

Answer: $m = 5.25$ g, $\Delta T = 132.5\,°\text{C} - 85.5\,°\text{C} = 47.0\,°\text{C}$, *SH* for titanium = 0.523 J/g °C

The known values are substituted into the heat equation making sure units cancel.

$$q = m \times \Delta T \times SH = 5.25 \text{ g} \times 47.0\,°\text{C} \times \frac{0.523 \text{ J}}{\text{g} \,°\text{C}}$$

$$= 129 \text{ J}$$

UNDERSTANDING THE CONCEPTS

The chapter sections to review are shown in parentheses at the end of each problem.

3.53 Identify each of the following as an element, a compound, or a mixture. Explain your choice. (3.1)

a.   **b.**

c.

3.54 Identify each of the following as a homogeneous or heterogeneous mixture. Explain your choice. (3.1)

a.   **b.**

c.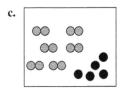

3.55 Classify each of the following as a homogeneous or heterogeneous mixture: (3.1)

a. lemon-flavored water
b. stuffed mushrooms
c. eye drops

3.56 Classify each of the following as a homogeneous or heterogeneous mixture: (3.1)

a. ketchup   **b.** hard-boiled egg   **c.** tortilla soup

3.57 State the temperature on the Celsius thermometer and convert to Fahrenheit. (3.3)

3.58 State the temperature on the Celsius thermometer and convert to Fahrenheit. (3.3)

3.59 Compost can be made at home from grass clippings, kitchen scraps, and dry leaves. As microbes break down organic matter, heat is generated, and the compost can reach a temperature of 155 °F, which kills most pathogens. What is this temperature in degrees Celsius? In kelvins? (3.3)

Compost produced from decayed plant material is used to enrich the soil.

3.60 After a week, biochemical reactions in compost slow, and the temperature drops to 45 °C. The dark brown organic-rich mixture is ready for use in the garden. What is this temperature in degrees Fahrenheit? In kelvins? (3.3)

3.61 Calculate the energy, in joules, to heat two cubes (gold and aluminum), each with a volume of 10.0 cm^3, from 15 °C to 25 °C. Refer to Tables 2.10 and 3.7. (3.5)

3.62 Calculate the energy, in joules, to heat two cubes (silver and copper), each with a volume of 10.0 cm^3, from 15 °C to 25 °C. Refer to Tables 2.10 and 3.7. (3.5)

Applications

3.63 A 70.0-kg person had a quarter-pound cheeseburger, french fries, and a chocolate shake. (3.6)

Item	Carbohydrate (g)	Fat (g)	Protein (g)
Cheeseburger	46	40.	47
French fries	47	16	4
Chocolate shake	76	10.	10.

a. Using Table 3.8, calculate the total kilocalories for each food type in this meal. (Round off the kilocalories to the tens place.)
b. Determine the total kilocalories for the meal.
c. Using Table 3.11, determine the number of hours of sleep needed to burn off the kilocalories in this meal.
d. Using Table 3.11, determine the number of hours of running needed to burn off the kilocalories in this meal.

3.64 Your friend, who has a mass of 70.0 kg, has a slice of pizza, a cola soft drink, and ice cream. (3.6)

Item	Carbohydrate (g)	Fat (g)	Protein (g)
Pizza	29	10.	13
Cola	51	0	0
Ice cream	44	28	8

a. Using Table 3.8, calculate the total kilocalories for each food type in this meal. (Round off the kilocalories to the tens place.)
b. Determine the total kilocalories for the meal.
c. Using Table 3.11, determine the number of hours of sitting needed to burn off the kilocalories in this meal.
d. Using Table 3.11, determine the number of hours of swimming needed to burn off the kilocalories in this meal.

ADDITIONAL PRACTICE PROBLEMS

3.65 Classify each of the following as an element, a compound, or a mixture: (3.1)
a. carbon in pencils
b. carbon monoxide (CO) in automobile exhaust
c. orange juice

3.66 Classify each of the following as an element, a compound, or a mixture: (3.1)
a. neon gas in lights
b. a salad dressing of oil and vinegar
c. sodium hypochlorite (NaClO) in bleach

3.67 Classify each of the following mixtures as homogeneous or heterogeneous: (3.1)
a. hot fudge sundae
b. herbal tea
c. vegetable oil

3.68 Classify each of the following mixtures as homogeneous or heterogeneous: (3.1)
a. water and sand
b. mustard
c. blue ink

3.69 Identify each of the following as solid, liquid, or gas: (3.2)
   a. vitamin tablets in a bottle
   b. helium in a balloon
   c. milk in a bottle
   d. the air you breathe
   e. charcoal briquettes on a barbecue

3.70 Identify each of the following as solid, liquid, or gas: (3.2)
   a. popcorn in a bag
   b. water in a garden hose
   c. a computer mouse
   d. air in a tire
   e. hot tea in a teacup

3.71 Identify each of the following as a physical or chemical property: (3.2)
   a. Gold is shiny.
   b. Gold melts at 1064 °C.
   c. Gold is a good conductor of electricity.
   d. When gold reacts with sulfur, a black sulfide compound forms.

3.72 Identify each of the following as a physical or chemical property: (3.2)
   a. A candle is 10 cm high and 2 cm in diameter.
   b. A candle burns.
   c. The wax of a candle softens on a hot day.
   d. A candle is blue.

3.73 Identify each of the following as a physical or chemical change: (3.2)
   a. A plant grows a new leaf.
   b. Chocolate is melted for a dessert.
   c. Wood is chopped for the fireplace.
   d. Wood burns in a woodstove.

3.74 Identify each of the following as a physical or chemical change: (3.2)
   a. Aspirin tablets are broken in half.
   b. Carrots are grated for use in a salad.
   c. Malt undergoes fermentation to make beer.
   d. A copper pipe reacts with air and turns green.

3.75 Calculate each of the following temperatures in degrees Celsius and kelvins: (3.3)
   a. The highest recorded temperature in the continental United States was 134 °F in Death Valley, California, on July 10, 1913.
   b. The lowest recorded temperature in the continental United States was −69.7 °F in Rogers Pass, Montana, on January 20, 1954.

3.76 Calculate each of the following temperatures in kelvins and degrees Fahrenheit: (3.3)

a. The highest recorded temperature in the world was 58.0 °C in El Azizia, Libya, on September 13, 1922.
b. The lowest recorded temperature in the world was −89.2 °C in Vostok, Antarctica, on July 21, 1983.

3.77 What is −15 °F in degrees Celsius and in kelvins? (3.3)

3.78 What is 56 °F in degrees Celsius and in kelvins? (3.3)

3.79 On a hot day, the beach sand gets hot but the water stays cool. Would you predict that the specific heat of sand is higher or lower than that of water? Explain. (3.5)

On a sunny day, the sand gets hot but the water stays cool.

3.80 On a hot sunny day, you get out of the swimming pool and sit in a metal chair, which is very hot. Would you predict that the specific heat of the metal is higher or lower than that of water? Explain. (3.5)

3.81 A 0.50-g sample of vegetable oil is placed in a calorimeter. When the sample is burned, 18.9 kJ is given off. What is the energy value (kcal/g) for the oil? (3.6)

3.82 A 1.3-g sample of rice is placed in a calorimeter. When the sample is burned, 22 kJ is given off. What is the energy value (kcal/g) for the rice? (3.6)

Applications

3.83 A hot-water bottle for a patient contains 725 g of water at 65 °C. If the water cools to body temperature (37 °C), how many kilojoules of heat could be transferred to sore muscles? (3.5)

3.84 The highest recorded body temperature that a person has survived is 46.5 °C. Calculate that temperature in degrees Fahrenheit and in kelvins. (3.3)

3.85 If you want to lose 1 lb of "body fat," which is 15% water, how many kilocalories do you need to expend? (3.6)

3.86 A young patient drinks whole milk as part of her diet. Calculate the total kilocalories if the glass of milk contains 12 g of carbohydrate, 9 g of fat, and 9 g of protein. (Round off the answer for each food type to the tens place.) (3.6)

CHALLENGE PROBLEMS

The following problems are related to the topics in this chapter. However, they do not all follow the chapter order, and they require you to combine concepts and skills from several sections. These problems will help you increase your critical thinking skills and prepare for your next exam.

3.87 When a 0.80-g sample of olive oil is burned in a calorimeter, the heat released increases the temperature of 450 g of water from 22.7 °C to 38.8 °C. What is the energy value (kcal/g) for the olive oil? (3.5, 3.6)

3.88 When 1.0 g of gasoline burns, it releases 11 kcal. The density of gasoline is 0.74 g/mL. (3.4)
   a. How many megajoules are released when 1.0 gal of gasoline burns?
   b. If a television requires 150 kJ/h to run, how many hours can the television run on the energy provided by 1.0 gal of gasoline?

3.89 A 70.-g piece of nickel metal at 54.0 °C is placed in 25 g of water at 21.9 °C. All the heat lost by the metal is used to heat the water, and the final temperature of the water and metal is 29.2 °C. What is the specific heat, in J/g °C, of nickel? (3.5)

3.90 A 125-g piece of metal is heated to 288 °C and dropped into 85.0 g of water at 12.0 °C. All the heat lost by the metal is used to heat the water, and the metal and water come to a final temperature of 24.0 °C. What is the specific heat, in J/g °C, of the metal? (3.5)

3.91 A metal is thought to be titanium or aluminum. When 4.7 g of the metal absorbs 11 J, its temperature rises by 4.5 °C. (3.5)

a. What is the specific heat, in J/g °C, of the metal?
b. Would you identify the metal as titanium or aluminum (see Table 3.7)?

3.92 A metal is thought to be copper or gold. When 18 g of the metal absorbs 58 cal, its temperature rises by 35 °C. (3.5)

a. What is the specific heat, in cal/g °C, of the metal?
b. Would you identify the metal as copper or gold (see Table 3.7)?

ANSWERS TO ENGAGE QUESTIONS

3.1 Both elements and compounds have a definite composition of one kind of atom or molecule.

3.2 A pizza is heterogeneous because the food items do not have a uniform distribution, whereas in vinegar, the components have a uniform distribution.

3.3 The components in a gas are moving rapidly in all directions until they take the shape and volume of the container.

3.4 The products of the decomposition of salt are new substances with different properties, which means that a chemical change has taken place.

3.5 The boiling point is a characteristic of iron, a physical property. Heating iron to form Fe_2O_3, rust, shows the ability of iron to from a new substance, which is a chemical property.

3.6 A degree Celsius is larger: It is 1.8 times the size of a degree Fahrenheit.

3.7 $T_C = \dfrac{T_F - 32}{1.8} = \dfrac{-40 - 32}{1.8} = \dfrac{-72}{1.8} = -40. \,°C$

3.8 A book on a high table has more potential energy because of its position above the floor.

3.9 Because silver has a smaller specific heat than iron, 1 g of silver will reach a higher temperature when it absorbs 400 J of heat than 1 g of iron.

3.10 $m = \dfrac{q}{\Delta T \times SH}$

3.11 Fat has the most energy per gram compared to carbohydrate or protein.

ANSWERS TO SELECTED PROBLEMS

3.1 **a.** element **b.** compound **c.** element
d. compound **e.** compound

3.3 **a.** pure substance **b.** mixture
c. pure substance **d.** pure substance
e. mixture

3.5 **a.** heterogeneous **b.** homogeneous
c. heterogeneous **d.** heterogeneous

3.7 **a.** gas **b.** gas **c.** solid

3.9 **a.** physical **b.** chemical **c.** physical
d. chemical **e.** chemical

3.11 **a.** physical **b.** chemical **c.** physical
d. physical **e.** physical

3.13 **a.** chemical **b.** physical **c.** physical
d. chemical **e.** physical

3.15 In the United States, we still use the Fahrenheit temperature scale. In °F, normal body temperature is 98.6. On the Celsius scale, her temperature would be 37.7 °C, a mild fever.

3.17 **a.** 98.6 °F **b.** 18.5 °C **c.** 246 K
d. 335 K **e.** 46 °C

3.19 **a.** 41 °C
b. No. The temperature is equivalent to 39 °C.

3.21 When the roller-coaster car stops at the top of the ramp, it has its maximum potential energy and a minimum of kinetic energy. As it descends, potential energy changes to kinetic energy. At the bottom, all the energy is kinetic.

3.23 **a.** potential **b.** kinetic
c. potential **d.** potential

3.25 **a.** 3.5 kcal **b.** 99.2 cal **c.** 120 J **d.** 1100 cal

3.27 **a.** 8.1×10^5 J **b.** 190 kcal

3.29 Copper, which has the lowest specific heat, would reach the highest temperature.

3.31 **a.** 0.389 J/g °C **b.** 0.313 J/g °C

3.33 **a.** 1380 J **b.** 1810 J

3.35 **a.** 54.5 g **b.** 216 g

3.37 **a.** 175 °C **b.** 11.3 °C

3.39 **a.** 0.84 J/g °C **b.** 0.50 J/g °C

3.41 **a.** 29.9 kcal **b.** 208 kcal

3.43 **a.** 470 kJ **b.** 18 g **c.** 130 kcal **d.** 38 g

3.45 210 kcal, 880 kJ

3.47 640 kcal

3.49 28.0 kJ/g

3.51 **a.** breakfast 270 kcal; lunch 420 kcal; dinner 440 kcal
b. 1130 kcal total
c. Yes. Charles should be losing weight.
d. 26 days

3.53 **a.** compound, the particles have a 2:1 ratio of atoms
b. mixture, has two different kinds of particles
c. element, has a single kind of atom

3.55 **a.** homogeneous **b.** heterogeneous
c. homogeneous

3.57 41.5 °C, 106.7 °F

3.59 68.3 °C, 341 K

3.61 gold, 250 J or 60. cal; aluminum, 240 J or 58 cal

3.63 **a.** carbohydrate, 680 kcal; fat, 590 kcal; protein, 240 kcal
b. 1510 kcal
c. 3×10^1 h
d. 2.0 h

3.65 **a.** element **b.** compound **c.** mixture

3.67 **a.** heterogeneous **b.** homogeneous
c. homogeneous

3.69 **a.** solid **b.** gas **c.** liquid
d. gas **e.** solid

3.71 **a.** physical **b.** physical
c. physical **d.** chemical

3.73 **a.** chemical **b.** physical
c. physical **d.** chemical

3.75 **a.** 56.7 °C, 330. K
b. −56.5 °C, 217 K

3.77 −26 °C, 247 K

3.79 The same amount of heat causes a greater temperature change in the sand than in the water; thus the sand must have a lower specific heat than that of water.

3.81 9.0 kcal/g

3.83 85 kJ

3.85 3500 kcal

3.87 9.0 kcal/g

3.89 0.44 J/g °C

3.91 **a.** 0.52 J/g °C **b.** titanium

CI.1 Gold, one of the most sought-after metals in the world, has a density of 19.3 g/cm^3, a melting point of 1064 °C, and a specific heat of 0.129 J/g °C. A gold nugget found in Alaska in 1998 weighed 20.17 lb. (2.4, 2.6, 2.7, 3.3, 3.5)

Gold nuggets, also called native gold, can be found in streams and mines.

a. How many significant figures are in the measurement of the weight of the nugget?
b. What is the mass of the nugget in kilograms?
c. If the nugget were pure gold, what would its volume be in cubic centimeters?
d. What is the melting point of gold in degrees Fahrenheit and kelvins?
e. How many kilocalories are required to raise the temperature of the nugget from 500. °C to 1064 °C?
f. If the price of gold is $42.06 per gram, what is the nugget worth in dollars?

CI.2 The mileage for a motorcycle with a fuel-tank capacity of 22 L is 35 mi/gal. (2.5, 2.6, 2.7, 3.4)

a. How long a trip, in kilometers, can be made on one full tank of gasoline?
b. If the price of gasoline is $2.82 per gallon, what would be the cost of fuel for the trip?
c. If the average speed during the trip is 44 mi/h, how many hours will it take to reach the destination?

When 1.00 g of gasoline burns, 47 kJ of energy is released.

d. If the density of gasoline is 0.74 g/mL, what is the mass, in grams, of the fuel in the tank?
e. When 1.00 g of gasoline burns, 47 kJ of energy is released. How many kilojoules are produced when the fuel in one full tank is burned?

CI.3 Answer the following for the water samples **A** and **B** shown in the diagrams: (3.1, 3.2, 3.3, 3.5)

A   B

a. In which sample (**A** or **B**) does the water have its own shape?
b. Which diagram (**1** or **2** or **3**) represents the arrangement of particles in water sample **A**?
c. Which diagram (**1** or **2** or **3**) represents the arrangement of particles in water sample **B**?

1   2   3

Answer the following for diagrams **1**, **2**, and **3**:
d. The state of matter indicated in diagram **1** is a _____; in diagram **2**, it is a _____; and in diagram **3**, it is a _____.
e. The motion of the particles is slowest in diagram _____.
f. The arrangement of particles is farthest apart in diagram _____.
g. The particles fill the volume of the container in diagram _____.

CI.4 The label of an energy bar with a mass of 68 g lists the nutrition facts as 39 g of carbohydrate, 5 g of fat, and 10. g of protein. (2.5, 2.6, 3.4, 3.6)

An energy bar contains carbohydrate, fat, and protein.

a. Using the energy values for carbohydrates, fats, and proteins (see Table 3.8), what are the total kilocalories for the energy bar? (Round off the answer for each food type to the tens place.)
b. What are the kilojoules for the energy bar? (Round off the answer for each food type to the tens place.)
c. If you obtain 160 kJ, how many grams of the energy bar did you eat?
d. If you are walking and using energy at a rate of 840 kJ/h, how many minutes will you need to walk to expend the energy from two energy bars?

CI.5 In one box of nails weighing 0.250 lb, there are 75 iron nails. The density of iron is 7.86 g/cm^3. The specific heat of iron is 0.452 J/g °C. The melting point of iron is 1535 °C. (2.5, 2.6, 2.7, 3.4, 3.5)

Nails made of iron have a density of 7.86 g/cm^3.

a. What is the volume, in cubic centimeters, of the iron nails in the box?
b. If 30 nails are added to a graduated cylinder containing 17.6 mL of water, what is the new level of water, in milliliters, in the cylinder?
c. How much heat, in joules, must be added to the nails in the box to raise their temperature from 16 °C to 125 °C?

CI.6 A hot tub is filled with 450 gal of water. (2.5, 2.6, 2.7, 3.3, 3.4, 3.5)

A hot tub filled with water is heated to 105 °F.

a. What is the volume of water, in liters, in the tub?
b. What is the mass, in kilograms, of water in the tub?
c. How many kilocalories are needed to heat the water from 62 °F to 105 °F?
d. If the hot-tub heater provides 5900 kJ/min, how long, in minutes, will it take to heat the water in the hot tub from 62 °F to 105 °F?

ANSWERS

CI.1 a. four SFs
   b. 9.17 kg
   c. 475 cm^3
   d. 1947 °F; 1337 K
   e. 159 kcal
   f. $386 000

CI.3 a. B
   b. A is represented by diagram **2**.
   c. B is represented by diagram **1**.

   d. solid; liquid; gas
   e. diagram **1**
   f. diagram **3**
   g. diagram **3**

CI.5 a. 14.4 cm^3
   b. 23.4 mL
   c. 5590 J

Atoms and Elements

John prepares for the next growing season by deciding how much of each crop to plant and where on his farm. The quality of the soil, including the pH, the amount of moisture, and the nutrient content in the soil helps John make his decisions. After performing chemical tests on the soil, John determines that several of his fields need additional fertilizer before the crops can be planted. John considers several different types of fertilizers, as each supplies different nutrients to the soil. Plants need three elements for growth: potassium, nitrogen, and phosphorus. Potassium (K on the periodic table) is a metal, whereas nitrogen (N) and phosphorus (P) are nonmetals. Fertilizers may also contain several other elements including calcium (Ca), magnesium (Mg), and sulfur (S). John applies a fertilizer containing a mixture of all of these elements to his soil and plans to recheck the soil nutrient content in a few days.

CAREER

Farmer

Farming involves much more than growing crops and raising animals. Farmers must understand how to perform chemical tests and how to apply fertilizer to soil and pesticides or herbicides to crops. Pesticides are chemicals used to kill insects that could destroy the crop, whereas herbicides are chemicals used to kill weeds that would compete with the crops for water and nutrients. Farmers must understand how to use these chemicals safely and effectively. In using this information, farmers are able to grow crops that produce a higher yield, greater nutritional value, and better taste.

UPDATE Improving Crop Production

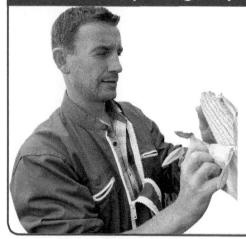

Last year, John noticed that the potatoes from one of his fields had brown spots on the leaves and were undersized. Last week, he obtained soil samples from that field to check the nutrient levels. You can see how John improved the soil and crop production in the **UPDATE** Improving **Crop Production**, page 117, and learn what kind of fertilizer John used and the quantity applied.

4.1 Elements and Symbols

LEARNING GOAL Given the name of an element, write its correct symbol; from the symbol, write the correct name.

All matter is composed of *elements*, of which there are 118 different kinds. Of these, 88 elements occur naturally and make up all the substances in our world. Many elements are already familiar to you. Perhaps you use aluminum in the form of foil or drink soft drinks from aluminum cans. You may have a ring or necklace made of gold, silver, or perhaps platinum. If you play tennis or golf, then you may have noticed that your racket or clubs may be made from the elements titanium or carbon. In our bodies, calcium and phosphorus form the structure of bones and teeth, iron and copper are needed in the formation of red blood cells, and iodine is required for thyroid function.

Elements are pure substances from which all other things are built. Elements cannot be broken down into simpler substances. Over the centuries, elements have been named for planets, mythological figures, colors, minerals, geographic locations, and famous people. Some sources of names of elements are listed in **TABLE 4.1**. A complete list of all the elements and their symbols are found on the inside front cover of this text.

TABLE 4.1 Some Elements, Symbols, and Source of Names

Element	Symbol	Source of Name
Uranium	U	The planet Uranus
Titanium	Ti	Titans (mythology)
Chlorine	Cl	*Chloros*: "greenish yellow" (Greek)
Iodine	I	*Ioeides*: "violet" (Greek)
Magnesium	Mg	Magnesia, a mineral
Tennessine	Ts	Tennessee
Curium	Cm	Marie and Pierre Curie
Copernicium	Cn	Nicolaus Copernicus

Chemical Symbols

Chemical symbols are one- or two-letter abbreviations for the names of the elements. Only the first letter of an element's symbol is capitalized. If the symbol has a second letter, it is lower-case so that we know when a different element is indicated. If two letters are capitalized, they

TABLE 4.2 Names and Symbols of Some Common Elements

Name*	Symbol	Name*	Symbol	Name*	Symbol
Aluminum	Al	Gallium	Ga	Oxygen	O
Argon	Ar	Gold (*aurum*)	Au	Phosphorus	P
Arsenic	As	Helium	He	Platinum	Pt
Barium	Ba	Hydrogen	H	Potassium (*kalium*)	K
Boron	B	Iodine	I	Radium	Ra
Bromine	Br	Iron (*ferrum*)	Fe	Silicon	Si
Cadmium	Cd	Lead (*plumbum*)	Pb	Silver (*argentum*)	Ag
Calcium	Ca	Lithium	Li	Sodium (*natrium*)	Na
Carbon	C	Magnesium	Mg	Strontium	Sr
Chlorine	Cl	Manganese	Mn	Sulfur	S
Chromium	Cr	Mercury (*hydrargyrum*)	Hg	Tin (*stannum*)	Sn
Cobalt	Co	Neon	Ne	Titanium	Ti
Copper (*cuprum*)	Cu	Nickel	Ni	Uranium	U
Fluorine	F	Nitrogen	N	Zinc	Zn

*Names given in parentheses are ancient Latin or Greek words from which the symbols are derived.

LOOKING AHEAD

4.1 Elements and Symbols 101
4.2 The Periodic Table 103
4.3 The Atom 108
4.4 Atomic Number and Mass Number 111
4.5 Isotopes and Atomic Mass 113

Aluminum (Al)

ENGAGE 4.1

What are the names and symbols of some elements that you encounter every day?

Gold (Au)

Carbon (C)

Silver (Ag)

Sulfur (S)

represent the symbols of two different elements. For example, the element cobalt has the symbol Co. However, the two capital letters CO specify two elements, carbon (C) and oxygen (O).

Although most of the symbols use letters from the current names, some are derived from their ancient names. For example, Na, the symbol for sodium, comes from the Latin word *natrium*. The symbol for silver, Ag, is derived from the Latin name *argentum*. TABLE 4.2 lists the names and symbols of some common elements. Learning their names and symbols will greatly help your learning of chemistry.

SAMPLE PROBLEM 4.1   Names and Symbols of Chemical Elements

TRY IT FIRST

Complete the following table with the correct name or symbol for each element:

Name	Symbol
nickel	
nitrogen	
	Zn
	K

SOLUTION

Name	Symbol
nickel	Ni
nitrogen	N
zinc	Zn
potassium	K

SELF TEST 4.1

Write the chemical symbols for the elements silicon, selenium, sodium, and seaborgium.

ANSWER

Si, Se, Na, and Sg

PRACTICE PROBLEMS

Try Practice Problems 4.1 to 4.8

PRACTICE PROBLEMS

4.1 Elements and Symbols

4.1 Write the symbols for the following elements:
a. copper   b. platinum   c. calcium   d. manganese
e. iron   f. barium   g. lead   h. strontium

4.2 Write the symbols for the following elements:
a. oxygen   b. lithium   c. uranium   d. titanium
e. hydrogen   f. chromium   g. tin   h. gold

4.3 Determine if each of the following symbols is correct. If incorrect, write the correct symbol.
a. silver, Si   b. silicon, SI   c. antimony, SB
d. mercury, Hg   e. fluorine, Fl

4.4 Determine if each of the following symbols is correct. If incorrect, write the correct symbol.
a. potassium, P   b. sodium, Na   c. chlorine, Ch
d. neon, N   e. nickel, Ni

Applications

4.5 Write the name for the symbol of each of the following elements found in the body:
a. C   b. Cl   c. I   d. Se
e. N   f. S   g. Zn   h. Co

4.6 Write the name for the symbol of each of the following elements found in the body:
a. V   b. P   c. Na   d. As
e. Ca   f. Mo   g. Mg   h. Si

4.7 Write the names for the elements in each of the following formulas of compounds used in medicine:
a. table salt, NaCl
b. plaster casts, $CaSO_4$
c. Demerol, $C_{15}H_{22}ClNO_2$
d. treatment of bipolar disorder, Li_2CO_3

4.8 Write the names for the elements in each of the following formulas of compounds used in medicine:
a. salt substitute, KCl
b. dental cement, $Zn_3(PO_4)_2$
c. antacid, $Mg(OH)_2$
d. contrast agent for X-ray, $BaSO_4$

4.2 The Periodic Table

LEARNING GOAL Use the periodic table to identify the group and the period of an element; identify an element as a metal, a nonmetal, or a metalloid.

By the late 1800s, scientists recognized that certain elements looked alike and behaved in similar ways. In 1869, a Russian chemist, Dmitri Mendeleev (1834–1907), arranged the 60 elements known at that time into groups with similar properties and placed them in order of increasing atomic masses. The current arrangement of the 118 elements we know about today is called the **periodic table** (see **FIGURE 4.1**).

Periodic Table of Elements

FIGURE 4.1 On the periodic table, groups are the elements arranged as vertical columns, and periods are the elements in each horizontal row.

Groups and Periods

Each vertical column on the periodic table contains a **group** (or family) of elements that have similar properties. A **group number** is written at the top of each vertical column (group) in the periodic table. For many years, the **representative elements** have had group numbers 1A to 8A. In the center of the periodic table is a block of elements known as the **transition elements**, which have numbers followed by the letter "B." A newer system assigns numbers 1 to 18 to the groups going left to right across the periodic table. Because both systems are in use, they are shown on the periodic table and are included in our discussions of elements and group numbers.

Each horizontal row in the periodic table is a **period**. The periods are counted down from the top of the table as Periods 1 to 7. The first period contains two elements: hydrogen (H) and helium (He). The second period contains eight elements: lithium (Li), beryllium (Be), boron (B), carbon (C), nitrogen (N), oxygen (O), fluorine (F), and neon (Ne). The third period also contains eight elements beginning with sodium (Na) and ending with argon (Ar). The fourth period, which begins with potassium (K), and the fifth period, which begins with rubidium (Rb), have 18 elements each. The sixth period, which begins with cesium (Cs), has 32 elements. The seventh period contains 32 elements, for a total of 118 elements. The elements called the *lanthanides* (atomic numbers 57 through 71) and the *actinides* (atomic numbers 89 through 103) are part of periods 6 and 7, but are usually moved to the bottom of the periodic table to allow them to fit on a page.

ENGAGE 4.2

What is the name and symbol of the element in Group 1A (1), Period 4?

Group
1A (1)

3
Li
11
Na
19
K
37
Rb
55
Cs
87
Fr

Potassium (K)

FIGURE 4.3 Lithium (Li), sodium (Na), and potassium (K) are alkali metals from Group 1A (1).

ENGAGE 4.3

What properties do the alkali metals have in common?

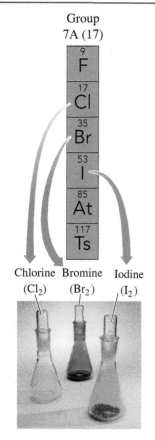

Group
7A (17)

9
F
17
Cl
35
Br
53
I
85
At
117
Ts

Chlorine Bromine Iodine
(Cl_2)   (Br_2)   (I_2)

FIGURE 4.4 Chlorine (Cl_2), bromine (Br_2), and iodine (I_2) are halogens from Group 7A (17).

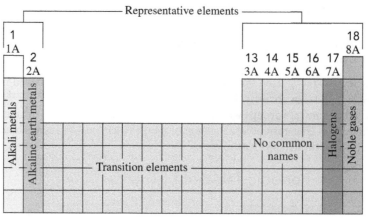

FIGURE 4.2 Certain groups on the periodic table have common names.

Names of Groups

Several groups in the periodic table have special names (see **FIGURE 4.2**). Group 1A (1) elements—lithium (Li), sodium (Na), potassium (K), rubidium (Rb), cesium (Cs), and francium (Fr)—are a family of elements known as the **alkali metals** (see **FIGURE 4.3**). The elements within this group are soft, shiny metals that are good conductors of heat and electricity and have relatively low melting points. Alkali metals react vigorously with water and form white products when they combine with oxygen.

Although hydrogen (H) is at the top of Group 1A (1), it is not an alkali metal and has very different properties than the rest of the elements in this group. Thus, hydrogen is not included in the alkali metals.

The **alkaline earth metals** are found in Group 2A (2). They include the elements beryllium (Be), magnesium (Mg), calcium (Ca), strontium (Sr), barium (Ba), and radium (Ra). The alkaline earth metals are shiny metals like those in Group 1A (1), but they are not as reactive.

The **halogens** are found on the right side of the periodic table in Group 7A (17). They include the elements fluorine (F), chlorine (Cl), bromine (Br), iodine (I), astatine (At), and tennessine (Ts) (see **FIGURE 4.4**). The halogens, especially fluorine and chlorine, are highly reactive and form compounds with most of the elements.

The **noble gases** are found in Group 8A (18). They include helium (He), neon (Ne), argon (Ar), krypton (Kr), xenon (Xe), radon (Rn), and oganesson (Og). They are quite unreactive and are seldom found in combination with other elements.

Metals, Nonmetals, and Metalloids

Another feature of the periodic table is the heavy zigzag line that separates the elements into the *metals* and the *nonmetals*. *Except for hydrogen*, the metals are to the left of the line with the nonmetals to the right.

In general, most **metals** are shiny solids, such as copper (Cu), gold (Au), and silver (Ag). Metals can be shaped into wires (ductile) or hammered into a flat sheet (malleable). Metals are good conductors of heat and electricity. They usually melt at higher temperatures than nonmetals. All the metals are solids at room temperature, except for mercury (Hg), which is a liquid.

Nonmetals are not especially shiny, ductile, or malleable, and they are often poor conductors of heat and electricity. They typically have low melting points and low densities. Some examples of nonmetals are hydrogen (H), carbon (C), nitrogen (N), oxygen (O), chlorine (Cl), and sulfur (S).

Except for aluminum and oganesson, the elements located along the heavy line are **metalloids**: B, Si, Ge, As, Sb, Te, Po, At, and Ts. Metalloids are elements that exhibit some properties that are typical of the metals and other properties that are characteristic of the nonmetals. For example, they are better conductors of heat and electricity than the nonmetals, but not as good as the metals. The metalloids are *semiconductors* because they can be modified to function as conductors or insulators. TABLE 4.3 compares some characteristics of silver, a metal, with those of antimony, a metalloid, and sulfur, a nonmetal.

TABLE 4.3 Some Characteristics of a Metal, a Metalloid, and a Nonmetal

	Silver (Ag)	Antimony (Sb)	Sulfur (S)
Type	Metal	Metalloid	Nonmetal
Luster	Shiny	Blue-gray, shiny	Dull, yellow
Ductility	Extremely ductile	Brittle	Brittle
Malleability	Can be hammered into sheets	Shatters when hammered	Shatters when hammered
Conductor	Good conductor	Poor conductor	Poor conductor
Uses	Manufacture of coins, jewelry, tableware	Hardens lead, colors glass and plastics	Manufacture of gunpowder, rubber, fungicides
Density	10.5 g/cm^3	6.7 g/cm^3	2.1 g/cm^3
Melting Point	962 °C	630 °C	113 °C

SAMPLE PROBLEM 4.2   Metals, Nonmetals, and Metalloids

TRY IT FIRST

Use the periodic table to classify each of the following elements by its group and period, group name (if any), and as a metal, a nonmetal, or a metalloid:

a. Na, important in nerve impulses, regulates blood pressure
b. I, needed to produce thyroid hormones
c. Si, needed for tendons and ligaments

SOLUTION

a. Na (sodium), Group 1A (1), Period 3, is an alkali metal.
b. I (iodine), Group 7A (17), Period 5, halogen, is a nonmetal.
c. Si (silicon), Group 4A (14), Period 3, is a metalloid.

SELF TEST 4.2

Strontium is an element that gives a brilliant red color to fireworks.

a. In what group is strontium found?
b. What is the name of this chemical family?
c. In what period is strontium found?
d. Is strontium a metal, a nonmetal, or a metalloid?

ANSWER

a. Group 2A (2)
b. alkaline earth metals
c. Period 5
d. a metal

ENGAGE 4.4

What elements are in Group 7A (17)?

Silver is a metal, antimony is a metalloid, and sulfur is a nonmetal.

The metalloids on the zigzag line exhibit characteristics of both metals and nonmetals.

Strontium provides the red color in fireworks.

PRACTICE PROBLEMS

Try Practice Problems 4.9 to 4.18

Chemistry Link to Health

Elements Essential to Health

Of all the elements, only about 20 are essential for the well-being and survival of the human body. Of those, four elements—oxygen, carbon, hydrogen, and nitrogen—which are representative elements in Period 1 and Period 2 on the periodic table—make up 96% of our body mass. Most of the food in our daily diet provides these elements to maintain a healthy body. These elements are found in carbohydrates, fats, and proteins. Most of the hydrogen and oxygen is found in water, which makes up 55 to 60% of our body mass.

The *macrominerals*—Ca, P, K, Cl, S, Na, and Mg—are located in Period 3 and Period 4 of the periodic table. They are involved in the formation of bones and teeth, maintenance of heart and blood vessels, muscle contraction, nerve impulses, acid–base balance of body fluids, and regulation of cellular metabolism. The macrominerals are

present in lower amounts than the major elements, so that smaller amounts are required in our daily diets.

The other essential elements, called *microminerals* or *trace elements*, are mostly transition elements in Period 4 along with Si in Period 3 and Mo and I in Period 5. They are present in the human body in very small amounts, some less than 100 mg. In recent years, the ability to detect small amounts has improved so that researchers can more easily identify the roles of trace elements. Some trace elements such as arsenic, chromium, and selenium are toxic at high levels in the body but are still required by the body. Other elements, such as tin and nickel, are thought to be essential, but their metabolic role has not yet been determined. Some examples and the amounts present in a 60.-kg person are listed in **TABLE 4.4**.

■ Major elements in the human body   ■ Macrominerals   ■ Microminerals (trace elements)

Elements essential to health include the major elements H, C, N, O (red), the macrominerals (blue), and the microminerals (purple).

TABLE 4.4 Typical Amounts of Essential Elements in a 60.-kg Adult		
Element	Quantity	Function
Major Elements		
Oxygen (O)	39 kg	Building block of biomolecules and water (H_2O)
Carbon (C)	11 kg	Building block of organic and biomolecules
Hydrogen (H)	6 kg	Component of biomolecules, water (H_2O), regulates pH of body fluids, stomach acid (HCl)
Nitrogen (N)	2 kg	Component of proteins and nucleic acids
Macrominerals		
Calcium (Ca)	1000 g	Needed for bones and teeth, muscle contraction, nerve impulses
Phosphorus (P)	600 g	Needed for bones and teeth, nucleic acids
Potassium (K)	120 g	Most common positive ion (K^+) in cells, muscle contraction, nerve impulses
Chlorine (Cl)	100 g	Most common negative ion (Cl^-) in fluids outside cells, stomach acid (HCl)
Sulfur (S)	86 g	Component of proteins, vitamin B_1, insulin
Sodium (Na)	60 g	Most common positive ion (Na^+) in fluids outside cells, involved in water balance, muscle contraction, nerve impulses
Magnesium (Mg)	36 g	Component of bones, required for metabolic reactions

Element	Quantity	Function
Microminerals (Trace Elements)		
Iron (Fe)	3600 mg	Component of oxygen carrier hemoglobin
Silicon (Si)	3000 mg	Needed for growth and maintenance of bones and teeth, tendons and ligaments, hair and skin
Zinc (Zn)	2000 mg	Needed for metabolic reactions in cells, DNA synthesis, growth of bones, teeth, connective tissue, immune system
Copper (Cu)	240 mg	Needed for blood vessels, blood pressure, immune system
Manganese (Mn)	60 mg	Needed for growth of bones, blood clotting, metabolic reactions
Iodine (I)	20 mg	Needed for proper thyroid function
Molybdenum (Mo)	12 mg	Needed to process Fe and N from food
Arsenic (As)	3 mg	Needed for growth and reproduction
Chromium (Cr)	3 mg	Needed for maintenance of blood sugar levels, synthesis of biomolecules
Cobalt (Co)	3 mg	Component of vitamin B_{12}, red blood cells
Selenium (Se)	2 mg	Needed for the immune system, heart, and pancreas
Vanadium (V)	2 mg	Needed in the formation of bones and teeth, and the extraction of energy from food

PRACTICE PROBLEMS

4.2 The Periodic Table

4.9 Identify the group or period number described by each of the following:
a. contains C, N, and O
b. begins with helium
c. contains the alkali metals
d. ends with neon

4.10 Identify the group or period number described by each of the following:
a. contains Na, K, and Rb
b. begins with Be
c. contains the noble gases
d. contains B, N, and F

4.11 Give the symbol of the element described by each of the following:
a. Group 4A (14), Period 2
b. the noble gas in Period 1
c. the alkali metal in Period 3
d. Group 2A (2), Period 4
e. Group 3A (13), Period 3

4.12 Give the symbol of the element described by each of the following:
a. the alkaline earth metal in Period 2
b. Group 5A (15), Period 3
c. the noble gas in Period 4
d. the halogen in Period 5
e. Group 4A (14), Period 4

4.13 Identify each of the following elements as a metal, a nonmetal, or a metalloid:
a. barium
b. sulfur
c. a shiny element
d. an element that is a gas at room temperature

e. located in Group 8A (18)
f. bromine
g. boron
h. silver

4.14 Identify each of the following elements as a metal, a nonmetal, or a metalloid:
a. located in Group 2A (2)
b. a good conductor of electricity
c. fluorine
d. bismuth
e. an element that is not shiny
f. scandium
g. polonium
h. tin

Applications

4.15 Using Table 4.4, identify the function of each of the following elements in the body and classify each as an alkali metal, an alkaline earth metal, a transition element, or a halogen:
a. Ca   b. Fe   c. K
d. Cl   e. V   f. Cr

4.16 Using Table 4.4, identify the function of each of the following elements in the body, and classify each as an alkali metal, an alkaline earth metal, a transition element, or a halogen:
a. Mg   b. Cu   c. I
d. Na   e. Co   f. Mn

4.17 Identify each of the elements in problem 4.15 as a metal, a nonmetal, or a metalloid.

4.18 Identify each of the elements in problem 4.16 as a metal, a nonmetal, or a metalloid.

Aluminum foil consists of atoms of aluminum.

ENGAGE 4.5

What does Dalton's atomic theory say about the atoms in a compound?

FIGURE 4.5 Images of gold atoms magnified 16 million times by a scanning tunneling microscope.

Positive charges repel

Negative charges repel

Unlike charges attract

FIGURE 4.6 Like charges repel and unlike charges attract.

4.3 The Atom

LEARNING GOAL Describe the electrical charge of a proton, neutron, and electron and identify their location within an atom.

All the elements listed on the periodic table are made up of atoms. An **atom** is the smallest particle of an element that retains the characteristics of that element. Imagine that you are tearing a piece of aluminum foil into smaller and smaller pieces. Now imagine that you have a microscopic piece so small that it cannot be divided any further. Then you would have a single atom of aluminum.

The concept of the atom is relatively recent. Although the Greek philosophers in 500 B.C.E. reasoned that everything must contain minute particles they called *atomos*, the idea of atoms did not become a scientific theory until 1808. Then, John Dalton (1766–1844) developed an atomic theory that proposed that atoms were responsible for the combinations of elements found in compounds.

Dalton's Atomic Theory

1. All matter is made up of tiny particles called atoms.
2. All atoms of a given element are the same and different from atoms of other elements.
3. Atoms of two or more different elements combine to form compounds. A particular compound is always made up of the same kinds of atoms and always has the same number of each kind of atom.
4. A chemical reaction involves the rearrangement, separation, or combination of atoms. Atoms are neither created nor destroyed during a chemical reaction.

Dalton's atomic theory formed the basis of current atomic theory, although we have modified some of Dalton's statements. We now know that atoms of the same element are not completely identical to each other and consist of even smaller particles. However, an atom is still the smallest particle that retains the properties of an element.

Although atoms are the building blocks of everything we see around us, we cannot see an atom or even a billion atoms with the naked eye. However, when billions and billions of atoms are packed together, the characteristics of each atom are added to those of the next until we can see the characteristics we associate with the element. For example, a small piece of the element gold consists of many, many gold atoms. A special kind of microscope called a *scanning tunneling microscope* (STM) produces images of individual atoms (see **FIGURE 4.5**).

Electrical Charges in an Atom

By the end of the 1800s, experiments showed that atoms were not solid spheres but were composed of smaller bits of matter called **subatomic particles**, three of which are the *proton*, *neutron*, and *electron*. Two of these subatomic particles were discovered because they have electrical charges.

An electrical charge can be positive or negative. Experiments show that like charges repel or push away from each other. When you brush your hair on a dry day, electrical charges that are alike build up on the brush and in your hair. As a result, your hair flies away from the brush. However, opposite or unlike charges attract. The crackle of clothes taken from the clothes dryer indicates the presence of electrical charges. The clinginess of the clothing results from the attraction of opposite, unlike, charges (see **FIGURE 4.6**).

Structure of the Atom

In 1897, J. J. Thomson (1856–1940), an English physicist, applied electricity to electrodes sealed in a glass tube, which produced streams of small particles called *cathode rays*. Because these rays were attracted to a positively charged electrode, Thomson realized that the particles in the rays must be negatively charged. In further experiments, these particles called **electrons** were found to be much smaller than the atom and to have extremely small masses. Because atoms are neutral, scientists soon discovered that atoms contained positively charged particles called **protons** that were much heavier than the electrons.

Positive electrode

+

+

Electron beam

−

−

Cathode ray tube

Negatively charged cathode rays (electrons) are attracted to the positive electrode.

Negatively charged electrons

Positively charged cloud

Thomson's "plum-pudding" model describes the electrons in an atom as scattered throughout a positively charged cloud.

Thomson proposed a model for the atom in which the electrons were distributed in a positively charged cloud. Thomson's model of the atom became know as the "plum-pudding" model because electrons reminded scientists at that time of "plums" in an English dessert, plum pudding. In 1911, Ernest Rutherford (1871–1937) worked with Thomson to test this model. In Rutherford's experiment, positively charged particles were aimed at a thin sheet of gold foil (see **FIGURE 4.7**). If the Thomson model were correct, the particles would travel in straight paths through the gold foil. Rutherford was greatly surprised to find that some of the particles were deflected as they passed through the gold foil, and a few particles were deflected so much that they went back in the opposite direction. According to Rutherford, it was as though he had shot a cannonball at a piece of tissue paper, and it bounced back at him.

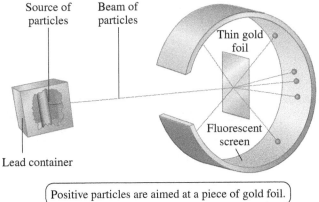

Source of particles

Beam of particles

Thin gold foil

Fluorescent screen

Lead container

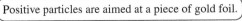

Positive particles are aimed at a piece of gold foil.

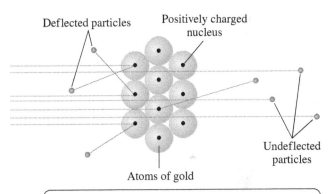

Deflected particles

Positively charged nucleus

Undeflected particles

Atoms of gold

Positive particles that come very close to atomic nuclei are deflected from their straight path.

FIGURE 4.7 Rutherford's gold-foil experiment

From his gold-foil experiments, Rutherford realized that the protons must be contained in a small, positively charged region at the center of the atom, which he called the **nucleus**. He proposed that the electrons in the atom occupy the space surrounding the nucleus through which most of the particles traveled undisturbed. Only the particles that came near this dense, positive center were deflected. The size of the nucleus is extremely small compared with the overall size of the atom. If an atom were the size of a football stadium, the nucleus would be about the size of a golf ball placed in the center of the field.

Later, when scientists realized that the nucleus was heavier than the mass of the protons, they looked for another subatomic particle. Eventually, they discovered that the nucleus also contained a particle that is neutral, which they called a **neutron**.

Mass of the Atom

All the subatomic particles are extremely small compared with the things you see around you (see **FIGURE 4.8**). One proton has a mass of 1.67×10^{-24} g, and the neutron is about the same. However, the electron has a mass 9.11×10^{-28} g, which is much less than the mass of either a proton or neutron. Because the masses of subatomic particles are so small, chemists use a very small unit of mass called an **atomic mass unit (amu)**. An amu is defined

INTERACTIVE VIDEO

Pearson eText

Rutherford's Gold-Foil Experiment

ENGAGE 4.6

Why are some particles deflected, whereas most pass through the gold foil undeflected?

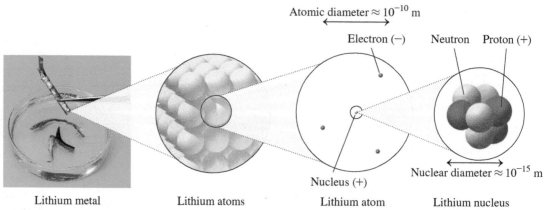

FIGURE 4.8 In an atom, the protons and neutrons that make up almost all the mass are packed into the tiny volume of the nucleus. The rapidly moving electrons (negative charge) surround the nucleus and account for the large volume of the atom.

ENGAGE 4.7

Why can we say that the atom is mostly empty space?

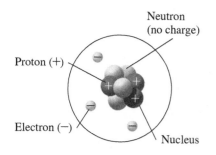

The nucleus of a typical lithium atom contains three protons and four neutrons.

as one-twelfth of the mass of a carbon atom, which has a nucleus containing six protons and six neutrons. In biology, the atomic mass unit is called a *Dalton* (Da) in honor of John Dalton. On the amu scale, the proton and neutron each have a mass of about 1 amu. Because the electron mass is so small, it is usually ignored in atomic mass calculations. TABLE 4.5 summarizes some information about the subatomic particles in an atom.

PRACTICE PROBLEMS

Try Practice Problems 4.19 to 4.26

TABLE 4.5 Subatomic Particles in the Atom

Particle	Symbol	Charge	Mass (amu)	Location in Atom
Proton	p or p^+	1+	1.007	Nucleus
Neutron	n or n^0	0	1.008	Nucleus
Electron	e^-	1−	0.000 55	Outside nucleus

PRACTICE PROBLEMS

4.3 The Atom

4.19 Identify each of the following as describing either a proton, a neutron, or an electron:
  a. has the smallest mass
  b. has a 1+ charge
  c. is found outside the nucleus
  d. is electrically neutral

4.20 Identify each of the following as describing either a proton, a neutron, or an electron:
  a. has a mass about the same as a proton
  b. is found in the nucleus
  c. is attracted to the protons
  d. has a 1− charge

4.21 What did Rutherford determine about the structure of the atom from his gold-foil experiment?

4.22 How did Thomson determine that the electrons have a negative charge?

4.23 Is each of the following statements *true* or *false*?
  a. A proton and an electron have opposite charges.
  b. The nucleus contains most of the mass of an atom.
  c. Electrons repel each other.
  d. A proton is attracted to a neutron.

4.24 Is each of the following statements *true* or *false*?
  a. A proton is attracted to an electron.
  b. A neutron has twice the mass of a proton.
  c. Neutrons repel each other.
  d. Electrons and neutrons have opposite charges.

4.25 On a dry day, your hair flies apart when you brush it. How would you explain this?

4.26 Sometimes clothes cling together when removed from a dryer. What kinds of charges are on the clothes?

4.4 Atomic Number and Mass Number

LEARNING GOAL Given the atomic number and the mass number of an atom, state the number of protons, neutrons, and electrons.

All the atoms of the same element always have the same number of protons. This feature distinguishes atoms of one element from atoms of all the other elements.

Atomic Number

The **atomic number** of an element is equal to the number of protons in every atom of that element. The atomic number is the whole number that appears above the symbol of each element on the periodic table.

Atomic number = number of protons in an atom

The periodic table on the inside front cover of this text shows the elements in order of atomic number from 1 to 118. We can use an atomic number to identify the number of protons in an atom of any element. For example, a lithium atom, with atomic number 3, has 3 protons. Any atom with 3 protons is always a lithium atom. In the same way, we determine that a carbon atom, with atomic number 6, has 6 protons. Any atom with 6 protons is carbon.

An atom is electrically neutral. That means that the number of protons in an atom is equal to the number of electrons, which gives every atom an overall charge of zero. Thus, the atomic number also gives the number of electrons in a neutral atom.

REVIEW

Using Positive and Negative Numbers in Calculations (1.4)

CORE CHEMISTRY SKILL

Counting Protons and Neutrons

ENGAGE 4.8

Why does every atom of barium have 56 protons and 56 electrons?

All atoms of lithium contain three protons and three electrons.

All atoms of carbon contain six protons and six electrons.

Mass Number

We now know that the protons and neutrons determine the mass of the nucleus. Thus, for a single atom, we assign a **mass number**, which is the total number of protons and neutrons in its nucleus. It is important to understand that the term *mass number* tells you how many protons and neutrons there are in the nucleus of an atom. Because these are the main subatomic particles that determine the mass of an atom, the count of protons and neutrons is called the "mass" number. The mass number does not appear on the periodic table because it applies to single atoms only.

Mass number = number of protons + number of neutrons

For example, the nucleus of an oxygen atom that contains 8 protons and 8 neutrons has a mass number of 16. An atom of iron that contains 26 protons and 32 neutrons has a mass number of 58.

ENGAGE 4.9

Which subatomic particles in an atom determine its mass number?

If we are given the mass number of an atom and its atomic number, we can calculate the number of neutrons in its nucleus.

Number of neutrons in a nucleus = mass number − number of protons

For example, if we are given an atom of chlorine (atomic number 17) that has a mass number of 37, we can calculate the number of neutrons in its nucleus.

Number of neutrons = 37 (mass number) − 17 (protons) = 20 neutrons

TABLE 4.6 illustrates the relationships between atomic number, mass number, and the number of protons, neutrons, and electrons in examples of single atoms for different elements.

TABLE 4.6 Composition of Some Atoms of Different Elements

Element	Symbol	Atomic Number	Mass Number	Number of Protons	Number of Neutrons	Number of Electrons
Hydrogen	H	1	1	1	0	1
Nitrogen	N	7	14	7	7	7
Oxygen	O	8	16	8	8	8
Chlorine	Cl	17	37	17	20	17
Iron	Fe	26	58	26	32	26
Gold	Au	79	197	79	118	79

SAMPLE PROBLEM 4.3   Calculating Numbers of Protons, Neutrons, and Electrons

TRY IT FIRST

Zinc, a micromineral, is needed for metabolic reactions in cells, DNA synthesis, the growth of bone, teeth, and connective tissue, and the proper functioning of the immune system. For an atom of zinc that has a mass number of 68, determine the number of:

a. protons   **b.** neutrons   **c.** electrons

SOLUTION

ANALYZE THE PROBLEM	Given	Need	Connect
	zinc (Zn), mass number 68	number of protons, neutrons, electrons	periodic table, atomic number

a. Zinc (Zn), with an atomic number of 30, has 30 protons.
b. The number of neutrons in this atom is found by subtracting the number of protons (atomic number) from the mass number.

Mass number	−	atomic number	=	number of neutrons
68	−	30	=	38

c. Because a zinc atom is neutral, the number of electrons is equal to the number of protons. A zinc atom has 30 electrons.

SELF TEST 4.3

a. How many neutrons are in the nucleus of a bromine atom that has a mass number of 80?
b. What is the mass number of a cesium atom that has 71 neutrons?

ANSWER

a. 45
b. 126

PRACTICE PROBLEMS

Try Practice Problems 4.27 to 4.34

PRACTICE PROBLEMS

4.4 Atomic Number and Mass Number

4.27 Would you use the atomic number, mass number, or both to determine each of the following?
a. number of protons in an atom
b. number of neutrons in an atom
c. number of particles in the nucleus
d. number of electrons in a neutral atom

4.28 Identify the type of subatomic particles described by each of the following:
a. atomic number
b. mass number
c. mass number − atomic number
d. mass number + atomic number

4.29 Write the names and symbols for the elements with the following atomic numbers:
a. 3   b. 9   c. 20   d. 30
e. 10   f. 14   g. 53   h. 8

4.30 Write the names and symbols for the elements with the following atomic numbers:
a. 1   b. 11   c. 19   d. 82
e. 35   f. 47   g. 15   h. 2

4.31 How many protons and electrons are there in a neutral atom of each of the following elements?
a. argon   b. manganese   c. iodine   d. cadmium

4.32 How many protons and electrons are there in a neutral atom of each of the following elements?
a. carbon   b. fluorine   c. tin   d. nickel

Applications

4.33 Complete the following table for atoms of essential elements in the body:

Name of the Element	Symbol	Atomic Number	Mass Number	Number of Protons	Number of Neutrons	Number of Electrons
	Zn		66			
		12			12	
Potassium					20	
				16	15	
			56			26

4.34 Complete the following table for atoms of essential elements in the body:

Name of the Element	Symbol	Atomic Number	Mass Number	Number of Protons	Number of Neutrons	Number of Electrons
	N		15			
Calcium			42			
				53	72	
		14			16	
		29	65			

4.5 Isotopes and Atomic Mass

LEARNING GOAL Determine the number of protons, neutrons, and electrons in one or more of the isotopes of an element; calculate the atomic mass of an element using the percentage abundance and mass of its naturally occurring isotopes.

We have seen that all atoms of the same element have the same number of protons and electrons. However, the atoms of any one element are not entirely identical because the atoms of most elements have different numbers of neutrons.

Atoms and Isotopes

Isotopes are atoms of the same element that have the same atomic number but different numbers of neutrons. For example, all atoms of the element magnesium (Mg) have an atomic number of 12. Thus every magnesium atom always has 12 protons. However, some naturally occurring magnesium atoms have 12 neutrons, others have 13 neutrons, and still others have 14 neutrons. The different numbers of neutrons give the magnesium atoms different mass numbers but do not change their chemical behavior.

To distinguish between the different isotopes of an element, we write an **atomic symbol** for a particular isotope that indicates the mass number in the upper left corner and the atomic number in the lower left corner.

Atomic symbol for an isotope of magnesium, Mg-24.

Atoms of Mg

Isotopes of Mg

$^{24}_{12}Mg$ $^{25}_{12}Mg$ $^{26}_{12}Mg$

The nuclei of three naturally occurring magnesium isotopes have the same number of protons but different numbers of neutrons.

An isotope may be referred to by its name or symbol, followed by its mass number, such as magnesium-24 or Mg-24. Magnesium has three naturally occurring isotopes, as shown in **TABLE 4.7**. In a large sample of naturally occurring magnesium atoms, each type of isotope can be present as a specific percentage. For example, the Mg-24 isotope makes up almost 80% of the total sample, whereas Mg-25 and Mg-26 each make up only about 10% of the total number of magnesium atoms.

TABLE 4.7 Isotopes of Magnesium

Atomic Symbol	$^{24}_{12}Mg$	$^{25}_{12}Mg$	$^{26}_{12}Mg$
Name	Mg-24	Mg-25	Mg-26
Number of Protons	12	12	12
Number of Electrons	12	12	12
Mass Number	24	25	26
Number of Neutrons	12	13	14
Mass of Isotope (amu)	23.99	24.99	25.98
Percentage Abundance	78.70	10.13	11.17

SAMPLE PROBLEM 4.4   Identifying Protons and Neutrons in Isotopes

TRY IT FIRST

Chromium, needed for maintenance of blood sugar levels, has four naturally occurring isotopes. Calculate the number of protons and number of neutrons in each of the following isotopes:

a. $^{50}_{24}Cr$ **b.** $^{52}_{24}Cr$ **c.** $^{53}_{24}Cr$ **d.** $^{54}_{24}Cr$

SOLUTION

ANALYZE THE PROBLEM	Given	Need	Connect
	atomic symbols for Cr isotopes	number of protons, number of neutrons	atomic number

In the atomic symbol, the mass number is shown in the upper left corner of the symbol, and the atomic number is shown in the lower left corner of the symbol. Thus, each isotope of Cr, atomic number 24, has 24 protons. The number of neutrons is found by subtracting the number of protons (24) from the mass number of each isotope.

Atomic Symbol	Atomic Number	Mass Number	Number of Protons	Number of Neutrons
a. $^{50}_{24}Cr$	24	50	24	26 (50 − 24)
b. $^{52}_{24}Cr$	24	52	24	28 (52 − 24)
c. $^{53}_{24}Cr$	24	53	24	29 (53 − 24)
d. $^{54}_{24}Cr$	24	54	24	30 (54 − 24)

ENGAGE 4.11

What is different and what is the same for an atom of Sn-112 and an atom of Sn-126?

SELF TEST 4.4

a. Vanadium is a micromineral needed in the formation of bones and teeth. Write the atomic symbol for the isotope of vanadium that has 27 neutrons.
b. Molybdenum, a micromineral needed to process Fe and N from food, has seven naturally occurring isotopes. Write the atomic symbols for two isotopes, Mo-97 and Mo-98.
c. Calculate the number of protons and number of neutrons in an atom of Mo-97.

ANSWER

a. $^{50}_{23}V$ **b.** $^{97}_{42}Mo$, $^{98}_{42}Mo$ **c.** 42 protons and 55 neutrons

PRACTICE PROBLEMS

Try Practice Problems 4.35 to 4.38

Atomic Mass

In laboratory work, a chemist generally uses samples with many atoms that contain all the different atoms or isotopes of an element. Because each isotope has a different mass, chemists have calculated an **atomic mass** for an "average atom," which is a *weighted average* of the masses of all the naturally occurring isotopes of that element. On the periodic table, the atomic mass is the number including decimal places that is given below the symbol of each element. Most elements consist of two or more isotopes, which is one reason that the atomic masses on the periodic table are seldom whole numbers.

ENGAGE 4.12

What is the difference between mass number and atomic mass?

Weighted Average Analogy

To understand how the atomic mass as a weighted average for a group of isotopes is calculated, we will use an analogy of bowling balls with different weights. Suppose that a bowling alley has ordered 5 bowling balls that weigh 8 lb each and 20 bowling balls that weigh 14 lb each. In this group of bowling balls, there are more 14-lb balls than 8-lb balls. The abundance of the 14-lb balls is 80.% (20/25), and the abundance of the 8-lb balls is 20.% (5/25). Now we can calculate a weighted average for the "average" bowling ball using the weight and percentage abundance of the two types of bowling balls.

The weighted average of 8-lb and 14-lb bowling balls is calculated using their percentage abundance.

Item	Weight (lb)	Percentage Abundance	Weight from Each Type
14-lb bowling ball	14 $\times$	$\dfrac{80.}{100}$ =	11.2 lb
8-lb bowling ball	8 $\times$	$\dfrac{20.}{100}$ =	1.6 lb
Weighted average mass of a bowling ball =			12.8 lb
"Atomic mass" of a bowling ball =			12.8 lb

CORE CHEMISTRY SKILL

Calculating Atomic Mass

Calculating Atomic Mass

To calculate the atomic mass of an element, we need to know the percentage abundance and the mass of each isotope, which must be determined experimentally. For example, a large sample of naturally occurring chlorine atoms is experimentally determined to contain 75.76% of $^{35}_{17}Cl$ atoms and 24.24% of $^{37}_{17}Cl$ atoms. The $^{35}_{17}Cl$ isotope has a mass of 34.97 amu, and the $^{37}_{17}Cl$ isotope has a mass of 36.97 amu.

$$\text{Atomic mass of Cl} = \underbrace{\text{mass of } ^{35}_{17}Cl \times \frac{^{35}_{17}Cl\%}{100\%}}_{\text{amu from } ^{35}_{17}Cl} + \underbrace{\text{mass of } ^{37}_{17}Cl \times \frac{^{37}_{17}Cl\%}{100\%}}_{\text{amu from } ^{37}_{17}Cl}$$

17 protons
Symbol for chlorine
Atomic mass 35.45 amu

Atomic Symbol	Mass (amu)	Percentage Abundance	Contribution to the Atomic Mass
$^{35}_{17}Cl$	34.97 $\times$	$\dfrac{75.76}{100}$ =	26.49 amu
$^{37}_{17}Cl$	36.97 $\times$	$\dfrac{24.24}{100}$ =	8.962 amu
Weighted average mass of Cl =			35.45 amu
Atomic mass of Cl =			35.45 amu

The atomic mass of 35.45 amu is the weighted average mass of a sample of Cl atoms, although no individual Cl atom actually has this mass. An atomic mass of 35.45, which is closer to the mass number of Cl-35, indicates there is a higher percentage of $^{35}_{17}Cl$ atoms in the chlorine sample. In fact, there are about three atoms of $^{35}_{17}Cl$ for every one atom of $^{37}_{17}Cl$ in a sample of chlorine atoms.

TABLE 4.8 lists the naturally occurring isotopes of some selected elements and their atomic masses.

Chlorine, with two naturally occurring isotopes, has an atomic mass of 35.45 amu.

TABLE 4.8 The Atomic Mass of Some Elements

Element	Naturally Occurring Isotopes	Atomic Mass (weighted average)
Lithium	$_3^6Li$, $_3^7Li$	6.941 amu
Carbon	$_6^{12}C$, $_6^{13}C$, $_6^{14}C$	12.01 amu
Oxygen	$_8^{16}O$, $_8^{17}O$, $_8^{18}O$	16.00 amu
Fluorine	$_9^{19}F$	19.00 amu
Sulfur	$_{16}^{32}S$, $_{16}^{33}S$, $_{16}^{34}S$, $_{16}^{36}S$	32.07 amu
Potassium	$_{19}^{39}K$, $_{19}^{40}K$, $_{19}^{41}K$	39.10 amu
Copper	$_{29}^{63}Cu$, $_{29}^{65}Cu$	63.55 amu

Magnesium, with three naturally occurring isotopes, has an atomic mass of 24.31 amu.

INTERACTIVE VIDEO

Isotopes and Atomic Mass

PRACTICE PROBLEMS

Try Practice Problems 4.39 to 4.48

SAMPLE PROBLEM 4.5   Calculating Atomic Mass

TRY IT FIRST

Magnesium is a macromineral needed in the contraction of muscles and metabolic reactions. Using Table 4.7, calculate the atomic mass for magnesium using the weighted average mass method.

SOLUTION

ANALYZE THE PROBLEM	Given	Need	Connect
	percentage abundance, atomic mass	atomic mass of Mg	weighted average mass

STEP 1 Multiply the mass of each isotope by its percentage abundance divided by 100.

Atomic Symbol	Mass (amu)	Percentage Abundance	Contribution to the Atomic Mass
$_{12}^{24}Mg$	23.99 $\times$	$\dfrac{78.70}{100}$ =	18.88 amu
$_{12}^{25}Mg$	24.99 $\times$	$\dfrac{10.13}{100}$ =	2.531 amu
$_{12}^{26}Mg$	25.98 $\times$	$\dfrac{11.17}{100}$ =	2.902 amu

STEP 2 Add the contribution of each isotope to obtain the atomic mass.

Atomic mass of Mg = 18.88 amu + 2.531 amu + 2.902 amu

= 24.31 amu (weighted average mass)

SELF TEST 4.5

There are two naturally occurring isotopes of boron. The isotope $_5^{10}B$ has a mass of 10.01 amu with an abundance of 19.80%, and the isotope $_5^{11}B$ has a mass of 11.01 amu with an abundance of 80.20%. Calculate the atomic mass for boron using the weighted average mass method.

ANSWER

10.81 amu

PRACTICE PROBLEMS

4.5 Isotopes and Atomic Mass

4.35 What are the number of protons, neutrons, and electrons in the following isotopes?

a. $_{38}^{89}Sr$ b. $_{24}^{52}Cr$ c. $_{16}^{34}S$ d. $_{35}^{81}Br$

4.36 What are the number of protons, neutrons, and electrons in the following isotopes?

a. $_1^2H$ b. $_7^{14}N$ c. $_{14}^{26}Si$ d. $_{30}^{70}Zn$

4.37 Write the atomic symbol for the isotope with each of the following characteristics:
  a. 15 protons and 16 neutrons
  b. 35 protons and 45 neutrons
  c. 50 electrons and 72 neutrons
  d. a chlorine atom with 18 neutrons
  e. a mercury atom with 122 neutrons

4.38 Write the atomic symbol for the isotope with each of the following characteristics:
  a. an oxygen atom with 10 neutrons
  b. 4 protons and 5 neutrons
  c. 25 electrons and 28 neutrons
  d. a mass number of 24 and 13 neutrons
  e. a nickel atom with 32 neutrons

4.39 Argon has three naturally occurring isotopes, with mass numbers 36, 38, and 40.
  a. Write the atomic symbol for each of these atoms.
  b. How are these isotopes alike?
  c. How are they different?
  d. Why is the atomic mass of argon listed on the periodic table not a whole number?
  e. Which isotope is the most abundant in a sample of argon?

4.40 Strontium has four naturally occurring isotopes, with mass numbers 84, 86, 87, and 88.
  a. Write the atomic symbol for each of these atoms.
  b. How are these isotopes alike?
  c. How are they different?

  d. Why is the atomic mass of strontium listed on the periodic table not a whole number?
  e. Which isotope is the most abundant in a sample of strontium?

4.41 What is the difference between the mass of an isotope and the atomic mass of an element?

4.42 What is the difference between the mass number and the atomic mass of an element?

4.43 Copper consists of two isotopes, $^{63}_{29}$Cu and $^{65}_{29}$Cu. If the atomic mass for copper on the periodic table is 63.55, are there more atoms of $^{63}_{29}$Cu or $^{65}_{29}$Cu in a sample of copper?

4.44 There are two naturally occurring isotopes of iridium: Ir-191 and Ir-193. Use the atomic mass of iridium listed on the periodic table to identify the more abundant isotope.

4.45 Cadmium consists of eight naturally occurring isotopes. Do you expect any of the isotopes to have the atomic mass listed on the periodic table for cadmium? Explain.

4.46 Zinc consists of five naturally occurring isotopes: $^{64}_{30}$Zn, $^{66}_{30}$Zn, $^{67}_{30}$Zn, $^{68}_{30}$Zn, and $^{70}_{30}$Zn. None of these isotopes has the atomic mass of 65.41 listed for zinc on the periodic table. Explain.

4.47 Two isotopes of gallium are naturally occurring, with $^{69}_{31}$Ga at 60.11% (68.93 amu) and $^{71}_{31}$Ga at 39.89% (70.92 amu). Calculate the atomic mass for gallium using the weighted average mass method.

4.48 Two isotopes of rubidium occur naturally, with $^{85}_{37}$Rb at 72.17% (84.91 amu) and $^{87}_{37}$Rb at 27.83% (86.91 amu). Calculate the atomic mass for rubidium using the weighted average mass method.

UPDATE Improving Crop Production

Plants need potassium (K) for many metabolic processes, including the regulation of growth, protein synthesis, photosynthesis, and ionic balance. Potassium-deficient potato plants show purple or brown spots and reduced plant, root, and seed growth. John notices that the leaves of his last crop of potatoes have brown spots, the potatoes are undersized, and the crop yield is low.

Tests on soil samples show that the potassium levels are below 100 ppm, indicating that the potato plants need supplemental potassium. John applies a fertilizer containing potassium chloride (KCl). To obtain the correct amount of potassium, John calculates that he needs to apply 170 kg of fertilizer per hectare.

A mineral deficiency of potassium causes brown spots on potato leaves.

Applications

4.49 **a.** What is the group number and name of the group that contains potassium?
  b. Is potassium a metal, a nonmetal, or a metalloid?
  c. How many protons are in an atom of potassium?

  d. Potassium has three naturally occurring isotopes. They are K-39, K-40, and K-41. Using the atomic mass of potassium, determine which isotope of potassium is the most abundant.

4.50 **a.** How many neutrons are in K-41?
  b. If John's potato field has an area of 34.5 hectares, how many pounds of fertilizer does John need to use?
  c. Potassium has three naturally occurring isotopes. They are K-39 (93.26%, 38.964 amu), K-40 (0.0117%, 39.964 amu), and K-41 (6.73%, 40.962 amu). Calculate the atomic mass for potassium from the naturally occurring isotopes and percentage abundance using the weighted average mass method.

CONCEPT MAP

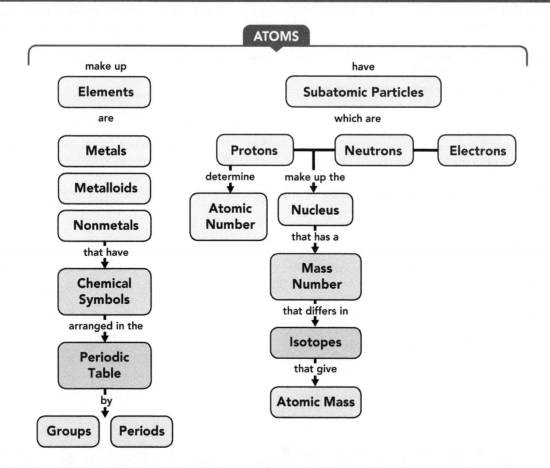

ATOMS

make up → **Elements**

have → **Subatomic Particles**

are → **Metals**, **Metalloids**, **Nonmetals**

which are → **Protons**, **Neutrons**, **Electrons**

Protons *determine* → **Atomic Number**

Neutrons *make up the* → **Nucleus**

that have → **Chemical Symbols**

arranged in the → **Periodic Table**

by → **Groups**, **Periods**

Nucleus *that has a* → **Mass Number**

that differs in → **Isotopes**

that give → **Atomic Mass**

CHAPTER REVIEW

4.1 Elements and Symbols

LEARNING GOAL Given the name of an element, write its correct symbol; from the symbol, write the correct name.

- Elements are the primary substances of matter.
- Chemical symbols are one- or two-letter abbreviations of the names of the elements.

4.2 The Periodic Table

LEARNING GOAL Use the periodic table to identify the group and the period of an element; identify an element as a metal, a nonmetal, or a metalloid.

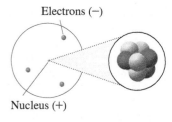

- The periodic table is an arrangement of the elements by increasing atomic number.
- A horizontal row is called a *period*.
- A vertical column on the periodic table containing elements with similar properties is called a *group*.

- Elements in Group 1A (1) are called the *alkali metals*; Group 2A (2), the *alkaline earth metals*; Group 7A (17), the *halogens*; and Group 8A (18), the *noble gases*.
- On the periodic table, *metals* are located on the left of the heavy zigzag line, and *nonmetals* are to the right of the heavy zigzag line.
- Except for aluminum and organesson, elements located along the heavy zigzag line are called *metalloids*.

4.3 The Atom

LEARNING GOAL Describe the electrical charge of a proton, neutron, and electron and identify their location within an atom.

- An atom is the smallest particle that retains the characteristics of an element.
- Atoms are composed of three types of subatomic particles.
- Protons have a positive charge (+), electrons carry a negative charge (−), and neutrons are electrically neutral.
- The protons and neutrons are found in the tiny, dense nucleus; electrons are located outside the nucleus.

Electrons (−)

Nucleus (+)

4.4 Atomic Number and Mass Number

LEARNING GOAL Given the atomic number and the mass number of an atom, state the number of protons, neutrons, and electrons.

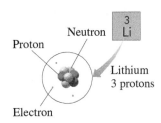

- The atomic number gives the number of protons in all the atoms of the same element.
- In a neutral atom, the number of protons and electrons is equal.
- The mass number is the total number of protons and neutrons in an atom.

4.5 Isotopes and Atomic Mass

LEARNING GOAL Determine the number of protons, neutrons, and electrons in one or more of the isotopes of an element; calculate the atomic mass of an element using the percentage abundance and mass of its naturally occurring isotopes.

- Atoms that have the same number of protons but different numbers of neutrons are called *isotopes*.
- The atomic mass of an element is the weighted average mass of all the isotopes in a naturally occurring sample of that element.

KEY TERMS

alkali metal An element in Group 1A (1), except hydrogen, that is a soft, shiny metal with one electron in its outermost energy level.

alkaline earth metal An element in Group 2A (2) that has two electrons in its outermost energy level.

atom The smallest particle of an element that retains the characteristics of the element.

atomic mass The weighted average mass of all the naturally occurring isotopes of an element.

atomic mass unit (amu) A small mass unit used to describe the mass of extremely small particles such as atoms and subatomic particles; 1 amu is equal to one-twelfth the mass of a $^{12}_{6}C$ atom.

atomic number A number that is equal to the number of protons in an atom.

atomic symbol An abbreviation used to indicate the mass number and atomic number of an isotope.

chemical symbol An abbreviation that represents the name of an element.

electron A negatively charged subatomic particle having a minute mass that is usually ignored in mass calculations; its symbol is e^-.

group A vertical column in the periodic table that contains elements having similar physical and chemical properties.

group number A number that appears at the top of each vertical column (group) in the periodic table and indicates the number of electrons in the outermost energy level.

halogen An element in Group 7A (17)—fluorine, chlorine, bromine, iodine, astatine, and tennessine—that has seven electrons in its outermost energy level.

isotope An atom that differs only in mass number from another atom of the same element. Isotopes have the same atomic number (number of protons), but different numbers of neutrons.

mass number The total number of protons and neutrons in the nucleus of an atom.

metal An element that is shiny, malleable, ductile, and a good conductor of heat and electricity. The metals are located to the left of the heavy zigzag line on the periodic table.

metalloid An element with properties of both metals and nonmetals located along the heavy zigzag line on the periodic table.

neutron A neutral subatomic particle having a mass of about 1 amu and found in the nucleus of an atom; its symbol is n or n^0.

noble gas An element in Group 8A (18) of the periodic table, generally unreactive and seldom found in combination with other elements, that has eight electrons (helium has two electrons) in its outermost energy level.

nonmetal An element with little or no luster that is a poor conductor of heat and electricity. The nonmetals are located to the right of the heavy zigzag line on the periodic table.

nucleus The compact, extremely dense center of an atom, containing the protons and neutrons of the atom.

period A horizontal row of elements in the periodic table.

periodic table An arrangement of elements by increasing atomic number such that elements having similar chemical behavior are grouped in vertical columns.

proton A positively charged subatomic particle having a mass of about 1 amu and found in the nucleus of an atom; its symbol is p or p^+.

representative element An element in the first two columns on the left of the periodic table and the last six columns on the right that has a group number of 1A through 8A or 1, 2, and 13 through 18.

subatomic particle A particle within an atom; protons, neutrons, and electrons are subatomic particles.

transition element An element in the center of the periodic table that is designated with the letter "B" or the group number of 3 through 12.

CORE CHEMISTRY SKILLS

The chapter section containing each Core Chemistry Skill is shown in parentheses at the end of each heading.

Counting Protons and Neutrons (4.4)

- The atomic number of an element is equal to the number of protons in every atom of that element. The atomic number is the whole number that appears above the symbol of each element on the periodic table.

  Atomic number = number of protons in an atom

- Because atoms are neutral, the number of electrons is equal to the number of protons. Thus, the atomic number gives the number of electrons.
- The mass number is the total number of protons and neutrons in the nucleus of an atom.

  Mass number = number of protons + number of neutrons

- The number of neutrons is calculated from the mass number and atomic number.

  Number of neutrons = mass number − number of protons

Example: Calculate the number of protons, neutrons, and electrons in a krypton atom with a mass number of 80.

Answer:

Element	Atomic Number	Mass Number	Number of Protons	Number of Neutrons	Number of Electrons
Kr	36	80	equal to atomic number 36	equal to mass number − number of protons 80 − 36 = 44	equal to number of protons 36

Example: Calculate the number of protons and neutrons in the cadmium isotope $^{112}_{48}\text{Cd}$.

Answer:

Atomic Symbol	Atomic Number	Mass Number	Number of Protons	Number of Neutrons
$^{112}_{48}\text{Cd}$	number in lower left corner 48	number in upper left corner 112	equal to atomic number 48	equal to mass number − number of protons 112 − 48 = 64

Writing Atomic Symbols for Isotopes (4.5)

- Isotopes are atoms of the same element that have the same atomic number but different numbers of neutrons.
- An atomic symbol is written for a particular isotope, with its mass number (protons and neutrons) shown in the upper left corner and its atomic number (protons) shown in the lower left corner.

Mass number ⟶ $^{24}_{12}\text{Mg}$ ⟵ Symbol of element

Atomic number ⟶

Calculating Atomic Mass (4.5)

- The atomic mass is calculated from the masses and percentage abundances of the isotopes of an element.

Atomic mass = %/100 × isotope (1) + %/100 × isotope (2) . . .

Example: Calculate the atomic mass for rhenium with two naturally occurring isotopes: 37.40% Re-185 (184.95 amu) and 62.60% Re-187 (186.96 amu).

Answer:
atomic mass Re = 37.40/100 × 184.95 amu + 62.60/100 × 186.96 amu

= 69.17 amu + 117.0 amu

= 186.2 amu

UNDERSTANDING THE CONCEPTS

The chapter sections to review are shown in parentheses at the end of each problem.

4.51 According to Dalton's atomic theory, which of the following are *true* or *false*? If *false*, correct the statement to make it *true*. (4.3)

a. Atoms of an element are identical to atoms of other elements.
b. Every element is made of atoms.
c. Atoms of different elements combine to form compounds.
d. In a chemical reaction, some atoms disappear and new atoms appear.

4.52 Use Rutherford's gold-foil experiment to answer each of the following: (4.3)

a. What did Rutherford expect to happen when he aimed particles at the gold foil?
b. How did the results differ from what he expected?
c. How did he use the results to propose a model of the atom?

4.53 Match the subatomic particles (**1** to **3**) to each of the descriptions below. (4.4)

1. protons   2. neutrons   3. electrons

a. atomic mass
b. atomic number
c. positive charge
d. negative charge
e. mass number − atomic number

4.54 Match the subatomic particles (**1** to **3**) to each of the descriptions below. (4.4)

1. protons   2. neutrons   3. electrons

a. mass number
b. surround the nucleus

c. in the nucleus
d. charge of 0
e. equal to number of electrons

4.55 Consider the following atoms in which X represents the chemical symbol of the element: (4.5)

$^{16}_{8}\text{X}$ $^{16}_{9}\text{X}$ $^{18}_{10}\text{X}$ $^{17}_{8}\text{X}$ $^{18}_{8}\text{X}$

a. What atoms have the same number of protons?
b. Which atoms are isotopes? Of what element?
c. Which atoms have the same mass number?

4.56 Consider the following atoms in which X represents the chemical symbol of the element: (4.5)

$^{124}_{47}\text{X}$ $^{116}_{49}\text{X}$ $^{116}_{50}\text{X}$ $^{124}_{50}\text{X}$ $^{116}_{48}\text{X}$

a. What atoms have the same number of protons?
b. Which atoms are isotopes? Of what element?
c. Which atoms have the same mass number?

4.57 For each representation of a nucleus **A** through **E**, write the atomic symbol, and identify which are isotopes. (4.5)

Proton ●

Neutron ●

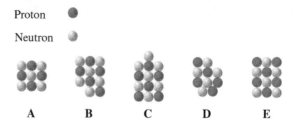

A   B   C   D   E

4.58 Identify the element represented by each nucleus **A** through **E** in problem 4.57 as a metal, a nonmetal, or a metalloid. (4.2)

Applications

4.59 Complete the following table for three of the naturally occurring isotopes of germanium, which is a metalloid used in semiconductors: (4.5)

	Atomic Symbol		
	$^{70}_{32}Ge$	$^{73}_{32}Ge$	$^{76}_{32}Ge$
Atomic Number			
Mass Number			
Number of Protons			
Number of Neutrons			
Number of Electrons			

4.60 Complete the following table for the three naturally occurring isotopes of silicon, the major component in computer chips: (4.5)

	Atomic Symbol		
	$^{28}_{14}Si$	$^{29}_{14}Si$	$^{30}_{14}Si$
Atomic Number			
Mass Number			
Number of Protons			
Number of Neutrons			
Number of Electrons			

Computer chips consist primarily of the element silicon.

ADDITIONAL PRACTICE PROBLEMS

4.61 Complete the following statements: (4.2, 4.4)
  a. The atomic number gives the number of _____ in the nucleus.
  b. In an atom, the number of electrons is equal to the number of _____.
  c. Sodium and potassium are examples of elements called _____.

4.62 Complete the following statements: (4.2, 4.4)
  a. The number of protons and neutrons in an atom is the _____ number.
  b. The elements in Group 7A (17) are called the _____.
  c. Elements that are shiny and conduct heat are called _____.

4.63 Write the chemical symbol and group and period number for each of the following elements: (4.2)
  a. bromine
  b. argon
  c. lithium
  d. radium

4.64 Write the chemical symbol and group and period number for each of the following elements: (4.2)
  a. radon
  b. tin
  c. carbon
  d. magnesium

4.65 The following trace elements have been found to be crucial to the functions of the body. Indicate each as a metal, a nonmetal, or a metalloid. (4.2)
  a. molybdenum
  b. vanadium
  c. silicon
  d. iodine

4.66 The following trace elements have been found to be crucial to the functions of the body. Indicate each as a metal, a nonmetal, or a metalloid. (4.2)
  a. copper
  b. selenium
  c. arsenic
  d. chromium

4.67 Indicate if each of the following statements is *true* or *false*: (4.3)
  a. The proton is a negatively charged particle.
  b. The neutron is 2000 times as heavy as a proton.
  c. The atomic mass unit is based on a carbon atom with six protons and six neutrons.
  d. The nucleus is the largest part of the atom.
  e. The electrons are located outside the nucleus.

4.68 Indicate if each of the following statements is *true* or *false*: (4.3)
  a. The neutron is electrically neutral.
  b. Most of the mass of an atom is because of the protons and neutrons.
  c. The charge of an electron is equal, but opposite, to the charge of a neutron.
  d. The proton and the electron have about the same mass.
  e. The mass number is the number of protons.

4.69 For the following atoms, give the number of protons, neutrons, and electrons: (4.4, 4.5)
  a. $^{114}_{48}Cd$ **b.** $^{98}_{43}Tc$
  c. $^{199}_{79}Au$ **d.** $^{222}_{86}Rn$
  e. $^{136}_{54}Xe$

4.70 For the following atoms, give the number of protons, neutrons, and electrons: (4.4, 4.5)
  a. $^{202}_{80}Hg$ **b.** $^{127}_{53}I$
  c. $^{75}_{35}Br$ **d.** $^{133}_{55}Cs$
  e. $^{195}_{78}Pt$

4.71 Complete the following table: (4.4, 4.5)

Name of the Element	Atomic Symbol	Number of Protons	Number of Neutrons	Number of Electrons
	$^{80}_{34}\text{Se}$			
		28	34	
Magnesium			14	
	$^{228}_{88}\text{Ra}$			

4.72 Complete the following table: (4.4, 4.5)

Name of the Element	Atomic Symbol	Number of Protons	Number of Neutrons	Number of Electrons
Potassium			22	
	$^{51}_{23}\text{V}$			
		48	64	
Barium			82	

4.73 Provide the following: (4.2, 4.4)
 a. the atomic number and symbol of the lightest alkali metal
 b. the atomic number and symbol of the heaviest noble gas
 c. the atomic mass and symbol of the alkaline earth metal in Period 3
 d. the atomic mass and symbol of the halogen with the fewest electrons

4.74 Provide the following: (4.2, 4.4)
 a. the atomic number and symbol of the heaviest metalloid in Group 4A (14)
 b. the atomic number and symbol of the element in Group 5A (15), Period 6
 c. the atomic mass and symbol of the alkali metal in Period 4
 d. the metalloid in Group 3A (13)

4.75 Write the names and symbols of the elements with the following atomic numbers: (4.4)
 a. 35   b. 57   c. 52
 d. 33   e. 50   f. 55

4.76 Write the names and symbols of the elements with the following atomic numbers: (4.4)
 a. 22   b. 49   c. 26
 d. 54   e. 78   f. 83

4.77 How many protons and electrons are there in a neutral atom of each of the following elements? (4.4)
 a. Pt   b. phosphorus   c. Sr
 d. Co   e. uranium

4.78 How many protons and electrons are there in a neutral atom of each of the following elements? (4.4)
 a. chromium   b. Cs   c. copper
 d. chlorine   e. Ba

4.79 Write the atomic symbol for each of the following: (4.5)
 a. an atom with 22 protons and 26 neutrons
 b. an atom with 12 protons and 14 neutrons
 c. a gallium atom with a mass number of 71
 d. an atom with 30 electrons and 40 neutrons

4.80 Write the atomic symbol for each of the following: (4.5)
 a. an aluminum atom with 14 neutrons
 b. an atom with atomic number 26 and 32 neutrons
 c. a lead atom with 125 neutrons
 d. an atom with a mass number of 72 and atomic number 33

4.81 Lead consists of four naturally occurring isotopes. Calculate the atomic mass for lead using the weighted average mass method. (4.5)

Isotope	Mass (amu)	Abundance (%)
$^{204}_{82}\text{Pb}$	204.0	1.40
$^{206}_{82}\text{Pb}$	206.0	24.10
$^{207}_{82}\text{Pb}$	207.0	22.10
$^{208}_{82}\text{Pb}$	208.0	52.40

4.82 Indium (In) has two naturally occurring isotopes: In-113 and In-115. In-113 has a 4.30% abundance and a mass of 112.9 amu, and In-115 has a 95.70% abundance and a mass of 114.9 amu. Calculate the atomic mass for indium using the weighted average mass method. (4.5)

CHALLENGE PROBLEMS

The following problems are related to the topics in this chapter. However, they do not all follow the chapter order, and they require you to combine concepts and skills from several sections. These problems will help you increase your critical thinking skills and prepare for your next exam.

4.83 The most abundant isotope of lead is $^{208}_{82}\text{Pb}$. (4.4, 4.5)
 a. How many protons, neutrons, and electrons are in $^{208}_{82}\text{Pb}$?
 b. What is the atomic symbol of another isotope of lead with 132 neutrons?
 c. What is the name and symbol of an atom with the same mass number as in part b and 131 neutrons?

4.84 The most abundant isotope of silver is $^{107}_{47}\text{Ag}$. (4.4, 4.5)
 a. How many protons, neutrons, and electrons are in $^{107}_{47}\text{Ag}$?
 b. What is the symbol of another isotope of silver with 62 neutrons?
 c. What is the name and symbol of an atom with the same mass number as in part b and 61 neutrons?

4.85 The most abundant isotope of gold is Au-197. (4.5)
 a. How many protons, neutrons, and electrons are in this isotope?
 b. What is the atomic symbol of another isotope of gold with 116 neutrons?
 c. What is the atomic symbol of an atom with an atomic number of 78 and 116 neutrons?

4.86 The most abundant isotope of calcium is Ca-40. (4.5)
 a. How many protons, neutrons, and electrons are in this isotope?
 b. What is the atomic symbol of another isotope of calcium with 24 neutrons?
 c. What is the atomic symbol of an atom with an atomic number of 21 and 24 neutrons?

4.87 Silicon has three naturally occurring isotopes: Si-28 that has a percentage abundance of 92.23% and a mass of 27.977 amu, Si-29 that has an abundance of 4.68% and a mass of 28.976 amu, and Si-30 that has a percentage abundance of 3.09% and a mass of 29.974 amu. Calculate the atomic mass for silicon using the weighted average mass method. (4.5)

4.88 Antimony (Sb) has two naturally occurring isotopes: Sb-121 that has a percentage abundance of 57.21% and a mass of 120.90 amu, and Sb-123 that has a percentage abundance of 42.79% and a mass of 122.90 amu. Calculate the atomic mass for antimony using the weighted average mass method. (4.5)

4.89 There are four naturally occurring isotopes of strontium: (4.4, 4.5)

$$^{84}_{38}Sr, \ ^{86}_{38}Sr, \ ^{87}_{38}Sr, \text{ and } ^{88}_{38}Sr$$

a. How many protons, neutrons, and electrons are in Sr-87?
b. What is the most abundant isotope in a strontium sample?

c. How many neutrons are in Sr-84?
d. Why don't any of the isotopes of strontium have the atomic mass of 87.62 amu listed on the periodic table?

4.90 There are four naturally occurring isotopes of iron: (4.4, 4.5)

$$^{54}_{26}Fe, \ ^{56}_{26}Fe, \ ^{57}_{26}Fe, \text{ and } ^{58}_{26}Fe$$

a. How many protons, neutrons, and electrons are in $^{58}_{26}Fe$?
b. What is the most abundant isotope in an iron sample?
c. How many neutrons are in $^{57}_{26}Fe$?
d. Why don't any of the isotopes of iron have the atomic mass of 55.85 amu listed on the periodic table?

ANSWERS TO ENGAGE QUESTIONS

4.1 Some elements we may encounter everyday are aluminum (Al), calcium (Ca), silver (Ag), carbon (C), and oxygen (O).

4.2 potassium, K

4.3 shiny, soft, good conductors of heat, and electricity, have relatively low melting points, react vigorously with water, and form white products when they combine with oxygen

4.4 fluorine, chlorine, bromine, iodine, astatine, and tennessine

4.5 Dalton said that compounds are always made up of the same kind and number of atoms.

4.6 Particles passing through the gold foil are deflected only if they go very close to the positively charged nucleus.

4.7 The neutrons and protons are contained in the very small nucleus with the electrons occupying the very large volume of the atom, which is mostly empty space.

4.8 Because atoms are neutral, 56 positively charged protons of every barium atom are balanced by 56 negatively charged electrons, which gives the atom an overall charge of zero.

4.9 The number of protons and neutrons determines the mass number of an atom.

4.10 An atom of tin with a mass number of 102 has 50 protons and 52 neutrons.

4.11 Atoms of Sn-112 and Sn-126 both have 50 protons; atoms of Sn-112 have 62 neutrons and atoms of Sn-126 have 76 neutrons.

4.12 The mass number is the number of protons and neutrons in an atom. The atomic mass is the weighted average of the masses of all the naturally occurring isotopes of an element.

ANSWERS TO SELECTED PROBLEMS

4.1 **a.** Cu   **b.** Pt   **c.** Ca   **d.** Mn
      e. Fe   **f.** Ba   **g.** Pb   **h.** Sr

4.3 **a.** Ag   **b.** Si   **c.** Sb   **d.** correct   **e.** F

4.5 **a.** carbon   **b.** chlorine   **c.** iodine   **d.** selenium
      e. nitrogen   **f.** sulfur   **g.** zinc   **h.** cobalt

4.7 **a.** sodium, chlorine
      b. calcium, sulfur, oxygen
      c. carbon, hydrogen, chlorine, nitrogen, oxygen
      d. lithium, carbon, oxygen

4.9 **a.** Period 2   **b.** Group 8A (18)
      c. Group 1A (1)   **d.** Period 2

4.11 **a.** C   **b.** He   **c.** Na   **d.** Ca   **e.** Al

4.13 **a.** metal   **b.** nonmetal   **c.** metal   **d.** nonmetal
       e. nonmetal   **f.** nonmetal   **g.** metalloid   **h.** metal

4.15 **a.** needed for bones and teeth, muscle contraction, nerve impulses; alkaline earth metal
       b. component of hemoglobin; transition element
       c. muscle contraction, nerve impulses; alkali metal
       d. found in fluids outside cells; halogen
       e. needed in the formation of bones and teeth, and the extraction of energy from food; transition element
       f. needed for maintenance of blood sugar levels; transition element

4.17 **a.** metal   **b.** metal   **c.** metal
       d. nonmetal   **e.** metal   **f.** metal

4.19 **a.** electron   **b.** proton
       c. electron   **d.** neutron

4.21 Rutherford determined that an atom contains a small, compact nucleus that is positively charged.

4.23 **a.** true
       b. true
       c. true
       d. false. A proton is attracted to an electron.

4.25 In the process of brushing hair, strands of hair become charged with like charges that repel each other.

4.27 **a.** atomic number   **b.** both
       c. mass number   **d.** atomic number

4.29 **a.** lithium, Li   **b.** fluorine, F   **c.** calcium, Ca
       d. zinc, Zn   **e.** neon, Ne   **f.** silicon, Si
       g. iodine, I   **h.** oxygen, O

4.31 **a.** 18 protons and 18 electrons
       b. 25 protons and 25 electrons
       c. 53 protons and 53 electrons
       d. 48 protons and 48 electrons

4.33

Name of the Element	Symbol	Atomic Number	Mass Number	Number of Protons	Number of Neutrons	Number of Electrons
Zinc	Zn	30	66	30	36	30
Magnesium	Mg	12	24	12	12	12
Potassium	K	19	39	19	20	19
Sulfur	S	16	31	16	15	16
Iron	Fe	26	56	26	30	26

4.35 **a.** 38 protons, 51 neutrons, 38 electrons
b. 24 protons, 28 neutrons, 24 electrons
c. 16 protons, 18 neutrons, 16 electrons
d. 35 protons, 46 neutrons, 35 electrons

4.37 **a.** $^{31}_{15}P$ **b.** $^{80}_{35}Br$ **c.** $^{122}_{50}Sn$ **d.** $^{35}_{17}Cl$ **e.** $^{202}_{80}Hg$

4.39 **a.** $^{36}_{18}Ar$ $^{38}_{18}Ar$ $^{40}_{18}Ar$
b. They all have the same number of protons and electrons.
c. They have different numbers of neutrons, which gives them different mass numbers.
d. The atomic mass of Ar listed on the periodic table is the weighted average atomic mass of all the naturally occurring isotopes.
e. The isotope Ar-40 is most abundant because its mass is closest to the atomic mass of Ar on the periodic table.

4.41 The mass of an isotope is the mass of an individual atom; the atomic mass of an element is the weighted average of the masses of all the naturally occurring isotopes of that element.

4.43 Because the atomic mass of copper is closer to 63 amu, there are more atoms of $^{63}_{29}Cu$.

4.45 No. The atomic mass is the weighted average of the masses of the eight naturally occurring isotopes of cadmium.

4.47 69.72 amu

4.49 **a.** Group 1A (1), alkali metals
b. metal
c. 19 protons
d. K-39

4.51 **a.** false. All atoms of a given element are different from atoms of other elements.
b. true
c. true
d. false. In a chemical reaction, atoms are neither created nor destroyed.

4.53 **a.** 1 + 2 **b.** 1 **c.** 1 **d.** 3 **e.** 2

4.55 **a.** $^{16}_{8}X$, $^{17}_{8}X$, $^{18}_{8}X$ all have eight protons.
b. $^{16}_{8}X$, $^{17}_{8}X$, $^{18}_{8}X$ are all isotopes of oxygen.
c. $^{16}_{8}X$ and $^{16}_{9}X$ have mass numbers of 16, and $^{18}_{10}X$ and $^{18}_{8}X$ have mass numbers of 18.

4.57 **a.** $^{9}_{4}Be$ **b.** $^{11}_{5}B$ **c.** $^{13}_{6}C$
d. $^{10}_{5}B$ **e.** $^{12}_{6}C$

B and D are isotopes of boron; C and E are isotopes of carbon.

4.59

	Atomic Symbol		
	$^{70}_{32}Ge$	$^{73}_{32}Ge$	$^{76}_{32}Ge$
Atomic Number	32	32	32
Mass Number	70	73	76
Number of Protons	32	32	32
Number of Neutrons	38	41	44
Number of Electrons	32	32	32

4.61 **a.** protons **b.** protons **c.** alkali metals

4.63 **a.** Group 7A (17), Period 4
b. Group 8A (18), Period 3
c. Group 1A (1), Period 2
d. Group 2A (2), Period 7

4.65 **a.** metal
b. metal
c. metalloid
d. nonmetal

4.67 **a.** false
b. false
c. true
d. false
e. true

4.69 **a.** 48 protons, 66 neutrons, 48 electrons
b. 43 protons, 55 neutrons, 43 electrons
c. 79 protons, 120 neutrons, 79 electrons
d. 86 protons, 136 neutrons, 86 electrons
e. 54 protons, 82 neutrons, 54 electrons

4.71

Name of the Element	Atomic Symbol	Number of Protons	Number of Neutrons	Number of Electrons
Selenium	$^{80}_{34}Se$	34	46	34
Nickel	$^{62}_{28}Ni$	28	34	28
Magnesium	$^{26}_{12}Mg$	12	14	12
Radium	$^{228}_{88}Ra$	88	140	88

4.73 **a.** 3, Li **b.** 118, Og **c.** 24.31 amu, Mg **d.** 19.00 amu, F

4.75 **a.** bromine, Br **b.** lanthanum, La **c.** tellurium, Te
d. arsenic, As **e.** tin, Sn **f.** cesium, Cs

4.77 **a.** 78 protons, 78 electrons **b.** 15 protons, 15 electrons
c. 38 protons, 38 electrons **d.** 27 protons, 27 electrons
e. 92 protons, 92 electrons

4.79 **a.** $^{48}_{22}Ti$ **b.** $^{26}_{12}Mg$ **c.** $^{71}_{31}Ga$ **d.** $^{70}_{30}Zn$

4.81 207.3 amu

4.83 **a.** 82 protons, 126 neutrons, 82 electrons
b. $^{214}_{82}Pb$
c. $^{214}_{83}Bi$

4.85 **a.** 79 protons, 118 neutrons, 79 electrons
b. $^{195}_{79}Au$
c. $^{194}_{78}Pt$

4.87 28.09 amu

4.89 **a.** 38 protons, 49 neutrons, 38 electrons
b. $^{88}_{38}Sr$
c. 46 neutrons
d. The atomic mass on the periodic table is the average of all the naturally occurring isotopes.

5

Electronic Structure of Atoms and Periodic Trends

Robert and Jennifer work in the materials science department of a research laboratory. As materials engineers, they develop and test various materials that are used in the manufacturing of consumer goods like computer chips, television sets, golf clubs, and snow skis. These engineers create new materials, which are needed for mechanical, chemical, and electrical industries. Often materials engineers study materials at an atomic level to learn how to improve the characteristics of materials.

In their research, Robert and Jennifer work with some of the elements in Groups 3A (13), 4A (14), and 5A (15) of the periodic table. These elements, such as silicon, have properties that make them good semiconductors. A microchip requires growing a single crystal of a semiconductor such as pure silicon. Microchips are manufactured for use in computers, cell phones, satellites, televisions, calculators, GPS, and many other devices. Robert and Jennifer are working on materials that may be used to develop more complex microchips that lead to new applications.

CAREER

Materials Engineer

Materials engineers work with metals, ceramics, plastics, semiconductors, and composites to develop new or improved products such as computer chips, aircraft material, and tennis racquets.

Engineers use mathematics and the principles of chemistry and other sciences to solve technical problems. They may also work in the development and testing of new materials. Materials engineers typically have bachelor's degrees in mechanical, electrical, or chemical engineering, or related fields.

UPDATE Developing New Materials for Computer Chips

Last week, Robert and Jennifer added impurities to silicon to evaluate changes in its properties as a semiconductor. You can see how Robert and Jennifer used the elements indium, in Group 3A (13), and tellurium, in Group 6A (16), to try to make new computer chips in the **UPDATE Developing New Materials for Computer Chips**, page 148, and learn some of the properties of indium and tellurium.

The wavelength is the distance between adjacent crests or troughs in a wave.

ENGAGE 5.1

How does the wavelength of red light compare to that of blue light?

ENGAGE 5.2

How many seconds will it take light to travel from Rome to Los Angeles, a distance of 10 200 km?

PRACTICE PROBLEMS

Try Practice Problems 5.1 to 5.8

5.1 Electromagnetic Radiation

LEARNING GOAL Compare the wavelength, frequency, and energy of electromagnetic radiation.

When we listen to a radio, use a microwave oven, turn on a light, see the colors of a rainbow, or have an X-ray taken, we are experiencing various forms of **electromagnetic radiation**. All of these types of electromagnetic radiation, including light, consist of particles that move as waves of energy.

Wavelength and Frequency

You are probably familiar with the action of waves in the ocean. When you go to a beach, you notice that the water in each wave rises and falls as the wave comes in to shore. The highest point on the wave is called a *crest*, whereas the lowest point is a *trough*. On a calm day, there might be long distances between crests or troughs. However, if there is a storm with a lot of energy, the crests or troughs are much closer together.

The waves of electromagnetic radiation also have crests and troughs. The **wavelength** (symbol λ, lambda) is the distance from a crest or trough in the wave to the next crest or trough in the wave (see **FIGURE 5.1**). In some types of radiation, the crests or troughs are far apart, while in others, they are close together.

FIGURE 5.1 Wavelengths for light

Light passing through a prism is separated into a spectrum of colors we see in a rainbow.

The wavelength is the distance from a crest or trough on a wave to the next crest or trough.

The **frequency** (symbol ν, nu) is the number of times the crests of a wave pass a point in 1 s. All electromagnetic radiation travels at the speed of light (c), which is equal to 3.00×10^8 m/s. Mathematically, the *wave equation* expresses the relationship of the speed of light (m/s) to wavelength (m) and frequency (s^{-1}).

$c = \lambda\nu$ Wave equation

Speed of light (c) = 3.00×10^8 m/s = wavelength (λ) × frequency (ν)

The speed of light is about a million times faster than the speed of sound, which is the reason we see lightning before we hear thunder during a storm.

Electromagnetic Spectrum

The **electromagnetic spectrum** is an arrangement of different types of electromagnetic radiation from longest wavelength to shortest wavelength. According to the wave equation, as the wavelength decreases, the frequency increases. Or as the wavelength increases, the frequency decreases. This type of relationship is called an *inverse relationship*.

Scientists have shown that the energy of electromagnetic radiation is directly related to frequency, which means that energy is inversely related to the wavelength. Thus, as the wavelength of radiation increases, the frequency and, therefore, the energy decrease.

At one end of the electromagnetic spectrum are radiations with long wavelengths such as *radio waves* that are used for AM and FM radio bands, cellular phones, and TV signals. The wavelength of a typical AM radio wave can be as long as a football field. *Microwaves*

Wavelength (m)

| | 10^5 | 10^4 | 10^3 | 10^2 | 10 | 1 | 10^{-1} | 10^{-2} | 10^{-3} | 10^{-4} | 10^{-5} | 10^{-6} | 10^{-7} | 10^{-8} | 10^{-9} | 10^{-10} | 10^{-11} |

| Radio waves | Microwave | Infrared | Ultraviolet | X-rays | Gamma rays |

Frequency (s^{-1})

| 10^4 | 10^5 | 10^6 | 10^7 | 10^8 | 10^9 | 10^{10} | 10^{11} | 10^{12} | 10^{13} | 10^{14} | 10^{15} | 10^{16} | 10^{17} | 10^{18} | 10^{19} |

Energy Increases →

Low energy   High energy

FIGURE 5.2 The electromagnetic spectrum shows the arrangement of wavelengths of electromagnetic radiation. The visible portion consists of wavelengths from 700 nm to 400 nm.

have shorter wavelengths and higher frequencies than radio waves. *Infrared radiation* (IR) is responsible for the heat we feel from sunlight and the infrared lamps used to warm food in restaurants. When we change the volume or the station on a TV set, we use a remote control to send infrared impulses to the receiver in the TV. Wireless technology uses radiation with higher frequencies than infrared to connect many electronic devices, including mobile and cell phones and laptops (see **FIGURE 5.2**).

Visible light, with wavelengths from 700 to 400 nm, is the only radiation our eyes can detect. Red light has the longest wavelength at 700 nm; orange is about 600 nm; green is about 500 nm; and violet at 400 nm has the shortest wavelength of visible light. We see objects as different colors because the objects reflect only certain wavelengths, which are absorbed by our eyes.

Ultraviolet (UV) *light* has shorter wavelengths and higher frequencies than violet light of the visible range. The UV radiation in sunlight can cause serious sunburn, which may lead to skin cancer. While some UV light from the Sun is blocked by the ozone layer, the cosmetic industry has developed sunscreens to prevent the absorption of UV light by the skin. *X-rays* have shorter wavelengths than ultraviolet light, which means they have some of the highest frequencies. X-rays can pass through soft substances but not metals or bone, which allow us to see images of the bones and teeth in the body.

Infrared radiation from heat lamps keeps food at proper serving temperature.

ENGAGE 5.3

Which type of electromagnetic radiation has lower frequency, ultraviolet or infrared?

SAMPLE PROBLEM 5.1   The Electromagnetic Spectrum

TRY IT FIRST

Arrange the following in order of decreasing wavelengths: X-rays, ultraviolet light, FM radio waves, and microwaves.

SOLUTION

The electromagnetic radiation with the longest wavelength is FM radio waves, then microwaves, followed by ultraviolet light, and then X-rays, which have the shortest wavelengths.

SELF TEST 5.1

Visible light contains colors from red to violet.

a. Does red light or violet light have the shorter wavelength?
b. Does red light or violet light have the lower frequency?

ANSWER

a. violet light   **b.** red light

Microwaves with wavelengths of about 1 cm heat water molecules in food.

PRACTICE PROBLEMS

Try Practice Problems 5.9 to 5.14

Chemistry Link to Health

Biological Reactions to UV Light

Our everyday life depends on sunlight, but exposure to sunlight can have damaging effects on living cells, and too much exposure can even cause their death. Light energy, especially ultraviolet (UV), excites electrons and may lead to unwanted chemical reactions. The list of damaging effects of sunlight includes sunburn; wrinkling; premature aging of the skin; changes in the DNA of the cells, which can lead to skin cancers; inflammation of the eyes; and perhaps cataracts. Some drugs, like the acne medications Accutane and Retin-A, as well as antibiotics, diuretics, sulfonamides, and estrogen, make the skin extremely sensitive to light.

Phototherapy uses light to treat certain skin conditions, including psoriasis, eczema, and dermatitis. In the treatment of psoriasis, for example, oral drugs are given to make the skin more photosensitive; then exposure to UV radiation follows. Low-energy radiation (blue light) with wavelengths from 390 to 470 nm is used to treat babies with *neonatal jaundice,* which converts high levels of bilirubin to water-soluble compounds that are excreted from the body. Sunlight is also a factor in stimulating the immune system.

In a disorder called *seasonal affective disorder* or SAD, people experience mood swings and depression during the winter. Some research suggests that SAD is the result of a decrease in serotonin, or an increase in melatonin, when there are fewer hours of sunlight. One treatment for SAD is therapy using bright light provided by a lamp called a light box. A daily exposure to blue light (460 nm) for 30 to 60 min seems to reduce symptoms of SAD.

Phototherapy is used to treat babies with neonatal jaundice.

A light box is used to provide light, which reduces symptoms of SAD.

PRACTICE PROBLEMS

5.1 Electromagnetic Radiation

5.1 What is meant by the wavelength of UV light?

5.2 How are the wavelength and frequency of light related?

5.3 What is the difference between "white" light and blue or red light?

5.4 Why can we use X-rays, but not radio waves or microwaves, to give an image of bones and teeth?

Applications

5.5 Ultraviolet radiation (UVB) used to treat psoriasis has a wavelength of 3×10^{-7} m, whereas infrared radiation used in

A thermal image of the hand of a patient with multiple sclerosis shows the warm areas in the palm of the hand as red and cool areas as blue in the fingers where blood flow is poor.

a thermal imaging camera has a wavelength of 1×10^{-5} m. Which has a higher frequency? Which has a higher energy?

5.6 AM radio waves have a frequency of $8 \times 10^5 \text{ s}^{-1}$, whereas TV Channel 2 has a frequency of $6 \times 10^7 \text{ s}^{-1}$. Which has a shorter wavelength? Which has a lower energy?

5.7 If orange light has a wavelength of 6.3×10^{-5} cm, what is its wavelength in meters and nanometers?

5.8 A wavelength of 850 nm is used for fiber-optic transmission. What is its wavelength in meters?

5.9 Which type of electromagnetic radiation, cell phones, AM radio, or infrared light has the longest wavelengths?

5.10 Of radio waves, infrared light, and UV light, which has the shortest wavelengths?

5.11 Place the following types of electromagnetic radiation in order of increasing wavelengths: the blue color in a rainbow, X-rays, and microwaves from an oven.

5.12 Place the following types of electromagnetic radiation in order of decreasing wavelengths: FM radio station, red color in neon lights, and cell phone.

5.13 Place the following types of electromagnetic radiation in order of increasing frequencies: TV signal, X-rays, and microwaves.

5.14 Place the following types of electromagnetic radiation in order of decreasing frequencies: AM music station, UV radiation from the Sun, and infrared radiation from a heat lamp.

5.2 Atomic Spectra and Energy Levels

LEARNING GOAL Explain how atomic spectra correlate with the energy levels in atoms.

When the white light from the Sun or a light bulb is passed through a prism or raindrops, it produces a *continuous spectrum*, like a rainbow. When atoms of elements are heated, they also produce light. At night, you may have seen the yellow color of sodium streetlamps or the red color of neon lights.

Photons

The light emitted from a streetlamp or by atoms that are heated is a stream of particles called *photons*. A **photon** is a packet of energy with both particle and wave characteristics that travels at the speed of light. High-energy photons have short wavelengths, whereas low-energy photons have long wavelengths.

Photons play an important role in our modern world, particularly in the use of lasers, which use a narrow range of wavelengths. For example, lasers use photons of a single frequency to scan bar codes on labels when we buy groceries. In hospitals, high-energy photons from X-rays and gamma rays are used in scans for diagnosis or for radiation treatments to reach tumors within the tissues without damaging the surrounding tissue.

Atomic Spectra

When the light emitted from heated elements is passed through a prism, it does not produce a continuous spectrum. Instead, an **atomic spectrum** is produced that consists of lines of different colors separated by dark areas (see **FIGURE 5.3**). This separation of colors indicates that only certain wavelengths of light are produced when an element is heated, which gives each element a unique atomic spectrum.

A rainbow forms when light passes through water droplets.

In gamma knife radiation therapy, gamma photons are used to kill cancer cells or reduce tumor size.

ENGAGE 5.4

Why does the light emitted by heated elements form atomic spectra whereas white light produces a continuous spectrum?

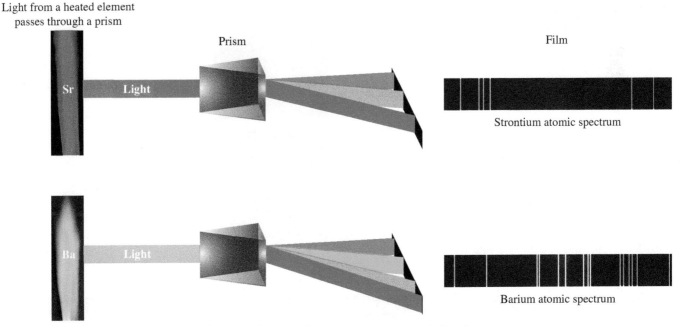

Light from a heated element passes through a prism

Prism

Film

Sr — Light

Strontium atomic spectrum

Ba — Light

Barium atomic spectrum

FIGURE 5.3 In an atomic spectrum, light from a heated element separates into distinct lines.

Electron Energy Levels

Scientists have now determined that the lines in atomic spectra are associated with changes in the energies of the electrons. In an atom, each electron has a specific energy known as its **energy level**, which is assigned a value called the **principal quantum number (*n*)**, ($n = 1, n = 2, \dots$). Generally, electrons in the lower energy levels are closer to the nucleus, while electrons in the higher energy levels are farther away. The energy of an

Energy Increases →

$n = 5$

$n = 4$

$n = 3$

$n = 2$

$n = 1$

Nucleus

An electron can have only the energy of one of the energy levels in an atom.

Colors are produced when electricity excites electrons in noble gases.

electron is *quantized*, which means that the energy of an electron can only have specific energy values, but cannot have values between these values.

Principal Quantum Number (*n*)

$1 < 2 < 3 < 4 < 5 < 6 < 7$

Lowest   ⟶   Highest
energy   energy

All the electrons with the same energy are grouped in the same energy level. As an analogy, we can think of the energy levels of an atom as similar to the shelves in a bookcase. The first shelf is the lowest energy level; the second shelf is the second energy level; and so on. If we are arranging books on the shelves, it would take less energy to fill the bottom shelf first, and then the second shelf, and so on. However, we could never get any book to stay in the space between any of the shelves. Similarly, the energy of an electron must be at a specific energy level, and not between.

Unlike standard bookcases, however, there is a large difference between the energy of the first and second energy levels, but then the higher energy levels are closer together. Another difference is that the lower electron energy levels hold fewer electrons than the higher energy levels.

Changes in Energy Levels

When an atom absorbs energy, an electron moves to a higher energy level called an *excited state*. An electron can change from one energy level to a higher energy level only if it absorbs the energy equal to the difference in energy levels. However, an electron in the excited state is less stable and falls to a lower-energy state by emitting a photon of a specific energy. If the energy emitted is in the visible range, we see one of the colors of visible light (see **FIGURE 5.4**). The yellow color of sodium streetlights and the red color of neon lights are examples of electrons emitting energy in the visible color range.

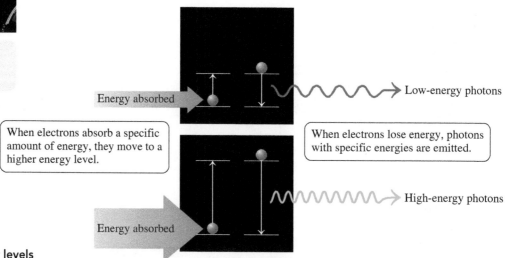

Energy absorbed → Low-energy photons

When electrons absorb a specific amount of energy, they move to a higher energy level.

When electrons lose energy, photons with specific energies are emitted.

Energy absorbed → High-energy photons

FIGURE 5.4 Changes in energy levels

ENGAGE 5.5

What causes electrons to move to higher energy levels?

SAMPLE PROBLEM 5.2   Changes in Energy Levels

TRY IT FIRST

a. When does an electron move to a higher energy level?
b. When an electron drops to a lower energy level, how is energy lost?

SOLUTION

a. An electron moves to a higher energy level when it absorbs an amount of energy equal to the difference between energy levels.
b. Energy, equal to the difference between energy levels, is emitted as a photon when an electron drops to a lower energy level.

SELF TEST 5.2

a. Why did scientists propose that electrons occupy specific energy levels in an atom?
b. How does the wavelength of a low-energy photon compare to that of a high-energy photon?

ANSWER

a. Because the atomic spectra of elements consist of discrete, separated lines, scientists concluded that electrons occupy only certain energy levels in the atom.
b. The wavelength of a low-energy photon would be longer than the wavelength of a high-energy photon.

PRACTICE PROBLEMS
Try Practice Problems 5.15 to 5.20

PRACTICE PROBLEMS

5.2 Atomic Spectra and Energy Levels

5.15 What feature of an atomic spectrum indicates that the energy emitted by heating an element is not continuous?

5.16 How can we explain the distinct lines that appear in an atomic spectrum?

5.17 Electrons can jump to higher energy levels when they _____ (absorb/emit) a photon.

5.18 Electrons drop to lower energy levels when they _____ (absorb/emit) a photon.

5.19 Identify the photon in each pair with the greater energy.
a. green light or yellow light
b. red light or blue light

5.20 Identify the photon in each pair with the greater energy.
a. orange light or violet light
b. infrared light or ultraviolet light

5.3 Sublevels and Orbitals

LEARNING GOAL Describe the sublevels and orbitals for the electrons in an atom.

We have seen that the protons and neutrons are contained in the small, dense nucleus of an atom. However, it is the electrons within the atoms that determine the physical and chemical properties of the elements. Therefore, we will look at the arrangement of electrons within the large volume of space surrounding the nucleus.

There is a limit to the number of electrons allowed in each energy level. Only a few electrons can occupy the lower energy levels, while more electrons can be accommodated in higher energy levels. The maximum number of electrons allowed in any energy level is calculated using the formula $2n^2$ (two times the square of the principal quantum number). **TABLE 5.1** shows the maximum number of electrons allowed in the first four energy levels.

TABLE 5.1 Maximum Number of Electrons Allowed in Energy Levels 1 to 4				
Energy Level (n)	1	2	3	4
$2n^2$	$2(1)^2$	$2(2)^2$	$2(3)^2$	$2(4)^2$
Maximum Number of Electrons	2	8	18	32

Sublevels

Each of the energy levels consists of one or more **sublevels**, in which electrons with identical energy are found. The sublevels are identified by the letters s, p, d, and f. The number of sublevels within an energy level is equal to the principal quantum number, n. For example, the first energy level ($n = 1$) has only one sublevel, $1s$. The second energy level ($n = 2$) has two sublevels, $2s$ and $2p$. The third energy level ($n = 3$) has three sublevels, $3s$, $3p$, and $3d$. The fourth energy level ($n = 4$) has four sublevels, $4s$, $4p$, $4d$, and $4f$. Energy levels $n = 5$, $n = 6$, and $n = 7$ also have as many sublevels as the value of n, but only s, p, d, and f sublevels are needed to hold the electrons in atoms of the 118 known elements (see **FIGURE 5.5**).

ENGAGE 5.6
How many sublevels are in energy level $n = 5$?

FIGURE 5.5 The number of sublevels in an energy level is the same as the principal quantum number, *n*.

Energy Level	Number of Sublevels	Types of Sublevels			
		s	*p*	*d*	*f*
n = 4	4				
n = 3	3				
n = 2	2				
n = 1	1				

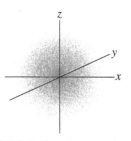

FIGURE 5.6 The electron cloud of an *s* orbital is spherical, which indicates a higher probability of finding an *s* electron near the nucleus.

Within each energy level, the *s* sublevel has the lowest energy. If there are additional sublevels, the *p* sublevel has the next lowest energy, then the *d* sublevel, and finally the *f* sublevel.

Order of Increasing Energy of Sublevels in an Energy Level

$$s < p < d < f$$

Lowest energy $\longrightarrow$ Highest energy

Orbitals

There is no way to know the exact location of an electron in an atom. Instead, scientists describe the location of an electron in terms of probability. The **orbital** is the three-dimensional volume in which electrons have the highest probability of being found.

As an analogy, imagine that you draw a circle with a 100-m radius around your chemistry classroom. There is a high probability of finding you within that circle when your chemistry class is in session. But once in a while, you may be outside that circle because you were sick or your car did not start.

3s

2s

1s

s orbitals

FIGURE 5.7 The sizes of the *s* orbitals increase because they contain electrons at higher energy levels.

Shapes of Orbitals

Each type of orbital has a unique three-dimensional shape. Electrons in an *s* orbital are most likely found in a region with a spherical shape. Imagine that you take a picture of the location of an electron in an *s* orbital every second for an hour. When all these pictures are overlaid, the result, called a *probability density*, would look like the electron cloud shown in **FIGURE 5.6**. For convenience, we draw this electron cloud as a sphere called an *s* orbital. There is one *s* orbital for every energy level starting with *n* = 1. For example, in the first, second, and third energy levels, there are *s* orbitals designated as 1*s*, 2*s*, and 3*s*. As the principal quantum number increases, there is an increase in the size of the *s* orbitals, although the shape is the same (see **FIGURE 5.7**).

The orbitals occupied by *p*, *d*, and *f* electrons have three-dimensional shapes different from those of the *s* electrons. There are three *p* orbitals, starting with *n* = 2. Each *p* orbital has two lobes like a balloon tied in the middle. The three *p* orbitals are arranged in three perpendicular directions, along the *x*, *y*, and *z* axes around the nucleus (see **FIGURE 5.8**). As with *s* orbitals, the shape of *p* orbitals is the same, but the volume increases at higher energy levels.

ENGAGE 5.7

What are some similarities and differences of the *p* orbitals in the *n* = 3 energy level?

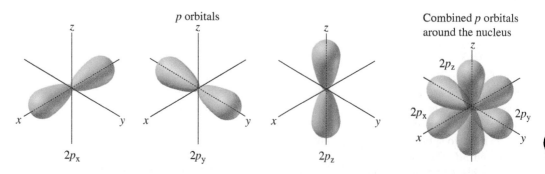

p orbitals

2*p*ₓ

2*p*ᵧ

2*p*ᵤ

Combined *p* orbitals around the nucleus

2*p*ᵤ

2*p*ₓ

2*p*ᵧ

FIGURE 5.8 Shapes of *p* orbitals

The electron cloud of a *p* orbital shows the highest probability of finding a *p* electron.

The *p* orbitals in the same energy level are arranged on their axes around the nucleus.

In summary, the $n = 2$ energy level, which has 2s and 2p sublevels, consists of one s orbital and three p orbitals.

Energy level $n = 2$ consists of one 2s orbital and three 2p orbitals.

Energy level $n = 3$ consists of three sublevels s, p, and d. The d sublevels contain five d orbitals (see **FIGURE 5.9**).

Energy level $n = 4$ consists of four sublevels s, p, d, and f. In the f sublevel, there are seven f orbitals. The shapes of f orbitals are complex, and we have not included them in this text.

ENGAGE 5.8

How many orbitals are there in the 5d sublevel?

FIGURE 5.9 Four of the five d orbitals consist of four lobes that are aligned along or between different axes. One d orbital consists of two lobes and a doughnut-shaped ring around its center.

SAMPLE PROBLEM 5.3   Energy Levels, Sublevels, and Orbitals

TRY IT FIRST

Indicate the type and number of orbitals in each of the following energy levels or sublevels:

a. 3p sublevel   **b.** $n = 2$ **c.** $n = 3$ **d.** 4d sublevel

SOLUTION

ANALYZE THE PROBLEM	Given	Need	Connect
	energy level, sublevel	orbitals	n, orbitals in s, p, d, f

a. The 3p sublevel contains three 3p orbitals.
b. The $n = 2$ energy level consists of one 2s and three 2p orbitals.
c. The $n = 3$ energy level consists of one 3s, three 3p, and five 3d orbitals.
d. The 4d sublevel contains five 4d orbitals.

SELF TEST 5.3

a. What is similar and what is different for 1s, 2s, and 3s orbitals?
b. How many orbitals would be in the 4f sublevel?

ANSWER

a. The 1s, 2s, and 3s orbitals are all spherical, but they increase in volume because the electron is most likely to be found farther from the nucleus for higher energy levels.
b. The 4f sublevel has seven orbitals.

PRACTICE PROBLEMS

Try Practice Problems 5.21 to 5.24

Electron spinning counterclockwise　　Electron spinning clockwise

Opposite spins of electrons in an orbital

An orbital can hold up to two electrons with opposite spins.

PRACTICE PROBLEMS

Try Practice Problems 5.25 to 5.28

Orbital Capacity and Electron Spin

The *Pauli exclusion principle* states that each orbital can hold a maximum of two electrons. According to a model for electron behavior, an electron is seen as spinning on its axis, which generates a magnetic field. When two electrons are in the same orbital, they will repel each other unless their magnetic fields cancel. This happens only when the two electrons spin in opposite directions. We can represent the spins of the electrons in the same orbital with one arrow pointing up and the other pointing down.

Number of Electrons in Sublevels

There is a maximum number of electrons that can occupy each sublevel. An *s* sublevel holds one or two electrons. Because each *p* orbital can hold up to two electrons, the three *p* orbitals in a *p* sublevel can accommodate six electrons. A *d* sublevel with five *d* orbitals can hold a maximum of 10 electrons. With seven *f* orbitals, an *f* sublevel can hold up to 14 electrons.

As mentioned earlier, higher energy levels such as $n = 5, 6,$ and 7 would have 5, 6, and 7 sublevels, but those beyond sublevel *f* are not utilized by the atoms of the elements known today. The total number of electrons in all the sublevels adds up to give the electrons allowed in an energy level. The number of sublevels, the number of orbitals, and the maximum number of electrons for energy levels 1 to 4 are shown in **TABLE 5.2**.

TABLE 5.2 Electron Capacity in Sublevels for Energy Levels 1 to 4

Energy Level (n)	Number of Sublevels	Type of Sublevel	Number of Orbitals	Maximum Number of Electrons	Total Electrons
1	1	1s	1	2	2
2	2	2s	1	2	
		2p	3	6	8
3	3	3s	1	2	
		3p	3	6	
		3d	5	10	18
4	4	4s	1	2	
		4p	3	6	
		4d	5	10	
		4f	7	14	32

PRACTICE PROBLEMS

5.3 Sublevels and Orbitals

5.21 Describe the shape of each of the following orbitals:
　a. 1s　　**b.** 2p　　**c.** 5s

5.22 Describe the shape of each of the following orbitals:
　a. 3p　　**b.** 6s　　**c.** 4p

5.23 Match statements **1** to **3** with **a** to **d**:
　1. They have the same shape.
　2. The maximum number of electrons is the same.
　3. They are in the same energy level.
　a. 1s and 2s orbitals　　**b.** 3s and 3p sublevels
　c. 3p and 4p sublevels　　**d.** three 3p orbitals

5.24 Match statements **1** to **3** with **a** to **d**:
　1. They have the same shape.
　2. The maximum number of electrons is the same.
　3. They are in the same energy level.
　a. 5s and 6s orbitals　　**b.** 3p and 4p orbitals
　c. 3s and 4s sublevels　　**d.** 2s and 2p orbitals

5.25 Indicate the number of each in the following:
　a. orbitals in the 3d sublevel
　b. sublevels in the $n = 1$ energy level
　c. orbitals in the 6s sublevel
　d. orbitals in the $n = 3$ energy level

5.26 Indicate the number of each in the following:
　a. orbitals in the $n = 2$ energy level
　b. sublevels in the $n = 4$ energy level
　c. orbitals in the 5f sublevel
　d. orbitals in the 6p sublevel

5.27 Indicate the maximum number of electrons in the following:
　a. 2p orbital　　　　**b.** 3p sublevel
　c. $n = 4$ energy level　　**d.** 5d sublevel

5.28 Indicate the maximum number of electrons in the following:
　a. 3s sublevel　　　　**b.** 4p orbital
　c. $n = 3$ energy level　　**d.** 4f sublevel

5.4 Orbital Diagrams and Electron Configurations

LEARNING GOAL Draw the orbital diagram and write the electron configuration for an element.

We can now look at how electrons are arranged in the orbitals within an atom. An **orbital diagram** shows the placement of the electrons in the orbitals in order of increasing energy (see **FIGURE 5.10**). In this energy diagram, we see that the electrons in the 1s orbital have the lowest energy level. The energy level is higher for the 2s orbital and is even higher for the 2p orbitals.

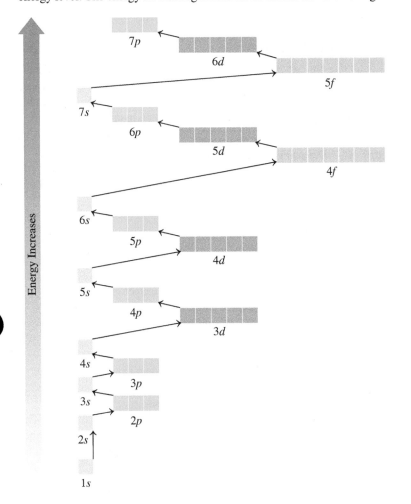

FIGURE 5.10 The orbitals in an atom fill in order of increasing energy beginning with 1s.

ENGAGE 5.9

Why do 3d orbitals fill after the 4s orbital?

We begin our discussion of electron configuration by using orbital diagrams in which boxes represent the orbitals. Any orbital can have a maximum of two electrons.

To draw an orbital diagram, the lowest energy orbitals are filled first. For example, we can draw the orbital diagram for carbon. The atomic number of carbon is 6, which means that a carbon atom has six electrons. The first two electrons go into the 1s orbital; the next two electrons go into the 2s orbital. In the orbital diagram, the two electrons in the 1s and 2s orbitals are shown with opposite spins; the first arrow is up and the second is down. The last two electrons in carbon begin to fill the 2p sublevel, which has the next lowest energy. However, there are three 2p orbitals of equal energy. Because the negatively charged electrons repel each other, they are placed in separate 2p orbitals. *Hund's rule* states that the lowest energy state is attained when electrons are placed singly in available orbitals of the same sublevel with the same spin. With few exceptions (which will be noted later in this chapter), lower energy sublevels are filled first, and then the "building" of electrons continues to the next lowest energy sublevel that is available until all the electrons are placed.

ENGAGE 5.10

When do we need to use Hund's rule in drawing orbital diagrams?

Orbital diagram for carbon

Half-filled   Empty

Electron Configurations

Chemists use a notation called the **electron configuration** to indicate the placement of the electrons of an atom in order of increasing energy. As in the orbital diagrams, an electron configuration is written with the lowest energy sublevel first, followed by the next lowest energy sublevel. The number of electrons in each sublevel is shown as a superscript.

Electron Configuration for Carbon

$$1s^2 2s^2 2p^2$$ Read as "one *s* two, two *s* two, two *p* two"

Period 1: Hydrogen and Helium

We will draw orbital diagrams and write the corresponding electron configurations for the elements H and He in Period 1. The 1*s* orbital (which is the 1*s* sublevel) is written first because it has the lowest energy. Hydrogen has one electron in the 1*s* sublevel; helium has two. In the orbital diagram, the electrons for helium are shown as arrows pointing in opposite directions.

Element	Atomic Number	Orbital Diagram	Electron Configuration
H	1	1*s* ↑	$1s^1$
He	2	↑↓	$1s^2$

Period 2: Lithium to Neon

Period 2 begins with lithium, which has three electrons. The first two electrons fill the 1*s* orbital, whereas the third electron goes into the 2*s* orbital, the sublevel with the next lowest energy. In beryllium, another electron is added to complete the 2*s* orbital. The next six electrons are used to fill the 2*p* orbitals. The electrons are added to separate *p* orbitals (Hund's rule) from boron to nitrogen, which gives three half-filled 2*p* orbitals.

From oxygen to neon, the remaining three electrons are paired up using opposite spins to complete the 2*p* sublevel. In writing the electron configurations for the elements in Period 2, begin with the 1*s* orbital followed by the 2*s* and then the 2*p* orbitals.

An electron configuration can also be written in an *abbreviated configuration*. The electron configuration of the preceding noble gas is replaced by writing its element symbol inside square brackets. For example, the electron configuration for lithium, $1s^2 2s^1$, can be abbreviated as [He]$2s^1$ where [He] replaces $1s^2$.

Element	Atomic Number	Orbital Diagram	Electron Configuration	Abbreviated Electron Configuration
Li	3	1*s* 2*s* ↑↓ ↑	$1s^2 2s^1$	[He]$2s^1$
Be	4	↑↓ ↑↓	$1s^2 2s^2$	[He]$2s^2$
B	5	↑↓ ↑↓ 2*p* ↑	$1s^2 2s^2 2p^1$	[He]$2s^2 2p^1$
C	6	↑↓ ↑↓ ↑ ↑	$1s^2 2s^2 2p^2$	[He]$2s^2 2p^2$
N	7	↑↓ ↑↓ ↑ ↑ ↑	$1s^2 2s^2 2p^3$	[He]$2s^2 2p^3$
O	8	↑↓ ↑↓ ↑↓ ↑ ↑	$1s^2 2s^2 2p^4$	[He]$2s^2 2p^4$
F	9	↑↓ ↑↓ ↑↓ ↑↓ ↑	$1s^2 2s^2 2p^5$	[He]$2s^2 2p^5$
Ne	10	↑↓ ↑↓ ↑↓ ↑↓ ↑↓	$1s^2 2s^2 2p^6$	[He]$2s^2 2p^6$

SAMPLE PROBLEM 5.4   Drawing Orbital Diagrams

TRY IT FIRST

Nitrogen is an element that is used in the formation of amino acids, proteins, and nucleic acids. Draw the orbital diagram for nitrogen.

SOLUTION

STEP 1   **Draw boxes to represent the orbitals.** Nitrogen has atomic number 7, which means it has seven electrons. For the orbital diagram, we draw boxes to represent the $1s$, $2s$, and $2p$ orbitals.

```
1s    2s     2p
□     □    □ □ □
```

STEP 2   **Place a pair of electrons with opposite spins in each filled orbital.** First, we place a pair of electrons with opposite spins in both the $1s$ and $2s$ orbitals.

```
1s    2s     2p
↑↓    ↑↓   □ □ □
```

STEP 3   **Place the remaining electrons in the last occupied sublevel in separate orbitals.** Then we place the three remaining electrons in three separate $2p$ orbitals with arrows drawn in the same direction.

```
1s    2s     2p
↑↓    ↑↓   ↑ ↑ ↑
```
Orbital diagram for nitrogen (N)

Fluorine is used to make nonstick coatings for cookware.

SELF TEST 5.4

Draw the orbital diagram for fluorine, which is used to make nonstick coatings for cookware.

ANSWER

```
1s    2s      2p
↑↓    ↑↓   ↑↓ ↑↓ ↑
```

Period 3: Sodium to Argon

In Period 3, electrons enter the orbitals of the $3s$ and $3p$ sublevels, but not the $3d$ sublevel. We notice that the elements sodium to argon, which are directly below the elements lithium to neon in Period 2, have a similar pattern of filling their s and p orbitals. In sodium and magnesium, one and two electrons go into the $3s$ orbital. The electrons for aluminum, silicon, and phosphorus go into separate $3p$ orbitals. The remaining electrons in sulfur, chlorine, and argon are paired up (opposite spins) with the electrons already in the $3p$ orbitals. For the abbreviated electron configurations of Period 3, the symbol [Ne] replaces $1s^2 2s^2 2p^6$.

Element	Atomic Number	Orbital Diagram (3s and 3p orbitals only)	Electron Configuration	Abbreviated Electron Configuration
		3s 3p		
Na	11	[Ne] ↑ ☐☐☐	$1s^2 2s^2 2p^6 3s^1$	[Ne]$3s^1$
Mg	12	[Ne] ↑↓ ☐☐☐	$1s^2 2s^2 2p^6 3s^2$	[Ne]$3s^2$
Al	13	[Ne] ↑↓ ↑☐☐	$1s^2 2s^2 2p^6 3s^2 3p^1$	[Ne]$3s^2 3p^1$
Si	14	[Ne] ↑↓ ↑↑☐	$1s^2 2s^2 2p^6 3s^2 3p^2$	[Ne]$3s^2 3p^2$
P	15	[Ne] ↑↓ ↑↑↑	$1s^2 2s^2 2p^6 3s^2 3p^3$	[Ne]$3s^2 3p^3$
S	16	[Ne] ↑↓ ↑↓↑↑	$1s^2 2s^2 2p^6 3s^2 3p^4$	[Ne]$3s^2 3p^4$
Cl	17	[Ne] ↑↓ ↑↓↑↓↑	$1s^2 2s^2 2p^6 3s^2 3p^5$	[Ne]$3s^2 3p^5$
Ar	18	[Ne] ↑↓ ↑↓↑↓↑↓	$1s^2 2s^2 2p^6 3s^2 3p^6$	[Ne]$3s^2 3p^6$

SAMPLE PROBLEM 5.5   Electron Configurations

TRY IT FIRST

Silicon is the basis of semiconductors. Write the complete and abbreviated electron configurations for silicon.

SOLUTION

	Given	Need	Connect
ANALYZE THE PROBLEM	silicon (Si)	electron configuration, abbreviated electron configuration	periodic table, atomic number, order of filling orbitals

STEP 1 **State the number of electrons from the atomic number on the periodic table.** Silicon has an atomic number of 14, which means it has 14 electrons.

STEP 2 **Write the number of electrons for each orbital in order of increasing energy until filling is complete.**

$1s^2 2s^2 2p^6 3s^2 3p^2$ Electron configuration for Si

STEP 3 **Write an abbreviated electron configuration by replacing the configuration of the preceding noble gas with its symbol.**

[Ne]$3s^2 3p^2$ Abbreviated electron configuration for Si

SELF TEST 5.5

Write the complete and abbreviated electron configurations for each of the following:

a. sulfur, a macromineral in proteins, vitamin B$_1$, and insulin
b. vanadium, a micromineral needed for the extraction of energy from foods

ANSWER

a. $1s^2 2s^2 2p^6 3s^2 3p^4$, [Ne]$3s^2 3p^4$
b. $1s^2 2s^2 2p^6 3s^2 3p^6 4s^2 3d^3$, [Ar]$4s^2 3d^3$

Sulfur atoms are part of the structure of human insulin.

PRACTICE PROBLEMS
Try Practice Problems 5.29 to 5.40

PRACTICE PROBLEMS

5.4 Orbital Diagrams and Electron Configurations

5.29 Compare the terms electron configuration and abbreviated electron configuration.

5.30 Compare the terms orbital diagram and electron configuration.

5.31 Draw the orbital diagram for each of the following:
  a. boron   **b.** aluminum
  c. phosphorus   **d.** argon

5.32 Draw the orbital diagram for each of the following:
  a. chlorine   **b.** sulfur
  c. magnesium   **d.** beryllium

5.33 Write the complete electron configuration for each of the following:
  a. nickel   **b.** sodium
  c. lithium   **d.** titanium

5.34 Write the complete electron configuration for each of the following:
  a. nitrogen   **b.** chlorine
  c. strontium   **d.** neon

5.35 Write the abbreviated electron configuration for each of the following:
  a. tin   **b.** cadmium
  c. selenium   **d.** fluorine

5.36 Write the abbreviated electron configuration for each of the following:
  a. barium   **b.** oxygen   **c.** manganese   **d.** arsenic

5.37 Give the symbol of the element with each of the following electron or abbreviated electron configurations:
  a. $1s^2 2s^1$ **b.** $1s^2 2s^2 2p^6 3s^2 3p^6 4s^2 3d^2$
  c. $[Ar]4s^2 3d^{10} 4p^2$ **d.** $[Ne]3s^2 3p^1$

5.38 Give the symbol of the element with each of the following electron or abbreviated electron configurations:
  a. $1s^2 2s^2 2p^4$ **b.** $[Ne]3s^2$
  c. $1s^2 2s^2 2p^6 3s^2 3p^6$ **d.** $[He]2s^2 2p^5$

5.39 Give the symbol of the element that meets the following conditions:
  a. has three electrons in the $n = 3$ energy level
  b. has two $2p$ electrons
  c. completes the $3p$ sublevel
  d. completes the $2s$ sublevel

5.40 Give the symbol of the element that meets the following conditions:
  a. has five electrons in the $3p$ sublevel
  b. has four $2p$ electrons
  c. completes the $3s$ sublevel
  d. has one electron in the $3s$ sublevel

5.5 Electron Configurations and the Periodic Table

LEARNING GOAL Write the electron configuration for an atom using the sublevel blocks on the periodic table.

Up to now, we have written electron configurations using their energy level diagrams. As configurations involve more energy levels, this can become tedious. However, the electron configurations of the elements are related to their position on the periodic table. Different sections or blocks within the periodic table correspond to the *s*, *p*, *d*, and *f* sublevels (see **FIGURE 5.11**). Therefore, we can "build" the electron configurations of atoms by reading the periodic table in order of increasing atomic number.

ENGAGE 5.11

How do you use the periodic table to determine the number of electrons in the 1s, 2s, and 2p sublevels of neon?

FIGURE 5.11 Electron configurations follow the order of occupied sublevels on the periodic table.

Blocks on the Periodic Table

1. The *s* **block** includes hydrogen and helium as well as the elements in Group 1A (1) and Group 2A (2). This means that the final one or two electrons in the elements of the *s* block are located in an *s* orbital. The period number indicates the particular *s* orbital that is filling: 1*s*, 2*s*, and so on.

2. The *p* **block** consists of the elements in Group 3A (13) to Group 8A (18). There are six *p* block elements in each period because three *p* orbitals can hold up to six electrons. The period number indicates the particular *p* sublevel that is filling: 2*p*, 3*p*, and so on.

3. The *d* **block**, containing the transition elements, first appears after calcium (atomic number 20). There are 10 elements in each period of the *d* block because five *d* orbitals can hold up to 10 electrons. The particular *d* sublevel is one less $(n - 1)$ than the period number. For example, in Period 4, the *d* block is the 3*d* sublevel. In Period 5, the *d* block is the 4*d* sublevel.

4. The *f* **block**, the inner transition elements, are the two rows at the bottom of the periodic table. There are 14 elements in each *f* block because seven *f* orbitals can hold up to 14 electrons. Elements that have atomic numbers higher than 57 (La) have electrons in the 4*f* block. The particular *f* sublevel is two less $(n - 2)$ than the period number. For example, in Period 6, the *f* block is the 4*f* sublevel. In Period 7, the *f* block is the 5*f* sublevel.

ENGAGE 5.12

On the periodic table, what are the group numbers that make up the *s* block? the *p* block? the *d* block?

Writing Electron Configurations Using Sublevel Blocks

CORE CHEMISTRY SKILL

Using the Periodic Table to Write Electron Configurations

Now we can write electron configurations using the sublevel blocks on the periodic table. As before, each configuration begins at H. But now we move across the table from left to right, writing down each sublevel block we come to until we reach the element for which we are writing an electron configuration. We will show how to write the electron configuration for chlorine (atomic number 17) from the sublevel blocks on the periodic table in Sample Problem 5.6.

SAMPLE PROBLEM 5.6 Using the Sublevel Blocks to Write Electron Configurations

TRY IT FIRST

Use the sublevel blocks on the periodic table to write the electron configuration for chlorine.

SOLUTION

ANALYZE THE PROBLEM	Given	Need	Connect
	element chlorine	electron configuration	sublevel blocks

STEP 1 **Locate the element on the periodic table.** Chlorine (atomic number 17) is in Period 3 and Group 7A (17).

STEP 2 **Write the filled sublevels in order, going across each period.**

Period	Sublevel Block Filling	Sublevel Block Notation (filled)
1	1*s* (H → He)	$1s^2$
2	2*s* (Li → Be) and 2*p* (B → Ne)	$2s^2 2p^6$
3	3*s* (Na → Mg)	$3s^2$

STEP 3 **Complete the configuration by counting the electrons in the last occupied sublevel block.** Because chlorine is the fifth element in the 3*p* block, there are five electrons in the 3*p* sublevel.

Period	Sublevel Block Filling	Sublevel Block Notation
3	3*p* (Al → Cl)	$3p^5$

The electron configuration for chlorine (Cl) is: $1s^2 2s^2 2p^6 3s^2 3p^5$.

Use the sublevel blocks on the periodic table to write the electron configuration for each of the following:

a. argon

b. cobalt

ANSWER

a. $1s^2 2s^2 2p^6 3s^2 3p^6$

b. $1s^2 2s^2 2p^6 3s^2 3p^6 4s^2 3d^7$

Electron Configurations for Period 4 and Above

Up to Period 4, the filling of the sublevels has progressed in order. However, if we look at the sublevel blocks in Period 4, we see that the $4s$ sublevel fills before the $3d$ sublevel. This occurs because the electrons in the $4s$ sublevel have slightly lower energy than the electrons in the $3d$ sublevel. This order occurs again in Period 5 when the $5s$ sublevel fills before the $4d$ sublevel, in Period 6 when the $6s$ fills before the $5d$, and in Period 7 when the $7s$ fills before the $6d$.

At the beginning of Period 4, the electrons in potassium (19) and calcium (20) go into the $4s$ sublevel. In scandium, the next electron added after the $4s$ sublevel is filled goes into the $3d$ block. The $3d$ block continues to fill until it is complete with 10 electrons at zinc (30). Once the $3d$ block is complete, the next six electrons, gallium to krypton, go into the $4p$ block.

ENGAGE 5.13

What sublevel blocks are used to add electrons to the elements from K to Zn?

Element	Atomic Number	Electron Configuration	Abbreviated Electron Configuration
4s Block			
K	19	$1s^2 2s^2 2p^6 3s^2 3p^6 4s^1$	$[Ar]4s^1$
Ca	20	$1s^2 2s^2 2p^6 3s^2 3p^6 4s^2$	$[Ar]4s^2$
3d Block			
Sc	21	$1s^2 2s^2 2p^6 3s^2 3p^6 4s^2 3d^1$	$[Ar]4s^2 3d^1$
Ti	22	$1s^2 2s^2 2p^6 3s^2 3p^6 4s^2 3d^2$	$[Ar]4s^2 3d^2$
V	23	$1s^2 2s^2 2p^6 3s^2 3p^6 4s^2 3d^3$	$[Ar]4s^2 3d^3$
Cr*	24	$1s^2 2s^2 2p^6 3s^2 3p^6 4s^1 3d^5$	$[Ar]4s^1 3d^5$ half-filled d sublevel is stable
Mn	25	$1s^2 2s^2 2p^6 3s^2 3p^6 4s^2 3d^5$	$[Ar]4s^2 3d^5$
Fe	26	$1s^2 2s^2 2p^6 3s^2 3p^6 4s^2 3d^6$	$[Ar]4s^2 3d^6$
Co	27	$1s^2 2s^2 2p^6 3s^2 3p^6 4s^2 3d^7$	$[Ar]4s^2 3d^7$
Ni	28	$1s^2 2s^2 2p^6 3s^2 3p^6 4s^2 3d^8$	$[Ar]4s^2 3d^8$
Cu*	29	$1s^2 2s^2 2p^6 3s^2 3p^6 4s^1 3d^{10}$	$[Ar]4s^1 3d^{10}$ filled d sublevel is stable
Zn	30	$1s^2 2s^2 2p^6 3s^2 3p^6 4s^2 3d^{10}$	$[Ar]4s^2 3d^{10}$
4p Block			
Ga	31	$1s^2 2s^2 2p^6 3s^2 3p^6 4s^2 3d^{10} 4p^1$	$[Ar]4s^2 3d^{10} 4p^1$
Ge	32	$1s^2 2s^2 2p^6 3s^2 3p^6 4s^2 3d^{10} 4p^2$	$[Ar]4s^2 3d^{10} 4p^2$
As	33	$1s^2 2s^2 2p^6 3s^2 3p^6 4s^2 3d^{10} 4p^3$	$[Ar]4s^2 3d^{10} 4p^3$
Se	34	$1s^2 2s^2 2p^6 3s^2 3p^6 4s^2 3d^{10} 4p^4$	$[Ar]4s^2 3d^{10} 4p^4$
Br	35	$1s^2 2s^2 2p^6 3s^2 3p^6 4s^2 3d^{10} 4p^5$	$[Ar]4s^2 3d^{10} 4p^5$
Kr	36	$1s^2 2s^2 2p^6 3s^2 3p^6 4s^2 3d^{10} 4p^6$	$[Ar]4s^2 3d^{10} 4p^6$

*Exceptions to the order of filling.

In Sample Problem 5.7, we use sublevel blocks on the periodic table to write the electron configuration for selenium (atomic number 34).

SAMPLE PROBLEM 5.7 **Using Sublevel Blocks to Write Electron Configurations**

TRY IT FIRST

Selenium is used in making glass and in pigments. Use the sublevel blocks on the periodic table to write the electron configuration for selenium.

SOLUTION

ANALYZE THE PROBLEM	Given	Need	Connect
	selenium	electron configuration	sublevel blocks

STEP 1 **Locate the element on the periodic table.** Selenium is in Period 4 and Group 6A (16).

STEP 2 **Write the filled sublevels in order, going across each period.**

Period	Sublevel Block Filling	Sublevel Block Notation (filled)
1	$1s$ (H → He)	$1s^2$
2	$2s$ (Li → Be) and $2p$ (B → Ne)	$2s^2 2p^6$
3	$3s$ (Na → Mg) and $3p$ (Al → Ar)	$3s^2 3p^6$
4	$4s$ (K → Ca) and $3d$ (Sc → Zn)	$4s^2 3d^{10}$

STEP 3 **Complete the configuration by counting the electrons in the last occupied sublevel block.** Because selenium is the fourth element in the $4p$ block, there are four electrons in the $4p$ sublevel.

Period	Sublevel Block Filling	Sublevel Block Notation
4	$4p$ (Ga → Se)	$4p^4$

The electron configuration for selenium (Se) is: $1s^2 2s^2 2p^6 3s^2 3p^6 4s^2 3d^{10} 4p^4$.

SELF TEST 5.7

a. Iodine is a micromineral needed for thyroid function. Use the sublevel blocks on the periodic table to write the electron configuration for iodine.
b. Manganese is a micromineral needed for metabolic function. Use the sublevel blocks on the periodic table to write the electron configuration for manganese.

ANSWER

a. $1s^2 2s^2 2p^6 3s^2 3p^6 4s^2 3d^{10} 4p^6 5s^2 4d^{10} 5p^5$ b. $1s^2 2s^2 2p^6 3s^2 3p^6 4s^2 3d^5$

Some dietary sources of iodine include seafood, eggs, and milk.

Some Exceptions in Sublevel Block Order

Within the filling of the $3d$ sublevel, exceptions occur for chromium and copper. In Cr and Cu, the $3d$ sublevel is close to being a half-filled or filled sublevel, which is particularly stable. Thus, the electron configuration for chromium has only one electron in the $4s$ and five electrons in the $3d$ sublevel to give the added stability of a half-filled d sublevel. This is shown in the abbreviated orbital diagram for chromium:

Abbreviated orbital diagram for chromium

A similar exception occurs when copper achieves a stable, filled $3d$ sublevel with 10 electrons and only one electron in the $4s$ orbital. This is shown in the abbreviated orbital diagram for copper:

After the $4s$ and $3d$ sublevels are completed, the $4p$ sublevel fills as expected from gallium to krypton, the noble gas that completes Period 4. There are also exceptions in filling for the higher d and f electron sublevels, some caused by the added stability of half-filled shells and others where the cause is not known.

PRACTICE PROBLEMS

Try Practice Problems 5.41 to 5.50

PRACTICE PROBLEMS

5.5 Electron Configurations and the Periodic Table

5.41 Use the sublevel blocks on the periodic table to write a complete electron configuration for an atom of each of the following:
 a. arsenic **b.** iron
 c. tin **d.** krypton

5.42 Use the sublevel blocks on the periodic table to write a complete electron configuration for an atom of each of the following:
 a. calcium **b.** nickel
 c. gallium **d.** cadmium

5.43 Use the sublevel blocks on the periodic table to write an abbreviated electron configuration for an atom of each of the following:
 a. titanium **b.** bromine
 c. barium **d.** lead

5.44 Use the sublevel blocks on the periodic table to write an abbreviated electron configuration for an atom of each of the following:
 a. vanadium **b.** palladium
 c. zinc **d.** cesium

5.45 Use the periodic table to give the symbol of the element with each of the following electron configurations:
 a. $1s^22s^22p^63s^23p^3$ **b.** $1s^22s^22p^63s^23p^64s^23d^7$
 c. $[Ar]4s^23d^{10}$ **d.** $[Xe]6s^24f^{14}5d^{10}6p^3$

5.46 Use the periodic table to give the symbol of the element with each of the following electron configurations:
 a. $1s^22s^22p^63s^23p^64s^23d^8$ **b.** $[Kr]5s^24d^{10}5p^4$
 c. $1s^22s^22p^63s^23p^64s^23d^{10}4p^2$ **d.** $[Ar]4s^23d^{10}4p^5$

5.47 Use the periodic table to give the symbol of the element that meets the following conditions:
 a. has three electrons in the $n = 4$ energy level
 b. has three $2p$ electrons
 c. completes the $5p$ sublevel
 d. has two electrons in the $4d$ sublevel

5.48 Use the periodic table to give the symbol of the element that meets the following conditions:
 a. has five electrons in the $n = 3$ energy level
 b. has one electron in the $6p$ sublevel
 c. completes the $7s$ sublevel
 d. has four $5p$ electrons

5.49 Use the periodic table to give the number of electrons in the indicated sublevels for the following:
 a. $3d$ in zinc **b.** $2p$ in sodium
 c. $4p$ in arsenic **d.** $5s$ in rubidium

5.50 Use the periodic table to give the number of electrons in the indicated sublevels for the following:
 a. $3d$ in manganese **b.** $5p$ in antimony
 c. $6p$ in lead **d.** $3s$ in magnesium

5.6 Trends in Periodic Properties

LEARNING GOAL Use the electron configurations of elements to explain the trends in periodic properties.

The electron configurations of atoms are an important factor in the physical and chemical properties of the elements and in the properties of the compounds that they form. In this section, we will look at the *valence electrons* in atoms, the trends in *atomic size*, *ionization energy*, and *metallic character*. Going across a period, there is a pattern of regular change in these properties from one group to the next. Known as *periodic properties*, each property increases or decreases across a period, and then the trend is repeated in each successive period. We can use the seasonal changes in temperatures as an analogy for periodic properties. In the winter, temperatures are cold and become warmer in the spring. By summer, the outdoor temperatures are hot but begin to cool in the fall. By winter, we expect cold temperatures again as the pattern of decreasing and increasing temperatures repeats for another year.

The change in temperature with the seasons is a periodic property.

Group Number and Valence Electrons

The chemical properties of representative elements are mostly due to the **valence electrons**, which are the electrons in the outermost energy level. These valence electrons occupy the s and p sublevels with the highest principal quantum number n. The group numbers indicate the number of valence (outer) electrons for the elements in each vertical column. For example, the elements in Group 1A (1), such as lithium, sodium, and potassium, all have one electron in an s orbital. Looking at the sublevel block, we can represent the valence electron in the alkali metals of Group 1A (1) as ns^1. All the elements in Group 2A (2), the alkaline earth metals, have two valence electrons, ns^2. The halogens in Group 7A (17) have seven valence electrons, ns^2np^5.

We can see the repetition of the outermost s and p electrons for the representative elements for Periods 1 to 4 in **TABLE 5.3**. Helium is included in Group 8A (18) because it is a noble gas, but it has only two electrons in its complete energy level.

ENGAGE 5.14

What do the group numbers indicate about the electron configurations of the elements in Groups 1A (1) to 8A (18)?

TABLE 5.3 Valence Electron Configuration for Representative Elements in Periods 1 to 4

1A (1)	2A (2)	3A (13)	4A (14)	5A (15)	6A (16)	7A (17)	8A (18)
1 H $1s^1$							2 He $1s^2$
3 Li $2s^1$	4 Be $2s^2$	5 B $2s^22p^1$	6 C $2s^22p^2$	7 N $2s^22p^3$	8 O $2s^22p^4$	9 F $2s^22p^5$	10 Ne $2s^22p^6$
11 Na $3s^1$	12 Mg $3s^2$	13 Al $3s^23p^1$	14 Si $3s^23p^2$	15 P $3s^23p^3$	16 S $3s^23p^4$	17 Cl $3s^23p^5$	18 Ar $3s^23p^6$
19 K $4s^1$	20 Ca $4s^2$	31 Ga $4s^24p^1$	32 Ge $4s^24p^2$	33 As $4s^24p^3$	34 Se $4s^24p^4$	35 Br $4s^24p^5$	36 Kr $4s^24p^6$

SAMPLE PROBLEM 5.8   Using Group Numbers

TRY IT FIRST

Using the periodic table, write the period, the group number, the number of valence electrons, and the valence electron configuration for each of the following:

a. calcium   **b.** iodine   **c.** lead

SOLUTION

The valence electrons are the outermost s and p electrons. Although there may be electrons in the d or f sublevels, they are not valence electrons.

a. Calcium is in Period 4, Group 2A (2), has two valence electrons, and has a valence electron configuration of $4s^2$.

b. Iodine is in Period 5, Group 7A (17), has seven valence electrons, and has a valence electron configuration of $5s^25p^5$.

c. Lead is in Period 6, Group 4A (14), has four valence electrons, and has a valence electron configuration of $6s^26p^2$.

SELF TEST 5.8

Using the periodic table, write the period, the group number, the number of valence electrons, and the valence electron configuration for each of the following:

a. sulfur   **b.** strontium

ANSWER

a. Sulfur is in Period 3, Group 6A (16), has six valence electrons, and a $3s^23p^4$ valence electron configuration.

b. Strontium is in Period 5, Group 2A (2), has two valence electrons, and a $5s^2$ valence electron configuration.

Atomic Size

The **atomic size** of an atom is determined by the distance of the valence electrons from the nucleus. For each group of representative elements, the atomic size *increases* going from the top to the bottom because the outermost electrons in each energy level are farther from the nucleus. For example, in Group 1A (1), Li has a valence electron in energy level 2; Na has a valence electron in energy level 3; and K has a valence electron in energy level 4. This means that a K atom is larger than a Na atom and a Na atom is larger than a Li atom (see FIGURE 5.12).

FIGURE 5.12 The atomic size increases going down a group but decreases going from left to right across a period.

The atomic size of representative elements is affected by the attractive forces of the protons in the nucleus on the electrons in the outermost level. For the elements going across a period, the increase in the number of protons in the nucleus increases the positive charge of the nucleus. As a result, the electrons are pulled closer to the nucleus, which means that the atomic size of representative elements decreases going from left to right across a period.

The size of atoms of transition elements within the same period changes only slightly because electrons are filling *d* orbitals rather than the outermost energy level. Because the increase in nuclear charge is canceled by an increase in *d* electrons, the attraction of the valence electrons by the nucleus remains about the same. Because there is little change in the nuclear attraction for the valence electrons, the atomic size remains relatively constant for the transition elements.

SAMPLE PROBLEM 5.9   Sizes of Atoms

TRY IT FIRST

Identify the smaller atom in each of the following pairs:

a. N or F   **b.** K or Kr   **c.** Ca or Sr

SOLUTION

a. The F atom has a greater positive charge on the nucleus, which pulls electrons closer, and makes the F atom smaller than the N atom. Atomic size decreases going from left to right across a period.

b. The Kr atom has a greater positive charge on the nucleus, which pulls electrons closer, and makes the Kr atom smaller than the K atom. Atomic size decreases going from left to right across a period.

c. The outer electrons in the Ca atom are closer to the nucleus than in the Sr atom, which makes the Ca atom smaller than the Sr atom. Atomic size increases going down a group.

ENGAGE 5.15

Why is a phosphorus atom larger than a nitrogen atom but smaller than a silicon atom?

PRACTICE PROBLEMS

Try Practice Problems 5.59 to 5.62

SELF TEST 5.9

Identify the largest atom in each of the following:

a. P, As, or Se   **b.** Mg, Si, or S

ANSWER

a. As   **b.** Mg

ENGAGE 5.16

When a magnesium atom ionizes to form a magnesium ion Mg^{2+}, which electrons are lost?

FIGURE 5.13 As the distance from the nucleus to the valence electron in Li, Na, and K atoms increases in Group 1A (1), the ionization energy decreases and less energy is required to remove the valence electron.

Ionization Energy

In an atom, negatively charged electrons are attracted to the positive charge of the protons in the nucleus. Therefore, energy is required to remove an electron from an atom. The **ionization energy** is the energy needed to remove one electron from an atom in the gaseous (*g*) state. When an electron is removed from a neutral atom, a cation with a 1+ charge is formed.

$$Na(g) + \text{energy (ionization)} \longrightarrow Na^+(g) + e^-$$

The attraction of a nucleus for the outermost electrons decreases as those electrons are farther from the nucleus. Thus the ionization energy decreases going down a group (see **FIGURE 5.13**). However, going across a period from left to right, the positive charge of the nucleus increases because there is an increase in the number of protons. Thus the ionization energy increases going from left to right across the periodic table.

In summary, the ionization energy is low for the metals and high for the nonmetals. The high ionization energies of the noble gases indicate that their electron configurations are especially stable.

ENGAGE 5.17

Why would Cs have a lower ionization energy than K?

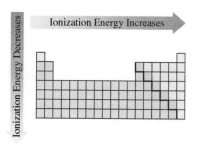

Ionization energy decreases going down a group and increases going from left to right across a period.

PRACTICE PROBLEMS

Try Practice Problems 5.63 to 5.66

SAMPLE PROBLEM 5.10   Ionization Energy

TRY IT FIRST

Indicate the element in each set that has the higher ionization energy and explain your choice.

a. S or Se

b. Mg or Cl

c. F, N, or C

SOLUTION

a. S. In S, an electron is removed from a sublevel closer to the nucleus, which requires a higher ionization energy for S compared with Se.

b. Cl. The increased nuclear charge of Cl increases the attraction for the valence electrons, which requires a higher ionization energy for Cl compared to Mg.

c. F. The increased nuclear charge of F increases the attraction for the valence electrons, which requires a higher ionization energy for F compared to C or N.

SELF TEST 5.10

a. Arrange Sr, I, and Sn in order of increasing ionization energy.

b. Arrange Ba, Mg, and Ca in order of decreasing ionization energy.

ANSWER

a. Sr, Sn, I. Ionization energy increases going from left to right across a period.

b. Ionization energy decreases going down a group: Mg, Ca, Ba.

Metallic Character

An element that has **metallic character** is an element that loses valence electrons easily. Metallic character is more prevalent in the elements on the left side of the periodic table (metals) and decreases going from left to right across a period. The elements on the right side of the periodic table (nonmetals) do not easily lose electrons, which means they are less metallic. Most of the metalloids between the metals and nonmetals tend to lose electrons, but not as easily as the metals. Thus, in Period 3, sodium, which loses electrons most easily, would be the most metallic. Going across from left to right in Period 3, metallic character decreases to argon, which has the least metallic character.

For elements in the same group of representative elements, metallic character increases going from top to bottom. Atoms at the bottom of any group have more electron levels, which makes it easier to lose electrons. Thus, the elements at the bottom of a group on the periodic table have lower ionization energy and are more metallic compared to the elements at the top.

A summary of the trends in periodic properties we have discussed is given in **TABLE 5.4**.

Metallic character increases going down a group and decreases going from left to right across a period.

ENGAGE 5.18

Why is magnesium more metallic than aluminum?

TABLE 5.4 Summary of Trends in Periodic Properties of Representative Elements

Periodic Property	Top to Bottom Within a Group	Left to Right Across a Period
Valence Electrons	Remains the same	Increases
Atomic Size	Increases because there is an increase in the number of energy levels	Decreases as the number of protons increases, which strengthens the attraction of the nucleus for the valence electrons, and pulls them closer to the nucleus
Ionization Energy	Decreases because the valence electrons are easier to remove when they are farther from the nucleus	Increases as the number of protons increases, which strengthens the attraction of the nucleus for the valence electrons, and more energy is needed to remove a valence electron
Metallic Character	Increases because the valence electrons are easier to remove when they are farther from the nucleus	Decreases as the number of protons increases, which strengthens the attraction of the nucleus for the valence electrons, and makes it more difficult to remove a valence electron

PRACTICE PROBLEMS

Try Practice Problems 5.67 to 5.74

PRACTICE PROBLEMS

5.6 Trends in Periodic Properties

5.51 What do the group numbers from 1A (1) to 8A (18) for the elements indicate about electron configurations of those elements?

5.52 What is similar and what is different about the valence electrons of the elements in a group?

5.53 Write the group number using both A/B and 1 to 18 notations for elements that have the following outer electron configuration:
a. $2s^2$ **b.** $3s^23p^3$ **c.** $4s^23d^5$ **d.** $5s^24d^{10}5p^4$

5.54 Write the group number using both A/B and 1 to 18 notations for elements that have the following outer electron configuration:
a. $4s^23d^{10}4p^5$ **b.** $4s^1$ **c.** $4s^23d^8$ **d.** $5s^24d^{10}5p^2$

5.55 Write the valence electron configuration for each of the following:
a. alkali metals   **b.** Group 4A (14)
c. Group 7A (17)   **d.** Group 5A (15)

5.56 Write the valence electron configuration for each of the following:
a. halogens   **b.** Group 6A (16)
c. Group 13   **d.** alkaline earth metals

5.57 Indicate the number of valence electrons in each of the following:
a. aluminum   **b.** Group 5A (15)
c. barium   **d.** F, Cl, Br, and I

5.58 Indicate the number of valence electrons in each of the following:
a. Li, Na, K, Rb, and Cs   **b.** Se
c. C, Si, Ge, Sn, and Pb   **d.** Group 8A (18)

5.59 Select the larger atom in each pair.
a. Na or Cl   **b.** Na or Rb
c. Na or Mg   **d.** Rb or I

5.60 Select the larger atom in each pair.
a. S or Ar   **b.** S or O
c. S or K   **d.** S or Mg

5.61 Place the elements in each set in order of decreasing atomic size.
    a. Al, Si, Mg   **b.** Cl, I, Br
    c. Sb, Sr, I   **d.** P, Si, Na

5.62 Place the elements in each set in order of decreasing atomic size.
    a. Cl, S, P   **b.** Ge, Si, C
    c. Ba, Ca, Sr   **d.** S, O, Se

5.63 Select the element in each pair with the higher ionization energy.
    a. Br or I   **b.** Mg or Sr
    c. Si or P   **d.** I or Xe

5.64 Select the element in each pair with the higher ionization energy.
    a. O or Ne   **b.** K or Br
    c. Ca or Ba   **d.** N or Ne

5.65 Arrange each set of elements in order of increasing ionization energy.
    a. F, Cl, Br   **b.** Na, Cl, Al
    c. Na, K, Cs   **d.** As, Ca, Br

5.66 Arrange each set of elements in order of increasing ionization energy.
    a. O, N, C   **b.** S, P, Cl
    c. P, As, N   **d.** Al, Si, P

5.67 Place the following in order of decreasing metallic character:
Br, Ge, Ca, Ga

5.68 Place the following in order of increasing metallic character:
Na, P, Al, Ar

5.69 Fill in each of the following blanks using *higher* or *lower*, *more* or *less*: Sr has a _____ ionization energy and is _____ metallic than Sb.

5.70 Fill in each of the following blanks using *higher* or *lower*, *more* or *less*: N has a _____ ionization energy and is _____ metallic than As.

5.71 Complete each of the following statements **a** to **d** using **1**, **2**, or **3**:
    1. decreases   **2.** increases   **3.** remains the same

Going down Group 6A (16),
    a. the ionization energy _____
    b. the atomic size _____
    c. the metallic character _____
    d. the number of valence electrons _____

5.72 Complete each of the following statements **a** to **d** using **1**, **2**, or **3**:
    1. decreases   **2.** increases   **3.** remains the same

Going from left to right across Period 3,
    a. the ionization energy _____
    b. the atomic size _____
    c. the metallic character _____
    d. the number of valence electrons _____

5.73 Which statements completed with **a** to **e** will be *true* and which will be *false*?

An atom of N compared to an atom of Li has a larger (greater)
    a. atomic size   **b.** ionization energy
    c. number of protons   **d.** metallic character
    e. number of valence electrons

5.74 Which statements completed with **a** to **e** will be *true* and which will be *false*?

An atom of C compared to an atom of Sn has a larger (greater)
    a. atomic size   **b.** ionization energy
    c. number of protons   **d.** metallic character
    e. number of valence electrons

UPDATE Developing New Materials for Computer Chips

As part of a new research project, Jennifer and Robert are assigned to investigate the properties of indium (In) and tellurium (Te). Both these elements are used to make computer chips.
As a result of their research, Jennifer and Robert hope to develop new computer chips that will be faster and less expensive to manufacture.

5.75   **a.** What is the atomic number of In?
       b. How many electrons are in an atom of In?
       c. Use the sublevel blocks on the periodic table to write the electron configuration and abbreviated electron configuration for In.
       d. Which is larger, an atom of indium or an atom of iodine?
       e. Which has a higher ionization energy, an atom of indium or an atom of iodine?

5.76   **a.** What is the atomic number of Te?
       b. How many electrons are in an atom of Te?
       c. Use the sublevel blocks on the periodic table to write the electron configuration and abbreviated electron configuration for Te.
       d. Which is smaller, an atom of selenium or an atom of tellurium?
       e. Which has a lower ionization energy, an atom of selenium or an atom of tellurium?

CONCEPT MAP

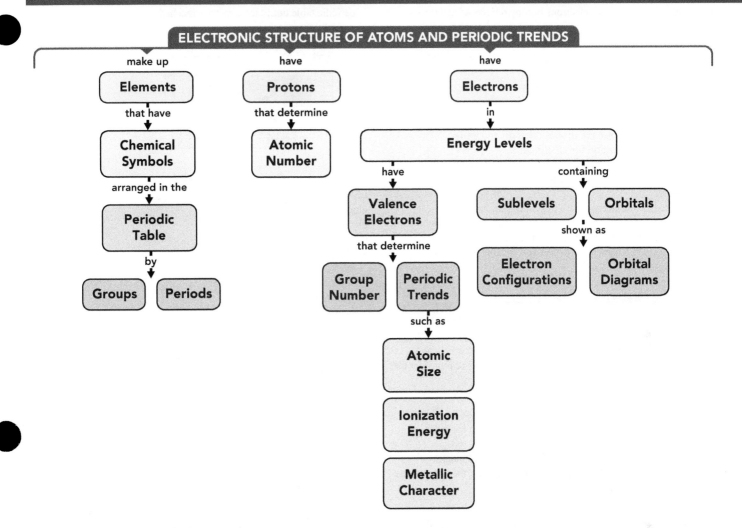

ELECTRONIC STRUCTURE OF ATOMS AND PERIODIC TRENDS

make up → **Elements** → that have → **Chemical Symbols** → arranged in the → **Periodic Table** → by → **Groups** **Periods**

have → **Protons** → that determine → **Atomic Number**

have → **Electrons** → in → **Energy Levels**

Energy Levels → have → **Valence Electrons** → that determine → **Group Number** **Periodic Trends**

Periodic Trends → such as → **Atomic Size**, **Ionization Energy**, **Metallic Character**

Energy Levels → containing → **Sublevels** **Orbitals** → shown as → **Electron Configurations** **Orbital Diagrams**

CHAPTER REVIEW

5.1 Electromagnetic Radiation

LEARNING GOAL Compare the wavelength, frequency, and energy of electromagnetic radiation.

- Electromagnetic radiation such as radio waves and visible light is energy that travels at the speed of light.
- Each particular type of radiation has a specific wavelength and frequency.
- A wavelength (symbol λ, lambda) is the distance between a crest or trough in a wave and the next crest or trough on that wave.
- The frequency (symbol ν, nu) is the number of waves that pass a certain point in 1 s.
- All electromagnetic radiation travels at the speed of light (c), which is 3.00×10^8 m/s.
- Mathematically, the relationship of the speed of light, wavelength, and frequency is expressed as $c = \lambda\nu$.
- Long-wavelength radiation has low frequencies, while short-wavelength radiation has high frequencies.
- Radiation with a high frequency has high energy.

5.2 Atomic Spectra and Energy Levels

LEARNING GOAL Explain how atomic spectra correlate with the energy levels in atoms.

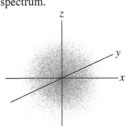

- The atomic spectra of elements are related to the specific energy levels occupied by electrons.
- Light consists of photons, which are particles of a specific energy.
- When an electron absorbs a photon of a particular energy, it attains a higher energy level. When an electron drops to a lower energy level, a photon of a particular energy is emitted.
- Each element has its own unique atomic spectrum.

5.3 Sublevels and Orbitals

LEARNING GOAL Describe the sublevels and orbitals for the electrons in an atom.

- An orbital is a region around the nucleus where an electron with a specific energy is most likely to be found.

- Each orbital holds a maximum of two electrons, which must have opposite spins.
- In each energy level (*n*), electrons occupy orbitals of identical energy within sublevels.
- An *s* sublevel contains one *s* orbital, a *p* sublevel contains three *p* orbitals, a *d* sublevel contains five *d* orbitals, and an *f* sublevel contains seven *f* orbitals.
- Each type of orbital has a unique shape.

5.4 Orbital Diagrams and Electron Configurations

LEARNING GOAL Draw the orbital diagram and write the electron configuration for an element.

- Within a sublevel, electrons enter orbitals in the same energy level one at a time until all the orbitals are half-filled.

- Additional electrons enter with opposite spins until the orbitals in that sublevel are filled with two electrons each.
- The orbital diagram for an element such as silicon shows the orbitals that are occupied by paired and unpaired electrons:

1*s*	2*s*	2*p*	3*s*	3*p*
↑↓	↑↓	↑↓ ↑↓ ↑↓	↑↓	↑ ↑

- The electron configuration for an element such as silicon shows the number of electrons in each sublevel: $1s^2 2s^2 2p^6 3s^2 3p^2$.
- An abbreviated electron configuration for an element such as silicon places the symbol of a noble gas in brackets to represent the filled sublevels: $[Ne]3s^2 3p^2$.

5.5 Electron Configurations and the Periodic Table

LEARNING GOAL Write the electron configuration for an atom using the sublevel blocks on the periodic table.

- The periodic table consists of *s*, *p*, *d*, and *f* sublevel blocks.
- An electron configuration can be written following the order of the sublevel blocks on the periodic table.

- Beginning with 1*s*, an electron configuration is obtained by writing the sublevel blocks in order going across each period on the periodic table until the element is reached.

5.6 Trends in Periodic Properties

LEARNING GOAL Use the electron configurations of elements to explain the trends in periodic properties.

Li atom

- The properties of elements are related to the valence electrons of the atoms.
- With only a few exceptions, each group of elements has the same arrangement of valence electrons differing only in the energy level.

Na atom

- The size of an atom increases going down a group and decreases going from left to right across a period.

Distance between the nucleus and valence electron

K atom

- The energy required to remove a valence electron is the ionization energy, which decreases going down a group, and increases going from left to right across a period.
- The metallic character of an element increases going down a group and decreases going from left to right across a period.

KEY TERMS

atomic size The distance between the outermost electrons and the nucleus.

atomic spectrum A series of lines specific for each element produced by photons emitted by electrons dropping to lower energy levels.

***d* block** The 10 elements in Groups 3B (3) to 2B (12) in which electrons fill the five *d* orbitals.

electromagnetic radiation Forms of energy such as visible light, microwaves, radio waves, infrared, ultraviolet light, and X-rays that travel as waves at the speed of light.

electromagnetic spectrum The arrangement of types of radiation from long wavelengths to short wavelengths.

electron configuration A list of the number of electrons in each sublevel within an atom, arranged by increasing energy.

energy level A group of electrons with similar energy.

***f* block** The 14 elements in the rows at the bottom of the periodic table in which electrons fill the seven 4*f* and 5*f* orbitals.

frequency The number of times the crests of a wave pass a point in 1 s.

ionization energy The energy needed to remove the least tightly bound electron from the outermost energy level of an atom.

metallic character A measure of how easily an element loses a valence electron.

orbital The region around the nucleus of an atom where electrons of certain energy are most likely to be found: *s* orbitals are spherical; *p* orbitals have two lobes.

orbital diagram A diagram that shows the distribution of electrons in the orbitals of the energy levels.

***p* block** The elements in Groups 3A (13) to 8A (18) in which electrons fill the *p* orbitals.

photon A packet of energy that has both particle and wave characteristics and travels at the speed of light.

principal quantum number (*n*) The number ($n = 1, n = 2, \ldots$) assigned to an energy level.

***s* block** The elements in Groups 1A (1) and 2A (2) in which electrons fill the *s* orbitals.

sublevel A group of orbitals of equal energy within energy levels. The number of sublevels in each energy level is the same as the principal quantum number (*n*).

valence electrons The electrons in the highest energy level of an atom.

wavelength The distance between adjacent crests or troughs in a wave.

CORE CHEMISTRY SKILLS

The chapter section containing each Core Chemistry Skill is shown in parentheses at the end of each heading.

Writing Electron Configurations (5.4)

- The electron configuration for an atom specifies the energy levels and sublevels occupied by the electrons of an atom.
- An electron configuration is written starting with the lowest energy sublevel, followed by the next lowest energy sublevel.
- The number of electrons in each sublevel is shown as a superscript.

Example: Write the electron configuration for palladium.

Answer: Palladium has atomic number 46, which means it has 46 protons and 46 electrons.

$$1s^2 2s^2 2p^6 3s^2 3p^6 4s^2 3d^{10} 4p^6 5s^2 4d^8$$

Using the Periodic Table to Write Electron Configurations (5.5)

- An electron configuration corresponds to the location of an element on the periodic table, where different blocks within the periodic table are identified as the *s*, *p*, *d*, and *f* sublevels.

Example: Use the periodic table to write the electron configuration for sulfur.

Answer: Sulfur (atomic number 16) is in Group 6A (16) and Period 3.

Period	Sublevel Block Filling	Sublevel Block Notation
1	1*s* (H → He)	$1s^2$
2	2*s* (Li → Be) and 2*p* (B → Ne)	$2s^2 2p^6$
3	3*s* (Na → Mg)	$3s^2$
	3*p* (Al → S)	$3p^4$

The electron configuration for sulfur (S) is: $1s^2 2s^2 2p^6 3s^2 3p^4$.

Identifying Trends in Periodic Properties (5.6)

- The size of an atom increases going down a group and decreases going from left to right across a period.
- The ionization energy decreases going down a group and increases going from left to right across a period.
- The metallic character of an element increases going down a group and decreases going from left to right across a period.

Example: For Mg, P, and Cl, identify which has the
     a. largest atomic size
     b. highest ionization energy
     c. greatest metallic character

Answer: **a.** Mg
         b. Cl
         c. Mg

UNDERSTANDING THE CONCEPTS

The chapter sections to review are shown in parentheses at the end of each problem.

Use the following diagram for problems 5.77 and 5.78:

A.

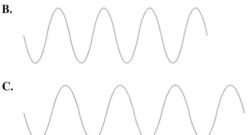

B.

C.

5.77 Select diagram **A**, **B**, or **C** that (5.1)
    a. has the longest wavelength
    b. has the shortest wavelength
    c. has the highest frequency
    d. has the lowest frequency

5.78 Select diagram **A**, **B**, or **C** that (5.1)
    a. has the highest energy
    b. has the lowest energy
    c. would represent blue light
    d. would represent red light

5.79 Match the following with an *s* or *p* orbital: (5.3)

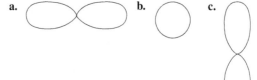
  a.   **b.**   **c.**

5.80 Match the following with *s* or *p* orbitals: (5.3)
    a. two lobes
    b. spherical shape
    c. found in $n = 1$
    d. found in $n = 3$

5.81 Indicate whether or not each of the following orbital diagrams is possible and explain. When possible, indicate the element it represents. (5.4)

a.
1s	2s	2p	3s
↑↓	↑↓	↑↓ ↑↓ ↑↓	↑↓

b.
1s	2s	2p	3s
↑↓	↑↓	↑↓ ↑↓ ↑	↑↑↓

5.82 Indicate whether or not each of the following abbreviated orbital diagrams is possible and explain. When possible, indicate the element it represents. (5.4)

a.
	2s	2p
[He]	↑↓	↑↓ ↑↑ ↑

b.
	4s	3d
[Ar]	↑	↑ ↑ ↑ ↑ ↑

5.83 Match the spheres **A** through **D** with atoms of Li, Na, K, and Rb. (5.6)

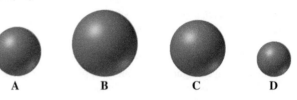

A   B   C   D

5.84 Match the spheres **A** through **D** with atoms of K, Ge, Ca, and Kr. (5.6)

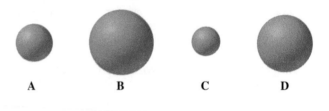

A   B   C   D

ADDITIONAL PRACTICE PROBLEMS

5.85 What is the difference between a continuous spectrum and an atomic spectrum? (5.1)

5.86 Why does a neon sign give off red light? (5.1)

5.87 What is the Pauli exclusion principle? (5.3)

5.88 Why would there be five unpaired electrons in a d sublevel but no paired electrons? (5.3)

5.89 Which of the following orbitals are possible in an atom: $4p$, $2d$, $3f$, and $5f$? (5.3)

5.90 Which of the following orbitals are possible in an atom: $1p$, $4f$, $6s$, and $4d$? (5.3)

5.91 a. What electron sublevel starts to fill after completion of the $3s$ sublevel? (5.4)
   b. What electron sublevel starts to fill after completion of the $4p$ sublevel?
   c. What electron sublevel starts to fill after completion of the $3d$ sublevel?
   d. What electron sublevel starts to fill after completion of the $3p$ sublevel?

5.92 a. What electron sublevel starts to fill after completion of the $5s$ sublevel? (5.4)
   b. What electron sublevel starts to fill after completion of the $4d$ sublevel?
   c. What electron sublevel starts to fill after completion of the $4f$ sublevel?
   d. What electron sublevel starts to fill after completion of the $5p$ sublevel?

5.93 a. How many $3d$ electrons are in Fe? (5.4)
   b. How many $5p$ electrons are in Ba?
   c. How many $4d$ electrons are in I?
   d. How many $7s$ electrons are in Ra?

5.94 a. How many $4d$ electrons are in Cd? (5.4)
   b. How many $4p$ electrons are in Br?
   c. How many $6p$ electrons are in Bi?
   d. How many $4s$ electrons are in Zn?

5.95 Write the abbreviated electron configuration and group number for each of the following elements: (5.4)
   a. Si   **b.** Se   **c.** Mn   **d.** Sb

5.96 Write the abbreviated electron configuration and group number for each of the following elements: (5.4)
   a. Br   **b.** Rh   **c.** Tc   **d.** Ra

5.97 What do the elements Ca, Sr, and Ba have in common in terms of their electron configuration? Where are they located on the periodic table? (5.4, 5.5)

5.98 What do the elements O, S, and Se have in common in terms of their electron configuration? Where are they located on the periodic table? (5.4, 5.5)

5.99 Name the element that corresponds to each of the following: (5.4, 5.5, 5.6)
   a. $1s^2 2s^2 2p^6 3s^2 3p^3$
   b. alkali metal with the smallest atomic size
   c. $[Kr]5s^2 4d^{10}$
   d. Group 5A (15) element with the highest ionization energy
   e. Period 3 element with the largest atomic size

5.100 Name the element that corresponds to each of the following: (5.4, 5.5, 5.6)
   a. $1s^2 2s^2 2p^6 3s^2 3p^6 4s^1 3d^5$
   b. $[Xe]6s^2 4f^{14} 5d^{10} 6p^5$
   c. halogen with the highest ionization energy
   d. Group 2A (2) element with the lowest ionization energy
   e. Period 4 element with the smallest atomic size

5.101 Why is the ionization energy of Ca higher than that of K but lower than that of Mg? (5.6)

5.102 Why is the ionization energy of Br lower than that of Cl but higher than that of Se? (5.6)

5.103 Select the more metallic element in each pair. (5.6)
   a. As or Sb   **b.** Sn or Sb   **c.** Cl or P   **d.** O or P

5.104 Select the more metallic element in each pair. (5.6)
   a. Sn or As   **b.** Cl or I
   c. Ca or Ba   **d.** Ba or Hg

5.105 Of the elements Na, P, Cl, and F, which (5.6)
   a. is a metal?
   b. is in Group 5A (15)?
   c. has the highest ionization energy?
   d. loses an electron most easily?
   e. is found in Group 7A (17), Period 3?

5.106 Of the elements K, Ca, Br, and Kr, which (5.6)
   a. is a halogen?
   b. has the smallest atomic size?
   c. has the lowest ionization energy?
   d. requires the most energy to remove an electron?
   e. is found in Group 2A (2), Period 4?

5.107 Consider three elements with the following abbreviated electron configurations: (5.4, 5.5, 5.6)
$$X = [Ar]4s^2 \quad Y = [Ne]3s^23p^4 \quad Z = [Ar]4s^23d^{10}4p^4$$
   a. Identify each element as a metal, nonmetal, or metalloid.
   b. Which element has the largest atomic size?
   c. Which element has the highest ionization energy?
   d. Which element has the smallest atomic size?

5.108 Consider three elements with the following abbreviated electron configurations: (5.4, 5.5, 5.6)
$$X = [Ar]4s^23d^5 \quad Y = [Ar]4s^23d^{10}4p^1 \quad Z = [Ar]4s^23d^{10}4p^6$$
   a. Identify each element as a metal, nonmetal, or metalloid.
   b. Which element has the smallest atomic size?
   c. Which element has the highest ionization energy?
   d. Which element has a half-filled sublevel?

CHALLENGE PROBLEMS

The following problems are related to the topics in this chapter. However, they do not all follow the chapter order, and they require you to combine concepts and skills from several sections. These problems will help you increase your critical thinking skills and prepare for your next exam.

5.109 How do scientists explain the colored lines observed in the spectra of heated atoms? (5.2)

5.110 Even though H has only one electron, there are many lines in the atomic spectrum of H. Explain. (5.2)

5.111 What is meant by an energy level, a sublevel, and an orbital? (5.3)

5.112 In some periodic tables, H is placed in Group 1A (1). In other periodic tables, H is also placed in Group 7A (17). Why? (5.4, 5.5)

5.113 Compare F, S, and Cl in terms of atomic size and ionization energy. (5.6)

5.114 Compare K, Mg, and Ca in terms of atomic size and ionization energy. (5.6)

5.115 Give the symbol of the element that has the (5.6)
   a. smallest atomic size in Group 6A (16)
   b. smallest atomic size in Period 3
   c. highest ionization energy in Group 3A (13)
   d. lowest ionization energy in Period 3
   e. abbreviated electron configuration $[Kr]5s^24d^6$

5.116 Give the symbol of the element that has the (5.6)
   a. largest atomic size in Period 5
   b. largest atomic size in Group 2A (2)
   c. highest ionization energy in Group 8A (18)
   d. lowest ionization energy in Period 2
   e. abbreviated electron configuration $[Kr]5s^24d^{10}5p^2$

ANSWERS TO ENGAGE QUESTIONS

5.1 The wavelength (distance from crest to crest) of red light is longer than that of blue light.

5.2 $10\,200\ \cancel{km} \times 1000\ \cancel{m}/1\ \cancel{km} \times 1\ s/3.00 \times 10^8\ \cancel{m} = 0.034\ s$

5.3 Infrared radiation has a lower frequency than ultraviolet light.

5.4 When electrons in a heated element absorb energy, they move to higher energy levels, then emit that energy by falling to lower energy levels. Only specific energies are emitted, not continuous energies.

5.5 Electrons absorb a specific amount of energy to move between their current energy level and a higher energy level.

5.6 Energy level $n = 5$ has five sublevels.

5.7 The p orbitals in the $n = 3$ energy level have the same shape and energy. However, they are oriented in different directions along the x, y, and z axes.

5.8 There are 5 orbitals in the $5d$ sublevel.

5.9 Because the $4s$ sublevel has a lower energy than the $3d$ sublevel, the $4s$ energy level fills before the $3d$ sublevel.

5.10 Hund's rule states there is less repulsion between electrons when they occupy orbitals of equal energy singly with the same spin. Thus, in an electron configuration, electrons are added singly to the orbitals in a sublevel, and then additional electrons for the same sublevel double up.

5.11 In neon, the $1s$ block and the $2s$ block each have two electrons. Neon is located at the end of Period 2; thus, neon has six electrons in the $2p$ sublevel.

5.12 The s block contains elements with group numbers 1 (1A) and 2 (2A), the p block contains elements with group numbers 13 (3A) through 18 (8A), and the d block contains elements with group numbers 3 (3B) through 12 (2B).

5.13 For elements K to Zn, the sublevel blocks used are $4s$ and $3d$.

5.14 The groups 1A to 8A indicate the number of s and p valence electrons for the elements in those groups.

5.15 Atomic size increases going from the top to the bottom of each group, which makes a phosphorus atom larger than a nitrogen atom. Atomic size also decreases going from left to right across a period as the number of protons in the nucleus increases. Thus, a phosphorus atom with more protons is smaller than a silicon atom.

5.16 In Cs, the outermost electron is farther from the nucleus, which means less energy is required to remove that electron.

5.17 The ionization of Mg to Mg^{2+} occurs with the loss of the two outermost electrons in the $3s$ sublevel.

5.18 The metallic character decreases going from left to right within a period. Thus, magnesium is more metallic than aluminum.

ANSWERS TO SELECTED PROBLEMS

5.1 The wavelength of UV light is the distance between crests of the wave.

5.3 White light has all the colors of the spectrum, including red and blue light.

5.5 Ultraviolet radiation has a higher frequency and higher energy.

5.7 6.3×10^{-7} m; 630 nm

5.9 AM radio has a longer wavelength than cell phones or infrared.

5.11 Order of increasing wavelengths: X-rays, blue light, microwaves

5.13 Order of increasing frequencies: TV, microwaves, X-rays

5.15 Atomic spectra consist of a series of lines separated by dark sections, indicating that the energy emitted by the elements is not continuous.

5.17 absorb

5.19 **a.** green light   **b.** blue light

5.21 **a.** spherical   **b.** two lobes   **c.** spherical

5.23 **a.** **1** and **2**   **b.** **3**
    c. **1** and **2**   **d.** **1, 2,** and **3**

5.25 **a.** There are five orbitals in the $3d$ sublevel.
    b. There is one sublevel in the $n = 1$ energy level.
    c. There is one orbital in the $6s$ sublevel.
    d. There are nine orbitals in the $n = 3$ energy level.

5.27 **a.** There is a maximum of two electrons in a $2p$ orbital.
    b. There is a maximum of six electrons in the $3p$ sublevel.
    c. There is a maximum of 32 electrons in the $n = 4$ energy level.
    d. There is a maximum of 10 electrons in the $5d$ sublevel.

5.29 The electron configuration shows the number of electrons in each sublevel of an atom. The abbreviated electron configuration uses the symbol of the preceding noble gas to show completed sublevels.

5.31 **a.**
$1s$	$2s$	$2p$		
↑↓	↑↓	↑		

    b.
$1s$	$2s$	$2p$			$3s$	$3p$		
↑↓	↑↓	↑↓	↑↓	↑↓	↑↓	↑		

    c.
$1s$	$2s$	$2p$			$3s$	$3p$		
↑↓	↑↓	↑↓	↑↓	↑↓	↑↓	↑	↑	↑

    d.
$1s$	$2s$	$2p$			$3s$	$3p$		
↑↓	↑↓	↑↓	↑↓	↑↓	↑↓	↑↓	↑↓	↑↓

5.33 **a.** $1s^22s^22p^63p^64s^23d^8$
    b. $1s^22s^22p^63s^1$
    c. $1s^22s^1$
    d. $1s^22s^22p^63s^23p^64s^23d^2$

5.35 **a.** $[Kr]5s^24d^{10}5p^2$
    b. $[Kr]5s^24d^{10}$
    c. $[Ar]4s^23d^{10}4p^4$
    d. $[He]2s^22p^5$

5.37 **a.** Li   **b.** Ti   **c.** Ge   **d.** Al

5.39 **a.** Al   **b.** C   **c.** Ar   **d.** Be

5.41 **a.** $1s^22s^22p^63s^23p^64s^23d^{10}4p^3$ **b.** $1s^22s^22p^63s^23p^64s^23d^6$
    c. $1s^22s^22p^63s^23p^64s^23d^{10}4p^65s^24d^{10}5p^2$
    d. $1s^22s^22p^63s^23p^64s^23d^{10}4p^6$

5.43 **a.** $[Ar]4s^23d^2$ **b.** $[Ar]4s^23d^{10}4p^5$
    c. $[Xe]6s^2$ **d.** $[Xe]6s^24f^{14}5d^{10}6p^2$

5.45 **a.** P   **b.** Co   **c.** Zn   **d.** Bi

5.47 **a.** Ga   **b.** N   **c.** Xe   **d.** Zr

5.49 **a.** 10   **b.** 6   **c.** 3   **d.** 1

5.51 The group numbers 1A to 8A indicate the number of valence electrons from 1 to 8.

5.53 **a.** 2A (2)   **b.** 5A (15)   **c.** 7B (7)   **d.** 6A (16)

5.55 **a.** ns^1 **b.** ns^2np^2 **c.** ns^2np^5 **d.** ns^2np^3

5.57 **a.** 3   **b.** 5   **c.** 2   **d.** 7

5.59 **a.** Na   **b.** Rb   **c.** Na   **d.** Rb

5.61 **a.** Mg, Al, Si   **b.** I, Br, Cl
    c. Sr, Sb, I   **d.** Na, Si, P

5.63 **a.** Br   **b.** Mg   **c.** P   **d.** Xe

5.65 **a.** Br, Cl, F   **b.** Na, Al, Cl
    c. Cs, K, Na   **d.** Ca, As, Br

5.67 Ca, Ga, Ge, Br

5.69 lower, more

5.71 **a.** **1.** decreases
    b. **2.** increases
    c. **2.** increases
    d. **3.** remains the same

5.73 **a.** false   **b.** true   **c.** true   **d.** false   **e.** true

5.75 **a.** 49
    b. 49
    c. $1s^22s^22p^63s^23p^64s^23d^{10}4p^65s^24d^{10}5p^1$; $[Kr]5s^24d^{10}5p^1$
    d. indium
    e. iodine

5.77 **a.** **C** has the longest wavelength.
    b. **A** has the shortest wavelength.
    c. **A** has the highest frequency.
    d. **C** has the lowest frequency.

5.79 **a.** p **b.** s **c.** p

5.81 **a.** This is possible. This element is magnesium.
    b. Not possible. The $2p$ sublevel would fill before the $3s$, and only two electrons are allowed in an s orbital.

5.83 Li is **D**, Na is **A**, K is **C**, and Rb is **B**.

5.85 A continuous spectrum from white light contains wavelengths of all energies. Atomic spectra are line spectra in which a series of lines corresponds to energy emitted when electrons drop from a higher energy level to a lower level.

5.87 The Pauli exclusion principle states that two electrons in the same orbital must have opposite spins.

5.89 A $4p$ orbital is possible because the $n = 4$ energy level has four sublevels, including a p sublevel. A $2d$ orbital is not possible because the $n = 2$ energy level has only s and p sublevels. There are no $3f$ orbitals because only s, p, and d sublevels are allowed for the $n = 3$ energy level. A $5f$ sublevel is possible in the $n = 5$ energy level because five sublevels are allowed, including an f sublevel.

5.91 **a.** *3p*   **b.** *5s*   **c.** *4p*   **d.** *4s*

5.93 **a.** 6   **b.** 6   **c.** 10   **d.** 2

5.95 **a.** [Ne]$3s^2 3p^2$; Group 4A (14)
 b. [Ar]$4s^2 3d^{10} 4p^4$; Group 6A (16)
 c. [Ar]$4s^2 3d^5$; Group 7B (7)
 d. [Kr]$5s^2 4d^{10} 5p^3$; Group 5A (15)

5.97 Ca, Sr, and Ba all have two valence electrons, ns^2, which place them in Group 2A (2).

5.99 **a.** phosphorus
 b. lithium (H is a nonmetal)
 c. cadmium
 d. nitrogen
 e. sodium

5.101 Calcium has a greater number of protons than K. The least tightly bound electron in Ca is farther from the nucleus than in Mg and needs less energy to remove.

5.103 **a.** Sb   **b.** Sn   **c.** P   **d.** P

5.105 **a.** Na   **b.** P   **c.** F   **d.** Na   **e.** Cl

5.107 **a.** X is a metal; Y and Z are nonmetals.
 b. X has the largest atomic size.
 c. Y has the highest ionization energy.
 d. Y has the smallest atomic size.

5.109 The series of lines separated by dark sections in atomic spectra indicate that the energy emitted by the elements is not continuous and that electrons are moving between discrete energy levels.

5.111 The energy level contains all the electrons with similar energy. A sublevel contains electrons with the same energy, while an orbital is the region around the nucleus where electrons of a certain energy are most likely to be found.

5.113 S has a larger atomic size than Cl; Cl is larger than F: S > Cl > F. F has a higher ionization energy than Cl; Cl has a higher ionization energy than S: F > Cl > S.

5.115 **a.** O   **b.** Ar   **c.** B   **d.** Na   **e.** Ru

Ionic and Molecular Compounds

Richard's doctor recommends that Richard take a low-dose aspirin (81 mg) every day to decrease the chance of a heart attack or stroke. Richard is concerned about taking aspirin and asks Dr. Chavez, a pharmacist working at a local pharmacy, about the effects of aspirin. Dr. Chavez, explains that aspirin is acetylsalicylic acid and has the chemical formula $C_9H_8O_4$. Aspirin is a molecular compound, often referred to as an organic molecule because it contains the nonmetal carbon (C) combined with the nonmetals hydrogen (H) and oxygen (O). She explains that aspirin is used to relieve minor pains, to reduce inflammation and fever, and to slow blood clotting. Aspirin is one of several nonsteroidal anti-inflammatory drugs (NSAIDs) that reduce pain and fever by blocking the formation of prostaglandins, which are chemical messengers that transmit pain signals to the brain and cause fever. Some potential side effects of aspirin may include heartburn, upset stomach, nausea, and an increased risk of a stomach ulcer.

CAREER

Pharmacist

Pharmacists work in hospitals, pharmacies, clinics, and long-term care facilities where they are responsible for the preparation and distribution of pharmaceutical medications based on a doctor's orders. They obtain the proper medication and also calculate, measure, and label the patients' medication. Pharmacists advise clients and health care practitioners on the selection of both prescription and over-the-counter drugs, proper dosages, and geriatric considerations, as well as possible side effects and interactions. They may also administer vaccinations; prepare sterile intravenous solutions; and advise clients about health, diet, and home medical equipment. Pharmacists also prepare insurance claims and create and maintain patient profiles.

UPDATE Compounds at the Pharmacy

Two weeks later, Richard returns to the pharmacy. Using his pharmacist's recommendations, he purchases Epsom salts for a sore toe, an antacid for an upset stomach, and an iron supplement. You can see the chemical formulas of these medications in the **UPDATE** Compounds at the Pharmacy, page 176.

6.1 Ions: Transfer of Electrons

LEARNING GOAL Write the symbols for the simple ions of the representative elements.

Most of the elements, except the noble gases, are found in nature combined as compounds. The noble gases are so stable that they form compounds only under extreme conditions. One explanation for the stability of noble gases is that they have a filled valence electron energy level.

Compounds form when electrons are transferred or shared between atoms to give stable electron configurations to the atoms. In the formation of either an *ionic bond* or a *covalent bond*, atoms lose, gain, or share valence electrons to acquire an octet of eight valence electrons. This tendency of atoms to attain a stable electron configuration is known as the **octet rule** and provides a key to our understanding of the ways in which atoms bond and form compounds. A few elements achieve the stability of helium with two valence electrons. However, the octet rule is not used with transition elements.

Ionic bonds occur when the valence electrons of atoms of a metal are transferred to atoms of nonmetals. For example, sodium atoms lose electrons and chlorine atoms gain electrons to form the ionic compound NaCl. **Covalent bonds** form when atoms of nonmetals share valence electrons. In the molecular compounds H_2O and C_3H_8, atoms share electrons (see **TABLE 6.1**).

TABLE 6.1 Types of Particles and Bonds in Compounds

Type	Ionic Compounds	Molecular Compounds	
Particles	Ions	Molecules	
Bonds	Ionic	Covalent	
Examples	Na^+ Cl^- ions	H_2O molecules	C_3H_8 molecules

M is a metal
Nm is a nonmetal

Transfer of electrons   Sharing electrons

Ionic bond   Covalent bond

Positive Ions: Loss of Electrons

In ionic bonding, **ions**, which have electrical charges, form when atoms lose or gain electrons to form a stable electron configuration. Because the ionization energies of metals of Groups 1A (1), 2A (2), and 3A (13) are low, metal atoms readily lose their valence electrons. In doing so, they form ions with positive charges. A metal atom obtains the same electron configuration as its nearest noble gas (usually eight valence electrons). For example, when a sodium atom loses its single valence electron, the remaining electrons have a stable configuration. By losing an electron, sodium has 10 negatively charged electrons instead of 11.

Name	Sodium atom Na		Sodium ion Na^+
		e^-	
		Loss of one valence electron	
Protons	$11\,p^+$		$11\,p^+$
Electrons	$11\,e^-$		$10\,e^-$
Electron Configuration	$1s^2 2s^2 2p^6 3s^1$		$1s^2 2s^2 2p^6$

Because there are still 11 positively charged protons in its nucleus, the sodium atom becomes a sodium ion, which is no longer neutral. It is now a sodium ion with a positive electrical charge (1+), called an **ionic charge**. The ionic charge of 1+ is written in the upper right corner Na^+, where the 1 is understood. A metal ion is named by its element name. Thus, Na^+ is named the *sodium ion*. The sodium ion is smaller than the sodium atom because the ion has lost its outermost electron from the third energy level. A positively charged ion of a metal is called a **cation** (pronounced *cat-eye-un*).

$$\text{Ionic charge} = \text{Charge of protons} + \text{Charge of electrons}$$
$$1+ = (11+) \qquad + (10-)$$

Magnesium, a metal in Group 2A (2), obtains a stable electron configuration by losing two valence electrons to form a magnesium ion with a 2+ ionic charge, Mg^{2+}. The magnesium ion is smaller than the magnesium atom because the outermost electrons in the third energy level were removed. The octet in the magnesium ion is made up of electrons that fill its second energy level.

Name	Magnesium atom Mg		Magnesium ion Mg^{2+}
Protons	$12 p^+$	Loss of two valence electrons	$12 p^+$
Electrons	$12 e^-$		$10 e^-$
Electron Configuration	$1s^2 2s^2 2p^6 3s^2$		$1s^2 2s^2 2p^6$

Negative Ions: Gain of Electrons

The ionization energy of a nonmetal atom in Groups 5A (15), 6A (16), or 7A (17) is high, which means that nonmetal atoms do not easily lose their valence electrons. Thus, a nonmetal atom gains one or more valence electrons to obtain a stable electron configuration. By gaining electrons, a nonmetal atom forms a negatively charged ion. For example, an atom of chlorine with seven valence electrons gains one electron to form an octet. Because it now has 18 electrons and 17 protons in its nucleus, chlorine is no longer a neutral atom. It is a *chloride ion* with an ionic charge of 1−, which is written as Cl^-, with the 1 understood. A negatively charged ion, called an **anion** (pronounced *an-eye-un*), is named by using the first syllable of its element name followed by *ide*. The chloride ion is larger than the chlorine atom because the ion has an additional electron, which completes its outermost energy level.

$$\text{Ionic charge} = \text{Charge of protons} + \text{Charge of electrons}$$
$$1- = (17+) \qquad + (18-)$$

Name	Chlorine atom Cl		Chloride ion Cl^-
Protons	$17 p^+$	Gain of one valence electron	$17 p^+$
Electrons	$17 e^-$		$18 e^-$

Electron Configuration	$1s^2 2s^2 2p^6 3s^2 3p^5$	$1s^2 2s^2 2p^6 3s^2 3p^6$

TABLE 6.2 lists the names of some important metal and nonmetal ions.

TABLE 6.2 Formulas and Names of Some Common Ions

Metals			Nonmetals		
Group Number	Cation	Name of Cation	Group Number	Anion	Name of Anion
1A (1)	Li^+	Lithium	5A (15)	N^{3-}	Nitride
	Na^+	Sodium		P^{3-}	Phosphide
	K^+	Potassium	6A (16)	O^{2-}	Oxide
2A (2)	Mg^{2+}	Magnesium		S^{2-}	Sulfide
	Ca^{2+}	Calcium	7A (17)	F^-	Fluoride
	Ba^{2+}	Barium		Cl^-	Chloride
3A (13)	Al^{3+}	Aluminum		Br^-	Bromide
				I^-	Iodide

SAMPLE PROBLEM 6.1   Ions

TRY IT FIRST

a. Write the symbol and name for the ion that has 7 protons and 10 electrons.
b. Write the symbol and name for the ion that has 20 protons and 18 electrons.

SOLUTION

a. The element with 7 protons is nitrogen. In an ion of nitrogen with 10 electrons, the ionic charge would be $3-$, $[(7+) + (10-) = 3-]$. The ion, written as N^{3-}, is the *nitride* ion.
b. The element with 20 protons is calcium. In an ion of calcium with 18 electrons, the ionic charge would be $2+$, $[(20+) + (18-) = 2+]$. The ion, written as Ca^{2+}, is the *calcium* ion.

SELF TEST 6.1

How many protons and electrons are in each of the following ions?

a. Sr^{2+} **b.** Cl^- **c.** As^{3+}

ANSWER

a. 38 protons, 36 electrons   **b.** 17 protons, 18 electrons   **c.** 33 protons, 30 electrons

Ionic Charges from Group Numbers

In ionic compounds, representative elements usually lose or gain electrons to give a stable electron arrangement like their nearest noble gas. We can use the group numbers in the periodic table to determine the charges for the ions of the representative elements. The elements in Group 1A (1) lose one electron to form ions with a 1+ charge. The elements in Group 2A (2) lose two electrons to form ions with a 2+ charge. The elements in Group 3A (13) lose three electrons to form ions with a 3+ charge. In this text, we do not use the group numbers of the transition elements to determine their ionic charges.

In ionic compounds, the elements in Group 7A (17) gain one electron to form ions with a 1− charge. The elements in Group 6A (16) gain two electrons to form ions with a 2− charge. The elements in Group 5A (15) gain three electrons to form ions with a 3− charge.

The nonmetals of Group 4A (14) do not typically form ions. However, the metals Sn and Pb in Group 4A (14) lose electrons to form positive ions. TABLE 6.3 lists the ionic charges for some common monatomic ions of representative elements.

PRACTICE PROBLEMS

Try Practice Problems 6.5 to 6.8

TABLE 6.3 Examples of Monatomic Ions and Their Nearest Noble Gases

Noble Gas	Metals Lose Valence Electrons			Nonmetals Gain Valence Electrons		
	1A (1)	2A (2)	3A (13)	5A (15)	6A (16)	7A (17)
He	Li^+					
Ne	Na^+	Mg^{2+}	Al^{3+}	N^{3-}	O^{2-}	F^-
Ar	K^+	Ca^{2+}		P^{3-}	S^{2-}	Cl^-
Kr	Rb^+	Sr^{2+}			Se^{2-}	Br^-
Xe	Cs^+	Ba^{2+}				I^-

ENGAGE 6.3

ENGAGE 6.3

Why do all the atoms in Group 2A (2) form ions with 2+ charges?

SAMPLE PROBLEM 6.2 Writing Symbols for Ions

TRY IT FIRST

Consider the elements aluminum and oxygen.

a. Identify each as a metal or a nonmetal.
b. State the number of valence electrons for each.
c. State the number of electrons that must be lost or gained for each to achieve an octet.
d. Write the symbol, including its ionic charge, and the name for each ion.

SOLUTION

Aluminum	Oxygen
a. metal	nonmetal
b. 3 valence electrons	6 valence electrons
c. loses 3 e^-	gains 2 e^-
d. Al^{3+}, $[(13+) + (10-)] = 3+$, aluminum ion	O^{2-}, $[(8+) + (10-)] = 2-$, oxide ion

SELF TEST 6.2

Consider the elements sulfur and potassium.

a. Identify each as a metal or a nonmetal.
b. State the number of valence electrons for each.
c. State the number of electrons that must be lost or gained for each to achieve an octet.
d. Write the symbol, including its ionic charge, and the name for each ion.

ANSWER

a. Sulfur is a nonmetal; potassium is a metal.
b. Sulfur has six valence electrons; potassium has one valence electron.
c. Sulfur will gain two electrons; potassium will lose one electron.
d. S^{2-}, sulfide ion; K^+, potassium ion

PRACTICE PROBLEMS

Try Practice Problems 6.9 to 6.14

PRACTICE PROBLEMS

6.1 Ions: Transfer of Electrons

6.1 State the number of electrons that must be lost by atoms of each of the following to achieve a stable electron configuration:
a. Li **b.** Ca **c.** Ga **d.** Cs **e.** Ba

6.2 State the number of electrons that must be gained by atoms of each of the following to achieve a stable electron configuration:
a. Cl **b.** Se **c.** N **d.** I **e.** S

6.3 State the number of electrons lost or gained when the following elements form ions:
a. Sr **b.** P **c.** Group 7A (17) **d.** Na **e.** Br

6.4 State the number of electrons lost or gained when the following elements form ions:
a. O **b.** Group 2A (2) **c.** F **d.** K **e.** Rb

6.5 Write the symbols for the ions with the following number of protons and electrons:
a. 3 protons, 2 electrons **b.** 9 protons, 10 electrons
c. 12 protons, 10 electrons **d.** 27 protons, 24 electrons

6.6 Write the symbols for the ions with the following number of protons and electrons:
a. 8 protons, 10 electrons **b.** 19 protons, 18 electrons
c. 35 protons, 36 electrons **d.** 50 protons, 46 electrons

6.7 State the number of protons and electrons in each of the following:
a. Cu^{2+} **b.** Se^{2-} **c.** Br^- **d.** Fe^{2+}

6.8 State the number of protons and electrons in each of the following:
a. S^{2-} **b.** Ni^{2+} **c.** Au^{3+} **d.** Ag^+

6.9 Write the symbol for the ion of each of the following:
a. chlorine **b.** cesium **c.** nitrogen **d.** radium

6.10 Write the symbol for the ion of each of the following:
a. fluorine **b.** barium **c.** sodium **d.** iodine

6.11 Write the names for each of the following ions:
a. Li^+ **b.** Ca^{2+} **c.** Ga^{3+} **d.** P^{3-}

6.12 Write the names for each of the following ions:
a. Rb^+ **b.** Sr^{2+} **c.** S^{2-} **d.** F^-

Applications

6.13 State the number of protons and electrons in each of the following ions:
a. O^{2-}, used to build biomolecules and water
b. K^+, most prevalent positive ion in cells; needed for muscle contraction, nerve impulses
c. I^-, needed for thyroid function
d. Ca^{2+}, needed for bones and teeth

6.14 State the number of protons and electrons in each of the following ions:
a. P^{3-}, needed for bones and teeth
b. F^-, used to strengthen tooth enamel
c. Mg^{2+}, needed for bones and teeth
d. Na^+, most prevalent positive ion in extracellular fluid

Chemistry Link to Health

Some Important Ions in the Body

Several ions in body fluids have important physiological and metabolic functions. Some are listed in TABLE 6.4.

Foods such as bananas, milk, cheese, and potatoes provide the body with ions that are important in regulating body functions.

Milk, cheese, bananas, cereal, and potatoes provide ions for the body.

TABLE 6.4 Ions in the Body

Ion	Occurrence	Function	Source	Result of Too Little	Result of Too Much
Na^+	Principal cation outside the cell	Regulates and controls body fluids	Salt, cheese, pickles	Hyponatremia, anxiety, diarrhea, circulatory failure, decrease in body fluid	Hypernatremia, little urine, thirst, edema
K^+	Principal cation inside the cell	Regulates body fluids and cellular functions	Bananas, orange juice, milk, prunes, potatoes	Hypokalemia (hypopotassemia), lethargy, muscle weakness, failure of neurological impulses	Hyperkalemia (hyperpotassemia), irritability, nausea, little urine, cardiac arrest
Ca^{2+}	Cation outside the cell; 90% of calcium in the body in bones	Major cation of bones; needed for muscle contraction	Milk, yogurt, cheese, greens, spinach	Hypocalcemia, tingling fingertips, muscle cramps, osteoporosis	Hypercalcemia, relaxed muscles, kidney stones, deep bone pain
Mg^{2+}	Cation outside the cell; 50% of magnesium in the body in bones	Essential for certain enzymes, muscles, nerve control	Widely distributed (part of chlorophyll of all green plants), nuts, whole grains	Disorientation, hypertension, tremors, slow pulse	Drowsiness
Cl^-	Principal anion outside the cell	Gastric juice, regulates body fluids	Salt	Same as for Na^+	Same as for Na^+

6.2 Ionic Compounds

LEARNING GOAL Using charge balance, write the correct formula for an ionic compound.

We use *ionic compounds* such as salt, NaCl, and baking soda, $NaHCO_3$, every day. Milk of magnesia, $Mg(OH)_2$, or calcium carbonate, $CaCO_3$, may be taken to settle an upset stomach. If you take a mineral supplement, iron may be present as iron(II) sulfate, $FeSO_4$, iodine as potassium iodide, KI, and manganese as manganese(II) sulfate, $MnSO_4$. Some sunscreens contain zinc oxide, ZnO, while tin(II) fluoride, SnF_2, in toothpaste provides fluoride to help prevent tooth decay. Gemstones are ionic compounds that are cut and polished to make jewelry. For example, sapphires and rubies are aluminum oxide, Al_2O_3. Impurities of chromium ions make rubies red, and iron and titanium ions make sapphires blue.

Rubies and sapphires are the ionic compound aluminum oxide, with chromium ions in rubies, and titanium and iron ions in sapphires.

Properties of Ionic Compounds

In an **ionic compound**, one or more electrons are transferred from metals to nonmetals, which form positive and negative ions. The attraction between these ions is called an ionic bond.

The physical and chemical properties of an ionic compound such as NaCl are very different from those of the original elements. For example, the original elements of NaCl were sodium, a soft, shiny metal, and chlorine, a yellow-green poisonous gas. However, when they react and form positive and negative ions, they produce NaCl, which is ordinary table salt, a hard, white, crystalline substance that is important in our diet.

In a crystal of NaCl, the larger Cl⁻ ions are arranged in a three-dimensional structure in which the smaller Na⁺ ions occupy the spaces between the Cl⁻ ions (see **FIGURE 6.1**). In this crystal, every Na⁺ ion is surrounded by six Cl⁻ ions, and every Cl⁻ ion is surrounded by six Na⁺ ions. Thus, there are many strong attractions between the positive and negative ions, which account for the high melting points of ionic compounds. For example, the melting point of NaCl is 801 °C. At room temperature, ionic compounds are solids.

PRACTICE PROBLEMS

Try Practice Problems 6.15 and 6.16

FIGURE 6.1 The elements sodium and chlorine react to form the ionic compound sodium chloride, the compound that makes up table salt. The magnification of NaCl crystals shows the arrangement of Na⁺ and Cl⁻ ions.

ENGAGE 6.4

What is the type of bonding between Na⁺ and Cl⁻ ions in NaCl?

Chemical Formulas of Ionic Compounds

The **chemical formula** of a compound represents the symbols and subscripts in the lowest whole-number ratio of the atoms or ions. In the formula of an ionic compound, the sum of the ionic charges in the formula is always zero. *Thus, the total amount of positive charge is equal to the total amount of negative charge.* For example, to achieve a stable electron configuration, one Na atom (metal) loses its one valence electron to form Na⁺, and one Cl atom (nonmetal) gains one electron to form a Cl⁻ ion. The formula NaCl indicates that the compound has charge balance because there is one sodium ion, Na⁺, for every chloride ion, Cl⁻. The ionic charges are not shown in the formula of the compound.

Subscripts in Formulas

Consider a compound of magnesium and chlorine. To achieve a stable electron configuration, one Mg atom (metal) loses its two valence electrons to form Mg^{2+}. Two Cl atoms (nonmetals) each gain one electron to form two Cl^- ions. The two Cl^- ions are needed to balance the positive charge of Mg^{2+}. This gives the formula $MgCl_2$, magnesium chloride, in which the subscript 2 shows that two Cl^- ions are needed for charge balance.

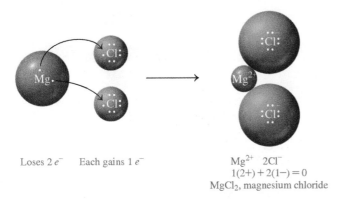

Loses 2 e^- Each gains 1 e^-

Mg^{2+} $2Cl^-$
$1(2+) + 2(1-) = 0$
$MgCl_2$, magnesium chloride

Writing Ionic Formulas from Ionic Charges

The subscripts in the formula of an ionic compound represent the number of positive and negative ions that give an overall charge of zero. Thus, we can now write a formula directly from the ionic charges of the positive and negative ions. Suppose we wish to write the formula for the ionic compound containing Na^+ and S^{2-} ions. To balance the ionic charge of the S^{2-} ion, we will need to place two Na^+ ions in the formula. This gives the formula Na_2S, which has an overall charge of zero. In the formula of an ionic compound, the cation is written first, followed by the anion. Appropriate subscripts are used to show the number of each of the ions. This formula is the lowest ratio of ions in the ionic compound. This lowest ratio of ions is called a *formula unit*.

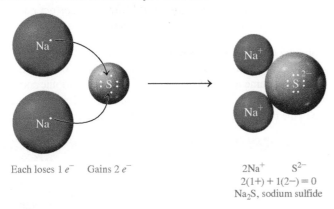

Each loses 1 e^- Gains 2 e^-

$2Na^+$ S^{2-}
$2(1+) + 1(2-) = 0$
Na_2S, sodium sulfide

SAMPLE PROBLEM 6.3   Writing Formulas from Ionic Charges

TRY IT FIRST

Write the symbols for the ions, and the correct formula for the ionic compound formed when lithium and nitrogen react.

SOLUTION

Lithium, which is a metal in Group 1A (1), forms Li^+; nitrogen, which is a nonmetal in Group 5A (15), forms N^{3-}. The charge of $3-$ is balanced by three Li^+ ions.

$$3(1+) + 1(3-) = 0$$

Writing the cation (positive ion) first and the anion (negative ion) second gives the formula Li_3N.

SELF TEST 6.3

Write the symbols for the ions, and the correct formula for the ionic compound that would form when each of the following react:
a. calcium and oxygen   **b.** magnesium and phosphorus

ANSWER
a. Ca^{2+}, O^{2-}, CaO   **b.** Mg^{2+}, P^{3-}, Mg_3P_2

PRACTICE PROBLEMS

Try Practice Problems 6.17 to 6.20

PRACTICE PROBLEMS

6.2 Ionic Compounds

6.15 Which of the following pairs of elements are likely to form an ionic compound?
  a. lithium and chlorine   **b.** oxygen and bromine
  c. potassium and oxygen   **d.** sodium and neon
  e. cesium and magnesium   **f.** nitrogen and fluorine

6.16 Which of the following pairs of elements are likely to form an ionic compound?
  a. helium and oxygen   **b.** magnesium and chlorine
  c. chlorine and bromine   **d.** potassium and sulfur
  e. sodium and potassium   **f.** nitrogen and iodine

6.17 Write the correct ionic formula for the compound formed between each of the following pairs of ions:
  a. Na^+ and O^{2-} **b.** Al^{3+} and Br^- **c.** Ba^{2+} and N^{3-}
  d. Mg^{2+} and F^- **e.** Al^{3+} and S^{2-}

6.18 Write the correct ionic formula for the compound formed between each of the following pairs of ions:
  a. Al^{3+} and Cl^- **b.** Ca^{2+} and S^{2-}
  c. Li^+ and S^{2-} **d.** Rb^+ and P^{3-}
  e. Cs^+ and I^-

6.19 Write the symbols for the ions, and the correct formula for the ionic compound formed by each of the following:
  a. potassium and sulfur   **b.** sodium and nitrogen
  c. aluminum and iodine   **d.** gallium and oxygen

6.20 Write the symbols for the ions, and the correct formula for the ionic compound formed by each of the following:
  a. calcium and chlorine   **b.** rubidium and bromine
  c. sodium and phosphorus   **d.** magnesium and oxygen

REVIEW

Solving Equations (1.4)

CORE CHEMISTRY SKILL

Naming Ionic Compounds

Iodized salt contains KI to prevent iodine deficiency.

ENGAGE 6.5

How are the positive and negative ions named in an ionic compound?

6.3 Naming and Writing Ionic Formulas

LEARNING GOAL Given the formula of an ionic compound, write the correct name; given the name of an ionic compound, write the correct formula.

In the name of an ionic compound made up of two elements, the name of the metal ion, which is written first, is the same as its element name. The name of the nonmetal ion is obtained by using the first syllable of its element name followed by *ide*. In the name of any ionic compound, a space separates the name of the cation from the name of the anion. Subscripts are not used; they are understood because of the charge balance of the ions in the compound (see **TABLE 6.5**).

TABLE 6.5 Names of Some Ionic Compounds

Compound	Metal Ion	Nonmetal Ion	Name
KI	K^+ Potassium	I^- Iodide	Potassium iodide
$MgBr_2$	Mg^{2+} Magnesium	Br^- Bromide	Magnesium bromide
Al_2O_3	Al^{3+} Aluminum	O^{2-} Oxide	Aluminum oxide

SAMPLE PROBLEM 6.4   Naming Ionic Compounds

TRY IT FIRST

Write the name for the ionic compound Mg_3N_2.

SOLUTION

ANALYZE THE PROBLEM	Given	Need	Connect
	Mg_3N_2	name	cation, anion

STEP 1 **Identify the cation and anion.** The cation is Mg^{2+} and the anion is N^{3-}.

STEP 2 **Name the cation by its element name.** The cation Mg^{2+} is magnesium.

STEP 3 **Name the anion by using the first syllable of its element name followed by *ide*.** The anion N^{3-} is nitride.

STEP 4 **Write the name for the cation first and the name for the anion second.** magnesium nitride

SELF TEST 6.4

Write the name for each of the following ionic compounds:
a. Ga_2S_3
b. Li_3P

ANSWER
a. gallium sulfide
b. lithium phosphide

PRACTICE PROBLEMS

Try Practice Problems 6.21 and 6.22

Metals with Variable Charges

We have seen that the charge of an ion of a representative element can be obtained from its group number. However, we cannot determine the charge of a transition element because it typically forms two or more positive ions. The transition elements lose electrons, but they are lost from the highest energy level and sometimes from a lower energy level as well. This is also true for metals of representative elements in Groups 4A (14) and 5A (15), such as Pb, Sn, and Bi.

In some ionic compounds, iron is in the Fe^{2+} form, but in other compounds, it has the Fe^{3+} form. Copper also forms two different ions, Cu^+ and Cu^{2+}. When a metal can form two or more types of ions, it has *variable charge*. Then we cannot predict the ionic charge from the group number.

For metals that form two or more ions, a naming system is used to identify the particular cation. A Roman numeral that is equal to the ionic charge is placed in parentheses immediately after the name of the metal. For example, Fe^{2+} is iron(II), and Fe^{3+} is iron(III). **TABLE 6.6** lists the ions of some metals that produce more than one ion.

The transition elements form more than one positive ion except for zinc (Zn^{2+}), cadmium (Cd^{2+}), and silver (Ag^+), which form only one ion. Thus, no Roman numerals are used with zinc, cadmium, and silver when naming their cations in ionic compounds. Metals in Groups 4A (14) and 5A (15) also form more than one type of positive ion. For example, lead and tin in Group 4A (14) form cations with charges of 2+ and 4+, and bismuth in Group 5A (15) forms cations with charges of 3+ and 5+.

Determination of Variable Charge

When you name an ionic compound, you need to determine if the metal is a representative element or a transition element. If it is a transition element, except for zinc, cadmium, or silver, you will need to use its ionic charge as a Roman numeral as part of its name. The calculation of ionic charge depends on the negative charge of the anions in the formula. For example, if we want to name the ionic compound $CuCl_2$, we use charge balance to determine the charge of the copper cation. Because there are two chloride ions, each with a $1-$ charge, the total negative charge is $2-$. To balance this $2-$ charge, the copper ion must have a charge of $2+$, or Cu^{2+}:

$CuCl_2$

$Cu^?$ Cl^-
          Cl^-
$$1(?) + 2(1-) = 0$$
$$? = 2+$$

To indicate the 2+ charge for the copper ion Cu^{2+}, we place the Roman numeral (II) immediately after copper when naming this compound: copper(II) chloride. Some ions and their location on the periodic table are seen in **FIGURE 6.2**.

ENGAGE 6.6

Why is a Roman numeral placed after the name of the cations of most transition elements?

TABLE 6.6 Some Metals That Form More Than One Positive Ion

Element	Positive Ions	Name of Ion
Bismuth	Bi^{3+}	Bismuth(III)
	Bi^{5+}	Bismuth(V)
Chromium	Cr^{2+}	Chromium(II)
	Cr^{3+}	Chromium(III)
Cobalt	Co^{2+}	Cobalt(II)
	Co^{3+}	Cobalt(III)
Copper	Cu^+	Copper(I)
	Cu^{2+}	Copper(II)
Gold	Au^+	Gold(I)
	Au^{3+}	Gold(III)
Iron	Fe^{2+}	Iron(II)
	Fe^{3+}	Iron(III)
Lead	Pb^{2+}	Lead(II)
	Pb^{4+}	Lead(IV)
Manganese	Mn^{2+}	Manganese(II)
	Mn^{3+}	Manganese(III)
Mercury	Hg_2^{2+}	Mercury(I)*
	Hg^{2+}	Mercury(II)
Nickel	Ni^{2+}	Nickel(II)
	Ni^{3+}	Nickel(III)
Tin	Sn^{2+}	Tin(II)
	Sn^{4+}	Tin(IV)

*Mercury(I) ions form an ion pair with a 2+ charge.

1 1A	2 2A	3 3B	4 4B	5 5B	6 6B	7 7B	8	9 8B	10	11 1B	12 2B	13 3A	14 4A	15 5A	16 6A	17 7A	18 8A
H^+																	
Li^+														N^{3-}	O^{2-}	F^-	
Na^+	Mg^{2+}											Al^{3+}		P^{3-}	S^{2-}	Cl^-	
K^+	Ca^{2+}				Cr^{2+} Cr^{3+}	Mn^{2+} Mn^{3+}	Fe^{2+} Fe^{3+}	Co^{2+} Co^{3+}	Ni^{2+} Ni^{3+}	Cu^+ Cu^{2+}	Zn^{2+}					Br^-	
Rb^+	Sr^{2+}									Ag^+	Cd^{2+}		Sn^{2+} Sn^{4+}			I^-	
Cs^+	Ba^{2+}									Au^+ Au^{3+}	Hg_2^{2+} Hg^{2+}		Pb^{2+} Pb^{4+}	Bi^{3+} Bi^{5+}			

Metals   Metalloids   Nonmetals

FIGURE 6.2 In ionic compounds, metals form positive ions, and nonmetals form negative ions.

TABLE 6.7 lists the names of some ionic compounds in which the transition elements and metals from Groups 4A (14) and 5A (15) have more than one positive ion.

ENGAGE 6.7

ENGAGE 6.7

What are the ions formed by copper and oxygen in ionic compounds?

TABLE 6.7 Some Ionic Compounds of Metals That Form Two Compounds

Compound	Systematic Name
$FeCl_2$	Iron(II) chloride
Fe_2O_3	Iron(III) oxide
Cu_3P	Copper(I) phosphide
$CrBr_2$	Chromium(II) bromide
$SnCl_2$	Tin(II) chloride
PbS_2	Lead(IV) sulfide
BiF_3	Bismuth(III) fluoride

SAMPLE PROBLEM 6.5   Naming Ionic Compounds with Variable Charge Metal Ions

TRY IT FIRST

Antifouling paint contains Cu_2O, which prevents the growth of barnacles and algae on the bottoms of boats. What is the name of Cu_2O?

SOLUTION

ANALYZE THE PROBLEM	Given	Need	Connect
	Cu_2O	name	cation, anion, charge balance

STEP 1   Determine the charge of the cation from the anion.

	Cation	Anion
Element	copper (Cu)	oxygen (O)
Group	transition element	6A (16)
Ion	$Cu^?$	O^{2-}
Calculation of Cation Charge	$2Cu^? + 2- = 0$ $2Cu^? = 2+$ $\dfrac{2Cu^?}{2} = \dfrac{2+}{2} = 1+$	
Ion	Cu^+	O^{2-}

STEP 2   Name the cation by its element name, and use a Roman numeral in parentheses for the charge. copper(I)

STEP 3   Name the anion by using the first syllable of its element name followed by *ide*. oxide

STEP 4   Write the name for the cation first and the name for the anion second. copper(I) oxide

The growth of barnacles is prevented by using a paint with Cu_2O on the bottom of a boat.

SELF TEST 6.5

Write the name for each of the following compounds:
a. Mn_2S_3 **b.** SnF_4

ANSWER

a. Mn_2S_3 with the variable charge metal ion Mn^{3+} is named manganese(III) sulfide.
b. SnF_4 with the variable charge metal ion Sn^{4+} is named tin(IV) fluoride.

PRACTICE PROBLEMS

Try Practice Problems 6.23 to 6.28

Writing Formulas from the Name of an Ionic Compound

The formula for an ionic compound is written from the first part of the name that describes the metal ion, including its charge, and the second part of the name that specifies the non-metal ion. Subscripts are added, as needed, to balance the charge. The steps for writing a formula from the name of an ionic compound are shown in Sample Problem 6.6.

INTERACTIVE VIDEO

Pearson eText
Naming and Writing
Ionic Formulas

SAMPLE PROBLEM 6.6   Writing Formulas for Ionic Compounds

TRY IT FIRST

Write the correct formula for iron(III) chloride.

SOLUTION

ANALYZE THE PROBLEM	Given	Need	Connect
	iron(III) chloride	formula	cation, anion, charge balance

STEP 1   Identify the cation and anion.

Type of Ion	Cation	Anion
Name	iron(III)	chloride
Symbol of Ion	Fe^{3+}	Cl^-

STEP 2   **Balance the charges.** The charge of 3+ is balanced by three Cl^- ions.
$$1(3+) + 3(1-) = 0$$

STEP 3   **Write the formula, cation first, using subscripts from the charge balance.** $FeCl_3$

SELF TEST 6.6

Write the correct formula for each of the following paint pigments:
a. chrome green, chromium(III) oxide   **b.** titanium white, titanium(IV) oxide

ANSWER

a. Cr_2O_3 **b.** TiO_2

The pigment chrome green contains chromium(III) oxide.

PRACTICE PROBLEMS

Try Practice Problems 6.29 to 6.34

PRACTICE PROBLEMS

6.3 Naming and Writing Ionic Formulas

6.21 Write the name for each of the following ionic compounds:
    a. AlF_3 **b.** $CaCl_2$
    c. Na_2O **d.** Mg_3P_2
    e. KI **f.** BaF_2

6.22 Write the name for each of the following ionic compounds:
    a. $MgCl_2$ **b.** K_3P
    c. Li_2S **d.** CsF
    e. MgO **f.** $SrBr_2$

6.23 Write the name for each of the following ions (include the Roman numeral when necessary):
    a. Fe^{2+} **b.** Cu^{2+} **c.** Zn^{2+}
    d. Pb^{4+} **e.** Cr^{3+} **f.** Mn^{2+}

6.24 Write the name for each of the following ions (include the Roman numeral when necessary):
    a. Ag^+ **b.** Cu^+ **c.** Bi^{3+}
    d. Sn^{2+} **e.** Au^{3+} **f.** Ni^{2+}

6.25 Write the name for each of the following ionic compounds:
    a. $SnCl_2$ **b.** FeO **c.** Cu_2S
    d. CuS **e.** $CdBr_2$ **f.** $HgCl_2$

6.26 Write the name for each of the following ionic compounds:
    a. Ag_3P **b.** PbS **c.** SnO_2
    d. $MnCl_3$ **e.** Bi_2O_3 **f.** $CoCl_2$

6.27 Write the symbol for the cation in each of the following ionic compounds:
    a. $AuCl_3$ **b.** Fe_2O_3 **c.** PbI_4 **d.** AlP

6.28 Write the symbol for the cation in each of the following ionic compounds:
    a. $FeCl_2$ **b.** CrO **c.** Ni_2S_3 **d.** $SnCl_2$

6.29 Write the formula for each of the following ionic compounds:
    a. magnesium chloride   **b.** sodium sulfide
    c. copper(I) oxide   **d.** zinc phosphide
    e. gold(III) nitride   **f.** cobalt(III) fluoride

6.30 Write the formula for each of the following ionic compounds:
    a. nickel(III) oxide   **b.** barium fluoride
    c. tin(IV) chloride   **d.** silver sulfide
    e. bismuth(V) chloride   **f.** potassium nitride

6.31 Write the formula for each of the following ionic compounds:
    a. cobalt(III) chloride   **b.** lead(IV) oxide
    c. silver iodide   **d.** calcium nitride
    e. copper(I) phosphide   **f.** chromium(II) chloride

6.32 Write the formula for each of the following ionic compounds:
    a. zinc bromide   **b.** iron(III) sulfide
    c. manganese(IV) oxide   **d.** chromium(III) iodide
    e. lithium nitride   **f.** gold(I) oxide

Applications

6.33 The following compounds contain ions that are required in small amounts by the body. Write the formula for each.
    a. potassium phosphide   **b.** copper(II) chloride
    c. iron(III) bromide   **d.** magnesium oxide

6.34 The following compounds contain ions that are required in small amounts by the body. Write the formula for each.
    a. calcium chloride   **b.** nickel(II) iodide
    c. manganese(II) oxide   **d.** zinc nitride

6.4 Polyatomic Ions

LEARNING GOAL Write the name and formula for an ionic compound containing a polyatomic ion.

An ionic compound may also contain a *polyatomic ion* as one of its cations or anions. A **polyatomic ion** is a group of covalently bonded atoms that has an overall ionic charge. Most polyatomic ions consist of a nonmetal such as phosphorus, sulfur, carbon, or nitrogen covalently bonded to oxygen atoms.

Almost all the polyatomic ions are anions with charges $1-$, $2-$, or $3-$. Only one common polyatomic ion, NH_4^+, has a positive charge. Some models of common polyatomic ions are shown in **FIGURE 6.3**.

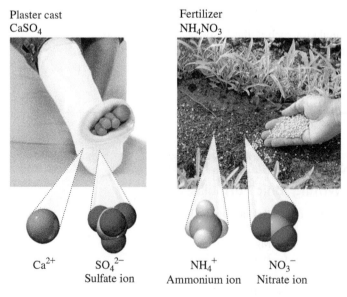

Plaster cast
$CaSO_4$

Fertilizer
NH_4NO_3

Ca^{2+} SO_4^{2-} NH_4^+ NO_3^-
Sulfate ion   Ammonium ion   Nitrate ion

FIGURE 6.3 Many products contain polyatomic ions, which are groups of atoms that have an ionic charge.

Names of Polyatomic Ions

The names of the most common polyatomic ions, which end in *ate*, are shown in bold in TABLE 6.8. When a related ion has one less O atom, the *ite* ending is used. For the same nonmetal, the *ate* ion and the *ite* ion have the same charge. For example, the sulfate ion is SO_4^{2-}, and the sulfite ion, which has one less oxygen atom, is SO_3^{2-}.

Formula	Charge	Name
SO_4^{2-}	2−	sul**fate**
SO_3^{2-}	2−	sul**fite**

Phosphate and phosphite ions each have a 3− charge.

Formula	Charge	Name
PO_4^{3-}	3−	phos**phate**
PO_3^{3-}	3−	phos**phite**

The formula of hydrogen carbonate, or *bicarbonate*, is written by placing a hydrogen in front of the polyatomic ion formula for carbonate (CO_3^{2-}), and the charge is decreased from 2− to 1− to give HCO_3^{-}.

Formula	Charge	Name
CO_3^{2-}	2−	carbon**ate**
HCO_3^{-}	1−	hydrogen carbon**ate**

The halogens form four different polyatomic ions with oxygen. The prefix *per* is used for one more O than the *ate* ion and the prefix *hypo* for one less O than the *ite* ion.

Formula	Charge	Name
ClO_4^{-}	1−	**per**chlor**ate**
ClO_3^{-}	1−	chlor**ate**
ClO_2^{-}	1−	chlor**ite**
ClO^{-}	1−	**hypo**chlor**ite**

Recognizing these prefixes and endings will help you identify polyatomic ions in the name of a compound. The hydroxide ion (OH^{-}) and cyanide ion (CN^{-}) are exceptions to this naming pattern.

ENGAGE 6.8

How do the number of O atoms in a phosphate ion compare to the number of O atoms in a phosphite ion?

TABLE 6.8 Names and Formulas of Some Common Polyatomic Ions

Nonmetal	Formula of Ion*	Name of Ion
Hydrogen	OH^{-}	Hydroxide
Nitrogen	NH_4^{+}	Ammonium
	NO_3^{-}	**Nitrate**
	NO_2^{-}	Nitrite
Chlorine	ClO_4^{-}	Perchlorate
	ClO_3^{-}	**Chlorate**
	ClO_2^{-}	Chlorite
	ClO^{-}	Hypochlorite
Carbon	CO_3^{2-}	**Carbonate**
	HCO_3^{-}	Hydrogen carbonate (or bicarbonate)
	CN^{-}	Cyanide
	$C_2H_3O_2^{-}$	Acetate
Sulfur	SO_4^{2-}	**Sulfate**
	HSO_4^{-}	Hydrogen sulfate (or bisulfate)
	SO_3^{2-}	Sulfite
	HSO_3^{-}	Hydrogen sulfite (or bisulfite)
Phosphorus	PO_4^{3-}	**Phosphate**
	HPO_4^{2-}	Hydrogen phosphate
	$H_2PO_4^{-}$	Dihydrogen phosphate
	PO_3^{3-}	Phosphite

*Formulas and names in bold type indicate the most common polyatomic ion for that element.

PRACTICE PROBLEMS

Try Practice Problems 6.35 to 6.38

Sodium chlorite is used as a disinfectant in contact lens cleaning solutions, mouthwashes, and as a medication to treat amyotrophic lateral sclerosis (ALS).

Writing Formulas for Compounds Containing Polyatomic Ions

No polyatomic ion exists by itself. Like any ion, a polyatomic ion must be associated with ions of opposite charge. The bonding between polyatomic ions and other ions is one of electrical attraction.

To write correct formulas for compounds containing polyatomic ions, we follow the same rules of charge balance that we used for writing the formulas for simple ionic compounds. The total negative and positive charges must equal zero. For example, consider the formula for a compound containing sodium ions and chlorite ions. The ions are written as

$$\underset{\text{Sodium ion}}{Na^+} \quad \underset{\text{Chlorite ion}}{ClO_2^-}$$
$$(1+) + (1-) = 0$$

Because one ion of each balances the charge, the formula is written as

$$\underset{\text{Sodium chlorite}}{NaClO_2}$$

When more than one polyatomic ion is needed for charge balance, parentheses are used to enclose the formula of the ion. A subscript is written outside the right parenthesis of the polyatomic ion to indicate the number needed for charge balance. Consider the formula for magnesium nitrate. The ions in this compound are the magnesium ion and the nitrate ion, a polyatomic ion.

$$\underset{\text{Magnesium ion}}{Mg^{2+}} \quad \underset{\text{Nitrate ion}}{NO_3^-}$$

To balance the positive charge of $2+$ of the magnesium ion, two nitrate ions are needed. In the formula of the compound, parentheses are placed around the nitrate ion, and the subscript 2 is written outside the right parenthesis.

ENGAGE 6.9

Why does the formula for magnesium nitrate contain two NO_3^- ions?

$$Mg^{2+} \quad \begin{matrix} NO_3^- \\ \\ NO_3^- \end{matrix} \qquad \underset{Mg(NO_3)_2}{\text{Magnesium nitrate}}$$
$$(2+) + 2(1-) = 0$$

Parentheses enclose the formula of the nitrate ion

Subscript outside the parenthesis indicates the use of two nitrate ions

---

SAMPLE PROBLEM 6.7 Writing Formulas Containing Polyatomic Ions

TRY IT FIRST

Antacid tablets contain aluminum hydroxide, which treats acid indigestion and heartburn. Write the formula for aluminum hydroxide.

SOLUTION

ANALYZE THE PROBLEM	Given	Need	Connect
	aluminum hydroxide	formula	cation, polyatomic anion, charge balance

Aluminum hydroxide is an antacid used to treat acid indigestion.

STEP 1 **Identify the cation and polyatomic ion (anion).**

Cation	Polyatomic Ion (anion)
aluminum	hydroxide
Al^{3+}	OH^-

STEP 2 **Balance the charges.** The charge of $3+$ is balanced by three OH^- ions.

$$\mathbf{1}(3+) + \mathbf{3}(1-) = 0$$

STEP 3   **Write the formula, cation first, using the subscripts from charge balance.**
The formula for the compound is written by enclosing the formula of the hydroxide ion, OH^-, in parentheses and writing the subscript 3 outside the right parenthesis.

$Al(OH)_3$

SELF TEST 6.7

Write the formula for a compound containing the following:

a. ammonium ions and phosphate ions   **b.** iron(III) ions and bicarbonate ions

ANSWER

a. $(NH_4)_3PO_4$ **b.** $Fe(HCO_3)_3$

Naming Ionic Compounds Containing Polyatomic Ions

When naming ionic compounds containing polyatomic ions, we first write the positive ion, usually a metal, and then we write the name for the polyatomic ion. It is important that you learn to recognize the polyatomic ion in the formula and name it correctly.

Na_2SO_4 $FePO_4$ $Al_2(CO_3)_3$

$Na_2\boxed{SO_4}$ $Fe\boxed{PO_4}$ $Al_2(\boxed{CO_3})_3$

Sodium sulfate   Iron(III) phosphate   Aluminum carbonate

TABLE 6.9 lists the formulas and names of some ionic compounds that include polyatomic ions and also gives their uses.

TABLE 6.9 Some Ionic Compounds That Contain Polyatomic Ions

Formula	Name	Medical Use
$AlPO_4$	Aluminum phosphate	Antacid
$Al_2(SO_4)_3$	Aluminum sulfate	Used as pH stabilizer in paper industry
$BaSO_4$	Barium sulfate	Contrast medium for X-rays
$CaCO_3$	Calcium carbonate	Antacid, calcium supplement
$Ca_3(PO_4)_2$	Calcium phosphate	Calcium dietary supplement
$Cr_2(SO_4)_3$	Chromium(III) sulfate	Green pigment for paints and inks
$MgSO_4$	Magnesium sulfate	Epsom salts, used to sooth sore muscles
K_2CO_3	Potassium carbonate	Alkalizer, diuretic
$AgNO_3$	Silver nitrate	Topical anti-infective
Na_2CO_3	Sodium carbonate	Manufacture of glass
$NaHCO_3$	Sodium hydrogen carbonate or sodium bicarbonate	Antacid, used to make baking powder
$Zn_3(PO_4)_2$	Zinc phosphate	Dental cement

A solution of Epsom salts, magnesium sulfate, $MgSO_4$, may be used to soothe sore muscles.

SAMPLE PROBLEM 6.8   Naming Compounds Containing Polyatomic Ions

TRY IT FIRST

Name the following ionic compounds:

a. $Cu(NO_3)_2$, used in fireworks

b. $KClO_3$, used to produce oxygen in aircraft and space stations

SOLUTION

	Given	Need	Connect
ANALYZE THE PROBLEM	formula	name	cation, polyatomic ion

Formula	Cation	Anion	Name of Cation	Name of Anion	Name of Compound
a. $Cu(NO_3)_2$	Cu^{2+}	NO_3^-	Copper(II) ion	Nitrate ion	Copper(II) nitrate
b. $KClO_3$	K^+	ClO_3^-	Potassium ion	Chlorate ion	Potassium chlorate

SELF TEST 6.8

Name each of the following compounds:
a. $Co_3(PO_4)_2$, a pigment called cobalt violet
b. $NaHCO_3$, corrects pH imbalance

PRACTICE PROBLEMS

Try Practice Problems 6.39 to 6.46

ANSWER
a. cobalt(II) phosphate   **b.**   sodium hydrogen carbonate or sodium bicarbonate

PRACTICE PROBLEMS

6.4 Polyatomic Ions

6.35 Write the formula including the charge for each of the following polyatomic ions:
a. hydrogen carbonate (bicarbonate)   **b.** ammonium
c. phosphite   **d.** chlorate

6.36 Write the formula including the charge for each of the following polyatomic ions:
a. nitrite   **b.** sulfite   **c.** hydroxide   **d.** acetate

6.37 Name the following polyatomic ions:
a. SO_4^{2-} **b.** CO_3^{2-} **c.** HSO_3^- **d.** NO_3^-

6.38 Name the following polyatomic ions:
a. OH^- **b.** PO_4^{3-} **c.** CN^- **d.** NO_2^-

6.39 Complete the following table with the formula and name of the compound that forms between each pair of ions:

	NO_2^-	CO_3^{2-}	HSO_4^-	PO_4^{3-}
Li^+				
Cu^{2+}				
Ba^{2+}				

6.40 Complete the following table with the formula and name of the compound that forms between each pair of ions:

	NO_3^-	HCO_3^-	SO_3^{2-}	HPO_4^{2-}
NH_4^+				
Al^{3+}				
Pb^{4+}				

6.41 Write the correct formula for the following ionic compounds:
a. barium hydroxide   **b.** sodium hydrogen sulfate
c. iron(II) nitrite   **d.** zinc phosphate
e. iron(III) carbonate

6.42 Write the correct formula for the following ionic compounds:
a. aluminum chlorate   **b.** ammonium oxide
c. magnesium bicarbonate   **d.** sodium nitrite
e. copper(I) sulfate

6.43 Write the formula for the polyatomic ion and name each of the following compounds:
a. Na_2CO_3 **b.** $(NH_4)_2S$
c. $Ca(OH)_2$ **d.** $Sn(NO_2)_2$

6.44 Write the formula for the polyatomic ion and name each of the following compounds:
a. $MnCO_3$ **b.** Au_2SO_4
c. $Ca_3(PO_4)_2$ **d.** $Fe(HCO_3)_3$

Applications

6.45 Name each of the following ionic compounds:
a. $Zn(C_2H_3O_2)_2$, cold remedy
b. $Mg_3(PO_4)_2$, antacid
c. NH_4Cl, expectorant
d. $Sr(NO_3)_2$, produces red color in fireworks
e. $NaNO_2$, meat preservative

6.46 Name each of the following ionic compounds:
a. Li_2CO_3, antidepressant
b. $MgSO_4$, Epsom salts
c. $NaClO$, disinfectant
d. Na_3PO_4, laxative
e. $Ba(OH)_2$, component of antacids

6.5 Molecular Compounds: Sharing Electrons

LEARNING GOAL Given the formula of a molecular compound, write its correct name; given the name of a molecular compound, write its formula.

A **molecular compound** consists of atoms of two or more nonmetals that share one or more valence electrons. The atoms are held together by **covalent bonds** that form a *molecule*. There are many more molecular compounds than there are ionic ones. For example, water (H_2O) and carbon dioxide (CO_2) are both molecular compounds. Molecular compounds consist of **molecules**, which are discrete groups of atoms in a definite proportion. A molecule of water (H_2O) consists of two atoms of hydrogen and one atom of oxygen. When you have iced tea, perhaps you add molecules of sugar ($C_{12}H_{22}O_{11}$), which is a molecular compound. Other familiar molecular compounds include propane (C_3H_8), alcohol (C_2H_6O), the antibiotic amoxicillin ($C_{16}H_{19}N_3O_5S$), and the antidepressant Prozac ($C_{17}H_{18}F_3NO$).

Names and Formulas of Molecular Compounds

When naming a molecular compound, the first nonmetal in the formula is named by its element name; the second nonmetal is named using the first syllable of its element name, followed by *ide*. When a subscript indicates two or more atoms of an element, a prefix is shown in front of its name. TABLE 6.10 lists prefixes used in naming molecular compounds.

The names of molecular compounds need prefixes because several different compounds can be formed from the same two nonmetals. For example, carbon and oxygen can form two different compounds, carbon monoxide, CO, and carbon dioxide, CO_2, in which the number of atoms of oxygen in each compound is indicated by the prefixes *mono* or *di* in their names.

When the vowels *o* and *o* or *a* and *o* appear together, the first vowel is omitted, as in carbon monoxide. In the name of a molecular compound, the prefix *mono* is usually omitted, as in NO, nitrogen oxide. Traditionally, however, CO is named carbon monoxide. TABLE 6.11 lists the formulas, names, and commercial uses of some molecular compounds.

TABLE 6.10 Prefixes Used in Naming Molecular Compounds

1 mono	6 hexa
2 di	7 hepta
3 tri	8 octa
4 tetra	9 nona
5 penta	10 deca

TABLE 6.11 Some Common Molecular Compounds

Formula	Name	Commercial Uses
CO_2	Carbon dioxide	Fire extinguishers, dry ice, propellant in aerosols, carbonation of beverages
CS_2	Carbon disulfide	Manufacture of rayon
NO	Nitrogen oxide	Stabilizer, biochemical messenger in cells
N_2O	Dinitrogen oxide	Inhalation anesthetic, "laughing gas"
SF_6	Sulfur hexafluoride	Electrical circuits
SO_2	Sulfur dioxide	Preserving fruits, vegetables; disinfectant in breweries; bleaching textiles
SO_3	Sulfur trioxide	Manufacture of explosives

SAMPLE PROBLEM 6.9 Naming Molecular Compounds

TRY IT FIRST

Name the molecular compound NCl_3.

SOLUTION

ANALYZE THE PROBLEM	Given	Need	Connect
	NCl_3	name	prefixes

STEP 1 Name the first nonmetal by its element name. In NCl_3, the first nonmetal (N) is nitrogen.

STEP 2 Name the second nonmetal by using the first syllable of its element name followed by *ide.* The second nonmetal (Cl) is chloride.

STEP 3 Add prefixes to indicate the number of atoms (subscripts). Because there is one nitrogen atom, no prefix is needed. The subscript 3 for the Cl atoms is shown as the prefix *tri*. The name of NCl_3 is nitrogen trichloride.

ENGAGE 6.10
Why are prefixes used to name molecular compounds?

SELF TEST 6.9

Name each of the following molecular compounds:

a. $SiBr_4$ b. Br_2O c. S_3N_2

PRACTICE PROBLEMS

Try Practice Problems 6.47 to 6.50

ANSWER

a. silicon tetrabromide   b. dibromine oxide   c. trisulfur dinitride

Writing Formulas from the Names of Molecular Compounds

In the name of a molecular compound, the names of two nonmetals are given along with prefixes for the number of atoms of each. To write the formula from the name, we use the symbol for each element and a subscript if a prefix indicates two or more atoms.

SAMPLE PROBLEM 6.10   Writing Formulas for Molecular Compounds

TRY IT FIRST

Write the formula for the molecular compound diboron trioxide.

SOLUTION

ANALYZE THE PROBLEM	Given	Need	Connect
	diboron trioxide	formula	subscripts from prefixes

STEP 1   **Write the symbols in the order of the elements in the name.**

Name of Element	boron	oxygen
Symbol of Element	B	O

STEP 2   **Write any prefixes as subscripts.** The prefix *di* in *di*boron indicates that there are two atoms of boron, shown as a subscript 2 in the formula. The prefix *tri* in *tri*oxide indicates that there are three atoms of oxygen, shown as a subscript 3 in the formula. B_2O_3

SELF TEST 6.10

Write the formula for each of the following molecular compounds:

a. iodine pentafluoride   b. carbon diselenide

PRACTICE PROBLEMS

Try Practice Problems 6.51 to 6.54

ANSWER

a. IF_5 b. CSe_2

Summary of Naming Ionic and Molecular Compounds

We have now examined strategies for naming ionic and molecular compounds. In general, compounds having two elements are named by stating the first element name followed by the name of the second element with an *ide* ending. If the first element is a metal, the compound is usually ionic; if the first element is a nonmetal, the compound is usually molecular. For ionic compounds, it is necessary to determine whether the metal can form more than one type of positive ion; if so, a Roman numeral following the name of the metal indicates the particular ionic charge. One exception is the ammonium ion, NH_4^+, which is also written first as a positively charged polyatomic ion. Ionic compounds having three or more elements include some type of polyatomic ion. They are named by ionic rules but have an *ate* or *ite* ending when the polyatomic ion has a negative charge.

In naming molecular compounds having two elements, prefixes are necessary to indicate two or more atoms of each nonmetal as shown in that particular formula (see **FIGURE 6.4**).

SAMPLE PROBLEM 6.11 Naming Ionic and Molecular Compounds

TRY IT FIRST

Identify each of the following compounds as ionic or molecular and write its name:
a. $NiSO_4$
b. SO_3

SOLUTION

a. $NiSO_4$, consisting of a cation of a transition element and a polyatomic ion SO_4^{2-}, is an ionic compound. As a transition element, Ni forms more than one type of ion. In this formula, the $2-$ charge of SO_4^{2-} is balanced by one nickel ion, Ni^{2+}. In the name, a Roman numeral written after the metal name, nickel(II), specifies the $2+$ charge. The anion SO_4^{2-} is a polyatomic ion named sulfate. The compound is named nickel(II) sulfate.

b. SO_3 consists of two nonmetals, which indicates that it is a molecular compound. The first element S is sulfur (no prefix is needed). The second element O, oxide, has subscript 3, which requires a prefix *tri* in the name. The compound is named sulfur trioxide.

ENGAGE 6.11

How do you know that sodium phosphate is an ionic compound and diphosphorus pentoxide a molecular compound?

SELF TEST 6.11

Name each of the following compounds:
a. IF_7
b. $Fe(NO_3)_3$

ANSWER

a. iodine heptafluoride
b. iron(III) nitrate

PRACTICE PROBLEMS

Try Practice Problems 6.55 and 6.56

ENGAGE 6.12

When are the names of some metal ions followed by a Roman numeral in the name of a compound?

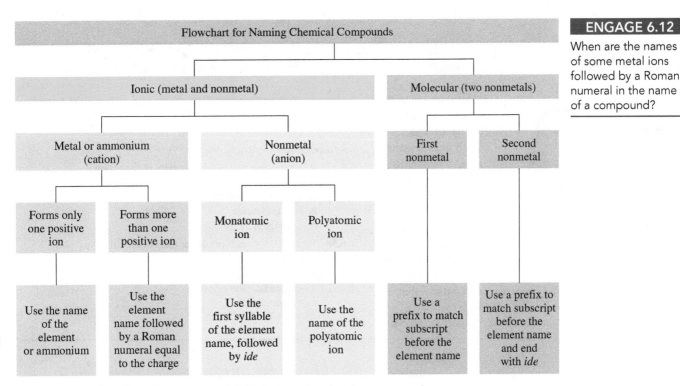

FIGURE 6.4 A flowchart illustrates naming for ionic and molecular compounds.

PRACTICE PROBLEMS

6.5 Molecular Compounds: Sharing Electrons

6.47 Name each of the following molecular compounds:
a. PBr_3
b. Cl_2O
c. CBr_4
d. HF
e. NF_3

6.48 Name each of the following molecular compounds:
a. IF_3
b. P_2O_5
c. SiO_2
d. PCl_3
e. CO

6.49 Name each of the following molecular compounds:
a. N_2O_3
b. Si_2Br_6
c. P_4S_3
d. PCl_5
e. SeF_6

6.50 Name each of the following molecular compounds:
a. SiF_4
b. IBr_3
c. CO_2
d. N_2F_2
e. N_2S_3

6.51 Write the formula for each of the following molecular compounds:
a. carbon tetrachloride
b. carbon monoxide
c. phosphorus trifluoride
d. dinitrogen tetroxide

6.52 Write the formula for each of the following molecular compounds:
a. sulfur dioxide
b. silicon tetrachloride
c. iodine trifluoride
d. dinitrogen oxide

6.53 Write the formula for each of the following molecular compounds:
a. oxygen difluoride
b. boron trichloride
c. dinitrogen trioxide
d. sulfur hexafluoride

6.54 Write the formula for each of the following molecular compounds:
a. sulfur dibromide
b. carbon disulfide
c. tetraphosphorus hexoxide
d. dinitrogen pentoxide

Applications

6.55 Name each of the following ionic or molecular compounds:
a. $Al_2(SO_4)_3$, antiperspirant
b. $Zn_3(PO_4)_2$, dental cement
c. N_2O, "laughing gas," inhaled anesthetic
d. $Mg(OH)_2$, laxative

6.56 Name each of the following ionic or molecular compounds:
a. $(NH_4)_3PO_4$, used as a fertilizer
b. $BaSO_4$, used to create images of the esophagus and stomach
c. NO, vasodilator
d. $Cu(OH)_2$, fungicide

UPDATE   Compounds at the Pharmacy

Richard returns to the pharmacy to talk to Dr. Chavez about a way to treat his sore toe. She recommends soaking his foot in a solution of Epsom salts, which is magnesium sulfate. Richard also asks Dr. Chavez to recommend an antacid for his upset stomach and an iron supplement. She suggests an antacid that contains calcium carbonate and aluminum hydroxide, and iron(II) sulfate as an iron supplement. Richard also picks up toothpaste containing tin(II) fluoride, and carbonated water, which contains carbon dioxide.

Applications

6.57 Write the correct formula for each of the following compounds:
a. magnesium sulfate
b. tin(II) fluoride
c. aluminum hydroxide

6.58 Write the correct formula for each of the following compounds:
a. iron(II) sulfate
b. calcium carbonate
c. carbon dioxide

CONCEPT MAP

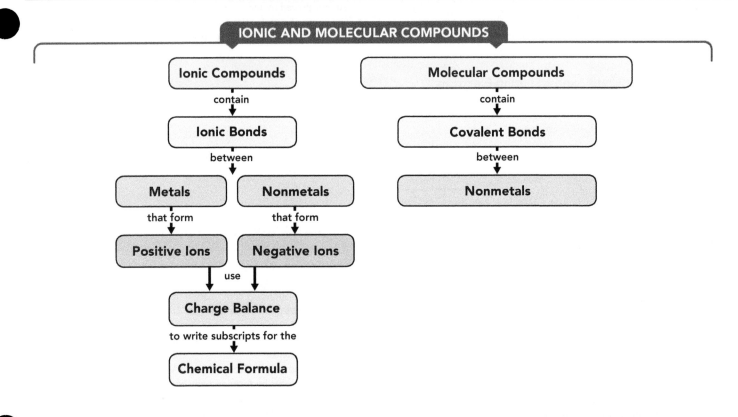

IONIC AND MOLECULAR COMPOUNDS

Ionic Compounds — contain → **Ionic Bonds** — between → **Metals** / **Nonmetals**

Metals — that form → **Positive Ions**

Nonmetals — that form → **Negative Ions**

Positive Ions / **Negative Ions** — use → **Charge Balance** — to write subscripts for the → **Chemical Formula**

Molecular Compounds — contain → **Covalent Bonds** — between → **Nonmetals**

CHAPTER REVIEW

6.1 Ions: Transfer of Electrons

LEARNING GOAL Write the symbols for the simple ions of the representative elements.

Transfer of electrons

Ionic bond

- The stability of the noble gases is associated with a stable electron configuration in the outermost energy level.
- With the exception of helium, which has two electrons, noble gases have eight valence electrons, which is an octet.
- Atoms of elements in Groups 1A to 7A (1, 2, 13 to 17) achieve stability by losing, gaining, or sharing their valence electrons in the formation of compounds.
- Metals of the representative elements lose valence electrons to form positively charged ions (cations): Groups 1A (1), 1+, 2A (2), 2+, and 3A (13), 3+.
- When reacting with metals, nonmetals gain electrons to form octets and form negatively charged ions (anions): Groups 5A (15), 3−, 6A (16), 2−, and 7A (17), 1−.

6.2 Ionic Compounds

LEARNING GOAL Using charge balance, write the correct formula for an ionic compound.

- The total positive and negative ionic charge is balanced in the formula of an ionic compound.
- Charge balance in a formula is achieved by using subscripts after each symbol so that the overall charge is zero.

Sodium chloride

6.3 Naming and Writing Ionic Formulas

LEARNING GOAL Given the formula of an ionic compound, write the correct name; given the name of an ionic compound, write the correct formula.

Metals   Metalloids   Nonmetals

- In naming ionic compounds, the positive ion is given first followed by the name of the negative ion.
- The names of ionic compounds containing two elements end with *ide*.
- Except for Ag, Cd, and Zn, transition elements form cations with two or more ionic charges.
- The charge of the cation is determined from the total negative charge in the formula and included as a Roman numeral immediately following the name of the metal that has a variable charge.

6.4 Polyatomic Ions

LEARNING GOAL Write the name and formula for an ionic compound containing a polyatomic ion.

- A polyatomic ion is a covalently bonded group of atoms with an electrical charge; for example, the carbonate ion has the formula CO_3^{2-}.

Ca^{2+} SO_4^{2-}
Sulfate ion

- Most polyatomic ions have names that end with *ate* or *ite*.
- Most polyatomic ions contain a nonmetal and one or more oxygen atoms.
- The ammonium ion, NH_4^+, is a positive polyatomic ion.
- When more than one polyatomic ion is used for charge balance, parentheses enclose the formula of the polyatomic ion.

- The first nonmetal in a molecular compound uses its element name; the second nonmetal uses the first syllable of its element name followed by *ide*.
- The name of a molecular compound with two different atoms uses prefixes to indicate the subscripts in the formula.

6.5 Molecular Compounds: Sharing Electrons

LEARNING GOAL Given the formula of a molecular compound, write its correct name; given the name of a molecular compound, write its formula.

- In a covalent bond, atoms of nonmetals share valence electrons such that each atom has a stable electron configuration.

1	mono
2	di
3	tri
4	tetra
5	penta

KEY TERMS

anion A negatively charged ion such as Cl^-, O^{2-}, or SO_4^{2-}.

cation A positively charged ion such as Na^+, Mg^{2+}, Al^{3+}, or NH_4^+.

chemical formula The group of symbols and subscripts that represents the atoms or ions in a compound.

covalent bond A sharing of valence electrons by atoms.

ion An atom or group of atoms having an electrical charge because of a loss or gain of electrons.

ionic bond The attraction between a positive ion and a negative ion when electrons are transferred from a metal to a nonmetal.

ionic charge The difference between the number of protons (positive) and the number of electrons (negative) written in the upper right corner of the symbol for the element or polyatomic ion.

ionic compound A compound of positive and negative ions held together by ionic bonds.

molecular compound A combination of atoms in which stable electron configurations are attained by sharing electrons.

molecule The smallest unit of two or more atoms held together by covalent bonds.

octet rule Elements in Groups 1A to 7A (1, 2, 13 to 17) react with other elements by forming ionic or covalent bonds to produce a stable electron configuration.

polyatomic ion A group of covalently bonded nonmetal atoms that has an overall electrical charge.

CORE CHEMISTRY SKILLS

The chapter section containing each Core Chemistry Skill is shown in parentheses at the end of each heading.

Writing Positive and Negative Ions (6.1)

- In the formation of an ionic bond, atoms of a metal lose and atoms of a nonmetal gain valence electrons to acquire a stable electron configuration, usually eight valence electrons.
- This tendency of atoms to attain a stable electron configuration is known as the octet rule.

Example: State the number of electrons lost or gained by atoms and the ion formed for each of the following to obtain a stable electron configuration:

   a. Br   **b.** Ca   **c.** S

Answer: **a.** Br atoms gain one electron to achieve a stable electron configuration, Br^-.
   b. Ca atoms lose two electrons to achieve a stable electron configuration, Ca^{2+}.
   c. S atoms gain two electrons to achieve a stable electron configuration, S^{2-}.

Writing Ionic Formulas (6.2)

- The chemical formula of a compound represents the lowest whole-number ratio of the atoms or ions.
- In the chemical formula of an ionic compound, the sum of the positive and negative charges is always zero.
- In a chemical formula of an ionic compound, the total positive charge is equal to the total negative charge.

Example: Write the formula for magnesium phosphide.

Answer: Magnesium phosphide is an ionic compound that contains the ions Mg^{2+} and P^{3-}.
Using charge balance, we determine the number(s) of each type of ion.
$3(2+) + 2(3-) = 0$
$3Mg^{2+}$ and $2P^{3-}$ give the formula Mg_3P_2.

Naming Ionic Compounds (6.3)

- In the name of an ionic compound made up of two elements, the name of the metal ion, which is written first, is the same as its element name.
- For metals that form two or more ions, a Roman numeral that is equal to the ionic charge is placed in parentheses immediately after the name of the metal.
- The name of a nonmetal ion is obtained by using the first syllable of its element name followed by *ide*.
- The name of a polyatomic ion ends in *ate* or *ite*.

Example: What is the name of $PbSO_4$?

Answer: This compound contains the SO_4^{2-} ion, which has a 2− charge. For charge balance, the positive ion must have a charge of 2+.
$Pb^? + (2-) = 0$ $Pb = 2+$
Because lead can form two different positive ions, a Roman numeral (II) is used in the name of the compound: lead(II) sulfate.

Writing the Names and Formulas for Molecular Compounds (6.5)

- When naming a molecular compound, the first nonmetal in the formula is named by its element name; the second nonmetal is named using the first syllable of its element name followed by *ide*.
- When a subscript indicates two or more atoms of an element, a prefix is shown in front of its name.

Example: Name the molecular compound BrF_5.

Answer: Two nonmetals share electrons and form a molecular compound. Br (first nonmetal) is bromine; F (second nonmetal) is fluoride. In the name for a molecular compound, prefixes indicate the subscripts in the formulas. The subscript 1 is understood for Br. The subscript 5 for fluoride is written with the prefix *penta*. The name is bromine pentafluoride.

UNDERSTANDING THE CONCEPTS

The chapter sections to review are shown in parentheses at the end of each problem.

6.59 a. How does the octet rule explain the formation of a magnesium ion? (6.1)
b. What noble gas has the same electron configuration as the magnesium ion?
c. Why are Group 1A (1) and Group 2A (2) elements found in many compounds, but not Group 8A (18) elements?

6.60 a. How does the octet rule explain the formation of a chloride ion? (6.1)
b. What noble gas has the same electron configuration as the chloride ion?
c. Why are Group 7A (17) elements found in many compounds, but not Group 8A (18) elements?

6.61 Identify each of the following atoms or ions: (6.1)

$18\,e^-$	$8\,e^-$	$28\,e^-$	$23\,e^-$
$15\,p^+$	$8\,p^+$	$30\,p^+$	$26\,p^+$
$16\,n$	$8\,n$	$35\,n$	$28\,n$
A	**B**	**C**	**D**

6.62 Identify each of the following atoms or ions: (6.1)

$2\,e^-$	$0\,e^-$	$3\,e^-$	$10\,e^-$
$3\,p^+$	$1\,p^+$	$3\,p^+$	$7\,p^+$
$4\,n$	$0\,n$	$4\,n$	$8\,n$
A	**B**	**C**	**D**

6.63 Element X has one valence electron and element Y has 6 valence electrons.

a. What are the group numbers of X and Y?
b. Will a compound of X and Y be ionic or molecular?
c. What ions would be formed by X and Y?
d. What would be the formula of a compound of X and Y?
e. What would be the formula of a compound of X and sulfur?
f. What would be the formula of a compound of Y and chlorine?
g. Is the compound in part **f** ionic or molecular?

6.64 Element X has two valence electrons and element Y has 5 valence electrons.

a. What are the group numbers of X and Y?
b. Will a compound of X and Y be ionic or molecular?
c. What ions would be formed by X and Y?
d. What would be the formula of a compound of X and Y?
e. What would be the formula of a compound of X and sulfur?
f. What would be the formula of a compound of Y and chlorine?
g. Is the compound in part **f** ionic or molecular?

6.65 Using each of the following electron configurations, write the formulas for the cation and anion that form, the formula for the compound they form, and its name. (6.2, 6.3)

Electron Configurations		Cation	Anion	Formula of Compound	Name of Compound
$1s^2 2s^2 2p^6 3s^2$	$1s^2 2s^2 2p^3$				
$1s^2 2s^2 2p^6 3s^2 3p^6 4s^1$	$1s^2 2s^2 2p^4$				
$1s^2 2s^2 2p^6 3s^2 3p^1$	$1s^2 2s^2 2p^6 3s^2 3p^5$				

6.66 Using each of the following electron configurations, write the formulas for the cation and anion that form, the formula for the compound they form, and its name. (6.2, 6.3)

Electron Configurations		Cation	Anion	Formula of Compound	Name of Compound
$1s^2 2s^2 2p^6 3s^1$	$1s^2 2s^2 2p^5$				
$1s^2 2s^2 2p^6 3s^2 3p^6 4s^2$	$1s^2 2s^2 2p^6 3s^2 3p^4$				
$1s^2 2s^1$	$1s^2 2s^2 2p^6 3s^2 3p^3$				

ADDITIONAL PRACTICE PROBLEMS

6.67 Write the name for each of the following ions: (6.1)
a. N^{3-} **b.** Mg^{2+}
c. O^{2-} **d.** Al^{3+}

6.68 Write the name for each of the following ions: (6.1)
a. K^+ **b.** Na^+
c. Ba^{2+} **d.** Cl^-

6.69 Consider an ion with the symbol X^{2+} formed from a representative element. (6.1, 6.2, 6.3)
a. What is the group number of the element?
b. If X is in Period 3, what is the element?
c. What is the formula of the compound formed from X and the nitride ion?

6.70 Consider an ion with the symbol Y^{3-} formed from a representative element. (6.1, 6.2, 6.3)
  a. What is the group number of the element?
  b. If Y is in Period 3, what is the element?
  c. What is the formula of the compound formed from the barium ion and Y?

6.71 One of the ions of tin is tin(IV). (6.1, 6.2, 6.3, 6.4)
  a. What is the symbol for this ion?
  b. How many protons and electrons are in the ion?
  c. What is the formula of tin(IV) oxide?
  d. What is the formula of tin(IV) phosphate?

6.72 One of the ions of gold is gold(III). (6.1, 6.2, 6.3, 6.4)
  a. What is the symbol for this ion?
  b. How many protons and electrons are in the ion?
  c. What is the formula of gold(III) sulfate?
  d. What is the formula of gold(III) nitrate?

6.73 Write the formula for each of the following ionic compounds: (6.2, 6.3)
  a. tin(II) sulfide   **b.** lead(IV) oxide
  c. silver chloride   **d.** calcium nitride
  e. copper(I) phosphide   **f.** chromium(II) bromide

6.74 Write the formula for each of the following ionic compounds: (6.2, 6.3)
  a. nickel(III) oxide   **b.** iron(III) sulfide
  c. lead(II) sulfate   **d.** chromium(III) iodide
  e. lithium nitride   **f.** gold(I) oxide

6.75 Name each of the following molecular compounds: (6.5)
  a. NCl_3 **b.** N_2S_3
  c. N_2O **d.** IF
  e. BF_3 **f.** P_2O_5

6.76 Name each of the following molecular compounds: (6.5)
  a. CF_4 **b.** SF_6
  c. BrCl   **d.** N_2O_4
  e. SO_2 **f.** CS_2

6.77 Write the formula for each of the following molecular compounds: (6.5)
  a. carbon sulfide   **b.** diphosphorus pentoxide
  c. dihydrogen sulfide   **d.** sulfur dichloride

6.78 Write the formula for each of the following molecular compounds: (6.5)
  a. silicon dioxide   **b.** carbon tetrabromide
  c. diphosphorus tetraiodide **d.** dinitrogen trioxide

6.79 Classify each of the following as ionic or molecular, and write its name: (6.3, 6.5)
  a. $FeCl_3$ **b.** Na_2SO_4
  c. NO_2 **d.** Rb_2S
  e. PF_5 **f.** CF_4

6.80 Classify each of the following as ionic or molecular, and write its name: (6.3, 6.5)
  a. $Al_2(CO_3)_3$ **b.** ClF_5
  c. BCl_3 **d.** Mg_3N_2
  e. ClO_2 **f.** $CrPO_4$

6.81 Write the formula for each of the following: (6.3, 6.4, 6.5)
  a. tin(II) carbonate   **b.** lithium phosphide
  c. silicon tetrachloride   **d.** manganese(III) oxide
  e. tetraphosphorus triselenide **f.** calcium bromide

6.82 Write the formula for each of the following: (6.3, 6.4, 6.5)
  a. sodium carbonate   **b.** nitrogen dioxide
  c. aluminum nitrate   **d.** copper(I) nitride
  e. potassium phosphate   **f.** cobalt(III) sulfate

CHALLENGE PROBLEMS

The following problems are related to the topics in this chapter. However, they do not all follow the chapter order, and they require you to combine concepts and skills from several sections. These problems will help you increase your critical thinking skills and prepare for your next exam.

6.83 Complete the following table for atoms or ions: (6.1)

Atom or Ion	Number of Protons	Number of Electrons	Electrons Lost/Gained
K^+			
	$12\ p^+$	$10\ e^-$	
	$8\ p^+$		$2\ e^-$ gained
		$10\ e^-$	$3\ e^-$ lost

6.84 Complete the following table for atoms or ions: (6.1)

Atom or Ion	Number of Protons	Number of Electrons	Electrons Lost/Gained
	$30\ p^+$		$2\ e^-$ lost
	$36\ p^+$	$36\ e^-$	
	$16\ p^+$		$2\ e^-$ gained
		$46\ e^-$	$4\ e^-$ lost

6.85 Identify the group number in the periodic table of X, a representative element, in each of the following ionic compounds: (6.2)
  a. XCl_3
  b. Al_2X_3
  c. XCO_3

6.86 Identify the group number in the periodic table of X, a representative element, in each of the following ionic compounds: (6.2)
  a. X_2O_3
  b. X_2SO_3
  c. Na_3X

6.87 Classify each of the following as ionic or molecular, and name each: (6.2, 6.3, 6.4, 6.5)
  a. Li_2HPO_4 **b.** ClF_3
  c. $Mg(ClO_2)_2$ **d.** NF_3
  e. $Ca(HSO_4)_2$ **f.** $KClO_4$
  g. $Au_2(SO_3)_3$

6.88 Classify each of the following as ionic or molecular, and name each: (6.2, 6.3, 6.4, 6.5)
  a. $FePO_3$ **b.** Cl_2O_7
  c. $Ca_3(PO_4)_2$ **d.** PCl_3
  e. $Al(ClO_2)_3$ **f.** $Pb(C_2H_3O_2)_2$
  g. $MgCO_3$

ANSWERS TO ENGAGE QUESTIONS

6.1 The net charge of the cesium ion with 55 protons (55+) and 54 electrons (54−) is 1+.

6.2 A Li atom loses one electron to form a stable Li^+ ion, whereas a Br atom gains one electron to form a stable Br^- ion.

6.3 The atoms in Group 2A (2) lose 2 electrons to form stable 2+ ions.

6.4 The bonding between Na^+ and Cl^- ions in NaCl is ionic.

6.5 In an ionic compound, the positive ion has the same name as the atom, and the negative ion uses the first syllable of the atom's name followed by *ide.*

6.6 Because most transition elements have more than one positive ion, a Roman numeral is used.

6.7 In ionic compounds, copper can form two different ions, Cu^+ and Cu^{2+}. Oxygen forms the ion O^{2-}.

6.8 In a phosphate ion, there are four O atoms; in a phosphite ion, there are three O atoms.

6.9 The 2+ charge of the magnesium ion must be balanced by the charges of two nitrate (NO_3^-) ions.

6.10 Because two nonmetals may form more than one molecular compound, prefixes are used.

6.11 Sodium phosphate is a combination of a metal ion and a polyatomic ion; it is an ionic compound. Diphosphorus pentoxide is a combination of two nonmetals; it is a molecular compound.

6.12 When a transition element that forms more than one positive ion is present in an ionic compound, a Roman numeral is used.

ANSWERS TO SELECTED PROBLEMS

6.1 **a.** 1 **b.** 2 **c.** 3 **d.** 1 **e.** 2

6.3 **a.** $2\,e^-$ lost **b.** $3\,e^-$ gained **c.** $1\,e^-$ gained
d. $1\,e^-$ lost **e.** $1\,e^-$ gained

6.5 **a.** Li^+ **b.** F^- **c.** Mg^{2+} **d.** Co^{3+}

6.7 **a.** 29 protons, 27 electrons **b.** 34 protons, 36 electrons
c. 35 protons, 36 electrons **d.** 26 protons, 24 electrons

6.9 **a.** Cl^- **b.** Cs^+ **c.** N^{3-} **d.** Ra^{2+}

6.11 **a.** lithium **b.** calcium
c. gallium **d.** phosphide

6.13 **a.** 8 protons, 10 electrons **b.** 19 protons, 18 electrons
c. 53 protons, 54 electrons **d.** 20 protons, 18 electrons

6.15 a and c

6.17 **a.** Na_2O **b.** $AlBr_3$ **c.** Ba_3N_2
d. MgF_2 **e.** Al_2S_3

6.19 **a.** K^+ and S^{2-}, K_2S **b.** Na^+ and N^{3-}, Na_3N
c. Al^{3+} and I^-, AlI_3 **d.** Ga^{3+} and O^{2-}, Ga_2O_3

6.21 **a.** aluminum fluoride **b.** calcium chloride
c. sodium oxide **d.** magnesium phosphide
e. potassium iodide **f.** barium fluoride

6.23 **a.** iron(II) **b.** copper(II)
c. zinc **d.** lead(IV)
e. chromium(III) **f.** manganese(II)

6.25 **a.** tin(II) chloride **b.** iron(II) oxide
c. copper(I) sulfide **d.** copper(II) sulfide
e. cadmium bromide **f.** mercury(II) chloride

6.27 **a.** Au^{3+} **b.** Fe^{3+} **c.** Pb^{4+} **d.** Al^{3+}

6.29 **a.** $MgCl_2$ **b.** Na_2S **c.** Cu_2O
d. Zn_3P_2 **e.** AuN **f.** CoF_3

6.31 **a.** $CoCl_3$ **b.** PbO_2 **c.** AgI
d. Ca_3N_2 **e.** Cu_3P **f.** $CrCl_2$

6.33 **a.** K_3P **b.** $CuCl_2$ **c.** $FeBr_3$ **d.** MgO

6.35 **a.** HCO_3^- **b.** NH_4^+ **c.** PO_3^{3-} **d.** ClO_3^-

6.37 **a.** sulfate **b.** carbonate
c. hydrogen sulfite (bisulfite) **d.** nitrate

6.39

	NO_2^-	CO_3^{2-}	HSO_4^-	PO_4^{3-}
Li^+	$LiNO_2$ Lithium nitrite	Li_2CO_3 Lithium carbonate	$LiHSO_4$ Lithium hydrogen sulfate	Li_3PO_4 Lithium phosphate
Cu^{2+}	$Cu(NO_2)_2$ Copper(II) nitrite	$CuCO_3$ Copper(II) carbonate	$Cu(HSO_4)_2$ Copper(II) hydrogen sulfate	$Cu_3(PO_4)_2$ Copper(II) phosphate
Ba^{2+}	$Ba(NO_2)_2$ Barium nitrite	$BaCO_3$ Barium carbonate	$Ba(HSO_4)_2$ Barium hydrogen sulfate	$Ba_3(PO_4)_2$ Barium phosphate

6.41 **a.** $Ba(OH)_2$ **b.** $NaHSO_4$ **c.** $Fe(NO_2)_2$
d. $Zn_3(PO_4)_2$ **e.** $Fe_2(CO_3)_3$

6.43 **a.** CO_3^{2-}, sodium carbonate
b. NH_4^+, ammonium sulfide
c. OH^-, calcium hydroxide
d. NO_2^-, tin(II) nitrite

6.45 **a.** zinc acetate
b. magnesium phosphate
c. ammonium chloride
d. strontium nitrate
e. sodium nitrite

6.47 **a.** phosphorus tribromide
b. dichlorine oxide
c. carbon tetrabromide
d. hydrogen fluoride
e. nitrogen trifluoride

6.49 **a.** dinitrogen trioxide
b. disilicon hexabromide
c. tetraphosphorus trisulfide
d. phosphorus pentachloride
e. selenium hexafluoride

6.51 **a.** CCl_4 **b.** CO **c.** PF_3 **d.** N_2O_4

6.53 **a.** OF_2 **b.** BCl_3 **c.** N_2O_3 **d.** SF_6

6.55 **a.** aluminum sulfate
    b. zinc phosphate
    c. dinitrogen oxide
    d. magnesium hydroxide

6.57 **a.** $MgSO_4$
    b. SnF_2
    c. $Al(OH)_3$

6.59 **a.** By losing two valence electrons from the third energy level, magnesium achieves an octet in the second energy level.
    b. The magnesium ion Mg^{2+} has the same electron configuration as Ne $(1s^2 2s^2 2p^6)$.
    c. Group 1A (1) and 2A (2) elements achieve octets by losing electrons to form compounds. Group 8A (18) elements are stable with octets (or two electrons for helium).

6.61 **a.** P^{3-} ion
    b. O atom
    c. Zn^{2+} ion
    d. Fe^{3+} ion

6.63 **a.** X = Group 1A (1), Y = Group 6A (16)
    b. ionic
    c. X^+ and Y^{2-}
    d. X_2Y
    e. X_2S
    f. YCl_2
    g. molecular

6.65

Electron Configurations		Cation	Anion	Formula of Compound	Name of Compound
$1s^2 2s^2 2p^6 3s^2$	$1s^2 2s^2 2p^3$	Mg^{2+}	N^{3-}	Mg_3N_2	Magnesium nitride
$1s^2 2s^2 2p^6 3s^2 3p^6 4s^1$	$1s^2 2s^2 2p^4$	K^+	O^{2-}	K_2O	Potassium oxide
$1s^2 2s^2 2p^6 3s^2 3p^1$	$1s^2 2s^2 2p^6 3s^2 3p^5$	Al^{3+}	Cl^-	$AlCl_3$	Aluminum chloride

6.67 **a.** nitride
    b. magnesium
    c. oxide
    d. aluminum

6.69 **a.** 2A (2)
    b. Mg
    c. X_3N_2

6.71 **a.** Sn^{4+}
    b. 50 protons and 46 electrons
    c. SnO_2
    d. $Sn_3(PO_4)_4$

6.73 **a.** SnS
    b. PbO_2
    c. AgCl
    d. Ca_3N_2
    e. Cu_3P
    f. $CrBr_2$

6.75 **a.** nitrogen trichloride
    b. dinitrogen trisulfide
    c. dinitrogen oxide
    d. iodine fluoride
    e. boron trifluoride
    f. diphosphorus pentoxide

6.77 **a.** CS
    b. P_2O_5
    c. H_2S
    d. SCl_2

6.79 **a.** ionic, iron(III) chloride
    b. ionic, sodium sulfate
    c. molecular, nitrogen dioxide
    d. ionic, rubidium sulfide
    e. molecular, phosphorus pentafluoride
    f. molecular, carbon tetrafluoride

6.81 **a.** $SnCO_3$
    b. Li_3P
    c. $SiCl_4$
    d. Mn_2O_3
    e. P_4Se_3
    f. $CaBr_2$

6.83

Atom or Ion	Number of Protons	Number of Electrons	Electrons Lost/ Gained
K^+	$19\ p^+$	$18\ e^-$	$1\ e^-$ lost
Mg^{2+}	$12\ p^+$	$10\ e^-$	$2\ e^-$ lost
O^{2-}	$8\ p^+$	$10\ e^-$	$2\ e^-$ gained
Al^{3+}	$13\ p^+$	$10\ e^-$	$3\ e^-$ lost

6.85 **a.** Group 3A (13)
    b. Group 6A (16)
    c. Group 2A (2)

6.87 **a.** ionic, lithium hydrogen phosphate
    b. molecular, chlorine trifluoride
    c. ionic, magnesium chlorite
    d. molecular, nitrogen trifluoride
    e. ionic, calcium bisulfate or calcium hydrogen sulfate
    f. ionic, potassium perchlorate
    g. ionic, gold(III) sulfite

Chemical Quantities

Max a six-year-old dog, is listless, drinking large amounts of water, and not eating his food. His owner takes Max to his veterinarian, Dr. Evans, for an examination. Dr. Evans weighs Max and obtains a blood sample for a blood chemistry profile, which determines the overall health, detects any metabolic disorders, and measures the concentration of electrolytes.

The results of the lab tests for Max indicate that the white blood count is elevated, which may indicate an infection or inflammation. His electrolytes, which are also indicators of good health, were all in the normal ranges. The electrolyte chloride (Cl^-) was in the normal range of 0.106 to 0.118 mol/L, and the electrolyte sodium (Na^+) has a normal range of 0.144 to 0.160 mol/L. Potassium (K^+), another important electrolyte, has a normal range of 0.0035 to 0.0058 mol/L.

CAREER

Veterinarian

Veterinarians care for domesticated pets such as dogs, cats, rats, and birds. Some veterinarians specialize in the treatment of large animals, such as horses and cattle. Veterinarians interact with pet owners as they give advice about feeding, behavior, and breeding. In the assessment of an ill animal, they record the animal's symptoms and medical history including dietary intake, medications, eating habits, weight, and any signs of disease.

To diagnose health problems, veterinarians perform laboratory tests on animals including a complete blood count and urinalysis. They also obtain tissue and blood samples, as well as vaccinate against distemper, rabies, and other diseases. Veterinarians prescribe medication if an animal has an infection or illness. If an animal is injured, a veterinarian treats wounds, and sets fractures. A veterinarian also performs surgery such as neutering and spaying, provides dental cleanings, removes tumors, and euthanizes animals.

UPDATE Prescriptions for Max

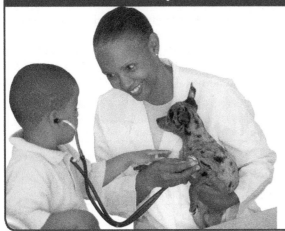

After Dr. Evans obtained the lab results from Max's tests, she determined that Max had a low-grade infection. You can see how Dr. Evans treated Max for the infection in the **UPDATE Prescription for Max**, page 204, and learn about the compounds that Max was given.

REVIEW

Writing Numbers in Scientific
   Notation (1.5)

Counting Significant Figures (2.2)

Using Significant Figures in
   Calculations (2.3)

Writing Conversion Factors
   from Equalities (2.5)

Using Conversion Factors (2.6)

CORE CHEMISTRY SKILL

Converting Particles to Moles

One mole of sulfur contains
6.022×10^{23} sulfur atoms.

7.1 The Mole

LEARNING GOAL Use Avogadro's number to calculate the number of particles in a given number of moles. Calculate the number of moles of an element in a given number of moles of a compound.

At the grocery store, you buy eggs by the dozen or soda by the case. In an office-supply store, pencils are ordered by the gross and paper by the ream. Common terms such as *dozen*, *case*, *gross*, and *ream* are used to count the number of items present. For example, when you buy a dozen eggs, you know you will get 12 eggs in the carton.

144 pencils = 1 gross

500 sheets = 1 ream

12 eggs = 1 dozen

Collections of items include dozen, gross, ream, and mole.

Avogadro's Number

In chemistry, particles such as atoms, molecules, and ions are counted by the **mole** (abbreviated *mol* in calculations), which contains 6.022×10^{23} items. This value, known as **Avogadro's number**, is a very big number because atoms are so small that it takes an extremely large number of atoms to provide a sufficient amount to weigh and use in chemical reactions. Avogadro's number is named for Amedeo Avogadro (1776–1856), an Italian physicist.

Avogadro's Number

$6.022 \times 10^{23} = 602\,200\,000\,000\,000\,000\,000\,000$

One mole of any element always contains Avogadro's number of atoms. For example, 1 mol of carbon contains 6.022×10^{23} carbon atoms; 1 mol of aluminum contains 6.022×10^{23} aluminum atoms; 1 mol of sulfur contains 6.022×10^{23} sulfur atoms.

1 mol of an element = 6.022×10^{23} atoms of that element

Avogadro's number tells us that 1 mol of a compound contains 6.022×10^{23} of the particular type of particles that make up that compound. One mole of a molecular compound contains Avogadro's number of molecules. For example, 1 mol of CO_2 contains 6.022×10^{23} molecules of CO_2. For an ionic compound, 1 mol contains Avogadro's number of formula units, which are the groups of ions represented by its formula. For the ionic formula, NaCl, 1 mol contains 6.022×10^{23} formula units of NaCl (Na^+, Cl^-). **TABLE 7.1** gives examples of the number of particles in some 1-mol quantities.

TABLE 7.1 Number of Particles in 1-Mol Quantities

Substance	Number and Type of Particles
1 mol of Al	6.022×10^{23} atoms of Al
1 mol of Fe	6.022×10^{23} atoms of Fe
1 mol of water (H_2O)	6.022×10^{23} molecules of H_2O
1 mol of vitamin C ($C_6H_8O_6$)	6.022×10^{23} molecules of vitamin C
1 mol of NaCl	6.022×10^{23} formula units of NaCl
1 mol of K_3PO_4	6.022×10^{23} formula units of K_3PO_4

Using Avogadro's Number as a Conversion Factor

We use Avogadro's number as a conversion factor to convert between the moles of a substance and the number of particles it contains.

$$\frac{6.022 \times 10^{23} \text{ particles}}{1 \text{ mol}} \quad \text{and} \quad \frac{1 \text{ mol}}{6.022 \times 10^{23} \text{ particles}}$$

For example, we use Avogadro's number to convert 4.00 mol of iron to atoms of iron.

$$4.00 \text{ mol Fe} \times \frac{6.022 \times 10^{23} \text{ atoms Fe}}{1 \text{ mol Fe}} = 2.41 \times 10^{24} \text{ atoms of Fe}$$

Avogadro's number as a conversion factor

We also use Avogadro's number to convert 3.01×10^{24} molecules of CO_2 to moles of CO_2.

$$3.01 \times 10^{24} \text{ molecules } CO_2 \times \frac{1 \text{ mol } CO_2}{6.022 \times 10^{23} \text{ molecules } CO_2} = 5.00 \text{ mol of } CO_2$$

Avogadro's number as a conversion factor

In calculations that convert between moles and particles, the number of moles will be a small number compared to the number of atoms or molecules, which will be large as shown in Sample Problem 7.1.

> **ENGAGE 7.1**
>
> Why is 0.20 mol of aluminum a small number, but the number 1.2×10^{23} atoms of aluminum in 0.20 mol is a large number?

SAMPLE PROBLEM 7.1 Calculating the Number of Molecules

TRY IT FIRST

How many molecules are present in 1.75 mol of carbon dioxide?

The solid form of carbon dioxide is known as "dry ice."

SOLUTION

STEP 1 State the given and needed quantities.

ANALYZE THE PROBLEM	Given	Need	Connect
	1.75 mol of CO_2	molecules of CO_2	Avogadro's number

STEP 2 Write a plan to convert moles to particles.

moles of CO_2 Avogadro's number   molecules of CO_2

STEP 3 Use Avogadro's number to write conversion factors.

$$1 \text{ mol of } CO_2 = 6.022 \times 10^{23} \text{ molecules of } CO_2$$

$$\frac{6.022 \times 10^{23} \text{ molecules } CO_2}{1 \text{ mol } CO_2} \quad \text{and} \quad \frac{1 \text{ mol } CO_2}{6.022 \times 10^{23} \text{ molecules } CO_2}$$

STEP 4   Set up the problem to calculate the number of particles.

$$1.75 \; \cancel{\text{mol CO}_2} \times \frac{6.022 \times 10^{23} \text{ molecules CO}_2}{1 \; \cancel{\text{mol CO}_2}} = 1.05 \times 10^{24} \text{ molecules of CO}_2$$

SELF TEST 7.1

a. How many moles of water, H_2O, contain 2.60×10^{23} molecules of water?

b. How many molecules are in 2.65 mol of dinitrogen oxide, N_2O, which is an anesthetic called "laughing gas"?

ANSWER

a. 0.432 mol of H_2O

b. 1.60×10^{24} molecules of N_2O

PRACTICE PROBLEMS

Try Practice Problems 7.1 to 7.4

Moles of Elements in a Chemical Compound

ENGAGE 7.2

Why does 1 mol of $Zn(C_2H_3O_2)_2$, a dietary supplement, contain 1 mol of Zn, 4 mol of C, 6 mol of H, and 4 mol of O?

We have seen that the subscripts in a chemical formula indicate the number of atoms of each type of element in a compound. For example, aspirin, $C_9H_8O_4$, is a drug used to reduce pain and inflammation in the body. Using the subscripts in the chemical formula of aspirin shows that there are 9 carbon atoms, 8 hydrogen atoms, and 4 oxygen atoms. The subscripts of the formula of aspirin, $C_9H_8O_4$, also tell us the number of moles of each element in 1 mol of aspirin: 9 mol of C atoms, 8 mol of H atoms, and 4 mol of O atoms.

Number of atoms in 1 molecule

Carbon (C)   Hydrogen (H)   Oxygen (O)

Aspirin $C_9H_8O_4$

Every molecule of aspirin $C_9H_8O_4$ contains 9 C atoms, 8 H atoms, and 4 O atoms.

The chemical formula subscripts specify the

$C_9H_8O_4$

	Carbon	**Hydrogen**	**Oxygen**
Atoms in 1 molecule	9 atoms of C	8 atoms of H	4 atoms of O
Moles of each element in 1 mol	9 mol of C	8 mol of H	4 mol of O

Using a Chemical Formula to Write Conversion Factors

Using the subscripts from the formula, $C_9H_8O_4$, we can write the conversion factors for each of the elements in 1 mol of aspirin:

$$\frac{9 \text{ mol C}}{1 \text{ mol } C_9H_8O_4} \qquad \frac{8 \text{ mol H}}{1 \text{ mol } C_9H_8O_4} \qquad \frac{4 \text{ mol O}}{1 \text{ mol } C_9H_8O_4}$$

$$\frac{1 \text{ mol } C_9H_8O_4}{9 \text{ mol C}} \qquad \frac{1 \text{ mol } C_9H_8O_4}{8 \text{ mol H}} \qquad \frac{1 \text{ mol } C_9H_8O_4}{4 \text{ mol O}}$$

SAMPLE PROBLEM 7.2 Calculating the Moles of an Element in a Compound

TRY IT FIRST

How many moles of carbon are present in 1.50 mol of aspirin, $C_9H_8O_4$?

SOLUTION

STEP 1 State the given and needed quantities.

ANALYZE THE PROBLEM	Given	Need	Connect
	1.50 mol of aspirin, $C_9H_8O_4$	moles of C	subscripts in formula

STEP 2 Write a plan to convert moles of a compound to moles of an element.

moles of $C_9H_8O_4$ ▸ Subscript ▸ moles of C

STEP 3 Write the equalities and conversion factors using subscripts.

$$1 \text{ mol of } C_9H_8O_4 = 9 \text{ mol of C}$$

$$\frac{9 \text{ mol C}}{1 \text{ mol } C_9H_8O_4} \quad \text{and} \quad \frac{1 \text{ mol } C_9H_8O_4}{9 \text{ mol C}}$$

STEP 4 Set up the problem to calculate the moles of an element.

$$1.50 \text{ mol } C_9H_8O_4 \times \frac{9 \text{ mol C}}{1 \text{ mol } C_9H_8O_4} = 13.5 \text{ mol of C}$$

SELF TEST 7.2

a. How many moles of aspirin, $C_9H_8O_4$, contain 0.480 mol of O?
b. How many moles of C are in 0.125 mol of aspirin?

ANSWER

a. 0.120 mol of aspirin   **b.** 1.13 mol of C

PRACTICE PROBLEMS

Try Practice Problems 7.5 to 7.10

PRACTICE PROBLEMS

7.1 The Mole

7.1 What is a mole?

7.2 What is Avogadro's number?

7.3 Calculate each of the following:
  a. number of C atoms in 0.500 mol of C
  b. number of SO_2 molecules in 1.28 mol of SO_2
  c. moles of Fe in 5.22×10^{22} atoms of Fe
  d. moles of C_2H_6O in 8.50×10^{24} molecules of C_2H_6O

7.4 Calculate each of the following:
  a. number of Li atoms in 4.5 mol of Li
  b. number of CO_2 molecules in 0.0180 mol of CO_2
  c. moles of Cu in 7.8×10^{21} atoms of Cu
  d. moles of C_2H_6 in 3.75×10^{23} molecules of C_2H_6

7.5 Calculate each of the following quantities in 2.00 mol of H_3PO_4:
  a. moles of H
  b. moles of O
  c. atoms of P
  d. atoms of O

7.6 Calculate each of the following quantities in 0.185 mol of $C_6H_{14}O$:
  a. moles of C
  b. moles of O
  c. atoms of H
  d. atoms of C

Applications

7.7 Quinine, $C_{20}H_{24}N_2O_2$, is a component of tonic water and bitter lemon.
  a. How many moles of H are in 1.5 mol of quinine?
  b. How many moles of C are in 5.0 mol of quinine?
  c. How many moles of N are in 0.020 mol of quinine?

7.8 Aluminum sulfate, $Al_2(SO_4)_3$, is used in some antiperspirants.
  a. How many moles of O are present in 3.0 mol of $Al_2(SO_4)_3$?
  b. How many moles of aluminum ions (Al^{3+}) are present in 0.40 mol of $Al_2(SO_4)_3$?
  c. How many moles of sulfate ions (SO_4^{2-}) are present in 1.5 mol of $Al_2(SO_4)_3$?

7.9   Naproxen is used to treat pain and inflammation caused by arthritis. Naproxen has a formula of $C_{14}H_{14}O_3$.
   a. How many moles of C are present in 2.30 mol of naproxen?
   b. How many moles of H are present in 0.444 mol of naproxen?
   c. How many moles of O are present in 0.0765 mol of naproxen?

7.10   Benadryl is an over-the-counter drug used to treat allergy symptoms. The formula of Benadryl is $C_{17}H_{21}NO$.
   a. How many moles of C are present in 0.733 mol of Benadryl?
   b. How many moles of H are present in 2.20 mol of Benadryl?
   c. How many moles of N are present in 1.54 mol of Benadryl?

7.2 Molar Mass

LEARNING GOAL Given the chemical formula of a substance, calculate its molar mass.

A single atom or molecule is much too small to weigh, even on the most accurate balance. In fact, it takes a huge number of atoms or molecules to make enough of a substance for you to see. An amount of water that contains Avogadro's number of water molecules is only a few sips. However, in the laboratory, we can use a balance to weigh out Avogadro's number of particles for 1 mol of substance.

For any element, the quantity called **molar mass** is the quantity in grams that equals the atomic mass of that element. We are counting 6.022×10^{23} atoms of an element when we weigh out the number of grams equal to its molar mass. For example, carbon has an atomic mass of 12.01 on the periodic table. This means 1 mol of carbon atoms has a mass of 12.01 g. Then to obtain 1 mol of carbon atoms, we would need to weigh out 12.01 g of carbon. Thus, the molar mass of carbon is found by looking at its atomic mass on the periodic table.

The molar mass of an element is obtained from the periodic table.

12.01 g of C atoms

1 mol of C atoms

6.022×10^{23} atoms of C

| 47 |
| Ag |
| 107.9 |

1 mol of silver atoms has a mass of 107.9 g

| 6 |
| C |
| 12.01 |

1 mol of carbon atoms has a mass of 12.01 g

| 16 |
| S |
| 32.07 |

1 mol of sulfur atoms has a mass of 32.07 g

Molar Mass of a Compound

To determine the molar mass of a compound, multiply the molar mass of each element by its subscript in the formula and add the results as shown in Sample Problem 7.3. **In this text, we round the molar mass of an element to the hundredths place (0.01) or use at least four significant figures for calculations.** FIGURE 7.1 shows some 1-mol quantities of substances.

1-Mol Quantities

| S | Fe | NaCl | $K_2Cr_2O_7$ | $C_{12}H_{22}O_{11}$ |
| (32.07 g) | (55.85 g) | (58.44 g) | (294.2 g) | (342.3 g) |

FIGURE 7.1 Each beaker contains a 1-mol sample of an element or compound.

ENGAGE 7.3

How is the molar mass for $K_2Cr_2O_7$ obtained?

SAMPLE PROBLEM 7.3 Calculating Molar Mass

TRY IT FIRST

Calculate the molar mass for lithium carbonate, Li_2CO_3, used to treat bipolar disorder.

SOLUTION

ANALYZE THE PROBLEM	Given	Need	Connect
	formula Li_2CO_3	molar mass of Li_2CO_3	periodic table

STEP 1 Obtain the molar mass of each element.

$$\frac{6.941 \text{ g Li}}{1 \text{ mol Li}} \quad \frac{12.01 \text{ g C}}{1 \text{ mol C}} \quad \frac{16.00 \text{ g O}}{1 \text{ mol O}}$$

Lithium carbonate is used to treat bipolar disorder.

STEP 2 Multiply each molar mass by the number of moles (subscript) in the formula.

Grams from 2 mol of Li:

$$2 \text{ mol Li} \times \frac{6.941 \text{ g Li}}{1 \text{ mol Li}} = 13.88 \text{ g of Li}$$

Grams from 1 mol of C:

$$1 \text{ mol C} \times \frac{12.01 \text{ g C}}{1 \text{ mol C}} = 12.01 \text{ g of C}$$

Grams from 3 mol of O:

$$3 \text{ mol O} \times \frac{16.00 \text{ g O}}{1 \text{ mol O}} = 48.00 \text{ g of O}$$

STEP 3   Calculate the molar mass by adding the masses of the elements.

2 mol of Li	= 13.88 g of Li
1 mol of C	= 12.01 g of C
3 mol of O	= 48.00 g of O
Molar mass of Li_2CO_3	= 73.89 g

SELF TEST 7.3

Calculate the molar mass for each of the following:
a. salicylic acid, $C_7H_6O_3$, used to treat skin conditions such as acne, psoriasis, and dandruff
b. vitamin D_3, $C_{27}H_{44}O$, a vitamin that prevents rickets

PRACTICE PROBLEMS

Try Practice Problems 7.11 to 7.20

ANSWER

a. 138.12 g   b. 384.6 g

PRACTICE PROBLEMS

7.2 Molar Mass

7.11 Calculate the molar mass for each of the following:
a. Cl_2 b. $C_3H_6O_3$ c. $Mg_3(PO_4)_2$

7.12 Calculate the molar mass for each of the following:
a. O_2 b. KH_2PO_4 c. $Fe(ClO_4)_3$

7.13 Calculate the molar mass for each of the following:
a. AlF_3 b. $C_2H_4Cl_2$ c. SnF_2

7.14 Calculate the molar mass for each of the following:
a. $C_4H_8O_4$ b. $Ga_2(CO_3)_3$ c. $KBrO_4$

Applications

7.15 Calculate the molar mass for each of the following:
a. KCl, salt substitute
b. Fe_2O_3, rust
c. $C_{19}H_{20}FNO_3$, Paxil, antidepressant

7.16 Calculate the molar mass for each of the following:
a. $FeSO_4$, iron supplement
b. Al_2O_3, absorbent and abrasive
c. $C_7H_5NO_3S$, saccharin, artificial sweetener

7.17 Calculate the molar mass for each of the following:
a. $Al_2(SO_4)_3$, antiperspirant
b. $KC_4H_5O_6$, cream of tartar
c. $C_{16}H_{19}N_3O_5S$, amoxicillin, antibiotic

7.18 Calculate the molar mass for each of the following:
a. C_3H_8O, rubbing alcohol
b. $(NH_4)_2CO_3$, baking powder
c. $Zn(C_2H_3O_2)_2$, zinc supplement

7.19 Calculate the molar mass for each of the following:
a. $C_8H_9NO_2$, acetaminophen used in Tylenol
b. $Ca_3(C_6H_5O_7)_2$, calcium supplement
c. $C_{17}H_{18}FN_3O_3$, Cipro, used to treat a range of bacterial infections

7.20 Calculate the molar mass for each of the following:
a. $CaSO_4$, calcium sulfate, used to make casts to protect broken bones
b. $C_6H_{12}N_2O_4Pt$, Carboplatin, used in chemotherapy
c. $C_{12}H_{18}O$, Propofol, used to induce anesthesia during surgery

---

7.3 Calculations Using Molar Mass

LEARNING GOAL Use molar mass to convert between grams and moles.

CORE CHEMISTRY SKILL

Using Molar Mass as a Conversion Factor

The molar mass of an element is one of the most useful conversion factors in chemistry because it converts moles of a substance to grams, or grams to moles. For example, 1 mol of silver has a mass of 107.9 g. To express molar mass of Ag as an equality, we write

1 mol of Ag = 107.9 g of Ag

From this equality for the molar mass, two conversion factors can be written as

$$\frac{107.9 \text{ g Ag}}{1 \text{ mol Ag}} \quad \text{and} \quad \frac{1 \text{ mol Ag}}{107.9 \text{ g Ag}}$$

Sample Problem 7.4 shows how the molar mass of silver is used as a conversion factor.

Silver metal is used to make jewelry.

SAMPLE PROBLEM 7.4 Converting Between Moles and Grams

TRY IT FIRST

Silver metal is used in the manufacture of tableware, mirrors, jewelry, and dental alloys. If the design for a piece of jewelry requires 0.750 mol of silver, how many grams of silver are needed?

SOLUTION

STEP 1 State the given and needed quantities.

ANALYZE THE PROBLEM	Given	Need	Connect
	0.750 mol of Ag	grams of Ag	molar mass

STEP 2 Write a plan to convert moles to grams.

moles of Ag   Molar mass   grams of Ag

STEP 3 Determine the molar mass and write conversion factors.

$$1 \text{ mol of Ag} = 107.9 \text{ g of Ag}$$
$$\frac{107.9 \text{ g Ag}}{1 \text{ mol Ag}} \quad \text{and} \quad \frac{1 \text{ mol Ag}}{107.9 \text{ g Ag}}$$

STEP 4 Set up the problem to convert moles to grams.

$$0.750 \text{ mol Ag} \times \frac{107.9 \text{ g Ag}}{1 \text{ mol Ag}} = 80.9 \text{ g of Ag}$$

SELF TEST 7.4

a. A dentist orders 24.4 g of gold (Au) to prepare dental crowns and fillings. Calculate the number of moles of gold in the order.
b. Pencil lead contains graphite, which is the element carbon. Calculate the number of grams in 0.520 mol of carbon.

ANSWER

a. 0.124 mol of Au
b. 6.25 g of C

Writing Conversion Factors for the Molar Mass of a Compound

The conversion factors for a compound are also written from the molar mass. For example, the molar mass of the compound H_2O is written

$$1 \text{ mol of } H_2O = 18.02 \text{ g of } H_2O$$

From this equality, conversion factors for the molar mass of H_2O are written as

$$\frac{18.02 \text{ g } H_2O}{1 \text{ mol } H_2O} \quad \text{and} \quad \frac{1 \text{ mol } H_2O}{18.02 \text{ g } H_2O}$$

We can now change from moles to grams, or grams to moles, using the conversion factors derived from the molar mass of a compound shown in Sample Problem 7.5. (Remember, you must determine the molar mass first.)

Table salt is sodium chloride, NaCl.

SAMPLE PROBLEM 7.5 Converting Between Mass and Moles of a Compound

TRY IT FIRST

A salt shaker contains 73.7 g of NaCl. How many moles of NaCl are present?

SOLUTION

STEP 1 State the given and needed quantities.

ANALYZE THE PROBLEM	Given	Need	Connect
	73.7 g of NaCl	moles of NaCl	molar mass

STEP 2 Write a plan to convert grams to moles.

grams of NaCl   Molar mass   moles of NaCl

STEP 3 Determine the molar mass and write conversion factors.

$$(1 \times 22.99) + (1 \times 35.45) = 58.44 \text{ g/mol}$$

$$1 \text{ mol of NaCl} = 58.44 \text{ g of NaCl}$$

$$\frac{58.44 \text{ g NaCl}}{1 \text{ mol NaCl}} \quad \text{and} \quad \frac{1 \text{ mol NaCl}}{58.44 \text{ g NaCl}}$$

STEP 4 Set up the problem to convert grams to moles.

$$73.7 \text{ g NaCl} \times \frac{1 \text{ mol NaCl}}{58.44 \text{ g NaCl}} = 1.26 \text{ mol of NaCl}$$

SELF TEST 7.5

a. One tablet of an antacid contains 680. mg of $CaCO_3$. How many moles of $CaCO_3$ are present?
b. A different brand of antacid contains 6.9×10^{-3} mol of $Mg(OH)_2$. How many grams of $Mg(OH)_2$ are present?

ANSWER

a. 0.006 79 or 6.79×10^{-3} mol of $CaCO_3$ **b.** 0.40 g of $Mg(OH)_2$

PRACTICE PROBLEMS

Try Practice Problems 7.21 to 7.30

ENGAGE 7.4

Which conversion steps are needed to calculate the number of atoms of H in 5.00 g of CH_4?

FIGURE 7.2 gives a summary of the calculations to show the connections between the moles of a compound, its mass in grams, the number of molecules (or formula units if ionic), and the moles and atoms of each element in that compound in the following flowchart:

FIGURE 7.2 The moles of a compound are related to its mass in grams by molar mass, to the number of molecules (or formula units) by Avogadro's number, and to the moles of each element by the subscripts in the formula.

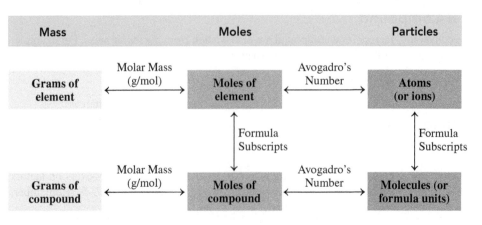

We can now convert the mass in grams of a compound to the mass of one of the elements in a compound as shown in Sample Problem 7.6.

SAMPLE PROBLEM 7.6 Converting Between Grams of Compound and Grams of Element

TRY IT FIRST

Hot packs are used to reduce muscle aches, inflammation, and muscle spasms. A hot pack consists of a bag of water and an inner bag containing 10.2 g of $CaCl_2$. When the bag is broken, the $CaCl_2$ dissolves in the water and heat is released. How many grams of Cl are in the $CaCl_2$ in the inner bag?

In some hot packs, heat is generated when $CaCl_2$ dissolves in water.

SOLUTION

STEP 1 State the given and needed quantities.

ANALYZE THE PROBLEM	Given	Need	Connect
	10.2 g of $CaCl_2$	grams of Cl	molar mass, subscript for Cl

STEP 2 Write a plan to convert grams of a compound to grams of an element.

grams of $CaCl_2$ Molar mass   moles of $CaCl_2$ Mole factor   moles of Cl

Molar mass   grams of Cl

STEP 3 Write the equalities and conversion factors for molar mass and mole factors.

ENGAGE 7.5

Why are there more grams of chlorine than grams of fluorine in 1 mol of Freon-12, CCl_2F_2?

$$1 \text{ mol of } CaCl_2 = 110.98 \text{ g of } CaCl_2$$
$$\frac{110.98 \text{ g } CaCl_2}{1 \text{ mol } CaCl_2} \quad \text{and} \quad \frac{1 \text{ mol } CaCl_2}{110.98 \text{ g } CaCl_2}$$

$$1 \text{ mol of } CaCl_2 = 2 \text{ mol of } Cl$$
$$\frac{2 \text{ mol } Cl}{1 \text{ mol } CaCl_2} \quad \text{and} \quad \frac{1 \text{ mol } CaCl_2}{2 \text{ mol } Cl}$$

$$1 \text{ mol of } Cl = 35.45 \text{ g of } Cl$$
$$\frac{35.45 \text{ g } Cl}{1 \text{ mol } Cl} \quad \text{and} \quad \frac{1 \text{ mol } Cl}{35.45 \text{ g } Cl}$$

STEP 4 Set up the problem to convert grams of a compound to grams of an element.

$$10.2 \text{ g } CaCl_2 \times \frac{1 \text{ mol } CaCl_2}{110.98 \text{ g } CaCl_2} \times \frac{2 \text{ mol } Cl}{1 \text{ mol } CaCl_2} \times \frac{35.45 \text{ g } Cl}{1 \text{ mol } Cl} = 6.52 \text{ g of Cl}$$

SELF TEST 7.6

a. Tin(II) fluoride (SnF_2) is added to toothpaste to strengthen tooth enamel. How many grams of tin(II) fluoride contain 1.46 g of F?

b. Sodium nitrite, $NaNO_2$, is used to give a longer storage life to ham and hot dogs. A package of ten hot dogs contains 110 milligrams of sodium nitrite. How many milligrams of N are in the package?

ANSWER

a. 6.02 g of SnF_2

b. 22 mg of N

PRACTICE PROBLEMS

Try Practice Problems 7.31 to 7.36

PRACTICE PROBLEMS

7.3 Calculations Using Molar Mass

7.21 Calculate the mass, in grams, for each of the following:
    a. 1.50 mol of Na   **b.** 2.80 mol of Ca
    c. 0.125 mol of CO_2 **d.** 0.0485 mol of Na_2CO_3
    e. 7.14×10^2 mol of PCl_3

7.22 Calculate the mass, in grams, for each of the following:
    a. 5.12 mol of Al   **b.** 0.75 mol of Cu
    c. 3.52 mol of $MgBr_2$ **d.** 0.145 mol of C_2H_6O
    e. 2.08 mol of $(NH_4)_2SO_4$

7.23 Calculate the mass, in grams, in 0.150 mol of each of the following:
    a. Ne   **b.** I_2 **c.** Na_2O
    d. $Ca(NO_3)_2$ **e.** C_6H_{14}

7.24 Calculate the mass, in grams, in 2.28 mol of each of the following:
    a. Pd   **b.** SO_3 **c.** $C_3H_6O_3$
    d. $Mg(HCO_3)_2$ **e.** SF_6

7.25 Calculate the number of moles in each of the following:
    a. 82.0 g of Ag   **b.** 0.288 g of C
    c. 15.0 g of ammonia, NH_3 **d.** 7.25 g of CH_4
    e. 245 g of Fe_2O_3

7.26 Calculate the number of moles in each of the following:
    a. 85.2 g of Ni   **b.** 144 g of K
    c. 6.4 g of H_2O **d.** 308 g of $BaSO_4$
    e. 252.8 g of fructose, $C_6H_{12}O_6$

7.27 Calculate the number of moles in 25.0 g of each of the following:
    a. He   **b.** O_2 **c.** $Al(OH)_3$
    d. Ga_2S_3 **e.** C_4H_{10}, butane

7.28 Calculate the number of moles in 4.00 g of each of the following:
    a. Au   **b.** SnO_2 **c.** CS_2
    d. Ca_3N_2 **e.** $C_6H_8O_6$, vitamin C

7.29 Calculate the mass, in grams, of C in each of the following:
    a. 0.688 g of CO_2
    b. 275 g of C_3H_6
    c. 1.84 g of C_2H_6O
    d. 73.4 g of $C_8H_{16}O_2$

7.30 Calculate the mass, in grams, of N in each of the following:
    a. 0.82 g of $NaNO_3$
    b. 40.0 g of $(NH_4)_3P$
    c. 0.464 g of N_2H_4
    d. 1.46 g of N_4O_6

Applications

7.31 Propane gas, C_3H_8, is used as a fuel for many barbecues.
    a. How many moles of H are in 34.0 g of propane?
    b. How many grams of C are in 1.50 mol of propane?
    c. How many grams of C are in 34.0 g of propane?
    d. How many grams of H are in 0.254 g of propane?

7.32 Allyl sulfide, $(C_3H_5)_2S$, gives garlic, onions, and leeks their characteristic odor.

The characteristic odor of garlic is due to a sulfur-containing compound.

    a. How many moles of S are in 23.2 g of allyl sulfide?
    b. How many grams of H are in 0.75 mol of allyl sulfide?
    c. How many grams of C are in 4.66 g of allyl sulfide?
    d. How many grams of S are in 15.0 g of allyl sulfide?

7.33 a. The compound $MgSO_4$, Epsom salts, is used to soothe sore feet and muscles. How many grams will you need to prepare a bath containing 5.00 mol of Epsom salts?
    b. Potassium iodide, KI, is used as an expectorant. How many grams of KI are in 0.450 mol of potassium iodide?

7.34 a. Cyclopropane, C_3H_6, is an anesthetic given by inhalation. How many grams are in 0.25 mol of cyclopropane?
    b. The sedative Demerol hydrochloride has the formula $C_{15}H_{22}ClNO_2$. How many grams are in 0.025 mol of Demerol hydrochloride?

7.35 Dinitrogen oxide (or nitrous oxide), N_2O, also known as laughing gas, is widely used as an anesthetic in dentistry.
    a. How many grams of the compound are in 1.50 mol of dinitrogen oxide?
    b. How many moles of the compound are in 34.0 g of dinitrogen oxide?
    c. How many grams of N are in 34.0 g of dinitrogen oxide?

7.36 Chloroform, $CHCl_3$, was formerly used as an anesthetic but its use was discontinued due to respiratory and cardiac failure.
    a. How many grams of the compound are in 0.122 mol of chloroform?
    b. How many moles of the compound are in 15.6 g of chloroform?
    c. How many grams of Cl are in 26.7 g of chloroform?

7.4 Mass Percent Composition

LEARNING GOAL Given the formula of a compound, calculate the mass percent composition.

Because the atoms of the elements in a compound are combined in a definite mole ratio, they are also combined in a definite proportion by mass. When we know the mass of an element in the mass of a sample of a compound, we can calculate its **mass percent composition** or **mass percent**, which is the mass of an element divided by the total mass of the compound and multiplied by 100%. For example, we can calculate the mass percent of N if we find from experiment that 7.64 g of N are present in 12.0 g of N_2O, "laughing gas," which is used as an anesthetic for surgery and in dentistry.

From the grams of N and the grams of N_2O, we calculate the mass percent of nitrogen as follows:

$$\text{Mass percent of an element} = \frac{\text{mass of an element}}{\text{total mass of the compound}} \times 100\%$$

$$\text{Mass percent of N} = \frac{7.64 \text{ g N}}{12.0 \text{ g } N_2O} \times 100\% = 63.7\% \text{ N}$$

Mass Percent Composition Using Molar Mass

We can also calculate mass percent composition of a compound from its molar mass. Then the total mass of each element is divided by the molar mass of the compound and multiplied by 100%.

$$\text{Mass percent composition} = \frac{\text{mass of each element}}{\text{molar mass of the compound}} \times 100\%$$

SAMPLE PROBLEM 7.7 Calculating Mass Percent Composition from Molar Mass

TRY IT FIRST

The odor of pears is due to the compound propyl acetate, which has a formula of $C_5H_{10}O_2$. What is the mass percent composition of propyl acetate?

SOLUTION

ANALYZE THE PROBLEM	Given	Need	Connect
	$C_5H_{10}O_2$	mass percent composition: %C, %H, %O	periodic table

STEP 1   Determine the total mass of each element in the molar mass of a formula.

$$5 \text{ mol C} \times \frac{12.01 \text{ g C}}{1 \text{ mol C}} = 60.05 \text{ g of C}$$

$$10 \text{ mol H} \times \frac{1.008 \text{ g H}}{1 \text{ mol H}} = 10.08 \text{ g of H}$$

$$2 \text{ mol O} \times \frac{16.00 \text{ g O}}{1 \text{ mol O}} = 32.00 \text{ g of O}$$

Molar mass of $C_5H_{10}O_2$ = 102.13 g of $C_5H_{10}O_2$

STEP 2   Divide the total mass of each element by the molar mass and multiply by 100%.

$$\text{Mass \% C} = \frac{60.05 \text{ g C}}{102.13 \text{ g } C_5H_{10}O_2} \times 100\% = 58.80\% \text{ C}$$

$$\text{Mass \% H} = \frac{10.08 \text{ g H}}{102.13 \text{ g } C_5H_{10}O_2} \times 100\% = 9.870\% \text{ H}$$

$$\text{Mass \% O} = \frac{32.00 \text{ g O}}{102.13 \text{ g } C_5H_{10}O_2} \times 100\% = 31.33\% \text{ O}$$

The total mass percent for all the elements in the compound should equal 100%. In some cases, because of rounding off, the sum of the mass percents may not total exactly 100%.

58.80% C + 9.870% H + 31.33% O = 100.00%

The odor of pears is due to propyl acetate, $C_5H_{10}O_2$.

SELF TEST 7.7

a. Ethylene glycol, $C_2H_6O_2$, used as automobile antifreeze, is a sweet-tasting liquid, which is toxic to humans and animals. What is the mass percent composition of ethylene glycol?

b. Iron(II) sulfate, $FeSO_4$, is a dietary supplement. What is the mass percent composition of iron(II) sulfate?

PRACTICE PROBLEMS

Try Practice Problems 7.37 to 7.42

ANSWER

a. 38.70% C; 9.744% H; 51.56% O   b. 36.76% Fe; 21.11% S; 42.13% O

Chemistry Link to the Environment

Fertilizers

Every year in the spring, homeowners and farmers add fertilizers to the soil to produce greener lawns and larger crops. Plants require several nutrients, but the major ones are nitrogen, phosphorus, and potassium. Nitrogen promotes green growth, phosphorus promotes strong root development for strong plants and abundant flowers, and potassium helps plants defend against diseases and weather extremes. The numbers on a package of fertilizer give the percentages each of N, P, and K by mass. For example, the set of numbers 30–3–4 describes a fertilizer that contains 30% N, 3% P, and 4% K.

The major nutrient, nitrogen, is present in huge quantities as N_2 in the atmosphere, but plants cannot utilize nitrogen in this form. Bacteria in the soil convert atmospheric N_2 to usable forms by *nitrogen fixation*. To provide additional nitrogen to plants, several types of nitrogen-containing chemicals, including ammonia, nitrates, and ammonium compounds, are added to soil. The nitrates are absorbed directly, but ammonia and ammonium salts are first converted to nitrates by soil bacteria.

The percent nitrogen depends on the type of nitrogen compound used in the fertilizer. The percent nitrogen by mass in each type is calculated using mass percent composition.

Nitrogen (N) 30%
Phosphorus (P) 3%
Potassium (K) 4%

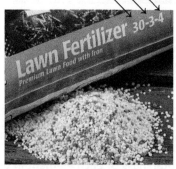

The label on a bag of fertilizer states the percentages of N, P, and K.

The choice of a fertilizer depends on its use and convenience. A fertilizer can be prepared as crystals or a powder, in a liquid solution, or as a gas such as ammonia. The ammonia and ammonium fertilizers are water soluble and quick-acting. Other forms may be made to slow-release by enclosing water-soluble ammonium salts in a thin plastic coating. The most commonly used fertilizer is NH_4NO_3 because it is easy to apply and has a high percent of N by mass.

Type of Fertilizer		Percent Nitrogen by Mass
NH_3	$\dfrac{14.01 \text{ g N}}{17.03 \text{ g NH}_3}$	$\times 100\% = 82.27\%$ N
NH_4NO_3	$\dfrac{28.02 \text{ g N}}{80.05 \text{ g NH}_4NO_3}$	$\times 100\% = 35.00\%$ N
$(NH_4)_2HPO_4$	$\dfrac{28.02 \text{ g N}}{132.06 \text{ g }(NH_4)_2HPO_4}$	$\times 100\% = 21.22\%$ N
$(NH_4)_2SO_4$	$\dfrac{28.02 \text{ g N}}{132.15 \text{ g }(NH_4)_2SO_4}$	$\times 100\% = 21.20\%$ N

PRACTICE PROBLEMS

7.4 Mass Percent Composition

7.37 Calculate the mass percent composition of each of the following:
   a. 4.68 g of Si and 5.32 g of O
   b. 5.72 g of C and 1.28 g of H
   c. 16.1 of Na, 22.5 g of S, and 11.3 g of O
   d. 6.22 g of C, 1.04 g of H, and 4.14 g of O

7.38 Calculate the mass percent composition of each of the following:
   a. 0.890 g of Ba and 1.04 g of Br
   b. 3.82 g of K and 1.18 g of I
   c. 3.29 of N, 0.946 g of H, and 3.76 g of S
   d. 4.14 g of C, 0.695 g of H, and 2.76 g of O

Applications

7.39 Calculate the mass percent composition of each of the following:
  a. MgF_2, magnesium fluoride
  b. $Ca(OH)_2$, calcium hydroxide
  c. $C_4H_8O_4$, erythrose, a carbohydrate
  d. $(NH_4)_3PO_4$, ammonium phosphate, fertilizer
  e. $C_{17}H_{19}NO_3$, morphine, a painkiller

7.40 Calculate the mass percent composition of each of the following:
  a. $CaCl_2$, calcium chloride
  b. $Na_2Cr_2O_7$, sodium dichromate
  c. $C_2H_3Cl_3$, trichloroethane, cleaning solvent
  d. $Ca_3(PO_4)_2$, calcium phosphate, found in bone and teeth
  e. $C_{18}H_{36}O_2$, stearic acid, from animal fat

7.41 Calculate the mass percent of N in each of the following:
  a. N_2O_5, dinitrogen pentoxide
  b. NH_4Cl, expectorant in cough medicine
  c. $C_2H_8N_2$, dimethylhydrazine, rocket fuel
  d. $C_9H_{15}N_5O$, Rogaine, stimulates hair growth
  e. $C_{14}H_{22}N_2O$, lidocaine, local anesthetic

7.42 Calculate the mass percent of S in each of the following:
  a. Na_2SO_4, sodium sulfate
  b. Al_2S_3, aluminum sulfide
  c. SO_3, sulfur trioxide
  d. C_2H_6SO, dimethylsulfoxide, topical anti-inflammatory
  e. $C_{10}H_{10}N_4O_2S$, sulfadiazine, antibacterial

7.5 Empirical Formulas

LEARNING GOAL From the mass percent composition, calculate the empirical formula for a compound.

Up to now, the formulas you have seen have been **molecular formulas**, which are the actual formulas of compounds. If we write a formula that represents the lowest whole-number ratio of the atoms in a compound, it is called the simplest or **empirical formula**. For example, the compound benzene, with molecular formula C_6H_6, has the empirical formula CH. Some molecular formulas and their empirical formulas are shown in **TABLE 7.2**.

TABLE 7.2 Examples of Molecular and Empirical Formulas

Name	Molecular (actual formula)	Empirical (simplest formula)
Acetylene	C_2H_2	CH
Benzene	C_6H_6	CH
Ammonia	NH_3	NH_3
Hydrazine	N_2H_4	NH_2
Ribose	$C_5H_{10}O_5$	CH_2O
Glucose	$C_6H_{12}O_6$	CH_2O

The empirical formula of a compound is determined by converting the number of grams of each element to moles and finding the lowest whole-number ratio to use as subscripts, as shown in Sample Problem 7.8.

CORE CHEMISTRY SKILL
Calculating an Empirical Formula

SAMPLE PROBLEM 7.8 Calculating an Empirical Formula

TRY IT FIRST

A compound of iron and chlorine is used to purify water in water-treatment plants. What is the empirical formula of the compound if experimental analysis shows that a sample of the compound contains 6.87 g of iron and 13.1 g of chlorine?

SOLUTION

ANALYZE THE PROBLEM	Given	Need	Connect
	6.87 g of Fe, 13.1 g of Cl	empirical formula	molar mass

Water-treatment plants use chemicals to purify sewage.

STEP 1　Calculate the moles of each element.

$$6.87 \ \cancel{g \ Fe} \times \frac{1 \ mol \ Fe}{55.85 \ \cancel{g \ Fe}} = 0.123 \ mol \ of \ Fe$$

$$13.1 \ \cancel{g \ Cl} \times \frac{1 \ mol \ Cl}{35.45 \ \cancel{g \ Cl}} = 0.370 \ mol \ of \ Cl$$

STEP 2　Divide by the smallest number of moles. The smallest number of moles is 0.123.

$$\frac{0.123 \ mol \ Fe}{0.123} = 1.00 \ mol \ of \ Fe$$

$$\frac{0.370 \ mol \ Cl}{0.123} = 3.01 \ mol \ of \ Cl$$

STEP 3　Use the lowest whole-number ratio of moles as subscripts. The relationship of moles of Fe to moles of Cl is 1 to 3, which we obtain by rounding off 3.01 to 3.

$$Fe_{1.00}Cl_{3.01} \longrightarrow Fe_1Cl_3, \ written \ as \ FeCl_3 \quad \text{Empirical formula}$$

SELF TEST 7.8

a. Phosphine is a highly toxic compound used for pest and rodent control. If a sample of phosphine contains 0.456 g of P and 0.0440 g of H, what is its empirical formula?
b. A compound of titanium and oxygen is used to make white pigment. If a sample of this compound contains 1.21 g of Ti and 0.806 g of O, what is its empirical formula?

ANSWER

a. PH_3　　　　　　　　　　　　　　b. TiO_2

Often, the mass percent of each element in a compound is given. The mass percent composition is true for any quantity of the compound. For example, methane, CH_4, always has a mass percent composition of 74.9% C and 25.1% H. Thus, if we assume that we have a sample of 100.0 g of the compound, we can determine the grams of each element in that sample and calculate the empirical formula as shown in Sample Problem 7.9.

Mass percent composition of methane, CH_4

Composition, in grams, for 50.0 g of methane, CH_4

Composition, in grams, for 100.0 g of methane, CH_4

The mass percent composition for any amount of methane is always the same.

SAMPLE PROBLEM 7.9	Calculating an Empirical Formula from the Mass Percent Composition

TRY IT FIRST

Tetrachloroethene is a colorless liquid used for dry cleaning. What is the empirical formula of tetrachloroethene if its mass percent composition is 14.5% C and 85.5% Cl?

SOLUTION

ANALYZE THE PROBLEM	Given	Need	Connect
	14.5% C, 85.5% Cl	empirical formula	molar mass

STEP 1 **Calculate the moles of each element.** In a sample size of 100. g of this compound, there are 14.5 g of C and 85.5 g of Cl.

$$14.5 \ \text{g C} \ \times \ \frac{1 \ \text{mol C}}{12.01 \ \text{g C}} = 1.21 \ \text{mol of C}$$

$$85.5 \ \text{g Cl} \ \times \ \frac{1 \ \text{mol Cl}}{35.45 \ \text{g Cl}} = 2.41 \ \text{mol of Cl}$$

STEP 2 **Divide by the smallest number of moles.** The smallest number of moles is 1.21.

$$\frac{1.21 \ \text{mol C}}{1.21} = 1.00 \ \text{mol of C}$$

$$\frac{2.41 \ \text{mol Cl}}{1.21} = 1.99 \ \text{mol of Cl}$$

STEP 3 **Use the lowest whole-number ratio of moles as subscripts.**

$$C_{1.00}Cl_{1.99} \longrightarrow C_1Cl_2 = CCl_2$$

SELF TEST 7.9

a. Sulfate of potash is the common name of a compound used in fertilizers to supply potassium and sulfur. What is the empirical formula of this compound if it has a mass percent composition of 44.9% K, 18.4% S, and 36.7% O?

b. Acrylonitrile is a compound used to make a variety of polymers. What is the empirical formula of acrylonitrile if it has a mass percent composition of 67.9% C, 5.7% H, and 26.4% N?

ANSWER

a. K_2SO_4 **b.** C_3H_3N

Sulfate of potash supplies sulfur and potassium.

Converting Decimal Numbers to Whole Numbers

Sometimes the result of dividing by the smallest number of moles gives a decimal instead of a whole number. Decimal values that are very close to whole numbers can be rounded off. For example, 2.04 rounds off to 2 and 6.98 rounds off to 7. *However, a decimal that is greater than 0.1 or less than 0.9 should not be rounded off.* Instead, we multiply by a small integer to obtain a whole number. Some multipliers that are typically used are listed in **TABLE 7.3**.

TABLE 7.3 Some Multipliers That Convert Decimals to Whole-Number Subscripts			
Decimal	Multiplier	Example		Whole Number
0.20	5	1.20×5	=	6
0.25	4	2.25×4	=	9
0.33	3	1.33×3	=	4
0.50	2	2.50×2	=	5
0.67	3	1.67×3	=	5

ENGAGE 7.6

If we obtain a ratio of 1.0 mol of P and 2.5 mol of O, why do we multiply by 2 to obtain an empirical formula of P_2O_5?

Let us suppose the numbers of moles we obtain give subscripts in the ratio of $C_{1.00}H_{2.33}O_{0.99}$. While 0.99 rounds off to 1, we cannot round off 2.33. If we multiply 2.33×2, we obtain 4.66, which is still not a whole number. If we multiply 2.33 by 3, the answer is 6.99, which rounds off to 7. To complete the empirical formula, *all the other subscripts must be multiplied by 3.*

$$C_{(1.00 \times 3)}H_{(2.33 \times 3)}O_{(0.99 \times 3)} = C_{3.00}H_{6.99}O_{2.97} \longrightarrow C_3H_7O_3$$

PRACTICE PROBLEMS

Try Practice Problems 7.43 to 7.46

Citrus fruits are a good source of vitamin C.

SAMPLE PROBLEM 7.10 Calculating an Empirical Formula Using Multipliers

TRY IT FIRST

Ascorbic acid (vitamin C), found in citrus fruits and vegetables, is important in metabolic reactions in the body, in the synthesis of collagen, and in the prevention of scurvy. If the mass percent composition of ascorbic acid is 40.9% C, 4.58% H, and 54.5% O, what is the empirical formula of ascorbic acid?

SOLUTION

ANALYZE THE PROBLEM	Given	Need	Connect
	40.9% C, 4.58% H, 54.5% O	empirical formula	molar mass

STEP 1 Calculate the moles of each element. In a sample size of 100. g, there are 40.9 g of C, 4.58 g of H, and 54.5 g of O.

$$40.9 \ \text{g C} \times \frac{1 \ \text{mol C}}{12.01 \ \text{g C}} = 3.41 \ \text{mol of C}$$

$$4.58 \ \text{g H} \times \frac{1 \ \text{mol H}}{1.008 \ \text{g H}} = 4.54 \ \text{mol of H}$$

$$54.5 \ \text{g O} \times \frac{1 \ \text{mol O}}{16.00 \ \text{g O}} = 3.41 \ \text{mol of O}$$

STEP 2 Divide by the smallest number of moles. The smallest number of moles is 3.41.

$$\frac{3.41 \ \text{mol C}}{3.41} = 1.00 \ \text{mol of C}$$

$$\frac{4.54 \ \text{mol H}}{3.41} = 1.33 \ \text{mol of H}$$

$$\frac{3.41 \ \text{mol O}}{3.41} = 1.00 \ \text{mol of O}$$

STEP 3 Use the lowest whole-number ratio of moles as subscripts. As calculated, the ratio of moles gives the formula

$$C_{1.00}H_{1.33}O_{1.00}$$

Because the subscript for H has a decimal that is greater than 0.1 and less than 0.9, it should not be rounded off. Then, we multiply each of the subscripts by 3 to obtain a whole number for H, which is 4. Thus, the empirical formula of ascorbic acid is $C_3H_4O_3$.

$$C_{(1.00 \times 3)}H_{(1.33 \times 3)}O_{(1.00 \times 3)} = C_{3.00}H_{3.99}O_{3.00} \longrightarrow C_3H_4O_3$$

SELF TEST 7.10

a. Glyoxylic acid is used by plants and bacteria to convert fats into glucose. What is the empirical formula of glyoxylic acid if it has a mass percent composition of 32.5% C, 2.70% H, and 64.8% O?

b. A compound of boron and oxygen is used to make glass that is resistant to thermal shocks. If this compound contains 31.1% B and 68.9% O, what is its empirical formula?

ANSWER

a. $C_2H_2O_3$ **b.** B_2O_3

PRACTICE PROBLEMS

7.5 Empirical Formulas

7.43 Calculate the empirical formula for each of the following:
   a. 3.57 g of N and 2.04 g of O
   b. 7.00 g of C and 1.75 g of H
   c. 0.175 g of H, 2.44 g of N, and 8.38 g of O
   d. 2.06 g of Ca, 2.66 g of Cr, and 3.28 g of O

7.44 Calculate the empirical formula for each of the following:
   a. 2.90 g of Ag and 0.430 g of S
   b. 2.22 g of Na and 0.774 g of O
   c. 2.11 g of Na, 0.0900 g of H, 2.94 g of S, and 5.86 g of O
   d. 5.52 g of K, 1.45 g of P, and 3.00 g of O

7.45 Calculate the empirical formula for each of the following:
   a. 70.9% K and 29.1% S
   b. 55.0% Ga and 45.0% F

   c. 69.6% Mn and 30.4% O
   d. 18.8% Li, 16.3% C, and 64.9% O
   e. 51.7% C, 6.95% H, and 41.3% O

7.46 Calculate the empirical formula for each of the following:
   a. 55.5% Ca and 44.5% S
   b. 78.3% Ba and 21.7% F
   c. 76.0% Zn and 24.0% P
   d. 29.1% Na, 40.6% S, and 30.3% O
   e. 19.8% C, 2.20% H, and 78.0% Cl

7.6 Molecular Formulas

LEARNING GOAL Determine the molecular formula of a substance from the empirical formula and molar mass.

An empirical formula represents the lowest whole-number ratio of atoms in a compound. However, empirical formulas do not necessarily represent the actual number of atoms in a molecule. A molecular formula is related to the empirical formula by a small integer such as 1, 2, or 3.

   Molecular formula = small integer × empirical formula

For example, in **TABLE 7.4**, we see several different compounds that have the same empirical formula, CH_2O. The molecular formulas are related to the empirical formulas by small whole numbers (integers). The same relationship is true for the molar mass and empirical formula mass. The molar mass of each of the different compounds is related to the mass of the empirical formula (30.03 g) by the same small integer.

Molecular formula		Molar mass	
	Small integer		
Empirical formula		Empirical formula mass	

TABLE 7.4 Comparing the Molar Mass of Some Compounds with the Empirical Formula of CH_2O

Compound	Empirical Formula	Molecular Formula	Molar Mass (g)	Integer × Empirical Formula	Integer × Empirical Formula Mass
Acetaldehyde	CH_2O	CH_2O	30.03	$1(CH_2O)$	1 × 30.03
Acetic acid	CH_2O	$C_2H_4O_2$	60.06	$2(CH_2O)$	2 × 30.03
Lactic acid	CH_2O	$C_3H_6O_3$	90.09	$3(CH_2O)$	3 × 30.03
Erythrose	CH_2O	$C_4H_8O_4$	120.12	$4(CH_2O)$	4 × 30.03
Ribose	CH_2O	$C_5H_{10}O_5$	150.15	$5(CH_2O)$	5 × 30.03

PRACTICE PROBLEMS

Try Practice Problems 7.47 and 7.48

Relating Empirical and Molecular Formulas

Once we determine the empirical formula, we can calculate the empirical formula mass in grams. If we are given the molar mass of the compound, we can calculate the value of the small integer.

$$\text{Small integer} = \frac{\text{molar mass of compound}}{\text{empirical formula mass}}$$

ENGAGE 7.7

Why do the molecular formulas $C_2H_4O_2$ and $C_5H_{10}O_5$ both have an empirical formula of CH_2O?

For example, when the molar mass of ribose is divided by the empirical formula mass, the integer is 5.

$$\text{Small integer} = \frac{\text{molar mass of ribose}}{\text{empirical formula mass of } CH_2O} = \frac{150.15 \text{ g}}{30.03 \text{ g}} = 5$$

Multiplying each of the subscripts in the empirical formula (CH_2O) by 5 gives the molecular formula of ribose, $C_5H_{10}O_5$.

$$5 \times \text{empirical formula } (CH_2O) = \text{molecular formula } (C_5H_{10}O_5)$$

Calculating a Molecular Formula

Earlier, in Sample Problem 7.10, we determined that the empirical formula of ascorbic acid (vitamin C) was $C_3H_4O_3$. From the molar mass for ascorbic acid, which is experimentally determined to be 176.12 g, we can calculate its molecular formula as follows:

The mass of the empirical formula $C_3H_4O_3$ is obtained in the same way as molar mass.

$$\text{Empirical formula} = 3 \text{ mol of C} + 4 \text{ mol of H} + 3 \text{ mol of O}$$

$$\text{Empirical formula mass} = (3 \times 12.01 \text{ g}) + (4 \times 1.008 \text{ g}) + (3 \times 16.00 \text{ g})$$

$$= 88.06 \text{ g}$$

$$\text{Small integer} = \frac{\text{molar mass of ascorbic acid}}{\text{empirical formula mass of } C_3H_4O_3} = \frac{176.12 \text{ g}}{88.06 \text{ g}} = 2$$

Multiplying all the subscripts in the empirical formula of ascorbic acid by 2 gives its molecular formula.

$$C_{(3 \times 2)}H_{(4 \times 2)}O_{(3 \times 2)} = C_6H_8O_6 \quad \text{Molecular formula}$$

SAMPLE PROBLEM 7.11   Determination of a Molecular Formula

TRY IT FIRST

Melamine, which is used to make plastic items such as dishes and toys, contains 28.57% C, 4.80% H, and 66.64% N. If the experimental molar mass is 125 g, what is the molecular formula of melamine?

SOLUTION

Some brightly colored dishes are made of melamine.

ANALYZE THE PROBLEM	Given	Need	Connect
	28.57% C, 4.80% H, 66.64% N, molar mass 125 g	molecular formula	empirical formula mass

STEP 1   Obtain the empirical formula and calculate the empirical formula mass.
In a sample size of 100. g of this compound, there are 28.57 g of C, 4.80 g of H, and 66.64 g of N.

$$28.57 \text{ g C} \times \frac{1 \text{ mol C}}{12.01 \text{ g C}} = 2.38 \text{ mol of C}$$

$$4.80 \text{ g H} \times \frac{1 \text{ mol H}}{1.008 \text{ g H}} = 4.76 \text{ mol of H}$$

$$66.64 \text{ g N} \times \frac{1 \text{ mol N}}{14.01 \text{ g N}} = 4.76 \text{ mol of N}$$

Divide the moles of each element by the smallest number of moles, 2.38, to obtain the subscripts of each element in the formula.

$$\frac{2.38 \text{ mol C}}{2.38} = 1.00 \text{ mol of C}$$

$$\frac{4.76 \text{ mol H}}{2.38} = 2.00 \text{ mol of H}$$

$$\frac{4.76 \text{ mol N}}{2.38} = 2.00 \text{ mol of N}$$

Using these values as subscripts, $C_{1.00}H_{2.00}N_{2.00}$, we write the empirical formula for melamine as CH_2N_2.

Now we calculate the molar mass for this empirical formula as follows:

Empirical formula = 1 mol of C + 2 mol of H + 2 mol of N

Empirical formula mass = $(1 \times 12.01) + (2 \times 1.008) + (2 \times 14.01) = 42.05$ g

STEP 2 **Divide the molar mass by the empirical formula mass to obtain a small integer.**

$$\text{Small integer} = \frac{\text{molar mass of melamine}}{\text{empirical formula mass of } CH_2N_2} = \frac{125 \text{ g}}{42.05 \text{ g}} = 2.97$$

STEP 3 **Multiply the empirical formula by the small integer to obtain the molecular formula.** Because the experimental molar mass is close to 3 times the empirical formula mass, the subscripts in the empirical formula are multiplied by 3 to give the molecular formula.

$C_{(1\times3)}H_{(2\times3)}N_{(2\times3)} = C_3H_6N_6$ Molecular formula

SELF TEST 7.11

a. The insecticide lindane has a mass percent composition of 24.78% C, 2.08% H, and 73.14% Cl. If its experimental molar mass is 290 g, what is the molecular formula?
b. Difluoroethane, used as a refrigerant, has a mass percent composition of 36.37% C, 6.10% H, and 57.53% F. If its experimental molar mass is 65 g, what is the molecular formula?

ANSWER

a. $C_6H_6Cl_6$ **b.** $C_2H_4F_2$

PRACTICE PROBLEMS
Try Practice Problems 7.49 to 7.58

PRACTICE PROBLEMS

7.6 Molecular Formulas

Applications

7.47 Write the empirical formula for each of the following substances:
a. H_2O_2, peroxide
b. $C_{18}H_{12}$, chrysene, used in the manufacture of dyes
c. $C_{10}H_{16}O_2$, chrysanthemic acid, in pyrethrum flowers
d. $C_9H_{18}N_6$, altretamine, anticancer medication
e. $C_2H_4N_2O_2$, oxamide, fertilizer

7.48 Write the empirical formula for each of the following substances:
a. $C_6H_6O_3$, pyrogallol, developer in photography
b. $C_6H_{12}O_6$, galactose, carbohydrate
c. $C_8H_6O_4$, terephthalic acid, used in the manufacture of plastic bottles
d. C_6Cl_6, hexachlorobenzene, fungicide
e. $C_{24}H_{16}O_{12}$, laccaic acid, crimson dye

7.49 The carbohydrate fructose found in honey and fruits has an empirical formula of CH_2O. If the experimental molar mass of fructose is 180 g, what is its molecular formula?

7.50 Caffeine has an empirical formula of $C_4H_5N_2O$. If it has an experimental molar mass of 194 g, what is the molecular formula of caffeine?

Coffee beans are a source of caffeine.

7.51 Benzene and acetylene have the same empirical formula, CH. However, benzene has an experimental molar mass of 78 g, and acetylene has an experimental molar mass of 26 g. What are the molecular formulas of benzene and acetylene?

7.52 Glyoxal, used in textiles; maleic acid, used to retard oxidation of fats and oils; and aconitic acid, a plasticizer, all have the same empirical formula, CHO. However, the experimental molar masses are glyoxal 58 g, maleic acid 117 g, and aconitic acid 174 g. What are the molecular formulas of glyoxal, maleic acid, and aconitic acid?

7.53 Mevalonic acid is involved in the biosynthesis of cholesterol. Mevalonic acid contains 48.64% C, 8.16% H, and 43.20% O. If mevalonic acid has an experimental molar mass of 148 g, what is its molecular formula?

7.54 Chloral hydrate, a sedative, contains 14.52% C, 1.83% H, 64.30% Cl, and 19.35% O. If it has an experimental molar mass of 165 g, what is the molecular formula of chloral hydrate?

7.55 Vanillic acid contains 57.14% C, 4.80% H, and 38.06% O, and has an experimental molar mass of 168 g. What is the molecular formula of vanillic acid?

7.56 Lactic acid, the substance that builds up in muscles during exercise, has a mass percent composition of 40.0% C, 6.71% H,

and 53.3% O, and an experimental molar mass of 90. g. What is the molecular formula of lactic acid?

7.57 A sample of nicotine, a poisonous compound found in tobacco leaves, contains 74.0% C, 8.70% H, and 17.3% N. If the experimental molar mass of nicotine is 162 g, what is its molecular formula?

7.58 Adenine, a nitrogen-containing compound found in DNA and RNA, contains 44.5% C, 3.70% H, and 51.8% N. If adenine has an experimental molar mass of 135 g, what is its molecular formula?

UPDATE Prescriptions for Max

After completing the examination and lab tests, Dr. Evans, the veterinarian, prescribes medications that contain clavulanic acid and phenylpropanolamine for Max.

7.60 Phenylpropanolamine (PPA) is used to treat urinary tract infections. The molecular formula of PPA is $C_9H_{13}NO$.
 a. What is the molar mass of PPA?
 b. What is the mass percent of N in PPA?
 c. Max weighs 12 kg. If the dose of PPA is 2.0 mg/kg, how many moles of PPA are given?

Applications

7.59 Clavulanic acid has a molecular formula of $C_8H_9NO_5$.
 a. What is the molar mass of clavulanic acid?
 b. What is the mass percent of C in clavulanic acid?
 c. Max weighs 12 kg. If the dose of clavulanic acid is 2.5 mg/kg, how many moles of clavulanic acid are given?

CONCEPT MAP

CHEMICAL QUANTITIES

Atoms
have an
Atomic Mass
on the
Periodic Table

used to calculate

Moles
contain
Avogadro's Number
(6.022 × 10²³)
and have
Molar Masses
that convert
Moles to Grams   **Grams to Moles**

Mass Percent Composition
give molar ratios for
Empirical Formula
and to find
Molecular Formula

CHAPTER REVIEW

7.1 The Mole

LEARNING GOAL Use Avogadro's
number to calculate the number of
particles in a given number of moles.
Calculate the number of moles of an
element in a given number of moles of
a compound.

- One mole of an element contains 6.022×10^{23} atoms.
- One mole of a compound contains 6.022×10^{23} molecules or
  formula units.

7.2 Molar Mass

LEARNING GOAL Given the chemical formula
of a substance, calculate its molar mass.

- The molar mass (g/mol) of a substance is the
  mass in grams equal numerically to its atomic
  mass, or the sum of the atomic masses, which
  have been multiplied by their subscripts in a
  formula.

7.3 Calculations Using Molar Mass

LEARNING GOAL Use molar
mass to convert between
grams and moles.

- The molar mass is used as a
  conversion factor to change a
  quantity in grams to moles or
  to change a given number of
  moles to grams.

7.4 Mass Percent Composition

LEARNING GOAL Given the formula of
a compound, calculate the mass percent
composition.

- The mass percent composition is
  obtained by dividing the mass in grams
  of each element in a compound by the
  mass of that compound.

7.5 Empirical Formulas

LEARNING GOAL From the mass
percent composition, calculate the
empirical formula for a compound.

- The empirical formula is calculated
  by determining the lowest whole-
  number mole ratio from the grams of
  the elements present in a sample.
- If mole ratios for an empirical formula are not all whole numbers,
  multiply all values by an integer to give whole numbers.

7.6 Molecular Formulas

LEARNING GOAL Determine the
molecular formula of a substance from
the empirical formula and molar mass.

- A molecular formula is equal to, or a
  multiple of, the empirical formula.
- The experimental molar mass, which must be known, is divided by
  the mass of the empirical formula to obtain the small integer used
  to convert the empirical formula to the molecular formula.

KEY TERMS

Avogadro's number The number of items in a mole, equal to
6.022×10^{23}.

empirical formula The simplest or smallest whole-number ratio of
the atoms in a formula.

mass percent composition The percent by mass of the elements in a
formula.

molar mass The mass in grams of 1 mol of an element equal
numerically to its atomic mass. The molar mass of a compound

is equal to the sum of the atomic masses of the elements in the
formula.

mole A group of atoms, molecules, or formula units that contains
6.022×10^{23} of these items.

molecular formula The actual formula that gives the number of
atoms of each type of element in the compound.

CORE CHEMISTRY SKILLS

*The chapter section containing each Core Chemistry Skill is shown in
parentheses at the end of each heading.*

Converting Particles to Moles (7.1)

- In chemistry, atoms, molecules, and ions are counted by the mole
  (abbreviated *mol* in calculations), a unit that contains 6.022×10^{23}
  items, which is Avogadro's number.
- For example, 1 mol of carbon contains 6.022×10^{23} atoms of
  carbon and 1 mol of H_2O contains 6.022×10^{23} molecules of H_2O.
- Avogadro's number is used to convert between particles and moles.

Example: How many moles of nickel contain 2.45×10^{24} Ni atoms?

Answer:

$$2.45 \times 10^{24} \text{ Ni atoms} \times \frac{1 \text{ mol Ni}}{6.022 \times 10^{23} \text{ Ni atoms}}$$

$$= 4.07 \text{ mol of Ni}$$

Calculating Molar Mass (7.2)

- The molar mass of an element is its mass in grams equal numerically to its atomic mass.
- The molar mass of a compound is the sum of the molar mass of
  each element in its chemical formula multiplied by its subscript in
  the formula.

Example: Calculate the molar mass for pinene, $C_{10}H_{16}$, a component of pine tree sap.

Pinene is a component of pine sap.

Answer: $10 \text{ mol C} \times \dfrac{12.01 \text{ g C}}{1 \text{ mol C}} = 120.1 \text{ g of C}$

$16 \text{ mol H} \times \dfrac{1.008 \text{ g H}}{1 \text{ mol H}} = \underline{16.13 \text{ g of H}}$

Molar mass of $C_{10}H_{16}$ = 136.2 g

Using Molar Mass as a Conversion Factor (7.3)

• Molar mass is used as a conversion factor to convert between the moles and grams of a substance.

Example: The frame of a bicycle contains 6500 g of aluminum. How many moles of aluminum are in the bicycle frame?

A racing bicycle has an aluminum frame.

Answer: Equality: 1 mol of Al = 26.98 g of Al

Conversion Factors: $\dfrac{26.98 \text{ g Al}}{1 \text{ mol Al}}$ and $\dfrac{1 \text{ mol Al}}{26.98 \text{ g Al}}$

$6500 \text{ g Al} \times \dfrac{1 \text{ mol Al}}{26.98 \text{ g Al}} = 240 \text{ mol of Al}$

Calculating Mass Percent Composition (7.4)

• The mass of a compound contains a definite proportion by mass of its elements.
• The mass percent of an element in a compound is calculated by dividing the mass of that element by the mass of the compound.

$$\text{Mass percent of an element} = \dfrac{\text{mass of an element}}{\text{mass of the compound}} \times 100\%$$

Example: Dinitrogen tetroxide, N_2O_4, is used in liquid fuels for rockets. If it has a molar mass of 92.02 g, what is the mass percent of nitrogen?

Answer: $\text{Mass \% N} = \dfrac{28.02 \text{ g N}}{92.02 \text{ g } N_2O_4} \times 100\% = 30.45\% \text{ N}$

Calculating an Empirical Formula (7.5)

• The empirical formula or simplest formula represents the lowest whole-number ratio of the atoms and therefore moles of elements in a compound.
• For example, dinitrogen tetroxide, N_2O_4, has the empirical formula NO_2.
• To calculate the empirical formula, the grams of each element are converted to moles and divided by the smallest number of moles to obtain the lowest whole-number ratio.

Example: Calculate the empirical formula for a compound that contains 3.28 g of Cr and 6.72 g of Cl.

Answer: Convert the grams of each element to moles.

$3.28 \text{ g Cr} \times \dfrac{1 \text{ mol Cr}}{52.00 \text{ g Cr}} = 0.0631 \text{ mol of Cr}$

$6.72 \text{ g Cl} \times \dfrac{1 \text{ mol Cl}}{35.45 \text{ g Cl}} = 0.190 \text{ mol of Cl}$

Divide by the smallest number of moles (0.0631) to obtain the empirical formula.

$\dfrac{0.0631 \text{ mol Cr}}{0.0631} = 1.00 \text{ mol of Cr}$

$\dfrac{0.190 \text{ mol Cl}}{0.0631} = 3.01 \text{ mol of Cl}$

Write the empirical formula using the whole-number ratios of moles.

$Cr_{1.00}Cl_{3.01} \longrightarrow CrCl_3$

Calculating a Molecular Formula (7.6)

• A molecular formula is related to the empirical formula by a small whole number (integer) such as 1, 2, or 3.

Molecular formula = small integer × empirical formula

• If the empirical formula mass and the molar mass are known for a compound, an integer can be calculated by dividing the molar mass by the empirical formula mass.

$$\text{Small integer} = \dfrac{\text{molar mass of compound}}{\text{empirical formula mass}}$$

Example: Cymene, a component in oil of thyme, has an empirical formula C_5H_7 and an experimental molar mass of 135 g. What is the molecular formula of cymene?

Answer: Empirical formula = 5 mol of C + 7 mol of H

Empirical formula mass = $(5 \times 12.01 \text{ g}) + (7 \times 1.008 \text{ g})$
= 67.11 g

$\dfrac{\text{Molar mass of cymene}}{\text{Empirical formula mass of } C_5H_7} = \dfrac{135 \text{ g}}{67.11 \text{ g}}$

= 2.01 (round off to 2)

The molecular formula of cymene is calculated by multiplying each of the subscripts in the empirical formula by 2.

$C_{(5 \times 2)}H_{(7 \times 2)} = C_{10}H_{14}$

UNDERSTANDING THE CONCEPTS

The chapter sections to review are shown in parentheses at the end of each problem.

7.61 Using the models of the molecules (black = C, white = H, yellow = S, green = Cl), determine each of the following for models of compounds **1** and **2**: (7.2, 7.4, 7.5, 7.6)

1.   **2.**

a. molecular formula   b. empirical formula
c. molar mass   d. mass percent composition

7.62 Using the models of the molecules (black = C, white = H, yellow = S, red = O), determine each of the following for models of compounds **1** and **2**: (7.2, 7.4, 7.5, 7.6)

1.   **2.**

a. molecular formula   b. empirical formula
c. molar mass   d. mass percent composition

Applications

7.63 A dandruff shampoo contains dipyrithione, $C_{10}H_8N_2O_2S_2$, an antibacterial and antifungal agent. (7.1, 7.2, 7.3, 7.4, 7.5)

Dandruff shampoo contains dipyrithione.

a. What is the empirical formula of dipyrithione?
b. What is the molar mass of dipyrithione?
c. What is the mass percent of O in dipyrithione?
d. How many grams of C are in 25.0 g of dipyrithione?
e. How many moles of dipyrithione are in 25.0 g of dipyrithione?

7.64 Ibuprofen, the anti-inflammatory drug in Advil, has the formula $C_{13}H_{18}O_2$. (7.1, 7.2, 7.3, 7.4, 7.5)

a. What is the empirical formula of ibuprofen?
b. What is the molar mass of ibuprofen?
c. What is the mass percent of O in ibuprofen?
d. How many grams of C are in 0.425 g of ibuprofen?
e. How many moles of ibuprofen are in 2.45 g of ibuprofen?

Ibuprofen is an anti-inflammatory drug.

ADDITIONAL PRACTICE PROBLEMS

7.65 Calculate the mass, in grams, of O in each of the following: (7.3)
a. 50.0 g of H_2O b. 17.5 g of CO_2
c. 48 g of $C_7H_6O_2$

7.66 Calculate the mass, in grams, of Cu in each of the following: (7.3)
a. 4.92 g of $CuCO_3$ b. 0.654 g of Cu_2S
c. 89 g of $Cu_3(PO_4)_2$

7.67 Calculate the mass percent composition for each of the following compounds: (7.4)
a. 3.85 g of Ca and 3.65 g of F
b. 0.389 g of Na and 0.271 g of O
c. 12.4 of K, 17.4 g of Mn, and 20.3 g of O

7.68 Calculate the mass percent composition for each of the following compounds: (7.4)
a. 0.457 g of C and 0.043 g of H
b. 3.65 g of Na, 2.54 g of S, and 3.81 g of O
c. 0.907 g of Na, 1.40 g of Cl, and 1.89 g of O

7.69 Calculate the mass percent composition for each of the following compounds: (7.4)
a. K_2CrO_4 b. $Al(HCO_3)_3$ c. $C_6H_{12}O_6$

7.70 Calculate the mass percent composition for each of the following compounds: (7.4)
a. $CaCO_3$ b. $NaC_2H_3O_2$ c. $Ba(NO_3)_2$

7.71 A mixture contains 0.250 mol of Mn_2O_3 and 20.0 g of MnO_2. (7.1, 7.2, 7.3)
a. How many moles of O are present in the mixture?
b. How many grams of Mn are in the mixture?

7.72 A mixture contains 4.00×10^{23} molecules of PCl_3 and 0.250 mol of PCl_5. (7.1, 7.2, 7.3)
a. How many grams of Cl are present in the mixture?
b. How many moles of P are in the mixture?

7.73 Calculate the empirical formula for each of the following compounds: (7.5)
  a. 2.20 g of S and 7.81 g of F
  b. 6.35 g of Ag, 0.825 g of N, and 2.83 g of O
  c. 43.6% P and 56.4% O
  d. 22.1% Al, 25.4% P, and 52.5% O

7.74 Calculate the empirical formula for each of the following compounds: (7.5)
  a. 5.13 g of Cr and 2.37 g of O
  b. 2.82 g of K, 0.870 g of C, and 2.31 g of O
  c. 61.0% Sn and 39.0% F
  d. 25.9% N and 74.1% O

7.75 Succinic acid is 40.7% C, 5.12% H, and 54.2% O. If it has an experimental molar mass of 118 g, what are the empirical and molecular formulas? (7.4, 7.5, 7.6)

7.76 A compound is 70.6% Hg, 12.5% Cl, and 16.9% O. If it has an experimental molar mass of 568 g, what are the empirical and molecular formulas? (7.4, 7.5, 7.6)

7.77 A sample of a compound contains 1.65×10^{23} atoms of C, 0.552 g of H, and 4.39 g of O. If 1 mol of the compound contains 4 mol of O, what is the molecular formula and molar mass of the compound? (7.4, 7.5, 7.6)

7.78 What is the molecular formula of a compound if 0.500 mol of the compound contains 0.500 mol of Sr, 1.81×10^{24} atoms of O, and 35.5 g of Cl? (7.4, 7.5, 7.6)

Applications

7.79 Calculate the molar mass for each of the following: (7.2)
  a. $ZnSO_4$, zinc sulfate, zinc supplement
  b. $Ca(IO_3)_2$, calcium iodate, iodine source in table salt
  c. $C_5H_8NNaO_4$, monosodium glutamate, flavor enhancer
  d. $C_6H_{12}O_2$, isoamyl formate, used to make artificial fruit syrups

7.80 Calculate the molar mass for each of the following: (7.2)
  a. $MgCO_3$, magnesium carbonate, an antacid
  b. $Au(OH)_3$, gold(III) hydroxide, used in gold plating
  c. $C_{18}H_{34}O_2$, oleic acid, from olive oil
  d. $C_{21}H_{26}O_5$, prednisone, anti-inflammatory

7.81 Aspirin, $C_9H_8O_4$, is used to reduce inflammation and reduce fever. (7.1, 7.2, 7.3, 7.4)
  a. What is the mass percent composition of aspirin?
  b. How many moles of aspirin contain 5.0×10^{24} atoms of C?
  c. How many grams of O are in 7.50 g of aspirin?
  d. How many molecules of aspirin contain 2.50 g of H?

7.82 Ammonium sulfate, $(NH_4)_2SO_4$, is used in fertilizers. (7.1, 7.2, 7.3, 7.4)
  a. What is the mass percent composition of $(NH_4)_2SO_4$?
  b. How many atoms of H are in 0.75 mol of $(NH_4)_2SO_4$?
  c. How many grams of O are in 4.50×10^{23} formula units of $(NH_4)_2SO_4$?
  d. What mass of $(NH_4)_2SO_4$ contains 2.50 g of S?

7.83 Write the empirical formula for each of the following: (7.5)
  a. $C_5H_5N_5$, adenine, found in RNA and DNA
  b. FeC_2O_4, iron(II) oxalate, photographic developer
  c. $C_{16}H_{16}N_4$, stilbamidine, antibiotic for animals
  d. $C_6H_{14}N_2O_2$, lysine, amino acid needed for growth

7.84 Write the empirical formula for each of the following: (7.5)
  a. $C_{12}H_{24}N_2O_4$, carisoprodol, skeletal muscle relaxant
  b. $C_{10}H_{10}O_5$, opianic acid, used to synthesize a drug to treat tuberculosis
  c. $CrCl_3$, chromium(III) chloride, used in chrome plating
  d. $C_{16}H_{16}N_2O_2$, lysergic acid, controlled substance from ergot

7.85 Oleic acid, a component of olive oil, is 76.54% C, 12.13% H, and 11.33% O. The experimental value of the molar mass is 282 g. (7.1, 7.2, 7.3, 7.4, 7.5, 7.6)
  a. What is the molecular formula of oleic acid?
  b. If oleic acid has a density of 0.895 g/mL, how many molecules of oleic acid are in 3.00 mL of oleic acid?

7.86 Iron pyrite, commonly known as "fool's gold," is 46.5% Fe and 53.5% S. (7.1, 7.2, 7.3, 7.4, 7.5, 7.6)
  a. If the empirical formula and the molecular formula are the same, what is the molecular formula of the compound?
  b. If the crystal contains 4.85 g of iron, how many grams of S are in the crystal?

Iron pyrite is commonly known as "fool's gold."

CHALLENGE PROBLEMS

The following problems are related to the topics in this chapter. However, they do not all follow the chapter order, and they require you to combine concepts and skills from several sections. These problems will help you increase your critical thinking skills and prepare for your next exam.

Applications

7.87 A toothpaste contains 0.240% by mass sodium fluoride used to prevent tooth decay and 5.0% by mass KNO_3, which decreases pain and sensitivity. One tube contains 119 g of toothpaste. (7.1, 7.2, 7.3, 7.4)
  a. How many moles of NaF are in the tube of toothpaste?
  b. How many fluoride ions, F^-, are in the tube of toothpaste?
  c. How many grams of sodium ion, Na^+, are in the tube of toothpaste?
  d. How many KNO_3 formula units are in the tube of toothpaste?

7.88 A gold bar is 5.50 cm long, 3.10 cm wide, and 0.300 cm thick. (7.1, 7.2, 7.3, 7.4, 7.5, 7.6)

A gold bar consists of gold atoms.

  a. If gold has a density of 19.3 g/cm³, what is the mass of the gold bar?
  b. How many atoms of gold are in the bar?
  c. When the same mass of gold combines with oxygen, the oxide product has a mass of 111 g. How many moles of O are combined with the gold?
  d. What is the molecular formula of the oxide product if it is the same as the empirical formula?

7.89 Iron(III) chromate, a yellow powder used as a pigment in paints, contains 24.3% Fe, 33.9% Cr, and 41.8% O. If it has an experimental molar mass of 460 g, what are its empirical and molecular formulas? (7.4, 7.5, 7.6)

Iron(III) chromate is a yellow pigment used in paints.

7.90 Sorbic acid, an inhibitor of mold in cheese, has a mass percent composition of 64.27% C, 7.19% H, and 28.54% O. If sorbic acid has an experimental molar mass of 112 g, what is its molecular formula? (7.4, 7.5, 7.6)

Cheese contains sorbic acid, which prevents the growth of mold.

ANSWERS TO ENGAGE QUESTIONS

7.1 There are a large number of atoms in 1 mol: 6.022×10^{23}. A small number of moles will contain a large number of atoms.

7.2 The formula of the dietary supplement zinc acetate, $Zn(C_2H_3O_2)_2$, shows that 1 mol contains 1 mol of Zn, 2×2 or 4 mol of C, 3×2 or 6 mol of H, and 2×2 or 4 mol of O.

7.3 The molar mass obtained by multiplying the molar mass of each element by its subscript in the formula. For $K_2Cr_2O_7$, the molar mass is 2×39.10 g $+ 2 \times 52.00$ g $+ 7 \times 16.00$ g or 294.2 g.

7.4 To calculate the number of atoms of H in 5.00 g of CH_4, first the moles of CH_4 are calculated, then the moles of H are calculated, and finally the atoms of H are calculated.

7.5 In 1 mol of Freon-12, there are 2 mol of fluorine and 2 mol of chlorine. Because the molar mass of chlorine is almost double the molar mass of fluorine, there are more grams of chlorine.

7.6 The atoms in a compound must be whole numbers, so multiplying both by 2 gives the whole-number ratio of 2 to 5.

7.7 By dividing the molecular formula $C_2H_4O_2$ by 2 and the molecular formula $C_5H_{10}O_5$ by 5, we see that they both have the empirical formula CH_2O.

ANSWERS TO SELECTED PROBLEMS

7.1 One mole contains 6.022×10^{23} atoms of an element, molecules of a molecular substance, or formula units of an ionic substance.

7.3
   a. 3.01×10^{23} atoms of C
   b. 7.71×10^{23} molecules of SO_2
   c. 0.0867 mol of Fe
   d. 14.1 mol of C_2H_6O

7.5
   a. 6.00 mol of H   **b.** 8.00 mol of O
   c. 1.20×10^{24} atoms of P   **d.** 4.82×10^{24} atoms of O

7.7
   a. 36 mol of H
   b. 1.0×10^2 mol of C
   c. 0.040 mol of N

7.9
   a. 32.2 mol of C   **b.** 6.22 mol of H
   c. 0.230 mol of O

7.11   **a.** 70.90 g   **b.** 90.08 g   **c.** 262.9 g

7.13   **a.** 83.98 g   **b.** 98.95 g   **c.** 156.7 g

7.15   **a.** 74.55 g   **b.** 159.7 g   **c.** 329.4 g

7.17   **a.** 342.2 g   **b.** 188.18 g   **c.** 365.5 g

7.19   **a.** 151.16 g   **b.** 498.4 g   **c.** 331.4 g

7.21   **a.** 34.5 g   **b.** 112 g   **c.** 5.50 g
             d. 5.14 g   **e.** 9.80×10^4 g

7.23   **a.** 3.03 g   **b.** 38.1 g   **c.** 9.30 g
             d. 24.6 g   **e.** 12.9 g

7.25
   a. 0.760 mol of Ag   **b.** 0.0240 mol of C
   c. 0.881 mol of NH_3 **d.** 0.452 mol of CH_4
   e. 1.53 mol of Fe_2O_3

7.27
   a. 6.25 mol of He   **b.** 0.781 mol of O_2
   c. 0.321 mol of $Al(OH)_3$ **d.** 0.106 mol of Ga_2S_3
   e. 0.430 mol of C_4H_{10}

7.29
   a. 0.188 g of C   **b.** 235 g of C
   c. 0.959 g of C   **d.** 48.9 g of C

7.31
   a. 6.17 mol of H
   b. 54.0 g of C
   c. 27.8 g of C
   d. 0.0465 g or 4.65×10^{-2} g of H

7.33   **a.** 602 g   **b.** 74.7 g

7.35
   a. 66.0 g of N_2O **b.** 0.772 mol of N_2O
   c. 21.6 g of N

7.37
   a. 46.8% Si; 53.2% O
   b. 81.7% C; 18.3% H
   c. 32.3% Na; 45.1% S; 22.6% O
   d. 54.6% C; 9.12% H; 36.3% O

7.39
   a. 39.01% Mg; 60.99% F
   b. 54.09% Ca; 43.18% O; 2.721% H
   c. 40.00% C; 6.714% H; 53.29% O
   d. 28.19% N; 8.115% H; 20.77% P; 42.92% O
   e. 71.55% C; 6.710% H; 4.909% N; 16.82% O

7.41 **a.** 25.94% N   **b.** 26.19% N   **c.** 46.62% N
d. 33.47% N   **e.** 11.96% N

7.43 **a.** N_2O **b.** CH_3 **c.** HNO_3 **d.** $CaCrO_4$

7.45 **a.** K_2S **b.** GaF_3 **c.** Mn_2O_3
d. Li_2CO_3 **e.** $C_5H_8O_3$

7.47 **a.** HO **b.** C_3H_2 **c.** C_5H_8O
d. $C_3H_6N_2$ **e.** CH_2NO

7.49 $C_6H_{12}O_6$

7.51 benzene C_6H_6; acetylene C_2H_2

7.53 $C_6H_{12}O_4$

7.55 $C_8H_8O_4$

7.57 $C_{10}H_{14}N_2$

7.59 **a.** 199.16 g   **b.** 48.24% C   **c.** 1.5×10^{-4} mol

7.61 **1.** **a.** S_2Cl_2 **b.** SCl
c. 135.04 g   **d.** 47.50% S; 52.50% Cl

2. **a.** C_6H_6 **b.** CH
c. 78.11 g   **d.** 92.25% C; 7.743% H

7.63 **a.** C_5H_4NOS **b.** 252.3 g
c. 12.68% O   **d.** 11.9 g of C
e. 0.0991 mol of dipyrithione

7.65 **a.** 44.4 g
b. 12.7 g
c. 13 g

7.67 **a.** 51.3% Ca; 48.7% F
b. 58.9% Na; 41.1% O
c. 24.8% K; 34.7% Mn; 40.5% O

7.69 **a.** 40.27% K; 26.78% Cr; 32.96% O
b. 12.85% Al; 1.440% H; 17.16% C; 68.57% O
c. 40.00% C; 6.716% H; 53.29% O

7.71 **a.** 1.210 mol of O
b. 40.1 g of Mn

7.73 **a.** SF_6
b. $AgNO_3$
c. P_2O_5
d. $AlPO_4$

7.75 The empirical formula is $C_2H_3O_2$; the molecular formula is $C_4H_6O_4$.

7.77 The molecular formula is $C_4H_8O_4$; the molar mass is 120.10 g.

7.79 **a.** 161.48 g
b. 389.9 g
c. 169.11 g
d. 116.16 g

7.81 **a.** 59.99% C; 4.475% H; 35.52% O
b. 0.92 mol of aspirin
c. 2.66 g of O
d. 1.87×10^{23} molecules of aspirin

7.83 **a.** CHN
b. FeC_2O_4
c. C_4H_4N
d. C_3H_7NO

7.85 **a.** $C_{18}H_{34}O_2$
b. 5.73×10^{21} molecules of oleic acid

7.87 **a.** 0.00680 mol of NaF
b. 4.10×10^{21} F^- ions
c. 0.156 g of Na^+
d. 3.5×10^{22} formula units of KNO_3

7.89 The empirical formula is $Fe_2Cr_3O_{12}$; the molecular formula is $Fe_2Cr_3O_{12}$.

CI.7 For parts **a** to **f**, consider the loss of electrons by atoms of the element X, and a gain of electrons by atoms of the element Y. Element X is in Group 2A (2), Period 3, and Y is in Group 7A (17), Period 3. (4.2, 5.4, 5.5, 6.2, 6.3)

$$X \qquad Y \qquad Y$$

   a. Which element is a metal, **X** or **Y**?
   b. Which element is a nonmetal, **X** or **Y**?
   c. What are the ionic charges of **X** and **Y**?
   d. Write the electron configurations for the atoms **X** and **Y**.
   e. Write the formula and name for the ionic compound formed by the ions of **X** and **Y**.

CI.8 A sterling silver bracelet, which is 92.5% silver by mass, has a volume of 25.6 cm^3 and a density of 10.2 g/cm^3. (2.6, 2.8, 4.4, 7.1, 7.5, 7.6)

Sterling silver is 92.5% silver by mass.

   a. What is the mass, in kilograms, of the bracelet?
   b. How many atoms of silver are in the bracelet?
   c. Determine the number of protons and neutrons in each of the two stable isotopes of silver:

$$^{107}_{47}\text{Ag} \qquad ^{109}_{47}\text{Ag}$$

   d. When silver combines with oxygen, a compound forms that contains 93.10% Ag by mass. What are the name and the molecular formula of the oxide product if the molecular formula is the same as the empirical formula?

CI.9 Oxalic acid, a compound found in plants and vegetables such as rhubarb, has a mass percent composition of 26.7% C, 2.24% H, and 71.1% O. Oxalic acid can interfere with respiration and cause kidney or bladder stones. If a large quantity of rhubarb leaves is ingested, the oxalic acid can be toxic. The lethal dose (LD$_{50}$) in rats for oxalic acid is 375 mg/kg. Rhubarb leaves contain about 0.5% by mass of oxalic acid. (2.7, 7.4, 7.5, 7.6)

Rhubarb leaves are a source of oxalic acid.

   a. What is the empirical formula of oxalic acid?
   b. If oxalic acid has an experimental molar mass of 90. g, what is its molecular formula?
   c. Using the LD$_{50}$, how many grams of oxalic acid would be toxic for a 160-lb person?
   d. How many kilograms of rhubarb leaves would the person in part **c** need to eat to reach the toxic level of oxalic acid?

CI.10 The active ingredient in an antacid tablet is calcium carbonate. One tablet contains 500. mg of calcium carbonate. (2.7, 6.3, 7.2, 7.3, 7.4)

   a. What is the chemical formula of calcium carbonate?
   b. What is the molar mass of calcium carbonate?
   c. What is the mass percent composition of calcium carbonate?
   d. How many moles of calcium carbonate are in 12 antacid tablets?
   e. If a person takes two antacid tablets, how many grams of calcium are obtained?
   f. If the Daily Value (DV) for Ca^{2+} to maintain bone strength in older women is 1500 mg, how many antacid tablets are needed each day?

CI.11 Oseltamivir, C$_{16}$H$_{28}$N$_2$O$_4$, is a drug that is used to treat influenza. The preparation of oseltamivir begins with the extraction of shikimic acid from the seedpods of star anise. From 2.6 g of star anise, 0.13 g of shikimic acid can be obtained and used to produce one capsule containing 75 mg of oseltamivir. The usual adult dosage for treatment of influenza is two capsules of oseltamivir daily for 5 days. (2.7, 6.5, 7.2, 7.3, 7.4, 7.5, 7.6)

The spice called star anise is a plant source of shikimic acid.

Shikimic acid is the basis for the antiviral drug oseltamivir.

   a. What is the empirical formula of oseltamivir?
   b. What is the mass percent composition of oseltamivir?
   c. If black spheres are carbon atoms, white spheres are hydrogen atoms, and red spheres are oxygen atoms, what is the molecular formula of shikimic acid?
   d. How many moles of shikimic acid are contained in 130 g of shikimic acid?
   e. How many capsules containing 75 mg of oseltamivir could be produced from 155 g of star anise?
   f. How many grams of C are in one dose (75 mg) of oseltamivir?
   g. How many kilograms of oseltamivir would be needed to treat all the people in a city with a population of 500 000 people if each person consumes two oseltamivir capsules a day for 5 days?

CI.12 The compound butyric acid gives rancid butter its characteristic odor. (2.7, 2.8, 6.5, 7.1, 7.2, 7.3, 7.4, 7.5)

    a. If black spheres are carbon atoms, white spheres are hydrogen atoms, and red spheres are oxygen atoms, what is the molecular formula of butyric acid?

    b. What is the empirical formula of butyric acid?

    c. What is the mass percent composition of butyric acid?

    d. How many grams of C are in 0.850 g of butyric acid?

    e. How many grams of butyric acid contain 3.28×10^{23} O atoms?

    f. Butyric acid has a density of 0.959 g/mL at 20 °C. How many moles of butyric acid are contained in 1.56 mL of butyric acid?

Butyric acid produces the characteristic odor of rancid butter.

ANSWERS

CI.7   **a.** X is a metal; elements in Group 2A (2) are metals.
      b. Y is a nonmetal; elements in Group 7A (17) are nonmetals.
      c. X^{2+}, Y^-
      d. $X = 1s^2 2s^2 2p^6 3s^2$ $Y = 1s^2 2s^2 2p^6 3s^2 3p^5$
      e. $MgCl_2$, magnesium chloride

CI.9   **a.** CHO_2
      b. $C_2H_2O_4$
      c. 27 g of oxalic acid
      d. 5 kg of rhubarb

CI.11   **a.** $C_8H_{14}NO_2$
      b. 61.52% C; 9.033% H; 8.969% N; 20.49% O
      c. $C_7H_{10}O_5$
      d. 0.75 mol of shikimic acid
      e. 59 capsules
      f. 0.046 g of C
      g. 4×10^2 kg

Chemical Reactions

Natalie was recently diagnosed with mild pulmonary emphysema due to secondhand cigarette smoke. She was referred to Angela, an exercise physiologist, who assesses Natalie's condition by connecting her to an electrocardiogram (ECG or EKG), a pulse oximeter, and a blood pressure cuff. The ECG records the electrical activity of Natalie's heart, which is used to measure her heart rate and rhythm and detect the possible presence of heart damage. The pulse oximeter measures her pulse and the saturation level of oxygen (the percentage of hemoglobin that is saturated with O_2) in her arterial blood. The blood pressure cuff determines the pressure exerted by the heart in pumping her blood.

To determine possible heart disease, Natalie has an exercise stress test on a treadmill to measure how her heart rate and blood pressure respond to exertion by walking faster as Angela increases the treadmill's slope. Angela measures the heart rate and blood pressure first at rest and then on the treadmill. Angela uses a face mask to collect expired air and measures Natalie's maximal volume of oxygen uptake, or $V_{O_2 \, max}$.

CAREER

Exercise Physiologist

Exercise physiologists work with athletes as well as patients diagnosed with diabetes, heart disease, pulmonary disease, or other chronic disabilities or diseases. Patients who have one of these diseases are often prescribed exercise as a form of treatment, and they are referred to an exercise physiologist. The exercise physiologist evaluates the patient's overall health and then creates a customized exercise program for that individual. The program for an athlete might focus on reducing the number of injuries, whereas a program for a cardiac patient would focus on strengthening the heart muscles. The exercise physiologist also monitors the patient for improvement and determines if the exercise is helping to reduce or reverse the progression of the disease.

UPDATE Improving Natalie's Overall Fitness

Natalie's test results indicate that her blood oxygen level is below normal. You can view Natalie's test results and lung function diagnosis in the **UPDATE Improving Natalie's Overall Fitness**, page 232. After reviewing the test results, Angela teaches Natalie how to improve her respiration and her overall fitness.

REVIEW

Writing Ionic Formulas (6.2)

Naming Ionic Compounds (6.3)

Writing the Names and Formulas for Molecular Compounds (6.5)

Ag   Ag₂S

FIGURE 8.1 The tarnishing of silver is a chemical change that produces a new substance with new properties.

ENGAGE 8.1

Why is the formation of rust a chemical change?

8.1 Equations for Chemical Reactions

LEARNING GOAL Identify a balanced chemical equation; determine the number of atoms in the reactants and products.

Chemical reactions occur everywhere. The fuel in our cars burns with oxygen to make the car move and run the air conditioner. When we cook our food or bleach our hair, chemical reactions take place. In our bodies, chemical reactions convert food into molecules that build muscles and move them. In the leaves of trees and plants, carbon dioxide and water are converted into carbohydrates. Some chemical reactions are simple, whereas others are quite complex. However, they can all be written with chemical equations that chemists use to describe chemical reactions. In every chemical reaction, the atoms in the reacting substances, called *reactants*, are rearranged to give new substances called *products*.

A *chemical change* occurs when a substance is converted into one or more new substances. For example, when silver tarnishes, the shiny silver metal (Ag) reacts with sulfur (S) to become the dull, black substance we call tarnish (Ag₂S) (see **FIGURE 8.1**).

A *chemical reaction* always involves chemical change because atoms of the reacting substances form new combinations with new properties. For example, a chemical reaction takes place when a piece of iron (Fe) combines with oxygen (O_2) in the air to produce a new substance, rust (Fe_2O_3), which has a reddish-brown color. During a chemical change, new properties become visible, which are an indication that a chemical reaction has taken place (see **TABLE 8.1**).

TABLE 8.1 Types of Evidence of a Chemical Reaction

1. Change in color	**2.** Formation of a gas (bubbles)
Fe Fe₂O₃ Iron nails change color when they react with oxygen to form rust.	Bubbles (gas) form when CaCO₃ reacts with acid.
3. Formation of a solid (precipitate)	**4.** Heat (or a flame) produced or heat absorbed
A yellow solid forms when potassium iodide is added to lead nitrate.	Methane gas burns in the air with a hot flame.

Writing a Chemical Equation

When you build a model airplane, prepare a new recipe, or mix a medication, you follow a set of directions. These directions tell you what materials to use and the products you will obtain. In chemistry, a *chemical equation* tells us the materials we need and the products that will form.

Suppose you work in a bicycle shop, assembling wheels and frames into bicycles. You could represent this process by a simple equation:

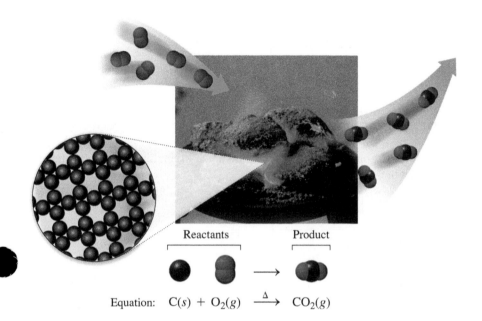

Equation: 2 Wheels + 1 Frame ⟶ 1 Bicycle

Reactants   Product

When you burn charcoal in a grill, the carbon in the charcoal combines with oxygen to form carbon dioxide. We can represent this reaction by a chemical equation.

Reactants   Product

Equation: $C(s) + O_2(g) \xrightarrow{\Delta} CO_2(g)$

In a chemical equation, the formulas of the reactants are written on the left of the arrow and the formulas of the products on the right.

In a **chemical equation**, the formulas of the **reactants** are written on the left of the arrow and the formulas of the **products** on the right. When there are two or more formulas on the same side, they are separated by plus (+) signs. The chemical equation for burning carbon is *balanced* because there is one carbon atom and two oxygen atoms in both the reactants and the products.

Generally, each formula in an equation is followed by an abbreviation, in parentheses, that gives the physical state of the substance: solid (*s*), liquid (*l*), or gas (*g*). If a substance is dissolved in water, it is in an aqueous (*aq*) solution. The delta sign (Δ) indicates that heat was used to start the reaction. **TABLE 8.2** summarizes some of the symbols used in equations.

Identifying a Balanced Chemical Equation

When a chemical reaction takes place, the bonds between the atoms of the reactants are broken, and new bonds are formed to give the products. All atoms are conserved, which means that atoms cannot be gained, lost, or changed into other types of atoms. Every chemical reaction must be written as a **balanced equation**, which shows the same number of atoms for each element in the reactants and in the products.

Now consider the balanced reaction in which hydrogen reacts with oxygen to form water, written as follows:

$H_2(g) + O_2(g) \longrightarrow H_2O(g)$ Not balanced

In the balanced equation, there are whole numbers called **coefficients** in front of the formulas. On the reactant side, the coefficient of 2 in front of the H_2 formula represents two molecules of hydrogen, which is 4 atoms of H. A coefficient of 1 is understood for O_2, which gives 2 atoms of O. On the product side, the coefficient of 2 in front of the H_2O formula

TABLE 8.2	Some Symbols Used in Writing Equations
Symbol	**Meaning**
+	Separates two or more formulas
⟶	Reacts to form products
(*s*)	Solid
(*l*)	Liquid
(*g*)	Gas
(*aq*)	Aqueous
$\xrightarrow{\Delta}$	Reactants are heated

$2H_2(g) + O_2(g) \longrightarrow 2H_2O(g)$

Balanced

H	O	H	O
Reactant atoms	=	Product atoms	

represents 2 molecules of water. Because the coefficient of 2 multiplies all the atoms in H_2O, there are 4 hydrogen atoms and 2 oxygen atoms in the products. Because there are the same number of hydrogen atoms and oxygen atoms in the reactants as in the products, we know that the equation is balanced. This illustrates the *Law of Conservation of Matter*, which states that matter cannot be created or destroyed during a chemical reaction.

ENGAGE 8.2

How can we determine that a chemical equation is balanced?

SAMPLE PROBLEM 8.1   Number of Atoms in Balanced Chemical Equations

TRY IT FIRST

Indicate the number of each type of atom in the following balanced chemical equation:

$$Fe_2S_3(s) + 6HCl(aq) \longrightarrow 2FeCl_3(aq) + 3H_2S(g)$$

	Reactants	Products
Fe		
S		
H		
Cl		

SOLUTION

The total number of atoms in each formula is obtained by multiplying the coefficient by each subscript in the chemical formula.

	Reactants	Products
Fe	2 (1 × 2)	2 (2 × 1)
S	3 (1 × 3)	3 (3 × 1)
H	6 (6 × 1)	6 (3 × 2)
Cl	6 (6 × 1)	6 (2 × 3)

SELF TEST 8.1

State the number of atoms of each element in the reactants and in the products for each of the following:

a. When ethane, C_2H_6, burns in oxygen, the products are carbon dioxide and water. The balanced equation is written as

$$2C_2H_6(g) + 7O_2(g) \xrightarrow{\Delta} 4CO_2(g) + 6H_2O(g)$$

b. Iron(III) oxide reacts with carbon monoxide to produce iron and carbon dioxide. The balanced chemical equation is written as

$$Fe_2O_3(s) + 3CO(g) \longrightarrow 2Fe(s) + 3CO_2(g)$$

ANSWER

a. In both the reactants and products, there are 4 C atoms, 12 H atoms, and 14 O atoms.
b. In both the reactants and products, there are 2 Fe atoms, 6 O atoms, and 3 C atoms.

PRACTICE PROBLEMS

Try Practice Problems 8.1 to 8.6

PRACTICE PROBLEMS

8.1 Equations for Chemical Reactions

8.1 State the number of atoms of oxygen in the reactants and in the products for each of the following equations:
 a. $3NO_2(g) + H_2O(l) \longrightarrow NO(g) + 2HNO_3(aq)$
 b. $5C(s) + 2SO_2(g) \longrightarrow CS_2(g) + 4CO(g)$
 c. $2C_2H_2(g) + 5O_2(g) \xrightarrow{\Delta} 4CO_2(g) + 2H_2O(g)$
 d. $N_2H_4(g) + 2H_2O_2(g) \longrightarrow N_2(g) + 4H_2O(g)$

8.2 State the number of atoms of oxygen in the reactants and in the products for each of the following equations:
 a. $CH_4(g) + 2O_2(g) \xrightarrow{\Delta} CO_2(g) + 2H_2O(g)$
 b. $4P(s) + 5O_2(g) \longrightarrow P_4O_{10}(s)$
 c. $4NH_3(g) + 6NO(g) \longrightarrow 5N_2(g) + 6H_2O(g)$
 d. $6CO_2(g) + 6H_2O(l) \longrightarrow C_6H_{12}O_6(aq) + 6O_2(g)$
 Glucose

8.3 Determine whether each of the following equations is balanced or not balanced:
a. $S(s) + O_2(g) \longrightarrow SO_3(g)$
b. $Al(s) + Cl_2(g) \longrightarrow AlCl_3(s)$
c. $H_2SO_4(aq) + 2NaOH(aq) \longrightarrow 2H_2O(l) + Na_2SO_4(aq)$
d. $C_3H_8(g) + 5O_2(g) \xrightarrow{\Delta} 3CO_2(g) + 4H_2O(g)$

8.4 Determine whether each of the following equations is balanced or not balanced:
a. $PCl_3(s) + Cl_2(g) \longrightarrow PCl_5(s)$
b. $CO(g) + 2H_2(g) \longrightarrow CH_4O(g)$
c. $2KClO_3(s) \longrightarrow 2KCl(s) + O_2(g)$
d. $Mg(s) + N_2(g) \longrightarrow Mg_3N_2(s)$

8.5 All of the following are balanced equations. State the number of atoms of each element in the reactants and in the products.
a. $2Na(s) + Cl_2(g) \longrightarrow 2NaCl(s)$
b. $PCl_3(l) + 3H_2(g) \longrightarrow PH_3(g) + 3HCl(g)$
c. $P_4O_{10}(s) + 6H_2O(l) \longrightarrow 4H_3PO_4(aq)$

8.6 All of the following are balanced equations. State the number of atoms of each element in the reactants and in the products.
a. $2N_2(g) + 3O_2(g) \longrightarrow 2N_2O_3(g)$
b. $Al_2O_3(s) + 6HCl(aq) \longrightarrow 3H_2O(l) + 2AlCl_3(aq)$
c. $C_5H_{12}(l) + 8O_2(g) \xrightarrow{\Delta} 5CO_2(g) + 6H_2O(g)$

8.2 Balancing a Chemical Equation

LEARNING GOAL Write a balanced chemical equation from the formulas of the reactants and products for a chemical reaction.

The chemical reaction that occurs in the flame of a gas burner you use in the laboratory or a gas cooktop is the reaction of methane gas, CH_4, and oxygen to produce carbon dioxide and water. We now show the process of writing and balancing the chemical equation for this reaction in Sample Problem 8.2.

SAMPLE PROBLEM 8.2   Writing and Balancing a Chemical Equation

TRY IT FIRST

The chemical reaction of methane gas (CH_4) and oxygen gas (O_2) produces the gases carbon dioxide (CO_2) and water (H_2O). Write a balanced chemical equation for this reaction.

SOLUTION

	Given	Need	Connect
ANALYZE THE PROBLEM	reactants, products	balanced equation	equal numbers of atoms in reactants and products

STEP 1   Write an equation using the correct formulas for the reactants and products.

$$CH_4(g) + O_2(g) \xrightarrow{\Delta} CO_2(g) + H_2O(g)$$

CH₄   O₂   CO₂   H₂O

In a balanced chemical equation, the number of each type of atom must be the same in the reactants and in the products.

STEP 2   **Count the atoms of each element in the reactants and products.** When we count the atoms on the reactant side and the atoms on the product side, we see that there are more H atoms in the reactants and more O atoms in the products.

$$CH_4(g) + O_2(g) \xrightarrow{\Delta} CO_2(g) + H_2O(g)$$

Reactants	Products	
1 C atom	1 C atom	Balanced
4 H atoms	2 H atoms	Not balanced
2 O atoms	3 O atoms	Not balanced

STEP 3   **Use coefficients to balance each element.** We will start by balancing the H atoms in CH_4 because it has the most atoms. By placing a coefficient of 2 in front of the formula for H_2O, a total of 4 H atoms in the products is obtained. *Only use coefficients to balance an equation. Do not change any of the subscripts: This would alter the chemical formula of a reactant or product.*

$$CH_4(g) + O_2(g) \xrightarrow{\Delta} CO_2(g) + 2H_2O(g)$$

Reactants	Products	
1 C atom	1 C atom	Balanced
4 H atoms	4 H atoms	Balanced
2 O atoms	4 O atoms	Not balanced

We can balance the O atoms on the reactant side by placing a coefficient of 2 in front of the formula O_2. There are now 4 O atoms in both the reactants and products.

$$CH_4(g) + 2O_2(g) \xrightarrow{\Delta} CO_2(g) + 2H_2O(g) \quad \text{Balanced}$$

STEP 4   **Check the final equation to confirm it is balanced.**

$$CH_4(g) + 2O_2(g) \xrightarrow{\Delta} CO_2(g) + 2H_2O(g) \quad \text{The equation is balanced.}$$

Reactants	Products	
1 C atom	1 C atom	Balanced
4 H atoms	4 H atoms	Balanced
4 O atoms	4 O atoms	Balanced

In a balanced chemical equation, the coefficients must be the *lowest possible whole numbers*. Suppose you had obtained the following for the balanced equation:

$$2CH_4(g) + 4O_2(g) \xrightarrow{\Delta} 2CO_2(g) + 4H_2O(g) \quad \text{Incorrect}$$

Although there are equal numbers of atoms on both sides of the equation, this is not written correctly. To obtain coefficients that are the lowest whole numbers, we divide all the coefficients by 2.

SELF TEST 8.2

Balance each of the following chemical equations:

a. $Al(s) + Cl_2(g) \longrightarrow AlCl_3(s)$
b. $Cr_2O_3(s) + CCl_4(l) \longrightarrow CrCl_3(s) + COCl_2(g)$

ANSWER

a. $2Al(s) + 3Cl_2(g) \longrightarrow 2AlCl_3(s)$
b. $Cr_2O_3(s) + 3CCl_4(l) \longrightarrow 2CrCl_3(s) + 3COCl_2(g)$

Whole-Number Coefficients

Sometimes the coefficients of the compounds in an equation need to be increased to give whole numbers for the coefficients. Then we need to adjust the coefficients and count the total atoms on both sides once again, as shown in Sample Problem 8.3.

SAMPLE PROBLEM 8.3   Balancing a Chemical Equation with Whole-Number Coefficients

TRY IT FIRST

Acetylene, C_2H_2, is used to produce high temperatures for welding by reacting it with O_2 to produce CO_2 and H_2O. All of the compounds are gases. Write a balanced chemical equation with whole-number coefficients for this reaction.

SOLUTION

	Given	Need	Connect
ANALYZE THE PROBLEM	reactants, products	balanced equation	equal numbers of atoms in reactants and products

STEP 1   Write an equation using the correct formulas of the reactants and products.

$$C_2H_2(g) + O_2(g) \xrightarrow{\Delta} CO_2(g) + H_2O(g) \quad \text{Not balanced}$$
Acetylene

STEP 2   Count the atoms of each element in the reactants and products. When we count the atoms on the reactant side and the product side, we see that there are more C atoms in the reactants and more O atoms in the products.

$$C_2H_2(g) + O_2(g) \xrightarrow{\Delta} CO_2(g) + H_2O(g)$$

Reactants	Products	
2 C atoms	1 C atom	Not balanced
2 H atoms	2 H atoms	Balanced
2 O atoms	3 O atoms	Not balanced

STEP 3   Use coefficients to balance each element. We start by balancing the C in C_2H_2 by placing a coefficient of 2 in front of the formula for CO_2. When we recheck all the atoms, there are 2 C atoms and 2 H atoms in both the reactants and products.

$$C_2H_2(g) + O_2(g) \xrightarrow{\Delta} 2CO_2(g) + H_2O(g)$$

Reactants	Products	
2 C atoms	2 C atoms	Balanced
2 H atoms	2 H atoms	Balanced
2 O atoms	5 O atoms	Not balanced

To balance the O atoms, we place a coefficient of 5/2 in front of the formula for O_2, which gives a total of 5 O atoms on each side of the equation.

$$C_2H_2(g) + \frac{5}{2}O_2(g) \xrightarrow{\Delta} 2CO_2(g) + H_2O(g)$$

Reactants	Products	
2 C atoms	2 C atoms	Balanced
2 H atoms	2 H atoms	Balanced
5 O atoms	5 O atoms	Balanced

Now the equation is balanced for atoms, but the coefficient for O_2 is a fraction. To obtain all whole-number coefficients, we multiply *all* the coefficients by 2.

$$2C_2H_2(g) + 5O_2(g) \xrightarrow{\Delta} 4CO_2(g) + 2H_2O(g)$$

STEP 4 **Check the final equation to confirm it is balanced.** A check of the total number of atoms indicates that the equation is now balanced with whole-number coefficients.

$$2C_2H_2(g) + 5O_2(g) \xrightarrow{\Delta} 4CO_2(g) + 2H_2O(g)$$

Reactants	Products	
4 C atoms	4 C atoms	Balanced
4 H atoms	4 H atoms	Balanced
10 O atoms	10 O atoms	Balanced

SELF TEST 8.3

Balance each of the following chemical equations:
a. $NO_2(g) + H_2(g) \longrightarrow NH_3(g) + H_2O(g)$
b. $Fe(aq) + F_2(g) \longrightarrow FeF_3(s)$

ANSWER

a. $2NO_2(g) + 7H_2(g) \longrightarrow 2NH_3(g) + 4H_2O(g)$
b. $2Fe(aq) + 3F_2(g) \longrightarrow 2FeF_3(s)$

Equations with Polyatomic Ions

Sometimes an equation contains the same polyatomic ion in both the reactants and the products. Then we can balance the polyatomic ions as a group on both sides of the equation, as shown in Sample Problem 8.4.

SAMPLE PROBLEM 8.4 **Balancing Chemical Equations with Polyatomic Ions**

TRY IT FIRST

Balance the following chemical equation:

$$Na_3PO_4(aq) + MgCl_2(aq) \longrightarrow Mg_3(PO_4)_2(s) + NaCl(aq)$$

SOLUTION

	Given	Need	Connect
ANALYZE THE PROBLEM	reactants, products	balanced equation	equal numbers of atoms in reactants and products

STEP 1 **Write an equation using the correct formulas for the reactants and products.**

$$Na_3PO_4(aq) + MgCl_2(aq) \longrightarrow Mg_3(PO_4)_2(s) + NaCl(aq) \quad \text{Not balanced}$$

STEP 2 **Count the atoms of each element in the reactants and products.** When we count the number of ions in the reactants and products, we find that the equation is not balanced. In this equation, we can balance the phosphate ion as a group of atoms because it appears on both sides of the equation.

Mg^{2+} Cl^- Na^+ PO_4^{3-}

ENGAGE 8.3

What is the evidence for a chemical reaction when $Na_3PO_4(aq)$ and $MgCl_2(aq)$ are mixed?

$$Na_3PO_4(aq) + MgCl_2(aq) \longrightarrow Mg_3(PO_4)_2(s) + NaCl(aq)$$

Reactants	Products	
3 Na^+	1 Na^+	Not balanced
1 PO_4^{3-}	2 PO_4^{3-}	Not balanced
1 Mg^{2+}	3 Mg^{2+}	Not balanced
2 Cl^-	1 Cl^-	Not balanced

STEP 3   **Use coefficients to balance each element.** We begin with the formula that has the highest subscript values, which in this equation is $Mg_3(PO_4)_2$. The subscript 3 in $Mg_3(PO_4)_2$ is used as a coefficient for $MgCl_2$ to balance magnesium. The subscript 2 in $Mg_3(PO_4)_2$ is used as a coefficient for Na_3PO_4 to balance the phosphate ion.

$$2Na_3PO_4(aq) + 3MgCl_2(aq) \longrightarrow Mg_3(PO_4)_2(s) + NaCl(aq)$$

Reactants	Products	
6 Na^+	1 Na^+	Not balanced
2 PO_4^{3-}	2 PO_4^{3-}	Balanced
3 Mg^{2+}	3 Mg^{2+}	Balanced
6 Cl^-	1 Cl^-	Not balanced

In the reactants and products, we see that the sodium and chloride ions are not yet balanced. A coefficient of 6 is placed in front of the NaCl to balance the equation.

$$2Na_3PO_4(aq) + 3MgCl_2(aq) \longrightarrow Mg_3(PO_4)_2(s) + 6NaCl(aq)$$

STEP 4   **Check the final equation to confirm it is balanced.** A check of the total number of ions confirms the equation is balanced. A coefficient of 1 is understood and not usually written.

$$2Na_3PO_4(aq) + 3MgCl_2(aq) \longrightarrow Mg_3(PO_4)_2(s) + 6NaCl(aq) \quad \text{Balanced}$$

Reactants	Products	
6 Na^+	6 Na^+	Balanced
2 PO_4^{3-}	2 PO_4^{3-}	Balanced
3 Mg^{2+}	3 Mg^{2+}	Balanced
6 Cl^-	6 Cl^-	Balanced

SELF TEST 8.4

Balance each of the following chemical equations:
a. $Pb(NO_3)_2(aq) + AlBr_3(aq) \longrightarrow PbBr_2(s) + Al(NO_3)_3(aq)$
b. $HNO_3(aq) + Fe_2(SO_4)_3(aq) \longrightarrow H_2SO_4(aq) + Fe(NO_3)_3(aq)$

ANSWER

a. $3Pb(NO_3)_2(aq) + 2AlBr_3(aq) \longrightarrow 3PbBr_2(s) + 2Al(NO_3)_3(aq)$
b. $6HNO_3(aq) + Fe_2(SO_4)_3(aq) \longrightarrow 3H_2SO_4(aq) + 2Fe(NO_3)_3(aq)$

PRACTICE PROBLEMS

Try Practice Problems 8.7 to 8.18

PRACTICE PROBLEMS

8.2 Balancing a Chemical Equation

8.7 Balance each of the following chemical equations:
a. $N_2(g) + O_2(g) \longrightarrow NO(g)$
b. $HgO(s) \xrightarrow{\Delta} Hg(l) + O_2(g)$
c. $Fe(s) + O_2(g) \longrightarrow Fe_2O_3(s)$
d. $Na(s) + Cl_2(g) \longrightarrow NaCl(s)$
e. $Cu_2O(s) + O_2(g) \longrightarrow CuO(s)$

8.8 Balance each of the following chemical equations:
a. $Ca(s) + Br_2(l) \longrightarrow CaBr_2(s)$
b. $P_4(s) + O_2(g) \longrightarrow P_4O_{10}(s)$
c. $C_4H_8(g) + O_2(g) \xrightarrow{\Delta} CO_2(g) + H_2O(g)$
d. $HNO_3(aq) + Ca(OH)_2(aq) \longrightarrow H_2O(l) + Ca(NO_3)_2(aq)$
e. $Fe_2O_3(s) + C(s) \longrightarrow Fe(s) + CO(g)$

8.9 Balance each of the following chemical equations:
a. $Mg(s) + AgNO_3(aq) \longrightarrow Ag(s) + Mg(NO_3)_2(aq)$
b. $CuCO_3(s) \longrightarrow CuO(s) + CO_2(g)$
c. $Al(s) + CuSO_4(aq) \longrightarrow Cu(s) + Al_2(SO_4)_3(aq)$
d. $Pb(NO_3)_2(aq) + NaCl(aq) \longrightarrow PbCl_2(s) + NaNO_3(aq)$
e. $HCl(aq) + Al(s) \longrightarrow H_2(g) + AlCl_3(aq)$

8.10 Balance each of the following chemical equations:
a. $HNO_3(aq) + Zn(s) \longrightarrow H_2(g) + Zn(NO_3)_2(aq)$
b. $H_2SO_4(aq) + Al(s) \longrightarrow H_2(g) + Al_2(SO_4)_3(aq)$
c. $K_2SO_4(aq) + BaCl_2(aq) \longrightarrow BaSO_4(s) + KCl(aq)$
d. $CaCO_3(s) \longrightarrow CaO(s) + CO_2(g)$
e. $AlCl_3(aq) + KOH(aq) \longrightarrow Al(OH)_3(s) + KCl(aq)$

8.11 Balance each of the following chemical equations:
a. $Fe_2O_3(s) + CO(g) \longrightarrow Fe(s) + CO_2(g)$
b. $Li_3N(s) \longrightarrow Li(s) + N_2(g)$
c. $HBr(aq) + Al(s) \longrightarrow H_2(g) + AlBr_3(aq)$
d. $Ba(OH)_2(aq) + Na_3PO_4(aq)$
$\longrightarrow Ba_3(PO_4)_2(s) + NaOH(aq)$
e. $As_4S_6(s) + O_2(g) \longrightarrow As_4O_6(s) + SO_2(g)$

8.12 Balance each of the following chemical equations:
a. $K(s) + H_2O(l) \longrightarrow H_2(g) + KOH(aq)$
b. $Cr(s) + S_8(s) \longrightarrow Cr_2S_3(s)$
c. $BCl_3(s) + H_2O(l) \longrightarrow H_3BO_3(aq) + HCl(aq)$
d. $H_2SO_4(aq) + Fe(OH)_3(s) \longrightarrow H_2O(l) + Fe_2(SO_4)_3(aq)$
e. $BaCl_2(aq) + Na_3PO_4(aq) \longrightarrow Ba_3(PO_4)_2(s) + NaCl(aq)$

8.13 Write a balanced equation using the correct formulas and include conditions (s, l, g, or aq) for each of the following chemical reactions:
a. Lithium metal reacts with liquid water to form hydrogen gas and aqueous lithium hydroxide.
b. Solid phosphorus reacts with chlorine gas to form solid phosphorus pentachloride.

c. Solid iron(II) oxide reacts with carbon monoxide gas to form solid iron and carbon dioxide gas.
d. Liquid pentene (C_5H_{10}) burns in oxygen gas to form carbon dioxide gas and water vapor.
e. Hydrogen sulfide gas and solid iron(III) chloride react to form solid iron(III) sulfide and hydrogen chloride gas.

8.14 Write a balanced equation using the correct formulas, and include conditions (s, l, g, or aq) for each of the following chemical reactions:
a. Solid sodium carbonate decomposes to produce solid sodium oxide and carbon dioxide gas.
b. Nitrogen oxide gas reacts with carbon monoxide gas to produce nitrogen gas and carbon dioxide gas.
c. Iron metal reacts with solid sulfur to produce solid iron(III) sulfide.
d. Solid calcium reacts with nitrogen gas to produce solid calcium nitride.
e. In the *Apollo* lunar module, hydrazine gas, N_2H_4, reacts with dinitrogen tetroxide gas to produce gaseous nitrogen and water vapor.

Applications

8.15 Dinitrogen oxide, also known as laughing gas, is a sweet-tasting gas used in dentistry. When solid ammonium nitrate is heated, the products are gaseous water and gaseous dinitrogen oxide. Write the balanced chemical equation for the reaction.

8.16 When ethanol $C_2H_6O(aq)$ is consumed, it reacts with oxygen gas in the body to produce gaseous carbon dioxide and liquid water. Write the balanced chemical equation for the reaction.

8.17 In the body, the amino acid alanine $C_3H_7NO_2(aq)$ reacts with oxygen gas to produce gaseous carbon dioxide, liquid water, and urea, $CH_4N_2O(aq)$. Write the balanced chemical equation for the reaction.

8.18 In the body, the amino acid asparagine $C_4H_8N_2O_3(aq)$ reacts with oxygen gas to produce gaseous carbon dioxide, liquid water, and urea, $CH_4N_2O(aq)$. Write the balanced chemical equation for the reaction.

8.3 Types of Chemical Reactions

LEARNING GOAL Identify a chemical reaction as a combination, decomposition, single replacement, double replacement, or combustion.

A great number of chemical reactions occur in nature, in biological systems, and in the laboratory. However, there are some general patterns that help us classify reactions into five general types.

CORE CHEMISTRY SKILL

Classifying Types of Chemical Reactions

Combination Reactions

In a **combination reaction**, two or more elements or compounds bond to form one product. For example, sulfur and oxygen combine to form the product sulfur dioxide.

$$S(s) + O_2(g) \longrightarrow SO_2(g)$$

In **FIGURE 8.2**, the elements magnesium and oxygen combine to form a single product, which is the ionic compound magnesium oxide formed from Mg^{2+} and O^{2-} ions.

$$2Mg(s) + O_2(g) \longrightarrow 2MgO(s)$$

Combination

Two or more reactants	combine to yield	a single product
A + B	$\longrightarrow$	A B

$$\underset{\text{Magnesium}}{2Mg(s)} \quad + \quad \underset{\text{Oxygen}}{O_2(g)} \quad \xrightarrow{\Delta} \quad \underset{\text{Magnesium oxide}}{2MgO(s)}$$

FIGURE 8.2 In a combination reaction, two or more substances combine to form one substance as product.

In other examples of combination reactions, elements or compounds combine to form a single product.

$$N_2(g) + 3H_2(g) \longrightarrow 2NH_3(g)$$
$$Cu(s) + S(s) \longrightarrow CuS(s)$$
$$MgO(s) + CO_2(g) \longrightarrow MgCO_3(s)$$

ENGAGE 8.4

What happens to the atoms in the reactants in a combination reaction?

Decomposition Reactions

In a **decomposition reaction**, a reactant splits into two or more simpler products. For example, when mercury(II) oxide is heated, the compound breaks apart into mercury atoms and oxygen (see **FIGURE 8.3**).

$$2HgO(s) \xrightarrow{\Delta} 2Hg(l) + O_2(g)$$

Decomposition

A reactant	splits into	two or more products
A B	$\longrightarrow$	A + B

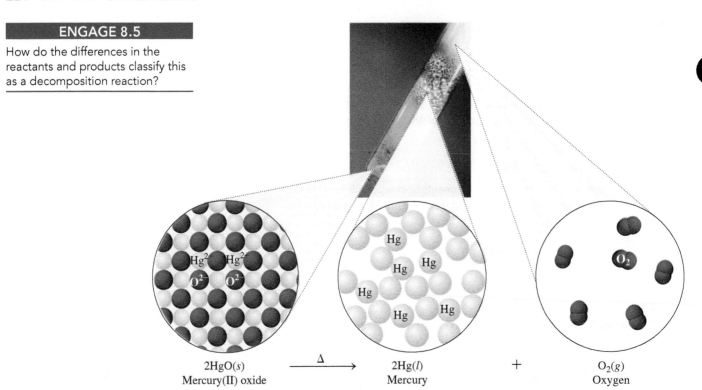

$$2HgO(s) \xrightarrow{\Delta} 2Hg(l) + O_2(g)$$

Mercury(II) oxide   Mercury   Oxygen

FIGURE 8.3 In a decomposition reaction, one reactant breaks down into two or more products.

In another example of a decomposition reaction, when calcium carbonate is heated, it breaks apart into simpler compounds of calcium oxide and carbon dioxide.

$$CaCO_3(s) \xrightarrow{\Delta} CaO(s) + CO_2(g)$$

Replacement Reactions

Single replacement

One element   replaces   another element

In a replacement reaction, elements in a compound are replaced by other elements. In a **single replacement reaction**, a reacting element switches place with an element in the other reacting compound.

In the single replacement reaction shown in **FIGURE 8.4**, zinc replaces hydrogen in hydrochloric acid, HCl(*aq*).

$$Zn(s) + 2HCl(aq) \longrightarrow H_2(g) + ZnCl_2(aq)$$

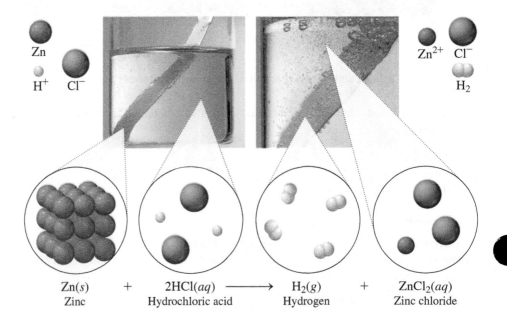

FIGURE 8.4 In a single replacement reaction, an atom or ion replaces an atom or ion in a compound.

$$Zn(s) + 2HCl(aq) \longrightarrow H_2(g) + ZnCl_2(aq)$$

Zinc   Hydrochloric acid   Hydrogen   Zinc chloride

In another single replacement reaction, chlorine replaces bromine in the compound potassium bromide.

$$Cl_2(g) + 2KBr(s) \longrightarrow 2KCl(s) + Br_2(l)$$

In a **double replacement reaction**, the positive ions in the reacting compounds switch places.

In the reaction shown in **FIGURE 8.5**, barium ions change places with sodium ions in the reactants to form sodium chloride and a white solid precipitate of barium sulfate. The formulas of the products depend on the charges of the ions.

$$BaCl_2(aq) + Na_2SO_4(aq) \longrightarrow BaSO_4(s) + 2NaCl(aq)$$

When sodium hydroxide and hydrochloric acid (HCl) react, sodium and hydrogen ions switch places, forming water and sodium chloride.

$$HCl(aq) + NaOH(aq) \longrightarrow H_2O(l) + NaCl(aq)$$

Double replacement

Two elements   replace   each other

$$SO_4^{2-} \quad Ba^{2+}$$
$$Na^+ \quad Cl^-$$

| Na$_2$SO$_4$(aq) | + | BaCl$_2$(aq) | $\longrightarrow$ | BaSO$_4$(s) | + | 2NaCl(aq) |
| Sodium sulfate | | Barium chloride | | Barium sulfate | | Sodium chloride |

FIGURE 8.5 In a double replacement reaction, the positive ions in the reactants replace each other.

ENGAGE 8.7

How do the formulas of the reactants change in a double replacement reaction?

Combustion Reactions

The burning of a candle and the burning of fuel in the engine of a car are examples of combustion reactions. In a **combustion reaction**, a carbon-containing compound, usually a fuel, burns in oxygen from the air to produce carbon dioxide (CO_2), water (H_2O), and energy in the form of heat or a flame. For example, methane gas (CH_4) undergoes combustion when used to cook our food on a gas cooktop and to heat our homes. In the equation for the combustion of methane, each element in the fuel (CH_4) forms a compound with oxygen.

$$CH_4(g) + 2O_2(g) \xrightarrow{\Delta} CO_2(g) + 2H_2O(g) + energy$$
Methane

The balanced equation for the combustion of propane (C_3H_8) is

$$C_3H_8(g) + 5O_2(g) \xrightarrow{\Delta} 3CO_2(g) + 4H_2O(g) + energy$$

In a combustion reaction, a candle burns by reacting with the oxygen in the air.

Propane is the fuel used in portable heaters and gas barbecues. Gasoline, a mixture of liquid hydrocarbons, is the fuel that powers our cars, lawn mowers, and snow blowers. TABLE 8.3 summarizes the reaction types and gives examples.

TABLE 8.3 Summary of Reaction Types

Reaction Type	Example
Combination $A + B \longrightarrow AB$	$Ca(s) + Cl_2(g) \longrightarrow CaCl_2(s)$
Decomposition $AB \longrightarrow A + B$	$Fe_2S_3(s) \longrightarrow 2Fe(s) + 3S(s)$
Single Replacement $A + BC \longrightarrow AC + B$	$Cu(s) + 2AgNO_3(aq) \longrightarrow 2Ag(s) + Cu(NO_3)_2(aq)$
Double Replacement $AB + CD \longrightarrow AD + CB$	$BaCl_2(aq) + K_2SO_4(aq) \longrightarrow BaSO_4(s) + 2KCl(aq)$
Combustion $C_XH_Y + ZO_2(g) \xrightarrow{\Delta} XCO_2(g) + \dfrac{Y}{2}H_2O(g) + energy$	$CH_4(g) + 2O_2(g) \xrightarrow{\Delta} CO_2(g) + 2H_2O(g) + energy$

SAMPLE PROBLEM 8.5   Identifying Reaction Types

TRY IT FIRST

Classify each of the following as a combination, decomposition, single replacement, double replacement, or combustion reaction:

a. $2Fe_2O_3(s) + 3C(s) \longrightarrow 3CO_2(g) + 4Fe(s)$

b. $2KClO_3(s) \xrightarrow{\Delta} 2KCl(s) + 3O_2(g)$

c. $2C_3H_6(g) + 9O_2(g) \xrightarrow{\Delta} 6CO_2(g) + 6H_2O(g) + energy$

SOLUTION

a. When a C atom replaces Fe in Fe_2O_3 to form the compound CO_2 and Fe atoms, the reaction is a single replacement.

b. When one reactant breaks down to produce two products, the reaction is decomposition.

c. The reaction of a carbon compound with oxygen to produce carbon dioxide, water, and energy makes this a combustion reaction.

SELF TEST 8.5

a. Nitrogen gas (N_2) and oxygen gas (O_2) react to form nitrogen dioxide gas. Write the balanced chemical equation using the correct chemical formulas for the reactants and product, and identify the reaction type.

b. Strontium metal and aqueous HCl react to give hydrogen gas and aqueous strontium chloride. Write the balanced chemical equation using the correct chemical formulas for the reactants and products, and identify the reaction type.

ANSWER

a. $N_2(g) + 2O_2(g) \longrightarrow 2NO_2(g)$ Combination

b. $2HCl(aq) + Sr(s) \longrightarrow H_2(g) + SrCl_2(aq)$ Single replacement

Chemistry Link to Health

Incomplete Combustion: Toxicity of Carbon Monoxide

When a propane heater, fireplace, or woodstove is used in a closed room, there must be adequate ventilation. If the supply of oxygen is limited, incomplete combustion from burning gas, oil, or wood produces carbon monoxide. The incomplete combustion of methane in natural gas is written

$$2CH_4(g) + 3O_2(g) \xrightarrow{\Delta} 2CO(g) + 4H_2O(g) + \text{energy}$$

Limited
oxygen
supply

Carbon
monoxide

Carbon monoxide (CO) is a colorless, odorless, poisonous gas. When inhaled, CO passes into the bloodstream, where it attaches to hemoglobin, which reduces the amount of oxygen (O_2) reaching the cells. As a result, a person can experience a reduction in exercise capability, visual perception, and manual dexterity.

Hemoglobin is the protein that transports O_2 in the blood. When the amount of hemoglobin bound to CO (COHb) is about 10%, a person may experience shortness of breath, mild headache, and drowsiness. Heavy smokers can have levels of COHb in their blood as high as 9%. When as much as 30% of the hemoglobin is bound to CO, a person may experience more severe symptoms, including dizziness, mental confusion, severe headache, and nausea. If 50% or more of the hemoglobin is bound to CO, a person could become unconscious and die if not treated immediately with oxygen.

SAMPLE PROBLEM 8.6   Writing an Equation for Combustion

TRY IT FIRST

A portable burner is fueled with butane gas, C_4H_{10}. Write the reactants and products for the complete combustion of butane, and balance the equation.

SOLUTION

	Given	Need	Connect
ANALYZE THE PROBLEM	C_4H_{10}	balanced equation for combustion	reactants: carbon compound + O_2, products: CO_2 + H_2O

When camping, a butane cartridge provides fuel for a portable burner.

In a combustion reaction, butane gas reacts with the gas O_2 to form the gases CO_2, H_2O, and energy. We write the unbalanced equation as

$$C_4H_{10}(g) + O_2(g) \xrightarrow{\Delta} CO_2(g) + H_2O(g) + \text{energy}$$

We can begin by using the subscripts in C_4H_{10} to balance the C atoms in CO_2 and the H atoms in H_2O. However, this gives a total of 13 O atoms in the products. This is balanced by placing a coefficient of 13/2 in front of the formula for O_2.

$$C_4H_{10}(g) + \frac{13}{2}O_2(g) \xrightarrow{\Delta} \underbrace{4CO_2(g) + 5H_2O(g)}_{\text{13 O atoms}} + \text{energy}$$

To obtain a whole number coefficient for O_2, we need to multiply all the coefficients by 2.

$$2C_4H_{10}(g) + 13O_2(g) \xrightarrow{\Delta} 8CO_2(g) + 10H_2O(g) + \text{energy}$$

SELF TEST 8.6

a. Ethene, used to ripen fruit, has the formula C_2H_4. Write a balanced chemical equation for the complete combustion of ethene.
b. Heptane, C_7H_{16}, is used in tests to determine the octane rating of gasoline. Write a balanced chemical equation for the complete combustion of heptane.

ANSWER

a. $C_2H_4(g) + 3O_2(g) \xrightarrow{\Delta} 2CO_2(g) + 2H_2O(g) + \text{energy}$
b. $C_7H_{16}(l) + 11O_2(g) \xrightarrow{\Delta} 7CO_2(g) + 8H_2O(g) + \text{energy}$

PRACTICE PROBLEMS

Try Practice Problems 8.19 to 8.24

PRACTICE PROBLEMS

8.3 Types of Chemical Reactions

8.19 Classify each of the following as a combination, decomposition, single replacement, double replacement, or combustion reaction:
a. $2Al_2O_3(s) \xrightarrow{\Delta} 4Al(s) + 3O_2(g)$
b. $Br_2(l) + BaI_2(s) \longrightarrow BaBr_2(s) + I_2(g)$
c. $2C_2H_2(g) + 5O_2(g) \xrightarrow{\Delta} 4CO_2(g) + 2H_2O(g)$
d. $BaCl_2(aq) + K_2CO_3(aq) \longrightarrow BaCO_3(s) + 2KCl(aq)$
e. $Pb(s) + O_2(g) \longrightarrow PbO_2(s)$

8.20 Classify each of the following as a combination, decomposition, single replacement, double replacement, or combustion reaction:
a. $H_2(g) + Br_2(l) \longrightarrow 2HBr(g)$
b. $AgNO_3(aq) + NaCl(aq) \longrightarrow AgCl(s) + NaNO_3(aq)$
c. $2H_2O_2(aq) \longrightarrow 2H_2O(l) + O_2(g)$
d. $Zn(s) + CuCl_2(aq) \longrightarrow Cu(s) + ZnCl_2(aq)$
e. $C_5H_8(g) + 7O_2(g) \xrightarrow{\Delta} 5CO_2(g) + 4H_2O(g)$

8.21 Classify each of the following as a combination, decomposition, single replacement, double replacement, or combustion reaction:
a. $4Fe(s) + 3O_2(g) \longrightarrow 2Fe_2O_3(s)$
b. $Mg(s) + 2AgNO_3(aq) \longrightarrow 2Ag(s) + Mg(NO_3)_2(aq)$
c. $CuCO_3(s) \xrightarrow{\Delta} CuO(s) + CO_2(g)$
d. $Al_2(SO_4)_3(aq) + 6KOH(aq) \longrightarrow$
$2Al(OH)_3(s) + 3K_2SO_4(aq)$
e. $C_4H_8(g) + 6O_2(g) \xrightarrow{\Delta} 4CO_2(g) + 4H_2O(g)$

8.22 Classify each of the following as a combination, decomposition, single replacement, double replacement, or combustion reaction:
a. $CuO(s) + 2HCl(aq) \longrightarrow CuCl_2(aq) + H_2O(l)$
b. $2Al(s) + 3Br_2(l) \longrightarrow 2AlBr_3(s)$
c. $C_6H_{12}(l) + 9O_2(g) \xrightarrow{\Delta} 6CO_2(g) + 6H_2O(g)$
d. $Fe_2O_3(s) + 3C(s) \longrightarrow 2Fe(s) + 3CO(g)$
e. $C_6H_{12}O_6(aq) \longrightarrow 2C_2H_6O(aq) + 2CO_2(g)$

8.23 Using Table 8.3, predict the products that would result from each of the following reactions and balance:
a. combination: $Mg(s) + Cl_2(g) \longrightarrow$
b. decomposition: $HBr(g) \longrightarrow$
c. single replacement: $Mg(s) + Zn(NO_3)_2(aq) \longrightarrow$
d. double replacement: $K_2S(aq) + Pb(NO_3)_2(aq) \longrightarrow$
e. combustion: $C_2H_6(g) + O_2(g) \xrightarrow{\Delta}$

8.24 Using Table 8.3, predict the products that would result from each of the following reactions and balance:
a. combination: $Ca(s) + O_2(g) \longrightarrow$
b. combustion: $C_6H_6(l) + O_2(g) \xrightarrow{\Delta}$
c. decomposition: $PbO_2(s) \xrightarrow{\Delta}$
d. single replacement: $KI(s) + Cl_2(g) \longrightarrow$
e. double replacement: $CuCl_2(aq) + Na_2S(aq) \longrightarrow$

8.4 Oxidation–Reduction Reactions

LEARNING GOAL Define the terms oxidation and reduction; identify the reactants oxidized and reduced.

Perhaps you have never heard of an oxidation and reduction reaction. However, this type of reaction has many important applications in your everyday life. When you see a rusty nail, tarnish on a silver spoon, or corrosion on metal, you are observing oxidation.

$$4Fe(s) + 3O_2(g) \longrightarrow 2Fe_2O_3(s) \quad \text{Fe is oxidized}$$
$$\text{Rust}$$

Rust forms when the oxygen in the air reacts with iron.

When we turn the lights on in our automobiles, an oxidation–reduction reaction within the car battery provides the electricity. On a cold, wintry day, we might build a fire. As the wood burns, oxygen combines with carbon and hydrogen to produce carbon dioxide, water, and heat. In the previous section, we called this a combustion reaction, but it is also an *oxidation–reduction reaction*. When we eat foods with starches in them, the starches break down to give glucose, which is oxidized in our cells to give us energy along with carbon dioxide and water. Every breath we take provides oxygen to carry out oxidation in our cells.

$$C_6H_{12}O_6(aq) + 6O_2(g) \longrightarrow 6CO_2(g) + 6H_2O(l) + \text{energy}$$
$$\text{Glucose}$$

Oxidation–Reduction Reactions

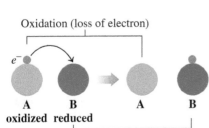

In an **oxidation–reduction reaction** (*redox*), electrons are transferred from one substance to another. If one substance loses electrons, another substance must gain electrons. **Oxidation** is defined as the *loss* of electrons; **reduction** is the *gain* of electrons.

One way to remember these definitions is to use the following acronym:

OIL RIG
Oxidation **I**s **L**oss of electrons
Reduction **I**s **G**ain of electrons

In general, atoms of metals lose electrons to form positive ions, whereas nonmetals gain electrons to form negative ions. Now we can say that metals are oxidized and nonmetals are reduced.

The green color that appears on copper surfaces from weathering, known as *patina*, is a mixture of $CuCO_3$ and CuO. We can now look at the oxidation and reduction reactions that take place when copper metal reacts with oxygen in the air to produce copper(II) oxide.

$$2Cu(s) + O_2(g) \longrightarrow 2CuO(s)$$

The element Cu in the reactants has a charge of 0, but in the CuO product, it is present as Cu^{2+}, which has a 2+ charge. Because the Cu atom lost two electrons, the Cu was oxidized in the reaction.

$$Cu^0(s) \longrightarrow Cu^{2+}(s) + 2\ e^- \quad \text{Oxidation: loss of electrons by Cu}$$

At the same time, the element O in the reactants has a charge of 0, but in the CuO product, it is present as O^{2-}, which has a 2− charge. Because the O atom has gained two electrons, the O was reduced in the reaction.

$$O_2{}^0(g) + 4\ e^- \longrightarrow 2O^{2-}(s) \quad \text{Reduction: gain of electrons by O}$$

Thus, the overall equation for the formation of CuO involves an oxidation and a reduction that occur simultaneously. In every oxidation and reduction, the number of electrons lost must be equal to the number of electrons gained. Therefore, we multiply the oxidation reaction of Cu by 2. Canceling the $4\ e^-$ on each side, we obtain the overall oxidation–reduction equation for the formation of CuO.

$$2Cu(s) \longrightarrow 2Cu^{2+}(s) + \cancel{4e^-} \quad \text{Oxidation}$$
$$\underline{O_2(g) + \cancel{4e^-} \longrightarrow 2O^{2-}(s)} \quad \text{Reduction}$$
$$2Cu(s) + O_2(g) \longrightarrow 2CuO(s) \quad \text{Oxidation–reduction equation}$$

As we see in the next reaction between zinc and copper(II) sulfate, there is always an oxidation with every reduction (see **FIGURE 8.6**).

$$Zn(s) + CuSO_4(aq) \longrightarrow ZnSO_4(aq) + Cu(s)$$

We rewrite the equation to show the atoms and ions.

$$\mathbf{Zn}(s) + \mathbf{Cu^{2+}}(aq) + SO_4{}^{2-}(aq) \longrightarrow \mathbf{Zn^{2+}}(aq) + SO_4{}^{2-}(aq) + \mathbf{Cu}(s)$$

In this reaction, Zn atoms lose two electrons to form Zn^{2+}, which indicates that Zn is oxidized. At the same time, Cu^{2+} gains two electrons, which indicates that Cu is reduced. The $SO_4{}^{2-}$ ions are *spectator ions*, which are present in both the reactants and products and do not change.

$$Zn(s) \longrightarrow Zn^{2+}(aq) + 2\ e^- \quad \text{Oxidation of Zn}$$
$$Cu^{2+}(aq) + 2\ e^- \longrightarrow Cu(s) \quad \text{Reduction of Cu}^{2+}$$

In this single replacement reaction, zinc was oxidized and copper(II) ion was reduced.

Oxidation and Reduction in Biological Systems

Oxidation may also involve the addition of oxygen or the loss of hydrogen, and reduction may involve the loss of oxygen or the gain of hydrogen. In the cells of the body, oxidation of organic (carbon) compounds involves the transfer of hydrogen atoms (H), which are composed of electrons and protons. For example, the oxidation of a typical biochemical

The green patina on copper is due to oxidation.

Reduced		Oxidized
Na	Oxidation: Lose e^-	$Na^+ + e^-$
Ca		$Ca^{2+} + 2\ e^-$
$2Br^-$	Reduction: Gain e^-	$Br_2 + 2\ e^-$
Fe^{2+}		$Fe^{3+} + e^-$

Oxidation is a loss of electrons; reduction is a gain of electrons.

CORE CHEMISTRY SKILL

Identifying Oxidized and Reduced Substances

PRACTICE PROBLEMS

Try Practice Problems 8.25 to 8.28

ENGAGE 8.8

How can you determine that Cu^{2+} is reduced in this reaction?

FIGURE 8.6 In this single replacement reaction, Zn(s) is oxidized to $Zn^{2+}(aq)$ when it provides two electrons to reduce $Cu^{2+}(aq)$ to Cu(s): $Zn(s) + Cu^{2+}(aq) \longrightarrow Zn^{2+}(aq) + Cu(s)$

molecule can involve the transfer of two hydrogen atoms (or $2H^+$ and $2\ e^-$) to a hydrogen ion acceptor such as the coenzyme FAD (flavin adenine dinucleotide). The coenzyme is reduced to $FADH_2$.

In many biochemical oxidation–reduction reactions, the transfer of hydrogen atoms is necessary for the production of energy in the cells. For example, methyl alcohol (CH_4O), a poisonous substance, is metabolized in the body by the following reactions:

$$CH_4O \longrightarrow CH_2O + 2H \quad \text{Oxidation: loss of H atoms}$$
Methyl   Formaldehyde
alcohol

The formaldehyde can be oxidized further, this time by the addition of oxygen, to produce formic acid.

$$2CH_2O + O_2 \longrightarrow 2CH_2O_2 \quad \text{Oxidation: addition of O atoms}$$
Formaldehyde   Formic acid

Finally, formic acid is oxidized to carbon dioxide and water.

$$2CH_2O_2 + O_2 \longrightarrow 2CO_2 + 2H_2O \quad \text{Oxidation: addition of O atoms}$$
Formic acid

The intermediate products of the oxidation of methyl alcohol are quite toxic, causing blindness and possibly death as they interfere with key reactions in the cells of the body.

In summary, we find that the particular definition of oxidation and reduction we use depends on the process that occurs in the reaction. All these definitions are summarized in **TABLE 8.4**. Oxidation always involves a loss of electrons, but it may also be seen as an addition of oxygen or the loss of hydrogen atoms. A reduction always involves a gain of electrons and may also be seen as the loss of oxygen or the gain of hydrogen.

TABLE 8.4 Characteristics of Oxidation and Reduction

Always Involves	May Involve
Oxidation	
Loss of electrons	Addition of oxygen
	Loss of hydrogen
Reduction	
Gain of electrons	Loss of oxygen
	Gain of hydrogen

SAMPLE PROBLEM 8.7   Oxidation and Reduction

TRY IT FIRST

For each of the following reactions, determine the number of electrons lost or gained and identify as an oxidation or a reduction:

a. $Mg^{2+}(aq) \longrightarrow Mg(s)$
b. $Fe(s) \longrightarrow Fe^{3+}(aq)$

SOLUTION

a. $Mg^{2+}(aq) + 2\ e^- \longrightarrow Mg(s)$ Reduction
b. $Fe(s) \longrightarrow Fe^{3+}(aq) + 3\ e^-$ Oxidation

SELF TEST 8.7

Identify each of the following as an oxidation or a reduction:

a. $Co^{3+}(aq) + e^- \longrightarrow Co^{2+}(aq)$
b. $2I^-(aq) \longrightarrow I_2(s) + 2\ e^-$

ANSWER

a. reduction
b. oxidation

PRACTICE PROBLEMS
Try Practice Problems 8.29 to 8.32

PRACTICE PROBLEMS

8.4 Oxidation–Reduction Reactions

8.25 Identify each of the following as an oxidation or a reduction:
  a. $Na^+(aq) + e^- \longrightarrow Na(s)$
  b. $Ni(s) \longrightarrow Ni^{2+}(aq) + 2\ e^-$
  c. $Cr^{3+}(aq) + 3\ e^- \longrightarrow Cr(s)$
  d. $2H^+(aq) + 2\ e^- \longrightarrow H_2(g)$

8.26 Identify each of the following as an oxidation or a reduction:
  a. $O_2(g) + 4\ e^- \longrightarrow 2O^{2-}(aq)$
  b. $Ag(s) \longrightarrow Ag^+(aq) + e^-$
  c. $Fe^{3+}(aq) + e^- \longrightarrow Fe^{2+}(aq)$
  d. $2Br^-(aq) \longrightarrow Br_2(l) + 2\ e^-$

8.27 In each of the following, identify the reactant that is oxidized and the reactant that is reduced:
  a. $Zn(s) + Cl_2(g) \longrightarrow ZnCl_2(s)$
  b. $Cl_2(g) + 2NaBr(aq) \longrightarrow 2NaCl(aq) + Br_2(l)$
  c. $2PbO(s) \longrightarrow 2Pb(s) + O_2(g)$
  d. $2Fe^{3+}(aq) + Sn^{2+}(aq) \longrightarrow 2Fe^{2+}(aq) + Sn^{4+}(aq)$

8.28 In each of the following, identify the reactant that is oxidized and the reactant that is reduced:
  a. $2Li(s) + F_2(g) \longrightarrow 2LiF(s)$
  b. $Cl_2(g) + 2KI(aq) \longrightarrow I_2(s) + 2KCl(aq)$
  c. $2Al(s) + 3Sn^{2+}(aq) \longrightarrow 2Al^{3+}(aq) + 3Sn(s)$
  d. $Fe(s) + CuSO_4(aq) \longrightarrow Cu(s) + FeSO_4(aq)$

Applications

8.29 In the mitochondria of human cells, energy is provided by the oxidation and reduction reactions of the iron ions in the cytochromes in electron transport. Identify each of the following as an oxidation or a reduction:
  a. $Fe^{3+} + e^- \longrightarrow Fe^{2+}$
  b. $Fe^{2+} \longrightarrow Fe^{3+} + e^-$

8.30 Chlorine (Cl_2) is a strong germicide used to disinfect drinking water and to kill microbes in swimming pools. If the product is Cl^-, was the elemental chlorine oxidized or reduced?

8.31 When linoleic acid, an unsaturated fatty acid, reacts with hydrogen, it forms a saturated fatty acid. Is linoleic acid oxidized or reduced in the hydrogenation reaction?

$$C_{18}H_{32}O_2 + 2H_2 \longrightarrow C_{18}H_{36}O_2$$
Linoleic acid

8.32 In one of the reactions in the citric acid cycle, which provides energy, succinic acid is converted to fumaric acid.

$$C_4H_6O_4 \longrightarrow C_4H_4O_4 + 2H$$
Succinic acid   Fumaric acid

The reaction is accompanied by a coenzyme, flavin adenine dinucleotide (FAD).

$$FAD + 2H \longrightarrow FADH_2$$

  a. Is succinic acid oxidized or reduced?
  b. Is FAD oxidized or reduced?
  c. Why would the two reactions occur together?

UPDATE Improving Natalie's Overall Fitness

Natalie's test results indicate that she has a blood oxygen level of 89%. The normal values for pulse oximeter readings are 95% to 100%, which means that Natalie's O_2 saturation is low. Thus, Natalie does not have an adequate amount of O_2 in her blood. This may be the reason she has noticed a shortness of breath and a dry cough. Her doctor diagnosed her with *interstitial lung disease*, which is scarring of the tissue of the lungs.

Angela teaches Natalie to inhale and exhale slower and deeper to fill the lungs with more air and thus more oxygen. Angela also develops a workout program with the goal of increasing Natalie's overall fitness level. During the exercises, Angela continues to monitor Natalie's heart rate, blood O_2 level, and blood pressure to ensure that Natalie is exercising at a level that will enable her to become stronger without breaking down muscle due to a lack of oxygen.

Low-intensity exercises are used at the beginning of Natalie's exercise program.

Applications

8.33 **a.** During cellular respiration, aqueous glucose ($C_6H_{12}O_6$) in the cells undergoes combustion. Write and balance the chemical equation for reaction of glucose.
b. In plants, carbon dioxide gas and liquid water are converted to aqueous glucose ($C_6H_{12}O_6$) and oxygen gas. Write and balance the chemical equation for production of glucose in plants.

8.34 Aqueous fatty acids undergo reaction with oxygen gas and form gaseous carbon dioxide and liquid water when utilized for energy in the body.
a. Write and balance the equation for the combustion of the fatty acid capric acid ($C_{10}H_{20}O_2$).
b. Write and balance the equation for the combustion of the fatty acid myristic acid ($C_{14}H_{28}O_2$).

CONCEPT MAP

CHEMICAL REACTIONS

show a chemical change between

Reactants and Products

are balanced with

Coefficients

to give

Equal Numbers of Atoms on Each Side

are classified as

Combination, Decomposition, Single Replacement, Double Replacement, and Combustion

and those with

Loss or Gain of Electrons

are also

Oxidation–Reduction Reactions

CHAPTER REVIEW

8.1 Equations for Chemical Reactions

LEARNING GOAL Identify a balanced chemical equation; determine the number of atoms in the reactants and products.

$$2H_2(g) + O_2(g) \longrightarrow 2H_2O(g)$$
Balanced

- A chemical change occurs when the atoms of the initial substances rearrange to form new substances.
- A chemical equation shows the formulas of the substances that react on the left side of a reaction arrow and the products that form on the right side of the reaction arrow.

8.2 Balancing a Chemical Equation

LEARNING GOAL Write a balanced chemical equation from the formulas of the reactants and products for a chemical reaction.

- A chemical equation is balanced by writing coefficients, small whole numbers, in front of formulas to equalize the atoms of each of the elements in the reactants and the products.

8.3 Types of Chemical Reactions

LEARNING GOAL Identify a chemical reaction as a combination, decomposition, single replacement, double replacement, or combustion.

Single replacement

- Many chemical reactions can be organized by reaction type: combination, decomposition, single replacement, double replacement, or combustion.

8.4 Oxidation–Reduction Reactions

LEARNING GOAL Define the terms oxidation and reduction; identify the reactants oxidized and reduced.

- When electrons are transferred in a reaction, it is an oxidation–reduction reaction.
- One reactant loses electrons, and another reactant gains electrons.
- Overall, the number of electrons lost and gained is equal.

KEY TERMS

balanced equation The final form of a chemical equation that shows the same number of atoms of each element in the reactants and products.

chemical equation A shorthand way to represent a chemical reaction using chemical formulas to indicate the reactants and products and coefficients to show reacting ratios.

coefficients Whole numbers placed in front of the formulas to balance the number of atoms or moles of atoms of each element on both sides of an equation.

combination reaction A chemical reaction in which reactants combine to form a single product.

combustion reaction A chemical reaction in which a fuel containing carbon reacts with oxygen to produce CO_2, H_2O, and energy.

decomposition reaction A reaction in which a single reactant splits into two or more simpler substances.

double replacement reaction A reaction in which the positive ions in the reacting compounds exchange places.

oxidation The loss of electrons by a substance. Biological oxidation may involve the addition of oxygen or the loss of hydrogen.

oxidation–reduction reaction A reaction in which the oxidation of one reactant is always accompanied by the reduction of another reactant.

products The substances formed as a result of a chemical reaction.

reactants The initial substances that undergo change in a chemical reaction.

reduction The gain of electrons by a substance. Biological reduction may involve the loss of oxygen or the gain of hydrogen.

single replacement reaction A reaction in which an element replaces a different element in a compound.

CORE CHEMISTRY SKILLS

The chapter section containing each Core Chemistry Skill is shown in parentheses at the end of each heading.

Balancing a Chemical Equation (8.2)

- In a *balanced* chemical equation, whole numbers called coefficients multiply each of the atoms in the chemical formulas so that the number of each type of atom in the reactants is equal to the number of the same type of atom in the products.

Example: Balance the following chemical equation:

$$SnCl_4(s) + H_2O(l) \longrightarrow Sn(OH)_4(s) + HCl(aq) \quad \text{Not balanced}$$

Answer: When we count the atoms on the reactant side and the product side, we see that there are more Cl atoms in the reactants and more O and H atoms in the products.

To balance the equation, we need to use coefficients in front of the formulas containing the Cl atoms, H atoms, and O atoms.

- Place a 4 in front of the formula HCl to give a total of 8 H atoms and 4 Cl atoms in the products.

$$SnCl_4(s) + H_2O(l) \longrightarrow Sn(OH)_4(s) + 4HCl(aq)$$

- Place a 4 in front of the formula H_2O to give 8 H atoms and 4 O atoms in the reactants.

$$SnCl_4(s) + 4H_2O(l) \longrightarrow Sn(OH)_4(s) + 4HCl(aq)$$

- The total number of Sn (1), Cl (4), H (8), and O (4) atoms is now equal on both sides of the equation. Thus, this equation is balanced.

$$SnCl_4(s) + 4H_2O(l) \longrightarrow Sn(OH)_4(s) + 4HCl(aq)$$

Classifying Types of Chemical Reactions (8.3)

- Chemical reactions are classified by identifying general patterns in their equations.
- In a combination reaction, two or more elements or compounds bond to form one product.
- In a decomposition reaction, a single reactant splits into two or more products.
- In a single replacement reaction, an uncombined element takes the place of an element in a compound.
- In a double replacement reaction, the positive ions in the reacting compounds switch places.
- In a combustion reaction, a fuel containing carbon and hydrogen reacts with oxygen from the air to produce carbon dioxide (CO_2), water (H_2O), and energy.

Example: Classify the type of the following reaction:

$$2Al(s) + Fe_2O_3(s) \xrightarrow{\Delta} Al_2O_3(s) + 2Fe(l)$$

Answer: The iron in iron(III) oxide is replaced by aluminum, which makes this a single replacement reaction.

Identifying Oxidized and Reduced Substances (8.4)

- In an oxidation–reduction reaction (abbreviated *redox*), one reactant is oxidized when it loses electrons, and another reactant is reduced when it gains electrons.
- Oxidation is the *loss* of electrons; reduction is the *gain* of electrons.

Example: For the following redox reaction, identify the reactant that is oxidized, and the reactant that is reduced:

$$Fe(s) + Cu^{2+}(aq) \longrightarrow Fe^{2+}(aq) + Cu(s)$$

Answer: $Fe^0(s) \longrightarrow Fe^{2+}(aq) + 2\,e^-$
Fe loses electrons; it is oxidized.
$Cu^{2+}(aq) + 2\,e^- \longrightarrow Cu^0(s)$
Cu^{2+} gains electrons; it is reduced.

UNDERSTANDING THE CONCEPTS

The chapter sections to review are shown in parentheses at the end of each problem.

8.35 Balance each of the following by adding coefficients, and identify the type of reaction for each: (8.1, 8.2, 8.3)

8.36 Balance each of the following by adding coefficients, and identify the type of reaction for each: (8.1, 8.2, 8.3)

8.37 If red spheres represent oxygen atoms, blue spheres represent nitrogen atoms, and all the molecules are gases, (8.1, 8.2, 8.3)

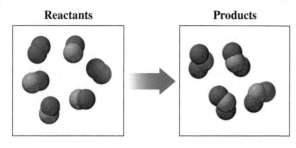

a. write the formula for each of the reactants and products.
b. write a balanced equation for the reaction.
c. indicate the type of reaction as combination, decomposition, single replacement, double replacement, or combustion.

8.38 If purple spheres represent iodine atoms, white spheres represent hydrogen atoms, and all the molecules are gases, (8.1, 8.2, 8.3)

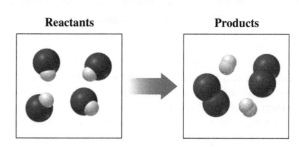

a. write the formula for each of the reactants and products.
b. write a balanced equation for the reaction.
c. indicate the type of reaction as combination, decomposition, single replacement, double replacement, or combustion.

8.39 If blue spheres represent nitrogen atoms, purple spheres represent iodine atoms, the reacting molecules are solid, and the products are gases, (8.1, 8.2, 8.3)

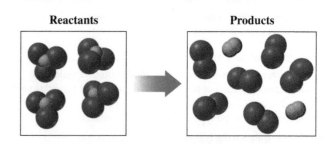

a. write the formula for each of the reactants and products.
b. write a balanced equation for the reaction.
c. indicate the type of reaction as combination, decomposition, single replacement, double replacement, or combustion.

8.40 If green spheres represent chlorine atoms, yellow-green spheres represent fluorine atoms, white spheres represent hydrogen atoms, and all the molecules are gases, (8.1, 8.2, 8.3)

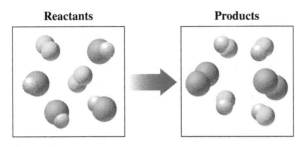

Reactants **Products**

a. write the formula for each of the reactants and products.
b. write a balanced equation for the reaction.
c. indicate the type of reaction as combination, decomposition, single replacement, double replacement, or combustion.

8.41 If green spheres represent chlorine atoms, red spheres represent oxygen atoms, and all the molecules are gases, (8.1, 8.2, 8.3)

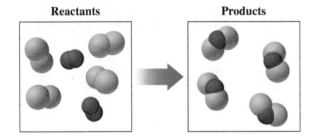

Reactants **Products**

a. write the formula for each of the reactants and products.
b. write a balanced equation for the reaction.
c. indicate the type of reaction as combination, decomposition, single replacement, double replacement, or combustion.

8.42 If blue spheres represent nitrogen atoms, purple spheres represent iodine atoms, the reacting molecules are gases, and the products are solid, (8.1, 8.2, 8.3)

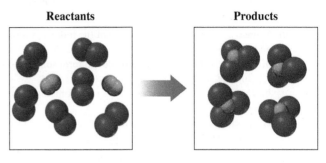

Reactants **Products**

a. write the formula for each of the reactants and products.
b. write a balanced equation for the reaction.
c. indicate the type of reaction as combination, decomposition, single replacement, double replacement, or combustion.

ADDITIONAL PRACTICE PROBLEMS

8.43 Identify the type of reaction for each of the following as combination, decomposition, single replacement, double replacement, or combustion: (8.3)
a. A metal and a nonmetal form an ionic compound.
b. A compound of hydrogen and carbon reacts with oxygen to produce carbon dioxide and water.
c. Heating magnesium carbonate produces magnesium oxide and carbon dioxide.
d. Zinc replaces copper in $Cu(NO_3)_2$.

8.44 Identify the type of reaction for each of the following as combination, decomposition, single replacement, double replacement, or combustion: (8.3)
a. A compound breaks apart into its elements.
b. Copper and bromine form copper(II) bromide.
c. Iron(II) sulfite breaks down to iron(II) oxide and sulfur dioxide.
d. Silver ion from $AgNO_3(aq)$ forms a solid with bromide ion from $KBr(aq)$.

8.45 Balance each of the following chemical equations, and identify the type of reaction: (8.1, 8.2, 8.3)
a. $NH_3(g) + HCl(g) \longrightarrow NH_4Cl(s)$
b. $C_4H_8(g) + O_2(g) \xrightarrow{\Delta} CO_2(g) + H_2O(g)$
c. $Sb(s) + Cl_2(g) \longrightarrow SbCl_3(s)$
d. $NI_3(s) \longrightarrow N_2(g) + I_2(g)$
e. $KBr(aq) + Cl_2(aq) \longrightarrow KCl(aq) + Br_2(l)$
f. $H_2SO_4(aq) + Fe(s) \longrightarrow H_2(g) + Fe_2(SO_4)_3(aq)$
g. $Al_2(SO_4)_3(aq) + NaOH(aq) \longrightarrow$
$Al(OH)_3(s) + Na_2SO_4(aq)$

8.46 Balance each of the following chemical equations, and identify the type of reaction: (8.1, 8.2, 8.3)
a. $Si_3N_4(s) \longrightarrow Si(s) + N_2(g)$
b. $Mg(s) + N_2(g) \longrightarrow Mg_3N_2(s)$
c. $H_3PO_4(aq) + Al(s) \longrightarrow H_2(g) + AlPO_4(aq)$
d. $C_3H_4(g) + O_2(g) \xrightarrow{\Delta} CO_2(g) + H_2O(g)$
e. $Cr_2O_3(s) + H_2(g) \longrightarrow Cr(s) + H_2O(g)$
f. $Al(s) + Cl_2(g) \longrightarrow AlCl_3(s)$
g. $MgCl_2(aq) + AgNO_3(aq) \longrightarrow AgCl(s) + Mg(NO_3)_2(aq)$

8.47 Predict the products and write a balanced equation for each of the following: (8.1, 8.2, 8.3)
a. single replacement:
$HCl(aq) + Zn(s) \longrightarrow \underline{\quad} + \underline{\quad}$
b. decomposition: $BaCO_3(s) \xrightarrow{\Delta} \underline{\quad} + \underline{\quad}$
c. double replacement:
$HCl(aq) + NaOH(aq) \longrightarrow \underline{\quad} + \underline{\quad}$
d. combination: $Al(s) + F_2(g) \longrightarrow \underline{\quad}$

8.48 Predict the products and write a balanced equation for each of the following: (8.1, 8.2, 8.3)
a. decomposition: $NaCl(s) \xrightarrow{\text{Electricity}} \underline{\quad} + \underline{\quad}$
b. combination: $Ca(s) + Br_2(l) \longrightarrow \underline{\quad}$
c. combustion: $C_2H_4(g) + O_2(g) \xrightarrow{\Delta} \underline{\quad} + \underline{\quad}$
d. double replacement:
$NiCl_2(aq) + NaOH(aq) \longrightarrow \underline{\quad} + \underline{\quad}$

8.49 Write a balanced equation for each of the following reactions and identify the type of reaction: (8.1, 8.2, 8.3)

   a. Sodium metal reacts with oxygen gas to form solid sodium oxide.

   b. Aqueous sodium chloride and aqueous silver nitrate react to form solid silver chloride and aqueous sodium nitrate.

   c. Gasohol is a fuel that contains liquid ethanol, C_2H_6O, which burns in oxygen gas to form two gases, carbon dioxide and water.

8.50 Write a balanced equation for each of the following reactions and identify the type of reaction: (8.1, 8.2, 8.3)

   a. Solid potassium chlorate is heated to form solid potassium chloride and oxygen gas.

   b. Carbon monoxide gas and oxygen gas combine to form carbon dioxide gas.

   c. Ethene gas, C_2H_4, reacts with chlorine gas, Cl_2, to form dichloroethane, $C_2H_4Cl_2$.

8.51 In each of the following reactions, identify the reactant that is oxidized and the reactant that is reduced: (8.4)

   a. $N_2(g) + 2O_2(g) \longrightarrow 2NO_2(g)$

   b. $CO(g) + 3H_2(g) \longrightarrow CH_4(g) + H_2O(g)$

   c. $Mg(s) + Br_2(l) \longrightarrow MgBr_2(s)$

8.52 In each of the following reactions, identify the reactant that is oxidized and the reactant that is reduced: (8.4)

   a. $2Al(s) + 3F_2(g) \longrightarrow 2AlF_3(s)$

   b. $ZnO(s) + H_2(g) \longrightarrow Zn(s) + H_2O(g)$

   c. $2CuS(s) + 3O_2(g) \longrightarrow 2CuO(s) + 2SO_2(g)$

CHALLENGE PROBLEMS

The following problems are related to the topics in this chapter. However, they do not all follow the chapter order, and they require you to combine concepts and skills from several sections. These problems will help you increase your critical thinking skills and prepare for your next exam.

8.53 Balance each of the following chemical equations, and identify the type of reaction: (8.1, 8.2, 8.3)

   a. $K_2O(s) + H_2O(g) \longrightarrow KOH(s)$

   b. $C_8H_{18}(l) + O_2(g) \xrightarrow{\Delta} CO_2(g) + H_2O(g)$

   c. $Fe(OH)_3(s) \longrightarrow Fe_2O_3(s) + H_2O(g)$

   d. $CuS(s) + HCl(aq) \longrightarrow CuCl_2(aq) + H_2S(g)$

8.54 Balance each of the following chemical equations, and identify the type of reaction: (8.1, 8.2, 8.3)

   a. $TiCl_4(s) + Mg(s) \longrightarrow MgCl_2(s) + Ti(s)$

   b. $P_4O_{10}(s) + H_2O(g) \xrightarrow{\Delta} H_3PO_4(aq)$

   c. $KClO_3(s) \xrightarrow{\Delta} KCl(s) + O_2(g)$

   d. $C_3H_6(g) + O_2(g) \xrightarrow{\Delta} CO_2(g) + H_2O(g)$

8.55 Complete and balance each of the following chemical equations: (8.1, 8.2, 8.3)

   a. single replacement: $Fe_3O_4(s) + H_2(g) \longrightarrow$

   b. combustion: $C_4H_{10}(g) + O_2(g) \xrightarrow{\Delta}$

   c. combination: $Al(s) + O_2(g) \longrightarrow$

   d. double replacement: $NaOH(aq) + ZnSO_4(aq) \longrightarrow$

8.56 Complete and balance each of the following chemical equations: (8.1, 8.2, 8.3)

   a. decomposition: $HgO(s) \xrightarrow{\Delta}$

   b. double replacement: $BaCl_2(aq) + AgNO_3(aq) \longrightarrow$

   c. single replacement: $Ca(s) + AlCl_3(s) \longrightarrow$

   d. combination: $Mg(s) + N_2(g) \longrightarrow$

8.57 Write the correct formulas for the reactants and products, the balanced equation for each of the following reaction descriptions, and identify each type of reaction: (8.1, 8.2, 8.3)

   a. An aqueous solution of lead(II) nitrate is mixed with aqueous sodium phosphate to produce solid lead(II) phosphate and aqueous sodium nitrate.

   b. Gallium metal heated in oxygen gas forms solid gallium(III) oxide.

   c. When solid sodium nitrate is heated, solid sodium nitrite and oxygen gas are produced.

8.58 Write the correct formulas for the reactants and products, the balanced equation for each of the following reaction descriptions, and identify each type of reaction: (8.1, 8.2, 8.3)

   a. Solid bismuth(III) oxide and solid carbon react to form bismuth metal and carbon monoxide gas.

   b. Solid sodium bicarbonate is heated and forms solid sodium carbonate, gaseous carbon dioxide, and liquid water.

   c. Liquid hexane, C_6H_{14}, reacts with oxygen gas to form two gaseous products: carbon dioxide and liquid water.

8.59 In the following diagram, if blue spheres are the element X and yellow spheres are the element Y: (8.1, 8.2, 8.3)

Reactants	Products

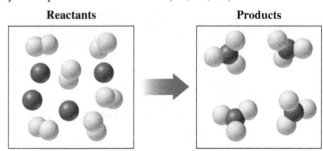

   a. Write the formula for each of the reactants and products.

   b. Write a balanced equation for the reaction.

   c. Indicate the type of reaction as combination, decomposition, single replacement, double replacement, or combustion.

8.60 In the following diagram, if red spheres are the element A, white spheres are the element B, and green spheres are the element C: (8.1, 8.2, 8.3)

Reactants	Products

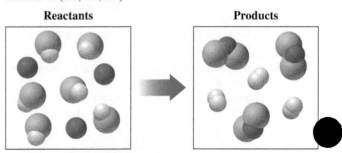

   a. Write the formula for each of the reactants and products.

   b. Write a balanced equation for the reaction.

   c. Indicate the type of reaction as combination, decomposition, single replacement, double replacement, or combustion.

ANSWERS TO ENGAGE QUESTIONS

8.1 A change in color in the formation of rust indicates a chemical change.

8.2 A chemical equation is balanced when there are the same numbers of atoms of each element in the reactants and the products, and the coefficients are the lowest possible whole numbers.

8.3 The formation of a white solid (precipitate) in the products, which has a different formula from the reactants, is evidence of a chemical reaction.

8.4 In a combination reaction, the atoms from the reactants combine to form one product.

8.5 In a decomposition reaction, a reactant splits into two or more simpler products.

8.6 In a single replacement reaction, one element in the reactants switches places with an element in the other reacting compound.

8.7 In a double replacement reaction, the positive ions in the reactants switch places.

8.8 When Cu^{2+} gains two electrons to form Cu^0, it is a reduction.

ANSWERS TO SELECTED PROBLEMS

8.1 **a.** reactants/products: 7 O atoms
   b. reactants/products: 4 O atoms
   c. reactants/products: 10 O atoms
   d. reactants/products: 4 O atoms

8.3 **a.** not balanced   **b.** not balanced
   c. balanced   **d.** balanced

8.5 **a.** 2 Na atoms, 2 Cl atoms
   b. 1 P atom, 3 Cl atoms, 6 H atoms
   c. 4 P atoms, 16 O atoms, 12 H atoms

8.7 **a.** $N_2(g) + O_2(g) \longrightarrow 2NO(g)$
   b. $2HgO(s) \xrightarrow{\Delta} 2Hg(l) + O_2(g)$
   c. $4Fe(s) + 3O_2(g) \longrightarrow 2Fe_2O_3(s)$
   d. $2Na(s) + Cl_2(g) \longrightarrow 2NaCl(s)$
   e. $2Cu_2O(s) + O_2(g) \longrightarrow 4CuO(s)$

8.9 **a.** $Mg(s) + 2AgNO_3(aq) \longrightarrow 2Ag(s) + Mg(NO_3)_2(aq)$
   b. $CuCO_3(s) \longrightarrow CuO(s) + CO_2(g)$
   c. $2Al(s) + 3CuSO_4(aq) \longrightarrow 3Cu(s) + Al_2(SO_4)_3(aq)$
   d. $Pb(NO_3)_2(aq) + 2NaCl(aq) \longrightarrow PbCl_2(s) + 2NaNO_3(aq)$
   e. $6HCl(aq) + 2Al(s) \longrightarrow 3H_2(g) + 2AlCl_3(aq)$

8.11 **a.** $Fe_2O_3(s) + 3CO(g) \longrightarrow 2Fe(s) + 3CO_2(g)$
   b. $2Li_3N(s) \longrightarrow 6Li(s) + N_2(g)$
   c. $6HBr(aq) + 2Al(s) \longrightarrow 3H_2(g) + 2AlBr_3(aq)$
   d. $3Ba(OH)_2(aq) + 2Na_3PO_4(aq) \longrightarrow$
   $Ba_3(PO_4)_2(s) + 6NaOH(aq)$
   e. $As_4S_6(s) + 9O_2(g) \longrightarrow As_4O_6(s) + 6SO_2(g)$

8.13 **a.** $2Li(s) + 2H_2O(l) \longrightarrow H_2(g) + 2LiOH(aq)$
   b. $2P(s) + 5Cl_2(g) \longrightarrow 2PCl_5(s)$
   c. $FeO(s) + CO(g) \longrightarrow Fe(s) + CO_2(g)$
   d. $2C_5H_{10}(l) + 15O_2(g) \xrightarrow{\Delta} 10CO_2(g) + 10H_2O(g)$
   e. $3H_2S(g) + 2FeCl_3(s) \longrightarrow Fe_2S_3(s) + 6HCl(g)$

8.15 $NH_4NO_3(s) \xrightarrow{\Delta} 2H_2O(g) + N_2O(g)$

8.17 $2C_3H_7NO_2(aq) + 6O_2(g) \longrightarrow$
   $5CO_2(g) + 5H_2O(l) + CH_4N_2O(aq)$

8.19 **a.** decomposition
   b. single replacement
   c. combustion
   d. double replacement
   e. combination

8.21 **a.** combination
   b. single replacement
   c. decomposition
   d. double replacement
   e. combustion

8.23 **a.** $Mg(s) + Cl_2(g) \longrightarrow MgCl_2(s)$
   b. $2HBr(g) \longrightarrow H_2(g) + Br_2(l)$
   c. $Mg(s) + Zn(NO_3)_2(aq) \longrightarrow Zn(s) + Mg(NO_3)_2(aq)$
   d. $K_2S(aq) + Pb(NO_3)_2(aq) \longrightarrow PbS(s) + 2KNO_3(aq)$
   e. $2C_2H_6(g) + 7O_2(g) \xrightarrow{\Delta} 4CO_2(g) + 6H_2O(g)$

8.25 **a.** reduction   **b.** oxidation
   c. reduction   **d.** reduction

8.27 **a.** Zn is oxidized; Cl_2 is reduced.
   b. The Br^- in NaBr is oxidized; Cl_2 is reduced.
   c. The O^{2-} in PbO is oxidized; the Pb^{2+} in PbO is reduced.
   d. Sn^{2+} is oxidized; Fe^{3+} is reduced.

8.29 **a.** reduction
   b. oxidation

8.31 Linoleic acid gains hydrogen atoms and is reduced.

8.33 **a.** $C_6H_{12}O_6(aq) + 6O_2(g) \longrightarrow 6CO_2(g) + 6H_2O(l)$
   b. $6CO_2(g) + 6H_2O(l) \longrightarrow C_6H_{12}O_6(aq) + 6O_2(g)$

8.35 **a.** 1,1,2 combination
   b. 2,2,1 decomposition

8.37 **a.** reactants NO and O_2; product NO_2
   b. $2NO(g) + O_2(g) \longrightarrow 2NO_2(g)$
   c. combination

8.39 **a.** reactant NI_3; products N_2 and I_2
   b. $2NI_3(s) \longrightarrow N_2(g) + 3I_2(g)$
   c. decomposition

8.41 **a.** reactants Cl_2 and O_2; product OCl_2
   b. $2Cl_2(g) + O_2(g) \longrightarrow 2OCl_2(g)$
   c. combination

8.43 **a.** combination   **b.** combustion
   c. decomposition   **d.** single replacement

8.45 **a.** $NH_3(g) + HCl(g) \longrightarrow NH_4Cl(s)$ combination
   b. $C_4H_8(g) + 6O_2(g) \xrightarrow{\Delta} 4CO_2(g) + 4H_2O(g)$ combustion
   c. $2Sb(s) + 3Cl_2(g) \longrightarrow 2SbCl_3(s)$ combination
   d. $2NI_3(s) \longrightarrow N_2(g) + 3I_2(g)$ decomposition
   e. $2KBr(aq) + Cl_2(aq) \longrightarrow 2KCl(aq) + Br_2(l)$
   single replacement
   f. $3H_2SO_4(aq) + 2Fe(s) \longrightarrow 3H_2(g) + Fe_2(SO_4)_3(aq)$
   single replacement
   g. $Al_2(SO_4)_3(aq) + 6NaOH(aq) \longrightarrow 2Al(OH)_3(s)$
   $+ 3Na_2SO_4(aq)$ double replacement

8.47 **a.** $2HCl(aq) + Zn(s) \longrightarrow H_2(g) + ZnCl_2(aq)$
   b. $BaCO_3(s) \xrightarrow{\Delta} BaO(s) + CO_2(g)$
   c. $HCl(aq) + NaOH(aq) \longrightarrow H_2O(l) + NaCl(aq)$
   d. $2Al(s) + 3F_2(g) \longrightarrow 2AlF_3(s)$

8.49 **a.** $4Na(s) + O_2(g) \longrightarrow 2Na_2O(s)$ combination
b. $NaCl(aq) + AgNO_3(aq) \longrightarrow AgCl(s) + NaNO_3(aq)$
double replacement
c. $C_2H_6O(l) + 3O_2(g) \xrightarrow{\Delta} 2CO_2(g) + 3H_2O(g)$ combustion

8.51 **a.** N_2 is oxidized; O_2 is reduced.
b. H_2 is oxidized; the C (in CO) is reduced.
c. Mg is oxidized; Br_2 is reduced.

8.53 **a.** $K_2O(s) + H_2O(g) \longrightarrow 2KOH(s)$ combination
b. $2C_8H_{18}(l) + 25O_2(g) \xrightarrow{\Delta} 16CO_2(g) + 18H_2O(g)$
combustion
c. $2Fe(OH)_3(s) \longrightarrow Fe_2O_3(s) + 3H_2O(g)$ decomposition
d. $CuS(s) + 2HCl(aq) \longrightarrow CuCl_2(aq) + H_2S(g)$
double replacement

8.55 **a.** $Fe_3O_4(s) + 4H_2(g) \longrightarrow 3Fe(s) + 4H_2O(g)$
b. $2C_4H_{10}(g) + 13O_2(g) \xrightarrow{\Delta} 8CO_2(g) + 10H_2O(g)$
c. $4Al(s) + 3O_2(g) \longrightarrow 2Al_2O_3(s)$
d. $2NaOH(aq) + ZnSO_4(aq) \longrightarrow$
$Zn(OH)_2(s) + Na_2SO_4(aq)$

8.57 **a.** $3Pb(NO_3)_2(aq) + 2Na_3PO_4(aq) \longrightarrow$
$Pb_3(PO_4)_2(s) + 6NaNO_3(aq)$ double replacement
b. $4Ga(s) + 3O_2(g) \xrightarrow{\Delta} 2Ga_2O_3(s)$ combination
c. $2NaNO_3(s) \xrightarrow{\Delta} 2NaNO_2(s) + O_2(g)$ decomposition

8.59 **a.** reactants: X and Y_2; product: XY_3
b. $2X + 3Y_2 \longrightarrow 2XY_3$
c. combination

Chemical Quantities in Reactions

Lance, an environmental scientist, collects soil and water samples at a nearby farm to test for the presence and concentration of insecticides and pharmaceuticals. Farmers use insecticides to increase food production and pharmaceuticals to treat and prevent animal-related diseases. Due to the common use of these chemicals, they may pass into the soil and water supply, potentially contaminating the environment and causing health problems.

Recently, a farmer treated his cotton and bean fields with an insecticide called Sevin (carbaryl), which is an acetylcholinesterase inhibitor that acts on the central nervous system. A few days later, Lance collects samples of soil and water and detects small amounts of Sevin. In humans, Sevin can cause headaches, nausea, and paralysis of the respiratory system. Because Sevin is very soluble, it is important that it does not contaminate the water supply. Lance advises the farmer to decrease the amount of Sevin he uses on his crops. Lance informs the farmer that he will return in a week to retest the soil and water for Sevin.

CAREER
Environmental Scientist

Environmental science is a multidisciplinary field that combines chemistry, biology, ecology, and geology to study environmental problems. Environmental scientists monitor environmental pollution to protect the health of the public. By using specialized equipment, environmental scientists measure pollution levels in soil, air, and water, as well as noise and radiation levels. They can specialize in a specific area, such as air quality or hazardous and solid waste. For instance, air-quality experts monitor indoor air for allergens, mold, and toxins; they measure outdoor air pollutants created by businesses, vehicles, and agriculture. Since environmental scientists obtain samples containing potentially hazardous materials, they must be knowledgeable about safety protocols and wear personal protective equipment. They may also recommend methods to diminish various pollutants and may assist in cleanup and remediation efforts.

UPDATE Testing Water Samples for Insecticides

Last week, Lance's measurements indicated that the insecticide level exceeded government guidelines. This week, Lance is at the farm rechecking the levels of insecticides in the soil and water, which he finds are much lower and now within government guidelines. You can view the information in the **UPDATE Testing Water Samples for Insecticides**, page 260.

9.1 Conservation of Mass

LEARNING GOAL Calculate the total mass of reactants and the total mass of products in a balanced chemical equation.

In any chemical reaction, the total amount of matter in the reactants is equal to the total amount of matter in the products. Thus, the total mass of all the reactants must be equal to the total mass of all the products. This is known as the **Law of Conservation of Mass**, which states that there is no change in the total mass of the substances reacting in a balanced chemical reaction. Thus, no material is lost or gained as original substances are changed to new substances. For example, tarnish (Ag_2S) forms when silver reacts with sulfur to form silver sulfide.

$$2Ag(s) + S(s) \longrightarrow Ag_2S(s)$$

$$2Ag(s) \qquad + \qquad S(s) \qquad \longrightarrow \qquad Ag_2S(s)$$

Mass of reactants   =   Mass of product

In the chemical reaction of Ag and S, the mass of the reactants is the same as the mass of the product, Ag_2S.

In this reaction, the number of silver atoms that reacts is twice the number of sulfur atoms. When 200 silver atoms react, 100 sulfur atoms are required. However, in the actual chemical reaction, many more atoms of both silver and sulfur would react. If we are dealing with molar amounts, then the coefficients in the equation can be interpreted in terms of moles. Thus, 2 mol of silver reacts with 1 mol of sulfur to produce 1 mol of Ag_2S. Because the molar mass of each can be determined, the moles of Ag, S, and Ag_2S can also be stated in terms of mass in grams of each. Thus, 215.8 g of Ag and 32.1 g of S react to form 247.9 g of Ag_2S. The total mass of the reactants (247.9 g) is equal to the mass of the product (247.9 g). The various ways in which a chemical equation can be interpreted are seen in **TABLE 9.1**.

ENGAGE 9.1

Why is the total mass of the reactants equal to the total mass of the products?

TABLE 9.1 Information Available from a Balanced Equation

	Reactants		Products
Equation	$2Ag(s)$	$+ S(s)$	$\longrightarrow Ag_2S(s)$
Atoms/Formula Units	2 Ag atoms	+ 1 S atom	$\longrightarrow$ 1 Ag_2S formula unit
	200 Ag atoms	+ 100 S atoms	$\longrightarrow$ 100 Ag_2S formula units
Avogadro's Number of Atoms	$2(6.022 \times 10^{23})$ Ag atoms	$+ \; 1(6.022 \times 10^{23})$ S atoms	$\longrightarrow 1(6.022 \times 10^{23})$ Ag_2S formula units
Moles	2 mol of Ag	+ 1 mol of S	$\longrightarrow$ 1 mol of Ag_2S
Mass (g)	2(107.9 g) of Ag	+ 1(32.1 g) of S	$\longrightarrow$ 1(247.9 g) of Ag_2S
Total Mass (g)	247.9 g		$\longrightarrow$ 247.9 g

SAMPLE PROBLEM 9.1 Conservation of Mass

TRY IT FIRST

The combustion of methane (CH_4) with oxygen produces carbon dioxide, water, and energy. Calculate the total mass of the reactants and the products for the following equation when 1 mol of CH_4 reacts:

$$CH_4(g) + 2O_2(g) \xrightarrow{\Delta} CO_2(g) + 2H_2O(g)$$

SOLUTION

Interpreting the coefficients in the equation as the number of moles of each substance and multiplying by its molar mass gives the total mass of reactants and products. The quantities of moles are exact because the coefficients in the balanced equation are exact.

	Reactants		Products
Equation	$CH_4(g) + 2O_2(g)$	$\xrightarrow{\Delta}$	$CO_2(g) + 2H_2O(g)$
Moles	1 mol of CH_4 + 2 mol of O_2	$\longrightarrow$	1 mol of CO_2 + 2 mol of H_2O
Mass	16.04 g of CH_4 + 64.00 g of O_2	$\longrightarrow$	44.01 g of CO_2 + 36.03 g of H_2O
Total Mass	80.04 g of reactants	=	80.04 g of products

SELF TEST 9.1

Calculate the total mass of the reactants and the products for each of the following equations:
a. $4K(s) + O_2(g) \longrightarrow 2K_2O(s)$
b. $6HCl(aq) + 2Al(s) \longrightarrow 3H_2(g) + 2AlCl_3(aq)$

ANSWER
a. 188.40 g of reactants and 188.40 g of products
b. 272.71 g of reactants and 272.71 g of products

PRACTICE PROBLEMS
Try Practice Problems 9.1 and 9.2

PRACTICE PROBLEMS

9.1 Conservation of Mass

9.1 Calculate the total mass of the reactants and the products for each of the following equations:
   a. $2SO_2(g) + O_2(g) \longrightarrow 2SO_3(g)$
   b. $4P(s) + 5O_2(g) \longrightarrow 2P_2O_5(s)$

9.2 Calculate the total mass of the reactants and the products for each of the following equations:
   a. $2Al(s) + 3Cl_2(g) \longrightarrow 2AlCl_3(s)$
   b. $4HCl(g) + O_2(g) \longrightarrow 2Cl_2(g) + 2H_2O(g)$

9.2 Mole Relationships in Chemical Equations

LEARNING GOAL Use a mole–mole factor from a balanced chemical equation to calculate the number of moles of another substance in the reaction.

When iron reacts with sulfur, the product is iron(III) sulfide.

$$2Fe(s) + 3S(s) \longrightarrow Fe_2S_3(s)$$

From the balanced equation, we see that 2 mol of iron reacts with 3 mol of sulfur to form 1 mol of iron(III) sulfide. Actually, any amount of iron or sulfur may be used, but the *ratio*

Iron (Fe)   Sulfur (S)   Iron(III) sulfide (Fe$_2$S$_3$)
2Fe(s)   +   3S(s) $\longrightarrow$ Fe$_2$S$_3$(s)

In the chemical reaction of Fe and S, the mass of the reactants is the same as the mass of the product, Fe$_2$S$_3$.

of iron reacting with sulfur will always be the same. From the coefficients, we can write **mole–mole factors** between reactants and between reactants and products. The coefficients used in the mole–mole factors are exact numbers; they do not limit the number of significant figures.

$$\text{Fe and S:} \quad \frac{2 \text{ mol Fe}}{3 \text{ mol S}} \quad \text{and} \quad \frac{3 \text{ mol S}}{2 \text{ mol Fe}}$$

$$\text{Fe and Fe}_2\text{S}_3: \quad \frac{2 \text{ mol Fe}}{1 \text{ mol Fe}_2\text{S}_3} \quad \text{and} \quad \frac{1 \text{ mol Fe}_2\text{S}_3}{2 \text{ mol Fe}}$$

PRACTICE PROBLEMS

Try Practice Problems 9.3 to 9.6

$$\text{S and Fe}_2\text{S}_3: \quad \frac{3 \text{ mol S}}{1 \text{ mol Fe}_2\text{S}_3} \quad \text{and} \quad \frac{1 \text{ mol Fe}_2\text{S}_3}{3 \text{ mol S}}$$

Using Mole–Mole Factors in Calculations

CORE CHEMISTRY SKILL

Using Mole–Mole Factors

Whenever you prepare a recipe, adjust an engine for the proper mixture of fuel and air, or prepare medicines in a pharmaceutical laboratory, you need to know the proper amounts of reactants to use and how much of the product will form. Now that we have written all the possible conversion factors for the balanced equation 2Fe(s) + 3S(s) $\longrightarrow$ Fe$_2$S$_3$(s), we will use those mole–mole factors in a chemical calculation in Sample Problem 9.2.

SAMPLE PROBLEM 9.2 Calculating Moles of a Reactant

TRY IT FIRST

In the chemical reaction of iron and sulfur, how many moles of sulfur are needed to react with 1.42 mol of iron?

2Fe(s) + 3S(s) $\longrightarrow$ Fe$_2$S$_3$(s)

SOLUTION

STEP 1 State the given and needed quantities (moles).

	Given	Need	Connect
ANALYZE THE PROBLEM	1.42 mol of Fe	moles of S	mole–mole factor
	Equation		
	2Fe(s) + 3S(s) $\longrightarrow$ Fe$_2$S$_3$(s)		

STEP 2 Write a plan to convert the given to the needed quantity (moles).

moles of Fe   Mole–mole factor   moles of S

STEP 3 Use coefficients to write mole–mole factors.

$$2 \text{ mol of Fe} = 3 \text{ mol of S}$$

$$\frac{2 \text{ mol Fe}}{3 \text{ mol S}} \quad \text{and} \quad \frac{3 \text{ mol S}}{2 \text{ mol Fe}}$$

STEP 4 Set up the problem to give the needed quantity (moles).

Exact

$$1.42 \text{ mol Fe} \times \frac{3 \text{ mol S}}{2 \text{ mol Fe}} = 2.13 \text{ mol of S}$$

Three SFs   Exact   Three SFs

SELF TEST 9.2

Using the equation in Sample Problem 9.2, calculate each of the following:
a. moles of iron needed to react with 2.75 mol of sulfur
b. moles of iron(III) sulfide produced by the reaction of 0.758 mol of sulfur

ANSWER

a. 1.83 mol of iron
b. 0.253 mol of iron(III) sulfide

SAMPLE PROBLEM 9.3   Calculating Moles of a Product

TRY IT FIRST

Propane gas (C_3H_8), a fuel used in camp stoves, soldering torches, and specially equipped automobiles, reacts with oxygen to produce carbon dioxide, water, and energy. How many moles of CO_2 can be produced when 2.25 mol of C_3H_8 reacts?

$$C_3H_8(g) + 5O_2(g) \xrightarrow{\Delta} 3CO_2(g) + 4H_2O(g)$$

SOLUTION

STEP 1 State the given and needed quantities (moles).

Propane fuel reacts with O_2 in the air to produce CO_2, H_2O, and energy.

	Given	Need	Connect
ANALYZE THE PROBLEM	2.25 mol of C_3H_8	moles of CO_2	mole–mole factor
	Equation		
	$C_3H_8(g) + 5O_2(g) \xrightarrow{\Delta} 3CO_2(g) + 4H_2O(g)$		

STEP 2 Write a plan to convert the given to the needed quantity (moles).

moles of C_3H_8 → Mole–mole factor → moles of CO_2

STEP 3 Use coefficients to write mole–mole factors.

$$1 \text{ mol of } C_3H_8 = 3 \text{ mol of } CO_2$$

$$\frac{3 \text{ mol } CO_2}{1 \text{ mol } C_3H_8} \quad \text{and} \quad \frac{1 \text{ mol } C_3H_8}{3 \text{ mol } CO_2}$$

STEP 4 Set up the problem to give the needed quantity (moles).

Exact

$$2.25 \ \cancel{\text{mol } C_3H_8} \times \frac{3 \ \text{mol CO}_2}{1 \ \cancel{\text{mol } C_3H_8}} = 6.75 \ \text{mol of CO}_2$$

Three SFs   Exact   Three SFs

SELF TEST 9.3

Using the equation in Sample Problem 9.3, calculate the moles of oxygen that must react to
a. produce 0.756 mol of water
b. react with 0.243 mol of C_3H_8

ANSWER
a. 0.945 mol of O_2
b. 1.22 mol of O_2

PRACTICE PROBLEMS

Try Practice Problems 9.7 to 9.10

PRACTICE PROBLEMS

9.2 Mole Relationships in Chemical Equations

9.3 Write all of the mole–mole factors for each of the following chemical equations:
a. $2SO_2(g) + O_2(g) \longrightarrow 2SO_3(g)$
b. $4P(s) + 5O_2(g) \longrightarrow 2P_2O_5(s)$

9.4 Write all of the mole–mole factors for each of the following chemical equations:
a. $2Al(s) + 3Cl_2(g) \longrightarrow 2AlCl_3(s)$
b. $4HCl(g) + O_2(g) \longrightarrow 2Cl_2(g) + 2H_2O(g)$

9.5 For the chemical equations in problem 9.3, write the problem setup with the correct mole–mole factor needed to find each of the following:
a. moles of SO_3 from the moles of SO_2
b. moles of O_2 needed to react with moles of P

9.6 For the chemical equations in problem 9.4, write the problem setup with the correct mole–mole factor needed to find each of the following:
a. moles of $AlCl_3$ from the moles of Cl_2
b. moles of O_2 needed to react with moles of HCl

9.7 The chemical reaction of hydrogen with oxygen produces water.

$$2H_2(g) + O_2(g) \longrightarrow 2H_2O(g)$$

a. How many moles of O_2 are required to react with 2.6 mol of H_2?
b. How many moles of H_2 are needed to react with 5.0 mol of O_2?
c. How many moles of H_2O form when 2.5 mol of O_2 reacts?

9.8 Ammonia is produced by the chemical reaction of nitrogen and hydrogen.

$$N_2(g) + 3H_2(g) \longrightarrow 2NH_3(g)$$

Ammonia

a. How many moles of H_2 are needed to react with 1.8 mol of N_2?
b. How many moles of N_2 reacted if 0.60 mol of NH_3 is produced?
c. How many moles of NH_3 are produced when 1.4 mol of H_2 reacts?

9.9 Carbon disulfide and carbon monoxide are produced when carbon is heated with sulfur dioxide.

$$5C(s) + 2SO_2(g) \xrightarrow{\Delta} CS_2(l) + 4CO(g)$$

a. How many moles of C are needed to react with 0.500 mol of SO_2?
b. How many moles of CO are produced when 1.2 mol of C reacts?
c. How many moles of SO_2 are required to produce 0.50 mol of CS_2?
d. How many moles of CS_2 are produced when 2.5 mol of C reacts?

9.10 In the acetylene torch, acetylene gas burns in oxygen to produce carbon dioxide, water, and energy.

$$2C_2H_2(g) + 5O_2(g) \xrightarrow{\Delta} 4CO_2(g) + 2H_2O(g)$$

Acetylene

a. How many moles of O_2 are needed to react with 2.40 mol of C_2H_2?
b. How many moles of CO_2 are produced when 3.5 mol of C_2H_2 reacts?
c. How many moles of C_2H_2 are required to produce 0.50 mol of H_2O?
d. How many moles of CO_2 are produced when 0.100 mol of O_2 reacts?

9.3 Mass Calculations for Chemical Reactions

LEARNING GOAL Given the mass in grams of a substance in a chemical reaction, calculate the mass in grams of another substance in the reaction.

When we have the balanced equation for a chemical reaction, we can use the mass of one of the substances (A) in the reaction to calculate the mass of another substance (B) in the reaction. However, the calculations require us to convert the mass of A to moles of A using the molar mass of A. Then we use the mole–mole factor that links substance A to substance B, which we obtain from the coefficients in the balanced equation. This mole–mole factor (B/A) will convert the moles of A to moles of B. Then the molar mass of B is used to calculate the grams of substance B.

REVIEW

Counting Significant Figures (2.2)

Using Significant Figures in Calculations (2.3)

Writing Conversion Factors from Equalities (2.5)

Using Conversion Factors (2.6)

Calculating Molar Mass (7.2)

Using Molar Mass as a Conversion Factor (7.3)

Substance A   **Substance B**

grams of A → [Molar mass A] → moles of A → [Mole–mole factor B/A] → moles of B → [Molar mass B] → grams of B

SAMPLE PROBLEM 9.4   Calculating Mass of a Product

TRY IT FIRST

When acetylene (C_2H_2) burns in oxygen, high temperatures are produced that are used for welding metals.

$$2C_2H_2(g) + 5O_2(g) \xrightarrow{\Delta} 4CO_2(g) + 2H_2O(g)$$

How many grams of CO_2 are produced when 54.6 g of C_2H_2 is burned?

SOLUTION

STEP 1   State the given and needed quantities (grams).

	Given	Need	Connect
ANALYZE THE PROBLEM	54.6 g of C_2H_2	grams of CO_2	molar masses, mole–mole factor
	Equation		
	$2C_2H_2(g) + 5O_2(g) \xrightarrow{\Delta} 4CO_2(g) + 2H_2O(g)$		

A mixture of acetylene and oxygen undergoes combustion during the welding of metals.

CORE CHEMISTRY SKILL

Converting Grams to Grams

STEP 2   Write a plan to convert the given to the needed quantity (grams).

grams of C_2H_2 → [Molar mass] → moles of C_2H_2 → [Mole–mole factor] → moles of CO_2 → [Molar mass] → grams of CO_2

STEP 3   Use coefficients to write mole–mole factors; write molar masses.

$$1 \text{ mol of } C_2H_2 = 26.04 \text{ g of } C_2H_2 \qquad 1 \text{ mol of } CO_2 = 44.01 \text{ g of } CO_2$$

$$\frac{26.04 \text{ g } C_2H_2}{1 \text{ mol } C_2H_2} \text{ and } \frac{1 \text{ mol } C_2H_2}{26.04 \text{ g } C_2H_2} \qquad \frac{44.01 \text{ g } CO_2}{1 \text{ mol } CO_2} \text{ and } \frac{1 \text{ mol } CO_2}{44.01 \text{ g } CO_2}$$

$$2 \text{ mol of } C_2H_2 = 4 \text{ mol of } CO_2$$

$$\frac{2 \text{ mol } C_2H_2}{4 \text{ mol } CO_2} \text{ and } \frac{4 \text{ mol } CO_2}{2 \text{ mol } C_2H_2}$$

Why is a mole–mole factor needed in the problem setup that converts grams of a reactant to grams of a product?

STEP 4 **Set up the problem to give the needed quantity (grams).**

	Exact		Exact		Four SFs		

$$54.6 \text{ g } C_2H_2 \times \frac{1 \text{ mol } C_2H_2}{26.04 \text{ g } C_2H_2} \times \frac{4 \text{ mol } CO_2}{2 \text{ mol } C_2H_2} \times \frac{44.01 \text{ g } CO_2}{1 \text{ mol } CO_2} = 185 \text{ g of } CO_2$$

Three SFs   Four SFs   Exact   Exact   Three SFs

SELF TEST 9.4

Using the equation in Sample Problem 9.4, calculate each of the following:
a. the grams of CO_2 produced when 25.0 g of O_2 reacts
b. the grams of C_2H_2 needed when 65.0 g of H_2O are produced

ANSWER

a. 27.5 g of CO_2 **b.** 93.9 g of C_2H_2

The octane rating is a measure of how well a fuel burns in an engine.

SAMPLE PROBLEM 9.5 **Calculating Mass of a Reactant**

TRY IT FIRST

The fuel heptane (C_7H_{16}) is designated as the zero point in the octane rating of gasoline. Heptane is an undesirable compound in gasoline because it burns rapidly and causes engine knocking. How many grams of O_2 are required to react with 22.5 g of C_7H_{16}?

$$C_7H_{16}(l) + 11O_2(g) \xrightarrow{\Delta} 7CO_2(g) + 8H_2O(g)$$

SOLUTION

STEP 1 **State the given and needed quantities (grams).**

ANALYZE THE PROBLEM	Given	Need	Connect
	22.5 g of C_7H_{16}	grams of O_2	molar masses, mole–mole factor
	Equation		
	$C_7H_{16}(l) + 11O_2(g) \xrightarrow{\Delta} 7CO_2(g) + 8H_2O(g)$		

STEP 2 **Write a plan to convert the given to the needed quantity (grams).**

grams of C_7H_{16} → Molar mass → moles of C_7H_{16} → Mole–mole factor → moles of O_2 → Molar mass → grams of O_2

STEP 3 **Use coefficients to write mole–mole factors; write molar masses.**

1 mol of C_7H_{16} = 100.2 g of C_7H_{16} 1 mol of O_2 = 32.00 g of O_2

$$\frac{100.2 \text{ g } C_7H_{16}}{1 \text{ mol } C_7H_{16}} \quad \text{and} \quad \frac{1 \text{ mol } C_7H_{16}}{100.2 \text{ g } C_7H_{16}} \qquad \frac{32.00 \text{ g } O_2}{1 \text{ mol } O_2} \quad \text{and} \quad \frac{1 \text{ mol } O_2}{32.00 \text{ g } O_2}$$

1 mol of C_7H_{16} = 11 mol of O_2

$$\frac{11 \text{ mol } O_2}{1 \text{ mol } C_7H_{16}} \quad \text{and} \quad \frac{1 \text{ mol } C_7H_{16}}{11 \text{ mol } O_2}$$

STEP 4 **Set up the problem to give the needed quantity (grams).**

	Exact		Exact		Four SFs		

$$22.5 \text{ g } C_7H_{16} \times \frac{1 \text{ mol } C_7H_{16}}{100.2 \text{ g } C_7H_{16}} \times \frac{11 \text{ mol } O_2}{1 \text{ mol } C_7H_{16}} \times \frac{32.00 \text{ g } O_2}{1 \text{ mol } O_2} = 79.1 \text{ g of } O_2$$

Three SFs   Four SFs   Exact   Exact   Three SFs

SELF TEST 9.5

Using the equation in Sample Problem 9.5, calculate each of the following:
a. the grams of C_7H_{16} that are needed to produce 15.0 g of H_2O
b. the grams of CO_2 produced from the reaction of 25.0 g of C_7H_{16}

ANSWER

a. 10.4 g of C_7H_{16} **b.** 76.9 g of CO_2

PRACTICE PROBLEMS
Try Practice Problems 9.11 to 9.18

PRACTICE PROBLEMS

9.3 Mass Calculations for Chemical Reactions

9.11 Sodium reacts with oxygen to produce sodium oxide.

$$4Na(s) + O_2(g) \longrightarrow 2Na_2O(s)$$

a. How many grams of Na_2O are produced when 57.5 g of Na reacts?
b. If you have 18.0 g of Na, how many grams of O_2 are required for reaction?
c. How many grams of O_2 are needed in a reaction that produces 75.0 g of Na_2O?

9.12 Nitrogen reacts with hydrogen to produce ammonia.

$$N_2(g) + 3H_2(g) \longrightarrow 2NH_3(g)$$

a. If you have 3.64 g of H_2, how many grams of NH_3 can be produced?
b. How many grams of H_2 are needed to react with 2.80 g of N_2?
c. How many grams of NH_3 can be produced from 12.0 g of H_2?

9.13 Ammonia and oxygen react to form nitrogen and water.

$$4NH_3(g) + 3O_2(g) \longrightarrow 2N_2(g) + 6H_2O(g)$$

a. How many grams of O_2 are needed to react with 13.6 g of NH_3?
b. How many grams of N_2 can be produced when 6.50 g of O_2 reacts?
c. How many grams of H_2O are formed from the reaction of 34.0 g of NH_3?

9.14 Iron(III) oxide reacts with carbon to give iron and carbon monoxide.

$$Fe_2O_3(s) + 3C(s) \longrightarrow 2Fe(s) + 3CO(g)$$

a. How many grams of C are required to react with 16.5 g of Fe_2O_3?
b. How many grams of CO are produced when 36.0 g of C reacts?
c. How many grams of Fe can be produced when 6.00 g of Fe_2O_3 reacts?

9.15 Nitrogen dioxide and water react to produce nitric acid, HNO_3, and nitrogen oxide.

$$3NO_2(g) + H_2O(l) \longrightarrow 2HNO_3(aq) + NO(g)$$

a. How many grams of H_2O are required to react with 28.0 g of NO_2?
b. How many grams of NO are produced from 15.8 g of H_2O?
c. How many grams of HNO_3 are produced from 8.25 g of NO_2?

9.16 Calcium cyanamide, $CaCN_2$, reacts with water to form calcium carbonate and ammonia.

$$CaCN_2(s) + 3H_2O(l) \longrightarrow CaCO_3(s) + 2NH_3(g)$$

a. How many grams of H_2O are needed to react with 75.0 g of $CaCN_2$?
b. How many grams of NH_3 are produced from 5.24 g of $CaCN_2$?
c. How many grams of $CaCO_3$ form if 155 g of H_2O reacts?

9.17 When solid lead(II) sulfide reacts with oxygen gas, the products are solid lead(II) oxide and sulfur dioxide gas.
a. Write the balanced chemical equation for the reaction.
b. How many grams of oxygen are required to react with 29.9 g of lead(II) sulfide?
c. How many grams of sulfur dioxide can be produced when 65.0 g of lead(II) sulfide reacts?
d. How many grams of lead(II) sulfide are used to produce 128 g of lead(II) oxide?

9.18 When the gases dihydrogen sulfide and oxygen react, they form the gases sulfur dioxide and water vapor.
a. Write the balanced chemical equation for the reaction.
b. How many grams of oxygen are required to react with 2.50 g of dihydrogen sulfide?
c. How many grams of sulfur dioxide can be produced when 38.5 g of oxygen reacts?
d. How many grams of oxygen are required to produce 55.8 g of water vapor?

9.4 Limiting Reactants

LEARNING GOAL Identify a limiting reactant when given the quantities of two reactants; calculate the amount of product formed from the limiting reactant.

When we make peanut butter sandwiches for lunch, we need 2 slices of bread and 1 tablespoon of peanut butter for each sandwich. As an equation, we could write:

2 slices of bread + 1 tablespoon of peanut butter $\longrightarrow$ 1 peanut butter sandwich

If we have 8 slices of bread and a full jar of peanut butter, we will run out of bread after we make 4 peanut butter sandwiches. We cannot make any more sandwiches once the bread is used up, even though there is a lot of peanut butter left in the jar. The number of slices of bread has limited the number of sandwiches we can make.

On a different day, we might have 8 slices of bread but only a tablespoon of peanut butter left in the peanut butter jar. We will run out of peanut butter after we make just 1 peanut butter sandwich and have 6 slices of bread left over. The small amount of peanut butter available has limited the number of sandwiches we can make.

8 slices of bread	+	1 jar of peanut butter	make	4 peanut butter sandwiches	+ peanut butter left over
8 slices of bread	+	1 tablespoon of peanut butter	make	1 peanut butter sandwich	+ 6 slices of bread left over

The reactant that is completely used up is the **limiting reactant**. The reactant that does not completely react and is left over is called the *excess reactant*.

Bread	Peanut Butter	Sandwiches	Limiting Reactant	Excess Reactant
1 loaf (20 slices)	1 tablespoon	1	peanut butter	bread
4 slices	1 full jar	2	bread	peanut butter
8 slices	1 full jar	4	bread	peanut butter

Calculating Moles of Product from a Limiting Reactant

In a similar way, the reactants in a chemical reaction do not always combine in quantities that allow each to be used up at exactly the same time. In many reactions, there is a limiting reactant that determines the amount of product that can be formed. When we know the quantities of the reactants of a chemical reaction, we calculate the amount of product that is possible from each reactant if it were completely consumed. We are looking for the *limiting reactant*, which is the one that runs out first, producing the smaller amount of product.

SAMPLE PROBLEM 9.6   Moles of Product from a Limiting Reactant

TRY IT FIRST

Carbon monoxide and hydrogen are used to produce methanol (CH_4O). The balanced chemical reaction is

$$CO(g) + 2H_2(g) \longrightarrow CH_4O(g)$$

If 3.00 mol of CO and 5.00 mol of H_2 are the initial reactants, what is the limiting reactant, and how many moles of methanol can be produced?

SOLUTION

STEP 1 State the given and needed quantity (moles).

	Given	Need	Connect
ANALYZE THE PROBLEM	3.00 mol of CO, 5.00 mol of H_2	limiting reactant, moles of CH_4O produced	mole–mole factors
	Equation		
	$CO(g) + 2H_2(g) \longrightarrow CH_4O(g)$		

STEP 2 Write a plan to convert the quantity (moles) of each reactant to quantity (moles) of product.

moles of CO | Mole–mole factor | moles of CH_4O

moles of H_2 | Mole–mole factor | moles of CH_4O

STEP 3 Use coefficients to write mole–mole factors.

$$1 \text{ mol of CO} = 1 \text{ mol of } CH_4O \qquad 2 \text{ mol of } H_2 = 1 \text{ mol of } CH_4O$$

$$\frac{1 \text{ mol CO}}{1 \text{ mol } CH_4O} \text{ and } \frac{1 \text{ mol } CH_4O}{1 \text{ mol CO}} \qquad \frac{2 \text{ mol } H_2}{1 \text{ mol } CH_4O} \text{ and } \frac{1 \text{ mol } CH_4O}{2 \text{ mol } H_2}$$

STEP 4 Calculate the quantity (moles) of product from each reactant, and select the smaller quantity (moles) as the limiting reactant.

Moles of CH_4O (product) from CO:

$$3.00 \text{ mol CO} \times \frac{1 \text{ mol } CH_4O}{1 \text{ mol CO}} = 3.00 \text{ mol of } CH_4O$$

Three SFs | Exact | Three SFs

Moles of CH_4O (product) from H_2:

$$5.00 \text{ mol } H_2 \times \frac{1 \text{ mol } CH_4O}{2 \text{ mol } H_2} = 2.50 \text{ mol of } CH_4O \quad \text{Smaller amount of product}$$

Limiting reactant | Exact | Three SFs

The smaller amount, 2.50 mol of CH_4O, is the maximum amount of methanol that can be produced from the limiting reactant, H_2, because it is completely consumed.

SELF TEST 9.6

a. If an initial mixture of reactants for Sample Problem 9.6 contains 4.00 mol of CO and 4.00 mol of H_2, what is the limiting reactant, and how many moles of methanol can be produced?

b. If an initial mixture of reactants for Sample Problem 9.6 contains 1.50 mol of CO and 6.00 mol of H_2, what is the limiting reactant, and how many moles of methanol can be produced?

ANSWER

a. H_2 is the limiting reactant; 2.00 mol of methanol can be produced.
b. CO is the limiting reactant; 1.50 mol of methanol can be produced.

PRACTICE PROBLEMS
Try Practice Problems 9.19 to 9.22

Calculating Mass of Product from a Limiting Reactant

The quantities of the reactants can also be given in grams. The calculations to identify the limiting reactant are the same as before, but the grams of each reactant must first be converted to moles, then to moles of product, and finally to grams of product. Then select the smaller mass of product, which is from complete use of the limiting reactant. This calculation is shown in Sample Problem 9.7.

A ceramic brake disc in a car withstands temperatures of 1400 °C.

SAMPLE PROBLEM 9.7   Mass of Product from a Limiting Reactant

TRY IT FIRST

When silicon dioxide (sand) and carbon are heated, the products are silicon carbide, SiC, and carbon monoxide. Silicon carbide is a ceramic material that tolerates extreme temperatures and is used as an abrasive and in the brake discs of cars. How many grams of CO are formed from 70.0 g of SiO_2 and 50.0 g of C?

$$SiO_2(s) + 3C(s) \xrightarrow{\Delta} SiC(s) + 2CO(g)$$

SOLUTION

STEP 1   State the given and needed quantity (grams).

	Given	Need	Connect
ANALYZE THE PROBLEM	70.0 g of SiO_2, 50.0 g of C	grams of CO from limiting reactant	molar masses, mole–mole factors
	Equation		
	$SiO_2(s) + 3C(s) \xrightarrow{\Delta} SiC(s) + 2CO(g)$		

STEP 2   Write a plan to convert the quantity (grams) of each reactant to quantity (grams) of product.

grams of SiO_2	Molar mass	moles of SiO_2	Mole–mole factor	moles of CO	Molar mass	grams of CO

grams of C	Molar mass	moles of C	Mole–mole factor	moles of CO	Molar mass	grams of CO

STEP 3   Use coefficients to write mole–mole factors; write molar masses.

1 mol of SiO_2 = 60.09 g of SiO_2 1 mol of C = 12.01 g of C   1 mol of CO = 28.01 g of CO

$$\frac{60.09 \text{ g } SiO_2}{1 \text{ mol } SiO_2} \text{ and } \frac{1 \text{ mol } SiO_2}{60.09 \text{ g } SiO_2} \qquad \frac{12.01 \text{ g C}}{1 \text{ mol C}} \text{ and } \frac{1 \text{ mol C}}{12.01 \text{ g C}} \qquad \frac{28.01 \text{ g CO}}{1 \text{ mol CO}} \text{ and } \frac{1 \text{ mol CO}}{28.01 \text{ g CO}}$$

2 mol of CO = 1 mol of SiO_2 2 mol of CO = 3 mol of C

$$\frac{2 \text{ mol CO}}{1 \text{ mol } SiO_2} \text{ and } \frac{1 \text{ mol } SiO_2}{2 \text{ mol CO}} \qquad \frac{2 \text{ mol CO}}{3 \text{ mol C}} \text{ and } \frac{3 \text{ mol C}}{2 \text{ mol CO}}$$

STEP 4 Calculate the quantity (grams) of product from each reactant, and select the smaller quantity (grams) as the limiting reactant.

Grams of CO (product) from SiO_2:

Three SFs	Exact	Exact	Four SFs	Three SFs

$$70.0 \text{ g } \cancel{SiO_2} \times \frac{1 \text{ mol } \cancel{SiO_2}}{60.09 \text{ g } \cancel{SiO_2}} \times \frac{2 \text{ mol } \cancel{CO}}{1 \text{ mol } \cancel{SiO_2}} \times \frac{28.01 \text{ g CO}}{1 \text{ mol } \cancel{CO}} = 65.3 \text{ g of CO}$$

Limiting reactant | Four SFs | Exact | Exact | Smaller amount of product

Grams of CO (product) from C:

Three SFs	Exact	Exact	Four SFs	Three SFs

$$50.0 \text{ g } \cancel{C} \times \frac{1 \text{ mol C}}{12.01 \text{ g } \cancel{C}} \times \frac{2 \text{ mol } \cancel{CO}}{3 \text{ mol } \cancel{C}} \times \frac{28.01 \text{ g CO}}{1 \text{ mol } \cancel{CO}} = 77.7 \text{ g of CO}$$

Four SFs | Exact | Exact

The smaller amount, 65.3 g of CO, is the most CO that can be produced. This also means that the SiO_2 is the limiting reactant.

SELF TEST 9.7

Hydrogen sulfide burns with oxygen to give sulfur dioxide and water.

$$2H_2S(g) + 3O_2(g) \xrightarrow{\Delta} 2SO_2(g) + 2H_2O(g)$$

a. How many grams of SO_2 are formed from the reaction of 8.52 g of H_2S and 9.60 g of O_2?

b. How many grams of H_2O are formed from the reaction of 15.0 g of H_2S and 25.0 g of O_2?

ANSWER

a. 12.8 g of SO_2 **b.** 7.93 g of H_2O

PRACTICE PROBLEMS
Try Practice Problems 9.23 to 9.26

PRACTICE PROBLEMS

9.4 Limiting Reactants

9.19 A taxi company has 10 taxis.
   a. On a certain day, only eight taxi drivers show up for work. How many taxis can be used to pick up passengers?
   b. On another day, 10 taxi drivers show up for work but three taxis are in the repair shop. How many taxis can be driven?

9.20 A clock maker has 15 clock faces. Each clock requires one face and two hands.
   a. If the clock maker has 42 hands, how many clocks can be produced?
   b. If the clock maker has only eight hands, how many clocks can be produced?

9.21 Nitrogen and hydrogen react to form ammonia.

$$N_2(g) + 3H_2(g) \longrightarrow 2NH_3(g)$$

Determine the limiting reactant in each of the following mixtures of reactants:
   a. 3.0 mol of N_2 and 5.0 mol of H_2
   b. 8.0 mol of N_2 and 4.0 mol of H_2
   c. 3.0 mol of N_2 and 12.0 mol of H_2

9.22 Iron and oxygen react to form iron(III) oxide.

$$4Fe(s) + 3O_2(g) \longrightarrow 2Fe_2O_3(s)$$

Determine the limiting reactant in each of the following mixtures of reactants:
   a. 2.0 mol of Fe and 6.0 mol of O_2
   b. 5.0 mol of Fe and 4.0 mol of O_2
   c. 16.0 mol of Fe and 20.0 mol of O_2

9.23 For each of the following reactions, 20.0 g of each reactant is present initially. Determine the limiting reactant, and calculate the grams of product in parentheses that would be produced.
   a. $2Al(s) + 3Cl_2(g) \longrightarrow 2AlCl_3(s)$ ($AlCl_3$)
   b. $4NH_3(g) + 5O_2(g) \longrightarrow 4NO(g) + 6H_2O(g)$ (H_2O)
   c. $CS_2(g) + 3O_2(g) \xrightarrow{\Delta} CO_2(g) + 2SO_2(g)$ (SO_2)

9.24 For each of the following reactions, 20.0 g of each reactant is present initially. Determine the limiting reactant, and calculate the grams of product in parentheses that would be produced.
   a. $4Al(s) + 3O_2(g) \longrightarrow 2Al_2O_3(s)$ (Al_2O_3)
   b. $3NO_2(g) + H_2O(l) \longrightarrow 2HNO_3(aq) + NO(g)$ (HNO_3)
   c. $C_2H_6O(l) + 3O_2(g) \xrightarrow{\Delta} 2CO_2(g) + 3H_2O(g)$ (H_2O)

9.25 For each of the following reactions, calculate the grams of indicated product when 25.0 g of the first reactant and 40.0 g of the second reactant are used:
   a. $2SO_2(g) + O_2(g) \longrightarrow 2SO_3(g)$ (SO_3)
   b. $3Fe(s) + 4H_2O(l) \longrightarrow Fe_3O_4(s) + 4H_2(g)$ (Fe_3O_4)
   c. $C_7H_{16}(l) + 11O_2(g) \xrightarrow{\Delta} 7CO_2(g) + 8H_2O(g)$ (CO_2)

9.26 For each of the following reactions, calculate the grams of indicated product when 15.0 g of the first reactant and 10.0 g of the second reactant are used:
   a. $4Li(s) + O_2(g) \longrightarrow 2Li_2O(s)$ (Li_2O)
   b. $Fe_2O_3(s) + 3H_2(g) \longrightarrow 2Fe(s) + 3H_2O(l)$ (Fe)
   c. $Al_2S_3(s) + 6H_2O(l) \longrightarrow 2Al(OH)_3(aq) + 3H_2S(g)$ (H_2S)

REVIEW

Calculating Percentages (1.4)

ENGAGE 9.4

For your chemistry class party, you have prepared cookie dough to make five dozen cookies. However, 12 of the cookies burned and you threw them away. Why is the percent yield of cookies for the chemistry party only 80%?

CORE CHEMISTRY SKILL

Calculating Percent Yield

On spacecraft, LiOH in canisters can be used to remove CO_2 from the air.

9.5 Percent Yield

LEARNING GOAL Given the actual quantity of product, determine the percent yield for a chemical reaction.

In our problems up to now, we assumed that all of the reactants changed completely to product. Thus, we have calculated the amount of product as the maximum quantity possible, or 100%. While this would be an ideal situation, it does not usually happen. As we carry out a reaction and transfer products from one container to another, some product is usually lost. In the lab as well as commercially, the starting materials may not be completely pure, and side reactions may use some of the reactants to give unwanted products. Thus, 100% of the desired product is not actually obtained.

When we do a chemical reaction in the laboratory, we measure out specific quantities of the reactants. We calculate the **theoretical yield** for the reaction, which is the amount of product (100%) we would expect if all the reactants were converted to the desired product. When the reaction ends, we collect and measure the mass of the product, which is the **actual yield** for the product. Because some product is usually lost, the actual yield is less than the theoretical yield. Using the actual yield and the theoretical yield for a product, we can calculate the **percent yield**.

$$\text{Percent yield (\%)} = \frac{\text{actual yield}}{\text{theoretical yield}} \times 100\%$$

SAMPLE PROBLEM 9.8   Calculating Percent Yield

TRY IT FIRST

On spacecraft, LiOH can be used to absorb exhaled CO_2 from breathing air to form $LiHCO_3$.

$$LiOH(s) + CO_2(g) \longrightarrow LiHCO_3(s)$$

What is the percent yield of $LiHCO_3$ for the reaction if 50.0 g of LiOH gives 72.8 g of $LiHCO_3$?

SOLUTION

STEP 1   State the given and needed quantities.

	Given	Need	Connect
ANALYZE THE PROBLEM	50.0 g of LiOH, 72.8 g of $LiHCO_3$ (actual yield)	percent yield of $LiHCO_3$	molar masses, mole–mole factor, percent yield expression
	Equation		
	$LiOH(s) + CO_2(g) \longrightarrow LiHCO_3(s)$		

STEP 2   Write a plan to calculate the theoretical yield and the percent yield.

grams of LiOH → Molar mass → moles of LiOH → Mole–mole factor → moles of $LiHCO_3$ → Molar mass → grams of $LiHCO_3$
Theoretical yield

$$\text{Percent yield } (\%) = \frac{\text{actual yield of LiHCO}_3}{\text{theoretical yield of LiHCO}_3} \times 100\%$$

STEP 3 **Use coefficients to write mole–mole factors; write molar masses.**

1 mol of LiOH = 23.95 g of LiOH   1 mol of LiHCO₃ = 67.96 g of LiHCO₃

$$\frac{23.95 \text{ g LiOH}}{1 \text{ mol LiOH}} \quad \text{and} \quad \frac{1 \text{ mol LiOH}}{23.95 \text{ g LiOH}} \qquad \frac{67.96 \text{ g LiHCO}_3}{1 \text{ mol LiHCO}_3} \quad \text{and} \quad \frac{1 \text{ mol LiHCO}_3}{67.96 \text{ g LiHCO}_3}$$

1 mol of LiHCO₃ = 1 mol of LiOH

$$\frac{1 \text{ mol LiHCO}_3}{1 \text{ mol LiOH}} \quad \text{and} \quad \frac{1 \text{ mol LiOH}}{1 \text{ mol LiHCO}_3}$$

STEP 4 **Calculate the percent yield by dividing the actual yield (given) by the theoretical yield and multiplying the result by 100%.**

$$50.0 \text{ g LiOH} \times \underset{\text{Four SFs}}{\frac{\overset{\text{Exact}}{1 \text{ mol LiOH}}}{23.95 \text{ g LiOH}}} \times \underset{\text{Exact}}{\frac{\overset{\text{Exact}}{1 \text{ mol LiHCO}_3}}{1 \text{ mol LiOH}}} \times \underset{\text{Exact}}{\frac{\overset{\text{Four SFs}}{67.96 \text{ g LiHCO}_3}}{1 \text{ mol LiHCO}_3}}$$

$$= 142 \text{ g of LiHCO}_3$$
Three SFs

Calculation of percent yield:

$$\frac{\text{actual yield (given)}}{\text{theoretical yield (calculated)}} \times 100\% = \frac{\overset{\text{Three SFs}}{72.8 \text{ g LiHCO}_3}}{142 \text{ g LiHCO}_3} \times 100\% = 51.3\%$$
                                                      Three SFs   Three SFs

A percent yield of 51.3% means that 72.8 g of the theoretical amount of 142 g of LiHCO₃ was actually produced by the reaction.

SELF TEST 9.8

Using the equation in Sample Problem 9.8, calculate each of the following:
a. the percent yield of LiHCO₃ if 8.00 g of CO₂ produces 10.5 g of LiHCO₃
b. the percent yield of LiHCO₃ if 35.0 g of LiOH produces 76.6 g of LiHCO₃

ANSWER
a. 84.7%   **b.** 77.1%

PRACTICE PROBLEMS
Try Practice Problems 9.27 to 9.32

PRACTICE PROBLEMS

9.5 Percent Yield

9.27 Carbon disulfide is produced by the reaction of carbon and sulfur dioxide.

$$5C(s) + 2SO_2(g) \longrightarrow CS_2(g) + 4CO(g)$$

a. What is the percent yield of carbon disulfide if the reaction of 40.0 g of carbon produces 36.0 g of carbon disulfide?
b. What is the percent yield of carbon disulfide if the reaction of 32.0 g of sulfur dioxide produces 12.0 g of carbon disulfide?

9.28 Iron(III) oxide reacts with carbon monoxide to produce iron and carbon dioxide.

$$Fe_2O_3(s) + 3CO(g) \longrightarrow 2Fe(s) + 3CO_2(g)$$

a. What is the percent yield of iron if the reaction of 65.0 g of iron(III) oxide produces 15.0 g of iron?
b. What is the percent yield of carbon dioxide if the reaction of 75.0 g of carbon monoxide produces 85.0 g of carbon dioxide?

9.29 Aluminum reacts with oxygen to produce aluminum oxide.

$$4Al(s) + 3O_2(g) \longrightarrow 2Al_2O_3(s)$$

Calculate the mass of Al₂O₃ that can be produced if the reaction of 50.0 g of aluminum and sufficient oxygen has a 75.0% yield.

9.30 Propane (C_3H_8) burns in oxygen to produce carbon dioxide and water.

$$C_3H_8(g) + 5O_2(g) \xrightarrow{\Delta} 3CO_2(g) + 4H_2O(g)$$

Calculate the mass of CO_2 that can be produced if the reaction of 45.0 g of propane and sufficient oxygen has a 60.0% yield.

9.31 When 30.0 g of carbon is heated with silicon dioxide, 28.2 g of carbon monoxide is produced. What is the percent yield of carbon monoxide for this reaction?

$$SiO_2(s) + 3C(s) \xrightarrow{\Delta} SiC(s) + 2CO(g)$$

9.32 When 56.6 g of calcium is reacted with nitrogen gas, 32.4 g of calcium nitride is produced. What is the percent yield of calcium nitride for this reaction?

$$3Ca(s) + N_2(g) \longrightarrow Ca_3N_2(s)$$

REVIEW

Using Energy Units (3.4)

9.6 Energy in Chemical Reactions

LEARNING GOAL Given the heat of reaction, calculate the loss or gain of heat for an exothermic or endothermic reaction.

Almost every chemical reaction involves a loss or gain of energy. To discuss energy change for a reaction, we look at the energy of the reactants before the reaction and the energy of the products after the reaction.

Energy Units for Chemical Reactions

The SI unit for energy is the *joule* (J). Often, the unit of *kilojoules* (kJ) is used to show the energy change in a reaction.

1 kilojoule (kJ) = 1000 joules (J)

Heat of Reaction

The **heat of reaction** is the amount of heat absorbed or released during a reaction that takes place at constant pressure. A change of energy occurs as reactants interact, bonds break apart, and products form. We determine a heat of reaction, symbol ΔH, as the difference in the energy of the products and the reactants.

$$\Delta H = H_{products} - H_{reactants}$$

Exothermic Reactions

In an **exothermic reaction** (*exo* means "out"), the energy of the products is lower than that of the reactants. This means that heat is released along with the products that form. Let us look at the equation for the exothermic reaction in which 185 kJ of heat is released when 1 mol of hydrogen and 1 mol of chlorine react to form 2 mol of hydrogen chloride. For an exothermic reaction, the heat of reaction can be written as one of the products. It can also be written as a ΔH value with a negative sign ($-$).

In an exothermic reaction, the energy of the products is lower than the energy of the reactants.

Exothermic, Heat Released	Heat Is a Product
$H_2(g) + Cl_2(g) \longrightarrow 2HCl(g) + 185$ kJ	$\Delta H = -185$ kJ Negative sign

Endothermic Reactions

In an **endothermic reaction** (*endo* means "within"), the energy of the products is higher than that of the reactants. Heat is required to convert the reactants to products. Let us look at the equation for the endothermic reaction in which 180 kJ of heat is needed to convert 1 mol of nitrogen and 1 mol of oxygen to 2 mol of nitrogen oxide. For an endothermic reaction, the heat of reaction can be written as one of the reactants. It can also be written as a ΔH value with a positive sign (+).

In an endothermic reaction, the energy of the products is higher than the energy of the reactants.

Endothermic, Heat Absorbed	Heat Is a Reactant
$N_2(g) + O_2(g) + 180\text{ kJ} \longrightarrow 2NO(g)$	$\Delta H = +180\text{ kJ}$ Positive sign

Reaction	Energy Change	Heat in the Equation	Sign of ΔH
Exothermic	Heat released	Product side	Negative sign (−)
Endothermic	Heat absorbed	Reactant side	Positive sign (+)

SAMPLE PROBLEM 9.9   Exothermic and Endothermic Reactions

TRY IT FIRST

In the reaction of 1 mol of solid carbon with 1 mol of oxygen gas, the energy of the carbon dioxide gas produced is 394 kJ lower than the energy of the reactants.
a. Is the reaction exothermic or endothermic?
b. Write the balanced chemical equation for the reaction, including the heat of reaction.
c. Is the sign of ΔH positive or negative?

SOLUTION

a. When the products have a lower energy than the reactants, the reaction is exothermic.
b. $C(s) + O_2(g) \longrightarrow CO_2(g) + 394\text{ kJ}$
c. negative

SELF TEST 9.9

In the reaction of 1 mol of solid carbon with 2 mol of solid sulfur, the energy of the liquid carbon disulfide produced is 92.0 kJ higher than the energy of the reactants.
a. Is the reaction exothermic or endothermic?
b. Write the balanced chemical equation for the reaction, including the heat of reaction.
c. Is the sign of ΔH positive or negative?

ANSWER

a. When the products have a higher energy than the reactants, the reaction is endothermic.
b. $C(s) + 2S(s) + 92.0\text{ kJ} \longrightarrow CS_2(l)$
c. positive

PRACTICE PROBLEMS

Try Practice Problems 9.33 to 9.38

Calculations of Heat in Reactions

The value of ΔH refers to the heat change, in kilojoules, for each substance in the balanced equation for the reaction. Consider the following decomposition reaction:

$$2H_2O(l) \longrightarrow 2H_2(g) + O_2(g) \quad \Delta H = +572 \text{ kJ}$$
$$2H_2O(l) + 572 \text{ kJ} \longrightarrow 2H_2(g) + O_2(g)$$

For this reaction, 572 kJ are absorbed by 2 mol of H_2O to produce 2 mol of H_2 and 1 mol of O_2. We can write two heat conversion factors for each substance in this reaction as:

$$\frac{+572 \text{ kJ}}{2 \text{ mol } H_2O} \text{ and } \frac{2 \text{ mol } H_2O}{+572 \text{ kJ}} \qquad \frac{+572 \text{ kJ}}{2 \text{ mol } H_2} \text{ and } \frac{2 \text{ mol } H_2}{+572 \text{ kJ}} \qquad \frac{+572 \text{ kJ}}{1 \text{ mol } O_2} \text{ and } \frac{1 \text{ mol } O_2}{+572 \text{ kJ}}$$

Suppose in this reaction that 9.00 g of H_2O undergoes reaction. We can calculate the kilojoules of heat absorbed as

$$9.00 \text{ g } H_2O \times \frac{1 \text{ mol } H_2O}{18.02 \text{ g } H_2O} \times \frac{+572 \text{ kJ}}{2 \text{ mol } H_2O} = +143 \text{ kJ}$$

SAMPLE PROBLEM 9.10   Calculating Heat in a Reaction

TRY IT FIRST

How much heat, in kilojoules, is released when nitrogen and hydrogen react to form 50.0 g of ammonia?

$$N_2(g) + 3H_2(g) \longrightarrow 2NH_3(g) \qquad \Delta H = -92.2 \text{ kJ}$$

SOLUTION

STEP 1   State the given and needed quantities.

ANALYZE THE PROBLEM	Given	Need	Connect
	50.0 g of NH_3	heat released, in kilojoules	molar mass, $\Delta H = -92.2$ kJ
	Equation		
	$N_2(g) + 3H_2(g) \longrightarrow 2NH_3(g)$		

STEP 2   Write a plan using the heat of reaction and any molar mass needed.

grams of NH_3 $\xrightarrow{\text{Molar mass}}$ moles of NH_3 $\xrightarrow{\text{Heat of reaction}}$ kilojoules

STEP 3   Write the conversion factors including heat of reaction.

$$1 \text{ mol of } NH_3 = 17.03 \text{ g of } NH_3 \qquad 2 \text{ mol of } NH_3 = -92.2 \text{ kJ}$$
$$\frac{17.03 \text{ g } NH_3}{1 \text{ mol } NH_3} \text{ and } \frac{1 \text{ mol } NH_3}{17.03 \text{ g } NH_3} \qquad \frac{-92.2 \text{ kJ}}{2 \text{ mol } NH_3} \text{ and } \frac{2 \text{ mol } NH_3}{-92.2 \text{ kJ}}$$

STEP 4   Set up the problem to calculate the heat.

$$\underset{\text{Three SFs}}{50.0 \text{ g } NH_3} \times \underset{\text{Four SFs}}{\frac{\overset{\text{Exact}}{1 \text{ mol } NH_3}}{17.03 \text{ g } NH_3}} \times \underset{\text{Exact}}{\frac{\overset{\text{Three SFs}}{-92.2 \text{ kJ}}}{2 \text{ mol } NH_3}} = \underset{\text{Three SFs}}{-135 \text{ kJ}}$$

Mercury(II) oxide decomposes to mercury and oxygen.

$$2HgO(s) \longrightarrow 2Hg(l) + O_2(g) \qquad \Delta H = +182 \text{ kJ}$$

a. Is the reaction exothermic or endothermic?
b. How many kilojoules are needed when 25.0 g of mercury(II) oxide reacts?

ANSWER

a. endothermic

b. 20.5 kJ

PRACTICE PROBLEMS

Try Practice Problems 9.39 and 9.40

Chemistry Link to Health

Cold Packs and Hot Packs

In a hospital, at a first-aid station, or at an athletic event, an instant cold pack may be used to reduce swelling from an injury, remove heat from inflammation, or decrease capillary size to lessen the effect of hemorrhaging. Inside the plastic container of a cold pack, there is a compartment containing solid ammonium nitrate (NH_4NO_3) that is separated from a compartment containing water.

Cold packs use an endothermic reaction.

The pack is activated when it is hit or squeezed hard enough to break the walls between the compartments and cause the ammonium nitrate to mix with the water (shown as H_2O over the reaction arrow). In an endothermic process, 1 mol of NH_4NO_3 that dissolves absorbs 26 kJ of heat. The temperature drops to about 4 to 5 °C to give a cold pack that is ready to use.

Endothermic Reaction in a Cold Pack

$$NH_4NO_3(s) + 26 \text{ kJ} \xrightarrow{H_2O} NH_4NO_3(aq)$$

Hot packs are used to relax muscles, lessen aches and cramps, and increase circulation by expanding capillary size. Constructed in the same way as cold packs, a hot pack contains a salt such as $CaCl_2$. When 1 mol of $CaCl_2$ dissolves in water, 82 kJ are released as heat. The temperature increases as much as 66 °C to give a hot pack that is ready to use.

Exothermic Reaction in a Hot Pack

$$CaCl_2(s) \xrightarrow{H_2O} CaCl_2(aq) + 82 \text{ kJ}$$

Hess's Law

According to **Hess's law**, heat can be absorbed or released in a single chemical reaction or in several steps. When there are two or more steps in the reaction, the overall energy change is the sum of the energy changes of those steps, provided they all occur at the same temperature.

Steps in solving problems involving Hess's law:

1. If you reverse a chemical equation, you must also reverse the sign of ΔH.
2. If a chemical equation is multiplied by some factor, then ΔH must be multiplied by the same factor.

We can see how the energy change for a specific reaction is the sum of two or more reactions in Sample Problem 9.11.

Calculate the ΔH value for the following chemical reaction:

$$C(s) + 2H_2O(g) \longrightarrow CO_2(g) + 2H_2(g)$$

First equation $C(s) + O_2(g) \longrightarrow CO_2(g) \qquad \Delta H = -394 \text{ kJ}$

Second equation $2H_2(g) + O_2(g) \longrightarrow 2H_2O(g) \qquad \Delta H = -484 \text{ kJ}$

SOLUTION

	Given	Need	Connect
ANALYZE THE PROBLEM	overall reaction, first and second equations	heat of reaction	combine equations, total heats of reactions

STEP 1 **Arrange the given equations to place reactants on the left and products on the right.** Use the first equation as written with $C(s)$ as a reactant and $CO_2(g)$ as a product.

$$C(s) + O_2(g) \longrightarrow CO_2(g) \qquad\qquad \Delta H = -394 \text{ kJ}$$

Reverse the second equation to give $H_2O(g)$ as a reactant and $H_2(g)$ as a product, and change the sign of the ΔH.

$$2H_2O(g) \longrightarrow 2H_2(g) + O_2(g) \qquad\qquad \Delta H = +484 \text{ kJ} \quad \text{Changed sign}$$

STEP 2 **If an equation is multiplied to balance coefficients, multiply the ΔH by the same number.** Now that $C(s)$ and $2H_2O(g)$ are reactants, and $CO_2(g)$ and $2H_2(g)$ are products, no further changes are needed.

STEP 3 **Combine the equations, cancel any substances that are common to both sides, and add the ΔH values.**

$$C(s) + \cancel{O_2(g)} \longrightarrow CO_2(g) \qquad\qquad \Delta H = -394 \text{ kJ}$$
$$2H_2O(g) \longrightarrow 2H_2(g) + \cancel{O_2(g)} \qquad\qquad \Delta H = +484 \text{ kJ}$$
$$\overline{C(s) + 2H_2O(g) \longrightarrow CO_2(g) + 2H_2(g) \qquad\quad \Delta H = +90.\text{ kJ}}$$

SELF TEST 9.11

Calculate the ΔH value for the following chemical reaction:

$$2NO(g) + O_2(g) \longrightarrow N_2O_4(g)$$

First equation	$N_2O_4(g) \longrightarrow 2NO_2(g)$	$\Delta H = +57.2 \text{ kJ}$
Second equation	$2NO(g) + O_2(g) \longrightarrow 2NO_2(g)$	$\Delta H = -114.0 \text{ kJ}$

ANSWER

$$\Delta H = -171.2 \text{ kJ}$$

PRACTICE PROBLEMS

Try Practice Problems 9.41 to 9.44

SAMPLE PROBLEM 9.12 **Calculating Heat of Reaction**

TRY IT FIRST

Calculate the ΔH value for the following chemical reaction:

$$CS_2(l) + 3O_2(g) \longrightarrow CO_2(g) + 2SO_2(g)$$

First equation	$C(s) + O_2(g) \longrightarrow CO_2(g)$	$\Delta H = -394 \text{ kJ}$
Second equation	$S(s) + O_2(g) \longrightarrow SO_2(g)$	$\Delta H = -297 \text{ kJ}$
Third equation	$C(s) + 2S(s) \longrightarrow CS_2(l)$	$\Delta H = +88 \text{ kJ}$

SOLUTION

	Given	Need	Connect
ANALYZE THE PROBLEM	overall reaction, first, second, third equations	heat of reaction	combine equations, total heats of reactions

STEP 1 **Arrange the given equations to place reactants on the left and products on the right.**

Use the first equation as written with $CO_2(g)$ as a product.

$$C(s) + O_2(g) \longrightarrow CO_2(g) \qquad\qquad \Delta H = -394 \text{ kJ}$$

Use the second equation as written with $SO_2(g)$ as a product.

$$S(s) + O_2(g) \longrightarrow SO_2(g) \qquad\qquad \Delta H = -297 \text{ kJ}$$

Reverse the third equation to give $CS_2(l)$ as a reactant, and change the sign of the ΔH.

$$CS_2(l) \longrightarrow C(s) + 2S(s) \qquad\qquad \Delta H = -88 \text{ kJ}$$

STEP 2 **If an equation is multiplied to balance coefficients, multiply the ΔH by the same number.** Obtain $2SO_2(g)$ on the product side by multiplying the second equation and the ΔH by 2.

$$2S(s) + 2O_2(g) \longrightarrow 2SO_2(g) \qquad\qquad \Delta H = -594 \text{ kJ}$$

STEP 3 **Combine the equations, cancel any substances that are common to both sides, and add the ΔH values.** The $C(s)$ and $2S(s)$ cancel.

$$\begin{array}{ll}
\cancel{C(s)} + O_2(g) \longrightarrow CO_2(g) & \Delta H = -394 \text{ kJ} \\
2\cancel{S(s)} + 2O_2(g) \longrightarrow 2SO_2(g) & \Delta H = -594 \text{ kJ} \\
CS_2(l) \longrightarrow \cancel{C(s)} + 2\cancel{S(s)} & \Delta H = -88 \text{ kJ} \\
\hline
CS_2(l) + 3O_2(g) \longrightarrow CO_2(g) + 2SO_2(g) & \Delta H = -1076 \text{ kJ}
\end{array}$$

SELF TEST 9.12

Calculate the ΔH value for the following chemical reaction:

$$C(s) + 2H_2(g) \longrightarrow CH_4(g)$$

First equation $\quad C(s) + O_2(g) \longrightarrow CO_2(g) \qquad\qquad \Delta H = -394 \text{ kJ}$

Second equation $\quad 2H_2(g) + O_2(g) \longrightarrow 2H_2O(g) \qquad\quad \Delta H = -484 \text{ kJ}$

Third equation $\quad CH_4(g) + 2O_2(g) \longrightarrow CO_2(g) + 2H_2O(g) \quad \Delta H = -802 \text{ kJ}$

ANSWER

-76 kJ

PRACTICE PROBLEMS

9.6 Energy in Chemical Reactions

9.33 In an exothermic reaction, is the energy of the products higher or lower than that of the reactants?

9.34 In an endothermic reaction, is the energy of the products higher or lower than that of the reactants?

9.35 Classify each of the following as exothermic or endothermic:
a. A reaction releases 550 kJ.
b. The energy level of the products is higher than that of the reactants.
c. The metabolism of glucose in the body provides energy.

9.36 Classify each of the following as exothermic or endothermic:
a. The energy level of the products is lower than that of the reactants.
b. In the body, the synthesis of proteins requires energy.
c. A reaction absorbs 125 kJ.

9.37 Classify each of the following as exothermic or endothermic, and give the ΔH for each:
a. $CH_4(g) + 2O_2(g) \xrightarrow{\Delta} CO_2(g) + 2H_2O(g) + 802 \text{ kJ}$
b. $Ca(OH)_2(s) + 65.3 \text{ kJ} \longrightarrow CaO(s) + H_2O(l)$
c. $2Al(s) + Fe_2O_3(s) \longrightarrow Al_2O_3(s) + 2Fe(l) + 850 \text{ kJ}$

The thermite reaction of aluminum and iron(III) oxide produces very high temperatures used to cut or weld railroad tracks.

9.38 Classify each of the following as exothermic or endothermic, and give the ΔH for each:

a. $C_3H_8(g) + 5O_2(g) \xrightarrow{\Delta} 3CO_2(g) + 4H_2O(g) + 2220$ kJ

b. $2Na(s) + Cl_2(g) \longrightarrow 2NaCl(s) + 819$ kJ

c. $PCl_5(g) + 67$ kJ $\longrightarrow PCl_3(g) + Cl_2(g)$

9.39 a. How many kilojoules are released when 125 g of Cl_2 reacts with silicon?

$$Si(s) + 2Cl_2(g) \longrightarrow SiCl_4(g) \quad \Delta H = -657 \text{ kJ}$$

b. How many kilojoules are absorbed when 278 g of PCl_5 reacts?

$$PCl_5(g) \longrightarrow PCl_3(g) + Cl_2(g) \quad \Delta H = +67 \text{ kJ}$$

9.40 a. How many kilojoules are released when 75.0 g of CH_4O reacts?

$$2CH_4O(l) + 3O_2(g) \longrightarrow 2CO_2(g) + 4H_2O(l)$$
$$\Delta H = -726 \text{ kJ}$$

b. How many kilojoules are absorbed when 315 g of $Ca(OH)_2$ reacts?

$$Ca(OH)_2(s) \longrightarrow CaO(s) + H_2O(l) \quad \Delta H = +65.3 \text{ kJ}$$

9.41 Calculate the energy change for the reaction $N_2(g) + 2O_2(g) \longrightarrow 2NO_2(g)$ from the following:

$$N_2(g) + O_2(g) \longrightarrow 2NO(g) \qquad \Delta H = +180 \text{ kJ}$$
$$2NO_2(g) \longrightarrow 2NO(g) + O_2(g) \qquad \Delta H = +114 \text{ kJ}$$

9.42 Calculate the energy change for the reaction $Fe_2O_3(s) + CO(g) \longrightarrow 2FeO(s) + CO_2(g)$ from the following:

$$Fe_2O_3(s) + 3CO(g) \longrightarrow 2Fe(s) + 3CO_2(g) \quad \Delta H = -23.4 \text{ kJ}$$
$$FeO(s) + CO(g) \longrightarrow Fe(s) + CO_2(g) \qquad \Delta H = -10.9 \text{ kJ}$$

9.43 Calculate the energy change for the reaction $S(s) + O_2(g) \longrightarrow SO_2(g)$ from the following:

$$2S(s) + 3O_2(g) \longrightarrow 2SO_3(g) \qquad \Delta H = -792 \text{ kJ}$$
$$2SO_2(g) + O_2(g) \longrightarrow 2SO_3(g) \qquad \Delta H = -198 \text{ kJ}$$

9.44 Calculate the energy change for the reaction $3C(s) + O_2(g) \longrightarrow C_3O_2(g)$ from the following:

$$2CO(g) + C(s) \longrightarrow C_3O_2(g) \qquad \Delta H = +127 \text{ kJ}$$
$$2C(s) + O_2(g) \longrightarrow 2CO(g) \qquad \Delta H = -222 \text{ kJ}$$

UPDATE   Testing Water Samples for Insecticides

One of the problems that Lance monitors is water pollution by insecticides. These insecticides are made by organic synthesis, in which smaller molecules are combined to form larger molecules, in a stepwise fashion.

Applications

9.45 In one step in the synthesis of the insecticide carbaryl (Sevin), naphthol reacts with phosgene as shown.

$$C_{10}H_8O + COCl_2 \longrightarrow C_{11}H_7O_2Cl + HCl$$
$\quad$ Naphthol $\quad$ Phosgene

a. How many kilograms of $C_{11}H_7O_2Cl$ form from 2.2×10^2 kg of naphthol?

b. If 100. g of naphthol and 100. g of phosgene react, what is the theoretical yield of $C_{11}H_7O_2Cl$?

c. If the actual yield of $C_{11}H_7O_2Cl$ in part **b** is 115 g, what is the percent yield of $C_{11}H_7O_2Cl$?

9.46 Another widely used insecticide is carbofuran (Furadan), an extremely toxic compound. In one step in the synthesis of carbofuran, the reaction shown is used.

$$C_6H_6O_2 + C_4H_7Cl \longrightarrow C_{10}H_{12}O_2 + HCl$$

a. How many grams of $C_6H_6O_2$ are needed to produce 3.8×10^3 g of $C_{10}H_{12}O_2$?

b. If 67.0 g of $C_6H_6O_2$ and 51.0 g of C_4H_7Cl react, what is the theoretical yield of $C_{10}H_{12}O_2$?

c. If the actual yield of $C_{10}H_{12}O_2$ in part **b** is 85.7 g, what is the percent yield of $C_{10}H_{12}O_2$?

CONCEPT MAP

CHEMICAL QUANTITIES IN REACTIONS

use a

Balanced Chemical Equation

have → **Coefficients** → that give ratios called → **Mole–Mole Factors**

use → **Molar Mass** → to calculate → **Grams of Reactants or Products**

- to calculate → **Limiting Reactant**
- to calculate → **Theoretical Yield** → to give → **Percent Yield**

have → **Heat of Reaction ΔH**

- that indicates an → **Exothermic Reaction** → when → **Heat Is Released**
- that indicates an → **Endothermic Reaction** → when → **Heat Is Absorbed**

→ **Hess's Law**

CHAPTER REVIEW

9.1 Conservation of Mass
LEARNING GOAL Calculate the total mass of reactants and the total mass of products in a balanced chemical equation.

$$2Ag(s) + S(s) \longrightarrow Ag_2S(s)$$
Mass of reactants = Mass of product

- In a balanced equation, the total mass of the reactants is equal to the total mass of the products.

9.2 Mole Relationships in Chemical Equations
LEARNING GOAL Use a mole–mole factor from a balanced chemical equation to calculate the number of moles of another substance in the reaction.

$$\text{Fe and S: } \frac{2 \text{ mol Fe}}{3 \text{ mol S}} \text{ and } \frac{3 \text{ mol S}}{2 \text{ mol Fe}}$$

- The coefficients in an equation describing the relationship between the moles of any two components are used to write mole–mole factors.
- When the number of moles for one substance is known, a mole–mole factor is used to find the moles of a different substance in the reaction.

9.3 Mass Calculations for Chemical Reactions
LEARNING GOAL Given the mass in grams of a substance in a chemical reaction, calculate the mass in grams of another substance in the reaction.

- In calculations using equations, the molar masses of the substances and their mole–mole factors are used to change the number of grams of one substance to the corresponding grams of a different substance.

9.4 Limiting Reactants
LEARNING GOAL Identify a limiting reactant when given the quantities of two reactants; calculate the amount of product formed from the limiting reactant.

- A limiting reactant is the reactant that produces the smaller amount of product while the other reactant is left over.
- When the masses of two reactants are given, the mass of a product is calculated from the limiting reactant.

9.5 Percent Yield

LEARNING GOAL Given the actual quantity of product, determine the percent yield for a chemical reaction.

- The percent yield for a chemical reaction indicates the percent of product actually produced during a reaction.
- The percent yield is calculated by dividing the actual yield in grams of a product by the theoretical yield in grams and multiplying by 100%.

9.6 Energy in Chemical Reactions

LEARNING GOAL Given the heat of reaction, calculate the loss or gain of heat for an exothermic or endothermic reaction.

- In chemical reactions, the heat of reaction (ΔH) is the energy difference between the products and the reactants.
- In an exothermic reaction, the energy of the products is lower than that of the reactants. Heat is released, and ΔH is negative.
- In an endothermic reaction, the energy of the products is higher than that of the reactants; heat is absorbed, and ΔH is positive.
- Hess's law states that heat can be absorbed or released in a single step or in several steps.

KEY TERMS

actual yield The actual amount of product produced by a reaction.

endothermic reaction A reaction in which the energy of the products is higher than that of the reactants.

exothermic reaction A reaction in which the energy of the products is lower than that of the reactants.

heat of reaction The heat (symbol ΔH) absorbed or released when a reaction takes place at constant pressure.

Hess's law Heat can be absorbed or released in a single chemical reaction or in several steps.

Law of Conservation of Mass In a chemical reaction, the total mass of the reactants is equal to the total mass of the products; matter is neither lost nor gained.

limiting reactant The reactant used up during a chemical reaction, which limits the amount of product that can form.

mole–mole factor A conversion factor that relates the number of moles of two compounds in an equation derived from their coefficients.

percent yield The ratio of the actual yield for a reaction to the theoretical yield possible for the reaction.

theoretical yield The maximum amount of product that a reaction can produce from a given amount of reactant.

CORE CHEMISTRY SKILLS

The chapter section containing each Core Chemistry Skill is shown in parentheses at the end of each heading.

Using Mole–Mole Factors (9.2)

Consider the balanced chemical equation

$$4Na(s) + O_2(g) \longrightarrow 2Na_2O(s)$$

- The coefficients in a balanced chemical equation represent the moles of reactants and the moles of products. Thus, 4 mol of Na react with 1 mol of O_2 to form 2 mol of Na_2O.
- From the coefficients, mole–mole factors can be written for any two substances as follows:

Na and O₂ $\dfrac{4 \text{ mol Na}}{1 \text{ mol O}_2}$ and $\dfrac{1 \text{ mol O}_2}{4 \text{ mol Na}}$

Na and Na₂O $\dfrac{4 \text{ mol Na}}{2 \text{ mol Na}_2\text{O}}$ and $\dfrac{2 \text{ mol Na}_2\text{O}}{4 \text{ mol Na}}$

O₂ and Na₂O $\dfrac{2 \text{ mol Na}_2\text{O}}{1 \text{ mol O}_2}$ and $\dfrac{1 \text{ mol O}_2}{2 \text{ mol Na}_2\text{O}}$

- A mole–mole factor is used to convert the number of moles of one substance in the reaction to the number of moles of another substance in the reaction.

Example: How many moles of sodium are needed to produce 3.5 mol of sodium oxide?

Answer:

Given	Need	Connect
3.5 mol of Na_2O	moles of Na	mole–mole factor

$$3.5 \text{ mol Na}_2\text{O} \times \frac{4 \text{ mol Na}}{2 \text{ mol Na}_2\text{O}} = 7.0 \text{ mol of Na}$$

Two SFs — Exact — Two SFs

Converting Grams to Grams (9.3)

When we have the balanced chemical equation for a reaction, we can use the mass of substance A and then calculate the mass of substance B. The process is as follows:

- Use the molar mass of A to convert the mass, in grams, of A to moles of A.
- Use the mole–mole factor that converts moles of A to moles of B.
- Use the molar mass of B to calculate the mass, in grams, of B.

grams of A $\xrightarrow{\text{Molar mass A}}$ mole of A $\xrightarrow{\text{Mole–mole factor}}$ moles of B $\xrightarrow{\text{Molar mass B}}$ grams of B

Example: How many grams of O_2 are needed to completely react with 14.6 g of Na?

$$4Na(s) + O_2(g) \longrightarrow 2Na_2O(s)$$

Answer:

$$14.6 \text{ g Na} \times \underset{\text{Three SFs}}{\frac{\overset{\text{Exact}}{1 \text{ mol Na}}}{22.99 \text{ g Na}}} \times \underset{\text{Four SFs}}{\frac{\overset{\text{Exact}}{1 \text{ mol } O_2}}{4 \text{ mol Na}}} \times \underset{\text{Exact}}{\frac{\overset{\text{Four SFs}}{32.00 \text{ g } O_2}}{1 \text{ mol } O_2}}$$

$= 5.08$ g of O_2
Three SFs

Calculating Quantity of Product from a Limiting Reactant (9.4)

Often in reactions, the reactants are not present in quantities that allow both reactants to be completely used up. Then one of the reactants, called the *limiting reactant*, determines the maximum amount of product that can form.

- To determine the limiting reactant, we calculate the amount of product that is possible from each reactant.
- The limiting reactant is the one that produces the smaller amount of product.

Example: If 12.5 g of S reacts with 17.2 g of O_2, what is the limiting reactant and the mass, in grams, of SO_3 produced?

$$2S(s) + 3O_2(g) \longrightarrow 2SO_3(g)$$

Answer:
Mass of SO_3 from S:

$$12.5 \text{ g S} \times \underset{\text{Three SFs}}{\frac{\overset{\text{Exact}}{1 \text{ mol S}}}{32.07 \text{ g S}}} \times \underset{\text{Four SFs}}{\frac{\overset{\text{Exact}}{2 \text{ mol } SO_3}}{2 \text{ mol S}}} \times \underset{\text{Exact}}{\frac{\overset{\text{Four SFs}}{80.07 \text{ g } SO_3}}{1 \text{ mol } SO_3}}$$

$= 31.2$ g of SO_3
Three SFs

Mass of SO_3 from O_2:

$$17.2 \text{ g } O_2 \times \underset{\text{Three SFs}}{\frac{\overset{\text{Exact}}{1 \text{ mol } O_2}}{32.00 \text{ g } O_2}} \times \underset{\text{Four SFs}}{\frac{\overset{\text{Exact}}{2 \text{ mol } SO_3}}{3 \text{ mol } O_2}} \times \underset{\text{Exact}}{\frac{\overset{\text{Four SFs}}{80.07 \text{ g } SO_3}}{1 \text{ mol } SO_3}}$$

$= 28.7$ g of SO_3
Three SFs

Therefore, O_2, is the limiting reactant, and 28.7 g of SO_3 can be produced.

Calculating Percent Yield (9.5)

- The *theoretical yield* for a reaction is the amount of product (100%) formed if all the reactants were converted to desired product.
- The *actual yield* for the reaction is the mass, in grams, of the product obtained at the end of the experiment. Because some product is usually lost, the actual yield is less than the theoretical yield.

- The *percent yield* is calculated from the actual yield divided by the theoretical yield and multiplied by 100%.

$$\text{Percent yield (\%)} = \frac{\text{actual yield}}{\text{theoretical yield}} \times 100\%$$

Example: If 22.6 g of Al reacts completely with O_2, and 37.8 g of Al_2O_3 is obtained, what is the percent yield of Al_2O_3 for the reaction?

$$4Al(s) + 3O_2(g) \longrightarrow 2Al_2O_3(s)$$

Answer:
Calculation of theoretical yield:

$$22.6 \text{ g Al} \times \underset{\text{Three SFs}}{\frac{\overset{\text{Exact}}{1 \text{ mol Al}}}{26.98 \text{ g Al}}} \times \underset{\text{Four SFs}}{\frac{\overset{\text{Exact}}{2 \text{ mol } Al_2O_3}}{4 \text{ mol Al}}} \times \underset{\text{Exact}}{\frac{\overset{\text{Five SFs}}{101.96 \text{ g } Al_2O_3}}{1 \text{ mol } Al_2O_3}}$$

$= 42.7$ g of Al_2O_3 Theoretical yield
Three SFs

Calculation of percent yield:

$$\text{Percent yield (\%)} = \frac{\text{actual yield (given)}}{\text{theoretical yield (calculated)}} \times 100\%$$

$$= \frac{\overset{\text{Three SFs}}{37.8 \text{ g } Al_2O_3}}{42.7 \text{ g } Al_2O_3} \times 100\% = 88.5\%$$
      Three SFs   Three SFs

Using the Heat of Reaction (9.6)

- The heat of reaction is the amount of heat, usually in kJ, that is absorbed or released during a reaction.
- The heat of reaction or *energy change*, symbol ΔH, is the difference in the energy of the products and the reactants.

$$\Delta H = H_{\text{products}} - H_{\text{reactants}}$$

- In an exothermic reaction, the energy of the products is lower than that of the reactants. This means that heat is released along with the products that form. The sign for the heat of reaction, ΔH, is negative.
- In an endothermic reaction, the energy of the products is higher than that of the reactants. The heat is required to convert the reactants to products. The sign for the heat of reaction, ΔH, is positive.

Example: How many kilojoules are released when 3.50 g of CH_4 undergoes combustion?

$$CH_4(g) + 2O_2(g) \longrightarrow CO_2(g) + 2H_2O(g) \quad \Delta H = -802 \text{ kJ}$$

Answer: $3.50 \text{ g } CH_4 \times \underset{\text{Three SFs}}{\frac{\overset{\text{Exact}}{1 \text{ mol } CH_4}}{16.04 \text{ g } CH_4}} \times \underset{\text{Exact}}{\frac{\overset{\text{Three SFs}}{-802 \text{ kJ}}}{1 \text{ mol } CH_4}}$
           Three SFs   Four SFs

$= -175$ kJ
Three SFs

Therefore, 175 kJ are released.

UNDERSTANDING THE CONCEPTS

The chapter sections to review are shown in parentheses at the end of each problem.

9.47 If red spheres represent oxygen atoms, blue spheres represent nitrogen atoms, and all the molecules are gases, (9.2, 9.4)

Reactants   **Products**

    a. write a balanced equation for the reaction.
    b. identify the limiting reactant.

9.48 If green spheres represent chlorine atoms, yellow-green spheres represent fluorine atoms, white spheres represent hydrogen atoms, and all the molecules are gases, (9.2, 9.4)

Reactants   **Products**

    a. write a balanced equation for the reaction.
    b. identify the limiting reactant.

9.49 If blue spheres represent nitrogen atoms, white spheres represent hydrogen atoms, and all the molecules are gases, (9.2, 9.4)

Reactants

Products

   A   B

 or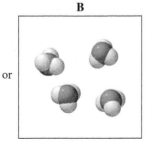

    a. write a balanced equation for the reaction.
    b. identify the diagram that shows the products.

9.50 If purple spheres represent iodine atoms, white spheres represent hydrogen atoms, and all the molecules are gases, (9.2, 9.4)

Reactants

Products

   A   B   C

 or or

    a. write a balanced equation for the reaction.
    b. identify the diagram that shows the products.

9.51 If blue spheres represent nitrogen atoms, purple spheres represent iodine atoms, and the reacting molecules are solid, and the products are gases, (9.2, 9.4, 9.5)

Reactants   **Actual products**

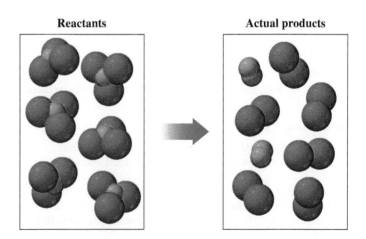

    a. write a balanced equation for the reaction.
    b. from the diagram of the actual products that result, calculate the percent yield for the reaction.

9.52 If green spheres represent chlorine atoms, red spheres represent oxygen atoms, and all the molecules are gases, (9.2, 9.4, 9.5)

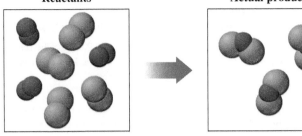

| Reactants | Actual products |

a. write a balanced equation for the reaction.
b. identify the limiting reactant.
c. from the diagram of the actual products that result, calculate the percent yield for the reaction.

9.53 Use the balanced chemical equation to complete the table: (9.1, 9.2)

$$2FeS(s) + 3O_2(g) \longrightarrow 2FeO(s) + 2SO_2(g)$$

FeS	O_2	FeO	SO_2
2.0 mol	_____ mol	_____ mol	_____ mol
_____ mol	_____ mol	_____ mol	4.6 mol

9.54 Use the balanced chemical equation to complete the table: (9.1, 9.2)

$$2C_2H_6(g) + 7O_2(g) \xrightarrow{\Delta} 4CO_2(g) + 6H_2O(g)$$

C_2H_6	O_2	CO_2	H_2O
_____ mol	_____ mol	3.2 mol	_____ mol
_____ mol	2.8 mol	_____ mol	_____ mol

ADDITIONAL PRACTICE PROBLEMS

9.55 When ammonia (NH_3) reacts with fluorine (F_2), the products are dinitrogen tetrafluoride and hydrogen fluoride. (9.2, 9.3)

$$2NH_3(g) + 5F_2(g) \longrightarrow N_2F_4(g) + 6HF(g)$$

a. How many moles of each reactant are needed to produce 4.00 mol of HF?
b. How many grams of F_2 are required to react with 25.5 g of NH_3?
c. How many grams of N_2F_4 can be produced when 3.40 g of NH_3 reacts?

9.56 Gasohol is a fuel that contains ethanol (C_2H_6O) that burns in oxygen (O_2) to give carbon dioxide and water. (9.2, 9.3)

$$C_2H_6O(g) + 3O_2(g) \xrightarrow{\Delta} 2CO_2(g) + 3H_2O(g)$$

a. How many moles of O_2 are needed to completely react with 4.0 mol of C_2H_6O?
b. If a car produces 88 g of CO_2, how many grams of O_2 are used up in the reaction?
c. If you add 125 g of C_2H_6O to your fuel, how many grams of CO_2 can be produced from the ethanol?

9.57 When hydrogen peroxide (H_2O_2) is used in rocket fuels, it produces water and oxygen. (9.2, 9.3)

$$2H_2O_2(l) \longrightarrow 2H_2O(l) + O_2(g)$$

a. How many moles of H_2O_2 are needed to produce 3.00 mol of H_2O?
b. How many grams of H_2O_2 are required to produce 36.5 g of O_2?
c. How many grams of H_2O can be produced when 12.2 g of H_2O_2 reacts?

9.58 Propane gas (C_3H_8) reacts with oxygen to produce carbon dioxide and water. (9.2, 9.3)

$$C_3H_8(g) + 5O_2(g) \xrightarrow{\Delta} 3CO_2(g) + 4H_2O(g)$$

a. How many moles of H_2O form when 5.00 mol of C_3H_8 completely reacts?
b. How many grams of CO_2 are produced from 18.5 g of O_2?
c. How many grams of H_2O can be produced when 46.3 g of C_3H_8 reacts?

9.59 When 12.8 g of Na and 10.2 g of Cl_2 react, what is the mass, in grams, of NaCl that is produced? (9.2, 9.3, 9.4)

$$2Na(s) + Cl_2(g) \longrightarrow 2NaCl(s)$$

9.60 If 35.8 g of CH_4 and 75.5 g of S react, how many grams of H_2S are produced? (9.2, 9.3, 9.4)

$$CH_4(g) + 4S(g) \longrightarrow CS_2(g) + 2H_2S(g)$$

9.61 Pentane gas (C_5H_{12}) reacts with oxygen to produce carbon dioxide and water. (9.2, 9.3, 9.4)

$$C_5H_{12}(g) + 8O_2(g) \xrightarrow{\Delta} 5CO_2(g) + 6H_2O(g)$$

a. How many moles of C_5H_{12} must react to produce 4.00 mol of H_2O?
b. How many grams of CO_2 are produced from 32.0 g of O_2?
c. How many grams of CO_2 are formed if 44.5 g of C_5H_{12} is mixed with 108 g of O_2?

9.62 When nitrogen dioxide (NO_2) from car exhaust combines with water in the air, it forms nitrogen oxide and nitric acid (HNO_3), which causes acid rain. (9.2, 9.3, 9.4)

$$3NO_2(g) + H_2O(l) \longrightarrow NO(g) + 2HNO_3(aq)$$

a. How many moles of NO_2 are needed to react with 0.250 mol of H_2O?
b. How many grams of HNO_3 are produced when 60.0 g of NO_2 completely reacts?
c. How many grams of HNO_3 can be produced if 225 g of NO_2 is mixed with 55.2 g of H_2O?

9.63 The acetylene (C_2H_2) used in welders' torches, burns according to the following equation: (9.2, 9.3, 9.5)

$$2C_2H_2(g) + 5O_2(g) \xrightarrow{\Delta} 4CO_2(g) + 2H_2O(g)$$

a. What is the theoretical yield, in grams, of CO_2, if 22.0 g of C_2H_2 completely reacts?
b. If the actual yield in part **a** is 64.0 g of CO_2, what is the percent yield of CO_2 for the reaction?

9.64 The equation for the decomposition of potassium chlorate is written as (9.2, 9.3, 9.5)

$$2KClO_3(s) \xrightarrow{\Delta} 2KCl(s) + 3O_2(g)$$

a. When 46.0 g of $KClO_3$ is completely decomposed, what is the theoretical yield, in grams, of O_2?
b. If the actual yield in part **a** is 12.1 g of O_2, what is the percent yield of O_2?

9.65 When 28.0 g of acetylene reacts with hydrogen, 24.5 g of ethane is produced. What is the percent yield of C_2H_6 for the reaction? (9.2, 9.3, 9.5)

$$C_2H_2(g) + 2H_2(g) \xrightarrow{Pt} C_2H_6(g)$$

9.66 When 50.0 g of iron(III) oxide reacts with carbon monoxide, 32.8 g of iron is produced. What is the percent yield of Fe for the reaction? (9.2, 9.3, 9.5)

$$Fe_2O_3(s) + 3CO(g) \longrightarrow 2Fe(s) + 3CO_2(g)$$

9.67 Nitrogen and hydrogen combine to form ammonia. (9.2, 9.3, 9.4, 9.5)

$$N_2(g) + 3H_2(g) \longrightarrow 2NH_3(g)$$

a. If 50.0 g of N_2 is mixed with 20.0 g of H_2, what is the theoretical yield, in grams, of NH_3?

b. If the reaction in part **a** has a percent yield of 62.0%, what is the actual yield, in grams, of NH_3?

9.68 Sodium and nitrogen combine to form sodium nitride. (9.2, 9.3, 9.4, 9.5)

$$6Na(s) + N_2(g) \longrightarrow 2Na_3N(s)$$

a. If 80.0 g of Na is mixed with 20.0 g of nitrogen gas, what is the theoretical yield, in grams, of Na_3N?

b. If the reaction in part **a** has a percent yield of 75.0%, what is the actual yield, in grams, of Na_3N?

9.69 The equation for the reaction of nitrogen and oxygen to form nitrogen oxide is written as (9.2, 9.6)

$$N_2(g) + O_2(g) \longrightarrow 2NO(g) \quad \Delta H = +90.2 \text{ kJ}$$

a. How many kilojoules are required to form 3.00 g of NO?

b. How many grams of NO form when 66.0 kJ are absorbed?

9.70 The equation for the reaction of iron and oxygen gas to form rust (Fe_2O_3) is written as (9.2, 9.6)

$$4Fe(s) + 3O_2(g) \longrightarrow 2Fe_2O_3(s) \quad \Delta H = -1.7 \times 10^3 \text{ kJ}$$

a. How many kilojoules are released when 2.00 g of Fe reacts?

b. How many grams of Fe_2O_3 form when 150 kJ are released?

Applications

9.71 Each of the following is a reaction that occurs in the cells of the body. Identify each as exothermic or endothermic. (9.6)

a. Succinyl CoA + H_2O $\longrightarrow$ succinate + CoA + 37 kJ

b. GDP + P_i + 34 kJ $\longrightarrow$ GTP + H_2O

9.72 Each of the following is a reaction that occurs in the cells of the body. Identify each as exothermic or endothermic. (9.6)

a. Phosphocreatine + H_2O $\longrightarrow$ creatine + P_i + 42.7 kJ

b. Fructose-6-phosphate + P_i + 16 kJ $\longrightarrow$ fructose-1,6-bisphosphate

CHALLENGE PROBLEMS

The following problems are related to the topics in this chapter. However, they do not all follow the chapter order, and they require you to combine concepts and skills from several sections. These problems will help you increase your critical thinking skills and prepare for your next exam.

9.73 Use the balanced chemical equation to complete the table: (9.1, 9.2, 9.3)

$$2FeS(s) + 3O_2(g) \longrightarrow 2FeO(s) + 2SO_2(g)$$

FeS	O_2	FeO	SO_2
26 g	_____ g	_____ g	_____ g
_____ g	_____ g	_____ g	7.94 g

9.74 Use the balanced chemical equation to complete the table: (9.1, 9.2, 9.3)

$$2C_2H_6(g) + 7O_2(g) \xrightarrow{\Delta} 4CO_2(g) + 6H_2O(g)$$

C_2H_6	O_2	CO_2	H_2O
_____ g	_____ g	39 g	_____ g
_____ g	73.3 g	_____ g	_____ g

9.75 Chromium and oxygen combine to form chromium(III) oxide. (9.2, 9.3, 9.4, 9.5)

$$4Cr(s) + 3O_2(g) \longrightarrow 2Cr_2O_3(s)$$

a. How many moles of O_2 react with 4.50 mol of Cr?

b. How many grams of Cr_2O_3 are produced when 24.8 g of Cr reacts?

c. When 26.0 g of Cr reacts with 8.00 g of O_2, how many grams of Cr_2O_3 can form?

d. If 74.0 g of Cr and 62.0 g of O_2 are mixed, and 87.3 g of Cr_2O_3 is actually obtained, what is the percent yield of Cr_2O_3 for the reaction?

9.76 Aluminum and chlorine combine to form aluminum chloride. (9.2, 9.3, 9.4, 9.5)

$$2Al(s) + 3Cl_2(g) \longrightarrow 2AlCl_3(s)$$

a. How many moles of Cl_2 are needed to react with 4.50 mol of Al?

b. How many grams of $AlCl_3$ are produced when 50.2 g of Al reacts?

c. When 13.5 g of Al reacts with 8.00 g of Cl_2, how many grams of $AlCl_3$ can form?

d. If 45.0 g of Al and 62.0 g of Cl_2 are mixed, and 66.5 g of $AlCl_3$ is actually obtained, what is the percent yield of $AlCl_3$ for the reaction?

9.77 The combustion of propyne (C_3H_4) releases heat when it burns according to the following equation: (9.2, 9.3, 9.5)

$$C_3H_4(g) + 4O_2(g) \xrightarrow{\Delta} 3CO_2(g) + 2H_2O(g)$$

a. How many moles of O_2 are needed to react completely with 0.225 mol of C_3H_4?

b. How many grams of H_2O are produced from the complete reaction of 64.0 g of O_2?

c. How many grams of CO_2 are produced from the complete reaction of 78.0 g of C_3H_4?

d. If the reaction in part **c** produces 186 g of CO_2, what is the percent yield of CO_2 for the reaction?

9.78 Butane gas (C_4H_{10}) burns according to the following equation: (9.2, 9.3, 9.5)

$$2C_4H_{10}(g) + 13O_2(g) \xrightarrow{\Delta} 8CO_2(g) + 10H_2O(g)$$

a. How many moles of H_2O are produced from the complete reaction of 2.50 mol of C_4H_{10}?

b. How many grams of O_2 are needed to react completely with 22.5 g of C_4H_{10}?

c. How many grams of CO_2 are produced from the complete reaction of 55.0 g of C_4H_{10}?

d. If the reaction in part **c** produces 145 g of CO_2, what is the percent yield of CO_2 for the reaction?

9.79 Sulfur trioxide decomposes to sulfur and oxygen. (9.2, 9.6)

$$2SO_3(g) \longrightarrow 2S(s) + 3O_2(g) \quad \Delta H = +790 \text{ kJ}$$

   a. Is the reaction exothermic or endothermic?

   b. How many kilojoules are required when 1.5 mol of SO_3 reacts?

   c. How many kilojoules are required when 150 g of O_2 is formed?

9.80 When hydrogen peroxide (H_2O_2) is used in rocket fuels, it produces water, oxygen, and heat. (9.2, 9.6)

$$2H_2O_2(l) \longrightarrow 2H_2O(l) + O_2(g) \quad \Delta H = -196 \text{ kJ}$$

   a. Is the reaction exothermic or endothermic?

   b. How many kilojoules are released when 2.50 mol of H_2O_2 reacts?

   c. How many kilojoules are released when 275 g of O_2 is produced?

9.81 Calculate the energy change for the reaction $NH_4Cl(s) \longrightarrow NH_3(g) + HCl(g)$ from the following: (9.6)

$$H_2(g) + Cl_2(g) \longrightarrow 2HCl(g) \qquad\qquad \Delta H = -184 \text{ kJ}$$
$$N_2(g) + 4H_2(g) + Cl_2(g) \longrightarrow 2NH_4Cl(s) \quad \Delta H = -631 \text{ kJ}$$
$$N_2(g) + 3H_2(g) \longrightarrow 2NH_3(g) \qquad\qquad \Delta H = -296 \text{ kJ}$$

9.82 Calculate the energy change for the reaction $Mg(s) + N_2(g) + 3O_2(g) \longrightarrow Mg(NO_3)_2(s)$ from the following: (9.6)

$$8Mg(s) + Mg(NO_3)_2(s) \longrightarrow Mg_3N_2(s) + 6MgO(s)$$
$$\qquad\qquad\qquad\qquad\qquad\qquad \Delta H = -3281 \text{ kJ}$$
$$Mg_3N_2(s) \longrightarrow 3Mg(s) + N_2(g) \qquad \Delta H = +461 \text{ kJ}$$
$$2MgO(s) \longrightarrow 2Mg(s) + O_2(g) \qquad\quad \Delta H = +1204 \text{ kJ}$$

Applications

9.83 In the body, the amino acid glycine ($C_2H_5NO_2$) reacts according to the following equation: (9.3)

$$2C_2H_5NO_2(aq) + 3O_2(g) \longrightarrow$$
$$\text{Glycine} \qquad\qquad 3CO_2(g) + 3H_2O(l) + CH_4N_2O(aq)$$
$$\text{Urea}$$

   a. How many grams of O_2 are needed to react with 15.0 g of glycine?

   b. How many grams of urea are produced from 15.0 g of glycine?

   c. How many grams of CO_2 are produced from 15.0 g of glycine?

9.84 In the body, ethanol (C_2H_6O) reacts according to the following equation: (9.3)

$$C_2H_6O(aq) + 3O_2(g) \longrightarrow 2CO_2(g) + 3H_2O(l)$$
$$\text{Ethanol}$$

   a. How many grams of O_2 are needed to react with 8.40 g of ethanol?

   b. How many grams of H_2O are produced from 8.40 g of ethanol?

   c. How many grams of CO_2 are produced from 8.40 g of ethanol?

9.85 In photosynthesis, glucose ($C_6H_{12}O_6$) and O_2 are produced from CO_2 and H_2O. Glucose from starches is the major fuel for the body. (9.3, 9.6)

$$6CO_2(g) + 6H_2O(l) + 680 \text{ kcal} \longrightarrow C_6H_{12}O_6(aq) + 6O_2(g)$$
$$\text{Glucose}$$

   a. Is the reaction exothermic or endothermic?

   b. How many grams of glucose are produced from 18.0 g of CO_2?

   c. How much heat, in kilojoules, is needed to produce 25.0 g of $C_6H_{12}O_6$?

9.86 Ethanol (C_2H_6O) reacts in the body using the following equation: (9.3, 9.5)

$$C_2H_6O(aq) + 3O_2(g) \longrightarrow 2CO_2(g) + 3H_2O(l) + 327 \text{ kcal}$$
$$\text{Ethanol}$$

   a. Is the reaction exothermic or endothermic?

   b. How much heat, in kilocalories, is produced when 5.00 g of ethanol reacts with O_2 gas?

   c. How many grams of ethanol react if 1250 kJ are produced?

ANSWERS TO ENGAGE QUESTIONS

9.1 The law of conservation of mass states that the total mass of reactants must be equal to the total mass of products in a chemical reaction. No material is lost or gained as original substances are changed to new substances.

9.2 The mole–mole factor relates the moles of reactants and products in a reaction. The grams must first be changed to moles and then the mole–mole factor is used.

9.3 When the 6 knives are used, no more people can be served. The spoons and forks are excess.

9.4 You prepared enough dough to make five dozen, or 60, cookies. If 12 were burned, 48 were made.

   $48/60 \times 100\% = 80\%$ yield

9.5 The heat of reaction can be related to each of the reactants and products in a reaction, making many heat conversion factors possible.

ANSWERS TO SELECTED PROBLEMS

9.1   **a.** 160.14 g of reactants = 160.14 g of products

     b. 283.88 g of reactants = 283.88 g of products

9.3   **a.** $\dfrac{2 \text{ mol } SO_2}{1 \text{ mol } O_2}$ and $\dfrac{1 \text{ mol } O_2}{2 \text{ mol } SO_2}$

          $\dfrac{2 \text{ mol } SO_2}{2 \text{ mol } SO_3}$ and $\dfrac{2 \text{ mol } SO_3}{2 \text{ mol } SO_2}$

          $\dfrac{2 \text{ mol } SO_3}{1 \text{ mol } O_2}$ and $\dfrac{1 \text{ mol } O_2}{2 \text{ mol } SO_3}$

     b. $\dfrac{4 \text{ mol } P}{5 \text{ mol } O_2}$ and $\dfrac{5 \text{ mol } O_2}{4 \text{ mol } P}$

          $\dfrac{4 \text{ mol } P}{2 \text{ mol } P_2O_5}$ and $\dfrac{2 \text{ mol } P_2O_5}{4 \text{ mol } P}$

          $\dfrac{5 \text{ mol } O_2}{2 \text{ mol } P_2O_5}$ and $\dfrac{2 \text{ mol } P_2O_5}{5 \text{ mol } O_2}$

9.5 a. $\text{mol SO}_2 \times \dfrac{2 \text{ mol SO}_3}{2 \text{ mol SO}_2} = \text{mol of SO}_3$

b. $\text{mol P} \times \dfrac{5 \text{ mol O}_2}{4 \text{ mol P}} = \text{mol of O}_2$

9.7 a. 1.3 mol of O_2
b. 10. mol of H_2
c. 5.0 mol of H_2O

9.9 a. 1.25 mol of C
b. 0.96 mol of CO
c. 1.0 mol of SO_2
d. 0.50 mol of CS_2

9.11 a. 77.5 g of Na_2O
b. 6.26 g of O_2
c. 19.4 g of O_2

9.13 a. 19.2 g of O_2
b. 3.79 g of N_2
c. 54.0 g of H_2O

9.15 a. 3.66 g of H_2O
b. 26.3 g of NO
c. 7.53 g of HNO_3

9.17 a. $2\text{PbS}(s) + 3\text{O}_2(g) \longrightarrow 2\text{PbO}(s) + 2\text{SO}_2(g)$
b. 6.00 g of O_2
c. 17.4 g of SO_2
d. 137 g of PbS

9.19 a. Eight taxis can be used to pick up passengers.
b. Seven taxis can be driven.

9.21 a. 5.0 mol of H_2
b. 4.0 mol of H_2
c. 3.0 mol of N_2

9.23 a. Cl_2 is the limiting reactant, which would produce 25.1 g of $AlCl_3$.
b. O_2 is the limiting reactant, which would produce 13.5 g of H_2O.
c. O_2 is the limiting reactant, which would produce 26.7 g of SO_2.

9.25 a. 31.2 g of SO_3
b. 34.6 g of Fe_3O_4
c. 35.0 g of CO_2

9.27 a. 71.0%
b. 63.2%

9.29 70.9 g of Al_2O_3

9.31 60.5%

9.33 In exothermic reactions, the energy of the products is lower than that of the reactants.

9.35 a. exothermic
b. endothermic
c. exothermic

9.37 a. Heat is released, exothermic, $\Delta H = -802$ kJ
b. Heat is absorbed, endothermic, $\Delta H = +65.3$ kJ
c. Heat is released, exothermic, $\Delta H = -850$ kJ

9.39 a. 579 kJ
b. 89 kJ

9.41 +66 kJ

9.43 −297 kJ

9.45 a. 3.2×10^2 kg
b. 143 g
c. 80.4%

9.47 a. $2\text{NO}(g) + \text{O}_2(g) \longrightarrow 2\text{NO}_2(g)$
b. NO is the limiting reactant.

9.49 a. $\text{N}_2(g) + 3\text{H}_2(g) \longrightarrow 2\text{NH}_3(g)$
b. A

9.51 a. $2\text{NI}_3(g) \longrightarrow \text{N}_2(g) + 3\text{I}_2(g)$
b. 67%

9.53

FeS	O₂	FeO	SO₂
2.0 mol	3.0 mol	2.0 mol	2.0 mol
4.6 mol	6.9 mol	4.6 mol	4.6 mol

9.55 a. 1.33 mol of NH_3 and 3.33 mol of F_2
b. 142 g of F_2
c. 10.4 g of N_2F_4

9.57 a. 3.00 mol of H_2O_2
b. 77.6 g of H_2O_2
c. 6.46 g of H_2O

9.59 16.8 g of NaCl

9.61 a. 0.667 mol of C_5H_{12}
b. 27.5 g of CO_2
c. 92.8 g of CO_2

9.63 a. 74.4 g of CO_2
b. 86.0%

9.65 75.9%

9.67 a. 60.8 g of NH_3
b. 37.7 g of NH_3

9.69 a. 4.51 kJ
b. 43.9 g of NO

9.71 a. exothermic
b. endothermic

9.73

FeS	O₂	FeO	SO₂
26 g	14 g	21 g	19 g
10.9 g	5.95 g	8.90 g	7.94 g

9.75 a. 3.38 mol of O_2
b. 36.2 g of Cr_2O_3
c. 25.3 g of Cr_2O_3
d. 80.8%

9.77 a. 0.900 mol of O_2
b. 18.0 g of O_2
c. 257 g of CO_2
d. 72.4%

9.79 a. endothermic
b. 590 kJ
c. 1200 kJ

9.81 76 kJ

9.83 a. 9.59 g of O_2
b. 6.00 g of urea
c. 13.2 g of CO_2

9.85 a. endothermic
b. 12.3 g of glucose
c. 390 kJ

Bonding and Properties of Solids and Liquids

Bill has been diagnosed with basal cell carcinoma, the most common form of skin cancer. He has an appointment to undergo Mohs surgery, a specialized procedure to remove the cancerous growth found on his shoulder. The surgeon begins by removing the abnormal growth, in addition to a thin layer of surrounding (margin) tissue, which he sends to Lisa, a histologist. Lisa prepares the tissue sample to be viewed by a pathologist. Tissue preparation requires Lisa to cut the tissue into a very thin section (normally about 0.001 cm), which is then mounted onto a microscope slide.

Lisa then treats the tissue with a dye to stain the cells, as this enables the pathologist to view any abnormal cells more easily. The pathologist examines the tissue sample and reports back to the surgeon. If no cancerous cells are detected, no more tissue will be removed. If cancer cells are still present, the surgeon will take more of the surrounding tissue.

CAREER

Histologist

Histologists study the microscopic makeup of tissues, cells, and bodily fluids in order to detect and identify the presence of a specific disease. They determine blood types and the concentrations of drugs and other substances in the blood. Sample preparation is a critical component of a histologist's job, as they prepare tissue samples from humans, animals, and plants. The tissue samples are cut into extremely thin sections, which are then mounted and stained using various chemical dyes. The dyes provide contrast for the cells to be viewed and help highlight any abnormalities that may exist. Utilization of various dyes requires the histologist to be familiar with solution preparation and the handling of potentially hazardous chemicals.

UPDATE   Histologist Stains Tissue with Dye

When Lisa obtains the tissue sample, she will freeze it to obtain thin slices of tissue for staining. You can learn more about how Lisa prepares the tissue from Bill's surgery for the pathologist by reading the **UPDATE Histologist Stains Tissue with Dye**, page 298.

REVIEW

Writing Electron Configurations (5.4)

Using the Periodic Table to Write
  Electron Configurations (5.5)

Writing Ionic Formulas (6.2)

10.1 Lewis Structures for Molecules and Polyatomic Ions

LEARNING GOAL Draw the Lewis structures for molecular compounds and polyatomic ions with single and multiple bonds.

Now we can investigate more complex chemical bonds and how they contribute to the structure of a molecule or a polyatomic ion. We will diagram the sharing of valence electrons in molecules and polyatomic ions, and determine the presence of single and multiple bonds.

Lewis Symbols for Atoms

A **Lewis symbol** is a convenient way to represent the valence electrons, which are shown as dots placed on the sides, top, or bottom of the symbol for the element. One to four valence electrons are arranged as single dots. When there are five to eight electrons, one or more electrons are paired. Any of the following would be an acceptable Lewis symbol for magnesium, which has two valence electrons:

Lewis Symbols for Magnesium

$$\overset{\bullet}{\text{Mg}}\bullet \qquad \overset{\bullet}{\underset{\bullet}{\text{Mg}}} \qquad \bullet\overset{\bullet}{\text{Mg}} \qquad \bullet\text{Mg}\bullet \qquad \text{Mg}\bullet \qquad \bullet\underset{\bullet}{\text{Mg}}$$

Lewis symbols for selected elements are given in **TABLE 10.1**.

TABLE 10.1 Lewis Symbols for Selected Elements in Periods 1 to 4

Group Number	1A (1)	2A (2)	3A (13)	4A (14)	5A (15)	6A (16)	7A (17)	8A (18)
Number of Valence Electrons	1	2	3	4	5	6	7	8
	Number of Valence Electrons Increases →							
Lewis Symbol	H•							He:*
	Li•	Be•	•B•	•C•	•N•	•O:	•F:	:Ne:
	Na•	Mg•	•Al•	•Si•	•P•	•S:	•Cl:	:Ar:
	K•	Ca•	•Ga•	•Ge•	•As•	•Se:	•Br:	:Kr:

*Helium (He) is stable with two valence electrons.

CORE CHEMISTRY SKILL

Drawing Lewis Symbols

SAMPLE PROBLEM 10.1   Drawing Lewis Symbols

TRY IT FIRST

Draw the Lewis symbol for each of the following:
a. bromine   **b.** aluminum

SOLUTION

a. The Lewis symbol for bromine, which is in Group 7A (17), has seven valence electrons. Thus, three pairs of dots and one single dot are drawn on the sides of the Br symbol.

$$\bullet\overset{\bullet\bullet}{\underset{\bullet\bullet}{\text{Br}}}\bullet\bullet$$

b. The Lewis symbol for aluminum, which is in Group 3A (13), has three valence electrons. They are drawn as single dots on the sides of the Al symbol.

$$\bullet\overset{\bullet}{\text{Al}}\bullet$$

ENGAGE 10.1

Why do nitrogen, phosphorus, and arsenic have five dots in their Lewis symbols?

SELF TEST 10.1

Draw the Lewis symbol for each of the following:

a. phosphorus, a macromineral needed for bones and teeth
b. selenium, a micromineral needed for the immune system

ANSWER

a. $\cdot \ddot{P} \cdot$ **b.** $\cdot \ddot{\underset{\cdot\cdot}{Se}}:$

Lewis Structures for Ionic Compounds

An ionic compound forms when valence electrons from a metal are transferred to a nonmetal. For example, the transfer of one valence electron from sodium to chlorine gives octets to both. The loss of an electron by sodium produces a sodium ion, and the gain of an electron by chlorine produces a chloride ion. We can use the Lewis symbols to show this. The *Lewis structure* of the compound NaCl shows the chloride ion in brackets with its negative charge in the upper right corner.

$$\text{Na}\cdot \ + \ \cdot\ddot{\underset{\cdot\cdot}{Cl}}: \ \longrightarrow \ \text{Na}^+ \left[:\ddot{\underset{\cdot\cdot}{Cl}}:\right]^- \qquad \text{Lewis structure}$$

Lewis Structures for Molecular Compounds

The simplest molecule is hydrogen, H_2. When two H atoms are far apart, there is no attraction between them. As the H atoms move closer, the positive charge of each nucleus attracts the electron of the other atom. This attraction, which is greater than the repulsion between the valence electrons, pulls the H atoms closer until they share a pair of valence electrons (see **FIGURE 10.1**). The result is called a *covalent bond*, in which the shared electrons give the stable electron configuration of He to *each* of the H atoms. When the H atoms form H_2, they are more stable than two individual H atoms.

CORE CHEMISTRY SKILL

Drawing Lewis Structures

FIGURE 10.1 A covalent bond forms as H atoms move close together to share electrons.

ENGAGE 10.2

What determines the attraction between two H atoms?

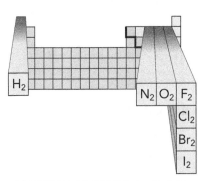

The elements hydrogen, nitrogen, oxygen, fluorine, chlorine, bromine, and iodine exist as diatomic molecules.

A molecule is represented by a **Lewis structure** in which the valence electrons of all the atoms are arranged to give octets, except for hydrogen, which has two electrons. The shared electrons, or *bonding pairs*, are shown as two dots or a single line between atoms. The nonbonding pairs of electrons, or *lone pairs*, are placed on the outside. For example, a fluorine molecule, F_2, consists of two fluorine atoms, which are in Group 7A (17), each with seven valence electrons. In the Lewis structure for the F_2 molecule, each F atom achieves an octet by sharing its unpaired valence electron.

TABLE 10.2 Elements That Exist as Diatomic Molecules	
Diatomic Molecule	Name
H_2	Hydrogen
N_2	Nitrogen
O_2	Oxygen
F_2	Fluorine
Cl_2	Chlorine
Br_2	Bromine
I_2	Iodine

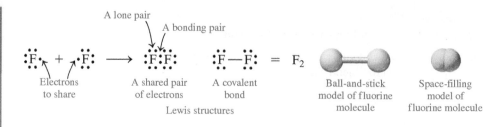

Hydrogen (H_2) and fluorine (F_2) are examples of nonmetal elements whose natural state is diatomic; that is, they contain two like atoms. The elements that exist as diatomic molecules are listed in TABLE 10.2.

Sharing Electrons Between Atoms of Different Elements

The number of electrons that a nonmetal atom shares and the number of covalent bonds it forms are usually equal to the number of electrons it needs to achieve a stable electron configuration. TABLE 10.3 gives the most typical bonding patterns for some nonmetals.

TABLE 10.3 Typical Bonding Patterns of Some Nonmetals					
1A (1)	3A (13)	4A (14)	5A (15)	6A (16)	7A (17)
*H 1 bond					
	*B 3 bonds	C 4 bonds	N 3 bonds	O 2 bonds	F 1 bond
		Si 4 bonds	P 3 bonds	S 2 bonds	Cl, Br, I 1 bond

*H and B do not form octets. H atoms share one electron pair; B atoms share three electron pairs for a set of six electrons.

TABLE 10.4 Representations for Some Molecular Compounds		
CH_4 Methane molecule	NH_3 Ammonia molecule	H_2O Water molecule
Lewis Structures		
(Lewis structures shown)		
Ball-and-Stick Models		
Space-Filling Models		

Drawing Lewis Structures

To draw the Lewis structure for CH_4, we first draw the Lewis symbols for carbon and hydrogen.

Then we determine the number of valence electrons needed for carbon and hydrogen. When a carbon atom shares its four electrons with four hydrogen atoms, carbon obtains an octet, and each hydrogen atom is complete with two shared electrons. The Lewis structure is drawn with the carbon atom as the central atom, with the hydrogen atoms on each of the sides. The bonding pairs of electrons, which are single covalent bonds, may also be shown as single lines between the carbon atom and each of the hydrogen atoms.

In addition to Lewis structures, three-dimensional models may be used to represent molecular compounds. A *ball-and-stick model* illustrates the shape of a molecule using spheres for atoms and sticks as bonding electron pairs. In a *space-filling model*, the atoms are shown as partial spheres that are proportional to the real size of the atoms. Because the bonds are within the filled spaces, they are not visible.

TABLE 10.4 gives examples of Lewis structures and molecular models for some molecules.

When we draw a Lewis structure for a molecule or polyatomic ion, we show the sequence of atoms, the bonding pairs of electrons shared between atoms, and the nonbonding (lone pairs) of electrons. From the formula, we identify the central atom, which is the element that has the fewer atoms. Then, the central atom is bonded to the other atoms, as shown in Sample Problem 10.2.

SAMPLE PROBLEM 10.2 Drawing Lewis Structures for Molecules

TRY IT FIRST

Draw the Lewis structures for PCl_3, phosphorus trichloride, used to prepare insecticides and flame retardants.

SOLUTION

ANALYZE THE PROBLEM	Given	Need	Connect
	PCl_3	Lewis structure	total valence electrons

STEP 1 **Determine the arrangement of atoms.** In PCl_3, the central atom is P because there is only one P atom.

Cl   P   Cl
    Cl

STEP 2 **Determine the total number of valence electrons.** We use the group number to determine the number of valence electrons for each of the atoms in the molecule.

Element	Group	Atoms		Valence Electrons		Total
P	5A (15)	1 P	×	$5\ e^-$	=	$5\ e^-$
Cl	7A (17)	3 Cl	×	$7\ e^-$	=	$21\ e^-$
				Total valence electrons for PCl_3	=	$26\ e^-$

STEP 3 **Attach each bonded atom to the central atom with a pair of electrons.** Each bonding pair can also be represented by a bond line.

Cl:P:Cl   or   Cl—P—Cl
  Cl   Cl

Six electrons ($3 \times 2\ e^-$) are used to bond the central P atom to three Cl atoms. Twenty valence electrons are left.

26 valence e^- − 6 bonding e^- = $20\ e^-$ remaining

STEP 4 **Use the remaining electrons to complete octets.** We use the remaining 20 electrons as lone pairs, which are placed around the outer Cl atoms and on the P atom, such that all the atoms have octets.

    or   :Cl—P—Cl:
                        :Cl:

SELF TEST 10.2

Draw the Lewis structures for Cl_2O, one with electron dots and the other with lines for bonding pairs.

ANSWER

  Cl_2O Ball-and-stick model

ENGAGE 10.3

How are the total number of valence electrons determined for a molecule?

PCl_3 Ball-and-stick model

PCl_3 Space-filling model

SAMPLE PROBLEM 10.3 Drawing Lewis Structures for Polyatomic Ions

TRY IT FIRST

Sodium chlorite, $NaClO_2$, is a component of mouthwashes, toothpastes, and contact lens cleaning solutions. Draw the Lewis structures for the chlorite ion, ClO_2^-.

SOLUTION

ANALYZE THE PROBLEM	Given	Need	Conment
	ClO_2^-	Lewis structure	total valence electrons

STEP 1 Determine the arrangement of atoms. For the polyatomic ion ClO_2^-, the central atom is Cl because there is only one Cl atom. For a polyatomic ion, the atoms and electrons are placed in brackets, and the charge is written outside at the upper right.

$$[O \quad Cl \quad O]^-$$

STEP 2 Determine the total number of valence electrons. We use the group numbers to determine the number of valence electrons for each of the atoms in the ion. Because the ion has a negative charge, one more electron is added to the valence electrons.

Element	Group	Atoms		Valence Electrons		Total
Cl	7A (17)	1 Cl	×	$7\,e^-$	=	$7\,e^-$
O	6A (16)	2 O	×	$6\,e^-$	=	$12\,e^-$
Ionic charge (negative) add				$1\,e^-$	=	$1\,e^-$
			Total valence electrons for ClO_2^-		=	$20\,e^-$

STEP 3 Attach each bonded atom to the central atom with a pair of electrons. Each bonding pair can also be represented by a line, which indicates a single bond.

$$[O\!:\!Cl\!:\!O]^- \quad or \quad [O-Cl-O]^-$$

Four electrons ($2 \times 2\,e^-$) are used to bond the O atoms to the central Cl atom, which leaves 16 valence electrons.

STEP 4 Use the remaining electrons to complete octets. Of the 16 remaining valence electrons, 12 are drawn as lone pairs to complete the octets of the O atoms.

$$\left[:\!\ddot{O}\!:\!Cl\!:\!\ddot{O}\!:\right]^- \quad or \quad \left[:\!\ddot{O}\!-\!Cl\!-\!\ddot{O}\!:\right]^-$$

The remaining four electrons are placed as two lone pairs on the central Cl atom.

$$\left[:\!\ddot{O}\!:\!\ddot{Cl}\!:\!\ddot{O}\!:\right]^- \quad or \quad \left[:\!\ddot{O}\!-\!\ddot{Cl}\!-\!\ddot{O}\!:\right]^-$$

ClO_2^- Ball-and-stick model

ClO_2^- Space-filling model

SELF TEST 10.3

Draw the Lewis structures for the polyatomic ion NH_2^-, one with electron dots and the other with lines for bonding pairs.

ANSWER

$$\left[H\!:\!\ddot{N}\!:\!H\right]^- \quad or \quad \left[H\!-\!\ddot{N}\!-\!H\right]^-$$

NH_2^- Ball-and-stick model

Double and Triple Bonds

 INTERACTIVE VIDEO

Drawing Lewis Structures with Multiple Bonds

Up to now, we have looked at bonding in molecules having only single bonds. In many molecular compounds, atoms share two or three pairs of electrons to complete their octets.

Double and triple bonds form when the number of valence electrons is not enough to complete the octets of all the atoms in the molecule. Then one or more lone pairs of electrons from the atoms attached to the central atom are shared with the central atom. A **double bond** occurs when two pairs of electrons are shared; in a **triple bond**, three pairs of electrons are shared. Atoms of carbon, oxygen, nitrogen, and sulfur are most likely to form multiple bonds.

Atoms of hydrogen and the halogens do not form double or triple bonds. The process of drawing a Lewis structure with multiple bonds is shown in Sample Problem 10.4.

SAMPLE PROBLEM 10.4 Drawing Lewis Structures with Multiple Bonds

TRY IT FIRST

Draw the Lewis structures for carbon dioxide, CO_2.

SOLUTION

ANALYZE THE PROBLEM	Given	Need	Connect
	CO_2	Lewis structure	total valence electrons

STEP 1 **Determine the arrangement of atoms.** In CO_2, the central atom is C because there is only one C atom.

O   C   O

STEP 2 **Determine the total number of valence electrons.** We use the group number to determine the number of valence electrons for each of the atoms in the molecule.

Element	Group	Atoms		Valence Electrons		Total
C	4A (14)	1 C	×	$4\,e^-$	=	$4\,e^-$
O	6A (16)	2 O	×	$6\,e^-$	=	$12\,e^-$
			Total valence electrons for CO_2 =			$16\,e^-$

STEP 3 **Attach each bonded atom to the central atom with a pair of electrons.**

O:C:O   or   O—C—O

We use four valence electrons to attach the central C atom to two O atoms, which leaves 12 valence electrons.

STEP 4 **Use the remaining electrons to complete octets, drawing multiple bonds if needed.**

The 12 remaining electrons are placed as six lone pairs of electrons on the outside O atoms. However, this does not complete the octet of the C atom.

:Ö:C:Ö: or   :Ö—C—Ö:

To obtain an octet, the C atom must share pairs of electrons from each of the O atoms. When two bonding pairs are drawn between two atoms, it is a double bond.

ENGAGE 10.4

How do you know when you need to draw one or more multiple bonds to complete a Lewis structure?

Lone pairs converted to bonding pairs

:Ö:C:Ö: or   :Ö⤸—C—⤸Ö:

Double bonds   Double bonds

    or

CO₂ Ball-and-stick model   **CO₂** Space-filling model

PRACTICE PROBLEMS

Try Practice Problems 10.1 to 10.8

SELF TEST 10.4

Draw the Lewis structures for HCN, which has a triple bond. Draw one Lewis structure with electron dots and the other with lines for bonding pairs.

ANSWER

H:C::N: or   H—C≡N:

HCN Ball-and-stick model   HCN Space-filling model

In BCl₃, the central B atom is bonded to three Cl atoms.

In SF₆, the central S atom is bonded to six F atoms.

Exceptions to the Octet Rule

Although the octet rule is useful for bonding in many compounds, there are exceptions. We have already seen that a hydrogen (H_2) molecule requires just two electrons or a single bond. Usually the nonmetals form octets. However, in BCl_3, the B atom has only three valence electrons to share. Boron compounds typically have six valence electrons on the central B atoms and form just three bonds. Although we will generally see compounds of P, S, Cl, Br, and I with octets, they can form molecules in which they share more of their valence electrons. This expands their valence electrons to 10, 12, or even 14 electrons. For example, we have seen that the P atom in PCl_3 has an octet, but in PCl_5, the P atom has five bonds with 10 valence electrons. In H_2S, the S atom has an octet, but in SF_6, there are six bonds to sulfur with 12 valence electrons.

PRACTICE PROBLEMS

10.1 Lewis Structures for Molecules and Polyatomic Ions

10.1 Determine the total number of valence electrons for each of the following:
   a. H_2S **b.** I_2 **c.** CCl_4 **d.** OH^-

10.2 Determine the total number of valence electrons for each of the following:
   a. SBr_2 **b.** NBr_3 **c.** CH_3OH **d.** NH_4^+

10.3 When is it necessary to draw a multiple bond in a Lewis structure?

10.4 If the available number of valence electrons for a molecule or polyatomic ion does not complete all of the octets in a Lewis structure, what should you do?

10.5 Draw the Lewis structures for each of the following molecules or ions:
   a. HF   **b.** SF_2 **c.** NBr_3 **d.** BH_4^-

10.6 Draw the Lewis structures for each of the following molecules or ions:
   a. H_2O **b.** CCl_4 **c.** H_3O^+ **d.** SiF_4

10.7 Draw the Lewis structures for each of the following molecules or ions:
   a. CO   **b.** CN^- **c.** H_2CO (C is the central atom)

10.8 Draw the Lewis structures for each of the following molecules or ions:
   a. HCCH (acetylene)   **b.** CS_2 **c.** NO^+

10.2 Resonance Structures

LEARNING GOAL Draw Lewis structures for molecules or polyatomic ions that have two or more resonance structures.

CORE CHEMISTRY SKILL

Drawing Resonance Structures

When a molecule or polyatomic ion contains multiple bonds, it may be possible to draw more than one Lewis structure. We can see how this happens when we draw the Lewis structure for ozone, O_3, a component in the stratosphere that protects us from the ultraviolet rays of the Sun.

To draw the Lewis structure for O_3, we determine the number of valence electrons for an O atom, and then the total number of valence electrons for O_3. Because O is in Group 6A (16), it has six valence electrons. Therefore, the compound O_3 would have a total of 18 valence electrons.

Element	Group	Atoms		Valence Electrons		Total
O	6A (16)	3 O	×	$6\,e^-$	=	$18\,e^-$

For the Lewis structure for O_3, we place three O atoms in a row. Using four of the valence electrons, we draw a bonding pair between each of the O atoms on the ends and the central O atom. Two bonding pairs require four valence electrons.

O—O—O

The remaining valence electrons (14) are placed as lone pairs of electrons around the O atoms on both ends of the Lewis structure, and one lone pair goes on the central O atom.

:Ö—Ö—Ö:

To complete an octet for the central O atom, one lone pair of electrons from an end O atom needs to be shared. But which lone pair should be used? One possibility is to form a double bond between the central O atom and the O on the left, and the other possibility is to form a double bond between the central O atom and the O on the right.

:Ö—Ö—Ö: or   :Ö—Ö—Ö:

Thus it is possible to draw two or more Lewis structures for a molecule such as O_3 or for a polyatomic ion. When this happens, all the Lewis structures are called **resonance structures**, and their relationship is shown by drawing a double-headed arrow between them.

:Ö=Ö—Ö: ⟷ :Ö—Ö=Ö:

Resonance structures

Experiments show that the actual bonds in ozone are equivalent to a molecule with "one-and-a-half" bonds between the central O atom and each outside O atom. In an actual ozone molecule, the electrons are spread equally over all the O atoms. When we draw resonance structures for molecules or polyatomic ions, the true structure is really an average of those structures.

Stratosphere

O_3 molecule

Ozone, O_3, is a component in the stratosphere that protects us from the ultraviolet rays of the Sun.

ENGAGE 10.5

Why are there two possible ways to form a double bond in ozone?

Representations of O_3 molecule

SAMPLE PROBLEM 10.5 Drawing Resonance Structures

TRY IT FIRST

Sulfur dioxide is produced by volcanic activity and the burning of sulfur-containing coal. Once in the atmosphere, the SO_2 is converted to SO_3, which combines with water, forming sulfuric acid, H_2SO_4, a component of acid rain. Draw two resonance structures for sulfur dioxide.

SOLUTION

ANALYZE THE PROBLEM	Given	Need	Connect
	SO_2	resonance structures	total valence electrons

STEP 1 **Determine the arrangement of atoms.** In SO_2, the S atom is the central atom because there is only one S atom.

O   S   O

STEP 2 **Determine the total number of valence electrons.** We use the group number to determine the number of valence electrons for each of the atoms in the molecule.

Element	Group	Atoms		Valence Electrons		Total
S	6A (16)	1 S	×	$6\,e^-$	=	$6\,e^-$
O	6A (16)	2 O	×	$6\,e^-$	=	$12\,e^-$
		Total valence electrons for SO_2			=	$18\,e^-$

SO₂ Ball-and-stick model

SO₂ Space-filling model

ENGAGE 10.6

Explain why SO₂ has resonance structures but SCl₂ does not.

PRACTICE PROBLEMS

Try Practice Problems 10.9 to 10.12

STEP 3   Attach each bonded atom to the central atom with a pair of electrons.

$$O—S—O$$

We use four electrons to form single bonds between the central S atom and the O atoms.

STEP 4   Place the remaining electrons using single or multiple bonds to complete octets. The remaining 14 electrons are drawn as lone pairs, which complete the octets for the O atoms but not the S atom.

$$:\ddot{O}—\ddot{S}—\ddot{O}:$$

To complete the octet for S, one lone pair of electrons from one of the O atoms is shared to form a double bond. One possibility is to form a double bond between the central S atom and the O on the left, and the other possibility is to form a double bond between the central S atom and the O on the right.

$$:O{=}\ddot{S}—\ddot{O}: \longleftrightarrow :\ddot{O}—\ddot{S}{=}O:$$

SELF TEST 10.5

Draw three resonance structures for SO₃.

ANSWER

$$:\ddot{O}—S{=}\ddot{O}: \longleftrightarrow :\ddot{O}—S—\ddot{O}: \longleftrightarrow :\ddot{O}{=}S—\ddot{O}:$$
$$\quad\;\;| \qquad\qquad\qquad || \qquad\qquad\qquad |$$
$$\quad\;\;:\ddot{O}: \qquad\qquad\; :\ddot{O}: \qquad\qquad\;\; :\ddot{O}:$$

TABLE 10.5 summarizes drawing Lewis structures for several molecules and ions.

TABLE 10.5 Using Valence Electrons to Draw Lewis Structures

Molecule or Polyatomic Ion	Total Valence Electrons	Form Single Bonds to Attach Atoms (electrons used)	Electrons Remaining	Completed Octets (or H:)
Cl_2	$2(7) = 14$	$Cl—Cl\ (2\ e^-)$	$14 - 2 = 12$	$:\ddot{Cl}—\ddot{Cl}:$
HCl	$1 + 7 = 8$	$H—Cl\ (2\ e^-)$	$8 - 2 = 6$	$H—\ddot{Cl}:$
H_2O	$2(1) + 6 = 8$	$H—O—H\ (4\ e^-)$	$8 - 4 = 4$	$H—\ddot{O}—H$
PCl_3	$5 + 3(7) = 26$	$Cl—P—Cl\ (6\ e^-)$ with Cl above	$26 - 6 = 20$	$:\ddot{Cl}—P—\ddot{Cl}:$ with $:\ddot{Cl}:$ above
ClO_3^-	$7 + 3(6) + 1 = 26$	$\left[O—Cl—O\right]^-\ (6\ e^-)$ with O above	$26 - 6 = 20$	$\left[:\ddot{O}—\ddot{Cl}—\ddot{O}:\right]^-$ with $:\ddot{O}:$ above
NO_2^-	$5 + 2(6) + 1 = 18$	$[O—N—O]^-\ (4\ e^-)$	$18 - 4 = 14$	$\left[:\ddot{O}—\ddot{N}{=}\ddot{O}:\right]^-$ $\updownarrow$ $\left[:\ddot{O}{=}\ddot{N}—\ddot{O}:\right]^-$

PRACTICE PROBLEMS

10.2 Resonance Structures

10.9 What is resonance?

10.10 When does a molecular compound have resonance?

10.11 Draw two resonance structures for each of the following molecules or ions:
   a. $ClNO_2$ (N is the central atom)
   b. OCN^- (C is the central atom)

10.12 Draw two resonance structures for each of the following molecules or ions:
   a. $H_2CNO_2^-$
   b. N_2O (N N O)

10.3 Shapes of Molecules and Polyatomic Ions (VSEPR Theory)

LEARNING GOAL Predict the three-dimensional structure of a molecule or a polyatomic ion.

Using the Lewis structures, we can predict the three-dimensional shapes of many molecules and polyatomic ions. The shape is important in our understanding of how molecules interact with enzymes or certain antibiotics or produce our sense of taste and smell.

> The three-dimensional shape of a molecule or polyatomic ion is determined by drawing its Lewis structure and identifying the number of electron groups (one or more electron pairs) around the central atom. We count lone pairs of electrons, single, double, and triple bonds as *one* electron group. In the **valence shell electron-pair repulsion (VSEPR) theory**, the electron groups are arranged as far apart as possible around the central atom to minimize the repulsion between the groups. Once we have counted the number of electron groups surrounding the central atom, we can determine its specific shape from the number of atoms bonded to the central atom.

Central Atoms with Two Electron Groups

In the Lewis structure for CO_2, there are two electron groups (two double bonds) attached to the central atom. According to VSEPR theory, minimal repulsion occurs when two electron groups are on opposite sides of the central C atom. This gives the CO_2 molecule a *linear* electron-group geometry and a shape that is **linear** with a bond angle of 180°.

$$:\ddot{O}=C=\ddot{O}:$$

Linear
electron-group
geometry

180°

Linear shape

Central Atoms with Three Electron Groups

In the Lewis structure for formaldehyde, H_2CO, the central atom C is attached to two H atoms by single bonds and to the O atom by a double bond. Minimal repulsion occurs when three electron groups are as far apart as possible around the central C atom, which gives 120° bond angles. This type of electron-group geometry is *trigonal planar* and gives a shape for H_2CO also called **trigonal planar**. When the central atom has the same number of electron groups as bonded atoms, the shape and the electron-group geometry are the same.

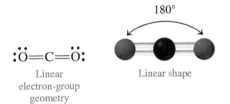

Lewis
structure

Trigonal planar
electron-group
geometry

120°

Trigonal planar shape

In the Lewis structure for SO_2, there are also three electron groups around the central S atom: a single bond to an O atom, a double bond to another O atom, and a lone pair of electrons. As in H_2CO, three electron groups have minimal repulsion when they form a trigonal planar electron-group geometry. However, in SO_2, one of the electron groups is a lone pair of electrons. Therefore, the shape of the SO_2 molecule is determined by the two O atoms bonded to the central S atom, which gives the SO_2 molecule a shape that is **bent** with a bond angle of 120°. When the central atom has more electron groups than bonded atoms, the shape and the electron-group geometry are different.

Lewis structure	Trigonal planar electron-group geometry	Bent shape

120°

Central Atom with Four Electron Groups

In a molecule of methane, CH_4, the central C atom is bonded to four H atoms. From the Lewis structure, you may think that CH_4 is planar with 90° bond angles. However, the best geometry for minimal repulsion is *tetrahedral*, giving bond angles of 109°. When there are four atoms attached to four electron groups, the shape of the molecule is also **tetrahedral**.

Lewis structure	Tetrahedral electron-group geometry	Tetrahedral shape	Tetrahedral wedge–dash notation

109°

A way to represent the three-dimensional structure of methane is to use the *wedge–dash notation*. In this representation, the two bonds connecting carbon to hydrogen by solid lines are in the plane of the paper. The wedge represents a carbon-to-hydrogen bond coming out of the page toward us, whereas the dash represents a carbon-to-hydrogen bond going into the page away from us.

Now we can look at molecules that have four electron groups, of which one or more are lone pairs of electrons. Then the central atom is attached to only two or three atoms. For example, in the Lewis structure for ammonia, NH_3, four electron groups have a tetrahedral electron-group geometry. However, in NH_3, one of the electron groups is a lone pair of electrons. Therefore, the shape of NH_3 is determined by the three H atoms bonded to the central N atom, which gives the NH_3 molecule a shape that is **trigonal pyramidal**, with a bond angle of 109°. The wedge–dash notation can also represent this three-dimensional structure of ammonia with one N—H bond in the plane, one N—H bond coming toward us, and one N—H bond going away from us.

Lone pair of electrons

Lewis structure	Tetrahedral electron-group geometry	Trigonal pyramidal shape	Trigonal pyramidal wedge–dash notation

In the Lewis structure for water, H_2O, there are also four electron groups, which have minimal repulsion when the electron-group geometry is tetrahedral. However, in H_2O, two

---

ENGAGE 10.7

If the four electron groups in a PH_3 molecule have a tetrahedral geometry, why does a PH_3 molecule have a trigonal pyramidal shape and not a tetrahedral shape?

of the electron groups are lone pairs of electrons. Because the shape of H_2O is determined by the two H atoms bonded to the central O atom, the H_2O molecule has a *bent shape* with a bond angle of 109°. **TABLE 10.6** gives the shapes of molecules with two, three, or four bonded atoms.

Lewis structure

Tetrahedral electron-group geometry

Bent shape

Bent wedge–dash notation

PRACTICE PROBLEMS

Try Practice Problems 10.13 and 10.14

TABLE 10.6 Molecular Shapes for a Central Atom with Two, Three, and Four Bonded Atoms

Electron Groups	Electron-Group Geometry	Bonded Atoms	Lone Pairs	Bond Angle*	Molecular Shape	Example	Three-Dimensional Models Ball-and-Stick	Space-Filling
2	Linear	2	0	180°	Linear	CO_2		
3	Trigonal planar	3	0	120°	Trigonal planar	H_2CO		
3	Trigonal planar	2	1	120°	Bent	SO_2		
4	Tetrahedral	4	0	109°	Tetrahedral	CH_4		
4	Tetrahedral	3	1	109°	Trigonal pyramidal	NH_3		
4	Tetrahedral	2	2	109°	Bent	H_2O		

*The bond angles in actual molecules may vary slightly.

SAMPLE PROBLEM 10.6 Shapes of Molecules

TRY IT FIRST

Use VSEPR theory to predict the shape of the molecule $SiCl_4$.

SOLUTION

	Given	Need	Connect
ANALYZE THE PROBLEM	$SiCl_4$	shape	Lewis structure, electron groups, bonded atoms

STEP 1 Draw the Lewis structure.

Element	Group	Atoms		Valence Electrons		Total
Si	4A (14)	1 Si	×	$4\,e^-$	=	$4\,e^-$
Cl	7A (17)	4 Cl	×	$7\,e^-$	=	$28\,e^-$
		Total valence electrons for SiCl₄			=	$32\,e^-$

Using $32\,e^-$, we draw the bonds and lone pairs for the Lewis structure of SiCl₄.

```
     ..
    :Cl:
  .. .. ..
 :Cl:Si:Cl:
  .. .. ..
    :Cl:
     ..
```

STEP 2 Arrange the electron groups around the central atom to minimize repulsion. In the Lewis structure of SiCl₄, there are four electron groups around the central Si atom. To minimize repulsion, the electron-group geometry would be tetrahedral.

STEP 3 Use the atoms bonded to the central atom to determine the shape. Because the central Si atom has four bonded atoms and no lone pairs of electrons, the SiCl₄ molecule has a tetrahedral shape.

SELF TEST 10.6

Use VSEPR theory to predict the shape of SCl₂.

ANSWER

The central atom S has four electron groups: two bonded atoms and two lone pairs of electrons. The shape of SCl₂ is bent, with a bond angle of 109°.

SAMPLE PROBLEM 10.7 Predicting Shape of an Ion

TRY IT FIRST

Use VSEPR theory to predict the shape of the polyatomic ion NO_3^-.

SOLUTION

	Given	Need	Connect
ANALYZE THE PROBLEM	NO_3^-	shape	Lewis structure, electron groups, bonded atoms

STEP 1 Draw the Lewis structure.

Element	Group	Atoms		Valence Electrons		Total
N	5A (15)	1 N	×	$5\,e^-$	=	$5\,e^-$
O	6A (16)	3 O	×	$6\,e^-$	=	$18\,e^-$
		Ionic charge (negative) add $1\,e^-$			=	$1\,e^-$
		Total valence electrons for NO_3^-			=	$24\,e^-$

The polyatomic ion NO_3^- contains three electron groups (two single bonds between the central N atom and O atoms, and one double bond between N and O). Note that the double bond can be drawn to any of the O atoms, which results in three resonance structures. However, we need just one of the structures to predict its shape.

$$\left[\ddot{O}-N=\ddot{O}\,\right]^-$$

STEP 2   **Arrange the electron groups around the central atom to minimize repulsion.** In the Lewis structure of NO_3^-, there are three electron groups around the central N atom. To minimize repulsion, the three electron groups have a trigonal planar geometry.

STEP 3   **Use the atoms bonded to the central atom to determine the shape.** Because NO_3^- has three bonded atoms and no lone pairs, it has a trigonal planar shape.

SELF TEST 10.7

Use VSEPR theory to predict the shape of each of the following:

a. ClO_2^- **b.** PO_2^+

ANSWER

a. With two bonded atoms and two lone pairs of electrons on the central Cl atom, the shape of ClO_2^- is bent, with a bond angle of 109°.

b. With two bonded atoms and no lone pairs on the central P atom, the shape of PO_2^+ is linear.

PRACTICE PROBLEMS
Try Practice Problems 10.15 to 10.22

PRACTICE PROBLEMS

10.3 Shapes of Molecules and Polyatomic Ions (VSEPR Theory)

10.13 Choose the shape (**1** to **6**) that matches each of the following descriptions (**a** to **c**):

     1. linear   **2.** bent (109°)   **3.** trigonal planar
     4. bent (120°)   **5.** trigonal pyramidal   **6.** tetrahedral

     a. a molecule with a central atom that has four electron groups and four bonded atoms
     b. a molecule with a central atom that has four electron groups and three bonded atoms
     c. a molecule with a central atom that has three electron groups and three bonded atoms

10.14 Choose the shape (**1** to **6**) that matches each of the following descriptions (**a** to **c**):

     1. linear   **2.** bent (109°)   **3.** trigonal planar
     4. bent (120°)   **5.** trigonal pyramidal   **6.** tetrahedral

     a. a molecule with a central atom that has four electron groups and two bonded atoms
     b. a molecule with a central atom that has two electron groups and two bonded atoms
     c. a molecule with a central atom that has three electron groups and two bonded atoms

10.15 Complete each of the following statements for a molecule of SeO_3:

     a. There are _____ electron groups around the central Se atom.
     b. The electron-group geometry is _____.
     c. The number of atoms attached to the central Se atom is _____.
     d. The shape of the molecule is _____.

10.16 Complete each of the following statements for a molecule of H_2S:

     a. There are _____ electron groups around the central S atom.
     b. The electron-group geometry is _____.
     c. The number of atoms attached to the central S atom is _____.
     d. The shape of the molecule is _____.

10.17 Compare the Lewis structures of CF_4 and NF_3. Why do these molecules have different shapes?

10.18 Compare the Lewis structures of CH_4 and H_2O. Why do these molecules have similar bond angles but different molecular shapes?

10.19 Use VSEPR theory to predict the shape of each of the following:

     a. GaH_3 **b.** OF_2 **c.** HCN   **d.** CCl_4

10.20 Use VSEPR theory to predict the shape of each of the following:

     a. CF_4 **b.** NCl_3 **c.** $SeBr_2$ **d.** CS_2

10.21 Draw the Lewis structure and predict the shape for each of the following:

     a. AlH_4^- **b.** SO_4^{2-} **c.** NH_4^+ **d.** NO_2^+

10.22 Draw the Lewis structure and predict the shape for each of the following:

     a. NO_2^- **b.** PO_4^{3-} **c.** ClO_4^- **d.** SF_3^+

10.4 Electronegativity and Bond Polarity

LEARNING GOAL Use electronegativity to determine the polarity of a bond.

We can learn more about the chemistry of compounds by looking at how bonding electrons are shared between atoms. The bonding electrons are shared equally in a bond between identical nonmetal atoms. However, when a bond is between atoms of different elements, the electron pairs are usually shared unequally. Then the shared pairs of electrons are attracted to one atom in the bond more than the other.

The **electronegativity** of an atom is its ability to attract the shared electrons in a chemical bond. Nonmetals have higher electronegativities than do metals, because nonmetals have a greater attraction for electrons than metals. On the electronegativity scale, the nonmetal fluorine, located in the upper right corner of the periodic table, was assigned a value of 4.0, and the electronegativities for all other elements were determined relative to the attraction of fluorine for shared electrons. The metal cesium, which has the lowest electronegativity (0.7), is located in the lower left corner of the periodic table. The electronegativities for the representative elements are shown in **FIGURE 10.2**. Note that there are no electronegativity values for the noble gases because they do not typically form bonds. The electronegativity values for transition elements are also low, but we have not included them in our discussion.

FIGURE 10.2 The electronegativity values of representative elements in Group 1A (1) to Group 7A (17), which indicate the ability of atoms to attract shared electrons, increase going across a period from left to right and decrease going down a group.

Polarity of Bonds

The difference in the electronegativity values of two atoms can be used to predict the type of chemical bond, ionic or covalent, that forms. For the H—H bond, the electronegativity difference is zero (2.1 − 2.1 = 0), which means the bonding electrons are shared equally. We illustrate this by drawing a symmetrical electron cloud around the H atoms. A bond between atoms with identical or very similar electronegativity values is a **nonpolar covalent bond**. However, when covalent bonds are between atoms with different electronegativity values, the electrons are shared unequally; the bond is a **polar covalent bond**. The electron cloud for a polar covalent bond is unsymmetrical. For the H—Cl bond, there is an electronegativity difference of 0.9 (3.0 − 2.1 = 0.9), which means that the H—Cl bond is polar covalent (see **FIGURE 10.3**). When finding the electronegativity difference, the smaller electronegativity is always subtracted from the larger; thus, the difference is always a positive number.

The **polarity** of a bond depends on the difference in the electronegativity values of its atoms. In a polar covalent bond, the shared electrons are attracted to the more electronegative atom, which makes it partially negative, because of the negatively charged electrons around that atom. At the other end of the bond, the atom with the lower electronegativity becomes partially positive because of the lack of electrons at that atom.

A bond becomes more *polar* as the electronegativity difference increases. A polar covalent bond that has a separation of charges is called a **dipole**. The positive and negative ends of the dipole are indicated by the lowercase Greek letter delta with a positive or negative sign, δ^+ and δ^-. Sometimes we use an arrow that points from the positive charge to the negative charge ⟼ to indicate the dipole.

$$C^{\delta+}\!\!-\!\!O^{\delta-} \quad N^{\delta+}\!\!-\!\!O^{\delta-} \quad Cl^{\delta+}\!\!-\!\!F^{\delta-}$$

Dipoles occur in polar covalent bonds containing N, O, or F.

H—H	$H^{\delta+}$—$Cl^{\delta-}$
Equal sharing of electrons in a nonpolar covalent bond	Unequal sharing of electrons in a polar covalent bond

FIGURE 10.3 In the nonpolar covalent bond of H_2, electrons are shared equally. In the polar covalent bond of HCl, electrons are shared unequally.

Variations in Bonding

The variations in bonding are continuous; there is no definite point at which one type of bond stops and the next starts. When the electronegativity difference is between 0.0 and 0.4, the electrons are considered to be shared equally in a *nonpolar covalent bond*. For example, the C—C bond (2.5 − 2.5 = 0.0) and the C—H bond (2.5 − 2.1 = 0.4) are classified as nonpolar covalent bonds.

As the electronegativity difference increases, the shared electrons are attracted more strongly to the more electronegative atom, which increases the polarity of the bond. When the electronegativity difference is from 0.5 to 1.8, the bond is a *polar covalent bond*. For example, the O—H bond (3.5 − 2.1 = 1.4) is classified as a polar covalent bond (see **TABLE 10.7**).

ENGAGE 10.9

Use electronegativity differences to explain why a Si—S bond is polar covalent and a Si—P bond is nonpolar covalent.

TABLE 10.7 Electronegativity Differences and Types of Bonds			
Electronegativity Difference	0.0 to 0.4	0.5 to 1.8	1.9 to 3.3
Bond Type	Nonpolar covalent	Polar covalent	Ionic
Electron Bonding	Electrons shared equally	Electrons shared unequally	Electrons transferred
		δ^+ δ^-	$+$ $-$

When the electronegativity difference is greater than 1.8, electrons are transferred from one atom to another, which results in an *ionic bond*. For example, the electronegativity difference for the ionic compound NaCl is 3.0 − 0.9 = 2.1. Thus, for large differences in electronegativity, we would predict an ionic bond (see **TABLE 10.8**).

PRACTICE PROBLEMS

Try Practice Problems 10.25 and 10.26

TABLE 10.8 Predicting Bond Type from Electronegativity Differences			
Formula	**Bond**	**Electronegativity Difference***	**Type of Bond**
H_2	H—H	2.1 − 2.1 = 0.0	Nonpolar covalent
BrCl	Br—Cl	3.0 − 2.8 = 0.2	Nonpolar covalent
HBr	$H^{\delta+}$—$Br^{\delta-}$	2.8 − 2.1 = 0.7	Polar covalent
HCl	$H^{\delta+}$—$Cl^{\delta-}$	3.0 − 2.1 = 0.9	Polar covalent
NaCl	Na^+Cl^-	3.0 − 0.9 = 2.1	Ionic
MgO	$Mg^{2+}O^{2-}$	3.5 − 1.2 = 2.3	Ionic
*Values are taken from Figure 10.2.			

SAMPLE PROBLEM 10.8   Bond Polarity

TRY IT FIRST

Using electronegativity values, classify the bond between each of the following pairs of atoms as nonpolar covalent, polar covalent, or ionic and label any polar covalent bond with δ^+ and δ^- and show the direction of the dipole:

a. K and O

b. As and Cl

c. N and N

d. P and Br

SOLUTION

ANALYZE THE PROBLEM	Given	Need	Connect
	pairs of atoms	type of bond	electronegativity values

For each pair of atoms, we obtain the electronegativity values and calculate the difference in electronegativity.

Atoms	Electronegativity Difference	Type of Bond	Dipole
a. K and O	$3.5 - 0.8 = 2.7$	Ionic	
b. As and Cl	$3.0 - 2.0 = 1.0$	Polar covalent	$\overset{\delta^+}{As}{-}\overset{\delta^-}{Cl}$ $\longmapsto$
c. N and N	$3.0 - 3.0 = 0.0$	Nonpolar covalent	
d. P and Br	$2.8 - 2.1 = 0.7$	Polar covalent	$\overset{\delta^+}{P}{-}\overset{\delta^-}{Br}$ $\longmapsto$

SELF TEST 10.8

Using electronegativity values, classify the bond between each of the following pairs of atoms as nonpolar covalent, polar covalent, or ionic and label any polar covalent bond with δ^+ and δ^- and show the direction of the dipole:

a. P and Cl

b. Br and Br

c. Na and O

d. O and I

ANSWER

a. polar covalent (0.9) $\overset{\delta^+}{P}{-}\overset{\delta^-}{Cl}$ $\longmapsto$

b. nonpolar covalent (0.0)

c. ionic (2.6)

d. polar covalent (1.0) $\overset{\delta^-}{O}{-}\overset{\delta^+}{I}$ $\longleftarrow\!\dashv$

PRACTICE PROBLEMS

Try Practice Problems 10.27 to 10.32

PRACTICE PROBLEMS

10.4 Electronegativity and Bond Polarity

10.23 Describe the trend in electronegativity as *increases* or *decreases* for each of the following:

a. from B to F

b. from Mg to Ba

c. from F to I

10.24 Describe the trend in electronegativity as *increases* or *decreases* for each of the following:

a. from Al to Cl

b. from Br to K

c. from Li to Cs

10.25 Which electronegativity difference (**a, b,** or **c**) would you expect for a nonpolar covalent bond?

a. from 0.0 to 0.4

b. from 0.5 to 1.8

c. from 1.9 to 3.3

10.26 Which electronegativity difference (**a, b,** or **c**) would you expect for a polar covalent bond?

a. from 0.0 to 0.4

b. from 0.5 to 1.8

c. from 1.9 to 3.3

10.27 Using the periodic table, arrange the atoms in each of the following sets in order of increasing electronegativity:

a. Li, Na, K

b. Na, Cl, P

c. Se, Ca, O

10.28 Using the periodic table, arrange the atoms in each of the following sets in order of increasing electronegativity:

a. Cl, F, Br

b. B, O, N

c. Mg, F, S

10.29 Predict whether the bond between each of the following pairs of atoms is nonpolar covalent, polar covalent, or ionic:
  a. Si and Br   **b.** Li and F
  c. Br and F   **d.** I and I
  e. N and P   **f.** C and P

10.30 Predict whether the bond between each of the following pairs of atoms is nonpolar covalent, polar covalent, or ionic:
  a. Si and O   **b.** K and Cl
  c. S and F   **d.** P and Br
  e. Li and O   **f.** N and S

10.31 For the bond between each of the following pairs of atoms, indicate the positive end with δ^+ and the negative end with δ^-. Draw an arrow to show the dipole for each.
  a. N and F   **b.** Si and Br   **c.** C and O
  d. P and Br   **e.** N and P

10.32 For the bond between each of the following pairs of atoms, indicate the positive end with δ^+ and the negative end with δ^-. Draw an arrow to show the dipole for each.
  a. P and Cl   **b.** Se and F   **c.** Br and F
  d. N and H   **e.** B and Cl

10.5 Polarity of Molecules

LEARNING GOAL Use the three-dimensional structure of a molecule to classify it as polar or nonpolar.

We have seen that covalent bonds in molecules can be polar or nonpolar. Now we will look at how the bonds in a molecule and its shape determine whether that molecule is classified as polar or nonpolar.

Identifying Polarity of Molecules

Nonpolar Molecules

In a **nonpolar molecule**, all the bonds are nonpolar or the polar bonds cancel each other out. Molecules such as H_2, Cl_2, and CH_4 are nonpolar because they contain only nonpolar covalent bonds. A *nonpolar molecule* also occurs when polar bonds (dipoles) cancel each other because they are in a symmetrical arrangement. For example, CO_2, a linear molecule, contains two equal polar covalent bonds whose dipoles point in opposite directions. As a result, the dipoles cancel, which makes a CO_2 molecule nonpolar.

Another example of a nonpolar molecule is the CCl_4 molecule, which has four polar bonds symmetrically arranged around the central C atom. Each of the C—Cl bonds has the same polarity, but because they have a tetrahedral arrangement, their opposing dipoles cancel. As a result, a molecule of CCl_4 is nonpolar.

The two C—O dipoles cancel; CO_2 is nonpolar.

The four C—Cl dipoles cancel; CCl_4 is nonpolar.

Polar Molecules

In a **polar molecule**, one end of the molecule is more negatively charged than the other end. Polarity in a molecule occurs when the dipoles from the individual polar bonds do not cancel each other. For example, HCl is a polar molecule because it has one covalent bond that is polar.

In molecules with two or more electron groups, the shape, such as bent or trigonal pyramidal, determines whether the dipoles cancel. For example, we have seen that H_2O has a bent shape. Thus, a water molecule is polar because the individual dipoles do not cancel.

A single dipole does not cancel; HCl is polar.

More negative end of molecule

More positive end of molecule

The dipoles do not cancel; H_2O is polar.

The NH_3 molecule has a tetrahedral electron-group geometry with three bonded atoms, which gives it a trigonal pyramidal shape. Thus, the NH_3 molecule is polar because the individual N—H dipoles do not cancel.

More negative end of molecule

More positive end of molecule

The dipoles do not cancel; NH_3 is polar.

The dipole does not cancel; CH_3F is polar.

In the molecule CH_3F, the C—F bond is polar covalent, but the three C—H bonds are nonpolar covalent. Because there is only one dipole in CH_3F, which does not cancel, CH_3F is a polar molecule.

SAMPLE PROBLEM 10.9 Polarity of Molecules

TRY IT FIRST

Determine whether a molecule of OF_2 is polar or nonpolar.

SOLUTION

ANALYZE THE PROBLEM	Given	Need	Connect
	OF_2	polarity	Lewis structure, bond polarity

STEP 1 **Determine if the bonds are polar covalent or nonpolar covalent.** From Figure 10.2, F and O have an electronegativity difference of 0.5 ($4.0 - 3.5 = 0.5$), which makes each of the O—F bonds polar covalent.

STEP 2 **If the bonds are polar covalent, draw the Lewis structure, determine the molecular shape, and determine if the dipoles cancel.** The Lewis structure for OF_2 has four electron groups and two bonded atoms. The molecule has a bent shape in which the dipoles of the O—F bonds do not cancel. The OF_2 molecule would be polar.

The dipoles do not cancel; OF_2 is polar.

SELF TEST 10.9

Determine if each of the following would be polar or nonpolar:

a. PCl_3 **b.** Cl_2

ANSWER

a. polar   **b.** nonpolar

PRACTICE PROBLEMS

Try Practice Problems 10.33 to 10.38

PRACTICE PROBLEMS

10.5 Polarity of Molecules

10.33 Why is F_2 a nonpolar molecule, but HF is a polar molecule?

10.34 Why is CCl_4 a nonpolar molecule, but PCl_3 is a polar molecule?

10.35 Identify each of the following molecules as polar or nonpolar:
    a. CS_2 **b.** NF_3 **c.** CHF_3 **d.** SO_3

10.36 Identify each of the following molecules as polar or nonpolar:
    a. SeF_2 **b.** PBr_3 **c.** SiF_4 **d.** SO_2

10.37 The molecule CO_2 is nonpolar, but CO is a polar molecule. Explain.

10.38 The molecules CH_4 and CH_3Cl both have tetrahedral shapes. Why is CH_4 nonpolar whereas CH_3Cl is polar?

10.6 Intermolecular Forces Between Atoms or Molecules

LEARNING GOAL Describe the intermolecular forces between ions, polar covalent molecules, and nonpolar covalent molecules.

CORE CHEMISTRY SKILL

Identifying Intermolecular Forces

The forces that keep the atoms or ions together in a compound are called *intramolecular forces*. In ionic compounds, we have seen how large differences in electronegativity produce positive and negative ions that are attracted to each other. Ionic bonds are the strongest of the forces in compounds. In molecules, we have seen that covalent bonds form by atoms sharing electrons. Ionic bonds and covalent bonds are involved in the shapes and polarity of compounds.

In solids and liquids, there are also *intermolecular forces* such as *dipole–dipole attractions*, *hydrogen bonding*, and *dispersion forces* between covalent molecules, which hold them close together. They are much weaker than the intramolecular forces, but important in the physical properties such as melting and boiling point.

Dipole–dipole
attraction

Dipole–Dipole Attractions

For polar molecules, intermolecular forces called **dipole–dipole attractions** occur between the positive end of one molecule and the negative end of another. For a polar molecule with a dipole such as HCl, the partially positive H atom of one HCl molecule attracts the partially negative Cl atom in another HCl molecule.

Hydrogen bond

Hydrogen Bonds

Polar molecules containing hydrogen atoms bonded to highly electronegative atoms of nitrogen, oxygen, or fluorine form especially strong dipole–dipole attractions. This type of intermolecular force, called a **hydrogen bond**, occurs between the partially positive hydrogen atom in one molecule and the partially negative nitrogen, oxygen, or fluorine atom in another molecule. Hydrogen bonds are the strongest type of intermolecular force between polar covalent molecules. They are a major factor in the formation and structure of biological molecules such as proteins and DNA.

Hydrogen bond

Dispersion Forces

Very weak attractions called **dispersion forces** are the only intermolecular forces that occur between nonpolar molecules. Usually, the electrons in a nonpolar covalent molecule are distributed symmetrically. However, the movement of the electrons may place more of them in one part of the molecule than another, which forms a *temporary dipole*. These momentary dipoles align the molecules so that the positive end of one molecule is attracted to the negative end of another molecule. Although dispersion forces are very weak, they make it possible for nonpolar molecules to form liquids and solids.

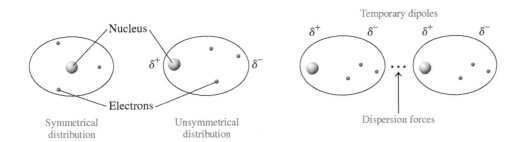

Nonpolar covalent molecules have weak attractions when they form temporary dipoles.

TABLE 10.9 compares the intramolecular and intermolecular types of bonding.

Intermolecular Forces and Melting Points

The melting point of a substance is related to the strength of the intermolecular forces between its particles. A compound with weak intermolecular forces, such as dispersion forces, has a low melting point because only a small amount of energy is needed to separate the molecules and form a liquid. A compound with dipole–dipole attractions requires more energy to break the intermolecular forces between the molecules. A compound with hydrogen bonds requires even more energy to overcome the intermolecular forces that exist between its molecules. **TABLE 10.10** compares the melting points of some substances.

TABLE 10.9 Comparison of Bonding Intermolecular Forces

Type of Force	Particle Arrangement	Example	Strength
Between Atoms or Ions (Intramolecular)			Strong
Ionic bonds		$Na^+ Cl^-$	
Covalent bonds (X = nonmetal)	X : X	$Cl-Cl$	
Between Molecules (Intermolecular)			
Hydrogen bonds (X = N, O, or F)	$\delta^+ \ \delta^-$ $\delta^+ \ \delta^-$ H X $\cdots$ H X	$H^{\delta^+}-F^{\delta^-} \cdots H^{\delta^+}-F^{\delta^-}$	
Dipole–dipole attractions (X and Y = nonmetals)	$\delta^+ \ \delta^-$ $\delta^+ \ \delta^-$ Y X $\cdots$ Y X	$H^{\delta^+}-Cl^{\delta^-} \cdots H^{\delta^+}-Cl^{\delta^-}$	
Dispersion forces (temporary shift of electrons in nonpolar bonds)	$\delta^+ \ \delta^-$ $\delta^+ \ \delta^-$ (temporary dipoles) X :X $\cdots$ X :X	$F^{\delta^+}-F^{\delta^-} \cdots F^{\delta^+}-F^{\delta^-}$	Weak

TABLE 10.10 Melting Points of Selected Substances

Substance	Melting Point (°C)
Hydrogen Bonds	
H_2O	0
NH_3	−78
Dipole–Dipole Attractions	
HI	−51
HBr	−89
HCl	−115
Dispersion Forces	
Br_2	−7
Cl_2	−101
F_2	−220

Size, Mass, and Melting and Boiling Points

As the size and mass of similar types of molecular compounds increase, there are more electrons that produce stronger temporary dipoles. As the molar mass of similar compounds increases, the dispersion forces also increase due to the increase in the number of electrons. In general, larger nonpolar molecules with increased molar masses also have higher melting and boiling points.

SAMPLE PROBLEM 10.10 Intermolecular Forces Between Particles

TRY IT FIRST

Indicate the major type of intermolecular forces—dipole–dipole attractions, hydrogen bonds, or dispersion forces—expected of each of the following:

a. HF   b. Br_2 c. PCl_3

SOLUTION

a. HF is a polar molecule that interacts with other HF molecules by hydrogen bonding.
b. Br_2 is nonpolar; only dispersion forces provide temporary intermolecular forces.
c. The polarity of PCl_3 molecules provides dipole–dipole attractions.

SELF TEST 10.10

Indicate the major type of intermolecular forces in each of the following:

a. H_2S b. H_2O

ANSWER

a. dipole–dipole attractions   b. hydrogen bonds

ENGAGE 10.10

Why does GeH_4 have a higher boiling point than CH_4?

PRACTICE PROBLEMS

Try Practice Problems 10.39 to 10.44

PRACTICE PROBLEMS

10.6 Intermolecular Forces Between Atoms or Molecules

10.39 Identify the major type of intermolecular forces between the particles of each of the following:
   a. BrF   **b.** H_2O_2 **c.** NF_3 **d.** Cl_2

10.40 Identify the major type of intermolecular forces between the particles of each of the following:
   a. HCl   **b.** Br_2 **c.** PBr_3 **d.** NH_3

10.41 Identify the strongest intermolecular forces between the particles of each of the following:
   a. CH_3OH **b.** CO   **c.** CF_4 **d.** CH_3CH_3

10.42 Identify the strongest intermolecular forces between the particles of each of the following:
   a. O_2 **b.** SiH_4 **c.** CH_3Cl **d.** $(CH_3)_2NH$

10.43 Identify the substance in each of the following pairs that would have the higher boiling point, and explain your choice:
   a. HF or HBr   **b.** HF or SF_2
   c. Br_2 or PBr_3 **d.** CH_4 or CH_3OH

10.44 Identify the substance in each of the following pairs that would have the higher boiling point, and explain your choice:
   a. H_2 or HCl   **b.** H_2O or H_2Se
   c. NH_3 or PH_3 **d.** F_2 or HF

10.7 Changes of State

LEARNING GOAL Describe the changes of state between solids, liquids, and gases; calculate the energy released or absorbed.

The states and properties of gases, liquids, and solids depend on the types of forces between their particles. Matter undergoes a **change of state** when it is converted from one state to another (see **FIGURE 10.4**).

REVIEW

Interpreting Graphs (1.4)
Using the Heat Equation (3.5)

Melting

Freezing

  Heat absorbed
  Heat released

ENGAGE 10.11

Is heat absorbed or released when liquid water freezes?

FIGURE 10.4 Changes of state include melting and freezing, vaporization and condensation, sublimation and deposition.

Solid   + Heat   Liquid

Melting

Freezing

− Heat

Melting and freezing are reversible processes.

Melting and Freezing

When heat is added to a solid, the particles move faster. At a temperature called the **melting point (mp)**, the particles of a solid gain sufficient energy to overcome the intermolecular forces that hold them together. The particles in the solid separate and move about in random patterns. The substance is **melting**, changing from a solid to a liquid.

If the temperature of a liquid is lowered, the reverse process takes place. Kinetic energy is lost, the particles slow down, and intermolecular forces pull the particles close together. The substance is **freezing**. A liquid changes to a solid at the **freezing point (fp)**, which is the same temperature as its melting point. Every substance has its own freezing (melting) point: Solid water (ice) melts at 0 °C when heat is added, and liquid water freezes at 0 °C when heat is removed. Gold melts at 1064 °C when heat is added, and freezes at 1064 °C when heat is removed.

During a change of state, the temperature of a substance remains constant. Suppose we have a glass containing ice and water. The ice melts when heat is added at 0 °C, forming liquid. When heat is removed at 0 °C, the liquid water freezes, forming solid. The process of melting requires heat; the process of freezing releases heat. Melting and freezing are reversible at 0 °C.

Heat of Fusion

During melting, the **heat of fusion** is the energy that must be added to convert exactly 1 g of solid to liquid at the melting point. For example, 334 J of heat is needed to melt exactly 1 g of ice at its melting point (0 °C).

$$H_2O(s) + 334 \text{ J/g} \longrightarrow H_2O(l) \qquad \text{Endothermic}$$

Heat of Fusion for Water

$$\frac{334 \text{ J}}{1 \text{ g } H_2O} \quad \text{and} \quad \frac{1 \text{ g } H_2O}{334 \text{ J}}$$

The heat of fusion is also the quantity of heat that must be removed to freeze exactly 1 g of water at its freezing point (0 °C).

Water is sometimes sprayed in fruit orchards during subfreezing weather. If the air temperature drops to 0 °C, the water begins to freeze. Because heat is released as the water molecules form solid ice, the air warms and protects the fruit.

$$H_2O(l) \longrightarrow H_2O(s) + 334 \text{ J/g} \qquad \text{Exothermic}$$

The heat of fusion can be used as a conversion factor in calculations. For example, to determine the heat needed to melt a sample of ice, the mass of ice, in grams, is multiplied by the heat of fusion.

Calculating Heat to Melt (or Freeze) Water

Heat = mass × heat of fusion

$$J = g \times \frac{334 \text{ J}}{g}$$

Sublimation and Deposition

In a process called **sublimation**, the particles on the surface of a solid change directly to a gas with no temperature change and without going through the liquid state. In the reverse process called **deposition**, gas particles change directly to a solid. For example, dry ice, which is solid carbon dioxide, sublimes at −78 °C. It is called "dry" because it does not form a liquid as it warms. In extremely cold areas, snow does not melt but sublimes directly to water vapor.

When frozen foods are left in the freezer for a long time, so much water sublimes that foods, especially meats, become dry and shrunken, a condition called *freezer burn*. Deposition occurs in a freezer when water vapor forms ice crystals on the surface of freezer bags and frozen food.

Freeze-dried foods prepared by sublimation are convenient for long-term storage and for camping and hiking. A food that has been frozen is placed in a vacuum chamber where it dries as the ice sublimes. The dried food retains all of its nutritional value and needs only

Solid   + Heat   Gas

Sublimation

Deposition

− Heat

Sublimation and deposition are reversible processes.

water to be edible. A food that is freeze-dried does not need refrigeration because bacteria cannot grow without moisture.

Evaporation, Boiling, and Condensation

Water in a mud puddle disappears, unwrapped food dries out, and clothes hung on a line dry. **Evaporation** is taking place as water molecules with sufficient energy escape from the liquid surface and enter the gas phase (see **FIGURE 10.5**). The loss of the "hot" water molecules removes heat, which cools the remaining liquid water. As heat is added, more and more water molecules gain sufficient energy to evaporate. At the **boiling point (bp)**, the molecules within a liquid have enough energy to overcome their intermolecular forces and become a gas. We observe the **boiling** of a liquid such as water as gas bubbles form throughout the liquid, rise to the surface, and escape.

Water vapor will change to solid on contact with a cold surface.

— 50 °C

Evaporation occurs at the surface of a liquid.

— 100 °C: the boiling point of water

Boiling occurs as bubbles of gas form throughout the liquid.

FIGURE 10.5 Evaporation and boiling

When heat is removed, a reverse process takes place. In **condensation**, water vapor is converted to liquid as the water molecules lose kinetic energy and slow down. Condensation occurs at the same temperature as boiling but differs because heat is removed. You may have noticed that condensation occurs when you take a hot shower and the water vapor forms water droplets on a mirror. Because a substance loses heat as it condenses, its surroundings become warmer. That is why, when a rainstorm is approaching, you may notice a warming of the air as gaseous water molecules condense to rain.

Heat of Vaporization and Condensation

The **heat of vaporization** is the energy that must be added to convert exactly 1 g of liquid to gas at its boiling point. For water, 2260 J is needed to convert 1 g of water to vapor at 100 °C.

$$H_2O(l) + 2260 \text{ J/g} \longrightarrow H_2O(g) \qquad \text{Endothermic}$$

This same amount of heat is released when 1 g of water vapor (gas) changes to liquid at 100 °C.

$$H_2O(g) \longrightarrow H_2O(l) + 2260 \text{ J/g} \qquad \text{Exothermic}$$

Therefore, 2260 J/g is also the *heat of condensation* of water.

Vaporization and condensation are reversible processes.

Heat of Vaporization for Water

$$\frac{2260 \text{ J}}{1 \text{ g H}_2\text{O}} \quad \text{and} \quad \frac{1 \text{ g H}_2\text{O}}{2260 \text{ J}}$$

Just as substances have different melting and boiling points, they also have different heats of fusion and vaporization. As seen in **FIGURE 10.6**, the heats of vaporization are larger than the heats of fusion.

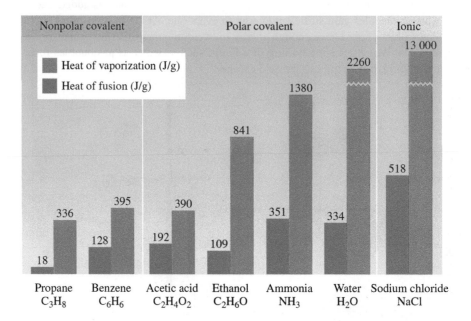

FIGURE 10.6 For any substance, the heat of vaporization is greater than the heat of fusion.

To determine the heat needed to boil a sample of water, the mass, in grams, is multiplied by the heat of vaporization. Because the temperature remains constant during a change of state (melting, freezing, boiling, or condensing), there is no temperature change used in the calculation, as shown in Sample Problem 10.11.

SAMPLE PROBLEM 10.11 Calculating Heat for Changes of State

TRY IT FIRST

In a sauna, 122 g of water is converted to steam at 100 °C. How many kilojoules are needed?

SOLUTION

STEP 1 State the given and needed quantities.

ANALYZE THE PROBLEM	Given	Need	Connect
	122 g of water at 100 °C	kilojoules to convert to steam at 100 °C	heat of vaporization, 2260 J/g

STEP 2 Write a plan to convert the given quantity to the needed quantity.

grams of water → Heat of vaporization → joules → Metric factor → kilojoules

STEP 3 Write the heat conversion factor and any metric factor.

$$1 \text{ g of H}_2\text{O}(l \rightarrow g) = 2260 \text{ J}$$

$$\frac{2260 \text{ J}}{1 \text{ g H}_2\text{O}} \quad \text{and} \quad \frac{1 \text{ g H}_2\text{O}}{2260 \text{ J}}$$

$$1 \text{ kJ} = 1000 \text{ J}$$

$$\frac{1000 \text{ J}}{1 \text{ kJ}} \quad \text{and} \quad \frac{1 \text{ kJ}}{1000 \text{ J}}$$

ENGAGE 10.12

Why is temperature not used in calculations involving changes of state?

STEP 4   **Set up the problem and calculate the needed quantity.**

Three SFs   Exact

$$\text{Heat} = 122 \text{ g H}_2\text{O} \times \frac{2260 \text{ J}}{1 \text{ g H}_2\text{O}} \times \frac{1 \text{ kJ}}{1000 \text{ J}} = 276 \text{ kJ}$$

Three SFs   Exact   Exact   Three SFs

SELF TEST 10.11

a. When steam from a pan of boiling water reaches a cool window, it condenses. How much heat, in kilojoules, is released when 25.0 g of steam condenses at 100 °C?
b. Ice bags are used by sports trainers to treat muscle injuries. If 260. g of ice is placed in an ice bag, how much heat, in kilojoules, will be released when all the ice melts at 0 °C?

ANSWER

a. 56.5 kJ   b. 86.8 kJ

PRACTICE PROBLEMS
Try Practice Problems 10.45 to 10.48

An ice bag is used to treat a sports injury.

Calculating Heat to Condense (or Boil) Water

$$\text{Heat} = \text{mass} \times \text{heat of vaporization}$$

$$\text{J} = \text{g} \times \frac{2260 \text{ J}}{\text{g}}$$

Heating and Cooling Curves

All the changes of state during the heating or cooling of a substance can be illustrated visually. On a **heating curve** or **cooling curve**, the temperature is shown on the vertical axis, and the loss or gain of heat is shown on the horizontal axis.

Steps on a Heating Curve

The first diagonal line indicates that the temperature of the solid increases as heat is added. When the melting temperature is reached, a horizontal line, or plateau, indicates that the solid is melting. As melting takes place, the solid is changing to liquid without any change in temperature (see **FIGURE 10.7**).

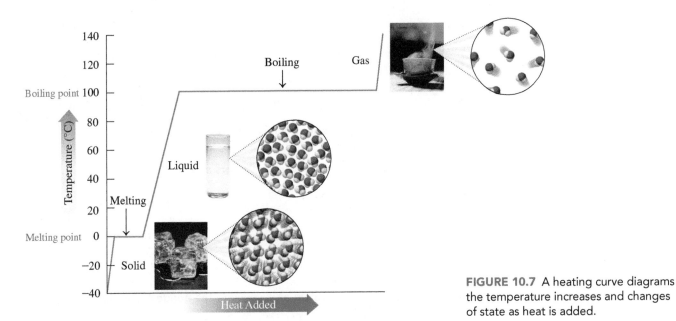

FIGURE 10.7 A heating curve diagrams the temperature increases and changes of state as heat is added.

When all the particles are in the liquid state, adding more heat will increase the temperature of the liquid. This increase is shown as a diagonal line from the melting point temperature to the boiling point temperature. Once the liquid reaches its boiling point, a horizontal line indicates that the liquid changes to gas at constant temperature. Because the heat of vaporization is greater than the heat of fusion, the horizontal line at the boiling point is longer than the line at the melting point. When all the liquid becomes gas, adding more heat increases the temperature of the gas.

Steps on a Cooling Curve

A cooling curve is a diagram in which the temperature decreases as heat is removed. Initially, a diagonal line is drawn to show that heat is removed from a gas until it begins to condense. At the condensation point, a horizontal line indicates a change of state as gas condenses to form a liquid. When all the gas has changed into liquid, further cooling lowers the temperature. The decrease in temperature is shown as a diagonal line from the condensation point to the freezing point, where another horizontal line indicates that liquid is changing to solid. When all the substance is frozen, the removal of more heat decreases the temperature of the solid below its freezing point, which is shown as a diagonal line.

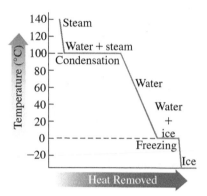

A cooling curve for water illustrates the temperature decreases and changes of state as heat is removed.

Combining Energy Calculations

Up to now, we have calculated one step in a heating or cooling curve. However, a problem may require a combination of steps that include a temperature change as well as a change of state. The heat is calculated for each step separately, and the results are added together to give the total energy, as seen in Sample Problem 10.12.

SAMPLE PROBLEM 10.12   Combining Heat Calculations

TRY IT FIRST

Charles has increased his activity by doing more exercise. After a session of using weights, he has a sore arm. An ice bag is filled with 125 g of ice at 0 °C. How much heat, in kilojoules, is absorbed to melt the ice and raise the temperature of the water to body temperature, 37.0 °C?

SOLUTION

STEP 1   State the given and needed quantities.

	Given	Need	Connect
ANALYZE THE PROBLEM	125 g of ice at 0 °C	total kilojoules to melt ice at 0 °C and to raise temperature of water to 37.0 °C	combine heat from change of state (heat of fusion) and temperature change (specific heat of water)

STEP 2   Write a plan to convert the given quantity to the needed quantity.

Total heat = kilojoules needed to melt the ice at 0 °C and heat the water from 0 °C to 37.0 °C

STEP 3   Write the heat conversion factors and any metric factor.

$$1 \text{ g of } H_2O \ (s \rightarrow l) = 334 \text{ J}$$

$$\frac{334 \text{ J}}{1 \text{ g } H_2O} \quad \text{and} \quad \frac{1 \text{ g } H_2O}{334 \text{ J}}$$

$$SH_{water} = \frac{4.184 \text{ J}}{\text{g °C}}$$

$$\frac{4.184 \text{ J}}{\text{g °C}} \quad \text{and} \quad \frac{\text{g °C}}{4.184 \text{ J}}$$

$$1 \text{ kJ} = 1000 \text{ J}$$

$$\frac{1000 \text{ J}}{1 \text{ kJ}} \quad \text{and} \quad \frac{1 \text{ kJ}}{1000 \text{ J}}$$

STEP 4   Set up the problem and calculate the needed quantity.

$$\Delta T = 37.0 \text{ °C} - 0 \text{ °C} = 37.0 \text{ °C}$$

Heat needed to change ice (solid) to water (liquid) at 0 °C:

$$\text{Heat} \;=\; \underset{\text{Three SFs}}{125 \; \text{g ice}} \;\times\; \underset{\text{Exact}}{\frac{334 \; J}{1 \; \text{g ice}}} \;\times\; \underset{\text{Exact}}{\frac{1 \; kJ}{1000 \; J}} \;=\; \underset{\text{Three SFs}}{41.8 \; kJ}$$

Heat needed to warm water (liquid) from 0 °C to water (liquid) at 37.0 °C:

$$\text{Heat} \;=\; \underset{\text{Three SFs}}{125 \; g} \;\times\; \underset{\text{Three SFs}}{37.0 \; °C} \;\times\; \underset{\text{Exact}}{\frac{4.184 \; J}{g \; °C}} \;\times\; \underset{\text{Exact}}{\frac{1 \; kJ}{1000 \; J}} \;=\; \underset{\text{Three SFs}}{19.4 \; kJ}$$

Calculate the total heat:

Melting ice at 0 °C	41.8 kJ
Heating water (0 °C to 37.0 °C)	19.4 kJ
Total heat needed	61.2 kJ

SELF TEST 10.12

a. How many kilojoules are released when 75.0 g of steam at 100 °C condenses, cools to 0 °C, and freezes at 0 °C? (*Hint*: The solution will require three energy calculations.)

b. How many kilocalories are needed to convert 18 g of ice at 0 °C to steam at 100 °C? (*Hint*: The solution will require three energy calculations.)

ANSWER

a. 226 kJ   **b.** 13 kcal

PRACTICE PROBLEMS

Try Practice Problems 10.49 to 10.52

Chemistry Link to Health

Steam Burns

Hot water at 100 °C will cause burns and damage to the skin. However, getting steam on the skin is even more dangerous. If 25 g of hot water at 100 °C falls on a person's skin, the temperature of the water will drop to body temperature, 37 °C. The heat released during cooling can cause severe burns. The amount of heat can be calculated from the mass, the temperature change, 100 °C − 37 °C = 63 °C, and the specific heat of water, 4.184 J/g °C.

$$25 \; g \times 63 \; °C \times \frac{4.184 \; J}{g \; °C} = 6600 \; J \text{ of heat released}$$
when water cools

The condensation of the same quantity of steam to liquid at 100 °C releases much more heat. The heat released when steam condenses can be calculated using the heat of vaporization, 2260 J/g.

$$25 \; g \times \frac{2260 \; J}{1 \; g} = 57\,000 \; J \text{ released when water (gas)}$$
condenses to water liquid) at 100 °C

The total heat released is calculated by combining the heat from the condensation at 100 °C and the heat from cooling of the steam from 100 °C to 37 °C (body temperature). We can see that most of the heat is from the condensation of steam.

Condensation (100 °C)	= 57 000 J
Cooling (100 °C to 37 °C)	= 6 600 J
Heat released	= 64 000 J (rounded off to thousands place)

The amount of heat released from steam is almost ten times greater than the heat from the same amount of hot water. This large amount of heat released on the skin is what causes the extensive damage from steam burns.

When 1 g of steam condenses, 2260 J or 540 cal is released.

PRACTICE PROBLEMS

10.7 Changes of State

10.45 Using Figure 10.6, calculate the heat change needed for each of the following at the melting/freezing point:
a. joules to melt 65.0 g of ice
b. joules to melt 17.0 g of benzene
c. kilojoules to freeze 225 g of benzene
d. kilojoules to freeze 0.0500 kg of water

10.46 Using Figure 10.6, calculate the heat change needed for each of the following at the melting/freezing point:
a. joules to freeze 35.2 g of acetic acid
b. joules to freeze 275 g of water
c. kilojoules to melt 145 g of ammonia
d. kilojoules to melt 5.00 kg of ice

10.47 Using Figure 10.6, calculate the heat change needed for each of the following at the boiling/condensation point:
a. joules to vaporize 10.0 g of water
b. kilojoules to vaporize 50.0 g of ethanol
c. joules to condense 8.00 g of acetic acid
d. kilojoules to condense 0.175 kg of ammonia

10.48 Using Figure 10.6, calculate the heat change needed for each of the following at the boiling/condensation point:
a. joules to condense 10.0 g of steam
b. kilojoules to condense 76.0 g of acetic acid
c. joules to vaporize 44.0 g of ammonia
d. kilojoules to vaporize 5.0 kg of water

10.49 Using Figure 10.6 and the specific heat of water, 4.184 J/g °C, calculate the total amount of heat for each of the following:
a. joules released when 2.00 g of water cools from 22.0 °C to 0 °C and freezes
b. joules needed to melt 50.0 g of ice at 0 °C and to warm the liquid to 65.0 °C

c. kilojoules released when 15.0 g of steam condenses at 100 °C and the liquid cools to 0 °C
d. kilojoules needed to melt 24.0 g of ice at 0 °C, warm the liquid to 100 °C, and change it to steam at 100 °C

10.50 Using Figure 10.6 and the specific heat of water, 4.184 J/g °C, calculate the total amount of heat for each of the following:
a. joules to condense 125 g of steam at 100 °C and to cool the liquid to 150 °C
b. joules needed to melt a 525-g ice sculpture at 0 °C and to warm the liquid to 15.0 °C
c. kilojoules released when 85.0 g of steam condenses at 100 °C, the liquid cools to 0 °C, and freezes
d. joules to warm 55.0 mL of water (density = 1.00 g/mL) from 10.0 °C to 100 °C and vaporize it at 100 °C

Applications

10.51 An ice bag containing 275 g of ice at 0 °C was used to treat sore muscles. When the bag was removed, the ice had melted and the liquid water had a temperature of 24.0 °C. How many kilojoules of heat were absorbed?

10.52 A patient arrives in the emergency room with a burn caused by steam. Calculate the heat, in kilojoules, that is released when 18.0 g of steam at 100 °C hits the skin, condenses, and cools to body temperature of 37.0 °C.

UPDATE Histologist Stains Tissue with Dye

When Lisa receives the sample of tissue from Bill's surgery, she places it in liquid nitrogen, which has a heat of vaporization of 198 J/g at the boiling point, which is −196 °C. When the tissue is frozen, Lisa places the tissue on a microtome, cuts thin slices, and mounts each on a glass slide. One of the stains that Lisa uses to allow the pathologist to see any abnormalities is Eosin, which has a formula, $C_{20}H_6Br_4Na_2O_5$. In preparing to stain the tissues, Lisa uses solutions of lithium carbonate and sodium bicarbonate.

The pathologist examines the tissue sample and reports back to the surgeon that no abnormal cells are present in the margin tissue. No further tissue removal is necessary.

Applications

10.53 At the boiling point of nitrogen, N_2, how many joules are needed to vaporize 15.8 g of N_2?

10.54 In the preparation of liquid nitrogen, how many kilojoules must be removed from 112 g of N_2 gas at its boiling point to form liquid N_2?

10.55 Using the electronegativity values in Figure 10.2, for each of the following bonds found in Eosin calculate the electronegativity difference and predict if the bond is nonpolar covalent, polar covalent, or ionic:
a. C and C   b. C and H   c. C and O

10.56 Using the electronegativity values in Figure 10.2, for each of the following bonds found in Eosin calculate the electronegativity difference and predict if the bond is nonpolar covalent, polar covalent, or ionic:
a. O and H   b. Na and O   c. C and Br

10.57 a. Draw three resonance structures for carbonate (CO_3^{2-}).
b. Determine the shape of the carbonate ion.

10.58 a. Draw two resonance structures for bicarbonate (HCO_3^-) with a central C atom and an H atom attached to one of the O atoms.
b. Determine the shape of the bicarbonate ion.

CONCEPT MAP

BONDING AND PROPERTIES OF COMPOUNDS

Ionic Compounds — contain → **Ionic Bonds**

Molecular Compounds — have → **Intermolecular Forces** — between → **Molecules** — that are → **Dipole–Dipole Attractions, Hydrogen Bonds, or Dispersion Forces**

Molecular Compounds — contain → **Covalent Bonds** — between → **Nonmetals** — can be → **Polar** / **Nonpolar**

Nonmetals — use → **VSEPR Theory** / **Electronegativity** — to determine → **Shape** / **Polarity**

Covalent Bonds — can be → **Single or Multiple Bonds**

Molecular Compounds — have → **Lewis Structures** — can be → **Single or Multiple Bonds**

States of Matter — are → **Solids, Liquids, Gases** — undergo → **Changes of State**

Changes of State — add/lose heat → **Melting/Freezing**, **Vaporization/Condensation**

Changes of State — are drawn as → **Heating/Cooling Curves**

CHAPTER REVIEW

10.1 Lewis Structures for Molecules and Polyatomic Ions

LEARNING GOAL Draw the Lewis structures for molecular compounds and polyatomic ions with single and multiple bonds.

- The Lewis symbol gives the symbol of the element with the valence electrons shown as dots around the symbol.
- The total number of valence electrons is determined for all the atoms in the molecule or polyatomic ion.
- Any negative charge is added to the total valence electrons, whereas any positive charge is subtracted.
- In the Lewis structure, a bonding pair of electrons is placed between the central atom and each of the attached atoms.
- Any remaining valence electrons are used as lone pairs to complete the octets of the surrounding atoms and then the central atom.
- If octets are not completed, one or more lone pairs of electrons are placed as bonding pairs forming double or triple bonds.

10.2 Resonance Structures

LEARNING GOAL Draw Lewis structures for molecules or polyatomic ions that have two or more resonance structures.

- Resonance structures can be drawn when there are two or more equivalent Lewis structures for a molecule or ion with multiple bonds.

10.3 Shapes of Molecules and Polyatomic Ions (VSEPR Theory)

LEARNING GOAL Predict the three-dimensional structure of a molecule or a polyatomic ion.

Tetrahedral shape

- The shape of a molecule is determined from the Lewis structure, the electron-group geometry, and the number of bonded atoms.
- The electron-group geometry around a central atom with two electron groups is linear; with three electron groups, the geometry is trigonal planar; and with four electron groups, the geometry is tetrahedral.
- When all the electron groups are bonded to atoms, the shape has the same name as the electron arrangement.
- A central atom with three electron groups and two bonded atoms has a bent shape, 120°.
- A central atom with four electron groups and three bonded atoms has a trigonal pyramidal shape.
- A central atom with four electron groups and two bonded atoms has a bent shape, 109°.

10.4 Electronegativity and Bond Polarity

LEARNING GOAL Use electronegativity to determine the polarity of a bond.

- Electronegativity is the ability of an atom to attract the electrons it shares with another atom. In general, the

electronegativities of metals are low, whereas nonmetals have high electronegativities.
- In a nonpolar covalent bond, atoms share electrons equally.
- In a polar covalent bond, the electrons are unequally shared because they are attracted to the more electronegative atom.
- In ionic bonds, atoms have large differences in electronegativites.
- The atom in a polar bond with the lower electronegativity is partially positive (δ^+), and the atom with the higher electronegativity is partially negative (δ^-).

10.5 Polarity of Molecules

LEARNING GOAL Use the three-dimensional structure of a molecule to classify it as polar or nonpolar.

- Nonpolar molecules contain nonpolar covalent bonds or have an arrangement of bonded atoms that causes the dipoles to cancel.
- In polar molecules, the dipoles do not cancel.

10.6 Intermolecular Forces Between Atoms or Molecules

LEARNING GOAL Describe the intermolecular forces between ions, polar covalent molecules, and nonpolar covalent molecules.

Dipole–dipole attraction

- In ionic solids, oppositely charged ions are held in a rigid structure by ionic bonds.
- Intermolecular forces called dipole–dipole attractions and hydrogen bonds hold the solid and liquid states of polar molecular compounds together.

- Nonpolar compounds form solids and liquids by weak intermolecular forces between temporary dipoles called dispersion forces.

10.7 Changes of State

LEARNING GOAL Describe the changes of state between solids, liquids, and gases; calculate the energy released or absorbed.

- Melting occurs when the particles in a solid absorb enough energy to break apart and form a liquid.
- The amount of energy required to convert exactly 1 g of solid to liquid is called the heat of fusion.
- For water, 334 J is needed to melt 1 g of ice or must be removed to freeze 1 g of water.
- Sublimation is a process whereby a solid changes directly to a gas.
- Evaporation occurs when particles in a liquid state absorb enough energy to break apart and form gaseous particles.
- Boiling is the vaporization of liquid at its boiling point. The heat of vaporization is the amount of heat needed to convert exactly 1 g of liquid to vapor.
- For water, 2260 J is needed to vaporize 1 g of water or must be removed to condense 1 g of steam.
- A heating or cooling curve illustrates the changes in temperature and state as heat is added to or removed from a substance. Plateaus on the graph indicate changes of state.
- The total heat absorbed or removed from a substance undergoing temperature changes and changes of state is the sum of energy calculations for changes of state and changes in temperature.

KEY TERMS

bent The shape of a molecule with two bonded atoms and one lone pair or two lone pairs.

boiling The formation of bubbles of gas throughout a liquid.

boiling point (bp) The temperature at which a liquid changes to gas (boils) and gas changes to liquid (condenses).

change of state The transformation of one state of matter to another, for example, solid to liquid, liquid to solid, liquid to gas.

condensation The change of state from a gas to a liquid.

cooling curve A diagram that illustrates temperature changes and changes of state for a substance as heat is removed.

deposition The change of a gas directly to a solid; the reverse of sublimation.

dipole The separation of positive and negative charge in a polar bond indicated by an arrow that is drawn from the more positive atom to the more negative atom.

dipole–dipole attractions Intermolecular forces between oppositely charged ends of polar molecules.

dispersion forces Weak dipole bonding that results from a momentary polarization of nonpolar molecules.

double bond A sharing of two pairs of electrons by two atoms.

electronegativity The relative ability of an element to attract electrons in a bond.

evaporation The formation of a gas (vapor) by the escape of high-energy molecules from the surface of a liquid.

freezing A change of state from liquid to solid.

freezing point (fp) The temperature at which a liquid changes to a solid (freezes) and a solid changes to a liquid (melts).

heat of fusion The energy required to melt exactly 1 g of a substance at its melting point. For water, 334 J is needed to melt 1 g of ice; 334 J is released when 1 g of water freezes.

heat of vaporization The energy required to vaporize 1 g of a substance at its boiling point. For water, 2260 J is needed to vaporize exactly 1 g of water; 1 g of steam gives off 2260 J when it condenses.

heating curve A diagram that illustrates the temperature changes and changes of state of a substance as it is heated.

hydrogen bond The attraction between a partially positive H atom and a strongly electronegative atom of N, O, or F.

Lewis structure A structure drawn in which the valence electrons of all the atoms are arranged to give octets except two electrons for hydrogen.

Lewis symbol The representation of an atom that shows the valence electrons as dots placed around the symbol of an element.

linear The shape of a molecule that has two bonded atoms and no lone pairs.

melting The change of state from a solid to a liquid.

melting point (mp) The temperature at which a solid becomes a liquid (melts). It is the same temperature as the freezing point.

nonpolar covalent bond A covalent bond in which the electrons are shared equally between atoms.

nonpolar molecule A molecule that has only nonpolar bonds or in which the bond dipoles cancel.

polar covalent bond A covalent bond in which the electrons are shared unequally between atoms.

polar molecule A molecule containing bond dipoles that do not cancel.

polarity A measure of the unequal sharing of electrons, indicated by the difference in electronegativities.

resonance structures Two or more Lewis structures that can be drawn for a molecule or polyatomic ion by placing a multiple bond between different atoms.

sublimation The change of state in which a solid is transformed directly to a gas without forming a liquid first.

tetrahedral The shape of a molecule with four bonded atoms.

trigonal planar The shape of a molecule with three bonded atoms and no lone pairs.

trigonal pyramidal The shape of a molecule that has three bonded atoms and one lone pair.

triple bond A sharing of three pairs of electrons by two atoms.

valence shell electron-pair repulsion (VSEPR) theory A theory that predicts the shape of a molecule by moving the electron groups on a central atom as far apart as possible to minimize the mutual repulsion of the electrons.

CORE CHEMISTRY SKILLS

The chapter section containing each Core Chemistry Skill is shown in parentheses at the end of each heading.

Drawing Lewis Symbols (10.1)

- The valence electrons are the electrons in the outermost energy level.
- The number of valence electrons is the same as the group number for the representative elements.
- A Lewis symbol represents the number of valence electrons shown as dots placed around the symbol for the element.

Example: Give the group number and number of valence electrons, and draw the Lewis symbol for each of the following:

    a. Rb   **b.** Se   **c.** Xe

Answer: **a.** Group 1A (1), one valence electron, Rb·

      b. Group 6A (16), six valence electrons, ·Se:

      c. Group 8A (18), eight valence electrons, :Xe:

Drawing Lewis Structures (10.1)

- The Lewis structure for a molecule or polyatomic ion shows the sequence of atoms, the bonding pairs of electrons shared between atoms, and the nonbonding or *lone pairs* of electrons.
- Double or triple bonds result when a second or third electron pair is shared between the same atoms to complete octets.

Example: Draw the Lewis structures for CS_2.

Answer: The central atom in the atom arrangement is C.

    S C S

Determine the total number of valence electrons.

$$2\,S \times 6\,e^- = 12\,e^-$$
$$1\,C \times 4\,e^- = \underline{4\,e^-}$$
$$\text{Total} = 16\,e^-$$

Attach each bonded atom to the central atom using a pair of electrons. Two bonding pairs use four electrons.

    S:C:S

Place the 12 remaining electrons as lone pairs around the S atoms.

    :S:C:S:

To complete the octet for C, a lone pair of electrons from each of the S atoms is shared with C, which forms two double bonds.

    :S::C::S: or :S=C=S:

Drawing Resonance Structures (10.2)

- When a molecule or polyatomic ion contains multiple bonds, it may be possible to draw multiple resonance structures.

Example: Draw two resonance structures for NO_2^-.

Answer: The central atom in the atom arrangement is N.

    O N O

Determine the total number of valence electrons.

$$1\,N \times 5\,e^- = 5\,e^-$$
$$2\,O \times 6\,e^- = 12\,e^-$$
$$\text{negative charge} = \underline{1\,e^-}$$
$$\text{Total} = 18\,e^-$$

Use electron pairs to attach each bonded atom to the central atom. Two bonding pairs use four electrons.

    O:N:O

Place the 14 remaining electrons as lone pairs around the O and N atoms.

$$\left[:\ddot{O}:\ddot{N}:\ddot{O}:\right]^-$$

To complete the octet for N, a lone pair from one O atom is shared with N, which forms one double bond. Because there are two O atoms, there are two resonance structures.

$$\left[:\ddot{O}=\ddot{N}-\ddot{O}:\right]^- \longleftrightarrow \left[:\ddot{O}-\ddot{N}=\ddot{O}:\right]^-$$

Predicting Shape (10.3)

- The three-dimensional shape of a molecule or polyatomic ion is determined by drawing a Lewis structure and identifying the number of electron groups (one or more electron pairs) around the central atom and the number of bonded atoms.
- In the valence shell electron-pair repulsion (VSEPR) theory, the electron groups are arranged as far apart as possible around a central atom to minimize the repulsion.
- A central atom with two electron groups bonded to two atoms is linear. A central atom with three electron groups bonded to three atoms is trigonal planar, and to two atoms is bent (120°). A central atom with four electron groups bonded to four atoms is tetrahedral, to three atoms is trigonal pyramidal, and to two atoms is bent (109°).

Example: Predict the shape of NO_2^-.

Answer: Using the resonance structures we drew for NO_2^-, we count three electron groups around the central N atom: one double bond, one single bond, and a lone pair of electrons.

$$\left[:\ddot{O}=\ddot{N}-\ddot{O}:\right]^- \text{ or } \left[:\ddot{O}-\ddot{N}=\ddot{O}:\right]^-$$

The electron-group geometry is trigonal planar, but with the central N atom bonded to two O atoms, the shape is bent, with a bond angle of 120°.

$$\left[\begin{array}{c}\ddot{N}\\ :\ddot{O} \qquad \ddot{O}:\end{array}\right]^-$$

Using Electronegativity (10.4)

- The electronegativity values indicate the ability of atoms to attract shared electrons.
- Electronegativity values increase going across a period from left to right, and decrease going down a group.
- A nonpolar covalent bond occurs between atoms with identical or very similar electronegativity values such that the electronegativity difference is 0.0 to 0.4.
- A polar covalent bond occurs between atoms with differing electronegativity values such that the electronegativity difference is 0.5 to 1.8.
- An ionic bond occurs when the difference in electronegativity for two atoms is greater than 1.8.

Example: Use electronegativity values to classify each of the following bonds as nonpolar covalent, polar covalent, or ionic:

    a. Sr and Cl   **b.** C and S   **c.** O and Br

Answer: **a.** An electronegativity difference of 2.0 (Cl 3.0 − Sr 1.0) makes this an ionic bond.

    b. An electronegativity difference of 0.0 (C 2.5 − S 2.5) makes this a nonpolar covalent bond.

    c. An electronegativity difference of 0.7 (O 3.5 − Br 2.8) makes this a polar covalent bond.

Identifying Polarity of Molecules (10.5)

- A molecule is nonpolar if all of its bonds are nonpolar or it has polar bonds that cancel. CCl_4 is a nonpolar molecule that has four polar bonds that cancel.

- A molecule is polar if it contains polar bonds that do not cancel. H_2O is a polar molecule that has polar bonds that do not cancel.

Example: Predict whether $AsCl_3$ is polar or nonpolar.

Answer: From its Lewis structure, we see that $AsCl_3$ has four electron groups with three bonded atoms.

:Cl:As:Cl:
:Cl:

The shape of a molecule of $AsCl_3$ would be trigonal pyramidal with three polar bonds (As and Cl = 3.0 − 2.0 = 1.0) that do not cancel. Thus, it is a polar molecule.

Identifying Intermolecular Forces (10.6)

- Dipole–dipole attractions occur between the dipoles in polar compounds because the positively charged end of one molecule is attracted to the negatively charged end of another molecule.

$\delta^+ \; \delta^- \qquad \delta^+ \; \delta^-$

H Cl ······ H Cl

Dipole–dipole
attraction

- Strong dipole–dipole attractions called hydrogen bonds occur in compounds in which H is bonded to N, O, or F. The partially positive H atom in one molecule has a strong attraction to the partially negative N, O, or F in another molecule.
- Dispersion forces are very weak intermolecular forces between nonpolar molecules that occur when temporary dipoles form as electrons are unsymmetrically distributed.

Example: Identify the strongest type of intermolecular forces in each of the following:

    a. HF   **b.** F_2 **c.** NF_3

Answer: **a.** HF molecules, which are polar with H bonded to F, have hydrogen bonding.

    b. Nonpolar F_2 molecules have only dispersion forces.

    c. NF_3 molecules, which are polar, have dipole–dipole attractions.

- Substances with hydrogen bonds have higher melting and boiling points than compounds with only dipole–dipole attractions. Substances with only dispersion forces would typically have the lowest melting and boiling points, which increase as molar mass increases.

Example: Identify the compound with the highest boiling point in each of the following:

    a. HI, HBr, HF
    b. F_2, Cl_2, I_2

Answer: **a.** HBr and HI have dipole–dipole attractions, but HF has hydrogen bonding, which gives HF the highest boiling point.

    b. Because F_2, Cl_2, and I_2 have only dispersion forces, I_2 with the greatest molar mass would have the highest boiling point.

Calculating Heat for Change of State (10.7)

- At the melting/freezing point, the heat of fusion is absorbed/released to convert 1 g of a solid to a liquid or 1 g of liquid to a solid.
- For example, 334 J of heat is needed to melt (freeze) exactly 1 g of ice at its melting (freezing) point (0 °C).
- At the boiling/condensation point, the heat of vaporization is absorbed/released to convert exactly 1 g of liquid to gas or 1 g of gas to liquid.
- For example, 2260 J of heat is needed to boil (condense) exactly 1 g of water/steam at its boiling (condensation) point, 100 °C.

Example: How many kilojoules are released when 45.8 g of steam (water) condenses at its boiling (condensation) point?

Answer:
$$45.8 \; \text{g steam} \times \frac{2260 \; J}{1 \; \text{g steam}} \times \frac{1 \; kJ}{1000 \; J} = 104 \; kJ$$

Three SFs   Exact   Three SFs
Three SFs   Exact   Exact   Three SFs

UNDERSTANDING THE CONCEPTS

The chapter sections to review are shown in parentheses at the end of each problem.

10.59 State the number of valence electrons, bonding pairs, and lone pairs in each of the following Lewis structures: (10.1)

   a. H:H   **b.** H:Br: **c.** :Br:Br:

10.60 State the number of valence electrons, bonding pairs, and lone pairs in each of the following Lewis structures: (10.1)

   a. H:Ö: **b.** H:N:H   **c.** :Br:O:Br:

10.61 Match each of the Lewis structures (**a** to **c**) with the correct diagram (**1** to **3**) of its shape, and name the shape; indicate if each molecule is polar or nonpolar. Assume X and Y are non-metals and all bonds are polar covalent. (10.1, 10.3, 10.5)

   a. X—Y—X   **b.** :Y—X   **c.** X—Y—X

10.62 Match each of the formulas (**a** to **c**) with the correct diagram (**1** to **3**) of its shape, and name the shape; indicate if each molecule is polar or nonpolar. (10.1, 10.3, 10.4, 10.5)

   a. PBr_3 **b.** $SiCl_4$ **c.** OF_2

10.63 Consider the following bonds: Ca and O, C and O, K and O, O and O, N and O. (10.4)
   a. Which bonds are polar covalent?
   b. Which bonds are nonpolar covalent?
   c. Which bonds are ionic?
   d. Arrange the covalent bonds in order of decreasing polarity.

10.64 Consider the following bonds: F and Cl, Cl and Cl, Cs and Cl, O and Cl, Ca and Cl. (10.4)
   a. Which bonds are polar covalent?
   b. Which bonds are nonpolar covalent?
   c. Which bonds are ionic?
   d. Arrange the covalent bonds in order of decreasing polarity.

10.65 Identify the major intermolecular forces between each of the following atoms or molecules: (10.6)
   a. PH_3 **b.** NO_2
   c. CH_3NH_2 **d.** Ar

10.66 Identify the major intermolecular forces between each of the following atoms or molecules: (10.6)
   a. He   **b.** HBr
   c. SnH_4 **d.** $CH_3CH_2CH_2OH$

10.67 Use your knowledge of changes of state to explain the following: (10.7)

Perspiration forms on the skin during heavy exercise.

   a. How does perspiration during heavy exercise cool the body?
   b. Why do towels dry more quickly on a hot summer day than on a cold winter day?
   c. Why do wet clothes stay wet in a plastic bag?

10.68 Use your knowledge of changes of state to explain the following: (10.7)

A spray is used to numb a sports injury.

   a. Why is a spray that evaporates quickly, such as ethyl chloride, used to numb a sports injury during a game?
   b. Why does water in a wide, flat, shallow dish evaporate more quickly than the same amount of water in a tall, narrow vase?
   c. Why does a sandwich on a plate dry out faster than a sandwich in plastic wrap?

10.69 Draw a heating curve for a sample of ice that is heated from $-20\,°C$ to $150\,°C$. Indicate the segment of the graph that corresponds to each of the following: (10.7)
   a. solid   **b.** melting   **c.** liquid
   d. boiling   **e.** gas

10.70 Draw a cooling curve for a sample of steam that cools from $110\,°C$ to $-10\,°C$. Indicate the segment of the graph that corresponds to each of the following: (10.7)
   a. solid   **b.** freezing   **c.** liquid
   d. condensing   **e.** gas

10.71 The following is a heating curve for chloroform, a solvent for fats, oils, and waxes: (10.5, 10.7)

a. What is the approximate melting point of chloroform?
b. What is the approximate boiling point of chloroform?
c. On the heating curve, identify the segments **A, B, C, D,** and **E** as solid, liquid, gas, melting, or boiling.
d. At the following temperatures, is chloroform a solid, liquid, or gas? $-80\ °C$; $-40\ °C$; $25\ °C$; $80\ °C$

10.72 Associate the contents of the beakers (**1** to **5**) with segments (**A** to **E**) on the heating curve for water. (10.7)

ADDITIONAL PRACTICE PROBLEMS

10.73 Determine the total number of valence electrons in each of the following: (10.1)
a. HNO_2 b. CH_3CHO c. PH_4^+ d. SO_3^{2-}

10.74 Determine the total number of valence electrons in each of the following: (10.1)
a. $COCl_2$ b. N_2O c. BrO_2^- d. $SeCl_2$

10.75 Draw the Lewis structures for each of the following: (10.1)
a. BF_4^- b. Cl_2O
c. H_2NOH (N is the central atom)   d. H_2CCCl_2

10.76 Draw the Lewis structures for each of the following: (10.1)
a. H_3COCH_3 (the atoms are in the order C O C)
b. HNO_2 (the atoms are in the order H O N O)
c. IO_3^- d. BrO^-

10.77 Draw resonance structures for each of the following: (10.2)
a. N_3^- b. NO_2^+ c. HCO_2^-

10.78 Draw resonance structures for each of the following: (10.2)
a. NO_3^- b. CO_3^{2-} c. SCN^-

10.79 Use the periodic table to arrange the following atoms in order of increasing electronegativity: (10.4)
a. I, F, Cl   b. Li, K, S, Cl   c. Mg, Sr, Ba, Be

10.80 Use the periodic table to arrange the following atoms in order of increasing electronegativity: (10.4)
a. Cl, Br, Se   b. Na, Cs, O, S   c. O, F, B, Li

10.81 Select the more polar bond in each of the following pairs: (10.4)
a. C and N or C and O   b. N and F or N and Br
c. Br and Cl or S and Cl   d. Br and Cl or Br and I
e. N and F or N and O

10.82 Select the more polar bond in each of the following pairs: (10.4)
a. C and C or C and O   b. P and Cl or P and Br
c. Si and S or Si and Cl   d. F and Cl or F and Br
e. P and O or P and S

10.83 Show the dipole arrow for each of the following bonds: (10.4)
a. Si and Cl   b. C and N   c. F and Cl
d. C and F   e. N and O

10.84 Show the dipole arrow for each of the following bonds: (10.4)
a. P and O   b. N and F   c. O and Cl
d. S and Cl   e. P and F

10.85 Calculate the electronegativity difference and classify each of the following bonds as nonpolar covalent, polar covalent, or ionic: (10.4)
a. Si and Cl   b. C and C   c. Na and Cl
d. C and H   e. F and F

10.86 Calculate the electronegativity difference and classify each of the following bonds as nonpolar covalent, polar covalent, or ionic: (10.4)
a. C and N   b. Cl and Cl   c. K and Br
d. H and H   e. N and F

10.87 For each of the following, draw the Lewis structure and determine the shape: (10.1, 10.3)
a. NF_3 b. $SiBr_4$ c. CSe_2 d. SO_2

10.88 For each of the following, draw the Lewis structure and determine the shape: (10.1, 10.3)
a. PCl_4^+ b. O_2^{2-}
c. $COCl_2$ (C is the central atom)
d. HCCH

10.89 For each of the following, draw the Lewis structure and determine the shape: (10.1, 10.3)
a. BrO_2^- b. H_2O c. CBr_4 d. PO_3^{3-}

10.90 For each of the following, draw the Lewis structure and determine the shape: (10.1, 10.3)
a. PH_3 b. NO_3^-
c. HCN   d. SO_3^{2-}

10.91 Predict the shape and polarity of each of the following molecules, which have polar covalent bonds: (10.3, 10.5)
a. A central atom with three identical bonded atoms and one lone pair.
b. A central atom with two bonded atoms and two lone pairs.

10.92 Predict the shape and polarity of each of the following molecules, which have polar covalent bonds: (10.3, 10.5)
a. A central atom with four identical bonded atoms and no lone pairs.
b. A central atom with four bonded atoms that are not identical and no lone pairs.

10.93 Classify each of the following molecules as polar or nonpolar: (10.3, 10.4, 10.5)
a. HBr   **b.** SiO_2 **c.** NCl_3
d. CH_3Cl **e.** NI_3 **f.** H_2O

10.94 Classify each of the following molecules as polar or nonpolar: (10.3, 10.4, 10.5)
a. GeH_4 **b.** I_2 **c.** CF_3Cl
d. PCl_3 **e.** BCl_3 **f.** SCl_2

10.95 Indicate the major type of intermolecular forces—(1) ionic bonds, (2) dipole–dipole attractions, (3) hydrogen bonds, (4) dispersion forces—that occurs between particles of the following: (10.6)
a. NF_3 **b.** ClF   **c.** Br_2
d. Cs_2O **e.** C_4H_{10} **f.** CH_3OH

10.96 Indicate the major type of intermolecular forces—(1) ionic bonds, (2) dipole–dipole attractions, (3) hydrogen bonds, (4) dispersion forces—that occurs between particles of the following: (10.6)
a. $CHCl_3$ **b.** H_2O **c.** LiCl
d. OBr_2 **e.** HBr   **f.** IBr

10.97 When it rains or snows, the air temperature seems warmer. Explain. (10.7)

10.98 Water is sprayed on the ground of an orchard when temperatures are near freezing to keep the fruit from freezing. Explain. (10.7)

10.99 Using Figure 10.6, calculate the grams of ice that will melt at 0 °C if 1540 J is absorbed. (10.7)

10.100 Using Figure 10.6, calculate the grams of ethanol that will vaporize at its boiling point if 4620 J is absorbed. (10.7)

10.101 Using Figure 10.6, calculate the grams of acetic acid that will freeze at its freezing point if 5.25 kJ is removed. (10.7)

10.102 Using Figure 10.6, calculate the grams of benzene that will condense at its boiling point if 8.46 kJ is removed. (10.7)

CHALLENGE PROBLEMS

The following problems are related to the topics in this chapter. However, they do not all follow the chapter order, and they require you to combine concepts and skills from several sections. These problems will help you increase your critical thinking skills and prepare for your next exam.

10.103 Complete the Lewis structure for each of the following: (10.1)

a.

b.

c. H—N—N—H

d.

10.104 Identify the errors in each of the following Lewis structures, and draw the correct formula: (10.1)

a. :Cl̈=O=C̈l:

b.

c. H—N̈=Ö—H
           |
           H

10.105 Predict the shape of each of the following molecules or ions: (10.3)
a. NH_2Cl (N is the central atom)   **b.** PH_4^+
c. SCN^-

10.106 Classify each of the following molecules as polar or nonpolar: (10.3, 10.4, 10.5)
a. N_2 **b.** TeO_2
c. NH_2Cl (N is the central atom)

10.107 The melting point of dibromomethane is −53 °C, and its boiling point is 97 °C. Sketch a heating curve for dibromomethane from −100 °C to 120 °C. (10.7)
a. What is the state of dibromomethane at −75 °C?
b. What happens on the curve at −53 °C?
c. What is the state of dibromomethane at −18 °C?
d. What is the state of dibromomethane at 110 °C?
e. At what temperature will both solid and liquid be present?

10.108 The melting point of benzene is 5.5 °C, and its boiling point is 80.1 °C. Draw a heating curve for benzene from 0 °C to 100 °C. (10.7)
a. What is the state of benzene at 15 °C?
b. What happens on the curve at 5.5 °C?
c. What is the state of benzene at 63 °C?
d. What is the state of benzene at 98 °C?
e. At what temperature will both liquid and gas be present?

10.109 A 45.0-g piece of ice at 0.0 °C is added to a sample of water at 8.0 °C. All of the ice melts, and the temperature of the water decreases to 0.0 °C. How many grams of water were in the sample? (10.7)

10.110 An ice cube at 0 °C with a mass of 115 g is added to water in a beaker that has a temperature of 64.0 °C. If the final temperature of the mixture is 24.0 °C, what was the initial mass of the warm water? (10.7)

10.111 An ice cube tray holds 325 g of water. If the water initially has a temperature of 25 °C, how many kilojoules of heat must be removed to cool and freeze the water at 0 °C? (10.7)

10.112 A 3.0-kg block of lead is taken from a furnace at 300. °C and placed on a large block of ice at 0 °C. The specific heat of lead is 0.13 J/g °C. If all the heat given up by the lead is used to melt ice, how much ice is melted when the temperature of the lead drops to 0 °C? (10.7)

ANSWERS TO ENGAGE QUESTIONS

10.1 Nitrogen, phosphorus, and arsenic are all in Group 5A (15) and have five valence electrons.

10.2 The positive charge of the proton in the nucleus of each H atom is attracted to the negative electron in the other H atom.

10.3 The total number of valence electrons in a molecule is the sum of the valence electrons of all the atoms in the molecule.

10.4 When octets are not complete for all atoms, some electrons are shared, which form multiple bonds.

10.5 Two pairs of electrons may be shared between the central O atom and one of the other two O atoms.

10.6 SO_2 can have a double bond from the S atom to one of the two O atoms; thus, it is possible to draw two Lewis structures for SO_2. SCl_2 has only single bonds between S and Cl; only one Lewis structure is possible.

10.7 There are four electron groups bonded to the central P in PH_3. However, one group is a lone pair of electrons, and the three H atoms determine the shape, which is trigonal pyramidal.

10.8 Chlorine is above iodine in Group 7A (17). Thus, the valence electrons are closer to the nucleus and have the greater attraction for shared electrons.

10.9 The difference in electronegativity between Si and S is 0.7, which makes the Si—S bond polar covalent. The difference in electronegativity between Si and P is 0.3, which makes the Si—P bond nonpolar covalent.

10.10 Both GeH_4 and CH_4 have nonpolar covalent bonds, but GeH_4 has a larger molar mass and more electrons, which produce more dispersion forces.

10.11 When liquid water freezes, heat is released.

10.12 When a substance changes state, there is no change in temperature. Thus, the temperature at which vaporization occurs is not used in the calculation.

ANSWERS TO SELECTED PROBLEMS

10.1 a. 8 valence electrons b. 14 valence electrons
  c. 32 valence electrons d. 8 valence electrons

10.3 If complete octets cannot be formed by using all the valence electrons, it is necessary to draw multiple bonds.

10.5 a. HF (8 e^-) $H\!:\!\ddot{F}\!:$ or $H\!-\!\ddot{F}\!:$

  b. SF_2 (20 e^-) $:\!\ddot{F}\!:\!\ddot{S}\!:\!\ddot{F}\!:$ or $:\!\ddot{F}\!-\!\ddot{S}\!-\!\ddot{F}\!:$

  c. NBr_3 (26 e^-) $:\!\ddot{Br}\!:\!\ddot{N}\!:\!\ddot{Br}\!:$ or $:\!\ddot{Br}\!-\!\ddot{N}\!-\!\ddot{Br}\!:$ with $:\!\ddot{Br}\!:$ above

  d. BH_4^- (8 e^-) $\left[H\!:\!\ddot{B}\!:\!H \right]^-$ with H above and below or $\left[H\!-\!B\!-\!H \right]^-$ with H above and below

10.7 a. CO (10 e^-) $:\!C\!:\!:\!O\!:$ or $:\!C\!\equiv\!O\!:$

  b. CN^- (10 e^-) $\left[:\!C\!:\!:\!N\!: \right]^-$ or $\left[:\!C\!\equiv\!N\!: \right]^-$

  c. H_2CO (12 e^-) $H\!:\!C\!:\!H$ with $:\!O\!:$ above or $H\!-\!C\!-\!H$ with $\overset{\ddot{O}}{\|}$ above

10.9 Resonance occurs when we can draw two or more Lewis structures for the same molecule or ion.

10.11 a. $ClNO_2$ (24 e^-) $:\!\ddot{Cl}\!-\!N\!-\!\ddot{O}\!:$ with $\overset{\ddot{O}}{\|}$ $\longleftrightarrow$ $:\!\ddot{Cl}\!-\!N\!=\!\ddot{O}\!:$ with $:\!\ddot{O}\!:$

  b. OCN^- (16 e^-) $\left[:\!O\!\equiv\!C\!-\!\ddot{N}\!: \right]^-$ $\longleftrightarrow$

  $\left[:\!\ddot{O}\!=\!C\!=\!\ddot{N}\!: \right]^-$ $\longleftrightarrow$ $\left[:\!\ddot{O}\!-\!C\!\equiv\!N\!: \right]^-$

10.13 a. 6, tetrahedral
  b. 5, trigonal pyramidal
  c. 3, trigonal planar

10.15 a. three
  b. trigonal planar
  c. three
  d. trigonal planar

10.17 In CF_4, the central atom C has four bonded atoms and no lone pairs of electrons, which gives it a tetrahedral shape. In NF_3, the central atom N has three bonded atoms and one lone pair of electrons, which gives NF_3 a trigonal pyramidal shape.

10.19 a. trigonal planar
  b. bent (109°)
  c. linear
  d. tetrahedral

10.21 a. AlH_4^- (8 e^-) $\left[H\!-\!Al\!-\!H \right]^-$ with H above and below tetrahedral

  b. SO_4^{2-} (32 e^-) $\left[:\!\ddot{O}\!-\!S\!-\!\ddot{O}\!: \right]^{2-}$ with $:\!\ddot{O}\!:$ above and below tetrahedral

  c. NH_4^+ (8 e^-) $\left[H\!-\!N\!-\!H \right]^+$ with H above and below tetrahedral

  d. NO_2^+ (16 e^-) $\left[:\!\ddot{O}\!=\!N\!=\!\ddot{O}\!: \right]^+$ linear

10.23 a. increases b. decreases c. decreases

10.25 a. K, Na, Li b. Na, P, Cl c. Ca, Se, O

10.27 **a.** between 0.0 and 0.4

10.29 **a.** polar covalent   **b.** ionic
    c. polar covalent   **d.** nonpolar covalent
    e. polar covalent   **f.** nonpolar covalent

10.31 **a.** $N^{\delta+}\!-\!F^{\delta-}$ **b.** $Si^{\delta+}\!-\!Br^{\delta-}$ **c.** $C^{\delta+}\!-\!O^{\delta-}$

      d. $P^{\delta+}\!-\!Br^{\delta-}$ **e.** $N^{\delta-}\!-\!P^{\delta+}$

10.33 Electrons are shared equally between two identical atoms and unequally between nonidentical atoms.

10.35 **a.** nonpolar   **b.** polar
    c. polar   **d.** nonpolar

10.37 In the molecule CO_2, the two $C\!-\!O$ dipoles cancel; in CO, there is only one dipole.

10.39 **a.** dipole–dipole attractions   **b.** hydrogen bonds
    c. dipole–dipole attractions   **d.** dispersion forces

10.41 **a.** hydrogen bonds   **b.** dipole–dipole attractions
    c. dispersion forces   **d.** dispersion forces

10.43 **a.** HF; hydrogen bonds are stronger than the dipole–dipole attractions in HBr.
    b. HF; hydrogen bonds are stronger than the dipole–dipole attractions in SF_2.
    c. PBr_3, the dipole–dipole attractions in PBr_3 are stronger than the dispersion forces in Br_2.
    d. CH_3OH; hydrogen bonds are stronger than the dispersion forces in CH_4.

10.45 **a.** 21 700 J   **b.** 2180 J
    c. 28.8 kJ   **d.** 16.7 kJ

10.47 **a.** 22 600 J   **b.** 42.1 kJ
    c. 3100 J   **d.** 242 kJ

10.49 **a.** 852 J   **b.** 30 300 J
    c. 40.2 kJ   **d.** 72.2 kJ

10.51 119.5 kJ

10.53 3130 J

10.55 **a.** 0.0, nonpolar covalent
    b. 0.4, nonpolar covalent
    c. 1.0, polar covalent

10.57 **a.**

$$\left[:\!\overset{..}{O}\!=\!C\!-\!\overset{..}{\underset{..}{O}}\!:\right]^{2-} \leftrightarrow \left[:\!\overset{..}{\underset{..}{O}}\!-\!C\!-\!\overset{..}{\underset{..}{O}}\!:\right]^{2-} \leftrightarrow \left[:\!\overset{..}{\underset{..}{O}}\!-\!C\!=\!\overset{..}{O}\!:\right]^{2-}$$

    b. Carbonate ion is trigonal planar.

10.59 **a.** two valence electrons, one bonding pair, no lone pairs
    b. eight valence electrons, one bonding pair, three lone pairs
    c. 14 valence electrons, one bonding pair, six lone pairs

10.61 **a.** 2, trigonal pyramidal, polar
    b. 1, bent (109°), polar
    c. 3, tetrahedral, nonpolar

10.63 **a.** C and O, N and O
    b. O and O
    c. Ca and O, K and O
    d. C and O, O and O, N and O

10.65 **a.** dispersion forces
    b. dipole–dipole attractions
    c. hydrogen bonds
    d. dispersion forces

10.67 **a.** The heat from the skin is used to evaporate the water (perspiration). Therefore, the skin is cooled.
    b. On a hot day, there are more molecules with sufficient energy to become water vapor.
    c. In a closed bag, some molecules evaporate, but they cannot escape and will condense back to liquid; the clothes will not dry.

10.69

10.71 **a.** about $-60\ °C$ **b.** about $60\ °C$
    c. The diagonal line **A** represents the solid state as temperature increases. The horizontal line **B** represents the change from solid to liquid or melting of the substance. The diagonal line **C** represents the liquid state as temperature increases. The horizontal line **D** represents the change from liquid to gas or boiling of the liquid. The diagonal line **E** represents the gas state as temperature increases.
    d. At $-80\ °C$, solid; at $-40\ °C$, liquid; at $25\ °C$, liquid; $80\ °C$, gas

10.73 **a.** $1 + 5 + 2(6) = 18$ valence electrons
    b. $2(4) + 4(1) + 6 = 18$ valence electrons
    c. $5 + 4(1) - 1 = 8$ valence electrons
    d. $6 + 3(6) + 2 = 26$ valence electrons

10.75 **a.** BF_4^- $(32\ e^-)$

$$\left[\overset{\;\;:\!\overset{..}{F}\!:}{\underset{:\!\overset{..}{F}\!:}{:\!\overset{..}{F}\!:\!B\!:\!\overset{..}{F}\!:}}\right]^{-} \text{ or } \left[\overset{\;\;:\!\overset{..}{F}\!:}{\underset{:\!\overset{..}{F}\!:}{:\!\overset{..}{F}\!-\!B\!-\!\overset{..}{F}\!:}}\right]^{-}$$

    b. Cl_2O $(20\ e^-)$ $:\!\overset{..}{\underset{..}{Cl}}\!:\!\overset{..}{\underset{..}{O}}\!:\!\overset{..}{\underset{..}{Cl}}\!:$ or $:\!\overset{..}{\underset{..}{Cl}}\!-\!\overset{..}{\underset{..}{O}}\!-\!\overset{..}{\underset{..}{Cl}}\!:$

    c. H_2NOH $(14\ e^-)$ $H\!:\!\overset{H}{\underset{..}{N}}\!:\!\overset{..}{\underset{..}{O}}\!:\!H$ or $H\!-\!\overset{\overset{\displaystyle H}{|}}{N}\!-\!\overset{..}{\underset{..}{O}}\!-\!H$

    d.

     H_2CCCl_2 $(24\ e^-)$ $H\!:\!\overset{..}{\underset{..}{Cl}}\!:\!\overset{H\;\;:\!\overset{..}{Cl}\!:}{C}\!:\!\overset{..}{\underset{..}{Cl}}\!:$ or $H\!-\!\overset{\overset{\displaystyle H}{|}}{C}\!=\!\overset{\overset{\displaystyle :\!\overset{..}{Cl}\!:}{|}}{C}\!-\!\overset{..}{\underset{..}{Cl}}\!:$

10.77 **a.** $(16\ e^-)$ $\left[:\!\overset{..}{N}\!:\!:\!N\!:\!:\!\overset{..}{N}\!:\right]^{-}$ or

$$\left[:\!\overset{..}{\underset{..}{N}}\!-\!N\!\equiv\!N\!:\right]^{-} \leftrightarrow \left[:\!\overset{..}{N}\!=\!N\!=\!\overset{..}{N}\!:\right]^{-} \leftrightarrow \left[:\!N\!\equiv\!N\!-\!\overset{..}{\underset{..}{N}}\!:\right]^{-}$$

    b. $(16\ e^-)$ $\left[:\!\overset{..}{O}\!:\!:\!N\!:\!:\!\overset{..}{O}\!:\right]^{+}$ or

$$\left[:\!\overset{..}{\underset{..}{O}}\!-\!N\!\equiv\!O\!:\right]^{+} \leftrightarrow \left[:\!\overset{..}{O}\!=\!N\!=\!\overset{..}{O}\!:\right]^{+} \leftrightarrow \left[:\!O\!\equiv\!N\!-\!\overset{..}{\underset{..}{O}}\!:\right]^{+}$$

    c. $(18\ e^-)$ $\left[\overset{\;\;:\!\overset{..}{O}\!:}{\underset{H\!:\!C\!:\!\overset{..}{\underset{..}{O}}\!:}{}}\right]^{-}$ or $\left[\overset{\;\;:\!\overset{..}{O}\!:}{\underset{H\!-\!C\!-\!\overset{..}{\underset{..}{O}}\!:}{\|}}\right]^{-} \leftrightarrow \left[\overset{\;\;:\!\overset{..}{\underset{..}{O}}\!:}{\underset{H\!-\!C\!=\!\overset{..}{O}\!:}{|}}\right]^{-}$

10.79 **a.** I, Cl, F **b.** K, Li, S, Cl **c.** Ba, Sr, Mg, Be

10.81 **a.** C and O **b.** N and F
  c. S and Cl **d.** Br and I
  e. N and F

10.83 **a.** Si—Cl
        →
    b. C—N
        →
  c. F—Cl
      ←+
    d. C—F
        +→
  e. N—O
      +→

10.85 **a.** polar covalent **b.** nonpolar covalent
  c. ionic **d.** nonpolar covalent
  e. nonpolar covalent

10.87 **a.** NF_3 $(26\ e^-)$:F̈—N̈—F̈: trigonal pyramidal
            :F̈:

  b. $SiBr_4$ $(32\ e^-)$
            :B̈r:
            |
        :B̈r—Si—B̈r: tetrahedral
            |
            :B̈r:

  c. CSe_2 $(16\ e^-)$:S̈e=C=S̈e: linear

  d. SO_2 $(18\ e^-)$:Ö=S̈—Ö: ⟷ :Ö—S̈=Ö:
                                bent (120°)

10.89 **a.** bent (109°) **b.** bent (109°)
  c. tetrahedral **d.** trigonal pyramidal

10.91 **a.** trigonal pyramidal, polar
  b. bent (109°), polar

10.93 **a.** polar **b.** nonpolar
  c. nonpolar **d.** polar
  e. polar **f.** polar

10.95 **a.** (2) dipole–dipole attractions
  b. (2) dipole–dipole attractions
  c. (4) dispersion forces
  d. (1) ionic bonds
  e. (4) dispersion forces
  f. (3) hydrogen bonds

10.97 When water vapor condenses or liquid water freezes, heat is released, which warms the air.

10.99 4.61 g of water

10.101 27.3 g of acetic acid

10.103 **a.** $(18\ e^-)$ H—N̈—C—H with H and :O: groups

  b. $(22\ e^-)$:C̈l—C—C≡N: with H groups

  c. $(12\ e^-)$ H—N̈=N̈—H

  d. $(30\ e^-)$:C̈l—C—Ö—C—H with :O: and H groups

10.105 **a.** trigonal pyramidal **b.** tetrahedral **c.** linear

10.107

$T\ °C$	Heat Added →

a. solid **b.** solid dibromomethane melts
c. liquid **d.** gas
e. −53 °C

10.109 450 g of water

10.111 143 kJ of heat is removed.

CI.13 In an experiment, the mass of a piece of copper is determined to be 8.56 g. Then the copper is reacted with sufficient oxygen gas to produce solid copper(II) oxide. (7.1, 8.1, 8.2, 8.3, 9.1, 9.2, 9.3, 9.5)

Cu
8.56 g

a. How many copper atoms are in the sample?
b. Write the balanced chemical equation for the reaction.
c. Classify the type of reaction.
d. How many grams of O_2 are required to completely react with the Cu?
e. How many grams of CuO will result from the reaction of 8.56 g of Cu and 3.72 g of oxygen?
f. How many grams of CuO will result in part **e**, if the yield for the reaction is 85.0%?

CI.14 One of the components in gasoline is octane, C_8H_{18}, which has a density of 0.803 g/cm³ and $\Delta H = -510$ kJ/mol. Suppose a hybrid car has a fuel tank with a capacity of 11.9 gal and has a gas mileage of 45 mi/gal. (7.10, 6.7, 8.3, 8.5, 9.6)

a. Write a balanced chemical equation for the complete combustion of octane including the heat of reaction.
b. What is the energy, in kilojoules, produced from one tank of fuel assuming it is all octane?
c. How many molecules of C_8H_{18} are present in one tank of fuel assuming it is all octane?

Octane is one of the components of motor fuel.

d. If this hybrid car is driven 24 500 miles in one year, how many kilograms of carbon dioxide will be produced from the combustion of the fuel assuming it is all octane?

CI.15 When clothes have stains, bleach may be added to the wash to react with the soil and make the stains colorless. The bleach solution is prepared by bubbling chlorine gas into a solution of sodium hydroxide to produce a solution of sodium hypochlorite, sodium chloride, and water. One brand of bleach contains 5.25% sodium hypochlorite by mass (active ingredient) with a density of 1.08 g/mL. (6.4, 7.1, 7.2, 8.2, 9.3, 9.4, 9.5, 10.1)

Bleach is a solution of sodium hypochlorite.

a. What is the formula and molar mass of sodium hypochlorite?
b. Draw the Lewis structure for the hypochlorite ion.
c. How many hypochlorite ions are present in 1.00 gal of bleach solution?
d. Write the balanced chemical equation for the preparation of bleach.

e. How many grams of NaOH are required to produce the sodium hypochlorite for 1.00 gal of bleach?
f. If 165 g of Cl_2 is passed through a solution containing 275 g of NaOH and 162 g of sodium hypochlorite is produced, what is the percent yield of sodium hypochlorite for the reaction?

CI.16 Ethanol, C_2H_6O, is obtained from renewable crops such as corn, which use the Sun as their source of energy. In the United States, automobiles can now use a fuel known as E85 that contains 85.0% ethanol and 15.0% gasoline by volume. Ethanol has a melting point of -115 °C, a boiling point of 78 °C, a heat of fusion of 109 J/g, and a heat of vaporization of 841 J/g. Liquid ethanol has a density of 0.796 g/mL and a specific heat of 2.46 J/g °C. (8.3, 8.5, 9.2, 9.3, 10.4, 10.5)

a. Draw a heating curve for ethanol from -150 °C to 100 °C.

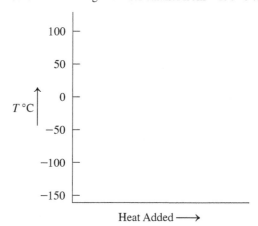

b. When 20.0 g of ethanol at -62 °C is heated and completely vaporized at 78 °C, how much energy, in kilojoules, is required?
c. If a 15.0-gal gas tank is filled with E85, how many liters of ethanol are in the gas tank?
d. Write the balanced chemical equation for the complete combustion of ethanol.
e. How many kilograms of CO_2 are produced by the complete combustion of the ethanol in a full 15.0-gal gas tank?
f. What would be the strongest intermolecular force between liquid ethanol molecules?

Applications

CI.17 Chloral hydrate, a sedative and hypnotic, was the first drug used to treat insomnia. Chloral hydrate has a melting point of 57 °C. At its boiling point of 98 °C, it breaks down to chloral and water. (7.4, 7.5, 8.4, 10.1)

Chloral hydrate   Chloral

a. Draw the Lewis structures for chloral hydrate and chloral.
b. What are the empirical formulas of chloral hydrate and chloral?
c. What is the mass percent of Cl in chloral hydrate?

CI.18 Ethylene glycol, $C_2H_6O_2$, used as a coolant and antifreeze, has a density of 1.11 g/mL. As a sweet-tasting liquid, it can be appealing to pets and small children, but it can be toxic, with an LD_{50} of 4700 mg/kg. Its accidental ingestion can cause kidney damage and difficulty with breathing. In the body, ethylene glycol is converted to another toxic substance, oxalic acid, $H_2C_2O_4$. (2.8, 7.4, 7.5, 8.2, 8.4, 10.1, 10.3, 10.4)

Antifreeze often contains ethylene glycol.

a. What are the empirical formulas of ethylene glycol and oxalic acid?
b. If ethylene glycol has a C—C single bond with two H atoms attached to each C atom, what is its Lewis structure?
c. Which bonds in ethylene glycol are polar covalent and which are nonpolar covalent?
d. How many milliliters of ethylene glycol could be toxic for an 11.0-lb cat?
e. What would be the strongest intermolecular force in ethylene glycol?
f. If oxalic acid has two carbon atoms attached by a C—C single bond with each carbon also attached to two oxygen atoms, what is its Lewis structure?
g. Write the balanced chemical equation for the reaction of ethylene glycol and oxygen (O_2) to give oxalic acid and water.

CI.19 Acetone (propanone), a clear liquid solvent with an acrid odor, is used to remove nail polish, paints, and resins. It has a low boiling point and is highly flammable. Acetone has a density of 0.786 g/mL and a heat of combustion of -1790 kJ/mol. (2.8, 7.2, 8.3, 8.4, 9.3, 9.6)

Acetone has carbon atoms (black), hydrogen atoms (white), and an oxygen atom (red).

a. Draw the Lewis structure for acetone.
b. What are the molecular formula and molar mass of acetone?
c. Write a balanced chemical equation for the complete combustion of acetone, including the heat of reaction.
d. Is the combustion of acetone an endothermic or exothermic reaction?
e. How much heat, in kilojoules, is released if 2.58 g of acetone reacts completely with oxygen?
f. How many grams of oxygen gas are needed to react with 15.0 mL of acetone?

CI.20 The compound dihydroxyacetone (DHA) is used in "sunless" tanning lotions, which darken the skin by reacting with the amino acids in the outer surface of the skin. A typical drugstore lotion contains 4.0% (mass/volume) DHA. (2.8, 7.2, 8.4)

Dihydroxyacetone has carbon atoms (black), hydrogen atoms (white), and oxygen atoms (red).

a. Draw the Lewis structure for DHA.
b. DHA has C—C, C—H, C—O, and O—H bonds. Which of these bonds are polar covalent and which are nonpolar covalent?
c. What are the molecular formula and molar mass of DHA?
d. A bottle of sunless tanning lotion contains 177 mL of lotion. How many milligrams of DHA are in a bottle?

ANSWERS

CI.13 a. 8.11×10^{22} atoms of copper
b. $2Cu(s) + O_2(g) \longrightarrow 2CuO(s)$
c. combination reaction
d. 2.16 g of O_2
e. 10.7 g of CuO
f. 9.10 g of CuO

CI.15 a. NaOCl, 74.44 g/mol
b. (14 e^-) $\left[:\!\overset{..}{\underset{..}{Cl}}\!-\!\overset{..}{\underset{..}{O}}\!: \right]^-$
c. 1.74×10^{24} ClO$^-$ ions
d. $2NaOH(aq) + Cl_2(g) \longrightarrow$
$\qquad\qquad NaClO(aq) + NaCl(aq) + H_2O(l)$
e. 231 g of NaOH
f. 93.6%

CI.17 a. (44 e^-)

$$:\overset{..}{\underset{..}{Cl}}-\overset{\overset{\displaystyle:\overset{..}{\underset{..}{Cl}}:}{|}}{\underset{\underset{\displaystyle:\overset{..}{\underset{..}{Cl}}:}{|}}{C}}-\overset{\overset{\displaystyle:\overset{..}{\underset{..}{O}}-H}{|}}{\underset{\underset{\displaystyle H}{|}}{C}}-\overset{..}{\underset{..}{O}}-H$$

CI.19 a. $CH_3-\overset{\overset{\displaystyle :\overset{..}{O}:}{||}}{C}-CH_3$
b. C_3H_6O; 58.08 g/mol

c. $C_3H_6O(l) + 4O_2(g) \overset{\Delta}{\longrightarrow}$
$\qquad\qquad 3CO_2(g) + 3H_2O(g) + 1790$ kJ
d. exothermic
e. 79.5 kJ
f. 26.0 g of O_2

(36 e^-)
$$:\overset{..}{\underset{..}{Cl}}-\overset{\overset{\displaystyle :\overset{..}{\underset{..}{Cl}}:}{|}}{\underset{\underset{\displaystyle :\overset{..}{\underset{..}{Cl}}:}{|}}{C}}-\overset{\overset{\displaystyle :\overset{..}{O}:}{||}}{C}-H$$

b. chloral hydrate: $C_2H_3O_2Cl_3$ chloral: C_2HOCl_3
c. 64.33% Cl (by mass)

Gases

11

After soccer practice, Louisa complains that she is having difficulty breathing. Her father takes her to the emergency room, where she is seen by Sam, a respiratory therapist, who listens to Louisa's chest and tests her breathing capacity using a spirometer. Based on her limited breathing capacity and the wheezing noise in her chest, Sam diagnoses Louisa with asthma.

Sam gives Louisa a nebulizer containing a bronchodilator that opens the airways and allows more air to go into the lungs. During the breathing treatment, he measures the amount of oxygen (O_2) in her blood and explains to Louisa and her father that air is a mixture of gases containing 78% nitrogen (N_2) gas and 21% O_2 gas. Because Louisa has difficulty obtaining sufficient oxygen, Sam gives her supplemental oxygen through a face mask. Within a short period of time, Louisa's breathing returns to normal. Sam explains that the lungs work according to Boyle's law: The volume of the lungs increases upon inhalation, and the pressure decreases to allow air to flow in. However, during an asthma attack, the airways become restricted, and it becomes more difficult to expand the volume of the lungs.

CAREER

Respiratory Therapist

Respiratory therapists assess and treat a range of patients, including premature infants whose lungs have not developed, asthmatics, and patients with emphysema or cystic fibrosis. In assessing patients, they perform a variety of diagnostic tests including breathing capacity and concentrations of oxygen and carbon dioxide in a patient's blood, as well as blood pH. In order to treat patients, therapists provide oxygen or aerosol medications to the patient, as well as chest physiotherapy to remove mucus from their lungs. Respiratory therapists also educate patients on how to use their inhalers correctly.

UPDATE Exercise-Induced Asthma

Louisa's doctor prescribes an inhaled medication that opens up her airways before she starts exercise. In the **UPDATE Exercise-Induced Asthma**, page 341, you can view the impact of Louisa's medication and learn about other treatments that help prevent exercise-induced asthma.

311

REVIEW

Using Significant Figures in
  Calculations (2.3)

Writing Conversion Factors from
  Equalities (2.5)

Using Conversion Factors (2.6)

INTERACTIVE VIDEO

Kinetic Molecular Theory

ENGAGE 11.1

Use the kinetic molecular theory to
explain why a gas completely fills a
container of any size and shape.

PRACTICE PROBLEMS

Try Practice Problems 11.1 and 11.2

FIGURE 11.1 Gas particles moving
in straight lines within a container
exert pressure when they collide
with the walls of the container.

11.1 Properties of Gases

LEARNING GOAL Describe the kinetic molecular theory of gases and the units of
measurement used for gases.

We all live at the bottom of a sea of gases called the atmosphere. The most important of these
gases is oxygen, which constitutes about 21% of the atmosphere. Without oxygen, life on
this planet would be impossible: Oxygen is vital to all life processes of plants and animals.
Ozone (O_3), formed in the upper atmosphere by the interaction of oxygen with ultraviolet
light, absorbs some of the harmful radiation before it can strike Earth's surface. The other
gases in the atmosphere include nitrogen (78%), argon, carbon dioxide (CO_2), and water
vapor. Carbon dioxide gas, a product of combustion and metabolism, is used by plants in
photosynthesis, which produces the oxygen that is essential for humans and animals.

The behavior of gases is quite different from that of liquids and solids. Gas particles
are far apart, whereas particles of both liquids and solids are held close together. Because
there are great distances between gas particles, a gas is less dense than a solid or liquid, and
easy to compress. A gas has no definite shape or volume and will completely fill any container. A model for the behavior of a gas, called the **kinetic molecular theory of gases**,
helps us understand gas behavior.

Kinetic Molecular Theory of Gases

1. **A gas consists of small particles (atoms or molecules) that move randomly with
   high velocities.** Gas molecules moving in random directions at high speeds cause a
   gas to fill the entire volume of a container.
2. **The attractive forces between the particles of a gas are usually very small.** Gas
   particles are far apart and fill a container of any size and shape.
3. **The actual volume occupied by gas molecules is extremely small compared to
   the volume that the gas occupies.** The volume of the gas is considered equal to the
   volume of the container. Most of the volume of a gas is empty space, which allows
   gases to be easily compressed.
4. **Gas particles are in constant motion, moving rapidly in straight paths.** When gas
   particles collide, they rebound and travel in new directions. Every time they hit the
   walls of the container, they exert pressure. An increase in the number or force of collisions against the walls of the container causes an increase in the pressure of the gas.
5. **The average kinetic energy of gas molecules is proportional to the Kelvin
   temperature.** Gas particles move faster as the temperature increases. At higher
   temperatures, gas particles hit the walls of the container more often and with more
   force, producing higher pressures.

The kinetic molecular theory helps explain some of the characteristics of gases. For
example, you can smell perfume when a bottle is opened on the other side of a room because
its particles move rapidly in all directions. At room temperature, the molecules in the air are
moving at about 450 m/s, which is 1000 mi/h. They move faster at higher temperatures and
more slowly at lower temperatures. Sometimes tires and gas-filled containers explode when
temperatures are too high. From the kinetic molecular theory, you know that gas particles
move faster when heated, hit the walls of a container with more force, and cause a buildup
of pressure inside a container.

When we talk about a gas, we describe it in terms of four properties: pressure, volume,
temperature, and the amount of gas.

Pressure (P)

Gas particles are extremely small and move rapidly. When they hit the walls of a container,
they exert a **pressure** (see **FIGURE 11.1**). If we heat the container, the molecules move faster
and smash into the walls of the container more often and with increased force, thus increasing the pressure. The gas particles in the air, mostly oxygen and nitrogen, exert a pressure
on us called **atmospheric pressure** (see **FIGURE 11.2**). As you go to higher altitudes, the

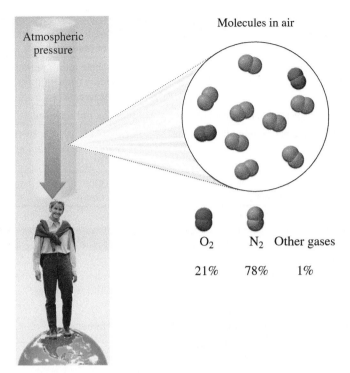

FIGURE 11.2 A column of air extending from the top of the atmosphere to the surface of Earth produces pressure on each of us of about 1 atm.

ENGAGE 11.2

Why is there less atmospheric pressure at a higher altitude?

atmospheric pressure is less because there are fewer gas particles. The most common units used to measure gas pressure are the *atmosphere* (atm) and *millimeters of mercury* (mmHg). On a TV weather report, the atmospheric pressure may be reported in inches of mercury, or in kilopascals in countries other than the United States. In a hospital, the unit torr or pounds per square inch (psi) may be used.

Volume (*V*)

The volume of gas equals the size of the container in which the gas is placed. When you inflate a balloon, you are adding more gas particles. The increase in the number of particles hitting the walls of the balloon increases the volume. The most common units for volume measurement are liters (L) and milliliters (mL).

Temperature (*T*)

The temperature of a gas is related to the kinetic energy of its particles. For example, if we have a gas at 200 K and heat it to a temperature of 400 K, the gas particles will have twice the kinetic energy that they did at 200 K. This also means that the gas at 400 K exerts twice the pressure of the gas at 200 K, if the volume and amount of gas do not change. Although we measure gas temperature using a Celsius thermometer, all comparisons of gas behavior and all calculations related to temperature must use the Kelvin temperature scale. No one has created the conditions for absolute zero (0 K), but scientists predict that the particles will have zero kinetic energy and exert zero pressure at absolute zero.

Amount of Gas (*n*)

When you add air to a bicycle tire, you increase the amount of gas, which results in a higher pressure in the tire. Usually, we measure the amount of gas by its mass, in grams. In gas law calculations, we need to change the grams of gas to moles.

A summary of the four properties of a gas is given in **TABLE 11.1**.

TABLE 11.1 Properties That Describe a Gas

Property	Description	Units of Measurement
Pressure (*P*)	The force exerted by a gas against the walls of the container	atmosphere (atm); millimeter of mercury (mmHg); torr (Torr); kilopascal (kPa)
Volume (*V*)	The space occupied by a gas	liter (L); milliliter (mL)
Temperature (*T*)	The determining factor of the kinetic energy and rate of motion of gas particles	degree Celsius (°C); kelvin (K) *is required in calculations*
Amount (*n*)	The quantity of gas present in a container	gram (g); mole (*n*) *is required in calculations*

SAMPLE PROBLEM 11.1   **Properties of Gases**

TRY IT FIRST

Identify the property of a gas that is described by each of the following:

a. increases the kinetic energy of gas particles
b. the force of the gas particles hitting the walls of the container
c. the space that is occupied by a gas

SOLUTION

a. temperature   **b.** pressure   **c.** volume

SELF TEST 11.1

What property of a gas is described in each of the following?

a. Helium gas is added to a balloon.
b. A balloon expands as it rises higher.

ANSWER

a. amount   **b.** volume

PRACTICE PROBLEMS
Try Practice Problems 11.3 and 11.4

FIGURE 11.3 A barometer: The pressure exerted by the gases in the atmosphere is equal to the downward pressure of a mercury column in a closed glass tube. The height of the mercury column measured in mmHg is called atmospheric pressure.

Measurement of Gas Pressure

When billions and billions of gas particles hit against the walls of a container, they exert pressure, which is a force acting on a certain area.

$$\text{Pressure } (P) = \frac{\text{force}}{\text{area}}$$

The atmospheric pressure can be measured using a barometer (see **FIGURE 11.3**). At a pressure of exactly 1 atmosphere (atm), a mercury column in an inverted tube would be *exactly* 760 mm high. One **atmosphere (atm)** is defined as *exactly* 760 mmHg (millimeters of mercury). One atmosphere is also 760 Torr, a pressure unit named to honor Evangelista Torricelli, the inventor of the barometer. Because the units of Torr and mmHg are equal, they are used interchangeably. One atmosphere is also equivalent to 29.9 in. of mercury (inHg).

1 atm = 760 mmHg = 760 Torr (exact)

1 atm = 29.9 inHg

1 mmHg = 1 Torr (exact)

In SI units, pressure is measured in pascals (Pa); 1 atm is equal to 101 325 Pa. Because a pascal is a very small unit, pressures are usually reported in kilopascals.

1 atm = 101 325 Pa = 101.325 kPa

The U.S. equivalent of 1 atm is 14.7 lb/in.^{2} When you use a pressure gauge to check the air pressure in the tires of a car, it may read 30 to 35 psi. This measurement is actually 30 to 35 psi above the pressure that the atmosphere exerts on the outside of the tire.

1 atm = 14.7 lb/in.^{2}

TABLE 11.2 summarizes the various units used in the measurement of pressure.

$P = 0.70$ atm $P = 1.0$ atm
(530 mmHg)   (760 mmHg)

TABLE 11.2 Units for Measuring Pressure

Unit	Abbreviation	Unit Equivalent to 1 atm
Atmosphere	atm	1 atm (exact)
Millimeters of Hg	mmHg	760 mmHg (exact)
Torr	Torr	760 Torr (exact)
Inches of Hg	inHg	29.9 inHg
Pounds per square inch	lb/in.^{2} (psi)	14.7 lb/in.^{2}
Pascal	Pa	101 325 Pa
Kilopascal	kPa	101.325 kPa

The atmospheric pressure decreases as the altitude increases.

Atmospheric pressure changes with variations in weather and altitude. On a hot, sunny day, the mercury column rises, indicating a higher atmospheric pressure. On a rainy day, the atmosphere exerts less pressure, which causes the mercury column to fall. In a weather report, this type of weather is called a *low-pressure system*. Above sea level, the density of the gases in the air decreases, which causes lower atmospheric pressures; the atmospheric pressure is greater than 760 mmHg at the Dead Sea because it is below sea level (see **TABLE 11.3**).

TABLE 11.3 Altitude and Atmospheric Pressure

Location	Altitude (km)	Atmospheric Pressure (mmHg)
Dead Sea	−0.40	800
Sea level	0.00	760
Los Angeles	0.09	752
Las Vegas	0.70	700
Denver	1.60	630
Mount Whitney	4.50	440
Mount Everest	8.90	253

Divers must be concerned about increasing pressures on their ears and lungs when they dive below the surface of the ocean. Because water is more dense than air, the pressure on a diver increases rapidly as the diver descends. At a depth of 33 ft below the surface of the ocean, an additional 1 atm of pressure is exerted by the water on a diver, which gives a total pressure of 2 atm. At 100 ft, there is a total pressure of 4 atm on a diver. The regulator that a diver uses continuously adjusts the pressure of the breathing mixture to match the increase in pressure.

SAMPLE PROBLEM 11.2   Units of Pressure

TRY IT FIRST

The oxygen in a tank in the hospital respiratory unit has a pressure of 4820 mmHg. Calculate the pressure, in atmospheres, of the oxygen gas.

SOLUTION

The equality 1 atm = 760 mmHg can be written as two conversion factors:

$$\frac{760 \text{ mmHg}}{1 \text{ atm}} \quad \text{and} \quad \frac{1 \text{ atm}}{760 \text{ mmHg}}$$

A patient with severe COPD obtains oxygen from an oxygen tank.

Using the conversion factor that cancels mmHg and gives atm, we can set up the problem as

$$4820 \text{ mmHg} \times \frac{1 \text{ atm}}{760 \text{ mmHg}} = 6.34 \text{ atm}$$

SELF TEST 11.2

A tank of nitrous oxide (N_2O) used as an anesthetic has a pressure of 48 psi.

a. What is that pressure in atmospheres?
b. What is the pressure in torr?

PRACTICE PROBLEMS

Try Practice Problems 11.5 to 11.8

ANSWER

a. 3.3 atm   **b.** 2500 (2.5×10^3) Torr

Chemistry Link to Health

Measuring Blood Pressure

Your blood pressure is one of the vital signs a doctor or nurse checks during a physical examination. It actually consists of two separate measurements. Acting like a pump, the heart contracts to create the pressure that pushes blood through the circulatory system. During contraction, the blood pressure is at its highest; this is your *systolic* pressure. When the heart muscles relax, the blood pressure falls; this is your *diastolic* pressure. The normal range for systolic pressure is 90 to 120 mmHg. For diastolic pressure, it is 60 to 80 mmHg. These two measurements are usually expressed as a ratio such as 100/80. These values are somewhat higher in older people. When blood pressures are elevated, such as 140/90, there is a greater risk of stroke, heart attack, or kidney damage. Low blood pressure prevents the brain from receiving adequate oxygen, causing dizziness and fainting.

The blood pressures are measured by a *sphygmomanometer*, an instrument consisting of a stethoscope and an inflatable cuff connected to a tube of mercury called a manometer. After the cuff is wrapped around the upper arm, it is pumped up with air until it cuts off the flow of blood. With the stethoscope over the artery, the air is slowly released from the cuff, decreasing the pressure on the artery. When the blood flow first starts again in the artery, a noise can be heard through the stethoscope signifying the systolic blood pressure as the pressure shown on the manometer. As air continues to be released, the cuff deflates until no sound is heard in the artery. A second pressure reading is taken at the moment of silence and denotes the diastolic pressure, the pressure when the heart is not contracting.

The use of digital blood pressure monitors is becoming more common. However, they have not been validated for use in all situations and can sometimes give inaccurate readings.

The measurement of blood pressure is part of a routine physical exam.

PRACTICE PROBLEMS

11.1 Properties of Gases

11.1 Use the kinetic molecular theory of gases to explain each of the following:
a. Gases move faster at higher temperatures.
b. Gases can be compressed much more easily than liquids or solids.
c. Gases have low densities.

11.2 Use the kinetic molecular theory of gases to explain each of the following:
a. A container of nonstick cooking spray explodes when thrown into a fire.
b. The air in a hot-air balloon is heated to make the balloon rise.
c. You can smell the odor of cooking onions from far away.

11.3 Identify the property of a gas that is measured in each of the following:
a. 350 K   **b.** 125 mL
c. 2.00 g of O_2 **d.** 755 mmHg

11.4 Identify the property of a gas that is measured in each of the following:
a. 425 K   **b.** 1.0 atm
c. 10.0 L   **d.** 0.50 mol of He

11.5 Which of the following statements describe the pressure of a gas?
a. the force of the gas particles on the walls of the container
b. the number of gas particles in a container
c. 4.5 L of helium gas
d. 750 Torr
e. 28.8 lb/in.^{2}

11.6 Which of the following statements describe the pressure of a gas?
a. the temperature of a gas
b. the volume of the container
c. 3.00 atm
d. 0.25 mol of O_2
e. 101 kPa

Applications

11.7 A tank contains oxygen (O_2) at a pressure of 2.00 atm. What is the pressure in the tank in terms of the following units?

    a. torr   **b.** lb/in.^{2} **c.** mmHg   **d.** kPa

11.8 On a climb up Mount Whitney, the atmospheric pressure drops to 467 mmHg. What is the pressure in terms of the following units?

    a. atm   **b.** torr   **c.** inHg   **d.** Pa

11.2 Pressure and Volume (Boyle's Law)

LEARNING GOAL Use the pressure–volume relationship (Boyle's law) to calculate the unknown pressure or volume when the temperature and amount of gas do not change.

Imagine that you can see air particles hitting the walls inside a bicycle tire pump. What happens to the pressure inside the pump as you push down on the handle? As the volume decreases, there is a decrease in the surface area of the container. The air particles are crowded together, more collisions occur, and the pressure increases within the container.

When a change in one property (in this case, volume) causes a change in another property (in this case, pressure), the properties are related. If the changes occur in opposite directions, the properties have an **inverse relationship**. The inverse relationship between the pressure and volume of a gas is known as **Boyle's law**. The law states that the volume (V) of a sample of gas changes inversely with the pressure (P) of the gas as long as there is no change in the temperature (T) and amount of gas (n), as illustrated in **FIGURE 11.4**.

If the volume or pressure of a gas changes without any change occurring in the temperature or in the amount of the gas, then the final pressure and volume will give the same PV product as the initial pressure and volume. Then we can set the initial and final PV products equal to each other. In the equation for Boyle's law, the initial pressure and volume are written as P_1 and V_1 and the final pressure and volume are written as P_2 and V_2.

REVIEW
Solving Equations (1.4)

Piston

$V_1 = 4\,\text{L}$ $V_2 = 2\,\text{L}$
$P_1 = 1\,\text{atm}$ $P_2 = 2\,\text{atm}$

FIGURE 11.4 Boyle's law: As volume decreases, gas molecules become more crowded, which causes the pressure to increase. Pressure and volume are inversely related.

Boyle's Law

$$P_1 V_1 = P_2 V_2 \quad \text{No change in temperature and amount of gas}$$

SAMPLE PROBLEM 11.3 Calculating Volume When Pressure Changes

TRY IT FIRST

When Louisa had her asthma attack, she was given oxygen through a face mask. The gauge on a 12-L tank of compressed oxygen reads 3800 mmHg. How many liters would this same gas occupy at a final pressure of 570 mmHg when temperature and amount of gas do not change?

CORE CHEMISTRY SKILL
Using the Gas Laws

SOLUTION

STEP 1 **State the given and needed quantities.** We place the gas data in a table by writing the initial pressure and volume as P_1 and V_1 and the final pressure and volume as P_2 and V_2. We see that the pressure decreases from 3800 mmHg to 570 mmHg. Using Boyle's law, we predict that the volume increases.

	Given	Need	Connect
ANALYZE THE PROBLEM	$P_1 = 3800\ \text{mmHg}$ $P_2 = 570\ \text{mmHg}$ $V_1 = 12\ \text{L}$ **Factors that do not change:** T and n	V_2	Boyle's law, $P_1 V_1 = P_2 V_2$ **Predict:** V increases

STEP 2 **Rearrange the gas law equation to solve for the unknown quantity.** For a PV relationship, we use Boyle's law and solve for V_2 by dividing both sides by P_2.

$$P_1 V_1 = P_2 V_2$$

Oxygen therapy increases the oxygen available to the tissues of the body.

If the volume of a gas increases, what will happen to its pressure, if its temperature and amount do not change?

$$\frac{P_1 V_1}{P_2} = \frac{P_2 V_2}{P_2}$$

$$V_2 = V_1 \times \frac{P_1}{P_2}$$

STEP 3 **Substitute values into the gas law equation and calculate.** When we substitute in the values with pressures in units of mmHg, the ratio of pressures (pressure factor) is greater than 1, which increases the volume as predicted.

$$V_2 = 12\ \text{L} \times \frac{3800\ \cancel{\text{mmHg}}}{570\ \cancel{\text{mmHg}}} = 80.\ \text{L}$$

Pressure factor
increases volume

SELF TEST 11.3

In an underground gas reserve, a bubble of methane gas (CH_4) has a volume of 45.0 mL at 1.60 atm pressure.

a. What volume, in milliliters, will the bubble occupy when it reaches the surface where the atmospheric pressure is 744 mmHg, if there is no change in the temperature and amount of gas?

b. What is the new pressure, in atmospheres, when the volume of the methane gas bubble expands to 125 mL, if there is no change in the temperature and amount of gas?

PRACTICE PROBLEMS

Try Practice Problems 11.9 to 11.24

ANSWER

a. 73.5 mL   **b.** 0.576 atm

Chemistry Link to Health

Pressure–Volume Relationship in Breathing

The importance of Boyle's law becomes apparent when you consider the mechanics of breathing. Our lungs are elastic, balloon-like structures contained within an airtight chamber called the thoracic cavity. The diaphragm, a muscle, forms the flexible floor of the cavity.

Inspiration

The process of taking a breath of air begins when the diaphragm contracts and the rib cage expands, causing an increase in the volume of the thoracic cavity. The elasticity of the lungs allows them to expand when the thoracic cavity expands. According to Boyle's law, the pressure inside the lungs decreases when their volume increases, causing the pressure inside the lungs to fall below the pressure of the atmosphere. This difference in pressures produces a *pressure gradient* between the lungs and the atmosphere. In a pressure gradient, molecules flow from an area of higher pressure to an area of lower pressure. During the inhalation phase of breathing, air flows into the lungs (*inspiration*), until the pressure within the lungs becomes equal to the pressure of the atmosphere.

Expiration

Expiration, the exhalation phase of breathing, occurs when the diaphragm relaxes and moves back up into the thoracic cavity to its

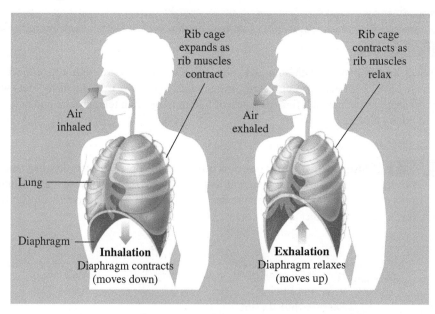

resting position. The volume of the thoracic cavity decreases, which squeezes the lungs and decreases their volume. Now the pressure in the lungs is higher than the pressure of the atmosphere, so air flows out of the lungs. Thus, breathing is a process in which pressure gradients are continuously created between the lungs and the environment because of the changes in the volume and pressure.

PRACTICE PROBLEMS

11.2 Pressure and Volume (Boyle's Law)

11.9 Why do scuba divers need to exhale air when they ascend to the surface of the water?

11.10 Why does a sealed bag of chips expand when you take it to a higher altitude?

11.11 The air in a cylinder with a piston has a volume of 220 mL and a pressure of 650 mmHg.
  a. To obtain a higher pressure inside the cylinder when the temperature and amount of gas do not change, would the cylinder change as shown in **A** or **B**? Explain your choice.

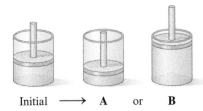

Initial $\longrightarrow$ **A**   or   **B**

  b. If the pressure inside the cylinder increases to 1.2 atm, what is the final volume, in milliliters, of the cylinder?

11.12 A balloon is filled with helium gas. When each of the following changes are made with no change in temperature and amount of gas, which of these diagrams (**A**, **B**, or **C**) shows the final volume of the balloon?

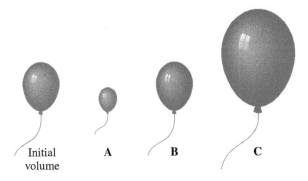

Initial volume   **A**   **B**   **C**

  a. The balloon floats to a higher altitude where the outside pressure is lower.
  b. The balloon is taken inside the house, but the atmospheric pressure does not change.
  c. The balloon is put in a hyperbaric chamber in which the pressure is increased.

11.13 A gas with a volume of 4.0 L is in a closed container. Indicate the changes (*increases, decreases, does not change*) in its pressure when the volume undergoes the following changes at the same temperature and amount of gas:
  a. The volume is compressed to 2.0 L.
  b. The volume expands to 12 L.
  c. The volume is compressed to 0.40 L.

11.14 A gas at a pressure of 2.0 atm is in a closed container. Indicate the changes (*increases, decreases, does not change*) in its volume when the pressure undergoes the following changes at the same temperature and amount of gas:
  a. The pressure increases to 6.0 atm.
  b. The pressure remains at 2.0 atm.
  c. The pressure drops to 0.40 atm.

11.15 A 10.0-L balloon contains helium gas at a pressure of 655 mmHg. What is the final pressure, in millimeters of mercury, when the helium is placed in tanks that have the following volumes, if there is no change in temperature and amount of gas?
  a. 20.0 L   **b.** 2.50 L
  c. 13 800 mL   **d.** 1250 mL

11.16 The air in a 5.00-L tank has a pressure of 1.20 atm. What is the final pressure, in atmospheres, when the air is placed in tanks that have the following volumes, if there is no change in temperature and amount of gas?
  a. 1.00 L   **b.** 2500. mL
  c. 750. mL   **d.** 8.00 L

11.17 A sample of nitrogen (N_2) has a volume of 50.0 L at a pressure of 760. mmHg. What is the final volume, in liters, of the gas at each of the following pressures, if there is no change in temperature and amount of gas?
  a. 725 mmHg   **b.** 1520 mmHg
  c. 0.500 atm   **d.** 850 Torr

11.18 A sample of methane (CH_4) has a volume of 25 mL at a pressure of 0.80 atm. What is the final volume, in milliliters, of the gas at each of the following pressures, if there is no change in temperature and amount of gas?
  a. 0.40 atm   **b.** 2.00 atm
  c. 2500 mmHg   **d.** 80.0 Torr

11.19 A sample of Ar gas has a volume of 5.40 L with an unknown pressure. The gas has a volume of 9.73 L when the pressure is 3.62 atm, with no change in temperature and amount of gas. What was the initial pressure, in atmospheres, of the gas?

11.20 A sample of Ne gas has a pressure of 654 mmHg with an unknown volume. The gas has a pressure of 345 mmHg when the volume is 495 mL, with no change in temperature and amount of gas. What was the initial volume, in milliliters, of the gas?

Applications

11.21 Cyclopropane, C_3H_6, is a general anesthetic. A 5.0-L sample has a pressure of 5.0 atm. What is the final volume, in liters, of this gas given to a patient at a pressure of 1.0 atm with no change in temperature and amount of gas?

11.22 A patient's oxygen tank holds 20.0 L of oxygen (O_2) at a pressure of 15.0 atm. What is the final volume, in liters, of this gas when it is released at a pressure of 1.00 atm with no change in temperature and amount of gas?

11.23 Use the words *inspiration* and *expiration* to describe the part of the breathing cycle that occurs as a result of each of the following:
  a. The diaphragm contracts.
  b. The volume of the lungs decreases.
  c. The pressure within the lungs is less than that of the atmosphere.

11.24 Use the words *inspiration* and *expiration* to describe the part of the breathing cycle that occurs as a result of each of the following:
  a. The diaphragm relaxes, moving up into the thoracic cavity.
  b. The volume of the lungs expands.
  c. The pressure within the lungs is higher than that of the atmosphere.

As the gas in a hot-air balloon is heated, it expands.

11.3 Temperature and Volume (Charles's Law)

LEARNING GOAL Use the temperature–volume relationship (Charles's law) to calculate the unknown temperature or volume when the pressure and amount of gas do not change.

Suppose that you are going to take a ride in a hot-air balloon. The captain turns on a propane burner to heat the air inside the balloon. As the air is heated, it expands and becomes less dense than the air outside, causing the balloon and its passengers to lift off. In 1787, Jacques Charles, a balloonist as well as a physicist, proposed that the volume of a gas is related to the temperature. This proposal became **Charles's law**, which states that the volume (V) of a gas is directly related to the temperature (T) when there is no change in the pressure (P) and amount (n) of gas. A **direct relationship** is one in which the related properties increase or decrease together. For two conditions, initial and final, we can write Charles's law as follows:

Charles's Law

$$\frac{V_1}{T_1} = \frac{V_2}{T_2} \quad \text{No change in pressure and amount of gas}$$

All temperatures used in gas law calculations must be converted to their corresponding Kelvin (K) temperatures.

To determine the effect of changing temperature on the volume of a gas, the pressure and the amount of gas are not changed. If we increase the temperature of a gas sample, the volume of the container must increase (see **FIGURE 11.5**). If the temperature of the gas is decreased, the volume of the container must also decrease when pressure and the amount of gas do not change.

$T_1 = 200 \text{ K}$ $T_2 = 400 \text{ K}$
$V_1 = 1 \text{ L}$ $V_2 = 2 \text{ L}$

FIGURE 11.5 Charles's law: The Kelvin temperature of a gas is directly related to the volume of the gas when there is no change in the pressure and amount of gas.

Why can we predict that the volume of the helium gas will increase when the temperature increases, if the pressure and amount of gas do not change?

SAMPLE PROBLEM 11.4 Calculating Volume When Temperature Changes

TRY IT FIRST

Helium gas is used to inflate the abdomen during laparoscopic surgery. A sample of helium gas has a volume of 5.40 L and a temperature of 15 °C. What is the final volume, in liters, of the gas after the temperature has been increased to 42 °C when the pressure and amount of gas do not change?

SOLUTION

STEP 1 **State the given and needed quantities.** We place the gas data in a table by writing the initial temperature and volume as T_1 and V_1 and the final temperature and volume as T_2 and V_2. We see that the temperature increases from 15 °C to 42 °C. Using Charles's law, we predict that the volume increases.

$$T_1 = 15 \text{ °C} + 273 = 288 \text{ K}$$

$$T_2 = 42 \text{ °C} + 273 = 315 \text{ K}$$

	Given	Need	Connect
ANALYZE THE PROBLEM	$T_1 = 288 \text{ K}$ $T_2 = 315 \text{ K}$ $V_1 = 5.40 \text{ L}$	V_2	Charles's law, $\dfrac{V_1}{T_1} = \dfrac{V_2}{T_2}$
	Factors that do not change: P and n		**Predict:** V increases

STEP 2 **Rearrange the gas law equation to solve for the unknown quantity.** In this problem, we want to know the final volume (V_2) when the temperature increases. Using Charles's law, we solve for V_2 by multiplying both sides by T_2.

$$\frac{V_1}{T_1} = \frac{V_2}{T_2}$$

$$\frac{V_1}{T_1} \times T_2 = \frac{V_2}{T_2} \times T_2$$

$$V_2 = V_1 \times \frac{T_2}{T_1}$$

STEP 3 **Substitute values into the gas law equation and calculate.** From the table, we see that the temperature has increased. Because temperature is directly related to volume, the volume must increase. When we substitute in the values, we see that the ratio of the temperatures (temperature factor) is greater than 1, which increases the volume as predicted.

$$V_2 = 5.40 \text{ L} \times \frac{315 \text{ K}}{288 \text{ K}} = 5.91 \text{ L}$$

Temperature factor
increases volume

SELF TEST 11.4

a. A mountain climber inhales air that has a temperature of −8 °C. If the final volume of air in the lungs is 569 mL at a body temperature of 37 °C, what was the initial volume of air, in milliliters, inhaled by the climber, if the pressure and amount of gas do not change?
b. A hiker inhales 598 mL of air. If the final volume of air in the lungs is 612 mL at a body temperature of 37 °C, what was the initial temperature of the air in degrees Celsius?

ANSWER
a. 486 mL   b. 30. °C

PRACTICE PROBLEMS
Try Practice Problems 11.25 to 11.32

PRACTICE PROBLEMS

11.3 Temperature and Volume (Charles's Law)

11.25 Select the diagram that shows the final volume of a balloon when each of the following changes are made when the pressure and amount of gas do not change:

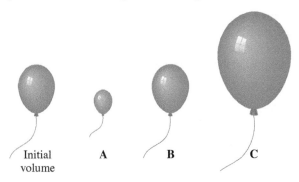

Initial volume   A   B   C

a. The temperature is changed from 100 K to 300 K.
b. The balloon is placed in a freezer.
c. The balloon is first warmed and then returned to its starting temperature.

11.26 Indicate whether the final volume of gas in each of the following is the *same, larger,* or *smaller* than the initial volume, if pressure and amount of gas do not change:
a. A volume of 505 mL of air on a cold winter day at −15 °C is breathed into the lungs, where body temperature is 37 °C.
b. The heater used to heat the air in a hot-air balloon is turned off.
c. A balloon filled with helium at the amusement park is left in a car on a hot day.

11.27 A sample of neon initially has a volume of 2.50 L at 15 °C. What final temperature, in degrees Celsius, is needed to change the volume of the gas to each of the following, if pressure and amount of gas do not change?
a. 5.00 L   b. 1250 mL   c. 7.50 L   d. 3550 mL

11.28 A gas has a volume of 4.00 L at 0 °C. What final temperature, in degrees Celsius, is needed to change the volume of the gas to each of the following, if pressure and amount of gas do not change?
a. 1.50 L   b. 1200 mL   c. 10.0 L   d. 50.0 mL

11.29 A balloon contains 2500 mL of helium gas at 75 °C. What is the final volume, in milliliters, of the gas when the temperature changes to each of the following, if pressure and amount of gas do not change?
a. 55 °C   b. 680. K   c. −25 °C   d. 240. K

11.30 An air bubble has a volume of 0.500 L at 18 °C. What is the final volume, in liters, of the gas when the temperature changes to each of the following, if pressure and amount of gas do not change?
a. 0 °C   b. 425 K   c. −12 °C   d. 575 K

11.31 A gas sample has a volume of 0.256 L with an unknown temperature. The same gas has a volume of 0.198 L when the temperature is 32 °C, with no change in the pressure and amount of gas. What was the initial temperature, in degrees Celsius, of the gas?

11.32 A gas sample has a temperature of 22 °C with an unknown volume. The same gas has a volume of 456 mL when the temperature is 86 °C, with no change in the pressure and amount of gas. What was the initial volume, in milliliters, of the gas?

$T_1 = 200$ K $T_2 = 400$ K
$P_1 = 1$ atm $P_2 = 2$ atm

FIGURE 11.6 Gay-Lussac's law:
When the Kelvin temperature of a gas is doubled and the volume and amount of gas do not change, the pressure also doubles.

11.4 Temperature and Pressure (Gay-Lussac's Law)

LEARNING GOAL Use the temperature–pressure relationship (Gay-Lussac's law) to calculate the unknown temperature or pressure when the volume and amount of gas do not change.

If we could observe the molecules of a gas as the temperature rises, we would notice that they move faster and hit the sides of the container more often and with greater force. If volume and amount of gas do not change, the pressure would increase. In the temperature–pressure relationship known as **Gay-Lussac's law**, the pressure of a gas is directly related to its Kelvin temperature. This means that an increase in temperature increases the pressure of a gas, and a decrease in temperature decreases the pressure of the gas as long as the volume and amount of gas do not change (see **FIGURE 11.6**).

Gay-Lussac's Law

$$\frac{P_1}{T_1} = \frac{P_2}{T_2} \quad \text{No change in volume and amount of gas}$$

All temperatures used in gas law calculations must be converted to their corresponding Kelvin (K) temperatures.

SAMPLE PROBLEM 11.5 Calculating Pressure When Temperature Changes

TRY IT FIRST

Home oxygen tanks can be dangerous if they are heated, because they can explode. Suppose an oxygen tank has a pressure of 120 atm at a room temperature of 25 °C. If a fire in the room causes the temperature of the gas inside the oxygen tank to reach 402 °C, what is its pressure, in atmospheres, if the volume and amount of gas do not change? The oxygen tank may rupture if the pressure inside exceeds 180 atm. Would you expect it to rupture?

SOLUTION

STEP 1 State the given and needed quantities. We place the gas data in a table by writing the initial temperature and pressure as T_1 and P_1 and the final temperature and pressure as T_2 and P_2. We see that the temperature increases from 25 °C to 402 °C. Using Gay-Lussac's law, we predict that the pressure increases.

$T_1 = 25\,°C + 273 = 298$ K
$T_2 = 402\,°C + 273 = 675$ K

ENGAGE 11.5

Why can we predict that the pressure in the oxygen tank will increase when the temperature increases, if the volume and amount of gas do not change?

	Given	Need	Connect
ANALYZE THE PROBLEM	$P_1 = 120$ atm $T_1 = 298$ K $T_2 = 675$ K **Factors that do not change:** V and n	P_2	Gay-Lussac's law, $\dfrac{P_1}{T_1} = \dfrac{P_2}{T_2}$ **Predict:** P increases

STEP 2 Rearrange the gas law equation to solve for the unknown quantity. Using Gay-Lussac's law, we solve for P_2 by multiplying both sides by T_2.

$$\frac{P_1}{T_1} = \frac{P_2}{T_2}$$

$$\frac{P_1}{T_1} \times T_2 = \frac{P_2}{\cancel{T_2}} \times \cancel{T_2}$$

$$P_2 = P_1 \times \frac{T_2}{T_1}$$

STEP 3   **Substitute values into the gas law equation and calculate.** When we substitute in the values, we see that the ratio of the temperatures (temperature factor) is greater than 1, which increases the pressure as predicted.

$$P_2 = 120 \text{ atm} \times \frac{675 \text{ K}}{298 \text{ K}} = 270 \text{ atm}$$

Temperature factor
increases pressure

Because the calculated pressure of 270 atm exceeds the limit of 180 atm, we would expect the oxygen tank to rupture.

SELF TEST 11.5

In a storage area of a hospital where the temperature has reached 55 °C, the pressure of oxygen gas in a 15.0-L steel cylinder is 965 Torr.

a. To what temperature, in degrees Celsius, would the gas have to be cooled to reduce the pressure to 850. Torr, when the volume and the amount of the gas do not change?
b. What is the final pressure, in millimeters of mercury, when the temperature of the oxygen gas drops to 24 °C, and the volume and the amount of the gas do not change?

ANSWER

a. 16 °C   **b.** 874 mmHg

Cylinders of oxygen gas are placed in a hospital storage room.

Vapor Pressure and Boiling Point

When liquid molecules with sufficient kinetic energy break away from the surface, they become gas particles or vapor. In an open container, all the liquid will eventually evaporate. In a closed container, the vapor accumulates and creates pressure called **vapor pressure**. Each liquid exerts its own vapor pressure at a given temperature. As temperature increases, more vapor forms, and vapor pressure increases. **TABLE 11.4** lists the vapor pressure of water at various temperatures.

A liquid reaches its boiling point when its vapor pressure becomes equal to the external pressure. As boiling occurs, bubbles of the gas form within the liquid and quickly rise to the

100 °C

Atmospheric pressure 760 mmHg

760 mmHg

Vapour pressure in the bubble equals the atmospheric pressure

TABLE 11.4 Vapor Pressure of Water	
Temperature (°C)	Vapor Pressure (mmHg)
0	5
10	9
20	18
30	32
37*	47
40	55
50	93
60	149
70	234
80	355
90	528
100	760
*Normal body temperature	

TABLE 11.5 Pressure and the Boiling Point of Water	
Pressure (mmHg)	Boiling Point (°C)
270	70
467	87
630	95
752	99
760	100
900	105
1075	110
1520 (2 atm)	120
3800 (5 atm)	160
7600 (10 atm)	180

surface. For example, at an atmospheric pressure of 760 mmHg, water will boil at 100 °C, the temperature at which its vapor pressure reaches 760 mmHg.

At high altitudes, where atmospheric pressures are lower than 760 mmHg, the boiling point of water is lower than 100 °C. For example, a typical atmospheric pressure in Denver is 630 mmHg. This means that water in Denver boils when the vapor pressure is 630 mmHg. From **TABLE 11.5**, we see that water has a vapor pressure of 630 mmHg at 95 °C, which means that water boils at 95 °C in Denver.

In a closed container such as a pressure cooker, a pressure greater than 1 atm can be obtained, which means that water boils at a temperature higher than 100 °C. Laboratories and hospitals use closed containers called *autoclaves* to sterilize laboratory and surgical equipment.

An autoclave used to sterilize equipment attains a temperature higher than 100 °C.

PRACTICE PROBLEMS

Try Practice Problems 11.33 to 11.42

PRACTICE PROBLEMS

11.4 Temperature and Pressure (Gay-Lussac's Law)

11.33 Calculate the final pressure, in millimeters of mercury, for each of the following, if volume and amount of gas do not change:
   a. A gas with an initial pressure of 1200 Torr at 155 °C is cooled to 0 °C.
   b. A gas in an aerosol can at an initial pressure of 1.40 atm at 12 °C is heated to 35 °C.

11.34 Calculate the final pressure, in atmospheres, for each of the following, if volume and amount of gas do not change:
   a. A gas with an initial pressure of 1.20 atm at 75 °C is cooled to −32 °C.
   b. A sample of N_2 with an initial pressure of 780. mmHg at −75 °C is heated to 28 °C.

11.35 Calculate the final temperature, in degrees Celsius, for each of the following, if volume and amount of gas do not change:
   a. A sample of xenon gas at 25 °C and 740. mmHg is cooled to give a pressure of 620. mmHg.
   b. A tank of argon gas with a pressure of 0.950 atm at −18 °C is heated to give a pressure of 1250 Torr.

11.36 Calculate the final temperature, in degrees Celsius, for each of the following, if volume and amount of gas do not change:
   a. A sample of helium gas with a pressure of 250 Torr at 0 °C is heated to give a pressure of 1500 Torr.
   b. A sample of air at 40 °C and 740. mmHg is cooled to give a pressure of 680. mmHg.

11.37 A gas sample has a pressure of 744 mmHg when the temperature is 22 °C. What is the final temperature, in degrees Celsius, when the pressure is 766 mmHg, with no change in the volume and amount of gas?

11.38 A gas sample has a pressure of 2.35 atm when the temperature is −15 °C. What is the final pressure, in atmospheres, when the temperature is 46 °C, with no change in the volume and amount of gas?

11.39 Explain each of the following observations:
   a. Water boils at 87 °C on the top of Mount Whitney.
   b. Food cooks more quickly in a pressure cooker than in an open pan.

11.40 Explain each of the following observations:
   a. Boiling water at sea level is hotter than boiling water in the mountains.
   b. Water used to sterilize surgical equipment is heated to 120 °C at 2.0 atm in an autoclave.

Applications

11.41 A tank contains isoflurane, an inhaled anesthetic, at a pressure of 1.8 atm and 5 °C. What is the pressure, in atmospheres, if the gas is warmed to a temperature of 22 °C, if volume and amount of gas do not change?

11.42 Bacteria and viruses are inactivated by temperatures above 135 °C. An autoclave contains steam at 1.00 atm and 100 °C. What is the pressure, in atmospheres, when the temperature of the steam in the autoclave reaches 135 °C, if volume and amount of gas do not change?

11.5 The Combined Gas Law

LEARNING GOAL Use the combined gas law to calculate the unknown pressure, volume, or temperature of a gas when changes in two of these properties are given and the amount of gas does not change.

All of the pressure–volume–temperature relationships for gases that we have studied may be combined into a single relationship called the **combined gas law**. This expression is useful for studying the effect of changes in two of these variables on the third as long as the amount of gas (number of moles) does not change.

Combined Gas Law

$$\frac{P_1V_1}{T_1} = \frac{P_2V_2}{T_2}$$ No change in number of moles of gas

By using the combined gas law, we can derive any of the gas laws by omitting those properties that do not change, as seen in **TABLE 11.6**.

TABLE 11.6 Summary of Gas Laws

Combined Gas Law	Properties That Do Not Change	Relationship	Name of Gas Law
$\dfrac{P_1V_1}{\cancel{T_1}} = \dfrac{P_2V_2}{\cancel{T_2}}$	T, n	$P_1V_1 = P_2V_2$	Boyle's law
$\dfrac{\cancel{P_1}V_1}{T_1} = \dfrac{\cancel{P_2}V_2}{T_2}$	P, n	$\dfrac{V_1}{T_1} = \dfrac{V_2}{T_2}$	Charles's law
$\dfrac{P_1\cancel{V_1}}{T_1} = \dfrac{P_2\cancel{V_2}}{T_2}$	V, n	$\dfrac{P_1}{T_1} = \dfrac{P_2}{T_2}$	Gay-Lussac's law

ENGAGE 11.6

Why does the pressure of a gas decrease to one-fourth of its initial pressure when the volume of the gas doubles and the Kelvin temperature decreases by half, if the amount of gas does not change?

SAMPLE PROBLEM 11.6 Using the Combined Gas Law

TRY IT FIRST

A 25.0-mL bubble is released from a diver's air tank at a pressure of 4.00 atm and a temperature of 11 °C. What is the volume, in milliliters, of the bubble when it reaches the ocean surface where the pressure is 1.00 atm and the temperature is 18 °C? (Assume the amount of gas in the bubble does not change.)

SOLUTION

STEP 1 **State the given and needed quantities.** We list the properties that change, which are the pressure, volume, and temperature. The temperatures in degrees Celsius must be changed to kelvins.

$$T_1 = 11\,°C + 273 = 284\ K$$
$$T_2 = 18\,°C + 273 = 291\ K$$

	Given	Need	Connect
ANALYZE THE PROBLEM	$P_1 = 4.00$ atm $P_2 = 1.00$ atm $V_1 = 25.0$ mL $T_1 = 284$ K $T_2 = 291$ K **Factor that does not change:** n	V_2	combined gas law, $\dfrac{P_1V_1}{T_1} = \dfrac{P_2V_2}{T_2}$

Under water, the pressure on a diver is greater than the atmospheric pressure.

STEP 2 **Rearrange the gas law equation to solve for the unknown quantity.** Using the combined gas law, we solve for V_2 by multiplying both sides by T_2 and dividing both sides by P_2.

$$\frac{P_1 V_1}{T_1} = \frac{P_2 V_2}{T_2}$$

$$\frac{P_1 V_1}{T_1} \times \frac{T_2}{P_2} = \frac{P_2 V_2}{T_2} \times \frac{T_2}{P_2}$$

$$V_2 = V_1 \times \frac{P_1}{P_2} \times \frac{T_2}{T_1}$$

ENGAGE 11.7

Rearrange the combined gas law to solve for T_1.

STEP 3 **Substitute values into the gas law equation and calculate.** From the data table, we determine that both the pressure decrease and the temperature increase will increase the volume.

$$V_2 = 25.0 \text{ mL} \times \underbrace{\frac{4.00 \text{ atm}}{1.00 \text{ atm}}}_{\substack{\text{Pressure} \\ \text{factor} \\ \text{increases} \\ \text{volume}}} \times \underbrace{\frac{291 \text{ K}}{284 \text{ K}}}_{\substack{\text{Temperature} \\ \text{factor} \\ \text{increases} \\ \text{volume}}} = 102 \text{ mL}$$

However, in situations where the unknown value is decreased by one change but increased by the second change, it is difficult to predict the overall change for the unknown.

SELF TEST 11.6

A weather balloon is filled with 15.0 L of helium at a temperature of 25 °C and a pressure of 685 mmHg.

a. What is the pressure, in millimeters of mercury, of the helium in the balloon in the upper atmosphere when the final temperature is −35 °C and the final volume becomes 34.0 L, if the amount of gas does not change?
b. What is the temperature, in degrees Celsius, if the helium in the balloon has a final volume of 22.0 L and a final pressure of 426 mmHg, if the amount of gas does not change?

ANSWER

a. 241 mmHg   **b.** −1 °C

PRACTICE PROBLEMS

Try Practice Problems 11.43 to 11.48

PRACTICE PROBLEMS

11.5 The Combined Gas Law

11.43 Rearrange the variables in the combined gas law to solve for T_2.

11.44 Rearrange the variables in the combined gas law to solve for P_2.

11.45 A sample of helium gas has a volume of 6.50 L at a pressure of 845 mmHg and a temperature of 25 °C. What is the final pressure of the gas, in atmospheres, when the volume and temperature of the gas sample are changed to the following, if the amount of gas does not change?
a. 1850 mL and 325 K   **b.** 2.25 L and 12 °C
c. 12.8 L and 47 °C

11.46 A sample of argon gas has a volume of 735 mL at a pressure of 1.20 atm and a temperature of 112 °C. What is the final volume of the gas, in milliliters, when the pressure and temperature of the gas sample are changed to the following, if the amount of gas does not change?
a. 658 mmHg and 281 K   **b.** 0.55 atm and 75 °C
c. 15.4 atm and −15 °C

Applications

11.47 A 124-mL bubble of hot gas initially at 212 °C and 1.80 atm is emitted from an active volcano. What is the final temperature, in degrees Celsius, of the gas in the bubble outside the volcano when the final volume of the bubble is 138 mL and the pressure is 0.800 atm, if the amount of gas does not change?

11.48 A scuba diver 60 ft below the ocean surface inhales 50.0 mL of compressed air from a scuba tank at a pressure of 3.00 atm and a temperature of 8 °C. What is the final pressure of air, in atmospheres, in the lungs when the gas expands to 150.0 mL at a body temperature of 37 °C, if the amount of gas does not change?

11.6 Volume and Moles (Avogadro's Law)

LEARNING GOAL Use Avogadro's law to calculate the unknown amount or volume of a gas when the pressure and temperature do not change.

In our study of the gas laws, we have looked at changes in properties for a specified amount (n) of gas. Now we consider how the properties of a gas change when there is a change in the number of moles or grams of the gas.

When you blow up a balloon, its volume increases because you add more air molecules. If the balloon has a small hole in it, air leaks out, causing its volume to decrease. In 1811, Amedeo Avogadro formulated **Avogadro's law**, which states that the volume of a gas is directly related to the number of moles of a gas when temperature and pressure do not change. For example, if the number of moles of a gas is doubled, then the volume will also double as long as we do not change the pressure or the temperature (see **FIGURE 11.7**). When pressure and temperature do not change, we can write Avogadro's law as follows:

Avogadro's Law

$$\frac{V_1}{n_1} = \frac{V_2}{n_2}$$ No change in pressure and temperature

REVIEW
Using Molar Mass as a Conversion Factor (7.3)

$n_1 = 1$ mol $n_2 = 2$ mol
$V_1 = 1$ L $V_2 = 2$ L

FIGURE 11.7 Avogadro's law: The volume of a gas is directly related to the number of moles of the gas. If the number of moles is doubled, the volume must double when temperature and pressure do not change.

SAMPLE PROBLEM 11.7 Calculating Volume for a Change in Moles

TRY IT FIRST

A weather balloon with a volume of 44 L is filled with 2.0 mol of helium. What is the final volume, in liters, if helium is added to give a total of 5.0 mol of helium, if the pressure and temperature do not change?

SOLUTION

STEP 1 **State the given and needed quantities.** We list those properties that change, which are volume and amount (moles). Because there is an increase in the number of moles of gas, we can predict that the volume increases.

	Given	Need	Connect
ANALYZE THE PROBLEM	$V_1 = 44$ L $n_1 = 2.0$ mol $n_2 = 5.0$ mol **Factors that do not change:** P and T	V_2	Avogadro's law, $\frac{V_1}{n_1} = \frac{V_2}{n_2}$ **Predict:** V increases

STEP 2 **Rearrange the gas law equation to solve for the unknown quantity.** Using Avogadro's law, we can solve for V_2 by multiplying both sides of the equation by n_2.

$$\frac{V_1}{n_1} = \frac{V_2}{n_2}$$

$$\frac{V_1}{n_1} \times n_2 = \frac{V_2}{n_2} \times n_2$$

$$V_2 = V_1 \times \frac{n_2}{n_1}$$

STEP 3 **Substitute values into the gas law equation and calculate.** When we substitute in the values, we see that the ratio of the moles (mole factor) is greater than 1, which increases the volume as predicted.

$$V_2 = 44 \text{ L} \times \frac{5.0 \text{ mol}}{2.0 \text{ mol}} = 110 \text{ L}$$

Mole factor
increases volume

ENGAGE 11.8
Why can we predict that the volume of a balloon increases when the number of moles of gas increases, if the pressure and temperature do not change?

A balloon containing 8.00 g of oxygen gas has a volume of 5.00 L.

a. What is the volume, in liters, after 4.00 g of oxygen gas is added to the 8.00 g of oxygen in the balloon, if the temperature and pressure do not change?

b. More oxygen gas is added to the balloon until the volume is 12.6 L. How many grams of oxygen gas are in the balloon, if the pressure and temperature do not change?

ANSWER

a. 7.50 L   **b.** 20.2 g of oxygen

PRACTICE PROBLEMS

Try Practice Problems 11.49 to 11.52

The molar volume of a gas at STP is about the same as the volume of three basketballs.

Molar Volume Conversion Factors

1 mol of gas $=$ 22.4 L (STP)

$$\frac{22.4\ \text{L (STP)}}{1\ \text{mol gas}} \quad \text{and} \quad \frac{1\ \text{mol gas}}{22.4\ \text{L (STP)}}$$

STP and Molar Volume

Using Avogadro's law, we can say that any two gases will have equal volumes if they contain the same number of moles of gas at the same temperature and pressure. To help us make comparisons between gases, arbitrary conditions called *standard temperature* (273 K) and *standard pressure* (1 atm), together abbreviated **STP**, were selected by scientists:

STP Conditions

Standard temperature is *exactly* 0 °C (273 K).
Standard pressure is *exactly* 1 atm (760 mmHg).

At STP, 1 mol of any gas occupies a volume of 22.4 L, which is about the same as the volume of three basketballs. This volume, 22.4 L, of any gas is called the **molar volume** (see **FIGURE 11.8**). When a gas is at STP conditions (0 °C and 1 atm), its molar volume can be used to write conversion factors between the number of moles of gas and its volume, in liters.

$V = 22.4$ L

1 mol of He
4.003 g of He
273 K
1 atm

1 mol of O_2
32.00 g of O_2
273 K
1 atm

1 mol of N_2
28.02 g of N_2
273 K
1 atm

FIGURE 11.8 Avogadro's law indicates that 1 mol of any gas at STP has a volume of 22.4 L.

What is the volume, in liters, of 64.0 g of O_2 gas at STP?

SOLUTION

Why would the molar volume of a gas be greater than 22.4 L if the pressure is 1 atm and the temperature is 100 °C?

STEP 1   State the given and needed quantities.

ANALYZE THE PROBLEM	Given	Need	Connect
	64.0 g of $O_2(g)$ at STP	liters of O_2 gas at STP	molar mass, molar volume (STP)

STEP 2 Write a plan to calculate the needed quantity.

grams of O_2 Molar mass moles of O_2 Molar volume liters of O_2

STEP 3 Write the equalities and conversion factors including 22.4 L/mol at STP.

$$1 \text{ mol of } O_2 = 32.00 \text{ g of } O_2$$
$$\frac{32.00 \text{ g } O_2}{1 \text{ mol } O_2} \quad \text{and} \quad \frac{1 \text{ mol } O_2}{32.00 \text{ g } O_2}$$

$$1 \text{ mol of } O_2 = 22.4 \text{ L of } O_2 \text{ (STP)}$$
$$\frac{22.4 \text{ L } O_2 \text{ (STP)}}{1 \text{ mol } O_2} \quad \text{and} \quad \frac{1 \text{ mol } O_2}{22.4 \text{ L } O_2 \text{ (STP)}}$$

STEP 4 Set up the problem with factors to cancel units.

$$64.0 \text{ g } O_2 \times \frac{1 \text{ mol } O_2}{32.00 \text{ g } O_2} \times \frac{22.4 \text{ L } O_2 \text{ (STP)}}{1 \text{ mol } O_2} = 44.8 \text{ L of } O_2 \text{ (STP)}$$

SELF TEST 11.8

a. How many grams of $Cl_2(g)$ are in 5.00 L of $Cl_2(g)$ at STP?
b. What is the volume, in liters, of 42.4 g of $O_2(g)$ at STP?

ANSWER
a. 15.8 g of Cl_2 **b.** 29.7 L of O_2

PRACTICE PROBLEMS
Try Practice Problems 11.53 and 11.54

PRACTICE PROBLEMS

11.6 Volume and Moles (Avogadro's Law)

11.49 What happens to the volume of a bicycle tire or a basketball when you use an air pump to add air?

11.50 Sometimes when you blow up a balloon and release it, it flies around the room. What is happening to the air in the balloon and its volume?

11.51 A sample containing 1.50 mol of Ne gas has an initial volume of 11.00 L. What is the final volume, in liters, when each of the following occurs and pressure and temperature do not change?
  a. A leak allows one-half of Ne atoms to escape.
  b. A sample of 3.50 mol of Ne is added to the 1.50 mol of Ne gas in the container.
  c. A sample of 25.0 g of Ne is added to the 1.50 mol of Ne gas in the container.

11.52 A sample containing 4.80 g of O_2 gas has an initial volume of 15.0 L. What is the final volume, in liters, when each of the following occurs and pressure and temperature do not change?

  a. A sample of 0.500 mol of O_2 is added to the 4.80 g of O_2 in the container.
  b. A sample of 2.00 g of O_2 is removed.
  c. A sample of 4.00 g of O_2 is added to the 4.80 g of O_2 gas in the container.

11.53 Use the molar volume to calculate each of the following at STP:
  a. the number of moles of O_2 in 2.24 L of O_2 gas
  b. the volume, in liters, occupied by 2.50 mol of N_2 gas
  c. the volume, in liters, occupied by 50.0 g of Ar gas
  d. the number of grams of H_2 in 1620 mL of H_2 gas

11.54 Use the molar volume to calculate each of the following at STP:
  a. the number of moles of CO_2 in 4.00 L of CO_2 gas
  b. the volume, in liters, occupied by 0.420 mol of He gas
  c. the volume, in liters, occupied by 6.40 g of O_2 gas
  d. the number of grams of Ne contained in 11.2 L of Ne gas

11.7 The Ideal Gas Law

LEARNING GOAL Use the ideal gas law equation to calculate the unknown P, V, T, or n of a gas when given three of the four values in the ideal gas law equation. Calculate the molar mass of a gas.

The **ideal gas law** is the relationship between the four properties used in the measurement of a gas—pressure (P), volume (V), temperature (T), and amount of a gas (n).

Ideal Gas Law

$$PV = nRT$$

Rearranging the ideal gas law equation shows that the four gas properties equal the **ideal gas constant, R**.

$$\frac{PV}{nT} = R$$

To calculate the value of R, we substitute the STP conditions for molar volume into the expression: 1.00 mol of any gas occupies 22.4 L at STP (273 K and 1.00 atm).

$$R = \frac{(1.00 \text{ atm})(22.4 \text{ L})}{(1.00 \text{ mol})(273 \text{ K})} = \frac{0.0821 \text{ L} \cdot \text{atm}}{\text{mol} \cdot \text{K}}$$

ENGAGE 11.10

When you use the value of 0.0821 for R, what unit must you use for pressure?

The value for R is 0.0821 L · atm per mol · K. If we use 760. mmHg for the pressure, we obtain another useful value for R of 62.4 L · mmHg per mol · K.

$$R = \frac{(760. \text{ mmHg})(22.4 \text{ L})}{(1.00 \text{ mol})(273 \text{ K})} = \frac{62.4 \text{ L} \cdot \text{mmHg}}{\text{mol} \cdot \text{K}}$$

The ideal gas law is a useful expression when you are given the quantities for any three of the four properties of a gas. Although real gases show some deviations in behavior, the ideal gas law closely approximates the behavior of real gases at typical conditions. In working problems using the ideal gas law, the units of each variable must match the units in the R you select.

Ideal Gas Constant (R)	$\dfrac{0.0821 \text{ L} \cdot \text{atm}}{\text{mol} \cdot \text{K}}$	$\dfrac{62.4 \text{ L} \cdot \text{mmHg}}{\text{mol} \cdot \text{K}}$
Pressure (P)	atm	mmHg
Volume (V)	L	L
Amount (n)	mol	mol
Temperature (T)	K	K

CORE CHEMISTRY SKILL

Using the Ideal Gas Law

Dinitrogen oxide is used as an anesthetic in dentistry.

SAMPLE PROBLEM 11.9   Using the Ideal Gas Law

TRY IT FIRST

Dinitrogen oxide, N_2O, which is used in dentistry, is an anesthetic called laughing gas. What is the pressure, in atmospheres, of 0.350 mol of N_2O gas at 22 °C in a 5.00-L container?

SOLUTION

STEP 1 **State the given and needed quantities.** When three of the four quantities (P, V, n, and T) are known, we use the ideal gas law equation to solve for the unknown quantity. It is helpful to organize the data in a table. The temperature is converted from degrees Celsius to kelvins so that the units of V, n, and T match the unit of the gas constant, R.

	Given	Need	Connect
ANALYZE THE PROBLEM	$V = 5.00$ L $n = 0.350$ mol $T = 22\ °C + 273 = 295$ K	P	ideal gas law, $PV = nRT$ $R = \dfrac{0.0821 \text{ L} \cdot \text{atm}}{\text{mol} \cdot \text{K}}$

STEP 2 **Rearrange the ideal gas law equation to solve for the needed quantity.** By dividing both sides of the ideal gas law equation by V, we solve for pressure, P.

$$PV = nRT \quad \text{Ideal gas law equation}$$

$$\frac{P\cancel{V}}{\cancel{V}} = \frac{nRT}{V}$$

$$P = \frac{nRT}{V}$$

STEP 3 Substitute the gas data into the equation, and calculate the needed quantity.

$$P = \dfrac{0.350 \ \cancel{mol} \times \dfrac{0.0821 \ \cancel{L} \cdot atm}{\cancel{mol} \cdot \cancel{K}} \times 295 \ \cancel{K}}{5.00 \ \cancel{L}} = 1.70 \ atm$$

SELF TEST 11.9

a. Chlorine gas, Cl_2, is used to purify water. How many moles of chlorine gas are in a 7.00-L tank if the gas has a pressure of 865 mmHg and a temperature of 24 °C?

b. Isobutane gas, C_4H_{10}, is used as a propellant in aerosol cans of whipping cream, shaving cream, and suntan spray. What is the pressure, in atmospheres, of 0.125 mol of isobutane in an aerosol can at 21 °C with a volume of 465 mL?

ANSWER

a. 0.327 mol of Cl_2 **b.** 6.49 atm

An aerosol spray can contains a gas the propels the contents.

Many times we need to know the amount of gas, in grams. Then the ideal gas law equation can be rearranged to solve for the amount (n) of gas, which is converted to mass in grams using its molar mass as shown in Sample Problem 11.10.

SAMPLE PROBLEM 11.10   Calculating Mass Using the Ideal Gas Law

TRY IT FIRST

Butane, C_4H_{10}, is used as a fuel for camping stoves. If you have 108 mL of butane gas at 715 mmHg and 25 °C, what is the mass, in grams, of butane?

When camping, butane is used as a fuel for a portable stove.

SOLUTION

STEP 1 **State the given and needed quantities.** When three of the quantities (P, V, and T) are known, we use the ideal gas law equation to solve for the unknown quantity moles (n). Because the pressure is given in mmHg, we will use R in mmHg. The volume given in milliliters (mL) is converted to a volume in liters (L). The temperature is converted from degrees Celsius to kelvins.

	Given	Need	Connect
ANALYZE THE PROBLEM	$P = 715 \ mmHg$ $V = 108 \ mL \ (0.108 \ L)$ $T = 25 \ °C + 273 = 298 \ K$	n	ideal gas law, $PV = nRT$ $R = \dfrac{62.4 \ L \cdot mmHg}{mol \cdot K}$ molar mass

STEP 2 **Rearrange the ideal gas law equation to solve for the needed quantity.** By dividing both sides of the ideal gas law equation by RT, we solve for moles, n.

$PV = n \ RT$ Ideal gas law equation

$$\dfrac{PV}{RT} = \dfrac{n \ \cancel{RT}}{\cancel{RT}}$$

$$n = \dfrac{PV}{RT}$$

STEP 3 Substitute the gas data into the equation and calculate the needed quantity.

$$n = \frac{715 \text{ mmHg} \times 0.108 \text{ L}}{\dfrac{62.4 \text{ L} \cdot \text{mmHg}}{\text{mol} \cdot \text{K}} \times 298 \text{ K}} = 0.004\ 15 \text{ mol } (4.15 \times 10^{-3} \text{ mol})$$

Now we can convert the moles of butane to grams using its molar mass of 58.12 g/mol:

$$0.004\ 15 \text{ mol } C_4H_{10} \times \frac{58.12 \text{ g } C_4H_{10}}{1 \text{ mol } C_4H_{10}} = 0.241 \text{ g of } C_4H_{10}$$

SELF TEST 11.10

a. What is the volume, in liters, of 1.20 g of carbon monoxide at 8 °C, if it has a pressure of 724 mmHg?
b. What is the pressure, in atmospheres, of 1.23 g of CH_4 gas in a 3.22-L container at 37 °C?

PRACTICE PROBLEMS

Try Practice Problems 11.55 to 11.60, 11.63 and 11.64

ANSWER

a. 1.04 L   **b.** 0.606 atm

Molar Mass of a Gas

Another use of the ideal gas law is to determine the molar mass of a gas. If the mass, in grams, of the gas is known, we calculate the number of moles of the gas using the ideal gas law equation. Then the molar mass (g/mol) can be determined.

SAMPLE PROBLEM 11.11   Molar Mass of a Gas Using the Ideal Gas Law

TRY IT FIRST

What is the molar mass, in grams per mole, of a gas if a 3.16-g sample of gas at 0.750 atm and 45 °C occupies a volume of 2.05 L?

SOLUTION

STEP 1   State the given and needed quantities.

	Given	Need	Connect
ANALYZE THE PROBLEM	$P = 0.750$ atm $V = 2.05$ L $T = 45\ °C + 273 = 318$ K mass $= 3.16$ g	n molar mass	ideal gas law, $PV = nRT$ $R = \dfrac{0.0821 \text{ L} \cdot \text{atm}}{\text{mol} \cdot \text{K}}$

STEP 2   Rearrange the ideal gas law equation to solve for the number of moles. To solve for moles, n, divide both sides of the ideal gas law equation by RT.

$$PV = n\,RT \quad \text{Ideal gas law equation}$$

$$\frac{PV}{RT} = \frac{n\,RT}{RT}$$

$$n = \frac{PV}{RT}$$

$$n = \frac{0.750 \text{ atm} \times 2.05 \text{ L}}{\dfrac{0.0821 \text{ L} \cdot \text{atm}}{\text{mol} \cdot \text{K}} \times 318 \text{ K}} = 0.0589 \text{ mol}$$

STEP 3 Obtain the molar mass by dividing the given number of grams by the number of moles.

$$\text{Molar mass} = \frac{\text{mass}}{\text{moles}} = \frac{3.16 \text{ g}}{0.0589 \text{ mol}} = 53.7 \text{ g/mol}$$

SELF TEST 11.11

a. What is the molar mass, in grams per mole, of an unknown gas in a 1.50-L container, if 0.488 g of the gas has a pressure of 0.0750 atm at 19 °C?

b. What is the molar mass, in grams per mole, of an unknown gas with a volume of 0.583 L, when 0.213 g of the gas has a pressure of 46.5 mmHg at 25 °C?

ANSWER

a. 104 g/mol   **b.** 146 g/mol

PRACTICE PROBLEMS

Try Practice Problems 11.61 and 11.62

PRACTICE PROBLEMS

11.7 The Ideal Gas Law

11.55 Calculate the pressure, in atmospheres, of 2.00 mol of helium gas in a 10.0-L container at 27 °C.

11.56 What is the volume, in liters, of 4.00 mol of methane gas, CH_4, at 18 °C and 1.40 atm?

11.57 An oxygen gas container has a volume of 20.0 L. How many grams of oxygen are in the container, if the gas has a pressure of 845 mmHg at 22 °C?

11.58 A 10.0-g sample of krypton has a temperature of 25 °C at 575 mmHg. What is the volume, in milliliters, of the krypton gas?

11.59 A 25.0-g sample of nitrogen, N_2, has a volume of 50.0 L and a pressure of 630. mmHg. What is the temperature, in kelvins and degrees Celsius, of the gas?

11.60 A 0.226-g sample of carbon dioxide, CO_2, has a volume of 525 mL and a pressure of 455 mmHg. What is the temperature, in kelvins and degrees Celsius, of the gas?

11.61 Determine the molar mass of each of the following gases:
   a. 0.84 g of a gas that occupies 450 mL at 0 °C and 1.00 atm (STP)
   b. 1.28 g of a gas that occupies 1.00 L at 0 °C and 760 mmHg (STP)

   c. 1.48 g of a gas that occupies 1.00 L at 685 mmHg and 22 °C
   d. 2.96 g of a gas that occupies 2.30 L at 0.95 atm and 24 °C

11.62 Determine the molar mass of each of the following gases:
   a. 2.90 g of a gas that occupies 0.500 L at 0 °C and 1.00 atm (STP)
   b. 1.43 g of a gas that occupies 2.00 L at 0 °C and 760 mmHg (STP)
   c. 0.726 g of a gas that occupies 855 mL at 1.20 atm and 18 °C
   d. 2.32 g of a gas that occupies 1.23 L at 685 mmHg and 25 °C

Applications

11.63 A single-patient hyperbaric chamber has a volume of 640 L. At a temperature of 24 °C, how many grams of oxygen are needed to give a pressure of 1.6 atm?

11.64 A multipatient hyperbaric chamber has a volume of 3400 L. At a temperature of 22 °C, how many grams of oxygen are needed to give a pressure of 2.4 atm?

11.8 Gas Laws and Chemical Reactions

LEARNING GOAL Calculate the mass or volume of a gas that reacts or forms in a chemical reaction.

Gases are involved as reactants and products in many chemical reactions. For example, we have seen that the combustion of organic fuels with oxygen gas produces carbon dioxide gas and water vapor. In combination reactions, we have seen that hydrogen gas and nitrogen gas react to form ammonia gas, and hydrogen gas and oxygen gas produce water. Typically, the information given for a gas in a reaction is its pressure (P), volume (V), and temperature (T). Then we can use the ideal gas law equation to determine the moles of a gas in a reaction. If we are given the number of moles for one of the gases in a reaction, we can use a mole–mole factor to determine the moles of any other substance.

SAMPLE PROBLEM 11.12   Gases in Chemical Reactions

TRY IT FIRST

Limestone ($CaCO_3$) reacts with HCl to produce carbon dioxide gas, water, and aqueous calcium chloride.

$$2HCl(aq) + CaCO_3(s) \longrightarrow CO_2(g) + H_2O(l) + CaCl_2(aq)$$

How many liters of CO_2 are produced at 752 mmHg and 24 °C from a 25.0-g sample of limestone?

SOLUTION

STEP 1   State the given and needed quantities.

	Given	Need	Connect
ANALYZE THE PROBLEM	25.0 g of $CaCO_3$ $P = 752$ mmHg $T = 24\ °C + 273 = 297$ K	V of $CO_2(g)$	ideal gas law, $PV = nRT$ $R = \dfrac{62.4\ L \cdot mmHg}{mol \cdot K}$ molar mass
	Equation		
	$2HCl(aq) + CaCO_3(s) \longrightarrow CO_2(g) + H_2O(l) + CaCl_2(aq)$		

STEP 2   Write a plan to convert the given quantity to the needed moles.

grams of $CaCO_3$ Molar mass   moles of $CaCO_3$ Mole–mole factor   moles of CO_2

STEP 3   Write the equalities and conversion factors for molar mass and mole–mole factors.

$$1 \text{ mol of } CaCO_3 = 100.09 \text{ g of } CaCO_3 \qquad 1 \text{ mol of } CaCO_3 = 1 \text{ mol of } CO_2$$

$$\frac{100.09 \text{ g } CaCO_3}{1 \text{ mol } CaCO_3} \text{ and } \frac{1 \text{ mol } CaCO_3}{100.09 \text{ g } CaCO_3} \qquad \frac{1 \text{ mol } CaCO_3}{1 \text{ mol } CO_2} \text{ and } \frac{1 \text{ mol } CO_2}{1 \text{ mol } CaCO_3}$$

STEP 4   Set up the problem to calculate moles of needed quantity.

$$25.0 \text{ g } CaCO_3 \times \frac{1 \text{ mol } CaCO_3}{100.09 \text{ g } CaCO_3} \times \frac{1 \text{ mol } CO_2}{1 \text{ mol } CaCO_3} = 0.250 \text{ mol of } CO_2$$

STEP 5 Convert the moles of needed quantity to volume using the ideal gas law equation.

$$V = \frac{nRT}{P}$$

$$V = \frac{0.250 \text{ mol} \times \dfrac{62.4 \text{ L} \cdot \text{mmHg}}{\text{mol} \cdot \text{K}} \times 297 \text{ K}}{752 \text{ mmHg}} = 6.16 \text{ L of } CO_2$$

SELF TEST 11.12

a. If 12.8 g of aluminum reacts with HCl, how many liters of H_2 would be formed at 715 mmHg and 19 °C?

$$6HCl(aq) + 2Al(s) \longrightarrow 3H_2(g) + 2AlCl_3(aq)$$

b. If 9.28 L of H_2 is produced at 24 °C and 1.20 atm, how many grams of aluminum react?

ANSWER

a. 18.1 L of H_2 b. 8.22 g of Al

When HCl reacts with aluminum, bubbles of H_2 gas form.

PRACTICE PROBLEMS

Try Practice Problems 11.65 to 11.70

PRACTICE PROBLEMS

11.8 Gas Laws and Chemical Reactions

11.65 HCl reacts with magnesium metal to produce hydrogen gas.

$$2HCl(aq) + Mg(s) \longrightarrow H_2(g) + MgCl_2(aq)$$

a. What volume, in liters, of hydrogen at 0 °C and 1.00 atm (STP) is released when 8.25 g of Mg reacts?

b. How many grams of magnesium are needed to prepare 5.00 L of H_2 at 735 mmHg and 18 °C?

11.66 When heated to 350 °C at 0.950 atm, ammonium nitrate decomposes to produce nitrogen, water, and oxygen gases.

$$2NH_4NO_3(s) \xrightarrow{\Delta} 2N_2(g) + 4H_2O(g) + O_2(g)$$

a. How many liters of water vapor are produced when 25.8 g of NH_4NO_3 decomposes?

b. How many grams of NH_4NO_3 are needed to produce 10.0 L of oxygen?

11.67 Butane undergoes combustion when it reacts with oxygen to produce carbon dioxide and water. What volume, in liters, of oxygen is needed to react with 55.2 g of butane at 0.850 atm and 25 °C?

$$2C_4H_{10}(g) + 13O_2(g) \xrightarrow{\Delta} 8CO_2(g) + 10H_2O(g)$$

11.68 Potassium nitrate decomposes to potassium nitrite and oxygen. What volume, in liters, of O_2 can be produced from the decomposition of 50.0 g of KNO_3 at 35 °C and 1.19 atm?

$$2KNO_3(s) \longrightarrow 2KNO_2(s) + O_2(g)$$

11.69 Aluminum and oxygen react to form aluminum oxide. How many liters of oxygen at 0 °C and 760 mmHg (STP) are required to completely react with 5.4 g of aluminum?

$$4Al(s) + 3O_2(g) \longrightarrow 2Al_2O_3(s)$$

11.70 Nitrogen dioxide reacts with water to produce oxygen and ammonia. How many grams of NH_3 can be produced when 4.00 L of NO_2 reacts at 415 °C and 725 mmHg?

$$4NO_2(g) + 6H_2O(g) \longrightarrow 7O_2(g) + 4NH_3(g)$$

11.9 Partial Pressures (Dalton's Law)

LEARNING GOAL Use Dalton's law of partial pressures to calculate the total pressure of a mixture of gases.

Many gas samples are a mixture of gases. For example, the air you breathe is a mixture of mostly oxygen and nitrogen gases. In gas mixtures, scientists observed that all gas particles behave in the same way. Therefore, the total pressure of the gases in a mixture is a result of the collisions of the gas particles regardless of what type of gas they are.

In a gas mixture, each gas exerts its **partial pressure**, which is the pressure it would exert if it were the only gas in the container. **Dalton's law** states that the total pressure of a gas mixture is the sum of the partial pressures of the gases in the mixture.

Dalton's Law

$$P_{total} = P_1 + P_2 + P_3 + \cdots$$

Total pressure of = Sum of the partial pressures
a gas mixture   of the gases in the mixture

Suppose we have two separate tanks, one filled with helium at a pressure of 2.0 atm and the other filled with argon at a pressure of 4.0 atm. When the gases are combined in a single tank with the same volume and temperature, the number of gas molecules, not the type of gas, determines the pressure in the container. The total pressure of the gas mixture would be 6.0 atm, which is the sum of their individual or partial pressures.

$$P_{total} = P_{He} + P_{Ar}$$
$$= 2.0 \text{ atm} + 4.0 \text{ atm}$$
$$= 6.0 \text{ atm}$$

$P_{He} = 2.0$ atm $P_{Ar} = 4.0$ atm

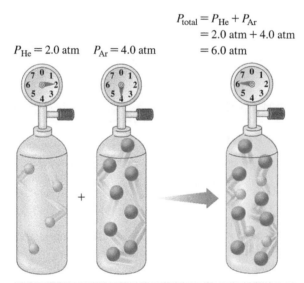

The total pressure of two gases is the sum of their partial pressures.

Air Is a Gas Mixture

The air you breathe is a mixture of gases. What we call the *atmospheric pressure* is actually the sum of the partial pressures of all the gases in the air. **TABLE 11.7** lists partial pressures for the gases in air on a typical day.

TABLE 11.7 Typical Composition of Air

Gas	Partial Pressure (mmHg)	Percentage (%)
Nitrogen, N_2	594	78.2
Oxygen, O_2	160.	21.0
Carbon dioxide, CO_2 Argon, Ar Water vapor, H_2O	6	0.8
Total air	760.	100

Chemistry Link to Health

Blood Gases

Our cells continuously use oxygen and produce carbon dioxide. Both gases move in and out of the lungs through the membranes of the alveoli, the tiny air sacs at the ends of the airways in the lungs. An exchange of gases occurs in which oxygen from the air diffuses into the lungs and into the blood, while carbon dioxide produced in the cells is carried to the lungs to be exhaled. In TABLE 11.8, partial pressures are given for the gases in air that we inhale (inspired air), air in the alveoli, and air that we exhale (expired air).

At sea level, oxygen normally has a partial pressure of 100 mmHg in the alveoli of the lungs. Because the partial pressure of oxygen in venous blood is 40 mmHg, oxygen diffuses from the alveoli into the bloodstream. The oxygen combines with hemoglobin, which carries it to the tissues of the body where the partial pressure of oxygen can be very low, less than 30 mmHg. Oxygen diffuses from the blood, where the partial pressure of O_2 is high, into the tissues where O_2 pressure is low.

As oxygen is used in the cells of the body during metabolic processes, carbon dioxide is produced, so the partial pressure of CO_2 may be as high as 50 mmHg or more. Carbon dioxide diffuses from the tissues into the bloodstream and is carried to the lungs. There it diffuses out of the blood, where CO_2 has a partial pressure of 46 mmHg, into the alveoli, where the CO_2 is at 40 mmHg and is exhaled. TABLE 11.9 gives the partial pressures of blood gases in the tissues and in oxygenated and deoxygenated blood.

TABLE 11.8 Partial Pressures of Gases During Breathing

Gas	Partial Pressure (mmHg)		
	Inspired Air	Alveolar Air	Expired Air
Nitrogen, N_2	594	573	569
Oxygen, O_2	160.	100.	116
Carbon dioxide, CO_2	0.3	40.	28
Water vapor, H_2O	5.7	47	47
Total	760.	760.	760.

TABLE 11.9 Partial Pressures of Oxygen and Carbon Dioxide in Blood and Tissues

Gas	Partial Pressure (mmHg)		
	Oxygenated Blood	Deoxygenated Blood	Tissues
O_2	100	40	30 or less
CO_2	40	46	50 or greater

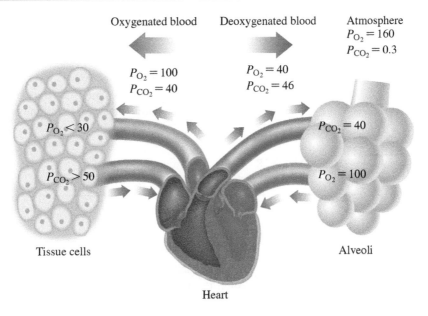

Oxygenated blood   Deoxygenated blood   Atmosphere
$P_{O_2} = 160$
$P_{CO_2} = 0.3$

$P_{O_2} = 100$ $P_{O_2} = 40$
$P_{CO_2} = 40$ $P_{CO_2} = 46$

$P_{O_2} < 30$ $P_{CO_2} = 40$

$P_{CO_2} > 50$ $P_{O_2} = 100$

Tissue cells   Alveoli

Heart

SAMPLE PROBLEM 11.13 Calculating the Partial Pressure of a Gas in a Mixture

TRY IT FIRST

A heliox breathing mixture of oxygen and helium is prepared for a patient with *chronic obstructive pulmonary disease* (COPD). The gas mixture has a total pressure of 7.00 atm. If the partial pressure of the oxygen in the tank is 1140 mmHg, what is the partial pressure, in atmospheres, of the helium in the breathing mixture?

SOLUTION

ANALYZE THE PROBLEM	Given	Need	Connect
	P_{total} = 7.00 atm P_{O_2} = 1140 mmHg	partial pressure of He	Dalton's law

STEP 1 Write the equation for the sum of the partial pressures.

$$P_{total} = P_{O_2} + P_{He} \quad \text{Dalton's law}$$

STEP 2 Rearrange the equation to solve for the unknown pressure. To solve for the partial pressure of helium (P_{He}), we rearrange the equation to give the following:

$$P_{He} = P_{total} - P_{O_2}$$

Convert units to match.

$$P_{O_2} = 1140 \; \text{mmHg} \times \frac{1 \; \text{atm}}{760 \; \text{mmHg}} = 1.50 \; \text{atm}$$

STEP 3 Substitute known pressures into the equation, and calculate the unknown pressure.

$$P_{He} = P_{total} - P_{O_2}$$
$$P_{He} = 7.00 \; \text{atm} - 1.50 \; \text{atm} = 5.50 \; \text{atm}$$

SELF TEST 11.13

a. An anesthetic consists of a mixture of cyclopropane gas, C_3H_6, and oxygen gas, O_2. If the mixture has a total pressure of 1.09 atm, and the partial pressure of the cyclopropane is 73 mmHg, what is the partial pressure, in millimeters of mercury, of the oxygen in the anesthetic?

b. Another anesthetic contains nitrous oxide gas, N_2O, and oxygen gas, O_2. If the mixture has a total pressure of 786 mmHg and the partial pressure of the nitrous oxide is 0.110 atm, what is the partial pressure, in millimeters of mercury, of the oxygen in the mixture?

PRACTICE PROBLEMS

Try Practice Problems 11.71 to 11.74, 11.77 to 11.80

ANSWER

a. 755 mmHg   b. 702 mmHg

Gases Collected Over Water

In the laboratory, gases are often collected by bubbling them through water into a container (see **FIGURE 11.9**). In a reaction, magnesium (Mg) reacts with HCl to form H_2 gas and $MgCl_2$.

$$2HCl(aq) + Mg(s) \longrightarrow H_2(g) + MgCl_2(aq)$$

As hydrogen is produced during the reaction, it displaces some of the water in the container. Because of the vapor pressure of water, the gas that is collected is a mixture of hydrogen and water vapor. For our calculation, we need the pressure of the dry hydrogen gas. We use the vapor pressure of water (see Table 11.4) at the experimental temperature, and subtract it from the total gas pressure. Then we can use the ideal gas law to determine the moles or grams of the hydrogen gas that were collected.

FIGURE 11.9 A gas from a reaction is collected by bubbling through water. Due to evaporation of water, the total pressure is equal to the partial pressure of the gas and the vapor pressure of water.

SAMPLE PROBLEM 11.14 Moles of Gas Collected Over Water

TRY IT FIRST

When magnesium reacts with HCl, a volume of 0.355 L of hydrogen gas is collected over water at 26 °C. The vapor pressure of water at 26 °C is 25 mmHg.

$$2HCl(aq) + Mg(s) \longrightarrow H_2(g) + MgCl_2(aq)$$

If the total pressure is 752 mmHg, how many moles of $H_2(g)$ were collected?

SOLUTION

	Given	Need	Connect
ANALYZE THE PROBLEM	$V = 0.355$ L of H_2 $P = 752$ mmHg $T = 26$ °C $+ 273 = 299$ K $P_{H_2O} = 25$ mmHg	moles of H_2 n	Dalton's law, ideal gas law, $PV = nRT$ $R = \dfrac{62.4 \text{ L} \cdot \text{mmHg}}{\text{mol} \cdot \text{K}}$

STEP 1 **Obtain the vapor pressure of water.** The vapor pressure of water at 26 °C is 25 mmHg.

STEP 2 **Subtract the vapor pressure from the total gas pressure to give the partial pressure of the needed gas.** Using Dalton's law of partial pressures, determine the partial pressure of H_2.

$$P_{total} = P_{H_2} + P_{H_2O}$$

Solving for the partial pressure of H_2 gives

$$P_{H_2} = P_{total} - P_{H_2O}$$
$$P_{H_2} = 752 \text{ mmHg} - 25 \text{ mmHg}$$
$$= 727 \text{ mmHg}$$

ENGAGE 11.12

Why is the vapor pressure of water subtracted from the total pressure of a gas collected over water?

STEP 3 **Use the ideal gas law to convert P_{gas} to moles of gas collected.** By dividing both sides of the ideal gas law equation by RT, we solve for moles, n, of gas.

$$PV = nRT \quad \text{Ideal gas law equation}$$

$$\frac{PV}{RT} = \frac{n\cancel{R}\cancel{T}}{\cancel{R}\cancel{T}}$$

$$n = \frac{PV}{RT}$$

Calculate the moles of H_2 gas by placing the partial pressure of H_2 (727 mmHg), volume of gas container (0.355 L), temperature (26 °C + 273 = 299 K), and R, using mmHg, into the ideal gas law equation.

$$n = \frac{727 \text{ } \cancel{mmHg} \times 0.355 \text{ } \cancel{L}}{\dfrac{62.4 \text{ } \cancel{L} \cdot \cancel{mmHg}}{mol \cdot \cancel{K}} \times 299 \text{ } \cancel{K}} = 0.0138 \text{ mol of } H_2 \, (1.38 \times 10^{-2} \text{ mol of } H_2)$$

SELF TEST 11.14

a. A 456-mL sample of oxygen gas (O_2) is collected over water at a pressure of 744 mmHg and a temperature of 20. °C. How many grams of dry O_2 are collected (see Table 11.4)?

b. A sample of nitrogen gas (N_2) with a volume of 376 mL is collected over water at a pressure of 766 mmHg and a temperature of 30. °C. How many grams of dry N_2 are collected (see Table 11.4)?

PRACTICE PROBLEMS

Try Practice Problems 11.75 and 11.76

ANSWER

a. 0.579 g of O_2 **b.** 0.409 g of N_2

Chemistry Link to Health

Hyperbaric Chambers

A burn patient may undergo treatment for burns and infections in a *hyperbaric chamber*, a device in which pressures can be obtained that are two to three times greater than atmospheric pressure. A greater oxygen pressure increases the level of dissolved oxygen in the blood and tissues, where it fights bacterial infections. High levels of oxygen are toxic to many strains of bacteria. The hyperbaric chamber may also be used during surgery, to help counteract carbon monoxide (CO) poisoning, and to treat some cancers.

The blood is normally capable of dissolving up to 95% of the oxygen. Thus, if the partial pressure of the oxygen in the hyperbaric chamber is 2280 mmHg (3 atm), about 2170 mmHg of oxygen can dissolve in the blood, saturating the tissues. In the treatment for carbon monoxide poisoning, oxygen at high pressure is used to displace the CO from the hemoglobin faster than breathing pure oxygen at 1 atm.

A patient undergoing treatment in a hyperbaric chamber must also undergo decompression (reduction of pressure) at a rate that slowly reduces the concentration of dissolved oxygen in the blood. If decompression is too rapid, the oxygen dissolved in the blood may form gas bubbles in the circulatory system.

Similarly, if a scuba diver does not decompress slowly, a condition called the "bends" may occur. While below the surface of the ocean, a diver uses a breathing mixture with higher pressures. If there is nitrogen in the mixture, higher quantities of nitrogen gas will dissolve in the blood. If the diver ascends to the surface too quickly, the dissolved nitrogen forms gas bubbles that can block a blood vessel and cut off the flow of blood in the joints and tissues of the body and be quite painful. A diver suffering from the bends is placed immediately into a hyperbaric chamber where pressure is first increased and then slowly decreased. The dissolved nitrogen can then diffuse through the lungs until atmospheric pressure is reached.

A hyperbaric chamber is used in the treatment of certain diseases.

PRACTICE PROBLEMS

11.9 Partial Pressures (Dalton's Law)

11.71 In a gas mixture, the partial pressures are nitrogen 425 Torr, oxygen 115 Torr, and helium 225 Torr. What is the total pressure, in torr, exerted by the gas mixture?

11.72 In a gas mixture, the partial pressures are argon 415 mmHg, neon 75 mmHg, and nitrogen 125 mmHg. What is the total pressure, in millimeters of mercury, exerted by the gas mixture?

11.73 A gas mixture containing oxygen, nitrogen, and helium exerts a total pressure of 925 Torr. If the partial pressures are oxygen 425 Torr and helium 75 Torr, what is the partial pressure, in torr, of the nitrogen in the mixture?

11.74 A gas mixture containing oxygen, nitrogen, and neon exerts a total pressure of 1.20 atm. If helium added to the mixture increases the pressure to 1.50 atm, what is the partial pressure, in atmospheres, of the helium?

11.75 When solid $KClO_3$ is heated, it decomposes to give solid KCl and O_2 gas. A volume of 256 mL of gas is collected over water at a total pressure of 765 mmHg and 24 °C. The vapor pressure of water at 24 °C is 22 mmHg.

$$2KClO_3(s) \xrightarrow{\Delta} 2KCl(s) + 3O_2(g)$$

a. What was the partial pressure of the O_2 gas?
b. How many moles of O_2 gas were produced in the reaction?

11.76 When Zn reacts with HCl solution, the products are H_2 gas and $ZnCl_2$. A volume of 425 mL of H_2 gas is collected over water at a total pressure of 758 mmHg and 16 °C. The vapor pressure of water at 16 °C is 14 mmHg.

$$2HCl(aq) + Zn(s) \longrightarrow H_2(g) + ZnCl_2(aq)$$

a. What was the partial pressure of the H_2 gas?
b. How many moles of H_2 gas were produced in the reaction?

Applications

11.77 An air sample in the lungs contains oxygen at 93 mmHg, nitrogen at 565 mmHg, carbon dioxide at 38 mmHg, and water vapor at 47 mmHg. What is the total pressure, in atmospheres, exerted by the gas mixture?

11.78 A nitrox II gas mixture for scuba diving contains oxygen gas at 53 atm and nitrogen gas at 94 atm. What is the total pressure, in torr, of the scuba gas mixture?

11.79 In certain lung ailments such as emphysema, there is a decrease in the ability of oxygen to diffuse into the blood.
a. How would the partial pressure of oxygen in the blood change?
b. Why does a person with severe emphysema sometimes use a portable oxygen tank?

11.80 Head trauma can affect the ability of a person to ventilate (breathe in and out).
a. What would happen to the partial pressures of oxygen and carbon dioxide in the blood if a person cannot properly ventilate?
b. When a person who cannot breathe properly is placed on a ventilator, an air mixture is delivered at pressures that are alternately above the air pressure in the person's lung, and then below. How will this move oxygen gas into the lungs, and carbon dioxide out?

UPDATE Exercise-Induced Asthma

Vigorous exercise can induce asthma, particularly in children. When Louisa had her asthma attack, her breathing became more rapid, the temperature within her airways increased, and the muscles around the bronchi contracted, causing a narrowing of the airways. Louisa's symptoms, which may occur within 5 to 20 min after the start of vigorous exercise, include shortness of breath, wheezing, and coughing.

Louisa now does several things to prevent exercise-induced asthma. She uses a pre-exercise inhaled medication before she starts her activity. The medication relaxes the muscles that surround the airways and opens up the airways. Then she does a warm-up set of exercises. If pollen counts are high, she avoids exercising outdoors.

Applications

11.81 Louisa's lung capacity was measured as 3.2 L at a body temperature of 37 °C and a pressure of 745 mmHg. What is her lung capacity, in liters, at STP?

11.82 Using the answer from problem 11.81, how many grams of nitrogen are in Louisa's lungs at STP if air contains 78% nitrogen?

CONCEPT MAP

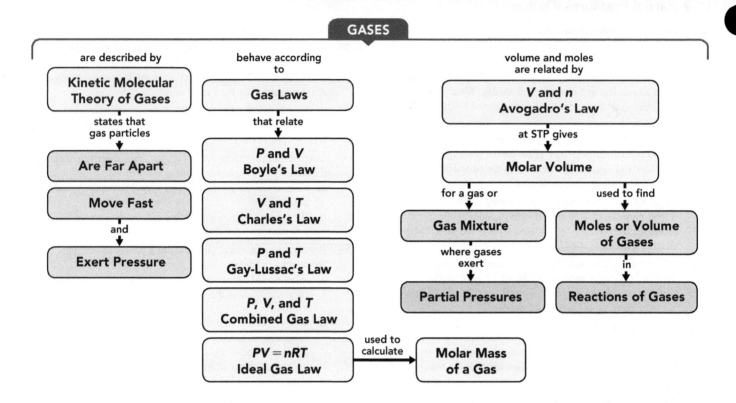

GASES

are described by
→ **Kinetic Molecular Theory of Gases**

states that gas particles
→ **Are Far Apart**
→ **Move Fast**
and
→ **Exert Pressure**

behave according to
→ **Gas Laws**

that relate
→ **P and V Boyle's Law**
→ **V and T Charles's Law**
→ **P and T Gay-Lussac's Law**
→ **P, V, and T Combined Gas Law**
→ **PV = nRT Ideal Gas Law**

used to calculate
→ **Molar Mass of a Gas**

volume and moles are related by
→ **V and n Avogadro's Law**

at STP gives
→ **Molar Volume**

for a gas or
→ **Gas Mixture**

where gases exert
→ **Partial Pressures**

used to find
→ **Moles or Volume of Gases**

in
→ **Reactions of Gases**

CHAPTER REVIEW

11.1 Properties of Gases

LEARNING GOAL Describe the kinetic molecular theory of gases and the units of measurement used for gases.

- In a gas, particles are so far apart and moving so fast that their attractions are negligible.
- A gas is described by the physical properties of pressure (P), volume (V), temperature (T), and amount in moles (n).
- A gas exerts pressure, the force of the gas particles striking the surface of a container.
- Gas pressure is measured in units such as torr, mmHg, atm, and Pa.

11.2 Pressure and Volume (Boyle's Law)

LEARNING GOAL Use the pressure–volume relationship (Boyle's law) to calculate the unknown pressure or volume when the temperature and amount of gas do not change.

Piston

$V_1 = 4 \text{ L}$ $V_2 = 2 \text{ L}$
$P_1 = 1 \text{ atm}$ $P_2 = 2 \text{ atm}$

- The volume (V) of a gas changes inversely with the pressure (P) of the gas, if there is no change in the temperature and the amount of gas.

$$P_1 V_1 = P_2 V_2$$

- The pressure increases if volume decreases; its pressure decreases if the volume increases.

11.3 Temperature and Volume (Charles's Law)

LEARNING GOAL Use the temperature–volume relationship (Charles's law) to calculate the unknown temperature or volume when the pressure and amount of gas do not change.

$T_1 = 200 \text{ K}$ $T_2 = 400 \text{ K}$
$V_1 = 1 \text{ L}$ $V_2 = 2 \text{ L}$

- The volume (V) of a gas is directly related to its Kelvin temperature (T) when there is no change in the pressure and the amount of gas.

$$\frac{V_1}{T_1} = \frac{V_2}{T_2}$$

- If the temperature of a gas increases, its volume increases; if its temperature decreases, the volume decreases.

11.4 Temperature and Pressure (Gay-Lussac's Law)

LEARNING GOAL Use the temperature–pressure relationship (Gay-Lussac's law) to calculate the unknown temperature or pressure when the volume and amount of gas do not change.

$T_1 = 200 \text{ K}$ $T_2 = 400 \text{ K}$
$P_1 = 1 \text{ atm}$ $P_2 = 2 \text{ atm}$

- The pressure (P) of a gas is directly related to its Kelvin temperature (T) when there is no change in the volume and the amount of the gas.

$$\frac{P_1}{T_1} = \frac{P_2}{T_2}$$

- As temperature of a gas increases, its pressure increases; if its temperature decreases, its pressure decreases.
- Vapor pressure is the pressure of the gas that forms when a liquid evaporates.
- At the boiling point of a liquid, the vapor pressure equals the external pressure.

11.5 The Combined Gas Law

LEARNING GOAL Use the combined gas law to calculate the unknown pressure, volume, or temperature of a gas when changes in two of these properties are given and the amount of gas does not change.

- The combined gas law is the relationship of pressure (P), volume (V), and temperature (T) when the amount of gas does not change.

$$\frac{P_1 V_1}{T_1} = \frac{P_2 V_2}{T_2}$$

- The combined gas law is used to determine the effect of changes in two of the variables on the third.

11.6 Volume and Moles (Avogadro's Law)

LEARNING GOAL Use Avogadro's law to calculate the unknown amount or volume of a gas when the pressure and temperature do not change.

$V = 22.4\ \text{L}$

1 mol of O_2
32.00 g of O_2
273 K
1 atm

- The volume (V) of a gas is directly related to the number of moles (n) of the gas when the pressure and temperature of the gas do not change.

$$\frac{V_1}{n_1} = \frac{V_2}{n_2}$$

- If the moles of gas increase, the volume must increase; if the moles of gas decrease, the volume must decrease.
- At standard temperature (273 K) and standard pressure (1 atm), abbreviated STP, 1 mol of any gas has a volume of 22.4 L.

11.7 The Ideal Gas Law

LEARNING GOAL Use the ideal gas law equation to solve for P, V, T, or n of a gas when given three of the four values in the ideal gas law equation. Calculate the molar mass of a gas.

- The ideal gas law gives the relationship of the quantities P, V, n, and T that describe and measure a gas.

$$PV = nRT$$

- Any of the four variables can be calculated if the values of the other three are known.
- The molar mass of a gas can be calculated using the ideal gas law equation.

11.8 Gas Laws and Chemical Reactions

LEARNING GOAL Calculate the mass or volume of a gas that reacts or forms in a chemical reaction.

- The ideal gas law equation is used to convert the quantities (P, V, and T) of gases to moles in a chemical reaction.
- The moles of gases can be used to determine the number of moles or grams of other substances in the reaction.

11.9 Partial Pressures (Dalton's Law)

$P_{total} = P_{He} + P_{Ar}$
$= 2.0\ \text{atm} + 4.0\ \text{atm}$
$= 6.0\ \text{atm}$

LEARNING GOAL Use Dalton's law of partial pressures to calculate the total pressure of a mixture of gases.

- In a mixture of two or more gases, the total pressure is the sum of the partial pressures of the individual gases.

$$P_{total} = P_1 + P_2 + P_3 + \cdots$$

- The partial pressure of a gas in a mixture is the pressure it would exert if it were the only gas in the container.
- For gases collected over water, the vapor pressure of water is subtracted from the total pressure of the gas mixture to obtain the partial pressure of the dry gas.

KEY TERMS

atmosphere (atm) A unit equal to the pressure exerted by a column of mercury 760 mm high.

atmospheric pressure The pressure exerted by the atmosphere.

Avogadro's law A gas law stating that the volume of a gas is directly related to the number of moles of gas when pressure and temperature do not change.

Boyle's law A gas law stating that the pressure of a gas is inversely related to the volume when temperature and moles of the gas do not change.

Charles's law A gas law stating that the volume of a gas is directly related to the Kelvin temperature when pressure and moles of the gas do not change.

combined gas law A relationship that combines several gas laws relating pressure, volume, and temperature when the amount of gas does not change.

$$\frac{P_1 V_1}{T_1} = \frac{P_2 V_2}{T_2}$$

Dalton's law A gas law stating that the total pressure exerted by a mixture of gases in a container is the sum of the partial pressures that each gas would exert alone.

direct relationship A relationship in which two properties increase or decrease together.

Gay-Lussac's law A gas law stating that the pressure of a gas is directly related to the Kelvin temperature when the number of moles of a gas and its volume do not change.

ideal gas constant, R A numerical value that relates the quantities P, V, n, and T in the ideal gas law, $PV = nRT$.

ideal gas law A law that combines the four measured properties of a gas.

$$PV = nRT$$

inverse relationship A relationship in which two properties change in opposite directions.

kinetic molecular theory of gases A model used to explain the behavior of gases.

molar volume A volume of 22.4 L occupied by 1 mol of a gas at STP conditions of 0 °C (273 K) and 1 atm.

partial pressure The pressure exerted by a single gas in a gas mixture.

pressure The force exerted by gas particles that hit the walls of a container.

STP Standard conditions of exactly 0 °C (273 K) temperature and 1 atm pressure used for the comparison of gases.

vapor pressure The pressure exerted by the particles of vapor above a liquid.

CORE CHEMISTRY SKILLS

The chapter section containing each Core Chemistry Skill is shown in parentheses at the end of each heading.

Using the Gas Laws (11.2)

- Boyle's, Charles's, Gay-Lussac's, and Avogadro's laws show the relationships between two properties of a gas.

$$P_1V_1 = P_2V_2 \quad \text{Boyle's law}$$

$$\frac{V_1}{T_1} = \frac{V_2}{T_2} \quad \text{Charles's law}$$

$$\frac{P_1}{T_1} = \frac{P_2}{T_2} \quad \text{Gay-Lussac's law}$$

$$\frac{V_1}{n_1} = \frac{V_2}{n_2} \quad \text{Avogadro's law}$$

- The combined gas law shows the relationship between P, V, and T for a gas.

$$\frac{P_1V_1}{T_1} = \frac{P_2V_2}{T_2}$$

- When two properties of a gas vary and the other two do not change, we list the initial and final conditions of each property in a table.

Example: A sample of helium gas (He) has a volume of 6.8 L and a pressure of 2.5 atm. What is the final volume, in liters, if it has a final pressure of 1.2 atm with no change in temperature and amount of gas?

Answer:

	Given	Need	Connect
ANALYZE THE PROBLEM	$P_1 = 2.5$ atm $P_2 = 1.2$ atm $V_1 = 6.8$ L	V_2	Boyle's law, $P_1V_1 = P_2V_2$
	Factors that do not change: T and n		**Predict:** V increases

Using Boyle's law, we can write the relationship for V_2, which we predict will increase.

$$V_2 = V_1 \times \frac{P_1}{P_2}$$

$$V_2 = 6.8 \text{ L} \times \frac{2.5 \text{ atm}}{1.2 \text{ atm}} = 14 \text{ L}$$

Using the Ideal Gas Law (11.7)

- The ideal gas law equation combines the relationships of the four properties of a gas into one equation.

$$PV = nRT$$

- When three of the four properties are given, we rearrange the ideal gas law equation for the needed quantity.

Example: What is the volume, in liters, of 0.750 mol of CO_2 at a pressure of 1340 mmHg and a temperature of 295 K?

Answer: $V = \dfrac{nRT}{P}$

$$= \frac{0.750 \text{ mol} \times \dfrac{62.4 \text{ L} \cdot \text{mmHg}}{\text{mol} \cdot \text{K}} \times 295 \text{ K}}{1340 \text{ mmHg}} = 10.3 \text{ L}$$

Calculating Mass or Volume of a Gas in a Chemical Reaction (11.8)

- The ideal gas law equation can be used to calculate the volume or mass of a gas in a chemical reaction.

Example: What is the volume, in liters, of N_2 required to react with 18.5 g of magnesium at a pressure of 1.20 atm and a temperature of 303 K?

$$3Mg(s) + N_2(g) \longrightarrow Mg_3N_2(s)$$

Answer: Initially, we convert the grams of Mg to moles and use a mole–mole factor from the balanced equation to calculate the moles of N_2 gas.

$$18.5 \text{ g Mg} \times \frac{1 \text{ mol Mg}}{24.31 \text{ g Mg}} \times \frac{1 \text{ mol N}_2}{3 \text{ mol Mg}} = 0.254 \text{ mol of N}_2$$

Now, we use the moles of N_2 in the ideal gas law equation and solve for liters, the needed quantity.

$$V = \frac{nRT}{P}$$

$$= \frac{0.254 \text{ mol N}_2 \times \dfrac{0.0821 \text{ L} \cdot \text{atm}}{\text{mol} \cdot \text{K}} \times 303 \text{ K}}{1.20 \text{ atm}} = 5.27 \text{ L}$$

Calculating Partial Pressure (11.9)

- In a gas mixture, each gas exerts its partial pressure, which is the pressure it would exert if it were the only gas in the container.
- Dalton's law states that the total pressure of a gas mixture is the sum of the partial pressures of the gases in the mixture.

$$P_{total} = P_1 + P_2 + P_3 + \cdots$$

Example: A gas mixture with a total pressure of 1.18 atm contains helium gas at a partial pressure of 465 mmHg and nitrogen gas. What is the partial pressure, in atmospheres, of the nitrogen gas?

Answer: Initially, we convert the partial pressure of helium gas from mmHg to atm.

$$465 \text{ mmHg} \times \frac{1 \text{ atm}}{760 \text{ mmHg}} = 0.612 \text{ atm of He gas}$$

Using Dalton's law, we solve for the needed quantity, P_{N_2}.

$$P_{total} = P_{N_2} + P_{He}$$

$$P_{N_2} = P_{total} - P_{He}$$

$$P_{N_2} = 1.18 \text{ atm} - 0.612 \text{ atm} = 0.57 \text{ atm}$$

UNDERSTANDING THE CONCEPTS

The chapter sections to review are shown in parentheses at the end of each problem.

11.83 Two flasks of equal volume and at the same temperature contain different gases. One flask contains 10.0 g of Ne, and the other flask contains 10.0 g of He. Is each of the following statements *true* or *false*? Explain. (11.1)
a. The flask that contains He has a higher pressure than the flask that contains Ne.
b. The densities of the gases are the same.

11.84 Two flasks of equal volume and at the same temperature contain different gases. One flask contains 5.0 g of O_2, and the other flask contains 5.0 g of H_2. Is each of the following statements *true* or *false*? Explain. (11.1)
a. Both flasks contain the same number of molecules.
b. The pressures in the flasks are the same.

11.85 At 100 °C, which of the following diagrams (**1**, **2**, or **3**) represents a gas sample that exerts the: (11.1)
a. lowest pressure?   **b.** highest pressure?

11.86 Indicate which diagram (**1**, **2**, or **3**) represents the volume of the gas sample in a flexible container when each of the following changes (**a** to **d**) takes place: (11.2, 11.3)

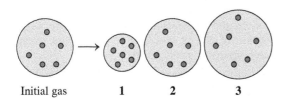

a. Temperature increases, if pressure does not change.
b. Temperature decreases, if pressure does not change.
c. Atmospheric pressure decreases, if temperature does not change.
d. Doubling the atmospheric pressure and doubling the Kelvin temperature.

11.87 A balloon is filled with helium gas with a partial pressure of 1.00 atm and neon gas with a partial pressure of 0.50 atm. For each of the following changes (**a** to **e**) of the initial balloon, select the diagram (**A**, **B**, or **C**) that shows the final volume of the balloon: (11.2, 11.3, 11.6)

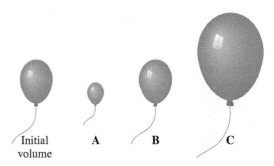

a. The balloon is put in a cold storage unit (pressure and amount of gas do not change).
b. The balloon floats to a higher altitude where the pressure is less (temperature and amount of gas do not change).
c. All of the neon gas is removed (temperature and pressure do not change).
d. The Kelvin temperature doubles and half of the gas atoms leak out (pressure does not change).
e. 2.0 mol of O_2 gas is added (temperature and pressure do not change).

11.88 Indicate if pressure *increases*, *decreases*, or *stays the same* in each of the following: (11.2, 11.4, 11.6)

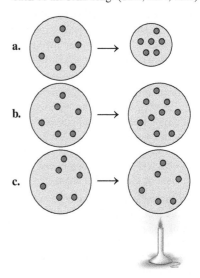

ADDITIONAL PRACTICE PROBLEMS

11.89 A gas sample has a volume of 4250 mL at 15 °C and 745 mmHg. What is the final temperature, in degrees Celsius, after the sample is transferred to a different container with a volume of 2.50 L and a pressure of 1.20 atm, when the amount of gas does not change? (11.5)

11.90 In the fermentation of glucose (wine making), 780 mL of CO_2 gas was produced at 37 °C and 1.00 atm. What is the final volume, in liters, of the gas when measured at 22 °C and 675 mmHg, when the amount of gas does not change? (11.5)

11.91 During laparoscopic surgery, carbon dioxide gas is used to expand the abdomen to help create a larger working space. If 4.80 L of CO_2 gas at 18 °C at 785 mmHg is used, what is the final volume, in liters, of the gas at 37 °C and a pressure of 745 mmHg, if the amount of CO_2 does not change? (11.5)

11.92 A weather balloon has a volume of 750 L when filled with helium at 8 °C at a pressure of 380 Torr. What is the final volume, in liters, of the balloon when the pressure is 0.20 atm, the temperature is −45 °C, and the amount of gas does not change? (11.5)

11.93 In 1783, Jacques Charles launched his first balloon filled with hydrogen gas (H_2), which he chose because it was lighter than air. The balloon had a volume of 31 000 L when it was filled at a pressure of 755 mmHg and a temperature of 22 °C. How many kilograms of hydrogen were needed to fill the balloon? (11.7)

Jacques Charles used hydrogen to launch his balloon in 1783.

11.94 When the balloon in problem 11.93 reached an altitude of 1000 m, the pressure was 658 mmHg and the temperature was −8 °C. What is the volume of the balloon at these conditions, if the amount of hydrogen does not change? (11.5)

11.95 You are doing research on planet X. The temperature inside the space station is a carefully controlled 24 °C, and the pressure is 755 mmHg. Suppose that a balloon, which has a volume of 850. mL inside the space station, is placed into the airlock and floats out to planet X. If planet X has an atmospheric pressure of 0.150 atm and the volume of the balloon changes to 3.22 L, what is the temperature, in degrees Celsius, on planet X (the amount of gas does not change)? (11.5)

11.96 Your spaceship has docked at a space station above Mars. The temperature inside the space station is a carefully controlled 24 °C at a pressure of 745 mmHg. A balloon with a volume of 425 mL drifts into the airlock where the temperature is −95 °C and the pressure is 0.115 atm. What is the final volume, in milliliters, of the balloon if the amount of gas does not change and the balloon is very elastic? (11.5)

11.97 A 2.00-L container is filled with methane gas, CH_4, at a pressure of 2500. mmHg and a temperature of 18 °C. How many grams of methane are in the container? (11.7)

11.98 A steel cylinder with a volume of 15.0 L is filled with 50.0 g of nitrogen gas at 25 °C. What is the pressure, in atmospheres, of the N_2 gas in the cylinder? (11.7)

11.99 A sample of gas with a mass of 1.62 g occupies a volume of 941 mL at a pressure of 748 Torr and a temperature of 20.0 °C. What is the molar mass of the gas? (11.7)

11.100 What is the molar mass of a gas if 1.15 g of the gas has a volume of 8 mL at 0 °C and 1.00 atm (STP)? (11.7)

11.101 How many grams of CO_2 are in 35.0 L of $CO_2(g)$ at 1.20 atm and 5 °C? (11.7)

11.102 A container is filled with 0.644 g of O_2 at 5 °C and 845 mmHg. What is the volume, in milliliters, of the container? (11.7)

11.103 How many liters of H_2 gas can be produced at 0 °C and 1.00 atm (STP) from 25.0 g of Zn? (11.7, 11.8)

$$2HCl(aq) + Zn(s) \longrightarrow H_2(g) + ZnCl_2(aq)$$

11.104 In the formation of smog, nitrogen and oxygen gas react to form nitrogen dioxide. How many grams of NO_2 will be produced when 2.0 L of nitrogen at 840 mmHg and 24 °C are completely reacted? (11.7, 11.8)

$$N_2(g) + 2O_2(g) \longrightarrow 2NO_2(g)$$

11.105 Nitrogen dioxide reacts with water to produce oxygen and ammonia. A 5.00-L sample of $H_2O(g)$ reacts at a temperature of 375 °C and a pressure of 725 mmHg. How many grams of NH_3 can be produced? (11.7, 11.8)

$$4NO_2(g) + 6H_2O(g) \longrightarrow 7O_2(g) + 4NH_3(g)$$

11.106 Hydrogen gas can be produced in the laboratory through the reaction of magnesium metal with hydrochloric acid. When 12.0 g of Mg reacts, what volume, in liters, of H_2 gas is produced at 24 °C and 835 mmHg? (11.7, 11.8)

$$2HCl(aq) + Mg(s) \longrightarrow H_2(g) + MgCl_2(aq)$$

11.107 A gas mixture with a total pressure of 2400 Torr is used by a scuba diver. If the mixture contains 2.0 mol of helium and 6.0 mol of oxygen, what is the partial pressure, in torr, of each gas in the sample? (11.9)

11.108 A gas mixture with a total pressure of 4.6 atm is used in a hospital. If the mixture contains 5.4 mol of nitrogen and 1.4 mol of oxygen, what is the partial pressure, in atmospheres, of each gas in the sample? (11.9)

11.109 A gas mixture contains oxygen and argon at partial pressures of 0.60 atm and 425 mmHg. If nitrogen gas added to the sample increases the total pressure to 1250 Torr, what is the partial pressure, in torr, of the nitrogen added? (11.9)

11.110 What is the total pressure, in millimeters of mercury, of a gas mixture containing argon gas at 0.25 atm, helium gas at 350 mmHg, and nitrogen gas at 360 Torr? (11.9)

CHALLENGE PROBLEMS

The following problems are related to the topics in this chapter. However, they do not all follow the chapter order, and they require you to combine concepts and skills from several sections. These problems will help you increase your critical thinking skills and prepare for your next exam.

11.111 Solid aluminum reacts with aqueous H_2SO_4 to form H_2 gas and aluminum sulfate. When a sample of Al is allowed to react, 415 mL of gas is collected over water at 23 °C, at a pressure of 755 mmHg. At 23 °C, the vapor pressure of water is 21 mmHg. (11.7, 11.8, 11.9)

$$3H_2SO_4(aq) + 2Al(s) \longrightarrow 3H_2(g) + Al_2(SO_4)_3(aq)$$

a. What is the pressure, in millimeters of mercury, of the dry H_2 gas?
b. How many moles of H_2 were produced?
c. How many grams of Al were reacted?

11.112 When heated, $KClO_3$ forms KCl and O_2. When a sample of $KClO_3$ is heated, 226 mL of gas with a pressure of 744 mmHg is collected over water at 26 °C. At 26 °C, the vapor pressure of water is 25 mmHg. (11.7, 11.8, 11.9)

$$2KClO_3(s) \xrightarrow{\Delta} 2KCl(s) + 3O_2(g)$$

a. What is the pressure, in millimeters of mercury, of the dry O_2 gas?
b. How many moles of O_2 were produced?
c. How many grams of $KClO_3$ were reacted?

11.113 A sample of gas with a mass of 1.020 g occupies a volume of 762 mL at 0 °C and 1.00 atm (STP). What is the molar mass of the gas? (11.7)

11.114 A sample of an unknown gas with a mass of 3.24 g occupies a volume of 1.88 L at a pressure of 748 mmHg and a temperature of 20. °C. What is the molar mass of the gas? (11.7)

Applications

11.115 The propane, C_3H_8, in a fuel cylinder, undergoes combustion with oxygen in the air. How many liters of CO_2 are produced at STP if the cylinder contains 881 g of propane? (11.8)

$$C_3H_8(g) + 5O_2(g) \xrightarrow{\Delta} 3CO_2(g) + 4H_2O(g)$$

11.116 When sensors in a car detect a collision, they cause the reaction of sodium azide, NaN_3, which generates nitrogen gas to fill the airbags within 0.03 s. How many liters of N_2 are produced at STP if the airbag contains 132 g of NaN_3? (11.8)

$$2NaN_3(s) \longrightarrow 2Na(s) + 3N_2(g)$$

11.117 Glucose, $C_6H_{12}O_6$, is metabolized in living systems. How many grams of water can be produced from the reaction of 18.0 g of glucose and 7.50 L of O_2 at 1.00 atm and 37 °C? (11.7, 11.8)

$$C_6H_{12}O_6(s) + 6O_2(g) \longrightarrow 6CO_2(g) + 6H_2O(l)$$

11.118 2.00 L of N_2, at 25 °C and 1.08 atm, is mixed with 4.00 L of O_2, at 25 °C and 0.118 atm, and the mixture is allowed to react. How much NO, in grams, is produced? (11.7, 11.8)

$$N_2(g) + O_2(g) \longrightarrow 2NO(g)$$

11.119 Hyperbaric therapy uses 100% oxygen at pressure to help heal wounds and infections, and to treat carbon monoxide poisoning. If the pressure inside a hyperbaric chamber is 3.0 atm, what is the volume, in liters, of the chamber containing 2400 g of O_2 at 28 °C? (11.7)

11.120 A hyperbaric chamber has a volume of 1510 L. How many kilograms of O_2 gas are needed to give an oxygen pressure of 2.04 atm at 25 °C? (11.7)

11.121 Laparoscopic surgery involves inflating the abdomen with carbon dioxide gas to separate the internal organs and the abdominal wall. If the CO_2 injected into the abdomen produces a pressure of 20. mmHg and a volume of 4.00 L at 32 °C, how many grams of CO_2 were used? (11.7)

11.122 In another laparoscopic surgery, helium is used to inflate the abdomen to separate the internal organs and the abdominal wall. If the He injected into the abdomen produces a pressure of 15 mmHg and a volume of 3.1 L at 28 °C, how many grams of He were used? (11.7)

11.123 At a restaurant, a customer chokes on a piece of food. You put your arms around the person's waist and use your fists to push up on the person's abdomen, an action called the Heimlich maneuver. (11.2)

a. How would this action change the volume of the chest and lungs?
b. Why does it cause the person to expel the food item from the airway?

11.124 An airplane is pressurized with air to 650. mmHg. (11.9)

a. If air is 21% oxygen, what is the partial pressure of oxygen on the plane?
b. If the partial pressure of oxygen drops below 100. mmHg, passengers become drowsy. If this happens, oxygen masks are released. What is the total cabin pressure at which oxygen masks are dropped?

ANSWERS TO ENGAGE QUESTIONS

11.1 According to the kinetic molecular theory, the molecules of a gas are moving rapidly in all directions to completely fill the size and shape of its container.

11.2 At a higher altitude there are less gas molecules, which means there is a lower pressure.

11.3 According to Boyle's law, the pressure of a gas decreases when the volume increases, if the temperature and amount of gas do not change.

11.4 According to Charles's law, the volume of a gas increases when the temperature increases, if the pressure and amount of gas do not change.

11.5 According to Gay-Lussac's law, the pressure of a gas increases when the temperature increases, if the volume and amount of gas do not change.

11.6 The pressure of a gas decreases to one-half when its volume doubles, if the amount of gas does not change. When the

Kelvin temperature of that gas also decreases by one-half, its pressure decreases to half. The total decrease in pressure is one-half times one-half or one-fourth of the initial pressure of the gas.

11.7 $T_1 = T_2 \times \dfrac{P_1}{P_2} \times \dfrac{V_1}{V_2}$

11.8 According to Avogadro's law, the volume of a balloon increases when the number of moles (or molecules) of gas in the balloon increases, if the pressure and temperature of the gas do not change.

11.9 Molar volume increases when the temperature increases from 0 °C to 100 °C.

11.10 If you use the value of 0.0821 for R, the pressure units must be atmospheres.

11.11 According to Dalton's law, partial pressures are additive. The total pressure in a container with 2.0 atm of helium and 4.0 atm of argon would be 6.0 atm.

11.12 According to Dalton's law, the total pressure of the gases collected over water includes the partial pressure of water vapor. To obtain the partial pressure of a gas, the partial pressure of water is subtracted from the total pressure.

ANSWERS TO SELECTED PROBLEMS

11.1 **a.** At a higher temperature, gas particles have greater kinetic energy, which makes them move faster.
  b. Because there are great distances between the particles of a gas, they can be pushed closer together and still remain a gas.
  c. Gas particles are very far apart, which means that the mass of a gas in a certain volume is very small, resulting in a low density.

11.3 **a.** temperature   **b.** volume
  c. amount   **d.** pressure

11.5 Statements **a**, **d**, and **e** describe the pressure of a gas.

11.7 **a.** 1520 Torr   **b.** 29.4 lb/in.^{2}
  c. 1520 mmHg   **d.** 203 kPa

11.9 As a diver ascends to the surface, external pressure decreases. If the air in the lungs were not exhaled, its volume would expand and severely damage the lungs. The pressure in the lungs must adjust to changes in the external pressure.

11.11 **a.** The pressure is greater in cylinder **A**. According to Boyle's law, a decrease in volume pushes the gas particles closer together, which will cause an increase in the pressure.
  b. 160 mL

11.13 **a.** increases   **b.** decreases
  c. increases

11.15 **a.** 328 mmHg   **b.** 2620 mmHg
  c. 475 mmHg   **d.** 5240 mmHg

11.17 **a.** 52.4 L   **b.** 25.0 L
  c. 100. L   **d.** 45 L

11.19 6.52 atm

11.21 25 L of cyclopropane

11.23 **a.** inspiration   **b.** expiration
  c. inspiration

11.25 **a.** C   **b.** A
  c. B

11.27 **a.** 303 °C   **b.** −129 °C
  c. 591 °C   **d.** 136 °C

11.29 **a.** 2400 mL   **b.** 4900 mL
  c. 1800 mL   **d.** 1700 mL

11.31 121 °C

11.33 **a.** 770 mmHg   **b.** 1150 mmHg

11.35 **a.** −23 °C   **b.** 168 °C

11.37 31 °C

11.39 **a.** On top of a mountain, water boils below 100 °C because the atmospheric (external) pressure is less than 1 atm.
  b. Because the pressure inside a pressure cooker is greater than 1 atm, water boils above 100 °C. At a higher temperature, food cooks faster.

11.41 1.9 atm

11.43 $T_2 = T_1 \times \dfrac{P_2}{P_1} \times \dfrac{V_2}{V_1}$

11.45 **a.** 4.26 atm   **b.** 3.07 atm
  c. 0.606 atm

11.47 −33 °C

11.49 The volume increases because the number of gas particles is increased.

11.51 **a.** 4.00 L   **b.** 26.7 L
  c. 14.6 L

11.53 **a.** 0.100 mol of O_2 **b.** 56.0 L
  c. 28.0 L   **d.** 0.146 g of H_2

11.55 4.93 atm

11.57 29.4 g of O_2

11.59 566 K (293 °C)

11.61 **a.** 42 g/mol   **b.** 28.7 g/mol
  c. 39.8 g/mol   **d.** 33 g/mol

11.63 1300 g of O_2

11.65 **a.** 7.60 L of H_2 **b.** 4.92 g of Mg

11.67 178 L of O_2

11.69 3.4 L of O_2

11.71 765 Torr

11.73 425 Torr

11.75 **a.** 743 mmHg   **b.** 0.0103 mol of O_2

11.77 0.978 atm

11.79 **a.** The partial pressure of oxygen will be lower than normal.
  b. Breathing a higher concentration of oxygen will help to increase the supply of oxygen in the lungs and blood and raise the partial pressure of oxygen in the blood.

11.81 2.8 L

11.83
  a. True. The flask containing helium has more moles of helium and thus more helium atoms.
  b. True. The mass and volume of each are the same, which means the mass/volume ratio or density is the same in both flasks.

11.85
  a. 2   **b.** 1

11.87
  a. A   **b.** C   **c.** A
  d. B   **e.** C

11.89 $-66\,°C$

11.91   5.39 L of CO_2

11.93   2.6 kg of H_2

11.95 $-103\,°C$

11.97   4.41 g of CH_4

11.99   42.1 g/mol

11.101   81.0 g of CO_2

11.103   8.56 L of H_2

11.105   1.02 g of NH_3

11.107   He 600 Torr, O_2 1800 Torr

11.109   370 Torr

11.111
  a. 734 mmHg
  b. 0.0165 mol of H_2
  c. 0.297 g of Al

11.113   30.0 g/mol

11.115   1340 L of CO_2

11.117   5.31 g of water

11.119   620 L

11.121   0.18 g of CO_2

11.123
  a. By pushing in on the abdomen, the volume of the chest and lungs is decreased.
  b. Because the volume is decreased, Boyle's law states that the pressure is increased, and this will expel the food blocking the airway.

Our kidneys produce urine, which carries waste products and excess fluid from the body. They also reabsorb electrolytes, such as potassium, and produce hormones that regulate blood pressure and the levels of calcium in the blood. Diseases such as diabetes and high blood pressure can cause a decrease in kidney function. Symptoms of kidney malfunction include protein in the urine, an abnormal level of urea in the blood, frequent urination, and swollen feet. If kidney failure occurs, it may be treated with dialysis or transplantation.

Michelle suffers from kidney disease because of severe strep throat she contracted as a child. When her kidneys stopped functioning, Michelle was placed on dialysis three times a week. As she enters the dialysis unit, her dialysis nurse, Amanda, asks Michelle how she is feeling. Michelle indicates that she feels tired today and has considerable swelling around her ankles. Amanda informs her that these side effects occur because of her body's inability to regulate the amount of water in her cells. Amanda explains that the retention of water is regulated by the concentration of electrolytes in her body fluids and the rate at which waste products are removed from her body. Amanda explains that although water is essential for the many chemical reactions that occur in the body, the amount of water can become too high or too low because of various diseases and conditions. Because Michelle's kidneys no longer function properly, she cannot regulate the amount of electrolytes or waste in her body fluids. As a result, she has an electrolyte imbalance and a buildup of waste products, so her body retains water. Amanda then explains that the dialysis machine does the work of her kidneys to reduce the high levels of electrolytes and waste products.

CAREER

Dialysis Nurse

A dialysis nurse specializes in assisting patients with kidney disease undergoing dialysis. This requires monitoring the patient before, during, and after dialysis for any complications such as a drop in blood pressure or cramping. The dialysis nurse connects the patient to the dialysis unit via a dialysis catheter that is inserted into the neck or chest. A dialysis nurse must have considerable knowledge about how the dialysis machine functions to ensure that it is operating correctly at all times.

UPDATE   Using Dialysis for Renal Failure

Michelle continues to have dialysis treatment three times a week. You can see more details of Michelle's dialysis in the **UPDATE Using Dialysis for Renal Failure**, page 388, and discover that dialysis requires 120 L of fluid containing electrolytes to adjust Michelle's blood level to that of normal serum.

12.1 Solutions

LEARNING GOAL Identify the solute and solvent in a solution; describe the formation of a solution.

Solutions are everywhere around us. Most of the gases, liquids, and solids we see are mixtures of at least one substance dissolved in another. There are different types of solutions. The air we breathe is a solution that is primarily oxygen and nitrogen gases. Carbon dioxide gas dissolved in water makes carbonated drinks. When we make solutions of coffee or tea, we use hot water to dissolve substances from coffee beans or tea leaves. The ocean is also a solution, consisting of many ionic compounds such as sodium chloride dissolved in water. In your medicine cabinet, the antiseptic tincture of iodine is a solution of iodine dissolved in ethanol.

A **solution** is a homogeneous mixture in which one substance, called the **solute**, is uniformly dispersed in another substance called the **solvent**. Because the solute and the solvent do not react with each other, they can be mixed in varying proportions. A solution with a small amount of salt dissolved in water tastes slightly salty. When a large amount of salt is dissolved in water, the solution tastes very salty. Usually, the solute (in this case, salt) is the substance present in the lesser amount, whereas the solvent (in this case, water) is present in the greater amount. For example, in a solution composed of 5.0 g of salt and 50. g of water, salt is the solute and water is the solvent. In a solution, the particles of the solute are evenly dispersed among the molecules within the solvent (see **FIGURE 12.1**).

FIGURE 12.1 A solution of copper(II) sulfate ($CuSO_4$) forms as particles of solute dissolve, move away from the crystal, and become evenly dispersed among the solvent (water) molecules.

Types of Solutes and Solvents

Solutes and solvents may be solids, liquids, or gases. The solution that forms has the same physical state as the solvent. When sugar crystals are dissolved in water, the resulting sugar solution is liquid. Sugar is the solute, and water is the solvent. Soda water and soft drinks are prepared by dissolving carbon dioxide gas in water. The carbon dioxide gas is the solute, and water is the solvent. **TABLE 12.1** lists some solutions and their solutes and solvents.

Water as a Solvent

Water is one of the most common solvents in nature. In the H_2O molecule, an oxygen atom shares electrons with two hydrogen atoms. Because oxygen is much more electronegative than hydrogen, the O — H bonds are polar. In each polar bond, the oxygen atom has a partial

REVIEW

Identifying Polarity of Molecules (10.5)

Identifying Intermolecular Forces (10.6)

Solute: The substance present in lesser amount

Salt

Water

Solvent: The substance present in greater amount

A solution has at least one solute dispersed in a solvent.

ENGAGE 12.1

What does the uniform blue color in the graduated cylinder on the right indicate?

TABLE 12.1 Some Examples of Solutions

Type	Example	Solute	Solvent
Gas Solutions			
Gas in a gas	Air	$O_2(g)$	$N_2(g)$
Liquid Solutions			
Gas in a liquid	Soda water	$CO_2(g)$	$H_2O(l)$
	Household ammonia	$NH_3(g)$	$H_2O(l)$
Liquid in a liquid	Vinegar	$HC_2H_3O_2(l)$	$H_2O(l)$
Solid in a liquid	Seawater	$NaCl(s)$	$H_2O(l)$
	Tincture of iodine	$I_2(s)$	$C_2H_6O(l)$
Solid Solutions			
Solid in a solid	Brass	$Zn(s)$	$Cu(s)$
	Steel	$C(s)$	$Fe(s)$

PRACTICE PROBLEMS

Try Practice Problems 12.1 and 12.2

negative (δ^-) charge, and the hydrogen atom has a partial positive (δ^+) charge. Because the shape of a water molecule is bent, its dipoles do not cancel out. Thus, water is polar and is a *polar solvent.*

Attractive forces known as *hydrogen bonds* occur between molecules where partially positive hydrogen atoms are attracted to the partially negative atoms N, O, or F. As seen in the diagram, the hydrogen bonds are shown as a series of dots. Although hydrogen bonds are much weaker than covalent or ionic bonds, there are many of them linking water molecules together. Hydrogen bonds are important in the properties of biological compounds such as proteins, carbohydrates, and DNA.

In water, hydrogen bonds form between an oxygen atom in one water molecule and the hydrogen atom in another.

Chemistry Link to Health

Water in the Body

The average adult is about 60% water by mass, and the average infant about 75%. About 60% of the body's water is contained within the cells as intracellular fluids; the other 40% makes up extracellular fluids, which include the interstitial fluid in tissue and the plasma in the blood. These external fluids carry nutrients and waste materials between the cells and the circulatory system.

Every day you lose between 1500 and 3000 mL of water from the kidneys as urine, from the skin as perspiration, from the lungs as you exhale, and from the gastrointestinal tract. Serious dehydration

Typical water gain and loss during 24 hours

Water Gain			Water Loss	
Liquid	1000 mL		Urine	1500 mL
Food	1200 mL		Perspiration	300 mL
Metabolism	300 mL		Breath	600 mL
			Feces	100 mL
Total	2500 mL		Total	2500 mL

The water lost from the body is replaced by the intake of fluids.

TABLE 12.2 Percentage of Water in Some Foods

Food	Water (% by mass)	Food	Water (% by mass)
Vegetables		**Meats/Fish**	
Carrot	88	Chicken, cooked	71
Celery	94	Hamburger, broiled	60
Cucumber	96	Salmon	71
Tomato	94		
Fruits		**Milk Products**	
Apple	85	Cottage cheese	78
Cantaloupe	91	Milk, whole	87
Orange	86	Yogurt	88
Strawberry	90		
Watermelon	93		

can occur in an adult if there is a 10% net loss in total body fluid; a 20% loss of fluid can be fatal. An infant suffers severe dehydration with only a 5 to 10% loss in body fluid.

Water loss is continually replaced by the liquids and foods in the diet and from metabolic processes that produce water in the cells of the body. **TABLE 12.2** lists the percentage by mass of water contained in some foods.

Formation of Solutions

The interactions between solute and solvent will determine whether a solution will form. Initially, energy is needed to separate the particles in the solute and the solvent particles. Then energy is released as solute particles move between the solvent particles to form a solution. However, there must be attractions between the solute and the solvent particles to provide the energy for the initial separation. These attractions occur when the solute and the solvent have similar polarities. The expression "like dissolves like" is a way of saying that the polarities of a solute and a solvent must be similar for a solution to form (see **FIGURE 12.2**). In the absence of attractions between a solute and a solvent, there is insufficient energy to form a solution (see **TABLE 12.3**).

Water (less dense)
CH_2Cl_2 (more dense)

CH_2Cl_2 is nonpolar and water is polar; the two layers do not mix.

The nonpolar solute I_2 (purple) is soluble in the nonpolar solvent CH_2Cl_2.

The ionic solute $Ni(NO_3)_2$ (green) is soluble in the polar solvent water.

FIGURE 12.2 Like dissolves like. In each test tube, the lower layer is CH_2Cl_2 (more dense), and the upper layer is water (less dense).

TABLE 12.3 Possible Combinations of Solutes and Solvents

Solutions Will Form		Solutions Will Not Form	
Solute	**Solvent**	**Solute**	**Solvent**
Polar	Polar	Polar	Nonpolar
Nonpolar	Nonpolar	Nonpolar	Polar

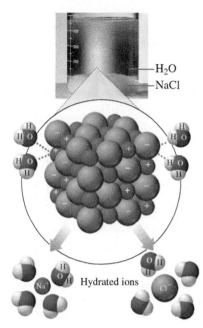

FIGURE 12.3 Ions on the surface of a crystal of NaCl dissolve in water as they are attracted to the polar water molecules that pull the ions into solution and surround them.

Solutions with Ionic and Polar Solutes

In ionic solutes such as sodium chloride, NaCl, there are strong ionic bonds between positively charged Na$^+$ ions and negatively charged Cl$^-$ ions. In water, a polar solvent, the hydrogen bonds provide strong solvent–solvent attractions. When NaCl crystals are placed in water, partially negative oxygen atoms in water molecules attract positive Na$^+$ ions, and the partially positive hydrogen atoms in other water molecules attract negative Cl$^-$ ions (see **FIGURE 12.3**). As soon as the Na$^+$ ions and the Cl$^-$ ions form a solution, they undergo **hydration** as water molecules surround each ion. Hydration of the ions diminishes their attraction to other ions and keeps them in solution.

In the equation for the formation of the NaCl solution, the solid and aqueous NaCl are shown with the formula H$_2$O over the arrow, which indicates that water is needed for the dissociation process but is not a reactant.

$$NaCl(s) \xrightarrow[\text{Dissociation}]{\text{H}_2\text{O}} Na^+(aq) + Cl^-(aq)$$

In another example, we find that a polar molecular compound such as methanol, CH$_4$O, is soluble in water because methanol has a polar —OH group that forms hydrogen bonds with water (see **FIGURE 12.4**). Polar solutes require polar solvents for a solution to form.

Methanol (CH$_4$O) solute

Water solvent

Methanol–water solution with hydrogen bonding

FIGURE 12.4 Polar molecules of methanol, CH$_4$O, form hydrogen bonds with polar water molecules to form a methanol–water solution.

ENGAGE 12.2

Why does KCl(s) form a solution with water, but nonpolar hexane (C$_6$H$_{14}$) does not form a solution with water?

PRACTICE PROBLEMS

Try Practice Problems 12.3 to 12.6

Solutions with Nonpolar Solutes

Compounds containing nonpolar molecules, such as iodine (I$_2$), oil, or grease, do not dissolve in water because there are no attractions between the particles of a nonpolar solute and the polar solvent. Nonpolar solutes require nonpolar solvents for a solution to form.

PRACTICE PROBLEMS

12.1 Solutions

12.1 Identify the solute and the solvent in each solution composed of the following:
 a. 10.0 g of NaCl and 100.0 g of H$_2$O
 b. 50.0 mL of ethanol, C$_2$H$_6$O, and 10.0 mL of H$_2$O
 c. 0.20 L of O$_2$ and 0.80 L of N$_2$

12.2 Identify the solute and the solvent in each solution composed of the following:
 a. 10.0 mL of acetic acid, HC$_2$H$_3$O$_2$, and 200. mL of H$_2$O
 b. 100.0 mL of H$_2$O and 5.0 g of sugar, C$_{12}$H$_{22}$O$_{11}$
 c. 1.0 g of Br$_2$ and 50.0 mL of methylene chloride, CH$_2$Cl$_2$

12.3 Describe the formation of an aqueous KI solution, when solid KI dissolves in water.

12.4 Describe the formation of an aqueous LiBr solution, when solid LiBr dissolves in water.

Applications

12.5 Water is a polar solvent and carbon tetrachloride (CCl$_4$) is a nonpolar solvent. In which solvent is each of the following, which is found or used in the body, more likely to be soluble?
 a. CaCO$_3$ (calcium supplement), ionic
 b. retinol (vitamin A), nonpolar
 c. sucrose (table sugar), polar
 d. cholesterol (lipid), nonpolar

12.6 Water is a polar solvent and hexane (C$_6$H$_{14}$) is a nonpolar solvent. In which solvent is each of the following, which is found or used in the body, more likely to be soluble?
 a. vegetable oil, nonpolar
 b. oleic acid (lipid), nonpolar
 c. niacin (vitamin B$_3$), polar
 d. FeSO$_4$ (iron supplement), ionic

12.2 Electrolytes and Nonelectrolytes

LEARNING GOAL Identify solutes as electrolytes or nonelectrolytes.

Solutes can be classified by their ability to conduct an electrical current. When **electrolytes** dissolve in water, the process of *dissociation* separates them into ions forming solutions that conduct electricity. When **nonelectrolytes** dissolve in water, they do not separate into ions, and their solutions do not conduct electricity.

To test solutions for the presence of ions, we can use an apparatus that consists of a battery and a pair of electrodes connected by wires to a light bulb. The light bulb glows when electricity can flow, which can only happen when electrolytes provide ions that move between the electrodes to complete the circuit.

Types of Electrolytes

Electrolytes can be further classified as *strong electrolytes* or *weak electrolytes*. For a **strong electrolyte**, such as sodium chloride (NaCl), there is 100% dissociation of the solute into ions. When the electrodes from the light bulb apparatus are placed in the NaCl solution, the light bulb glows very brightly.

In an equation for dissociation of a compound in water, the charges must balance. For example, magnesium nitrate dissociates to give one magnesium ion for every two nitrate ions. However, only the ionic bonds between Mg^{2+} and NO_3^- are broken, not the covalent bonds within the polyatomic ion. The equation for the dissociation of $Mg(NO_3)_2$ is written as follows:

$$Mg(NO_3)_2(s) \xrightarrow[\text{Dissociation}]{H_2O} Mg^{2+}(aq) + 2NO_3^-(aq)$$

A **weak electrolyte** is a compound that dissolves in water mostly as molecules. Only a few of the dissolved solute molecules undergo dissociation, producing a small number of ions in solution. Thus, solutions of weak electrolytes do not conduct electrical current as well as solutions of strong electrolytes. When the electrodes are placed in a solution of a weak electrolyte, the glow of the light bulb is very dim. In an aqueous solution of the weak electrolyte HF, a few HF molecules dissociate to produce H^+ and F^- ions. As more H^+ and F^- ions form, some recombine to give HF molecules. These forward and reverse reactions of molecules to ions and back again are indicated by two arrows between reactant and products that point in opposite directions:

$$HF(aq) \underset{\text{Recombination}}{\overset{\text{Dissociation}}{\rightleftharpoons}} H^+(aq) + F^-(aq)$$

A nonelectrolyte such as methanol (CH_4O) dissolves in water only as molecules, which do not dissociate. When electrodes of the light bulb apparatus are placed in a solution of a nonelectrolyte, the light bulb does not glow because the solution does not contain ions and cannot conduct electricity.

$$CH_4O(l) \xrightarrow{H_2O} CH_4O(aq)$$

TABLE 12.4 summarizes the classification of solutes in aqueous solutions.

REVIEW

Writing Conversion Factors from Equalities (2.5)

Using Conversion Factors (2.6)

Writing Positive and Negative Ions (6.1)

Strong electrolyte

Weak electrolyte

TABLE 12.4 Classification of Solutes in Aqueous Solutions				
Type of Solute	In Solution	Type(s) of Particles in Solution	Conducts Electricity?	Examples
Strong electrolyte	Dissociates completely	Only ions	Yes	Ionic compounds such as NaCl, KBr, $MgCl_2$, $NaNO_3$; bases such as NaOH, KOH; acids such as HCl, HBr, HI, HNO_3, $HClO_4$, H_2SO_4
Weak electrolyte	Dissociates partially	Mostly molecules and a few ions	Weakly	HF, H_2O, NH_3, $HC_2H_3O_2$ (acetic acid)
Nonelectrolyte	No dissociation	Only molecules	No	Carbon compounds such as CH_4O (methanol), C_2H_6O (ethanol), $C_{12}H_{22}O_{11}$ (sucrose), CH_4N_2O (urea)

Nonelectrolyte

SAMPLE PROBLEM 12.1   Solutions of Electrolytes and Nonelectrolytes

TRY IT FIRST

Indicate whether solutions of each of the following contain only ions, only molecules, or mostly molecules and a few ions. Write the equation for the formation of a solution for each of the following:

a. $Na_2SO_4(s)$, a strong electrolyte
b. sucrose, $C_{12}H_{22}O_{11}(s)$, a nonelectrolyte
c. acetic acid, $HC_2H_3O_2(l)$, a weak electrolyte

SOLUTION

a. An aqueous solution of $Na_2SO_4(s)$ contains only the ions Na^+ and SO_4^{2-}.

$$Na_2SO_4(s) \xrightarrow{H_2O} 2Na^+(aq) + SO_4^{2-}(aq)$$

b. A nonelectrolyte such as sucrose, $C_{12}H_{22}O_{11}(s)$, produces only molecules when it dissolves in water.

$$C_{12}H_{22}O_{11}(s) \xrightarrow{H_2O} C_{12}H_{22}O_{11}(aq)$$

c. A weak electrolyte such as $HC_2H_3O_2(l)$ produces mostly molecules and a few ions when it dissolves in water.

$$HC_2H_3O_2(l) \underset{}{\overset{H_2O}{\rightleftarrows}} H^+(aq) + C_2H_3O_2^-(aq)$$

SELF TEST 12.1

a. Boric acid, $H_3BO_3(s)$, is a weak electrolyte. Would you expect a boric acid solution to contain only ions, only molecules, or mostly molecules and a few ions?
b. Hydrochloric acid, HCl, is a strong electrolyte. Would you expect a solution of hydrochloric acid to contain only ions, only molecules, or mostly molecules and a few ions?

ANSWER

a. A solution of a weak electrolyte would contain mostly molecules and a few ions.
b. A solution of a strong electrolyte would contain only ions.

ENGAGE 12.3

Why does a solution of $LiNO_3$, a strong electrolyte, contain only ions, whereas a solution of urea, CH_4N_2O, a nonelectrolyte, contains only molecules?

PRACTICE PROBLEMS

Try Practice Problems 12.7 to 12.14

Chemistry Link to Health

Electrolytes in Body Fluids

Electrolytes in the body play an important role in maintaining the proper function of the cells and organs in the body. Typically, the electrolytes sodium, potassium, chloride, and bicarbonate are measured in a blood test. Sodium ions regulate the water content in the body and are important in carrying electrical impulses through the nervous system. Potassium ions are also involved in the transmission of electrical impulses and play a role in the maintenance of a regular heartbeat. Chloride ions balance the charges of the positive ions and also control the balance of fluids in the body. Bicarbonate is important in maintaining the proper pH of the blood. Sometimes when vomiting, diarrhea, or sweating is excessive, the concentrations of certain electrolytes may decrease. Then fluids such as Pedialyte may be given to return electrolyte levels to normal.

An intravenous solution is used to replace electrolytes in the body.

12.2 Electrolytes and Nonelectrolytes

12.7 KF is a strong electrolyte, and HF is a weak electrolyte. How is the solution of KF different from that of HF?

12.8 NaOH is a strong electrolyte, and CH_4O is a nonelectrolyte. How is the solution of NaOH different from that of CH_4O?

12.9 Write a balanced equation for the dissociation of each of the following strong electrolytes in water:
- **a.** KCl
- **b.** $CaCl_2$
- **c.** K_3PO_4
- **d.** $Fe(NO_3)_3$

12.10 Write a balanced equation for the dissociation of each of the following strong electrolytes in water:
- **a.** LiBr
- **b.** $NaNO_3$
- **c.** $CuCl_2$
- **d.** K_2CO_3

12.11 Indicate whether aqueous solutions of each of the following solutes contain only ions, only molecules, or mostly molecules and a few ions:
- **a.** acetic acid, $HC_2H_3O_2$, a weak electrolyte
- **b.** NaBr, a strong electrolyte
- **c.** fructose, $C_6H_{12}O_6$, a nonelectrolyte

12.12 Indicate whether aqueous solutions of each of the following solutes contain only ions, only molecules, or mostly molecules and a few ions:
- **a.** NH_4Cl, a strong electrolyte
- **b.** ethanol, C_2H_6O, a nonelectrolyte
- **c.** hydrocyanic acid, HCN, a weak electrolyte

12.13 Classify the solute represented in each of the following equations as a strong, weak, or nonelectrolyte:

- **a.** $K_2SO_4(s) \xrightarrow{H_2O} 2K^+(aq) + SO_4^{2-}(aq)$

- **b.** $NH_3(g) + H_2O(l) \rightleftarrows NH_4^+(aq) + OH^-(aq)$

- **c.** $C_6H_{12}O_6(s) \xrightarrow{H_2O} C_6H_{12}O_6(aq)$

12.14 Classify the solute represented in each of the following equations as a strong, weak, or nonelectrolyte:

- **a.** $C_4H_{10}O(l) \xrightarrow{H_2O} C_4H_{10}O(aq)$

- **b.** $MgCl_2(s) \xrightarrow{H_2O} Mg^{2+}(aq) + 2Cl^-(aq)$

- **c.** $HClO(aq) \rightleftarrows H^+(aq) + ClO^-(aq)$

12.3 Solubility

LEARNING GOAL Define solubility; distinguish between an unsaturated and a saturated solution. Identify an ionic compound as soluble or insoluble.

The term *solubility* is used to describe the amount of a solute that can dissolve in a given amount of solvent. Many factors, such as the type of solute, the type of solvent, and the temperature, affect the solubility of a solute. **Solubility**, usually expressed in grams of solute in 100. g of solvent, is the maximum amount of solute that can dissolve at a certain temperature. If a solute readily dissolves when added to the solvent, the solution does not contain the maximum amount of solute. We call this solution an **unsaturated solution**.

A solution that contains all the solute that can dissolve is a **saturated solution**. When a solution is saturated, the rate at which the solute dissolves becomes equal to the rate at which solid forms, a process known as *crystallization*. Then there is no further change in the amount of dissolved solute in solution.

$$\text{Solute + solvent} \underset{\text{Solute crystallizes}}{\overset{\text{Solute dissolves}}{\rightleftarrows}} \text{saturated solution}$$

We can prepare a saturated solution by adding an amount of solute greater than that needed to reach solubility. Stirring the solution will dissolve the maximum amount of solute and leave the excess on the bottom of the container. The addition of more solute to the saturated solution will only increase the amount of undissolved solute.

REVIEW

Interpreting Graphs (1.4)

Unsaturated solution   Saturated solution

Additional solute can dissolve in an unsaturated solution, but not in a saturated solution.

SAMPLE PROBLEM 12.2 Saturated Solutions

TRY IT FIRST

At 20 °C, the solubility of KCl is 34 g/100. g of H_2O. In the laboratory, a student mixes 75 g of KCl with 200. g of H_2O at a temperature of 20 °C.

a. How much of the KCl will dissolve?

b. Is the solution saturated or unsaturated?

c. What is the mass, in grams, of any solid KCl left undissolved on the bottom of the container?

SOLUTION

a. At 20 °C, KCl has a solubility of 34 g of KCl in 100. g of water. Using the solubility as a conversion factor, we can calculate the maximum amount of KCl that can dissolve in 200. g of water as follows:

$$200. \; g \, H_2O \times \frac{34 \; g \; KCl}{100. \; g \, H_2O} = 68 \; g \; of \; KCl$$

b. Because 75 g of KCl exceeds the maximum amount (68 g) that can dissolve in 200. g of water, the KCl solution is saturated.

c. If we add 75 g of KCl to 200. g of water and only 68 g of KCl can dissolve, there is 7 g (75 g − 68 g) of solid (undissolved) KCl on the bottom of the container.

SELF TEST 12.2

At 40 °C, the solubility of KNO_3 is 65 g/100. g of H_2O.

a. How many grams of KNO_3 will dissolve in 120 g of H_2O at 40 °C?

b. How many grams of H_2O are needed to completely dissolve 110 g of KNO_3 at 20 °C?

ANSWER

a. 78 g of KNO_3 **b.** 170 g of H_2O

PRACTICE PROBLEMS

Try Practice Problems 12.15 to 12.20

Chemistry Link to Health

Gout and Kidney Stones: Saturation in Body Fluids

The conditions of gout and kidney stones involve compounds in the body that exceed their solubility levels and form solid products. Gout affects adults, primarily men, over the age of 40. Attacks of gout may occur when the concentration of uric acid in blood plasma exceeds its solubility, which is 7 mg/100 mL of plasma at 37 °C. Insoluble deposits of needle-like crystals of uric acid can form in the cartilage, tendons, and soft tissues, where they cause painful gout attacks. They may also form in the tissues of the kidneys, where they can cause renal damage. High levels of uric acid in the body can be caused by an increase in uric acid production, failure of the kidneys to remove uric acid, or a diet with an overabundance of foods containing purines, which are metabolized to uric acid in the body. Foods in the diet that contribute to high levels of uric acid include certain meats, sardines, mushrooms, asparagus, and beans. Drinking alcoholic beverages may also significantly increase uric acid levels and bring about gout attacks.

Treatment for gout involves diet changes and drugs. Medications, such as probenecid, which helps the kidneys eliminate uric acid, or allopurinol, which blocks the production of uric acid by the body, may be useful.

Kidney stones are solid materials that form in the urinary tract. Most kidney stones are composed of calcium phosphate and calcium oxalate, although they can be solid uric acid. Insufficient water intake and high levels of calcium, oxalate, and phosphate in the urine can lead to the formation of kidney stones. When a kidney stone passes through the urinary tract, it causes considerable pain and discomfort, necessitating the use of painkillers and surgery. Sometimes ultrasound is used to break up kidney stones. Persons prone to kidney stones are advised to drink six to eight glasses of water every day to prevent saturation levels of minerals in the urine.

Gout occurs when uric acid exceeds its solubility in blood plasma.

Kidney stones form when calcium phosphate exceeds its solubility.

Effect of Temperature on Solubility

The solubility of most solids is greater as temperature increases, which means that solutions usually contain more dissolved solute at higher temperature. A few substances show little change in solubility at higher temperatures, and a few are less soluble (see **FIGURE 12.5**). For example, when you add sugar to iced tea, some undissolved sugar may form on the bottom of the glass. But if you add sugar to hot tea, many teaspoons of sugar are needed before solid sugar appears. Hot tea dissolves more sugar than does cold tea because the solubility of sugar is much greater at a higher temperature.

When a saturated solution is carefully cooled, it becomes a *supersaturated solution* because it contains more solute than the solubility allows. Such a solution is unstable, and if the solution is agitated or if a solute crystal is added, the excess solute will crystallize to give a saturated solution again.

Conversely, the solubility of a gas in water decreases as the temperature increases. At higher temperatures, more gas molecules have the energy to escape from the solution. Perhaps you have observed the bubbles escaping from a carbonated soft drink as it warms. At high temperatures, bottles containing carbonated solutions may burst as more gas molecules leave the solution and increase the gas pressure inside the bottle. Biologists have found that increased temperatures in rivers and lakes cause the amount of dissolved oxygen to decrease until the warm water can no longer support a biological community. Electricity-generating plants are required to have their own ponds to use with their cooling towers to lessen the threat of thermal pollution in surrounding waterways.

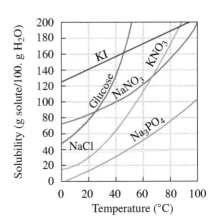

FIGURE 12.5 In water, most common solids are more soluble as the temperature increases.

ENGAGE 12.4

Compare the solubility of $NaNO_3$ at 20 °C and 60 °C.

Henry's Law

Henry's law states that the solubility of gas in a liquid is directly related to the pressure of that gas above the liquid. At higher pressures, there are more gas molecules available to enter and dissolve in the liquid. A can of soda is carbonated by using CO_2 gas at high pressure to increase the solubility of the CO_2 in the beverage. When you open the can at atmospheric pressure, the pressure on the CO_2 drops, which decreases the solubility of CO_2. As a result, bubbles of CO_2 rapidly escape from the solution. The burst of bubbles is even more noticeable when you open a warm can of soda.

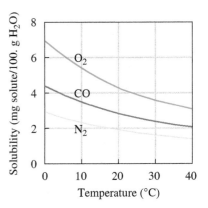

In water, gases are less soluble as the temperature increases.

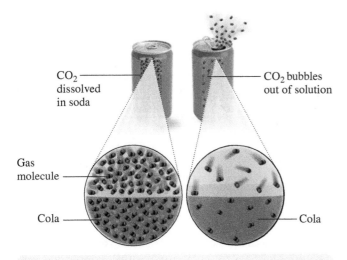

When the pressure of a gas above a solution decreases, the solubility of that gas in the solution also decreases.

PRACTICE PROBLEMS

Try Practice Problems 12.21 and 12.22

Soluble and Insoluble Ionic Compounds

In our discussion up to now, we have considered ionic compounds that dissolve in water. However, some ionic compounds do not dissociate into ions and remain as solids even in contact with water. The **solubility rules** give some guidelines about the solubility of ionic compounds in water.

ENGAGE 12.5

Why is K_2S soluble in water whereas $PbCl_2$ is not soluble?

ENGAGE 12.6

Why are each of the ionic compounds in Figure 12.6 insoluble in water?

Ionic compounds that are soluble in water typically contain at least one of the ions in **TABLE 12.5**. *Only an ionic compound containing a soluble cation or anion will dissolve in water.* Most ionic compounds containing Cl^- are soluble, but $AgCl$, $PbCl_2$, and Hg_2Cl_2 are insoluble. Similarly, most ionic compounds containing SO_4^{2-} are soluble, but a few are insoluble. Most other ionic compounds are insoluble (see **FIGURE 12.6**). In an insoluble ionic compound, the ionic bonds between its positive and negative ions are too strong for the polar water molecules to break. We can use the solubility rules to predict whether a solid ionic compound would be soluble or not. **TABLE 12.6** illustrates the use of these rules.

TABLE 12.5 Solubility Rules for Ionic Compounds in Water

An ionic compound is soluble in water if it contains one of the following:	
Positive Ions	Li^+, Na^+, K^+, Rb^+, Cs^+, NH_4^+
Negative Ions	NO_3^-, $C_2H_3O_2^-$ Cl^-, Br^-, I^- except when combined with Ag^+, Pb^{2+}, or Hg_2^{2+} SO_4^{2-} except when combined with Ba^{2+}, Pb^{2+}, Ca^{2+}, Sr^{2+}, or Hg_2^{2+}
Ionic compounds that do not contain at least one of these ions are usually insoluble.	

CdS   FeS PbI_2 $Ni(OH)_2$

FIGURE 12.6 If an ionic compound contains a combination of a cation and an anion that are not soluble, that ionic compound is insoluble. For example, combinations of cadmium and sulfide, iron and sulfide, lead and iodide, and nickel and hydroxide do not contain any soluble ions. Thus, they form insoluble ionic compounds.

FIGURE 12.7 A barium sulfate-enhanced X-ray of the abdomen shows the lower gastrointestinal (GI) tract.

TABLE 12.6 Using Solubility Rules

Ionic Compound	Solubility in Water	Reasoning
K_2S	Soluble	Contains K^+
$Ca(NO_3)_2$	Soluble	Contains NO_3^-
$PbCl_2$	Insoluble	Is an insoluble chloride
$NaOH$	Soluble	Contains Na^+
$AlPO_4$	Insoluble	Contains no soluble ions

In medicine, insoluble $BaSO_4$ is used as an opaque substance to enhance X-rays of the gastrointestinal tract (see **FIGURE 12.7**). $BaSO_4$ is so insoluble that it does not dissolve in gastric fluids. Other ionic barium compounds cannot be used because they would dissolve in water, releasing Ba^{2+}, which is poisonous.

SAMPLE PROBLEM 12.3   Soluble and Insoluble Ionic Compounds

TRY IT FIRST

Predict whether each of the following ionic compounds is soluble in water or not and explain your answer:

a. Na_3PO_4 **b.** $CaCO_3$

SOLUTION

a. The ionic compound Na_3PO_4 is soluble in water because any compound that contains Na^+ is soluble.
b. The ionic compound $CaCO_3$ is not soluble because it does not contain a soluble positive or negative ion.

SELF TEST 12.3

a. In some electrolyte drinks, $MgCl_2$ is added to provide magnesium. Why would you expect $MgCl_2$ to be soluble in water?
b. Even though barium ions are toxic if ingested, why can barium sulfate be given to a patient for an X-ray of the lower gastrointestinal tract?

ANSWER

a. $MgCl_2$ is soluble in water because ionic compounds that contain chloride are soluble unless they contain Ag^+, Pb^{2+}, or Hg_2^{2+}.
b. Barium sulfate is insoluble in water; it does not produce barium ions as it passes through the GI tract during an X-ray.

PRACTICE PROBLEMS

Try Practice Problems 12.23 and 12.24

Formation of a Solid

We can use solubility rules to predict whether a solid, called a *precipitate*, forms when two solutions containing soluble reactants are mixed, as shown in Sample Problem 12.4.

SAMPLE PROBLEM 12.4 Writing Equations for the Formation of an Insoluble Ionic Compound

TRY IT FIRST

When solutions of NaCl and $AgNO_3$ are mixed, a white solid forms. Write the ionic and net ionic equations for the reaction.

SOLUTION

STEP 1 Write the ions of the reactants.

Reactants (initial combinations)

$Ag^+(aq) + NO_3^-(aq)$

$Na^+(aq) + Cl^-(aq)$

STEP 2 Write the combinations of ions, and determine if any are insoluble. When we look at the ions of each solution, we see that the combination of Ag^+ and Cl^- forms an insoluble ionic compound.

Mixture (new combinations)	Product	Soluble
$Ag^+(aq) + Cl^-(aq)$	AgCl	No
$Na^+(aq) + NO_3^-(aq)$	$NaNO_3$	Yes

ENGAGE 12.7

When mixing solutions of $Pb(NO_3)_2$ and NaBr, how do we know that $PbBr_2(s)$ is the solid that forms?

STEP 3 Write the ionic equation including any solid. In the *ionic equation*, we show all the ions of the reactants. The products include the solid AgCl that forms along with the remaining ions Na^+ and NO_3^-.

$$Ag^+(aq) + NO_3^-(aq) + Na^+(aq) + Cl^-(aq) \longrightarrow AgCl(s) + Na^+(aq) + NO_3^-(aq)$$

STEP 4 **Write the net ionic equation.** We remove the Na^+ and NO_3^- ions, known as *spectator ions*, which are unchanged. This gives the *net ionic equation*, which only shows the ions that form a solid precipitate.

$$Ag^+(aq) + \underbrace{NO_3^-(aq) + Na^+(aq)}_{\text{Spectator ions}} + Cl^-(aq) \longrightarrow AgCl(s) + \underbrace{Na^+(aq) + NO_3^-(aq)}_{\text{Spectator ions}}$$

$$Ag^+(aq) + Cl^-(aq) \longrightarrow AgCl(s) \quad \text{Net ionic equation}$$

Ag^+ Cl^-

Na^+ NO_3^-

Insoluble ionic compound

Type of Equation	
Chemical	$AgNO_3(aq) \quad\quad + NaCl(aq) \longrightarrow AgCl(s) + NaNO_3(aq)$
Ionic	$Ag^+(aq) + NO_3^-(aq) + Na^+(aq) + Cl^-(aq) \rightarrow AgCl(s) + Na^+(aq) + NO_3^-(aq)$
Net Ionic	$Ag^+(aq) \quad\quad + Cl^-(aq) \longrightarrow AgCl(s)$

SELF TEST 12.4

Predict whether a solid might form in each of the following mixtures of solutions. If so, write the net ionic equation for the reaction.
a. $NH_4Cl(aq) + Ca(NO_3)_2(aq)$ b. $Pb(NO_3)_2(aq) + KCl(aq)$

ANSWER

a. No solid forms because the products, $NH_4NO_3(aq)$ and $CaCl_2(aq)$, are soluble.
b. $Pb^{2+}(aq) + 2Cl^-(aq) \longrightarrow PbCl_2(s)$

PRACTICE PROBLEMS

Try Practice Problems 12.25 and 12.26

PRACTICE PROBLEMS

12.3 Solubility

12.15 State whether each of the following refers to a saturated or an unsaturated solution:
a. A crystal added to a solution does not change in size.
b. A sugar cube completely dissolves when added to a cup of coffee.
c. A uric acid concentration of 4.6 mg/100 mL in the kidney does not cause gout.

12.16 State whether each of the following refers to a saturated or an unsaturated solution:
a. A spoonful of salt added to boiling water dissolves.
b. A layer of sugar forms on the bottom of a glass of tea as ice is added.
c. A kidney stone of calcium phosphate forms in the kidneys when urine becomes concentrated.

Use the following table for problems 12.17 to 12.20:

	Solubility (g/100. g H_2O)	
Substance	20 °C	50 °C
KCl	34	43
$NaNO_3$	88	110
$C_{12}H_{22}O_{11}$ (sugar)	204	260

12.17 Determine whether each of the following solutions will be saturated or unsaturated at 20 °C:
a. adding 25 g of KCl to 100. g of H_2O
b. adding 11 g of $NaNO_3$ to 25 g of H_2O
c. adding 400. g of sugar to 125 g of H_2O

12.18 Determine whether each of the following solutions will be saturated or unsaturated at 50 °C:
a. adding 25 g of KCl to 50. g of H_2O
b. adding 150. g of $NaNO_3$ to 75 g of H_2O
c. adding 80. g of sugar to 25 g of H_2O

12.19 A solution containing 80. g of KCl in 200. g of H_2O at 50 °C is cooled to 20 °C.
a. How many grams of KCl remain in solution at 20 °C?
b. How many grams of solid KCl crystallized after cooling?

12.20 A solution containing 80. g of $NaNO_3$ in 75 g of H_2O at 50 °C is cooled to 20 °C.
a. How many grams of $NaNO_3$ remain in solution at 20 °C?
b. How many grams of solid $NaNO_3$ crystallized after cooling?

12.21 Explain the following observations:
a. More sugar dissolves in hot tea than in iced tea.
b. Champagne in a warm room goes flat.
c. A warm can of soda has more spray when opened than a cold one.

12.22 Explain the following observations:
a. An open can of soda loses its "fizz" faster at room temperature than in the refrigerator.
b. Chlorine gas in tap water escapes as the sample warms to room temperature.
c. Less sugar dissolves in iced coffee than in hot coffee.

12.23 Predict whether each of the following ionic compounds is soluble in water:
a. LiCl **b.** PbS **c.** $BaCO_3$
d. K_2O **e.** $Fe(NO_3)_3$

12.24 Predict whether each of the following ionic compounds is soluble in water:
a. AgCl **b.** KI **c.** Na_2S
d. Ag_2O **e.** $CaSO_4$

12.25 Determine whether a solid forms when solutions containing the following ionic compounds are mixed. If so, write the ionic equation and the net ionic equation.
a. KCl(aq) and Na_2S(aq)
b. $AgNO_3$(aq) and K_2S(aq)
c. $CaCl_2$(aq) and Na_2SO_4(aq)
d. $CuCl_2$(aq) and Li_3PO_4(aq)

12.26 Determine whether a solid forms when solutions containing the following ionic compounds are mixed. If so, write the ionic equation and the net ionic equation.
a. Na_3PO_4(aq) and $AgNO_3$(aq)
b. K_2SO_4(aq) and Na_2CO_3(aq)
c. $Pb(NO_3)_2$(aq) and Na_2CO_3(aq)
d. $BaCl_2$(aq) and KOH(aq)

12.4 Solution Concentrations

LEARNING GOAL Calculate the concentration of a solute in a solution; use concentration units to calculate the amount of solute or solution.

Our body fluids contain water and dissolved substances including glucose, urea, and electrolytes such as K^+, Na^+, Cl^-, Mg^{2+}, HCO_3^-, and HPO_4^{2-}. Proper amounts of each of these dissolved substances and water must be maintained in the body fluids. Small changes in electrolyte levels can seriously disrupt cellular processes and endanger our health. Solutions can be described by their **concentration**, which is the amount of solute in a specific amount of that solution.

We will look at different ways to express a concentration as a ratio of a certain amount of solute in a given amount of solution as shown in **TABLE 12.7**. The amount of a solute may be expressed in units of grams, milliliters, or moles. The amount of a solution may be expressed in units of grams, milliliters, or liters.

$$\text{Concentration of a solution} = \frac{\text{amount of solute}}{\text{amount of solution}}$$

REVIEW
Calculating Percentages (1.4)
Using Molar Mass as a Conversion Factor (7.3)
Using Mole–Mole Factors (9.2)

CORE CHEMISTRY SKILL
Calculating Concentration

TABLE 12.7 Summary of Types of Concentration Expressions and Their Units

	Mass Percent (m/m)	Volume Percent (v/v)	Mass/Volume Percent (m/v)	Molarity (M)
Solute Unit	g	mL	g	mole
Solvent Unit	g	mL	mL	L

When water is added to 8.00 g of KCl to form 50.00 g of a KCl solution, the mass percent concentration is 16.0% (m/m).

Mass Percent (m/m) Concentration

Mass percent (m/m) describes the mass of the solute in grams for 100. g of solution. The mass percent is calculated by dividing the mass of a solute by the mass of the solution multiplied by 100% to give the percentage. In the calculation of mass percent (m/m), the units of mass of the solute and solution must be the same. If the mass of the solute is given as grams, then the mass of the solution must also be grams. The mass of the solution is the sum of the mass of the solute and the mass of the solvent.

$$\text{Mass percent (m/m)} = \frac{\text{mass of solute (g)}}{\text{mass of solute (g)} + \text{mass of solvent (g)}} \times 100\%$$

$$= \frac{\text{mass of solute (g)}}{\text{mass of solution (g)}} \times 100\%$$

Suppose we prepared a solution by mixing 8.00 g of KCl (solute) with 42.00 g of water (solvent). Together, the mass of the solute and mass of solvent give the mass of the solution (8.00 g + 42.00 g = 50.00 g). Mass percent is calculated by substituting the mass of the solute and the mass of the solution into the mass percent expression.

$$\frac{8.00 \text{ g KCl}}{50.00 \text{ g solution}} \times 100\% = 16.0\% \text{ (m/m) KCl solution}$$

$$\underbrace{8.00 \text{ g KCl}}_{\text{Solute}} + \underbrace{42.00 \text{ g H}_2\text{O}}_{\text{Solvent}}$$

SAMPLE PROBLEM 12.5 Calculating Mass Percent (m/m) Concentration

TRY IT FIRST

What is the mass percent of NaOH in a solution prepared by dissolving 30.0 g of NaOH in 120.0 g of H_2O?

SOLUTION

STEP 1 State the given and needed quantities.

	Given	Need	Connect
ANALYZE THE PROBLEM	30.0 g of NaOH, 120.0 g of H_2O	mass percent (m/m)	$\frac{\text{mass of solute}}{\text{mass of solution}} \times 100\%$

STEP 2 Write the concentration expression.

$$\text{Mass percent (m/m)} = \frac{\text{grams of solute}}{\text{grams of solution}} \times 100\%$$

STEP 3 Substitute solute and solution quantities into the expression and calculate. The mass of the solution is obtained by adding the mass of the solute and the mass of the solution.

$$\text{mass of solution} = 30.0 \text{ g NaOH} + 120.0 \text{ g H}_2\text{O} = 150.0 \text{ g of NaOH solution}$$

$$\text{Mass percent (m/m)} = \frac{\overset{\text{Three SFs}}{30.0 \text{ g NaOH}}}{\underset{\text{Four SFs}}{150.0 \text{ g solution}}} \times 100\%$$

$$= \underset{\text{Three SFs}}{20.0\% \text{ (m/m) NaOH solution}}$$

SELF TEST 12.5

a. What is the mass percent (m/m) of NaCl in a solution made by dissolving 2.0 g of NaCl in 56.0 g of H_2O?

b. What is the mass percent (m/m) of $MgCl_2$ in a solution prepared by dissolving 1.8 g of $MgCl_2$ in 18.5 g of H_2O?

ANSWER

a. 3.4% (m/m) NaCl solution **b.** 8.9% (m/m) $MgCl_2$ solution

PRACTICE PROBLEMS

Try Practice Problems 12.29 and 12.30

Using Mass Percent Concentration as a Conversion Factor

In the preparation of solutions, we often need to calculate the amount of solute or solution. Then the concentration of a solution is useful as a conversion factor as shown in Sample Problem 12.6.

SAMPLE PROBLEM 12.6   Using Mass Percent to Calculate Mass of Solute

TRY IT FIRST

The topical antibiotic ointment Neosporin is 3.5% (m/m) neomycin solution. How many grams of neomycin are in a tube containing 64 g of ointment?

SOLUTION

STEP 1   State the given and needed quantities.

	Given	Need	Connect
ANALYZE THE PROBLEM	64 g of 3.5% (m/m) neomycin solution	grams of neomycin	mass percent factor $\dfrac{\text{g of solute}}{100.\ \text{g of solution}}$

STEP 2   Write a plan to calculate the mass.

grams of ointment % (m/m) factor grams of neomycin

STEP 3   Write equalities and conversion factors. The mass percent (m/m) indicates the grams of a solute in every 100. g of a solution. The mass percent (3.5% m/m) can be written as two conversion factors.

3.5 g of neomycin = 100. g of ointment

$$\frac{3.5\ \text{g neomycin}}{100.\ \text{g ointment}} \quad \text{and} \quad \frac{100.\ \text{g ointment}}{3.5\ \text{g neomycin}}$$

STEP 4   Set up the problem to calculate the mass.

Two SFs

$$64\ \text{g ointment} \times \frac{3.5\ \text{g neomycin}}{100.\ \text{g ointment}} = 2.2\ \text{g of neomycin}$$

Two SFs   Exact   Two SFs

SELF TEST 12.6

a. Calculate the grams of KCl in 225 g of an 8.00% (m/m) KCl solution.
b. Calculate the grams of KNO_3 in 45.0 g of a 1.85% (m/m) KNO_3 solution.

ANSWER

a. 18.0 g of KCl   **b.** 0.833 g of KNO_3

Volume Percent (v/v) Concentration

Because the volumes of liquids or gases are easily measured, the concentrations of their solutions are often expressed as **volume percent (v/v)**. The units of volume used in the ratio must be the same, for example, both in milliliters or both in liters.

$$\text{Volume percent (v/v)} = \frac{\text{volume of solute}}{\text{volume of solution}} \times 100\%$$

We interpret a volume percent as the volume of solute in 100. mL of solution. On a bottle of extract of vanilla, a label that reads alcohol 35% (v/v) means 35 mL of alcohol solute in 100. mL of vanilla solution.

The label indicates that vanilla extract contains 35% (v/v) alcohol.

Lemon extract is a solution of lemon flavor and alcohol.

SAMPLE PROBLEM 12.7 Calculating Volume Percent (v/v) Concentration

TRY IT FIRST

A bottle contains 59 mL of lemon extract solution. If the extract contains 49 mL of alcohol, what is the volume percent (v/v) of the alcohol in the solution?

SOLUTION

STEP 1 State the given and needed quantities.

ANALYZE THE PROBLEM	Given	Need	Connect
	49 mL of alcohol, 59 mL of solution	volume percent (v/v)	$\dfrac{\text{volume of solute}}{\text{volume of solution}} \times 100\%$

STEP 2 Write the concentration expression.

$$\text{Volume percent (v/v)} = \frac{\text{volume of solute}}{\text{volume of solution}} \times 100\%$$

STEP 3 Substitute solute and solution quantities into the expression and calculate.

$$\text{Volume percent (v/v)} = \frac{\overset{\text{Two SFs}}{49 \text{ mL alcohol}}}{\underset{\text{Two SFs}}{59 \text{ mL solution}}} \times 100\% = \underset{\text{Two SFs}}{83\%} \text{ (v/v) alcohol solution}$$

SELF TEST 12.7

a. What is the volume percent (v/v) of Br_2 in a solution prepared by dissolving 12 mL of liquid bromine (Br_2) in the solvent carbon tetrachloride (CCl_4) to make 250 mL of solution?
b. What is the volume percent (v/v) of a "rubbing alcohol" solution that contains 14 mL of isopropyl alcohol in 20. mL of solution?

ANSWER

a. 4.8% (v/v) Br_2 solution   **b.** 70.% (v/v) isopropyl alcohol solution

---

PRACTICE PROBLEMS

Try Practice Problems 12.35 and 12.36

Mass/Volume Percent (m/v) Concentration

Mass/volume percent (m/v) describes the mass of the solute in grams for 100. mL of solution. In the calculation of mass/volume percent, the unit of mass of the solute is grams, and the unit of the solution volume is milliliters.

$$\text{Mass/volume percent (m/v)} = \frac{\text{grams of solute}}{\text{milliliters of solution}} \times 100\%$$

The mass/volume percent is widely used in hospitals and pharmacies for the preparation of intravenous solutions and medicines. For example, a 5% (m/v) glucose solution contains 5 g of glucose in 100. mL of solution. The volume of solution represents the combined volumes of the glucose and H_2O.

ENGAGE 12.9

What units should be used for a 3.0% (m/v) solution?

SAMPLE PROBLEM 12.8 Calculating Mass/Volume Percent (m/v) Concentration

TRY IT FIRST

A potassium iodide solution may be used in a diet that is low in iodine. A KI solution is prepared by dissolving 5.0 g of KI in enough water to give a final volume of 250 mL. What is the mass/volume percent (m/v) of KI in the solution?

SOLUTION

STEP 1 State the given and needed quantities.

ANALYZE THE PROBLEM	Given	Need	Connect
	5.0 g of KI solute, 250 mL of KI solution	mass/volume percent (m/v)	$\dfrac{\text{mass of solute}}{\text{volume of solution}} \times 100\%$

STEP 2 **Write the concentration expression.**

$$\text{Mass/volume percent (m/v)} = \frac{\text{mass of solute}}{\text{volume of solution}} \times 100\%$$

STEP 3 **Substitute solute and solution quantities into the expression and calculate.**

$$\text{Mass/volume percent (m/v)} = \frac{\overset{\text{Two SFs}}{5.0 \text{ g KI}}}{\underset{\text{Two SFs}}{250 \text{ mL solution}}} \times 100\% = \underset{\text{Two SFs}}{2.0\% \text{ (m/v) KI solution}}$$

Water added to make a solution   250 mL

SELF TEST 12.8

a. What is the mass/volume percent (m/v) of NaOH in a solution prepared by dissolving 12 g of NaOH in enough water to make 220 mL of solution?

b. What is the mass/volume percent (m/v) of $MgCl_2$ in a solution prepared by dissolving 3.25 g of $MgCl_2$ in enough water to make 125 mL of solution?

5.0 g of KI   250 mL of
KI solution

ANSWER

a. 5.5% (m/v) NaOH solution

b. 2.60% (m/v) $MgCl_2$ solution

SAMPLE PROBLEM 12.9 Using Mass/Volume Percent to Calculate Mass of Solute

TRY IT FIRST

A topical antibiotic is 1.0% (m/v) clindamycin. How many grams of clindamycin are in 60. mL of the 1.0% (m/v) clindamycin solution?

SOLUTION

STEP 1 **State the given and needed quantities.**

	Given	Need	Connect
ANALYZE THE PROBLEM	60. mL of 1.0% (m/v) clindamycin solution	grams of clindamycin	% (m/v) factor

STEP 2 **Write a plan to calculate the mass.**

milliliters of solution   % (m/v) factor   grams of clindamycin

STEP 3 **Write equalities and conversion factors.** The percent (m/v) indicates the grams of a solute in every 100. mL of a solution. The 1.0% (m/v) can be written as two conversion factors.

$$1.0 \text{ g of clindamycin} = 100. \text{ mL of solution}$$

$$\frac{1.0 \text{ g clindamycin}}{100. \text{ mL solution}} \quad \text{and} \quad \frac{100. \text{ mL solution}}{1.0 \text{ g clindamycin}}$$

STEP 4 **Set up the problem to calculate the mass.** The volume of the solution is converted to mass of solute using the conversion factor that cancels mL.

$$60. \text{ mL solution} \times \frac{\overset{\text{Two SFs}}{1.0 \text{ g clindamycin}}}{100. \text{ mL solution}} = 0.60 \text{ g of clindamycin}$$

Two SFs   Exact   Two SFs

SELF TEST 12.9

a. In 2010, the FDA approved a 2.0% (m/v) morphine oral solution to treat severe or chronic pain. How many grams of morphine does a patient receive if 0.60 mL of 2.0% (m/v) morphine solution was ordered?

PRACTICE PROBLEMS

Try Practice Problems 12.27 and 12.28, 12.31 to 12.34, 12.37 and 12.38, 12.47 to 12.50

b. For a skin infection, the doctor orders 300 mg of ampicillin. How many milliliters of 5.0% (m/v) ampicillin should be given?

ANSWER

a. 0.012 g of morphine   **b.** 6 mL

Volumetric flask

1.0 mol of NaCl

Add water until 1-liter mark is reached.

Mix

A 1.0 molar (M) NaCl solution

Molarity (M) Concentration

When chemists work with solutions, they often use **molarity (M)**, a concentration that states the number of moles of solute in exactly 1 L of solution.

$$\text{Molarity (M)} = \frac{\text{moles of solute}}{\text{liters of solution}}$$

The molarity of a solution can be calculated when we know the moles of solute and the volume of solution in liters. For example, if 1.0 mol of NaCl were dissolved in enough water to prepare 1.0 L of solution, the resulting NaCl solution has a molarity of 1.0 M. The abbreviation M indicates the units of mole per liter (mol/L).

$$M = \frac{\text{moles of solute}}{\text{liters of solution}} = \frac{1.0 \text{ mol NaCl}}{1 \text{ L solution}} = 1.0 \text{ M NaCl solution}$$

SAMPLE PROBLEM 12.10   Calculating Molarity

TRY IT FIRST

What is the molarity (M) of 60.0 g of NaOH in 0.250 L of NaOH solution?

SOLUTION

STEP 1   State the given and needed quantities.

	Given	Need	Connect
ANALYZE THE PROBLEM	60.0 g of NaOH, 0.250 L of NaOH solution	molarity (mol/L)	molar mass of NaOH, $\dfrac{\text{moles of solute}}{\text{liters of solution}}$

To calculate the moles of NaOH, we need to write the equality and conversion factors for the molar mass of NaOH. Then the moles in 60.0 g of NaOH can be determined.

$$1 \text{ mol of NaOH} = 40.00 \text{ g of NaOH}$$

$$\frac{40.00 \text{ g NaOH}}{1 \text{ mol NaOH}} \quad \text{and} \quad \frac{1 \text{ mol NaOH}}{40.00 \text{ g NaOH}}$$

$$\text{moles of NaOH} = 60.0 \text{ g NaOH} \times \frac{1 \text{ mol NaOH}}{40.00 \text{ g NaOH}}$$

$$= 1.50 \text{ mol of NaOH}$$

$$\text{volume of solution} = 0.250 \text{ L of NaOH solution}$$

STEP 2   Write the concentration expression.

$$\text{Molarity (M)} = \frac{\text{moles of solute}}{\text{liters of solution}}$$

STEP 3   Substitute solute and solution quantities into the expression and calculate.

$$M = \frac{1.50 \text{ mol NaOH}}{0.250 \text{ L solution}} = \frac{6.00 \text{ mol NaOH}}{1 \text{ L solution}} = 6.00 \text{ M NaOH solution}$$

ENGAGE 12.10

Why does a solution containing 2.0 mol of KCl in 4.0 L of KCl solution have a molarity of 0.50 M?

a. What is the molarity of a solution that contains 75.0 g of KNO_3 dissolved in 0.350 L of solution?

b. What is the molarity of a mouthwash solution that contains 6.40 g of the disinfectant thymol, $C_{10}H_{14}O$, in 2.00 L of solution?

ANSWER

a. 2.12 M KNO_3 solution

b. 0.0213 M thymol solution

SAMPLE PROBLEM 12.11   Using Molarity to Calculate Volume of Solution

TRY IT FIRST

How many liters of a 2.00 M NaCl solution are needed to provide 67.3 g of NaCl?

SOLUTION

STEP 1   State the given and needed quantities.

ANALYZE THE PROBLEM	Given	Need	Connect
	67.3 g of NaCl, 2.00 M NaCl solution	liters of NaCl solution	molar mass of NaCl, molarity

STEP 2   Write a plan to calculate the volume.

grams of NaCl   Molar mass   moles of NaCl   Molarity   liters of NaCl solution

STEP 3   Write equalities and conversion factors.

1 mol of NaCl = 58.44 g of NaCl

$$\frac{58.44 \text{ g NaCl}}{1 \text{ mol NaCl}} \quad \text{and} \quad \frac{1 \text{ mol NaCl}}{58.44 \text{ g NaCl}}$$

1 L of NaCl solution = 2.00 mol of NaCl

$$\frac{2.00 \text{ mol NaCl}}{1 \text{ L NaCl solution}} \quad \text{and} \quad \frac{1 \text{ L NaCl solution}}{2.00 \text{ mol NaCl}}$$

STEP 4   Set up the problem to calculate the volume.

$$67.3 \text{ g NaCl} \times \frac{1 \text{ mol NaCl}}{58.44 \text{ g NaCl}} \times \frac{1 \text{ L NaCl solution}}{2.00 \text{ mol NaCl}} = 0.576 \text{ L of NaCl solution}$$

a. How many milliliters of a 6.0 M HCl solution will provide 164 g of HCl?

b. What volume, in milliliters, of a 2.50 M NaOH solution contains 12.5 g of NaOH?

ANSWER

a. 750 mL of HCl solution

b. 125 mL of NaOH solution

PRACTICE PROBLEMS

Try Practice Problems 12.39 to 12.46

A summary of percent concentrations and molarity, their meanings, and conversion factors are given in **TABLE 12.8**.

TABLE 12.8 Conversion Factors from Concentrations

Percent Concentration	Meaning	Conversion Factors
10% (m/m) KCl solution	10 g of KCl in 100. g of KCl solution	$\dfrac{10\text{ g KCl}}{100.\text{ g solution}}$ and $\dfrac{100.\text{ g solution}}{10\text{ g KCl}}$
12% (v/v) ethanol solution	12 mL of ethanol in 100. mL of ethanol solution	$\dfrac{12\text{ mL ethanol}}{100.\text{ mL solution}}$ and $\dfrac{100.\text{ mL solution}}{12\text{ mL ethanol}}$
5% (m/v) glucose solution	5 g of glucose in 100. mL of glucose solution	$\dfrac{5\text{ g glucose}}{100.\text{ mL solution}}$ and $\dfrac{100.\text{ mL solution}}{5\text{ g glucose}}$

Molarity	Meaning	Conversion Factors
6.0 M HCl solution	6.0 mol of HCl in 1 L of HCl solution	$\dfrac{6.0\text{ mol HCl}}{1\text{ L solution}}$ and $\dfrac{1\text{ L solution}}{6.0\text{ mol HCl}}$

PRACTICE PROBLEMS

12.4 Solution Concentrations

12.27 What is the difference between a 5.00% (m/m) glucose solution and a 5.00% (m/v) glucose solution?

12.28 What is the difference between a 10.0% (v/v) methanol solution and a 10.0% (m/m) methanol solution?

12.29 Calculate the mass percent (m/m) for the solute in each of the following:
a. 25 g of KCl and 125 g of H_2O
b. 12 g of sucrose in 225 g of tea solution
c. 8.0 g of $CaCl_2$ in 80.0 g of $CaCl_2$ solution

12.30 Calculate the mass percent (m/m) for the solute in each of the following:
a. 75 g of NaOH in 325 g of NaOH solution
b. 2.0 g of KOH and 20.0 g of H_2O
c. 48.5 g of Na_2CO_3 in 250.0 g of Na_2CO_3 solution

12.31 Calculate the mass/volume percent (m/v) for the solute in each of the following:
a. 75 g of Na_2SO_4 in 250 mL of Na_2SO_4 solution
b. 39 g of sucrose in 355 mL of a carbonated drink

12.32 Calculate the mass/volume percent (m/v) for the solute in each of the following:
a. 2.50 g of LiCl in 40.0 mL of LiCl solution
b. 7.5 g of casein in 120 mL of low-fat milk

12.33 Calculate the grams or milliliters of solute needed to prepare the following:
a. 50. g of a 5.0% (m/m) KCl solution
b. 1250 mL of a 4.0% (m/v) NH_4Cl solution
c. 250. mL of a 10.0% (v/v) acetic acid solution

12.34 Calculate the grams or milliliters of solute needed to prepare the following:
a. 150. g of a 40.0% (m/m) $LiNO_3$ solution
b. 450 mL of a 2.0% (m/v) KOH solution
c. 225 mL of a 15% (v/v) isopropyl alcohol solution

12.35 A mouthwash contains 22.5% (v/v) alcohol. If the bottle of mouthwash contains 355 mL, what is the volume, in milliliters, of alcohol?

12.36 A bottle of champagne is 11% (v/v) alcohol. If there are 750 mL of champagne in the bottle, what is the volume, in milliliters, of alcohol?

12.37 For each of the following solutions, calculate the:
a. grams of a 25% (m/m) $LiNO_3$ solution that contains 5.0 g of $LiNO_3$
b. milliliters of a 10.0% (m/v) KOH solution that contains 40.0 g of KOH
c. milliliters of a 10.0% (v/v) formic acid solution that contains 2.0 mL of formic acid

12.38 For each of the following solutions, calculate the:
a. grams of a 2.0% (m/m) NaCl solution that contains 7.50 g of NaCl
b. milliliters of a 25% (m/v) NaF solution that contains 4.0 g of NaF
c. milliliters of an 8.0% (v/v) ethanol solution that contains 20.0 mL of ethanol

12.39 Calculate the molarity of each of the following:
a. 2.00 mol of glucose in 4.00 L of a glucose solution
b. 4.00 g of KOH in 2.00 L of a KOH solution
c. 5.85 g of NaCl in 400. mL of a NaCl solution

12.40 Calculate the molarity of each of the following:
a. 0.500 mol of glucose in 0.200 L of a glucose solution
b. 73.0 g of HCl in 2.00 L of a HCl solution
c. 30.0 g of NaOH in 350. mL of a NaOH solution

12.41 Calculate the grams of solute needed to prepare each of the following:
a. 2.00 L of a 1.50 M NaOH solution
b. 4.00 L of a 0.200 M KCl solution
c. 25.0 mL of a 6.00 M HCl solution

12.42 Calculate the grams of solute needed to prepare each of the following:
a. 2.00 L of a 6.00 M NaOH solution
b. 5.00 L of a 0.100 M $CaCl_2$ solution
c. 175 mL of a 3.00 M $NaNO_3$ solution

12.43 For each of the following solutions, calculate the:
a. liters of a 2.00 M KBr solution to obtain 3.00 mol of KBr
b. liters of a 1.50 M NaCl solution to obtain 15.0 mol of NaCl
c. milliliters of a 0.800 M $Ca(NO_3)_2$ solution to obtain 0.0500 mol of $Ca(NO_3)_2$

12.44 For each of the following solutions, calculate the:
    a. liters of a 4.00 M KCl solution to obtain 0.100 mol of KCl
    b. liters of a 6.00 M HCl solution to obtain 5.00 mol of HCl
    c. milliliters of a 2.50 M K_2SO_4 solution to obtain 1.20 mol of K_2SO_4

12.45 Calculate the volume, in milliliters, for each of the following that provides the given amount of solute:
    a. 12.5 g of Na_2CO_3 from a 0.120 M Na_2CO_3 solution
    b. 0.850 mol of $NaNO_3$ from a 0.500 M $NaNO_3$ solution
    c. 30.0 g of LiOH from a 2.70 M LiOH solution

12.46 Calculate the volume, in liters, for each of the following that provides the given amount of solute:
    a. 5.00 mol of NaOH from a 12.0 M NaOH solution
    b. 15.0 g of Na_2SO_4 from a 4.00 M Na_2SO_4 solution
    c. 28.0 g of $NaHCO_3$ from a 1.50 M $NaHCO_3$ solution

Applications

12.47 A patient receives 100. mL of 20.% (m/v) mannitol solution every hour.
    a. How many grams of mannitol are given in 1 h?
    b. How many grams of mannitol does the patient receive in 12 h?

12.48 A patient receives 250 mL of a 4.0% (m/v) amino acid solution twice a day.
    a. How many grams of amino acids are in 250 mL of solution?
    b. How many grams of amino acids does the patient receive in 1 day?

12.49 A patient needs 100. g of glucose in the next 12 h. How many liters of a 5% (m/v) glucose solution must be given?

12.50 A patient received 2.0 g of NaCl in 8 h. How many milliliters of a 0.90% (m/v) NaCl (saline) solution were delivered?

12.5 Dilution of Solutions

LEARNING GOAL Describe the dilution of a solution; calculate the unknown concentration or volume when a solution is diluted.

REVIEW
Solving Equations (1.4)

In chemistry and biology, we often prepare diluted solutions from more concentrated solutions. In a process called **dilution**, a solvent, usually water, is added to a solution, which increases the volume. As a result, the concentration of the solution decreases. In an everyday example, you are making a dilution when you add three cans of water to a can of concentrated orange juice.

1 can of orange juice concentrate   +   3 cans of water   =   4 cans of orange juice

In the dilution of a solution, solvent is added to increase the volume and decrease the concentration.

Although the addition of solvent increases the volume, the amount of solute does not change; it is the same in the concentrated solution and the diluted solution (see **FIGURE 12.8**).

Grams or moles of solute = grams or moles of solute
    Concentrated solution   Diluted solution

FIGURE 12.8 When water is added to a concentrated solution, there is no change in the number of particles. However, the solute particles spread out as the volume of the diluted solution increases.

We can write this equality in terms of the concentration, C, and the volume, V. The concentration, C, may be percent concentration or molarity.

$$C_1V_1 = C_2V_2$$
Concentrated solution   Diluted solution

If we are given any three of the four variables (C_1, C_2, V_1, or V_2) we can rearrange the dilution expression to solve for the unknown quantity as seen in Sample Problems 12.12 and 12.13.

PRACTICE PROBLEMS

Try Practice Problems 12.51 and 12.52

SAMPLE PROBLEM 12.12   Calculations Involving Concentration of Solutions

TRY IT FIRST

What is the molarity of a solution when 75.0 mL of a 4.00 M KCl solution is diluted to a volume of 500. mL?

SOLUTION

STEP 1   **Prepare a table of the concentrations and volumes of the solutions.**

ANALYZE THE PROBLEM	Given	Need	Connect
	$C_1 = 4.00$ M $V_1 = 75.0$ mL $V_2 = 500.$ mL	C_2	$C_1V_1 = C_2V_2$ **Predict:** C_2 decreases

STEP 2   **Rearrange the dilution expression to solve for the unknown quantity.**

$$C_1V_1 = C_2V_2$$

$$\frac{C_1V_1}{V_2} = \frac{C_2V_2}{V_2} \quad \text{Divide both sides by } V_2$$

$$C_2 = C_1 \times \frac{V_1}{V_2}$$

ENGAGE 12.11

How do we know that the dilution of this KCl solution has a lower molarity?

STEP 3   **Substitute the known quantities into the dilution expression and calculate.**

$$C_2 = 4.00 \text{ M} \times \frac{\overset{\text{Three SFs}}{75.0 \text{ mL}}}{500. \text{ mL}} = 0.600 \text{ M (diluted KCl solution)}$$

Three SFs   Three SFs   Three SFs

Volume factor decreases concentration

When the initial molarity (C_1) is multiplied by a ratio of the volumes (volume factor) that is less than 1, the molarity of the diluted solution (C_2) decreases as predicted in Step 1.

SELF TEST 12.12

a. What is the molarity of a solution when 50.0 mL of a 4.00 M KOH solution is diluted to 200. mL?

b. What is the final concentration (m/v) when 18 mL of a 12% (m/v) NaCl solution is diluted to a volume of 85 mL?

PRACTICE PROBLEMS

Try Practice Problems 12.53 to 12.56

ANSWER

a. 1.00 M KOH solution   **b.** 2.5% (m/v) NaCl solution

SAMPLE PROBLEM 12.13   Calculations Involving Volume of Solutions

TRY IT FIRST

A doctor orders 1000. mL of a 35.0% (m/v) dextrose solution. If you have a 50.0% (m/v) dextrose solution, how many milliliters would you use to prepare 1000. mL of 35.0% (m/v) dextrose solution?

SOLUTION

STEP 1   Prepare a table of the concentrations and volumes of the solutions.

ANALYZE THE PROBLEM	Given	Need	Connect
	$C_1 = 50.0\%$ (m/v) $C_2 = 35.0\%$ (m/v) $V_2 = 1000.$ mL	V_1	$C_1V_1 = C_2V_2$ Predict: V_1 decreases

STEP 2   Rearrange the dilution expression to solve for the unknown quantity.

$$C_1V_1 = C_2V_2$$

$$\frac{C_1V_1}{C_1} = \frac{C_2V_2}{C_1} \quad \text{Divide both sides by } C_1$$

$$V_1 = V_2 \times \frac{C_2}{C_1}$$

STEP 3   Substitute the known quantities into the dilution expression and calculate.

$$V_1 = 1000.\text{ mL} \times \overset{\text{Three SFs}}{\frac{35.0\%}{50.0\%}} = 700.\text{ mL of dextrose solution}$$

Four SFs   Three SFs   Three SFs
Concentration factor
decreases volume

When the final volume (V_2) is multiplied by a ratio of the percent concentrations (concentration factor) that is less than 1, the initial volume (V_1) is less than the final volume (V_2) as predicted in Step 1.

SELF TEST 12.13

a. What initial volume, in milliliters, of a 15% (m/v) mannose solution is needed to prepare 125 mL of a 3.0% (m/v) mannose solution?
b. What initial volume, in milliliters, of a 2.40 M NaCl solution is needed to prepare 80.0 mL of a 0.600 M NaCl solution?

ANSWER

a. 25 mL of a 15% (m/v) mannose solution
b. 20.0 mL of a 2.40 M NaCl solution

PRACTICE PROBLEMS
Try Practice Problems 12.57 to 12.60

PRACTICE PROBLEMS

12.5 Dilution of Solutions

12.51 To make tomato soup, you add one can of water to the condensed soup. Why is this a dilution?

12.52 A can of frozen lemonade calls for the addition of three cans of water to make a pitcher of the beverage. Why is this a dilution?

12.53 Calculate the final concentration of each of the following:
a. 2.0 L of a 6.0 M HCl solution is added to water so that the final volume is 6.0 L.
b. Water is added to 0.50 L of a 12 M NaOH solution to make 3.0 L of a diluted NaOH solution.
c. A 10.0-mL sample of a 25% (m/v) KOH solution is diluted with water so that the final volume is 100.0 mL.
d. A 50.0-mL sample of a 15% (m/v) H_2SO_4 solution is added to water to give a final volume of 250 mL.

12.54 Calculate the final concentration of each of the following:
a. 1.0 L of a 4.0 M HNO_3 solution is added to water so that the final volume is 8.0 L.
b. Water is added to 0.25 L of a 6.0 M NaF solution to make 2.0 L of a diluted NaF solution.
c. A 50.0-mL sample of an 8.0% (m/v) KBr solution is diluted with water so that the final volume is 200.0 mL.
d. A 5.0-mL sample of a 50.0% (m/v) acetic acid ($HC_2H_3O_2$) solution is added to water to give a final volume of 25 mL.

12.55 Determine the final volume, in milliliters, of each of the following:

a. a 1.5 M HCl solution prepared from 20.0 mL of a 6.0 M HCl solution

b. a 2.0% (m/v) LiCl solution prepared from 50.0 mL of a 10.0% (m/v) LiCl solution

c. a 0.500 M H_3PO_4 solution prepared from 50.0 mL of a 6.00 M H_3PO_4 solution

d. a 5.0% (m/v) glucose solution prepared from 75 mL of a 12% (m/v) glucose solution

12.56 Determine the final volume, in milliliters, of each of the following:

a. a 1.00% (m/v) H_2SO_4 solution prepared from 10.0 mL of a 20.0% H_2SO_4 solution

b. a 0.10 M HCl solution prepared from 25 mL of a 6.0 M HCl solution

c. a 1.0 M NaOH solution prepared from 50.0 mL of a 12 M NaOH solution

d. a 1.0% (m/v) $CaCl_2$ solution prepared from 18 mL of a 4.0% (m/v) $CaCl_2$ solution

12.57 Determine the initial volume, in milliliters, required to prepare each of the following:

a. 255 mL of a 0.200 M HNO_3 solution using a 4.00 M HNO_3 solution

b. 715 mL of a 0.100 M $MgCl_2$ solution using a 6.00 M $MgCl_2$ solution

c. 0.100 L of a 0.150 M KCl solution using an 8.00 M KCl solution

12.58 Determine the initial volume, in milliliters, required to prepare each of the following:

a. 20.0 mL of a 0.250 M KNO_3 solution using a 6.00 M KNO_3 solution

b. 25.0 mL of a 2.50 M H_2SO_4 solution using a 12.0 M H_2SO_4 solution

c. 0.500 L of a 1.50 M NH_4Cl solution using a 10.0 M NH_4Cl solution

Applications

12.59 You need 500. mL of a 5.0% (m/v) glucose solution. If you have a 25% (m/v) glucose solution on hand, how many milliliters do you need?

12.60 A doctor orders 100. mL of 2.0% (m/v) ibuprofen. If you have 8.0% (m/v) ibuprofen on hand, how many milliliters do you need?

12.6 Chemical Reactions in Solution

LEARNING GOAL Given the volume and concentration of a solution in a chemical reaction, calculate the amount of a reactant or product in the reaction.

INTERACTIVE VIDEO

Solutions

When chemical reactions involve aqueous solutions, we use the balanced chemical equation, the molarity, and the volume to determine the moles or grams of a reactant or product. For example, we can determine the volume of a solution from the molarity and the grams of reactant as seen in Sample Problem 12.14.

Zinc reacts when placed in a solution of HCl.

SAMPLE PROBLEM 12.14 Calculations Involving Solutions in Reactions

TRY IT FIRST

HCl reacts with zinc to produce hydrogen gas, H_2, and $ZnCl_2$.

$$2HCl(aq) + Zn(s) \longrightarrow H_2(g) + ZnCl_2(aq)$$

How many liters of a 1.50 M HCl solution completely react with 5.32 g of zinc?

SOLUTION

STEP 1 State the given and needed quantities.

	Given	Need	Connect
ANALYZE THE PROBLEM	5.32 g of Zn, 1.50 M HCl solution	liters of HCl solution	molar mass of Zn, molarity, mole–mole factor
	Equation		
	$2HCl(aq) + Zn(s) \longrightarrow H_2(g) + ZnCl_2(aq)$		

STEP 2 Write a plan to calculate the needed quantity.

grams of Zn | Molar mass | moles of Zn | Mole–mole factor | moles of HCl | Molarity | liters of HCl solution

STEP 3 Write equalities and conversion factors including mole–mole and concentration factors.

$$1 \text{ mol of Zn} = 65.41 \text{ g of Zn}$$
$$\frac{65.41 \text{ g Zn}}{1 \text{ mol Zn}} \quad \text{and} \quad \frac{1 \text{ mol Zn}}{65.41 \text{ g Zn}}$$

$$1 \text{ mol of Zn} = 2 \text{ mol of HCl}$$
$$\frac{2 \text{ mol HCl}}{1 \text{ mol Zn}} \quad \text{and} \quad \frac{1 \text{ mol Zn}}{2 \text{ mol HCl}}$$

$$1 \text{ L of solution} = 1.50 \text{ mol of HCl}$$
$$\frac{1.50 \text{ mol HCl}}{1 \text{ L solution}} \quad \text{and} \quad \frac{1 \text{ L solution}}{1.50 \text{ mol HCl}}$$

STEP 4 Set up the problem to calculate the needed quantity.

$$5.32 \text{ g Zn} \times \frac{1 \text{ mol Zn}}{65.41 \text{ g Zn}} \times \frac{2 \text{ mol HCl}}{1 \text{ mol Zn}} \times \frac{1 \text{ L solution}}{1.50 \text{ mol HCl}} = 0.108 \text{ L of HCl solution}$$

> **CORE CHEMISTRY SKILL**
> Calculating the Quantity of a Reactant or Product for a Chemical Reaction in Solution

SELF TEST 12.14

a. Using the reaction in Sample Problem 12.14, how many grams of zinc can react with 225 mL of a 0.200 M HCl solution?

b. Using the reaction in Sample Problem 12.14, how many milliliters of a 0.235 M HCl solution will react with 0.426 g of zinc ?

ANSWER

a. 1.47 g of Zn   **b.** 28.6 mL of HCl solution

SAMPLE PROBLEM 12.15   Calculations Involving Solutions in Reactions

TRY IT FIRST

How many milliliters of a 0.250 M $BaCl_2$ solution are needed to react with 0.0325 L of a 0.160 M Na_2SO_4 solution?

$$Na_2SO_4(aq) + BaCl_2(aq) \longrightarrow BaSO_4(s) + 2NaCl(aq)$$

SOLUTION

STEP 1 State the given and needed quantities.

When a $BaCl_2$ solution is added to a Na_2SO_4 solution, $BaSO_4$, a white solid, forms.

ANALYZE THE PROBLEM	Given	Need	Connect
	0.0325 L of 0.160 M Na_2SO_4 solution, 0.250 M $BaCl_2$ solution	milliliters of $BaCl_2$ solution	mole–mole factor, metric factor
	Equation		
	$Na_2SO_4(aq) + BaCl_2(aq) \longrightarrow BaSO_4(s) + 2NaCl(aq)$		

STEP 2 Write a plan to calculate the needed quantity.

liters of Na_2SO_4 solution → Molarity → moles of Na_2SO_4 → Mole–mole factor → moles of $BaCl_2$

Molarity → liters of $BaCl_2$ solution → Metric factor → milliliters of $BaCl_2$ solution

> **INTERACTIVE VIDEO**
> Calculations Involving Solutions in Reaction

STEP 3 Write equalities and conversion factors including mole–mole and concentration factors.

$$1 \text{ L of solution} = 0.160 \text{ mol of } Na_2SO_4$$

$$\frac{0.160 \text{ mol } Na_2SO_4}{1 \text{ L solution}} \quad \text{and} \quad \frac{1 \text{ L solution}}{0.160 \text{ mol } Na_2SO_4}$$

$$1 \text{ mol of } Na_2SO_4 = 1 \text{ mol of } BaCl_2$$

$$\frac{1 \text{ mol } BaCl_2}{1 \text{ mol } Na_2SO_4} \quad \text{and} \quad \frac{1 \text{ mol } Na_2SO_4}{1 \text{ mol } BaCl_2}$$

$$1 \text{ L of solution} = 0.250 \text{ mol of } BaCl_2$$

$$\frac{0.250 \text{ mol } BaCl_2}{1 \text{ L solution}} \quad \text{and} \quad \frac{1 \text{ L solution}}{0.250 \text{ mol } BaCl_2}$$

$$1 \text{ L} = 1000 \text{ mL}$$

$$\frac{1000 \text{ mL}}{1 \text{ L}} \quad \text{and} \quad \frac{1 \text{ L}}{1000 \text{ mL}}$$

STEP 4 Set up the problem to calculate the needed quantity.

$$0.0325 \text{ L solution} \times \frac{0.160 \text{ mol } Na_2SO_4}{1 \text{ L solution}} \times \frac{1 \text{ mol } BaCl_2}{1 \text{ mol } Na_2SO_4} \times \frac{1 \text{ L solution}}{0.250 \text{ mol } BaCl_2} \times \frac{1000 \text{ mL } BaCl_2 \text{ solution}}{1 \text{ L solution}}$$

$$= 20.8 \text{ mL of } BaCl_2 \text{ solution}$$

SELF TEST 12.15

a. For the reaction in Sample Problem 12.15, how many milliliters of a 0.330 M Na_2SO_4 solution are needed to react with 26.8 mL of a 0.216 M $BaCl_2$ solution?

b. Using the reaction in Sample Problem 12.15, what is the molarity of a $BaCl_2$ solution if 19.4 mL of the $BaCl_2$ solution is needed to react with 15.0 mL of a 0.225 M Na_2SO_4 solution?

ANSWER

a. 17.5 mL of Na_2SO_4 solution   b. 0.174 M $BaCl_2$ solution

---

SAMPLE PROBLEM 12.16   Volume of a Gas from a Reaction in Solution

TRY IT FIRST

Acid rain results from the reaction of nitrogen dioxide with water in the air.

$$3NO_2(g) + H_2O(l) \longrightarrow 2HNO_3(aq) + NO(g)$$

At STP, how many liters of NO_2 gas are required to produce 0.275 L of a 0.400 M HNO_3 solution?

SOLUTION

STEP 1 State the given and needed quantities.

	Given	Need	Connect
ANALYZE THE PROBLEM	0.275 L of 0.400 M HNO_3 solution	liters of $NO_2(g)$ at STP	mole–mole factor, molar volume
	Equation		
	$3NO_2(g) + H_2O(l) \longrightarrow 2HNO_3(aq) + NO(g)$		

STEP 2 Write a plan to calculate the needed quantity. We start the problem with the volume and molarity of the HNO_3 solution to calculate moles. Then we can use the mole–mole factor and the molar volume to calculate the liters of NO_2 gas.

liters of solution ▸ Molarity ▸ moles of HNO_3 ▸ Mole–mole factor ▸ moles of NO_2 ▸ Molar volume ▸ liters of NO_2 (at STP)

STEP 3 Write equalities and conversion factors including mole–mole and concentration factors.

$$1 \text{ L of solution} = 0.400 \text{ mol of HNO}_3$$
$$\frac{0.400 \text{ mol HNO}_3}{1 \text{ L solution}} \quad \text{and} \quad \frac{1 \text{ L solution}}{0.400 \text{ mol HNO}_3}$$

$$3 \text{ mol of NO}_2 = 2 \text{ mol of HNO}_3$$
$$\frac{3 \text{ mol NO}_2}{2 \text{ mol HNO}_3} \quad \text{and} \quad \frac{2 \text{ mol HNO}_3}{3 \text{ mol NO}_2}$$

$$1 \text{ mol of NO}_2 = 22.4 \text{ L of NO}_2 \text{ (STP)}$$
$$\frac{22.4 \text{ L NO}_2 \text{ (STP)}}{1 \text{ mol NO}_2} \quad \text{and} \quad \frac{1 \text{ mol NO}_2}{22.4 \text{ L NO}_2 \text{ (STP)}}$$

STEP 4 Set up the problem to calculate the needed quantity.

$$0.275 \text{ L solution} \times \frac{0.400 \text{ mol HNO}_3}{1 \text{ L solution}} \times \frac{3 \text{ mol NO}_2}{2 \text{ mol HNO}_3} \times \frac{22.4 \text{ L NO}_2 \text{ (STP)}}{1 \text{ mol NO}_2}$$

$$= 3.70 \text{ L of NO}_2 \text{ (STP)}$$

SELF TEST 12.16

a. Using the equation in Sample Problem 12.16, determine the volume of NO produced at 100 °C and 1.20 atm, when 2.20 L of a 1.50 M HNO$_3$ solution is produced.
b. Using the reaction in Sample Problem 12.16, how many milliliters of 0.344 M HNO$_3$ solution are produced when 1.40 L of NO$_2$ gas at 0.800 atm and 20. °C reacts?

ANSWER
a. 42.1 L of NO
b. 90.3 mL of HNO$_3$ solution

ENGAGE 12.12

What sequence of conversion factors would you use to calculate the number of grams of CaCO$_3$ needed to react with 1.50 L of a 2.00 M HCl solution in the reaction $2HCl(aq) + CaCO_3(s) \longrightarrow CaCl_2(aq) + CO_2(g) + H_2O(l)$?

PRACTICE PROBLEMS

Try Practice Problems 12.61 to 12.66

FIGURE 12.9 gives a summary of the pathways and conversion factors needed for substances including solutions involved in chemical reactions.

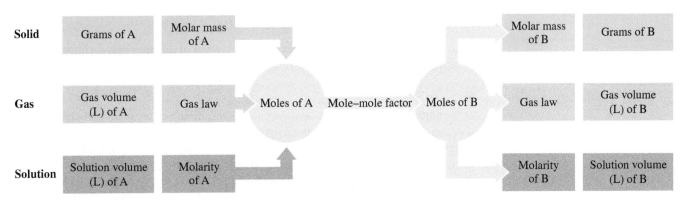

FIGURE 12.9 In calculations involving chemical reactions, substance A is converted to moles of A using molar mass (if solid), gas laws (if gas), or molarity (if solution). Then moles of A are converted to moles of substance B, which are converted to grams of solid, liters of gas, or liters of solution, as needed.

12.6 Chemical Reactions in Solution

12.61 Answer the following for the reaction:

$$Pb(NO_3)_2(aq) + 2KCl(aq) \longrightarrow PbCl_2(s) + 2KNO_3(aq)$$

a. How many grams of $PbCl_2$ will be formed from 50.0 mL of a 1.50 M KCl solution?
b. How many milliliters of a 2.00 M $Pb(NO_3)_2$ solution will react with 50.0 mL of a 1.50 M KCl solution?
c. What is the molarity of 20.0 mL of a KCl solution that reacts completely with 30.0 mL of a 0.400 M $Pb(NO_3)_2$ solution?

12.62 Answer the following for the reaction:

$$NiCl_2(aq) + 2NaOH(aq) \longrightarrow Ni(OH)_2(s) + 2NaCl(aq)$$

a. How many milliliters of a 0.200 M NaOH solution are needed to react with 18.0 mL of a 0.500 M $NiCl_2$ solution?
b. How many grams of $Ni(OH)_2$ are produced from the reaction of 35.0 mL of a 1.75 M NaOH solution and excess $NiCl_2$?
c. What is the molarity of 30.0 mL of a $NiCl_2$ solution that reacts completely with 10.0 mL of a 0.250 M NaOH solution?

12.63 Answer the following for the reaction:

$$2HCl(aq) + Mg(s) \longrightarrow H_2(g) + MgCl_2(aq)$$

a. How many milliliters of a 6.00 M HCl solution are required to react with 15.0 g of magnesium?
b. How many liters of hydrogen gas can form at STP when 0.500 L of a 2.00 M HCl solution reacts with excess magnesium?
c. What is the molarity of a HCl solution if the reaction of 45.2 mL of the HCl solution with excess magnesium produces 5.20 L of H_2 gas at 735 mmHg and 25 °C?

12.64 Answer the following for the reaction:

$$2HCl(aq) + CaCO_3(s) \longrightarrow CO_2(g) + H_2O(l) + CaCl_2(aq)$$

a. How many milliliters of a 0.200 M HCl solution can react with 8.25 g of $CaCO_3$?
b. How many liters of CO_2 gas can form at STP when 15.5 mL of a 3.00 M HCl solution reacts with excess $CaCO_3$?
c. What is the molarity of a HCl solution if the reaction of 200. mL of the HCl solution with excess $CaCO_3$ produces 12.0 L of CO_2 gas at 725 mmHg and 18 °C?

12.65 Answer the following for the reaction:

$$2HBr(aq) + Zn(s) \longrightarrow H_2(g) + ZnBr_2(aq)$$

a. How many milliliters of a 3.50 M HBr solution are required to react with 8.56 g of zinc?
b. How many liters of hydrogen gas can form at STP when 0.750 L of a 6.00 M HBr solution reacts with excess zinc?
c. What is the molarity of a HBr solution if the reaction of 28.7 mL of the HBr solution with excess zinc produces 0.620 L of H_2 gas at 725 mmHg and 24 °C?

12.66 Answer the following for the reaction:

$$3AgNO_3(aq) + Na_3PO_4(aq) \longrightarrow Ag_3PO_4(s) + 3NaNO_3(aq)$$

a. How many milliliters of a 0.265 M $AgNO_3$ solution are required to react with 31.2 mL of 0.154 M Na_3PO_4 solution?
b. How many grams of silver phosphate are produced from the reaction of 25.0 mL of a 0.165 M $AgNO_3$ solution and excess Na_3PO_4?
c. What is the molarity of 20.0 mL of a Na_3PO_4 solution that reacts completely with 15.0 mL of a 0.576 M $AgNO_3$ solution?

12.7 Molality and Freezing Point Lowering/Boiling Point Elevation

LEARNING GOAL Identify a mixture as a solution, a colloid, or a suspension. Using the molality, calculate the freezing point and boiling point for a solution.

The size and number of solute particles in different types of mixtures play an important role in determining the properties of that mixture.

Solutions

In the solutions discussed up to now, the solute was dissolved as small particles that are uniformly dispersed throughout the solvent to give a homogeneous solution. When you observe a solution, such as salt water, you cannot visually distinguish the solute from the solvent. The solution appears transparent, although it may have a color. The particles are so small that they go through filters and through *semipermeable membranes*. A semipermeable membrane allows solvent molecules such as water and very small solute particles to pass through, but does not allow the passage of large solute molecules.

Colloids

The particles in a **colloid** are much larger than solute particles in a solution. Colloidal particles are large molecules, such as proteins, or groups of molecules or ions. Colloids, similar to solutions, are homogeneous mixtures that do not separate or settle out. Colloidal particles are small enough to pass through filters, but too large to pass through semipermeable membranes. **TABLE 12.9** lists several examples of colloids.

TABLE 12.9 Examples of Colloids

Colloid	Substance Dispersed	Dispersing Medium
Fog, clouds, hair sprays	Liquid	Gas
Dust, smoke	Solid	Gas
Shaving cream, whipped cream, soapsuds	Gas	Liquid
Styrofoam, marshmallows	Gas	Solid
Mayonnaise, homogenized milk	Liquid	Liquid
Cheese, butter	Liquid	Solid
Blood plasma, paints (latex), gelatin	Solid	Liquid

Suspensions

Suspensions are heterogeneous, nonuniform mixtures that are very different from solutions or colloids. The particles of a suspension are so large that they can often be seen with the naked eye. They are trapped by filters and semipermeable membranes.

The weight of the suspended solute particles causes them to settle out soon after mixing. If you stir muddy water, it mixes but then quickly separates as the suspended particles settle to the bottom and leave clear liquid at the top. You can find suspensions among the medications in a hospital or in your medicine cabinet. These include Kaopectate, calamine lotion, antacid mixtures, and liquid penicillin. It is important to follow the instructions on the label that state "shake well before using" so that the particles form a suspension.

Water-treatment plants make use of the properties of suspensions to purify water. When chemicals such as aluminum sulfate or iron(III) sulfate are added to untreated water, they react with impurities to form large suspended particles called *floc*. In the water-treatment plant, a system of filters traps the suspended particles, but clean water passes through.

TABLE 12.10 compares the different types of mixtures, and **FIGURE 12.10** illustrates some properties of solutions, colloids, and suspensions.

Suspensions separate into solids and water that can be removed by filtration.

TABLE 12.10 Comparison of Solutions, Colloids, and Suspensions

Type of Mixture	Type of Particle	Settling	Separation
Solution	Small particles such as atoms, ions, or small molecules	Particles do not settle	Particles cannot be separated by filters or semipermeable membranes
Colloid	Larger molecules or groups of molecules or ions	Particles do not settle	Particles can be separated by semipermeable membranes but not by filters
Suspension	Very large particles that may be visible	Particles settle rapidly	Particles can be separated by filters

PRACTICE PROBLEMS

Try Practice Problems 12.67 and 12.68

ENGAGE 12.13

A filter can be used to separate suspension particles from a solution, but a semipermeable membrane is needed to separate colloids from a solution. Explain.

● Solution
▲ Colloid
■ Suspension

FIGURE 12.10 Properties of different types of mixtures.

Vapor pressure

Lower vapor pressure

Pure solvent   Solute and solvent

The Alaska Upis beetle produces biological antifreeze to survive subfreezing temperatures.

Ethylene glycol is added to a radiator to form an aqueous solution that has a lower freezing point than water and a higher boiling point.

Freezing Point Lowering and Boiling Point Elevation

When we add a solute to water, it changes the vapor pressure, freezing point, and boiling point of the solvent pure water. These types of changes in physical properties, known as *colligative properties*, depend only on the concentration of solute particles in the solution.

We can illustrate how these changes in physical properties occur by comparing the number of evaporating solvent molecules in pure solvent with those in a solution with a nonvolatile solute in the solvent. In the solution, there are fewer solvent molecules at the surface because the solute that has been added takes up some of the space at the surface. As a result, fewer solvent molecules can evaporate compared to the pure solvent. Vapor pressure is lowered for the solution. If we add more solute molecules, the vapor pressure will be lowered even more.

The freezing point of a solvent is lowered when a nonvolatile solute is added. In this case, the solute particles prevent the organization of solvent molecules needed to form the solid state. Thus, a lower temperature is required before the molecules of the solvent can become organized enough to freeze. When you spread salt on an icy sidewalk when the temperature is below freezing, the particles from the salt combine with water to lower the freezing point, which causes the ice to melt.

Insects and fish in climates with subfreezing temperatures control ice formation by producing biological antifreezes made of glycerol, proteins, and sugars, such as glucose, within their bodies. Some insects can survive temperatures below $-60\ °C$. These forms of biological antifreezes may one day be applied to the long-term preservation of human organs.

The boiling point of a solvent is raised when a nonvolatile solute is added. The vapor pressure of a solvent must reach atmospheric pressure before it begins boiling. However, because the solute lowers the vapor pressure of the solvent, a temperature higher than the normal boiling point of the pure solvent is needed to cause the solution to boil. Antifreeze, which is a compound such as ethylene glycol, $C_2H_6O_2$, is added to the water in a car radiator. If an ethylene glycol and water mixture is about 50–50% by mass, it does not freeze until the temperature drops to about $-30\ °F$, and does not boil unless the temperature reaches about 225 °F. The solution in the radiator prevents the water in the radiator from forming ice in cold weather or boiling over on a hot desert highway.

Particles in Solution

A solute that is a nonelectrolyte dissolves as molecules, whereas a solute that is a strong electrolyte dissolves entirely as ions. The antifreeze ethylene glycol, $C_2H_6O_2$, a nonelectrolyte, dissolves as molecules.

Nonelectrolyte: 1 mol of $C_2H_6O_2(l)$ = 1 mol of $C_2H_6O_2(aq)$

However, when 1 mol of a strong electrolyte, such as NaCl or $CaCl_2$, dissolves in water, the NaCl solution will contain 2 mol of particles, and the $CaCl_2$ solution will contain 3 mol of particles.

Strong electrolytes:

$$\text{1 mol of NaCl}(s) = \underbrace{\text{1 mol of Na}^+(aq) + \text{1 mol of Cl}^-(aq)}_{\text{2 mol of particles }(aq)}$$

$$\text{1 mol of CaCl}_2(s) = \underbrace{\text{1 mol of Ca}^{2+}(aq) + \text{2 mol of Cl}^-(aq)}_{\text{3 mol of particles }(aq)}$$

Molality (m)

The calculation for freezing point lowering or boiling point elevation uses a concentration unit of *molality*. The **molality**, abbreviation **m**, of a solution is the number of moles of solute particles per kilogram of solvent. This may seem similar to molarity, but the denominator for molality refers to the mass of the solvent, not the volume of the solution.

$$\text{Molality }(m) = \frac{\text{moles of solute particles}}{\text{kilograms of solvent}}$$

SAMPLE PROBLEM 12.17 Calculating Molality

TRY IT FIRST

Calculate the molality of a solution containing 35.5 g of the nonelectrolyte glucose ($C_6H_{12}O_6$) in 0.400 kg of water.

SOLUTION

STEP 1 **State the given and needed quantities.**

	Given	Need	Connect
ANALYZE THE PROBLEM	35.5 g of glucose ($C_6H_{12}O_6$), 0.400 kg of water	molality (m)	molar mass of glucose

STEP 2 **Write the molality expression.**

$$\text{Molality } (m) \text{ of glucose solution} = \frac{\text{moles of glucose}}{\text{kilograms of water}}$$

STEP 3 **Substitute solute and solvent quantities into the expression and calculate.** To match the unit of moles in the molality expression, we convert the grams of glucose to moles of glucose, using its molar mass.

$$1 \text{ mol of glucose} = 180.16 \text{ g of glucose}$$

$$\frac{180.16 \text{ g glucose}}{1 \text{ mol glucose}} \quad \text{and} \quad \frac{1 \text{ mol glucose}}{180.16 \text{ g glucose}}$$

$$\text{moles of glucose} = 35.5 \text{ g glucose} \times \frac{1 \text{ mol glucose}}{180.16 \text{ g glucose}} = 0.197 \text{ mol of glucose}$$

$$\text{Molality } (m) = \frac{0.197 \text{ mol glucose}}{0.400 \text{ kg water}} = 0.493 \ m$$

SELF TEST 12.17

Calculate the molality of each of the following solutions:
a. 15.8 g of urea, CH_4N_2O, a nonelectrolyte, in 250. g of water
b. 44.6 g of ethanol, C_2H_6O, a nonelectrolyte, in 340. g of water

ANSWER

a. 1.05 *m*   **b.** 2.84 *m*

PRACTICE PROBLEMS

Try Practice Problems 12.69 and 12.70

Freezing Point Lowering

The change in the freezing point temperature of a solvent (ΔT_f) is determined from the molality (*m*) of the particles in the solution and the freezing point constant, K_f, for the solvent.

$$\Delta T_f = m \times K_f$$

The freezing point constant (K_f) for water, which is determined experimentally, is $\frac{1.86 \,^\circ C}{m}$.

For a 1 *m* solution, we can calculate the change in the freezing point temperature as

$$\Delta T_f = m \times K_f = 1 \ m \times \frac{1.86 \,^\circ C}{m} = 1.86 \,^\circ C$$

Then we can calculate the new, lower freezing point of the solution:

$$T_{solution} = T_{water} - \Delta T_f$$
$$= 0.00 \,^\circ C - 1.86 \,^\circ C$$
$$= -1.86 \,^\circ C$$

CORE CHEMISTRY SKILL

Calculating the Freezing Point/ Boiling Point of a Solution

ENGAGE 12.14

Why does a 1 *m* solution of KBr, a strong electrolyte, lower the freezing point of water more than a 1 *m* solution of urea, a nonelectrolyte?

A truck spreads calcium chloride on the road to melt ice and snow.

SAMPLE PROBLEM 12.18   Calculating the Freezing Point of a Solution

TRY IT FIRST

In the northeastern United States during freezing weather, $CaCl_2$ is spread on highways to melt the ice. Calculate the freezing point lowering and freezing point of a solution containing 225 g of $CaCl_2$ in 500. g of water.

SOLUTION

STEP 1   State the given and needed quantities.

	Given	Need	Connect
ANALYZE THE PROBLEM	225 g of $CaCl_2$, 500. g of water = 0.500 kg of water	ΔT_f, freezing point	molar mass, $\Delta T_f = m \times K_f$

$$CaCl_2(s) \longrightarrow Ca^{2+}(aq) + 2Cl^-(aq) = 3 \text{ mol of solute particles}$$

1 mol of $CaCl_2$ = 110.98 g of $CaCl_2$

$\dfrac{110.98 \text{ g } CaCl_2}{1 \text{ mol } CaCl_2}$ and $\dfrac{1 \text{ mol } CaCl_2}{110.98 \text{ g NaCl}}$

1 mol of $CaCl_2$ = 3 mol of solute particles

$\dfrac{3 \text{ mol solute particles}}{1 \text{ mol } CaCl_2}$ and $\dfrac{1 \text{ mol } CaCl_2}{3 \text{ mol solute particles}}$

STEP 2   Determine the number of moles of solute particles and calculate the molality.

We use molar mass to calculate the moles of $CaCl_2$. Then we multiply by three to obtain the number of moles of ions (particles) produced by 1 mol of $CaCl_2$ in solution.

$$\text{moles of particles} = 225 \text{ g } CaCl_2 \times \frac{1 \text{ mol } CaCl_2}{110.98 \text{ g } CaCl_2} \times \frac{3 \text{ mol particles}}{1 \text{ mol } CaCl_2}$$

$$= 6.08 \text{ mol of particles}$$

The molality (m) of the particles in solution is obtained by dividing the moles of particles by the number of kilograms of water in the solution.

$$m = \frac{\text{moles of particles}}{\text{kilograms of water}}$$

$$\text{Molality } (m) = \frac{6.08 \text{ mol particles}}{0.500 \text{ kg water}} = 12.2 \, m$$

STEP 3   Calculate the temperature change and subtract from the freezing point. The freezing point lowering is calculated using the molality and the freezing point constant. The freezing point lowering is subtracted from 0.00 °C to obtain the new freezing point of the $CaCl_2$ solution.

$$\Delta T_f = m \times K_f$$

$$\Delta T_f = 12.2 \, m \times \frac{1.86 \, °C}{m} = 22.7 \, °C$$

$$T_{solution} = T_{water} - \Delta T_f$$

$$= 0.00 \, °C - 22.7 \, °C = -22.7 \, °C$$

SELF TEST 12.18

Calculate the freezing point of each of the following solutions:
a. ethylene glycol, $C_2H_6O_2$, a nonelectrolyte, is added to the water in a radiator to give a solution containing 315 g of ethylene glycol in 565 g of water
b. sodium nitrate, $NaNO_3$, a strong electrolyte, is added to water to give a solution containing 315 g of sodium nitrate in 565 g of water

ANSWER

a. $-16.7 \, °C$ b. $-24.4 \, °C$

Boiling Point Elevation

A similar change occurs with the boiling point of water. The boiling point elevation (ΔT_b) is determined from the molality (m) of the particles in the solution and the boiling point constant, K_b, $\dfrac{0.51\ °C}{m}$ for water, which is determined experimentally.

$$\Delta T_b = m \times K_b$$

The effect of some solutions on freezing and boiling point is summarized in **TABLE 12.11**.

TABLE 12.11 Effect of Solute Concentration on Freezing and Boiling Points of 1 kg of Water

Substance/kg water	Type of Solute	Molality of Particles	Freezing Point	Boiling Point
Pure water	None	0	0.00 °C	100.00 °C
1 mol of $C_2H_6O_2$	Nonelectrolyte	1 m	−1.86 °C	100.51 °C
1 mol of NaCl	Strong electrolyte	2 m	−3.72 °C	101.02 °C
1 mol of $CaCl_2$	Strong electrolyte	3 m	−5.58 °C	101.53 °C

PRACTICE PROBLEMS

Try Practice Problems 12.71 to 12.74

As the concentration of a solution increases, the freezing point is lowered and the boiling point is raised.

SAMPLE PROBLEM 12.19   Calculating the Boiling Point of a Solution

TRY IT FIRST

Propylene glycol, $C_3H_8O_2$, is a nonelectrolyte that is added to the water in a radiator to increase the boiling point. If 4.6 mol of propylene glycol is added to 1.55 kg of water (solvent) in a radiator, what is the boiling point of the solution?

SOLUTION

STEP 1 State the given and needed quantities.

ANALYZE THE PROBLEM	Given	Need	Connect
	4.6 mol of propylene glycol, 1.55 kg of water	boiling point	$\Delta T_b = m \times K_b$

STEP 2 **Determine the number of moles of solute particles and calculate the molality.** Since propylene glycol is a nonelectrolyte, the moles of propylene glycol is equal to the moles of particles.

$$m = \frac{\text{moles of particles}}{\text{kilograms of water}} = \frac{4.6 \text{ mol}}{1.55 \text{ kg}} = 3.0 \; m$$

STEP 3 **Calculate the temperature change and add to the boiling point.**

$$\Delta T_b = m \times K_b = 3.0 \; \cancel{m} \times \frac{0.51 \,^\circ C}{\cancel{m}} = 1.5 \,^\circ C$$

$$T_{solution} = T_{water} + \Delta T_b$$
$$= 100.0 \,^\circ C + 1.5 \,^\circ C$$
$$= 101.5 \,^\circ C$$

SELF TEST 12.19

Calculate the boiling point of each of the following solutions:
a. 1.2-mol of potassium phosphate, K_3PO_4, a strong electrolyte, added to 0.26 kg of water
b. 36 g of ethylene glycol, $C_2H_6O_2$, a nonelectrolyte, added to 130 g of water

ANSWER
a. 109.6 °C   **b.** 102.3 °C

PRACTICE PROBLEMS

12.7 Molality and Freezing Point Lowering/Boiling Point Elevation

12.67 Identify each of the following as characteristic of a solution, colloid, or suspension:
 a. a mixture that cannot be separated by a semipermeable membrane
 b. a mixture that settles out upon standing

12.68 Identify each of the following as characteristic of a solution, colloid, or suspension:
 a. particles of this mixture remain inside a semipermeable membrane but pass through filters
 b. particles of solute in this mixture are very large and visible

12.69 Calculate the molality (m) of the following solutions:
 a. 325 g of methanol, CH_4O, a nonelectrolyte, added to 455 g of water
 b. 640. g of the antifreeze propylene glycol, $C_3H_8O_2$, a nonelectrolyte, dissolved in 1.22 kg of water

12.70 Calculate the molality (m) of the following solutions:
 a. 65.0 g of glucose, $C_6H_{12}O_6$, a nonelectrolyte, dissolved in 112 g of water
 b. 110. g of sucrose, $C_{12}H_{22}O_{11}$, a nonelectrolyte, dissolved in 1.50 kg of water

12.71 Calculate the freezing point and boiling point of each solution in problem 12.69.

12.72 Calculate the freezing point and boiling point of each solution in problem 12.70.

12.73 In each pair, identify the solution that will have a lower freezing point. Explain.
 a. 1.0 mol of glycerol (nonelectrolyte) and 2.0 mol of ethylene glycol (nonelectrolyte) each in 1.0 kg of water
 b. 0.50 mol of KCl (strong electrolyte) and 0.50 mol of $MgCl_2$ (strong electrolyte) each in 1.0 kg of water

12.74 In each pair, identify the solution that will have a higher boiling point. Explain.
 a. 1.50 mol of LiOH (strong electrolyte) and 3.00 mol of KOH (strong electrolyte) each in 1.0 kg of water
 b. 0.40 mol of $Al(NO_3)_3$ (strong electrolyte) and 0.40 mol of CsCl (strong electrolyte) each in 1.0 kg of water

12.8 Properties of Solutions: Osmosis

LEARNING GOAL Describe how the number of particles in solution affects osmosis.

The movement of water into and out of the cells of plants as well as the cells of our bodies is an important biological process that also depends on the solute concentration. In a process called **osmosis**, water molecules move through a semipermeable membrane from the solution with the lower concentration of solute into a solution with the higher solute concentration. In an osmosis apparatus, water is placed on one side of a semipermeable membrane and a sucrose (sugar) solution on the other side. The semipermeable membrane allows water molecules to flow back and forth but blocks the sucrose molecules because they cannot pass through the membrane. Because the sucrose solution has a higher solute concentration, more water molecules flow into the sucrose solution than out of the sucrose solution. The volume level of the sucrose solution rises as the volume level on the water side falls. The increase of water dilutes the sucrose solution to equalize (or attempt to equalize) the concentrations on both sides of the membrane.

Eventually the height of the sucrose solution creates sufficient pressure to equalize the flow of water between the two compartments. This pressure, called **osmotic pressure**, prevents the flow of additional water into the more concentrated solution. Then there is no further change in the volumes of the two solutions. The osmotic pressure depends on the

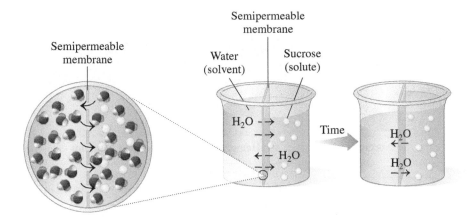

Semipermeable membrane

Semipermeable membrane

Water (solvent)   Sucrose (solute)

$H_2O \rightarrow$

$\leftarrow H_2O$

Time

$H_2O \leftarrow$

$H_2O \rightarrow$

Water flows into the solution with a higher solute concentration until the flow of water becomes equal in both directions.

concentration of solute particles in the solution. The greater the number of particles dissolved, the higher its osmotic pressure. In this example, the sucrose solution has a higher osmotic pressure than pure water, which has an osmotic pressure of zero.

In a process called *reverse osmosis*, a pressure greater than the osmotic pressure is applied to a solution so that it is forced through a purification membrane. The flow of water is reversed because water flows from an area of lower water concentration to an area of higher water concentration. The molecules and ions in solution stay behind, trapped by the membrane, while water passes through the membrane. This process of reverse osmosis is used in a few desalination plants to obtain pure water from sea (salt) water. However, the pressure that must be applied requires so much energy that reverse osmosis is not yet an economical method for obtaining pure water in most parts of the world.

PRACTICE PROBLEMS

Try Practice Problems 12.75 to 12.78

Isotonic Solutions

Because the cell membranes in biological systems are semipermeable, osmosis is an ongoing process. The solutes in body solutions such as blood, tissue fluids, lymph, and plasma all exert osmotic pressure. Most intravenous (IV) solutions used in a hospital are *isotonic solutions*, which exert the same osmotic pressure as body fluids such as blood. The percent concentration typically used in IV solutions is mass/volume percent (m/v), which is a type of percent concentration we have already discussed. The most typical isotonic solutions are 0.9% (m/v) NaCl solution, or 0.9 g of NaCl/100. mL of solution, and 5% (m/v) glucose, or 5 g of glucose/100. mL of solution. Although they do not contain the same kinds of particles, a 0.9% (m/v) NaCl

A 0.9% NaCl solution is isotonic with the solute concentration of the blood cells of the body.

solution as well as a 5% (m/v) glucose solution both have the same osmotic pressure. A red blood cell placed in an isotonic solution retains its volume because there is an equal flow of water into and out of the cell (see **FIGURE 12.11a**).

Hypotonic and Hypertonic Solutions

If a red blood cell is placed in a solution that is not isotonic, the differences in osmotic pressure inside and outside the cell can drastically alter the volume of the cell. When a red blood cell is placed in a *hypotonic solution*, which has a lower solute concentration (*hypo* means "lower than"), water flows into the cell by osmosis. The increase in fluid causes the cell to swell, and possibly burst—a process called *hemolysis* (see **FIGURE 12.11b**). A similar process occurs when you place dehydrated food, such as raisins or dried fruit, in water. The water enters the cells, and the food becomes plump and smooth.

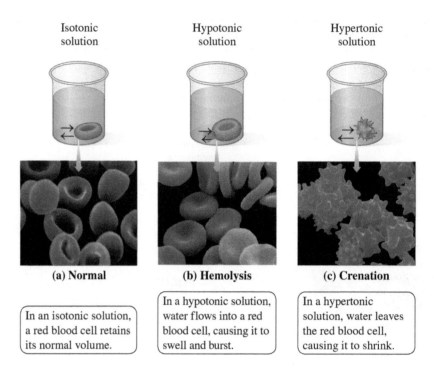

| Isotonic solution | Hypotonic solution | Hypertonic solution |

(a) Normal | **(b) Hemolysis** | **(c) Crenation**

| In an isotonic solution, a red blood cell retains its normal volume. | In a hypotonic solution, water flows into a red blood cell, causing it to swell and burst. | In a hypertonic solution, water leaves the red blood cell, causing it to shrink. |

FIGURE 12.11 Red blood cells in isotonic, hypotonic, and hypertonic solutions

If a red blood cell is placed in a *hypertonic solution*, which has a higher solute concentration (*hyper* means "greater than"), water flows out of the cell into the hypertonic solution by osmosis. Suppose a red blood cell is placed in a 10% (m/v) NaCl solution. Because the osmotic pressure in the red blood cell is the same as a 0.9% (m/v) NaCl solution, the 10% (m/v) NaCl solution has a much greater osmotic pressure. As water leaves the cell, it shrinks, a process called *crenation* (see **FIGURE 12.11c**). A similar process occurs when making pickles, which uses a hypertonic salt solution that causes the cucumbers to shrivel as they lose water.

ENGAGE 12.15

What happens to a red blood cell placed in a 4% NaCl solution?

PRACTICE PROBLEMS

Try Practice Problems 12.79 and 12.80

SAMPLE PROBLEM 12.20 Isotonic, Hypotonic, and Hypertonic Solutions

TRY IT FIRST

Describe each of the following solutions as isotonic, hypotonic, or hypertonic. Indicate whether a red blood cell placed in each solution will undergo hemolysis, crenation, or no change.

a. a 5% (m/v) glucose solution   **b.** a 0.2% (m/v) NaCl solution

SOLUTION

a. A 5% (m/v) glucose solution is isotonic. A red blood cell will not undergo any change.
b. A 0.2% (m/v) NaCl solution is hypotonic. A red blood cell will undergo hemolysis.

a. What will happen to a red blood cell placed in a 10% (m/v) glucose solution?

b. What will happen to a red blood cell placed in a 1.0% (m/v) glucose solution?

ANSWER

a. The red blood cell will shrink (crenate).

b. The red blood cell will undergo hemolysis.

Initial Final

○ Solution particles such as Na^+, Cl^-, glucose

● Colloidal particles such as protein, starch

Solution particles pass through a dialyzing membrane, but colloidal particles are retained.

PRACTICE PROBLEMS

Try Practice Problems 12.81 and 12.82

Dialysis

Dialysis is a process that is similar to osmosis. In dialysis, a semipermeable membrane, called a dialyzing membrane, permits small solute molecules and ions as well as solvent water molecules to pass through, but it retains large particles, such as colloids.

Suppose we fill a cellophane bag with a solution containing NaCl, glucose, starch, and protein and place it in pure water. Cellophane is a dialyzing membrane, and the sodium ions, chloride ions, and glucose molecules will pass through it into the surrounding water. However, large colloidal particles, like starch and protein, remain inside. Water molecules will flow into the cellophane bag. Eventually the concentrations of sodium ions, chloride ions, and glucose molecules inside and outside the dialysis bag become equal. To remove more NaCl or glucose, the cellophane bag must be placed in a fresh sample of pure water.

Chemistry Link to Health

Hemodialysis and the Artificial Kidney

In an adult, each kidney contains about 2 million nephrons. At the top of each nephron, there is a network of arterial capillaries called the *glomerulus*. As blood flows into the glomerulus, small particles, such as amino acids, glucose, urea, water, and certain ions dialyze through the capillary membranes into the nephron. As this solution moves through the nephron, substances still of value to the body (such as amino acids, glucose, certain ions, and 99% of the water) are reabsorbed. The major waste product, urea, is excreted in the urine. If the kidneys fail to remove waste products, increased levels of urea can become life-threatening in a relatively short time.

A person with kidney failure must use an artificial kidney, which cleanses the blood by *hemodialysis*. A typical artificial kidney machine contains a large tank filled with water containing selected electrolytes. In the dialyzing bath (dialysate), there is a dialyzing membrane made of cellulose tubing. As the patient's blood flows through the dialyzing membrane, the concentrated waste products dialyze out of the blood. No blood is lost because the membrane is not permeable to red blood cells. Dialysis patients do not produce much urine. As a result, they retain large amounts of water between dialysis treatments, which produces a strain on the heart. For some

Blood out (filtered)

Blood in

Glomerulus

Collecting duct

Urine to ureter

In the kidneys, each nephron contains a glomerulus where urea and waste products are removed to form urine.

dialysis patients, 2 to 10 L of water may be removed during one treatment. At a treatment center, dialysis patients typically have three treatments a week for about 4 h each. For many patients, dialysis is done at home using a home dialysis unit for 1.5 to 2 h every day.

During dialysis, waste products and excess water are removed from the blood.

Dialyzed blood

Blood

Pump

Dialysate

○ = Urea and other waste products

Dialyzing coil

PRACTICE PROBLEMS

12.8 Properties of Solutions: Osmosis

12.75 A 10% (m/v) starch solution is separated from a 1% (m/v) starch solution by a semipermeable membrane. (Starch is a colloid.)
a. Which compartment has the higher osmotic pressure?
b. In which direction will water flow initially?
c. In which compartment will the volume level rise?

12.76 A 0.1% (m/v) albumin solution is separated from a 2% (m/v) albumin solution by a semipermeable membrane. (Albumin is a colloid.)
a. Which compartment has the higher osmotic pressure?
b. In which direction will water flow initially?
c. In which compartment will the volume level rise?

12.77 Indicate the compartment (**A** or **B**) that will increase in volume for each of the following pairs of solutions separated by a semipermeable membrane:

A	B
a. 5% (m/v) sucrose	10% (m/v) sucrose
b. 8% (m/v) albumin	4% (m/v) albumin
c. 0.1% (m/v) starch	10% (m/v) starch

12.78 Indicate the compartment (**A** or **B**) that will increase in volume for each of the following pairs of solutions separated by a semipermeable membrane:

A	B
a. 20% (m/v) starch	10% (m/v) starch
b. 10% (m/v) albumin	2% (m/v) albumin
c. 0.5% (m/v) sucrose	5% (m/v) sucrose

Applications

12.79 Are the following solutions isotonic, hypotonic, or hypertonic compared with a red blood cell?
a. distilled H_2O
b. 1% (m/v) glucose
c. 0.9% (m/v) NaCl
d. 15% (m/v) glucose

12.80 Will a red blood cell undergo crenation, hemolysis, or no change in each of the following solutions?
a. 1% (m/v) glucose
b. 2% (m/v) NaCl
c. 5% (m/v) glucose
d. 0.1% (m/v) NaCl

12.81 Each of the following mixtures is placed in a dialyzing bag and immersed in distilled water. Which substances will be found outside the bag in the distilled water?
a. NaCl solution
b. starch solution (colloid) and alanine (an amino acid) solution
c. NaCl solution and starch solution (colloid)
d. urea solution

12.82 Each of the following mixtures is placed in a dialyzing bag and immersed in distilled water. Which substances will be found outside the bag in the distilled water?
a. KCl solution and glucose solution
b. albumin solution (colloid)
c. an albumin solution (colloid), KCl solution, and glucose solution
d. urea solution and NaCl solution

UPDATE   Using Dialysis for Renal Failure

As a dialysis patient, Michelle has a 4-h dialysis treatment three times a week. When she arrives at the dialysis clinic, her weight, temperature, and blood pressure are taken and blood tests are done to determine the level of electrolytes and urea in her blood. In the dialysis center, tubes to the dialyzer are connected to the catheter she has had implanted. Blood is then pumped out of her body, through the dialyzer where it is filtered, and returned to her body. As Michelle's blood flows through the dialyzer, electrolytes from the dialysate move into her blood, and waste products in her blood move into the dialysate, which is continually renewed.

To achieve normal serum electrolyte levels, dialysate fluid contains sodium, chloride, and magnesium levels that are equal to serum concentrations. These electrolytes are removed from the blood only if their concentrations are higher than normal. Typically, in dialysis patients, the potassium ion level is higher than normal. Therefore, initial dialysis may start with a low concentration of potassium ion in the dialysate. During dialysis excess fluid is removed

by osmosis. A 4-h dialysis session requires at least 120 L of dialysis fluid. During dialysis, the electrolytes in the dialysate are adjusted until the electrolytes have the same levels as normal serum. Initially the dialysate solution prepared for Michelle contains the following: HCO_3^-, K^+, Na^+, Ca^{2+}, Mg^{2+}, Cl^-, and glucose.

Applications

12.83 A doctor orders 0.075 g of chlorpromazine, which is used to treat schizophrenia. If the stock solution is 2.5% (m/v), how many milliliters are administered to the patient?

12.84 A doctor orders 5.0 mg of compazine, which is used to treat nausea, vertigo, and migraine headaches. If the stock solution is 2.5% (m/v), how many milliliters are administered to the patient?

12.85 A $CaCl_2$ solution is given to increase blood levels of calcium. If a patient receives 5.0 mL of a 10.% (m/v) $CaCl_2$ solution, how many grams of $CaCl_2$ were given?

12.86 An intravenous solution of mannitol is used as a diuretic to increase the loss of sodium and chloride by a patient. If a patient receives 30.0 mL of a 25% (m/v) mannitol solution, how many grams of mannitol were given?

CONCEPT MAP

SOLUTIONS

consist of a

Solute — **Solvent**

Concentration — of — **Solute Particles**

amount dissolved is

Solubility

the maximum amount that dissolves is

Saturated

as

Electrolytes

dissociate

100% → **Strong**

slightly → **Weak**

Nonelectrolytes

do not dissociate

m/m, v/v, m/v → **Percent**

add water for → **Dilutions**

mol/L → **Molarity**

is used to calculate → **Moles** / **Volume**

mol/kg → **Molality**

is used to calculate → **Freezing and Boiling Point Change**

change the → **Vapor Pressure, Freezing Point, Boiling Point, Osmotic Pressure**

CHAPTER REVIEW

12.1 Solutions

LEARNING GOAL Identify the solute and solvent in a solution; describe the formation of a solution.

Solute: The substance present in lesser amount

— Salt

— Water

Solvent: The substance present in greater amount

- A solution forms when a solute dissolves in a solvent.
- In a solution, the particles of solute are evenly dispersed in the solvent.
- The solute and solvent may be solid, liquid, or gas.
- The polar O—H bond leads to hydrogen bonding between water molecules.
- An ionic solute dissolves in water, a polar solvent, because the polar water molecules attract and pull the ions into solution, where they become hydrated.
- The expression *like dissolves like* means that a polar or an ionic solute dissolves in a polar solvent while a nonpolar solute dissolves in a nonpolar solvent.

12.2 Electrolytes and Nonelectrolytes

LEARNING GOAL Identify solutes as electrolytes or nonelectrolytes.

Na^+ Cl^-
Cl^-
Na^+
Na^+ Cl^-

Strong electrolyte

- Substances that produce ions in water are called electrolytes because their solutions will conduct an electrical current.
- Strong electrolytes are completely dissociated, whereas weak electrolytes are only partially dissociated.
- Nonelectrolytes are substances that dissolve in water to produce only molecules and cannot conduct electrical currents.

12.3 Solubility

LEARNING GOAL Define solubility; distinguish between an unsaturated and a saturated solution. Identify an ionic compound as soluble or insoluble.

Dissolved solute

Undissolved solute

Dissolving / Crystallization

Saturated solution

- The solubility of a solute is the maximum amount of a solute that can dissolve in 100. g of solvent.
- A solution that contains the maximum amount of dissolved solute is a saturated solution.
- A solution containing less than the maximum amount of dissolved solute is unsaturated.
- An increase in temperature increases the solubility of most solids in water, but decreases the solubility of gases in water.
- Ionic compounds that are soluble in water usually contain Li^+, Na^+, K^+, NH_4^+, NO_3^-, or acetate, $C_2H_3O_2^-$.
- When two solutions are mixed, solubility rules can be used to predict whether a precipitate will form.

12.4 Solution Concentrations

LEARNING GOAL Calculate the concentration of a solute in a solution; use concentration units to calculate the amount of solute or solution.

Water added to make a solution   250 mL

5.0 g of KI   250 mL of KI solution

- Mass percent expresses the mass/mass (m/m) ratio of the mass of solute to the mass of solution multiplied by 100%.
- Percent concentration can also be expressed as volume/volume (v/v) and mass/volume (m/v) ratios.
- Molarity is the moles of solute per liter of solution.
- In calculations of grams or milliliters of solute or solution, the concentration is used as a conversion factor.
- Molarity (or mol/L) is written as conversion factors to solve for moles of solute or volume of solution.

12.5 Dilution of Solutions

LEARNING GOAL Describe the dilution of a solution; calculate the unknown concentration or volume when a solution is diluted.

- In dilution, a solvent such as water is added to a solution, which increases its volume and decreases its concentration.

12.6 Chemical Reactions in Solution

LEARNING GOAL Given the volume and concentration of a solution in a chemical reaction, calculate the amount of a reactant or product in the reaction.

- When solutions are involved in chemical reactions, the moles of a substance in solution can be determined from the volume and molarity of the solution.
- When mass, volume, and molarities of substances in a reaction are given, the balanced equation is used to determine the quantities or concentrations of other substances in the reaction.

12.7 Molality and Freezing Point Lowering/Boiling Point Elevation

LEARNING GOAL Identify a mixture as a solution, a colloid, or a suspension. Using the molality, calculate the freezing point and boiling point for a solution.

- Colloids contain particles that pass through most filters but do not settle out or pass through semipermeable membranes.
- Suspensions have very large particles that settle out.
- The particles in a solution lower the vapor pressure, lower the freezing point, and raise the boiling point.
- Molality is the moles of solute per kilogram of solvent, usually water.

12.8 Properties of Solutions: Osmosis

LEARNING GOAL Describe how the number of particles in solution affects osmosis.

Semipermeable membrane

- The particles in a solution increase the osmotic pressure.
- In osmosis, solvent (water) passes through a semipermeable membrane from a solution with a lower osmotic pressure (lower solute concentration) to a solution with a higher osmotic pressure (higher solute concentration).
- Isotonic solutions have osmotic pressures equal to that of body fluids.
- A red blood cell maintains its volume in an isotonic solution but swells in a hypotonic solution, and shrinks in a hypertonic solution.
- In dialysis, water and small solute particles pass through a dialyzing membrane, while larger particles are retained.

KEY TERMS

colloid A mixture having particles that are moderately large. Colloids pass through filters but cannot pass through semipermeable membranes.

concentration A measure of the amount of solute that is dissolved in a specified amount of solution.

dialysis A process in which water and small solute particles pass through a semipermeable membrane.

dilution A process by which water (solvent) is added to a solution to increase the volume and decrease (dilute) the solute concentration.

electrolyte A substance that produces ions when dissolved in water; its solution conducts electricity.

Henry's law The solubility of a gas in a liquid is directly related to the pressure of that gas above the liquid.

hydration The process of surrounding dissolved ions by water molecules.

mass percent (m/m) The grams of solute in 100. g of solution.

mass/volume percent (m/v) The grams of solute in 100. mL of solution.

molality (m) The number of moles of solute particles in exactly 1 kg of solvent.

molarity (M) The number of moles of solute in exactly 1 L of solution.

nonelectrolyte A substance that dissolves in water as molecules; its solution does not conduct an electrical current.

osmosis The flow of a solvent, usually water, through a semipermeable membrane into a solution of higher solute concentration.

osmotic pressure The pressure that prevents the flow of water into the more concentrated solution.

saturated solution A solution containing the maximum amount of solute that can dissolve at a given temperature. Any additional solute will remain undissolved in the container.

solubility The maximum amount of solute that can dissolve in 100. g of solvent, usually water, at a given temperature.

solubility rules A set of guidelines that states whether an ionic compound is soluble or insoluble in water.

solute The component in a solution that is present in the lesser amount.

solution A homogeneous mixture in which the solute is made up of small particles (ions or molecules) that can pass through filters and semipermeable membranes.

solvent The substance in which the solute dissolves; usually the component present in greater amount.

strong electrolyte A compound that ionizes completely when it dissolves in water; its solution is a good conductor of electricity.

suspension A mixture in which the solute particles are large enough and heavy enough to settle out and be retained by both filters and semipermeable membranes.

unsaturated solution A solution that contains less solute than can be dissolved.

volume percent (v/v) The milliliters of solute in 100. mL of solution.

weak electrolyte A substance that produces only a few ions along with many molecules when it dissolves in water; its solution is a weak conductor of electricity.

CORE CHEMISTRY SKILLS

The chapter section containing each Core Chemistry Skill is shown in parentheses at the end of each heading.

Using Solubility Rules (12.3)

- Ionic compounds that are soluble in water contain Li^+, Na^+, K^+, NH_4^+, NO_3^-, or $C_2H_3O_2^-$ (acetate).
- Ionic compounds containing Cl^-, Br^-, or I^- are soluble, but if they contain Ag^+, Pb^{2+}, or Hg_2^{2+}, they are insoluble.
- Most ionic compounds containing SO_4^{2-} are soluble, but if they contain Ba^{2+}, Pb^{2+}, Ca^{2+}, Sr^{2+}, or Hg_2^{2+}, they are insoluble.
- Most other ionic compounds, including those containing the anions CO_3^{2-}, S^{2-}, PO_4^{3-}, or OH^-, are insoluble.
- To write an equation or ionic equation for the formation of an insoluble ionic compound, we write the cations and anions to identify any combination that would form an insoluble ionic compound.

Example: Determine if an ionic compound forms when solutions of $CaCl_2$ and K_2CO_3 are mixed. If so, write the net ionic equation for the reaction.

Answer: In the ionic equation, we show all the ions of the reactants and show the ionic compound $CaCO_3$ as a solid.

$$Ca^{2+}(aq) + 2Cl^-(aq) + 2K^+(aq) + CO_3^{2-}(aq) \longrightarrow$$
$$CaCO_3(s) + 2K^+(aq) + 2Cl^-(aq)$$

For the net ionic equation, spectator ions that appear on both sides of the equation are removed.

$$Ca^{2+}(aq) + CO_3^{2-}(aq) \longrightarrow CaCO_3(s)$$

Calculating Concentration (12.4)

The amount of solute dissolved in a certain amount of solution is called the concentration of the solution.

- Mass percent $(m/m) = \dfrac{\text{mass of solute}}{\text{mass of solution}} \times 100\%$

- Volume percent $(v/v) = \dfrac{\text{volume of solute}}{\text{volume of solution}} \times 100\%$

- Mass/volume percent $(m/v) = \dfrac{\text{grams of solute}}{\text{milliliters of solution}} \times 100\%$

- Molarity $(M) = \dfrac{\text{moles of solute}}{\text{liters of solution}}$

Example: What is the mass/volume percent (m/v) and the molarity (M) of 225 mL (0.225 L) of a LiCl solution that contains 17.1 g of LiCl?

Answer: Mass/volume % $(m/v) = \dfrac{\text{grams of solute}}{\text{milliliters of solution}} \times 100\%$

$$= \dfrac{17.1 \text{ g LiCl}}{225 \text{ mL solution}} \times 100\%$$

$$= 7.60\% \text{ (m/v) LiCl solution}$$

$$\text{moles of LiCl} = 17.1 \text{ g LiCl} \times \dfrac{1 \text{ mol LiCl}}{42.39 \text{ g LiCl}}$$

$$= 0.403 \text{ mol of LiCl}$$

$$\text{Molarity (M)} = \dfrac{\text{moles of solute}}{\text{liters of solution}} = \dfrac{0.403 \text{ mol LiCl}}{0.225 \text{ L solution}}$$

$$= 1.79 \text{ M LiCl solution}$$

Using Concentration as a Conversion Factor (12.4)

- When we need to calculate the amount of solute or solution, we use the concentration as a conversion factor.
- For example, the concentration of a 4.50 M HCl solution means there are 4.50 mol of HCl in 1 L of HCl solution, which gives two conversion factors written as

$$\dfrac{4.50 \text{ mol HCl}}{1 \text{ L solution}} \quad \text{and} \quad \dfrac{1 \text{ L solution}}{4.50 \text{ mol HCl}}$$

Example: How many milliliters of a 4.50 M HCl solution will provide 41.2 g of HCl?

Answer:

$$41.2 \text{ g HCl} \times \dfrac{1 \text{ mol HCl}}{36.46 \text{ g HCl}} \times \dfrac{1 \text{ L solution}}{4.50 \text{ mol HCl}} \times \dfrac{1000 \text{ mL solution}}{1 \text{ L solution}}$$

$$= 251 \text{ mL of HCl solution}$$

Calculating the Quantity of a Reactant or Product for a Chemical Reaction in Solution (12.6)

- When chemical reactions involve aqueous solutions of reactants or products, we use the balanced chemical equation, the molarity, and the volume to determine the moles or grams of the reactants or products.

Example: How many grams of zinc metal will react with 0.315 L of a 1.20 M HCl solution?

$$2HCl(aq) + Zn(s) \longrightarrow H_2(g) + ZnCl_2(aq)$$

Answer:

$$0.315 \text{ L solution} \times \dfrac{1.20 \text{ mol HCl}}{1 \text{ L solution}} \times \dfrac{1 \text{ mol Zn}}{2 \text{ mol HCl}} \times \dfrac{65.41 \text{ g Zn}}{1 \text{ mol Zn}}$$

$$= 12.4 \text{ g of Zn}$$

Calculating the Freezing Point/Boiling Point of a Solution (12.7)

- The particles in a solution lower the freezing point, raise the boiling point, and increase the osmotic pressure.
- The freezing point lowering (ΔT_f) is determined from the molality (m) of the particles in the solution and the freezing point constant, K_f.

$$\Delta T_f = m \times K_f$$

- The boiling point elevation (ΔT_b) is determined from the molality (m) of the particles in the solution and the boiling point constant, K_b.

$$\Delta T_b = m \times K_b$$

Example: What is the boiling point of a solution that contains 1.5 mol of the strong electrolyte KCl in 1 kg of water?

Answer: A solution of 1.5 mol of KCl in 1 kg of water, which contains 3.0 mol of particles (1.5 mol of K^+ and 1.5 mol of Cl^-), is a 3.0-m solution.

$$\Delta T_b = m \times K_b = 3.0 \text{ } m \times \dfrac{0.51 \text{ °C}}{m} = 1.5 \text{ °C}$$

$$T_{\text{solution}} = T_{\text{water}} + \Delta T_b$$

$$= 100.0 \text{ °C} + 1.5 \text{ °C}$$

$$= 101.5 \text{ °C}$$

UNDERSTANDING THE CONCEPTS

The chapter sections to review are shown in parentheses at the end of each problem.

12.87 Match the diagrams with the following: (12.1)
 a. a polar solute and a polar solvent
 b. a nonpolar solute and a polar solvent
 c. a nonpolar solute and a nonpolar solvent

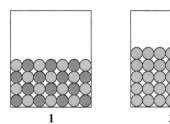

12.88 If all the solute is dissolved in diagram **1**, how would heating or cooling the solution cause each of the following changes? (12.3)
 a. **2** to **3**
 b. **2** to **1**

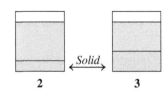

12.89 Select the diagram that represents the solution formed by a solute that is a (12.2)
 a. nonelectrolyte
 b. weak electrolyte
 c. strong electrolyte

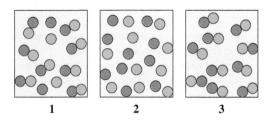

12.90 Select the container that represents the dilution of a 4% (m/v) KCl solution to give each of the following: (12.5)
 a. a 2% (m/v) KCl solution
 b. a 1% (m/v) KCl solution

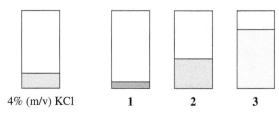

Use the following illustration of beakers and solutions for problems 12.91 and 12.92:

12.91 Use the following ions: (12.3)

Na^+ ● Cl^- ● Ag^+ ● NO_3^- ●

 a. Select the beaker (**1**, **2**, **3**, or **4**) that contains the products after the solutions in beakers **A** and **B** are mixed.
 b. If an insoluble ionic compound forms, write the ionic equation.
 c. If a reaction occurs, write the net ionic equation.

12.92 Use the following ions: (12.3)

K^+ ● NO_3^- ● NH_4^+ ● Br^- ●

 a. Select the beaker (**1**, **2**, **3**, or **4**) that contains the products after the solutions in beakers **A** and **B** are mixed.
 b. If an insoluble ionic compound forms, write the ionic equation.
 c. If a reaction occurs, write the net ionic equation.

12.93 A pickle is made by soaking a cucumber in brine, a salt-water solution. What makes the smooth cucumber become wrinkled like a prune? (12.8)

12.94 Why do lettuce leaves in a salad wilt after a vinaigrette dressing containing salt is added? (12.8)

12.95 A semipermeable membrane separates two compartments, **A** and **B**. If the levels in **A** and **B** are equal initially, select the diagram that illustrates the final levels in **a** to **d**: (12.8)

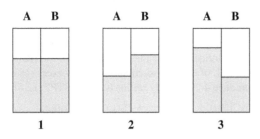

Solution in A	Solution in B
a. 2% (m/v) starch	8% (m/v) starch
b. 1% (m/v) starch	1% (m/v) starch
c. 5% (m/v) sucrose	1% (m/v) sucrose
d. 0.1% (m/v) sucrose	1% (m/v) sucrose

12.96 Select the diagram that represents the shape of a red blood cell when placed in each of the following **a** to **e**: (12.8)

1   2   3

Normal red blood cell

a. 0.9% (m/v) NaCl solution
b. 10% (m/v) glucose solution
c. 0.01% (m/v) NaCl solution
d. 5% (m/v) glucose solution
e. 1% (m/v) glucose solution

ADDITIONAL PRACTICE PROBLEMS

12.97 Why does iodine dissolve in hexane, but not in water? (12.1)

12.98 How do temperature and pressure affect the solubility of solids and gases in water? (12.3)

12.99 Potassium nitrate has a solubility of 32 g of KNO_3 in 100. g of H_2O at 20 °C. Determine if each of the following forms an unsaturated or saturated solution at 20 °C: (12.3)
a. adding 32 g of KNO_3 to 200. g of H_2O
b. adding 19 g of KNO_3 to 50. g of H_2O
c. adding 68 g of KNO_3 to 150. g of H_2O

12.100 Potassium chloride has a solubility of 43 g of KCl in 100. g of H_2O at 50 °C. Determine if each of the following forms an unsaturated or saturated solution at 50 °C: (12.3)
a. adding 25 g of KCl to 100. g of H_2O
b. adding 15 g of KCl to 25 g of H_2O
c. adding 86 g of KCl to 150. g of H_2O

12.101 Indicate whether each of the following ionic compounds is soluble or insoluble in water: (12.3)
a. KCl   **b.** $MgSO_4$ **c.** CuS
d. $AgNO_3$ **e.** $Ca(OH)_2$

12.102 Indicate whether each of the following ionic compounds is soluble or insoluble in water: (12.3)
a. $CuCO_3$ **b.** FeO   **c.** $Mg_3(PO_4)_2$
d. $(NH_4)_2SO_4$ **e.** $NaHCO_3$

12.103 Write the net ionic equation to show the formation of a solid (insoluble ionic compound) when the following solutions are mixed. Write *none* if no solid forms. (12.3)
a. $AgNO_3(aq)$ and $LiCl(aq)$
b. $NaCl(aq)$ and $KNO_3(aq)$
c. $Na_2SO_4(aq)$ and $BaCl_2(aq)$

12.104 Write the net ionic equation to show the formation of a solid (insoluble ionic compound) when the following solutions are mixed. Write *none* if no solid forms. (12.3)
a. $Ca(NO_3)_2(aq)$ and $Na_2S(aq)$
b. $Na_3PO_4(aq)$ and $Pb(NO_3)_2(aq)$
c. $FeCl_3(aq)$ and $NH_4NO_3(aq)$

12.105 If NaCl has a solubility of 36.0 g in 100. g of H_2O at 20 °C, how many grams of water are needed to prepare a saturated solution containing 80.0 g of NaCl? (12.3)

12.106 If the solid NaCl in a saturated solution of NaCl continues to dissolve, why is there no change in the concentration of the NaCl solution? (12.3)

12.107 Calculate the mass percent (m/m) of a solution containing 15.5 g of Na_2SO_4 and 75.5 g of H_2O. (12.4)

12.108 Calculate the mass percent (m/m) of a solution containing 26 g of K_2CO_3 and 724 g of H_2O. (12.4)

12.109 How many milliliters of a 12% (v/v) propyl alcohol solution would you need to obtain 4.5 mL of propyl alcohol? (12.4)

12.110 An 80-proof brandy is a 40.% (v/v) ethanol solution. The "proof" is twice the percent concentration of alcohol in the beverage. How many milliliters of alcohol are present in 750 mL of brandy? (12.4)

12.111 How many liters of a 12% (m/v) KOH solution would you need to obtain 86.0 g of KOH? (12.4)

12.112 How many liters of a 5.0% (m/v) glucose solution would you need to obtain 75 g of glucose? (12.4)

12.113 What is the molarity of a solution containing 8.0 g of NaOH in 400. mL of NaOH solution? (12.4)

12.114 What is the molarity of a solution containing 15.6 g of KCl in 274 mL of KCl solution? (12.4)

12.115 How many milliliters of a 1.75 M LiCl solution contain 15.2 g of LiCl? (12.4)

12.116 How many milliliters of a 1.50 M NaBr solution contain 75.0 g of NaBr? (12.4)

12.117 How many liters of a 2.50 M KNO_3 solution can be prepared from 60.0 g of KNO_3? (12.4)

12.118 How many liters of a 4.00 M NaCl solution will provide 25.0 g of NaCl? (12.4)

12.119 How many grams of solute are in each of the following solutions? (12.4)
  a. 2.5 L of a 3.0 M $Al(NO_3)_3$ solution
  b. 75 mL of a 0.50 M $C_6H_{12}O_6$ solution
  c. 235 mL of a 1.80 M LiCl solution

12.120 How many grams of solute are in each of the following solutions? (12.4)
  a. 0.428 L of a 0.450 M K_2CO_3 solution
  b. 10.5 mL of a 2.50 M $AgNO_3$ solution
  c. 28.4 mL of a 6.00 M H_3PO_4 solution

12.121 Calculate the final concentration of the solution when water is added to prepare each of the following: (12.5)
  a. 25.0 mL of a 0.200 M NaBr solution is diluted to 50.0 mL
  b. 15.0 mL of a 12.0% (m/v) K_2SO_4 solution is diluted to 40.0 mL
  c. 75.0 mL of a 6.00 M NaOH solution is diluted to 255 mL

12.122 Calculate the final concentration of the solution when water is added to prepare each of the following: (12.5)
  a. 25.0 mL of an 18.0 M HCl solution is diluted to 500. mL
  b. 50.0 mL of a 15.0% (m/v) NH_4Cl solution is diluted to 125 mL
  c. 4.50 mL of an 8.50 M KOH solution is diluted to 75.0 mL

12.123 What is the initial volume, in milliliters, needed to prepare each of the following diluted solutions? (12.5)
  a. 250 mL of a 3.0% (m/v) HCl solution using a 10.0% (m/v) HCl solution
  b. 500. mL of a 0.90% (m/v) NaCl solution using a 5.0% (m/v) NaCl solution
  c. 350. mL of a 2.00 M NaOH solution using a 6.00 M NaOH solution

12.124 What is the initial volume, in milliliters, needed to prepare each of the following diluted solutions? (12.5)
  a. 250 mL of a 5.0% (m/v) glucose solution using a 20.% (m/v) glucose solution
  b. 45.0 mL of a 1.0% (m/v) $CaCl_2$ solution using a 5.0% (m/v) $CaCl_2$ solution
  c. 100. mL of a 6.00 M H_2SO_4 solution using an 18.0 M H_2SO_4 solution

12.125 What is the final volume, in milliliters, when 25.0 mL of each of the following solutions is diluted to provide the given concentration? (12.5)
  a. 10.0% (m/v) HCl solution to give a 2.50% (m/v) HCl solution
  b. 5.00 M HCl solution to give a 1.00 M HCl solution
  c. 6.00 M HCl solution to give a 0.500 M HCl solution

12.126 What is the final volume, in milliliters, when 5.00 mL of each of the following solutions is diluted to provide the given concentration? (12.5)
  a. 20.0% (m/v) NaOH solution to give a 4.00% (m/v) NaOH solution
  b. 0.600 M NaOH solution to give a 0.100 M NaOH solution
  c. 16.0% (m/v) NaOH solution to give a 2.00% (m/v) NaOH solution

12.127 The antacid Amphojel contains aluminum hydroxide $Al(OH)_3$. How many milliliters of a 6.00 M HCl solution are required to react with 60.0 mL of a 2.00 M $Al(OH)_3$ solution? (12.6)

$$3HCl(aq) + Al(OH)_3(s) \longrightarrow 3H_2O(l) + AlCl_3(aq)$$

12.128 Cadmium reacts with HCl to produce hydrogen gas and cadmium chloride. What is the molarity (M) of the HCl solution if 250. mL of the HCl solution reacts with excess cadmium to produce 4.20 L of H_2 gas measured at STP? (12.6)

$$2HCl(aq) + Cd(s) \longrightarrow H_2(g) + CdCl_2(aq)$$

12.129 Calculate the freezing point of each of the following solutions: (12.7)
  a. 0.580 mol of lactose, a nonelectrolyte, added to 1.00 kg of water
  b. 45.0 g of KCl, a strong electrolyte, dissolved in 1.00 kg of water
  c. 1.5 mol of K_3PO_4, a strong electrolyte, dissolved in 1.00 kg of water

12.130 Calculate the boiling point of each of the following solutions: (12.7)
  a. 175 g of glucose, $C_6H_{12}O_6$, a nonelectrolyte, added to 1.00 kg of water
  b. 1.8 mol of $CaCl_2$, a strong electrolyte, dissolved in 1.00 kg of water
  c. 50.0 g of $LiNO_3$, a strong electrolyte, dissolved in 1.00 kg of water

Applications

12.131 An antacid tablet, such as Amphojel, may be taken to reduce excess stomach acid, which is a 0.20 M HCl solution. If one dose of Amphojel contains 450 mg of $Al(OH)_3$, what volume, in milliliters, of stomach acid will be neutralized? (12.6)

$$3HCl(aq) + Al(OH)_3(s) \longrightarrow 3H_2O(l) + AlCl_3(aq)$$

12.132 Calcium carbonate, $CaCO_3$, reacts with stomach acid, (HCl, hydrochloric acid) according to the following equation: (12.6)

$$2HCl(aq) + CaCO_3(s) \longrightarrow$$
$$CO_2(g) + H_2O(l) + CaCl_2(aq)$$

Tums, an antacid, contains $CaCO_3$. If Tums is added to 20.0 mL of a 0.400 M HCl solution, how many grams of CO_2 gas are produced?

CHALLENGE PROBLEMS

The following problems are related to the topics in this chapter. However, they do not all follow the chapter order, and they require you to combine concepts and skills from several sections. These problems will help you increase your critical thinking skills and prepare for your next exam.

12.133 Write the net ionic equation to show the formation of a solid when the following solutions are mixed. Write *none* if no solid forms. (12.3)
   a. $AgNO_3(aq) + Na_2SO_4(aq)$
   b. $KCl(aq) + Pb(NO_3)_2(aq)$
   c. $CaCl_2(aq) + (NH_4)_3PO_4(aq)$
   d. $K_2SO_4(aq) + BaCl_2(aq)$

12.134 Write the net ionic equation to show the formation of a solid when the following solutions are mixed. Write *none* if no solid forms. (12.3)
   a. $Pb(NO_3)_2(aq) + NaBr(aq)$
   b. $AgNO_3(aq) + (NH_4)_2CO_3(aq)$
   c. $Na_3PO_4(aq) + Al(NO_3)_3(aq)$
   d. $NaOH(aq) + CuCl_2(aq)$

12.135 In a laboratory experiment, a 10.0-mL sample of NaCl solution is poured into an evaporating dish with a mass of 24.10 g. The combined mass of the evaporating dish and NaCl solution is 36.15 g. After heating, the evaporating dish and dry NaCl have a combined mass of 25.50 g. (12.4)
   a. What is the mass percent (m/m) of the NaCl solution?
   b. What is the molarity (M) of the NaCl solution?
   c. If water is added to 10.0 mL of the initial NaCl solution to give a final volume of 60.0 mL, what is the molarity of the diluted NaCl solution?

12.136 In a laboratory experiment, a 15.0-mL sample of KCl solution is poured into an evaporating dish with a mass of 24.10 g. The combined mass of the evaporating dish and KCl solution is 41.50 g. After heating, the evaporating dish and dry KCl have a combined mass of 28.28 g. (12.4)
   a. What is the mass percent (m/m) of the KCl solution?
   b. What is the molarity (M) of the KCl solution?
   c. If water is added to 10.0 mL of the initial KCl solution to give a final volume of 60.0 mL, what is the molarity of the diluted KCl solution?

12.137 Potassium fluoride has a solubility of 92 g of KF in 100. g of H_2O at 18 °C. Determine if each of the following mixtures forms an unsaturated or saturated solution at 18 °C: (12.3)
   a. adding 35 g of KF to 25 g of H_2O
   b. adding 42 g of KF to 50. g of H_2O
   c. adding 145 g of KF to 150. g of H_2O

12.138 Lithium chloride has a solubility of 55 g of LiCl in 100. g of H_2O at 25 °C. Determine if each of the following mixtures forms an unsaturated or saturated solution at 25 °C: (12.3)
   a. adding 10 g of LiCl to 15 g of H_2O
   b. adding 25 g of LiCl to 50. g of H_2O
   c. adding 75 g of LiCl to 150. g of H_2O

12.139 A solution is prepared with 70.0 g of HNO_3 and 130.0 g of H_2O. The HNO_3 solution has a density of 1.21 g/mL. (12.4)
   a. What is the mass percent (m/m) of the HNO_3 solution?
   b. What is the total volume, in milliliters, of the solution?
   c. What is the mass/volume percent (m/v) of the solution?
   d. What is the molarity (M) of the solution?

12.140 A solution is prepared by dissolving 22.0 g of NaOH in 118.0 g of water. The NaOH solution has a density of 1.15 g/mL. (12.4)
   a. What is the mass percent (m/m) of the NaOH solution?
   b. What is the total volume, in milliliters, of the solution?
   c. What is the mass/volume percent (m/v) of the solution?
   d. What is the molarity (M) of the solution?

12.141 A 355-mL sample of a HCl solution reacts with excess Mg to produce 4.20 L of H_2 gas measured at 745 mmHg and 35 °C. What is the molarity (M) of the HCl solution? (12.6)

$$2HCl(aq) + Mg(s) \longrightarrow H_2(g) + MgCl_2(aq)$$

12.142 A 255-mL sample of a HCl solution reacts with excess Ca to produce 14.0 L of H_2 gas measured at 732 mmHg and 22 °C. What is the molarity (M) of the HCl solution? (12.6)

$$2HCl(aq) + Ca(s) \longrightarrow H_2(g) + CaCl_2(aq)$$

12.143 How many moles of each of the following strong electrolytes are needed to give the same freezing point lowering as 1.2 mol of the nonelectrolyte ethylene glycol in 1 kg of water? (12.7)
   a. NaCl   **b.** K_3PO_4

12.144 How many moles of each of the following are needed to give the same freezing point lowering as 3.0 mol of the nonelectrolyte ethylene glycol in 1 kg of water? (12.7)
   a. CH_4O, a nonelectrolyte
   b. KNO_3, a strong electrolyte

12.145 The boiling point of a NaCl solution is 101.04 °C. (12.7)
   a. What is the molality (*m*) of the solute particles in the NaCl solution?
   b. What is the freezing point of the NaCl solution?

12.146 The freezing point of a $CaCl_2$ solution is −25 °C. (12.7)
   a. What is the molality (*m*) of the solute particles in the $CaCl_2$ solution?
   b. What is the boiling point of the $CaCl_2$ solution?

12.147 Calculate the freezing point of each of the following solutions:
   a. 1.36 mol of methanol, CH_4O, a nonelectrolyte, added to 1.00 kg of water
   b. 640. g of the antifreeze propylene glycol, $C_3H_8O_2$, a nonelectrolyte, dissolved in 1.00 kg of water
   c. 111 g of KCl, a strong electrolyte, dissolved in 1.00 kg of water

12.148 Calculate the boiling point of each of the following solutions:
   a. 2.12 mol of glucose, $C_6H_{12}O_6$, a nonelectrolyte, added to 1.00 kg of water
   b. 110. g of sucrose, $C_{12}H_{22}O_{11}$, a nonelectrolyte, dissolved in 1.00 kg of water
   c. 146 g of $NaNO_3$, a strong electrolyte, dissolved in 1.00 kg of water

ANSWERS TO ENGAGE QUESTIONS

12.1 The uniform blue color indicates that the solute particles are evenly dispersed among the solvent particles.

12.2 KCl is a polar solute, and water is a polar solvent.

12.3 In an aqueous solution, a strong electrolyte such as $LiNO_3$ dissociates into ions Li^+ and NO_3^-, whereas a polar nonelectrolyte such as CH_4N_2O dissolves as molecules only.

12.4 At 20 °C, the solubility of $NaNO_3$ is 85 g/100 g H_2O, and at 60 °C, it is 120 g/100 g H_2O.

12.5 K_2S is soluble in water because the K^+ is a soluble cation, which makes any compound soluble.

12.6 Each of the test tubes contains a cation or an anion or both that form a compound that is not soluble.

12.7 The combination of Pb^{2+} and Br^- forms an insoluble compound.

12.8 The mass percent (m/m) of a solution is an equality of mass of solute in exactly 100 g of solution. By multiplying the mass of a solution and the percent factor (g solute/100. g solution), the mass in grams of the solute is obtained.

12.9 The units for a mass/volume percent solution are g of solute in 100. mL of solution. A 3.0% solution contains 3.0 g of solute in 100. mL of solution.

12.10 2.0 mol of KCl in 4.00 L of solution is 2.0 mol/4.0 L = 0.50 mol/L = 0.50 M.

12.11 Increasing the volume during dilution decreases the mol/L concentration of KCl.

12.12 $\dfrac{2.00 \text{ mol HCl}}{1 \text{ L solution}}, \dfrac{1 \text{ mol CaCO}_3}{2 \text{ mol HCl}}, \dfrac{100.09 \text{ g CaCO}_3}{1 \text{ mol CaCO}_3}$

12.13 Suspension particles are very large and can be trapped by filters. Because colloids are smaller, a semipermeable membrane is needed to separate colloids from the solute.

12.14 A 1 m solution of KBr contains 2 mol of particles (K^+ and Br^-) in 1 L of water, whereas a 1 m solution of the nonelectrolyte urea contains 1 mol of molecules in 1 L of water. Thus, 2 mol of particles (KBr) lowers the freezing point of water more than 1 mol of particles (urea).

12.15 A 4% NaCl solution is hypertonic; a red blood cell will crenate (shrivel up) in this solution.

ANSWERS TO SELECTED PROBLEMS

12.1
 a. NaCl, solute; water, solvent
 b. water, solute; ethanol, solvent
 c. oxygen, solute; nitrogen, solvent

12.3 The polar water molecules pull the K^+ and I^- ions away from the solid and into solution, where they are hydrated.

12.5 a. water b. CCl_4 c. water d. CCl_4

12.7 In a solution of KF, only the ions of K^+ and F^- are present in the solvent. In a HF solution, there are a few ions of H^+ and F^- present but mostly dissolved HF molecules.

12.9
 a. $KCl(s) \xrightarrow{H_2O} K^+(aq) + Cl^-(aq)$

 b. $CaCl_2(s) \xrightarrow{H_2O} Ca^{2+}(aq) + 2Cl^-(aq)$

 c. $K_3PO_4(s) \xrightarrow{H_2O} 3K^+(aq) + PO_4^{3-}(aq)$

 d. $Fe(NO_3)_3(s) \xrightarrow{H_2O} Fe^{3+}(aq) + 3NO_3^-(aq)$

12.11
 a. mostly molecules and a few ions
 b. only ions c. only molecules

12.13 a. strong electrolyte b. weak electrolyte
 c. nonelectrolyte

12.15 a. saturated b. unsaturated

12.17 a. unsaturated b. unsaturated c. saturated

12.19 a. 68 g of KCl b. 12 g of KCl

12.21
 a. The solubility of solid solutes typically increases as temperature increases.
 b. The solubility of a gas is less at a higher temperature.
 c. Gas solubility is less at a higher temperature and the CO_2 pressure in the can is increased.

12.23 a. soluble b. insoluble c. insoluble
 d. soluble e. soluble

12.25 a. No solid forms.
 b. $2Ag^+(aq) + 2NO_3^-(aq) + 2K^+(aq) + S^{2-}(aq) \longrightarrow$
$Ag_2S(s) + 2K^+(aq) + 2NO_3^-(aq)$
$2Ag^+(aq) + S^{2-}(aq) \longrightarrow Ag_2S(s)$

 c. $Ca^{2+}(aq) + 2Cl^-(aq) + 2Na^+(aq) + SO_4^{2-}(aq) \longrightarrow$
$CaSO_4(s) + 2Na^+(aq) + 2Cl^-(aq)$
$Ca^{2+}(aq) + SO_4^{2-}(aq) \longrightarrow CaSO_4(s)$

 d. $3Cu^{2+}(aq) + 6Cl^-(aq) + 6Li^+(aq) + 2PO_4^{3-}(aq) \longrightarrow$
$Cu_3(PO_4)_2(s) + 6Li^+(aq) + 6Cl^-(aq)$
$3Cu^{2+}(aq) + 2PO_4^{3-}(aq) \longrightarrow Cu_3(PO_4)_2(s)$

12.27 A 5.00% (m/m) glucose solution can be made by adding 5.00 g of glucose to 95.00 g of water, while a 5.00% (m/v) glucose solution can be made by adding 5.00 g of glucose to enough water to make 100.0 mL of solution.

12.29 a. 17% (m/m) KCl solution
 b. 5.3% (m/m) sucrose solution
 c. 10.% (m/m) $CaCl_2$ solution

12.31 a. 30.% (m/v) Na_2SO_4 solution
 b. 11% (m/v) sucrose solution

12.33 a. 2.5 g of KCl b. 50. g of NH_4Cl
 c. 25.0 mL of acetic acid

12.35 79.9 mL of alcohol

12.37 a. 20. g of $LiNO_3$ solution
 b. 400. mL of KOH solution
 c. 20. mL of formic acid solution

12.39 a. 0.500 M glucose solution
 b. 0.0356 M KOH solution
 c. 0.250 M NaCl solution

12.41 a. 120. g of NaOH b. 59.6 g of KCl
 c. 5.47 g of HCl

12.43 a. 1.50 L of KBr solution
 b. 10.0 L of NaCl solution
 c. 62.5 mL of $Ca(NO_3)_2$ solution

12.45 a. 983 mL b. 1.70×10^3 mL c. 464 mL

12.47 a. 20. g of mannitol b. 240 g of mannitol

12.49 2 L of glucose solution

12.51 Adding water (solvent) to the soup increases the volume and dilutes the tomato soup concentration.

12.53 **a.** 2.0 M HCl solution
b. 2.0 M NaOH solution
c. 2.5% (m/v) KOH solution
d. 3.0% (m/v) H_2SO_4 solution

12.55 **a.** 80. mL of HCl solution
b. 250 mL of LiCl solution
c. 600. mL of H_3PO_4 solution
d. 180 mL of glucose solution

12.57 **a.** 12.8 mL of 4.00 M HNO_3 solution
b. 11.9 mL of 6.00 M $MgCl_2$ solution
c. 1.88 mL of 8.00 M KCl solution

12.59 1.0×10^2 mL

12.61 **a.** 10.4 g of $PbCl_2$
b. 18.8 mL of $Pb(NO_3)_2$ solution
c. 1.20 M KCl solution

12.63 **a.** 206 mL of HCl solution
b. 11.2 L of H_2 gas
c. 9.12 M HCl solution

12.65 **a.** 74.8 mL of 3.50 M HBr solution
b. 50.4 L of H_2
c. 1.69 M HBr

12.67 **a.** solution   **b.** suspension

12.69 **a.** 22.3 m **b.** 6.89 m

12.71 **a.** freezing point: $-41.5\ °C$; boiling point: $111.6\ °C$
b. freezing point: $-12.8\ °C$; boiling point: $103.6\ °C$

12.73 **a.** 2.0 mol of ethylene glycol in 1.0 kg of water will have a lower freezing point because it has more particles in solution.
b. 0.50 mol of $MgCl_2$ in 1.0 kg of water has a lower freezing point because each formula unit of $MgCl_2$ dissociates in water to give three particles, whereas each formula unit of KCl dissociates to give only two particles.

12.75 **a.** 10% (m/v) starch solution
b. from the 1% (m/v) starch solution into the 10% (m/v) starch solution
c. 10% (m/v) starch solution

12.77 **a.** B 10% (m/v) sucrose solution
b. A 8% (m/v) albumin solution
c. B 10% (m/v) starch solution

12.79 **a.** hypotonic   **b.** hypotonic
c. isotonic   **d.** hypertonic

12.81 **a.** Na^+, Cl^- **b.** alanine
c. Na^+, Cl^- **d.** urea

12.83 3.0 mL of chlorpromazine solution

12.85 0.50 g of $CaCl_2$

12.87 **a.** 1   **b.** 2
c. 1

12.89 **a.** 3   **b.** 1
c. 2

12.91 **a.** beaker 3
b. $Na^+(aq) + Cl^-(aq) + Ag^+(aq) + NO_3^-(aq) \longrightarrow$
$AgCl(s) + Na^+(aq) + NO_3^-(aq)$
c. $Ag^+(aq) + Cl^-(aq) \longrightarrow AgCl(s)$

12.93 The skin of the cucumber acts like a semipermeable membrane, and the more dilute solution inside flows into the brine solution.

12.95 **a.** 2   **b.** 1
c. 3   **d.** 2

12.97 Because iodine is a nonpolar molecule, it will dissolve in hexane, a nonpolar solvent. Iodine does not dissolve in water because water is a polar solvent.

12.99 **a.** unsaturated solution   **b.** saturated solution
c. saturated solution

12.101 **a.** soluble   **b.** soluble
c. insoluble   **d.** soluble
e. insoluble

12.103 **a.** $Ag^+(aq) + Cl^-(aq) \rightarrow AgCl(s)$
b. none
c. $Ba^{2+}(aq) + SO_4^{2-}(aq) \longrightarrow BaSO_4(s)$

12.105 222 g of water

12.107 17.0% (m/m) Na_2SO_4 solution

12.109 38 mL of propyl alcohol solution

12.111 0.72 L of KOH solution

12.113 0.500 M NaOH solution

12.115 205 mL of LiCl solution

12.117 0.237 L of KNO_3 solution

12.119 **a.** 1600 g of $Al(NO_3)_3$ **b.** 6.8 g of $C_6H_{12}O_6$
c. 17.9 g of LiCl

12.121 **a.** 0.100 M NaBr solution
b. 4.50% (m/v) K_2SO_4 solution
c. 1.76 M NaOH solution

12.123 **a.** 75 mL of 10.0% (m/v) HCl solution
b. 90. mL of 5.0% (m/v) NaCl solution
c. 117 mL of 6.00 M NaOH solution

12.125 **a.** 100. mL
b. 125 mL
c. 300. mL

12.127 60.0 mL of HCl solution

12.129 **a.** $-1.08\ °C$
b. $-2.25\ °C$
c. $-11\ °C$

12.131 87 mL of HCl solution

12.133 **a.** $2Ag^+(aq) + SO_4^{2-}(aq) \longrightarrow Ag_2SO_4(s)$
b. $Pb^{2+}(aq) + 2Cl^-(aq) \longrightarrow PbCl_2(s)$
c. $3Ca^{2+}(aq) + 2PO_4^{3-}(aq) \longrightarrow Ca_3(PO_4)_2(s)$
d. $Ba^{2+}(aq) + SO_4^{2-}(aq) \longrightarrow BaSO_4(s)$

12.135 **a.** 11.6% (m/m) NaCl solution
b. 2.40 M NaCl solution
c. 0.400 M NaCl solution

12.137 **a.** saturated
b. unsaturated
c. saturated

12.139 **a.** 35.0% (m/m) HNO_3 solution
b. 165 mL
c. 42.4% (m/v) HNO_3 solution
d. 6.73 M HNO_3 solution

12.141 0.917 M HCl solution

12.143 **a.** 0.60 mol of NaCl   **b.** 0.30 mol of K_3PO_4

12.145 **a.** 2.0 m **b.** $-3.7\ °C$

12.147 **a.** $-2.53\ °C$ **b.** $-15.6\ °C$ **c.** $-5.54\ °C$

Reaction Rates and Chemical Equilibrium

Peter, a chemical oceanographer, is collecting data concerning the amount of dissolved gases, specifically carbon dioxide (CO_2) in the Atlantic Ocean. Studies indicate that CO_2 in the atmosphere has increased as much as 45% since the eighteenth century. Peter's research involves measuring the amount of dissolved CO_2 in the oceans and trying to determine its impact. The oceans are a complex mixture of many different chemicals including gases, elements and minerals, and organic and particulate matter. Peter understands that CO_2 is absorbed in the ocean through a series of equilibrium reactions that convert CO_2 to HCO_3^-.

$$CO_2(g) + H_2O(l) \rightleftharpoons HCO_3^-(aq) + H^+(aq)$$

An equilibrium reaction is a reversible reaction in which both the products and the reactants are present. Because the equilibrium reaction shifts according to Le Châtelier's principle, an increase in the CO_2 concentration could eventually increase the amount of dissolved calcium carbonate, $CaCO_3$, which makes up coral reefs and shells.

CAREER

Chemical Oceanographer

A chemical oceanographer, also called a marine chemist, studies the chemistry of the ocean. One area of study includes how chemicals or pollutants enter into and affect the ocean. These include sewage, oil or fuels, chemical fertilizers, plastics, and storm drain overflows. Oceanographers analyze how these chemicals interact with seawater, marine life, and sediments, as they can behave differently due to the ocean's varied environmental conditions. Chemical oceanographers also study how the various elements are cycled within the ocean. For instance, oceanographers quantify the amount and rate at which carbon dioxide is absorbed at the ocean's surface and eventually transferred to deep waters. Chemical oceanographers also aid ocean engineers in the development of instruments and vessels that enable researchers to collect data and discover previously unknown marine life.

UPDATE   Equilibrium of CO₂ in the Ocean

Peter studies the effect of changes in the acidity of ocean water on coral. Read how CO_2 dissolved in oceans changes the pH in the **UPDATE Equilibrium of CO₂ in the Ocean**, page 424.

13.1 Rates of Reactions

LEARNING GOAL Describe how temperature, concentration, and catalysts affect the rate of a reaction.

Earlier we looked at chemical reactions and determined the amounts of substances that react and the products that form. Now we are interested in how fast a reaction goes. If we know how fast a medication acts on the body, we can adjust the time over which the medication is taken. Some reactions such as explosions or the formation of precipitates in a solution are very fast. When we roast a turkey or bake a cake, the reaction is slower. Some reactions such as the tarnishing of silver and the aging of the body are much slower (see **FIGURE 13.1**). We have seen that some reactions need energy while other reactions produce energy. In this chapter, we will also look at the effect of changing the concentrations of reactants or products on the rate of reaction.

FIGURE 13.1 Reaction rates vary greatly for everyday processes. A banana ripens in a few days, silver tarnishes in a few months, while the aging process of humans takes many years.

For a chemical reaction to take place, the molecules of the reactants must come in contact with each other. The **collision theory** indicates that a reaction takes place only when molecules collide with the proper orientation and sufficient energy. Many collisions can occur, but only a few actually lead to the formation of product. For example, consider the reaction of nitrogen and oxygen molecules (see **FIGURE 13.2**). To form nitrogen oxide (NO), the collisions between N_2 and O_2 molecules must place the atoms in the proper alignment. If the molecules are not aligned properly, no reaction takes place.

Activation Energy

Even when a collision has the proper orientation, there still must be sufficient energy to break the bonds between the atoms of the reactants. The **activation energy** is the minimum amount of energy required to break the bonds between atoms of the reactants. The concept of activation energy is analogous to climbing a hill. To reach a destination on the other side, we must have the energy needed to climb to the top of the hill. Once we are at the top, we can run down the other side. The energy needed to get us from our starting point to the top of the hill would be our activation energy.

In the same way, a collision must provide enough energy to push the reactants to the top of the energy hill. Then the reactants may be converted to products. If the energy

Collision that forms products

FIGURE 13.2 Reacting molecules must collide, have a minimum amount of energy, and have the proper orientation to form product.

What happens in a collision of reacting molecules that have the proper orientation, but not the minimum energy of activation?

provided by the collision is less than the activation energy, the molecules simply bounce apart, and no reaction occurs. The features that lead to a successful reaction are summarized next (see **FIGURE 13.3**).

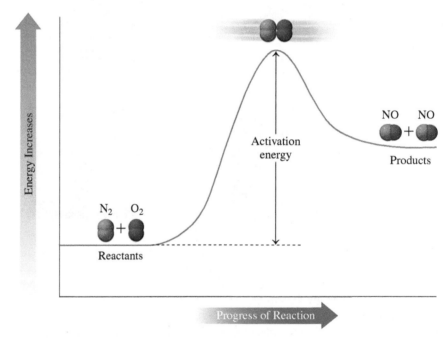

FIGURE 13.3 The activation energy is the minimum energy needed to convert the colliding molecules into product.

Three Conditions Required for a Reaction to Occur

1. **Collision**   The reactants must collide.
2. **Orientation** The reactants must align properly to break and form bonds.
3. **Energy**   The collision must provide the energy of activation.

Rate of Reaction

The **rate** (or speed) **of reaction** is determined by measuring the amount of a reactant used up, or the amount of a product formed, in a certain period of time.

$$\text{Rate of reaction} = \frac{\text{change in concentration of reactant or product}}{\text{change in time}}$$

We can describe the rate of reaction with the analogy of eating a pizza. When we start to eat, we have a whole pizza. As time goes by, there are fewer slices of pizza left. If we know how long it took to eat the pizza, we could determine the rate at which the pizza was consumed. Let's assume 4 slices are eaten every 8 minutes. That gives a rate of $\frac{1}{2}$ slice per minute. After 16 minutes, all 8 slices are gone.

Rate at Which Pizza Slices Are Eaten				
Slices Eaten	0	4 slices	6 slices	8 slices
Time (min)	0	8 min	12 min	16 min

Factors That Affect the Rate of a Reaction

Reactions with low activation energies go faster than reactions with high activation energies. Some reactions go very fast, while others are very slow. For any reaction, the rate is affected by changes in temperature, changes in the concentration of the reactants, and the addition of catalysts.

Temperature

At higher temperatures, the increase in kinetic energy of the reactants makes them move faster and collide more often, and it provides more collisions with the required energy of activation. Reactions almost always go faster at higher temperatures. For every 10 °C increase in temperature, most reaction rates approximately double. If we want food to cook faster, we increase the temperature. When body temperature rises, there is an increase in the pulse rate, rate of breathing, and metabolic rate. If we are exposed to extreme heat, we may experience *heat stroke*, which is a condition that occurs when body temperature goes above 40.5 °C (105 °F). If the body loses its ability to regulate temperature, body temperature continues to rise and may cause damage to the brain and internal organs. On the other hand, we slow down a reaction by decreasing the temperature. For example, we refrigerate perishable foods to make them last longer. In some cardiac surgeries, body temperature is decreased to 28 °C so the heart can be stopped and less oxygen is required by the brain. This is also the reason why some people have survived submersion in icy lakes for long periods of time. Cool water or an ice blanket may also be used to decrease the body temperature of a person with hyperthermia or heat stroke.

Concentrations of Reactants

For virtually all reactions, the rate of a reaction increases when the concentration of the reactants increases. When there are more reacting molecules, more collisions that form products can occur, and the reaction goes faster (see **FIGURE 13.4**). For example, a patient having difficulty breathing may be given a breathing mixture with a higher oxygen content than the atmosphere. The increase in the number of oxygen molecules in the lungs increases the rate at which oxygen combines with hemoglobin. The increased rate of oxygenation of the blood means that the patient can breathe more easily.

$$\underset{\text{Hemoglobin}}{\text{Hb}(aq)} \quad + \quad \underset{\text{Oxygen}}{O_2(g)} \longrightarrow \underset{\text{Oxyhemoglobin}}{\text{HbO}_2(aq)}$$

Catalysts

Another way to speed up a reaction is to lower the *energy of activation*. The energy of activation is the minimum energy needed to break apart the bonds of the reacting molecules. If a collision provides less than the activation energy, the bonds do not break, and the reactant molecules bounce apart. A **catalyst** speeds up a reaction by providing an alternative

ENGAGE 13.2

Why would an increase in temperature increase the rate of a reaction?

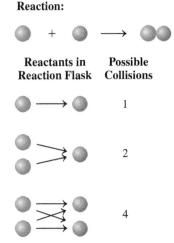

FIGURE 13.4 Increasing the concentration of a reactant increases the number of collisions that are possible.

pathway that has a lower energy of activation. When activation energy is lowered, more collisions provide sufficient energy for reactants to form product. During a reaction, a catalyst is not changed or consumed.

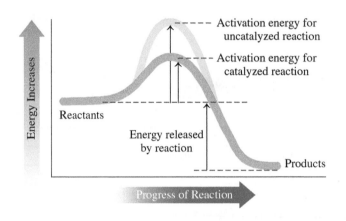

When a catalyst lowers the activation energy, the reaction occurs at a faster rate.

Catalysts have many uses in industry. In the manufacturing of margarine, hydrogen (H_2) is added to vegetable oils. Normally, the reaction is very slow because it has a high activation energy. However, when platinum (Pt) is used as a catalyst, the reaction occurs rapidly. In the body, biocatalysts called *enzymes* make metabolic reactions proceed at rates necessary for proper cellular activity. Enzymes are added to laundry detergents to break down proteins (proteases), starches (amylases), or greases (lipases) that have stained clothes. Such enzymes function at the low temperatures that are used in home washing machines, and they are biodegradable as well.

The factors affecting reaction rates are summarized in **TABLE 13.1**.

TABLE 13.1 Factors That Increase Reaction Rate

Factor	Reason
Increasing temperature	More collisions, more collisions with energy of activation
Increasing reactant concentration	More collisions
Adding a catalyst	Lowers energy of activation

SAMPLE PROBLEM 13.1 Factors That Affect the Rate of Reaction

TRY IT FIRST

Explain why each of the following changes will increase, decrease, or have no effect on the rate of reaction:

a. increasing the temperature
b. decreasing the number of reacting molecules
c. adding a catalyst

SOLUTION

a. A higher temperature increases the kinetic energy of the particles, which increases the number of collisions and makes more collisions effective, causing an increase in the rate of reaction.
b. Decreasing the number of reacting molecules decreases the number of collisions and the rate of the reaction.
c. Adding a catalyst increases the rate of reaction by lowering the activation energy, which increases the number of collisions that form product.

SELF TEST 13.1

a. How does using an ice blanket on a patient affect the rate of metabolism in the body?
b. How does adding more reacting molecules affect the rate of the reaction?

ANSWER

a. Lowering the temperature will decrease the rate of metabolism.
b. Adding more reacting molecules will increase the rate of the reaction.

PRACTICE PROBLEMS

Try Practice Problems 13.1 to 13.6

PRACTICE PROBLEMS

13.1 Rates of Reactions

13.1 **a.** What is meant by the rate of a reaction?
b. Why does bread grow mold more quickly at room temperature than in the refrigerator?

13.2 **a.** How does a catalyst affect the activation energy?
b. Why is pure oxygen used in respiratory distress?

13.3 In the following reaction, what happens to the number of collisions when more $Br_2(g)$ molecules are added?

$$H_2(g) + Br_2(g) \longrightarrow 2HBr(g)$$

13.4 In the following reaction, what happens to the number of collisions when the temperature of the reaction is decreased?

$$2H_2(g) + CO(g) \longrightarrow CH_4O(g)$$

13.5 How would each of the following change the rate of the reaction shown here?

$$2SO_2(g) + O_2(g) \longrightarrow 2SO_3(g)$$

a. adding more $SO_2(g)$ **b.** increasing the temperature
c. adding a catalyst   **d.** removing some $O_2(g)$

13.6 How would each of the following change the rate of the reaction shown here?

$$2NO(g) + 2H_2(g) \longrightarrow N_2(g) + 2H_2O(g)$$

a. adding more $NO(g)$ **b.** decreasing the temperature
c. removing some $H_2(g)$ **d.** adding a catalyst

13.2 Chemical Equilibrium

LEARNING GOAL Use the concept of reversible reactions to explain chemical equilibrium.

Previously we assumed that all of the reactants in a chemical reaction were converted to products. However, most of the time reactants are not completely converted to products because a *reverse reaction* takes place in which products collide to form the reactants. When a reaction proceeds in both a forward and reverse direction, it is said to be *reversible*. We have looked at other reversible processes. For example, the melting of solids to form liquids and the freezing of liquids to solids is a reversible physical change. Even in our daily life we have reversible events. We go from home to school and we return from school to home. We go up an escalator and we come back down. We put money in our bank account and we take money out.

An analogy for a forward and reverse reaction can be found in the phrase "We are going to the grocery store." Although we mention our trip in one direction, we know that we will also return home from the store. Because our trip has both a forward and reverse direction, we can say the trip is reversible. It is not very likely that we would stay at the store forever.

A trip to the grocery store can be used to illustrate another aspect of reversible reactions. Perhaps the grocery store is nearby and we usually walk. However, we can change our rate. Suppose that one day we drive to the store, which increases our rate and gets us to the store faster. Correspondingly, a car also increases the rate at which we return home.

As a reaction progresses, the rate of the forward reaction decreases and that of the reverse reaction increases. At equilibrium, the rates of the forward and reverse reactions are equal.

Reversible Chemical Reactions

A **reversible reaction** proceeds in both the forward and reverse direction. That means there are two reaction rates: one is the rate of the forward reaction, and the other is the rate of the reverse reaction. When molecules begin to react, the rate of the forward reaction is faster than the rate of the reverse reaction. As reactants are consumed and products accumulate, the rate of the forward reaction decreases, and the rate of the reverse reaction increases.

Equilibrium is reached when there are no further changes in the concentrations of reactants and products.

Equilibrium

Eventually, the rates of the forward and reverse reactions become equal; the reactants form products at the same rate that the products form reactants. A reaction reaches **chemical equilibrium** when no further change takes place in the concentrations of the reactants and products, even though the two reactions continue at equal but opposite rates.

At Equilibrium:

- The rate of the forward reaction is equal to the rate of the reverse reaction.
- No further changes occur in the concentrations of reactants and products, even though the two reactions continue at equal but opposite rates.

ENGAGE 13.4

Why do the concentrations of the reactants decrease before equilibrium is reached?

ENGAGE 13.5

How do the rates of the forward and reverse reactions compare once a chemical reaction reaches equilibrium?

Let us look at the process as the reaction of H_2 and I_2 proceeds to equilibrium. Initially, only the reactants H_2 and I_2 are present. Soon, a few molecules of HI are produced by the forward reaction. With more time, additional HI molecules are produced. As the concentration of HI increases, more HI molecules collide and react in the reverse direction. As HI product builds up, the rate of the reverse reaction increases, while the rate of the forward reaction decreases. Eventually, the rates become equal, which means the reaction has reached equilibrium. Even though the concentrations remain constant at equilibrium, the forward and reverse reactions continue to occur. The forward and reverse reactions are usually shown together in a single equation by using a double arrow. A reversible reaction is two opposing reactions that occur at the same time (see **FIGURE 13.5**).

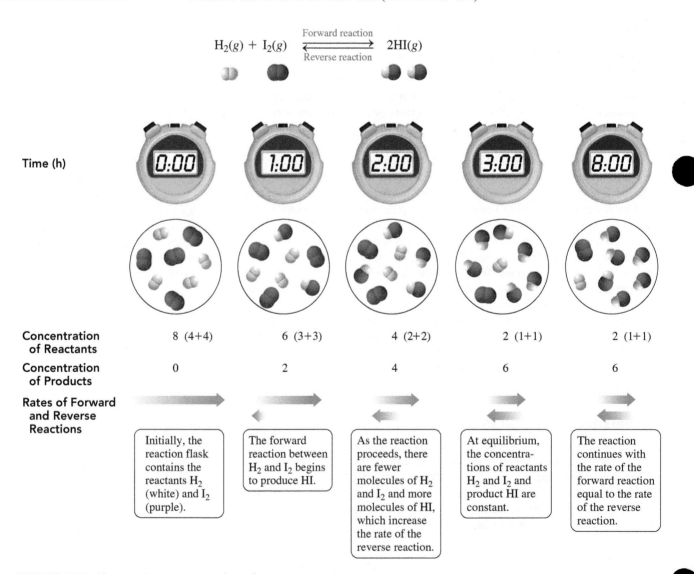

FIGURE 13.5 Changes in reactant and product concentrations

SAMPLE PROBLEM 13.2 Reaction Rates and Equilibrium

TRY IT FIRST

Complete each of the following with *change* or *do not change*, *faster* or *slower*, *equal* or *not equal*:
a. Before equilibrium is reached, the concentrations of the reactants and products
_____.
b. Initially, reactants placed in a container have a _____ rate of reaction than the rate of reaction of the products.
c. At equilibrium, the rate of the forward reaction is _____ to the rate of the reverse reaction.

SOLUTION

a. Before equilibrium is reached, the concentrations of the reactants and products *change*.
b. Initially, reactants placed in a container have a *faster* rate of reaction than the rate of reaction of the products.
c. At equilibrium, the rate of the forward reaction is *equal* to the rate of the reverse reaction.

SELF TEST 13.2

Complete the following statements with *changes* or *does not change*:
a. At equilibrium, the concentration of the reactant _____.
b. The rate of the forward reaction _____ as the reaction proceeds toward equilibrium.

ANSWER

a. At equilibrium, the concentration of the reactant *does not change*.
b. The rate of the forward reaction *changes* as the reaction proceeds toward equilibrium.

PRACTICE PROBLEMS

Try Practice Problems 13.7 to 13.12

We can set up a reaction starting with only reactants or with only products. Let's look at the initial reactions in each, the forward and reverse reactions, and the equilibrium mixture that forms (see **FIGURE 13.6**).

$$2SO_2(g) + O_2(g) \rightleftharpoons 2SO_3(g)$$

If we start with only the reactants SO_2 and O_2 in the container, the forward reaction to form SO_3 takes place until equilibrium is reached. However, if we start with only the product SO_3 in the container, the reverse reaction to form SO_2 and O_2 takes place until equilibrium is reached. In both containers, the equilibrium mixture contains the same concentrations of SO_2, O_2, and SO_3.

Forward Reaction

Reverse Reaction

Initially, the reaction flask contains $SO_2(g)$ and $O_2(g)$.

At equilibrium, the reaction flask contains mostly $SO_3(g)$ and small amounts of $SO_2(g)$ and $O_2(g)$.

Initially, the reaction flask contains $SO_3(g)$.

FIGURE 13.6 Reactions of $SO_2(g)$ and $O_2(g)$ or $SO_3(g)$ reach the same equilibrium mixture

PRACTICE PROBLEMS

13.2 Chemical Equilibrium

13.7 What is meant by the term reversible reaction?

13.8 When does a reversible reaction reach equilibrium?

13.9 Which of the following are at equilibrium?
   a. The rate of the forward reaction is twice as fast as the rate of the reverse reaction.
   b. The concentrations of the reactants and the products do not change.
   c. The rate of the reverse reaction does not change.

13.10 Which of the following are *not* at equilibrium?
   a. The rates of the forward and reverse reactions are equal.
   b. The rate of the forward reaction does not change.
   c. The concentrations of reactants and the products are not constant.

13.11 The following diagrams show the chemical reaction with time:

   A ⇌ B

   If A is blue and B is orange, state whether or not the reaction has reached equilibrium in this time period and explain why.

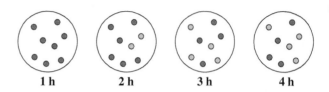

1 h   2 h   3 h   4 h

13.12 The following diagrams show the chemical reaction with time:

   C ⇌ D

   If C is blue and D is yellow, state whether or not the reaction has reached equilibrium in this time period and explain why.

1 h   2 h   3 h   4 h

REVIEW

Using Significant Figures in Calculations (2.3)

Balancing a Chemical Equation (8.2)

Calculating Concentration (12.4)

At equilibrium, the number of people riding up the lift and the number of people skiing down the slope are constant.

CORE CHEMISTRY SKILL

Writing the Equilibrium Expression

13.3 Equilibrium Constants

LEARNING GOAL Calculate the equilibrium constant for a reversible reaction given the concentrations of reactants and products at equilibrium.

At equilibrium, the concentrations of the reactants and products are constant. We can use a ski lift as an analogy. Early in the morning, skiers at the bottom of the mountain begin to ride the ski lift up to the slopes. After the skiers reach the top of the mountain, they ski down. Eventually, the number of people riding up the ski lift becomes equal to the number of people skiing down the mountain. There is no further change in the number of skiers on the slopes; the system is at equilibrium.

Equilibrium Expression

At equilibrium, the concentrations can be used to set up a relationship between the products and the reactants. Suppose we write a general equation for reactants A and B that form products C and D. The small italic letters are the coefficients in the balanced equation.

$$aA + bB \rightleftharpoons cC + dD$$

An **equilibrium expression** for a reversible chemical reaction multiplies the concentrations of the products together and divides by the concentrations of the reactants. Each concentration is raised to a power that is equal to its coefficient in the balanced chemical equation. The square bracket around each substance indicates that the concentration is expressed in moles per liter (M). For our general reaction, this is written as:

$$K_c = \frac{[\text{Products}]}{[\text{Reactants}]} = \frac{[C]^c [D]^d}{[A]^a [B]^b}$$

⟩ Coefficients

Equilibrium expression

We can now describe how to write the equilibrium expression for the reaction of H_2 and I_2 that forms HI. The balanced chemical equation is written with a double arrow between the reactants and the products.

$$H_2(g) + I_2(g) \rightleftharpoons 2HI(g)$$

We show the concentration of the products using brackets in the numerator and the concentrations of the reactants in brackets in the denominator and write any coefficient as an exponent of its concentration (a coefficient 1 is understood).

$$K_c = \frac{[HI]^2 \longleftarrow \text{Coefficient of HI}}{[H_2][I_2]}$$

ENGAGE 13.6

How is a coefficient in a chemical equation indicated in the equilibrium expression?

SAMPLE PROBLEM 13.3 Writing Equilibrium Expressions

TRY IT FIRST

Write the equilibrium expression for the following reaction:

$$2SO_2(g) + O_2(g) \rightleftharpoons 2SO_3(g)$$

SOLUTION

	Given	Need	Connect
ANALYZE THE PROBLEM	equation	equilibrium expression	$\frac{[products]}{[reactants]}$

STEP 1 Write the balanced chemical equation.

$$2SO_2(g) + O_2(g) \rightleftharpoons 2SO_3(g)$$

STEP 2 Write the concentrations of the products as the numerator and the reactants as the denominator.

$$\frac{[Products]}{[Reactants]} \longrightarrow \frac{[SO_3]}{[SO_2][O_2]}$$

STEP 3 Write any coefficient in the equation as an exponent.

$$K_c = \frac{[SO_3]^2}{[SO_2]^2[O_2]}$$

SELF TEST 13.3

Write the equilibrium expression for each of the following reactions:

a. $2NO(g) + O_2(g) \rightleftharpoons 2NO_2(g)$ **b.** $2N_2O_5(g) \rightleftharpoons 4NO_2(g) + O_2(g)$

ANSWER

a. $K_c = \dfrac{[NO_2]^2}{[NO]^2[O_2]}$ **b.** $K_c = \dfrac{[NO_2]^4[O_2]}{[N_2O_5]^2}$

PRACTICE PROBLEMS

Try Practice Problems 13.13 and 13.14

Calculating Equilibrium Constants

The **equilibrium constant, K_c,** is the numerical value obtained by substituting experimentally measured molar concentrations at equilibrium into the equilibrium expression. For example, the equilibrium expression for the reaction of H_2 and I_2 is written

CORE CHEMISTRY SKILL

Calculating an Equilibrium Constant

$$H_2(g) + I_2(g) \rightleftharpoons 2HI(g) \qquad K_c = \frac{[HI]^2}{[H_2][I_2]}$$

In the first experiment, the molar concentrations for the reactants and products at equilibrium are found to be $[H_2] = 0.10$ M, $[I_2] = 0.20$ M, and $[HI] = 1.04$ M. When we substitute these values into the equilibrium expression, we obtain the numerical value of K_c.

In additional experiments 2 and 3, the mixtures have different equilibrium concentrations for the system at equilibrium at the same temperature. However, when these concentrations are used to calculate the equilibrium constant, we obtain the same value of K_c for each (see **TABLE 13.2**). *Thus, a reaction at a specific temperature can have only one value for the equilibrium constant.*

TABLE 13.2 Equilibrium Constant for $H_2(g) + I_2(g) \rightleftharpoons 2HI(g)$ at 427 °C				
Experiment	Concentrations at Equilibrium			Equilibrium Constant
	$[H_2]$	$[I_2]$	$[HI]$	$K_c = \dfrac{[HI]^2}{[H_2][I_2]}$
1	0.10 M	0.20 M	1.04 M	$K_c = \dfrac{[1.04]^2}{[0.10][0.20]} = 54$
2	0.20 M	0.20 M	1.47 M	$K_c = \dfrac{[1.47]^2}{[0.20][0.20]} = 54$
3	0.30 M	0.17 M	1.66 M	$K_c = \dfrac{[1.66]^2}{[0.30][0.17]} = 54$

The units of K_c depend on the specific equation. In this example, the units of $[M]^2/[M]^2$ cancel to give a value of 54. In other equations, the concentration units do not cancel. However, in this text, the numerical value will be given without any units as shown in Sample Problem 13.4.

SAMPLE PROBLEM 13.4   Calculating an Equilibrium Constant

TRY IT FIRST

The decomposition of dinitrogen tetroxide forms nitrogen dioxide.

$$N_2O_4(g) \rightleftharpoons 2NO_2(g)$$

What is the numerical value of K_c at 100 °C if a reaction mixture at equilibrium contains 0.45 M N_2O_4 and 0.31 M NO_2?

SOLUTION

STEP 1   State the given and needed quantities.

	Given	Need	Connect
ANALYZE THE PROBLEM	0.45 M N_2O_4, 0.31 M NO_2	K_c	equilibrium expression
	Equation		
	$N_2O_4(g) \rightleftharpoons 2NO_2(g)$		

STEP 2   Write the equilibrium expression.

$$K_c = \frac{[NO_2]^2}{[N_2O_4]}$$

STEP 3   Substitute equilibrium (molar) concentrations and calculate K_c.

$$K_c = \frac{[0.31]^2}{[0.45]} = 0.21$$

SELF TEST 13.4

a. Calculate the numerical value of K_c if an equilibrium mixture contains 0.040 M NH_3, 0.60 M H_2, and 0.20 M N_2.

$$2NH_3(g) \rightleftharpoons 3H_2(g) + N_2(g)$$

b. Calculate the numerical value of K_c if an equilibrium mixture contains 0.20 M ICl, 0.010 M I_2, and 0.048 M Cl_2.

$$2ICl(g) \rightleftharpoons I_2(g) + Cl_2(g)$$

ANSWER

a. $K_c = 27$ **b.** $K_c = 0.012$

PRACTICE PROBLEMS

Try Practice Problems 13.15 to 13.20

Heterogeneous Equilibrium

Up to now, our examples have been reactions that involve only gases. A reaction in which all the reactants and products are in the same state reaches **homogeneous equilibrium**. When the reactants and products are in two or more states, the equilibrium is termed a **heterogeneous equilibrium**. In the following reaction, solid calcium carbonate reaches heterogeneous equilibrium with solid calcium oxide and carbon dioxide gas (see **FIGURE 13.7**).

$$CaCO_3(s) \rightleftharpoons CaO(s) + CO_2(g)$$

In contrast to gases, the concentrations of pure solids and pure liquids are constant; they do not change. Therefore, pure solids and liquids are not included in the equilibrium expression. For this heterogeneous equilibrium, the K_c expression does not include the concentration of $CaCO_3(s)$ or $CaO(s)$. It is written as $K_c = [CO_2]$.

$$CaCO_3(s) \rightleftharpoons CaO(s) + CO_2(g)$$

$T = 800\ °C$

$T = 800\ °C$

FIGURE 13.7 At equilibrium at constant temperature, the concentration of CO_2 is the same regardless of the amounts of $CaCO_3(s)$ and $CaO(s)$ in the container.

SAMPLE PROBLEM 13.5   Heterogeneous Equilibrium Expression

TRY IT FIRST

Write the equilibrium expression for the following reaction at equilibrium:

$$4HCl(g) + O_2(g) \rightleftharpoons 2H_2O(l) + 2Cl_2(g)$$

SOLUTION

STEP 1   Write the balanced chemical equation.

$$4HCl(g) + O_2(g) \rightleftharpoons 2H_2O(l) + 2Cl_2(g)$$

STEP 2   Write the concentrations of the products as the numerator and the reactants as the denominator. In this heterogeneous reaction, the concentration of the H_2O, which is a pure liquid, is not included in the equilibrium expression.

$$\frac{[Products]}{[Reactants]} \longrightarrow \frac{[Cl_2]}{[HCl][O_2]}$$

STEP 3   Write any coefficient in the equation as an exponent.

$$K_c = \frac{[Cl_2]^2}{[HCl]^4[O_2]}$$

SELF TEST 13.5

a. Solid iron(II) oxide and carbon monoxide gas react to produce solid iron and carbon dioxide gas. Write the balanced chemical equation and the equilibrium expression for this reaction at equilibrium.
b. When heated, solid sodium hydrogen carbonate decomposes to form solid sodium carbonate, gaseous water, and carbon dioxide gas. Write the balanced chemical equation and the equilibrium expression for this reaction at equilibrium.

ANSWER

a. $FeO(s) + CO(g) \rightleftharpoons Fe(s) + CO_2(g)$ $K_c = \dfrac{[CO_2]}{[CO]}$

b. $2NaHCO_3(s) \rightleftharpoons Na_2CO_3(s) + H_2O(g) + CO_2(g)$ $K_c = [H_2O][CO_2]$

ENGAGE 13.7

Why are the concentrations of $CaO(s)$ and $CaCO_3(s)$ not included in the equilibrium expression for K_c for the decomposition of $CaCO_3(s)$?

PRACTICE PROBLEMS

Try Practice Problems 13.21 to 13.26

PRACTICE PROBLEMS

13.3 Equilibrium Constants

13.13 Write the equilibrium expression for each of the following reactions:

   a. $CH_4(g) + 2H_2S(g) \rightleftharpoons CS_2(g) + 4H_2(g)$
   b. $2NO(g) \rightleftharpoons N_2(g) + O_2(g)$
   c. $2SO_3(g) + CO_2(g) \rightleftharpoons CS_2(g) + 4O_2(g)$
   d. $CH_4(g) + H_2O(g) \rightleftharpoons 3H_2(g) + CO(g)$

13.14 Write the equilibrium expression for each of the following reactions:

   a. $2HBr(g) \rightleftharpoons H_2(g) + Br_2(g)$
   b. $2BrNO(g) \rightleftharpoons Br_2(g) + 2NO(g)$
   c. $CH_4(g) + Cl_2(g) \rightleftharpoons CH_3Cl(g) + HCl(g)$
   d. $Br_2(g) + Cl_2(g) \rightleftharpoons 2BrCl(g)$

13.15 Write the equilibrium expression for the reaction in the diagram, and calculate the numerical value of K_c. In the diagram, X atoms are orange and Y atoms are blue; the volume of the container is 1.0 L.

$$X_2(g) + Y_2(g) \rightleftharpoons 2XY(g)$$

Equilibrium
mixture

13.16 Write the equilibrium expression for the reaction in the diagram, and calculate the numerical value of K_c. In the diagram, A atoms are red and B atoms are green; the volume of the container is 1.0 L.

$$2AB(g) \rightleftharpoons A_2(g) + B_2(g)$$

Equilibrium
mixture

13.17 What is the numerical value of K_c for the following reaction if the equilibrium mixture contains 0.030 M N_2O_4 and 0.21 M NO_2?

$$N_2O_4(g) \rightleftharpoons 2NO_2(g)$$

13.18 What is the numerical value of K_c for the following reaction if the equilibrium mixture contains 0.30 M CO_2, 0.033 M H_2, 0.20 M CO, and 0.30 M H_2O?

$$CO_2(g) + H_2(g) \rightleftharpoons CO(g) + H_2O(g)$$

13.19 What is the numerical value of K_c for the following reaction if the equilibrium mixture contains 0.51 M CO, 0.30 M H_2, 1.8 M CH_4, and 2.0 M H_2O?

$$CO(g) + 3H_2(g) \rightleftharpoons CH_4(g) + H_2O(g)$$

13.20 What is the numerical value of K_c for the following reaction if the equilibrium mixture contains 0.44 M N_2, 0.40 M H_2, and 2.2 M NH_3?

$$N_2(g) + 3H_2(g) \rightleftharpoons 2NH_3(g)$$

13.21 Identify each of the following as a homogeneous or heterogeneous equilibrium:

   a. $2O_3(g) \rightleftharpoons 3O_2(g)$
   b. $2NaHCO_3(s) \rightleftharpoons Na_2CO_3(s) + CO_2(g) + H_2O(g)$
   c. $C_6H_6(g) + 3H_2(g) \rightleftharpoons C_6H_{12}(g)$
   d. $4HCl(g) + Si(s) \rightleftharpoons SiCl_4(l) + 2H_2(g)$

13.22 Identify each of the following as a homogeneous or heterogeneous equilibrium:

   a. $CO(g) + H_2(g) \rightleftharpoons C(s) + H_2O(g)$
   b. $NH_4Cl(s) \rightleftharpoons NH_3(g) + HCl(g)$
   c. $CS_2(g) + 4H_2(g) \rightleftharpoons CH_4(g) + 2H_2S(g)$
   d. $Ti(s) + 2Cl_2(g) \rightleftharpoons TiCl_4(g)$

13.23 Write the equilibrium expression for each of the reactions in problem 13.21.

13.24 Write the equilibrium expression for each of the reactions in problem 13.22.

13.25 What is the numerical value of K_c for the following reaction if the equilibrium mixture at 750 °C contains 0.20 M CO and 0.052 M CO_2?

$$FeO(s) + CO(g) \rightleftharpoons Fe(s) + CO_2(g)$$

13.26 What is the numerical value of K_c for the following reaction if the equilibrium mixture at 800 °C contains 0.030 M CO_2?

$$CaCO_3(s) \rightleftharpoons CaO(s) + CO_2(g)$$

REVIEW

Solving Equations (1.4)

13.4 Using Equilibrium Constants

LEARNING GOAL Use an equilibrium constant to predict the extent of reaction and to calculate equilibrium concentrations.

The values of K_c can be large or small. The size of the equilibrium constant depends on whether equilibrium is reached with more products than reactants, or more reactants than products. However, the size of an equilibrium constant does not affect how fast equilibrium is reached.

Equilibrium with a Large K_c

When a reaction has a large equilibrium constant, it means that the forward reaction produced a large amount of products when equilibrium was reached. Then the equilibrium mixture contains mostly products, which makes the concentrations of the products in the numerator higher than the concentrations of the reactants in the denominator. Thus at equilibrium, this reaction has a large K_c. Consider the reaction of SO_2 and O_2, which has a large K_c. At equilibrium, the reaction mixture contains mostly product and few reactants (see **FIGURE 13.8**).

ENGAGE 13.8

Why does an equilibrium mixture containing mostly products have a large K_c?

$$2SO_2(g) + O_2(g) \rightleftharpoons 2SO_3(g)$$

$$K_c = \frac{[SO_3]^2}{[SO_2]^2[O_2]} \qquad \frac{\text{Mostly product}}{\text{Few reactants}} = 3.4 \times 10^2$$

$$2SO_2(g) + O_2(g) \rightleftharpoons 2SO_3(g)$$

FIGURE 13.8 In the reaction of $SO_2(g)$ and $O_2(g)$, the equilibrium mixture contains mostly product $SO_3(g)$, which results in a large K_c.

Equilibrium with a Small K_c

When a reaction has a small equilibrium constant, the equilibrium mixture contains a high concentration of reactants and a low concentration of products. Then the equilibrium expression has a small number in the numerator and a large number in the denominator. Thus at equilibrium, this reaction has a small K_c. Consider the reaction for the formation of $NO(g)$ from $N_2(g)$ and $O_2(g)$, which has a small K_c (see **FIGURE 13.9**).

$$N_2(g) + O_2(g) \rightleftharpoons 2NO(g)$$

$$K_c = \frac{[NO]^2}{[N_2][O_2]} \qquad \frac{\text{Few products}}{\text{Mostly reactants}} = 2 \times 10^{-9}$$

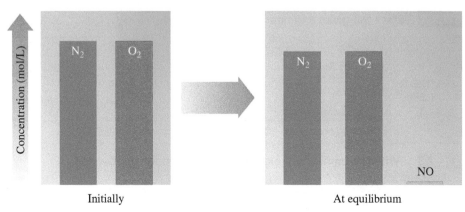

$$N_2(g) + O_2(g) \rightleftharpoons 2NO(g)$$

FIGURE 13.9 The equilibrium mixture contains a very small amount of the product NO and a large amount of the reactants N_2 and O_2, which results in a small K_c.

A few reactions have equilibrium constants close to 1, which means they have about equal concentrations of reactants and products. Moderate amounts of reactants have been converted to products upon reaching equilibrium (see **FIGURE 13.10**).

| Small K_c | $K_c \approx 1$ | Large K_c |

Mostly reactants ← → Mostly products

| Products $<<$ Reactants
Little reaction takes place | Reactants $\approx$ Products
Moderate reaction | Products $>>$ Reactants
Reaction essentially
complete |

FIGURE 13.10 At equilibrium, a reaction with a large K_c contains mostly products, whereas a reaction with a small K_c contains mostly reactants.

TABLE 13.3 lists some equilibrium constants and the extent of their reaction.

TABLE 13.3 Examples of Reactions with Large and Small K_c Values

Reactants	Products	K_c	Equilibrium Mixture Contains
$2CO(g) + O_2(g) \rightleftharpoons$	$2CO_2(g)$	2×10^{11}	Mostly product
$2H_2(g) + S_2(g) \rightleftharpoons$	$2H_2S(g)$	1.1×10^7	Mostly product
$N_2(g) + 3H_2(g) \rightleftharpoons$	$2NH_3(g)$	1.6×10^2	Mostly product
$PCl_5(g) \rightleftharpoons$	$PCl_3(g) + Cl_2(g)$	1.2×10^{-2}	Mostly reactant
$N_2(g) + O_2(g) \rightleftharpoons$	$2NO(g)$	2×10^{-9}	Mostly reactants

Calculating Concentrations at Equilibrium

When we know the numerical value of the equilibrium constant and all the equilibrium concentrations except one, we can calculate the unknown concentration as shown in Sample Problem 13.6.

SAMPLE PROBLEM 13.6 Calculating Concentration Using an Equilibrium Constant

TRY IT FIRST

For the reaction of carbon dioxide and hydrogen, the equilibrium concentrations are 0.25 M CO_2, 0.80 M H_2, and 0.50 M H_2O. What is the equilibrium concentration of $CO(g)$?

$$CO_2(g) + H_2(g) \rightleftharpoons CO(g) + H_2O(g) \quad K_c = 0.11$$

SOLUTION

STEP 1 State the given and needed quantities.

ANALYZE THE PROBLEM	Given	Need	Connect
	0.25 M CO_2, 0.80 M H_2, 0.50 M H_2O	[CO]	equilibrium expression, value of K_c
	Equation		
	$CO_2(g) + H_2(g) \rightleftharpoons CO(g) + H_2O(g) \quad K_c = 0.11$		

STEP 2 Write the equilibrium expression and solve for the needed concentration.

$$K_c = \frac{[CO][H_2O]}{[CO_2][H_2]}$$

We rearrange the equilibrium expression to solve for the unknown [CO] as follows:

Multiply both sides by $[CO_2][H_2]$.

$$K_c \times [CO_2][H_2] = \frac{[CO][H_2O]}{[\cancel{CO_2}][\cancel{H_2}]} \times [\cancel{CO_2}][\cancel{H_2}]$$

$$K_c \times [CO_2][H_2] = [CO][H_2O]$$

Divide both sides by $[H_2O]$.

$$K_c \times \frac{[CO_2][H_2]}{[H_2O]} = \frac{[CO][\cancel{H_2O}]}{[\cancel{H_2O}]}$$

$$[CO] = K_c \times \frac{[CO_2][H_2]}{[H_2O]}$$

STEP 3 **Substitute the equilibrium (molar) concentrations and calculate the needed concentration.**

$$[CO] = K_c \times \frac{[CO_2][H_2]}{[H_2O]} = 0.11 \times \frac{[0.25][0.80]}{[0.50]} = 0.044 \ M$$

SELF TEST 13.6

a. When ethene (C_2H_4) reacts with water vapor, ethanol (C_2H_6O) is produced. If an equilibrium mixture contains 0.020 M C_2H_4 and 0.015 M H_2O, what is the equilibrium concentration of C_2H_6O? At 327 °C, the K_c is 9.0×10^3.

$$C_2H_4(g) + H_2O(g) \rightleftharpoons C_2H_6O(g)$$

b. $NO(g)$ and $Br_2(g)$ react to form $NOBr(g)$.

$$2NO(g) + Br_2(g) \rightleftharpoons 2NOBr(g)$$

If an equilibrium mixture contains 0.70 M NO and 0.060 M NOBr, what is the equilibrium concentration of Br_2? At 1000 K, the K_c is 1.3×10^{-2}.

ANSWER

a. $[C_2H_6O] = 2.7 \ M$ **b.** $[Br_2] = 0.57 \ M$

PRACTICE PROBLEMS

Try Practice Problems 13.31 to 13.34

PRACTICE PROBLEMS

13.4 Using Equilibrium Constants

13.27 If the K_c for this reaction is 4, which of the following diagrams represents the molecules in an equilibrium mixture? In the diagrams, X atoms are orange and Y atoms are blue.

$$X_2(g) + Y_2(g) \rightleftharpoons 2XY(g)$$

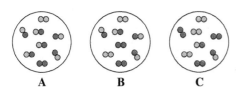

     A   B   C

13.28 If the K_c for this reaction is 2, which of the following diagrams represents the molecules in an equilibrium mixture? In the diagrams, A atoms are orange and B atoms are green.

$$2AB(g) \rightleftharpoons A_2(g) + B_2(g)$$

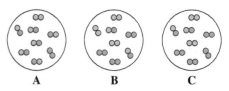

     A   B   C

13.29 Indicate whether each of the following equilibrium mixtures contains mostly products or mostly reactants:
a. $Cl_2(g) + NO(g) \rightleftharpoons 2NOCl(g)$ $K_c = 3.7 \times 10^8$
b. $2H_2(g) + S_2(g) \rightleftharpoons 2H_2S(g)$ $K_c = 1.1 \times 10^7$
c. $3O_2(g) \rightleftharpoons 2O_3(g)$ $K_c = 1.7 \times 10^{-56}$

13.30 Indicate whether each of the following equilibrium mixtures contains mostly products or mostly reactants:
a. $CO(g) + Cl_2(g) \rightleftharpoons COCl_2(g)$ $K_c = 5.0 \times 10^{-9}$
b. $2HF(g) \rightleftharpoons H_2(g) + F_2(g)$ $K_c = 1.0 \times 10^{-95}$
c. $2NO(g) + O_2(g) \rightleftharpoons 2NO_2(g)$ $K_c = 6.0 \times 10^{13}$

13.31 The numerical value of the equilibrium constant, K_c, for this reaction is 54.

$$H_2(g) + I_2(g) \rightleftharpoons 2HI(g)$$

If the equilibrium mixture contains 0.015 M I_2 and 0.030 M HI, what is the molar concentration of H_2?

13.32 The numerical value of the equilibrium constant, K_c, for the following reaction is 4.6×10^{-3}. If the equilibrium mixture contains 0.050 M NO_2, what is the molar concentration of N_2O_4?

$$N_2O_4(g) \rightleftharpoons 2NO_2(g)$$

13.33 The numerical value of the equilibrium constant, K_c, for the following reaction is 2.0. If the equilibrium mixture contains 2.0 M NO and 1.0 M Br_2, what is the molar concentration of NOBr?

$$2NOBr(g) \rightleftharpoons 2NO(g) + Br_2(g)$$

13.34 The numerical value of the equilibrium constant, K_c, for the following reaction is 1.7×10^2. If the equilibrium mixture contains 0.18 M H_2 and 0.020 M N_2, what is the molar concentration of NH_3?

$$3H_2(g) + N_2(g) \rightleftharpoons 2NH_3(g)$$

13.5 Changing Equilibrium Conditions: Le Châtelier's Principle

LEARNING GOAL Use Le Châtelier's principle to describe the changes made in equilibrium concentrations when reaction conditions change.

We have seen that when a reaction reaches equilibrium, the rates of the forward and reverse reactions are equal and the concentrations remain constant. Now we will look at what happens to a system at equilibrium when changes occur in reaction conditions, such as changes in concentration, volume, or temperature.

Le Châtelier's Principle

When we alter any of the conditions of a system at equilibrium, the rates of the forward and reverse reactions may no longer be equal. We say that a *stress* is placed on the equilibrium. Then the system responds by changing the rate of the forward or reverse reaction in the direction that relieves that stress to reestablish equilibrium. We can use **Le Châtelier's principle**, which states that when a system at equilibrium is disturbed, the system will shift in the direction that will reduce that stress.

> ### Le Châtelier's Principle
> When a stress (change in conditions) is placed on a reaction at equilibrium, the equilibrium will shift in the direction that relieves the stress.

Suppose we have two water tanks connected by a pipe. When the water levels in the tanks are equal, water flows in the forward direction from Tank A to Tank B at the same rate as it flows in the reverse direction from Tank B to Tank A. Suppose we add more water

At equilibrium, the water levels are equal.

Tank A → Tank B

Water added to Tank A increases the rate of flow in the forward direction.

Tank A → Tank B

Equilibrium is reached again when the water levels are equal.

Tank A → Tank B

to Tank A. With a higher level of water in Tank A, more water flows in the forward direction from Tank A to Tank B than in the reverse direction from Tank B to Tank A, which is shown with a longer arrow. Eventually, equilibrium is reached as the levels in both tanks become equal, but higher than before. Then the rate of water flows equally between Tank A and Tank B.

Effect of Concentration Changes on Equilibrium

We will now use the reaction of H_2 and I_2 to illustrate how a change in concentration disturbs the equilibrium and how the system responds to that stress.

$$H_2(g) + I_2(g) \rightleftharpoons 2HI(g)$$

Suppose that more of the reactant H_2 is added to the equilibrium mixture, which increases the concentration of H_2. Because K_c cannot change for a reaction at a given temperature, adding more H_2 places a stress on the system (see **FIGURE 13.11**). Then the system

CORE CHEMISTRY SKILL
Using Le Châtelier's Principle

ENGAGE 13.11

If more product HI is added, will the equilibrium shift in the direction of the product or reactants?

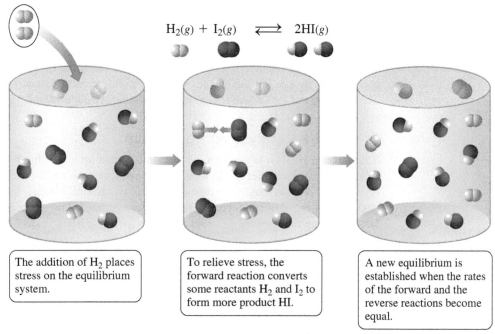

$$H_2(g) + I_2(g) \rightleftharpoons 2HI(g)$$

The addition of H_2 places stress on the equilibrium system.

To relieve stress, the forward reaction converts some reactants H_2 and I_2 to form more product HI.

A new equilibrium is established when the rates of the forward and the reverse reactions become equal.

FIGURE 13.11 Effect of adding reactant to a system at equilibrium

relieves this stress by increasing the rate of the forward reaction, which is indicated by the direction of the large arrow. Thus, more product is formed until the system is again at equilibrium. According to Le Châtelier's principle, adding more reactant causes the system to *shift* in the direction of the product until equilibrium is reestablished.

Add H_2

$$H_2(g) + I_2(g) \rightleftharpoons 2HI(g)$$

Suppose now that some H_2 is removed from the reaction mixture at equilibrium, which lowers the concentration of H_2 and slows the rate of the forward reaction. Using Le Châtelier's principle, we know that when some of the reactants are removed, the system will *shift* in the direction of the reactants until equilibrium is reestablished.

Remove H_2

$$H_2(g) + I_2(g) \rightleftharpoons 2HI(g)$$

The concentrations of the products of an equilibrium mixture can also increase or decrease. For example, if more HI is added, there is an increase in the rate of the reaction in the reverse direction, which converts some of the product to reactants. The concentration of the products decreases and the concentration of the reactants increases until equilibrium is reestablished. Using Le Châtelier's principle, we see that the addition of a product causes the system to *shift* in the direction of the reactants.

Add HI

$$H_2(g) + I_2(g) \rightleftharpoons 2HI(g)$$

In another example, some HI is removed from an equilibrium mixture, which decreases the concentration of the product. Then there is a *shift* in the direction of the product to reestablish equilibrium.

Remove HI

$$H_2(g) + I_2(g) \rightleftharpoons 2HI(g)$$

In summary, Le Châtelier's principle indicates that a stress caused by adding a substance at equilibrium is relieved when the equilibrium system shifts the reaction away from that substance. Adding more reactant causes an increase in the forward reaction to products. Adding more product causes an increase in the reverse reaction to reactants. When some of a substance is removed, the equilibrium system shifts in the direction of that substance. These features of Le Châtelier's principle are summarized in **TABLE 13.4**.

Effect of a Catalyst on Equilibrium

Sometimes a catalyst is added to a reaction to speed up a reaction by lowering the activation energy. As a result, the rates of both the forward and reverse reactions increase. The time required to reach equilibrium is shorter, but the same ratios of products and reactants are attained. Therefore, a catalyst speeds up the forward and reverse reactions, but it has no effect on the equilibrium mixture.

Effect of Volume Change on Equilibrium

If there is a change in the volume of a gas mixture at equilibrium, there will also be a change in the concentrations of those gases. Decreasing the volume will increase the concentration of gases, whereas increasing the volume will decrease their concentration. Then the system responds to reestablish equilibrium.

Let's look at the effect of decreasing the volume of the equilibrium mixture of the following reaction:

$$2CO(g) + O_2(g) \rightleftharpoons 2CO_2(g)$$

If we decrease the volume, all the concentrations increase. According to Le Châtelier's principle, the increase in concentration is relieved when the system shifts in the direction of the smaller number of moles.

Decrease V

$$2CO(g) + O_2(g) \rightleftharpoons 2CO_2(g)$$
3 mol of gas   2 mol of gas

On the other hand, when the volume of the equilibrium gas mixture increases, the concentrations of all the gases decrease. Then the system shifts in the direction of the greater number of moles to reestablish equilibrium.

Increase V

$$2CO(g) + O_2(g) \rightleftharpoons 2CO_2(g)$$
3 mol of gas   2 mol of gas

TABLE 13.4 Effect of Concentration Changes on Equilibrium
$H_2(g) + I_2(g) \rightleftharpoons 2HI(g)$

Stress	Shift in the Direction of
Increasing [H₂]	Product
Decreasing [H₂]	Reactants
Increasing [I₂]	Product
Decreasing [I₂]	Reactants
Increasing [HI]	Reactants
Decreasing [HI]	Product

$$2CO(g) + O_2(g) \rightleftharpoons 2CO_2(g)$$

Decreasing the volume of the container shifts the equilibrium in the direction of fewer moles of gas.

Increasing the volume of the container shifts the equilibrium in the direction of more moles of gas.

When a reaction has the same number of moles of reactants as products, a volume change does not affect the equilibrium mixture because the concentrations of the reactants and products change in the same way.

$$H_2(g) + I_2(g) \rightleftharpoons 2HI(g)$$
2 mol of gas   2 mol of gas

Chemistry Link to Health

Oxygen–Hemoglobin Equilibrium and Hypoxia

The transport of oxygen in the blood involves an equilibrium between hemoglobin (Hb), oxygen, and oxyhemoglobin (HbO_2). Hemoglobin, present in our red blood cells, is an iron-containing protein that carries oxygen from our lungs to the muscles and tissues of our body. Each hemoglobin molecule contains four iron-containing heme groups. Because each heme group can attach to one O_2 in the lungs, one hemoglobin can carry as many as four O_2 molecules. The transport of oxygen involves an equilibrium reaction between the reactants Hb and O_2 and the product HbO_2.

$$Hb(aq) + O_2(g) \rightleftharpoons HbO_2(aq)$$

The equilibrium expression can be written

$$K_c = \frac{[HbO_2]}{[Hb][O_2]}$$

According to Le Châtelier's principle, we see that a decrease in oxygen will shift the equilibrium in the direction of the reactants. Such a shift depletes the concentration of HbO_2 and causes the hypoxia.

Remove O_2
$$Hb(aq) + O_2(g) \rightleftharpoons HbO_2(aq)$$

Immediate treatment of altitude sickness includes hydration, rest, and if necessary, descending to a lower altitude. The body's adaptation to lowered oxygen levels requires about 10 days. During this time, the bone marrow increases red blood cell production, providing more red blood cells and more hemoglobin. A person living at a high altitude can have 50% more red blood cells than

In the lungs,
Hb binds O_2.

Red cells carry O_2 as HbO_2 from the lungs to the tissues and muscles.

In the tissues and muscles, O_2 is released from HbO_2.

In the alveoli of the lungs where the O_2 concentration is high, the forward reaction is faster, which shifts the equilibrium in the direction of the product HbO_2. As a result, O_2 binds to hemoglobin in the lungs.

$$Hb(aq) + O_2(g) \longrightarrow HbO_2(aq)$$

In the tissues and the muscles where the O_2 concentration is low, the reverse reaction is faster, which shifts the equilibrium in the direction of the reactant Hb and releases the oxygen from the hemoglobin.

$$Hb(aq) + O_2(g) \longleftarrow HbO_2(aq)$$

At normal atmospheric pressure, oxygen diffuses into the blood because the partial pressure of oxygen in the alveoli is higher than that in the blood. At an altitude above 8000 ft, a decrease in the atmospheric pressure results in a significant reduction in the partial pressure of oxygen, which means that less oxygen is available for the blood and body tissues. At an altitude of 18 000 ft, a person will obtain 29% less oxygen. When oxygen levels are lowered, a person may experience *hypoxia*, characterized by increased respiratory rate, headache, decreased mental acuteness, fatigue, decreased physical coordination, nausea, vomiting, and cyanosis. A similar problem occurs in persons with a history of lung disease that impairs gas diffusion in the alveoli or in persons with a reduced number of red blood cells, such as in smokers.

someone at sea level. This increase in hemoglobin causes a shift in the equilibrium back in the direction of HbO_2 product. Eventually, the higher concentration of HbO_2 will provide more oxygen to the tissues and the symptoms of hypoxia will lessen.

Add Hb
$$Hb(aq) + O_2(g) \rightleftharpoons HbO_2(aq)$$

For some who climb high mountains, it is important to stop and acclimatize for several days at increasing altitudes. At very high altitudes, it may be necessary to use an oxygen tank.

Hypoxia may occur at high altitudes where the oxygen concentration is lower.

Effect of a Change in Temperature on Equilibrium

We can think of heat as a reactant or a product in a reaction. For example, in the equation for an endothermic reaction, heat is written on the reactant side. When the temperature of an endothermic reaction increases, the system responds by shifting in the direction of the products to remove heat.

Increase T

$$N_2(g) + O_2(g) + heat \rightleftharpoons 2NO(g)$$

If the temperature is decreased for an endothermic reaction, there is a decrease in heat. Then the system shifts in the direction of the reactants to add heat.

Decrease T

$$N_2(g) + O_2(g) + heat \rightleftharpoons 2NO(g)$$

In the equation for an exothermic reaction, heat is written on the product side. When the temperature of an exothermic reaction increases, the system responds by shifting in the direction of the reactants to remove heat.

Increase T

$$2SO_2(g) + O_2(g) \rightleftharpoons 2SO_3(g) + heat$$

If the temperature is decreased for an exothermic reaction, there is a decrease in heat. Then the system shifts in the direction of the products to add heat.

Decrease T

$$2SO_2(g) + O_2(g) \rightleftharpoons 2SO_3(g) + heat$$

ENGAGE 13.12

If you want to decrease the amount of product in an exothermic reaction, would you increase or decrease the temperature?

TABLE 13.5 summarizes the ways we can use Le Châtelier's principle to determine the shift in equilibrium that relieves a stress caused by the change in a condition.

TABLE 13.5 Effects of Condition Changes on Equilibrium

Condition	Change (Stress)	Shift in the Direction of
Concentration	Adding a reactant	Products (forward reaction)
	Removing a reactant	Reactants (reverse reaction)
	Adding a product	Reactants (reverse reaction)
	Removing a product	Products (forward reaction)
Volume (container)	Decreasing the volume	Fewer moles of gas
	Increasing the volume	More moles of gas
Temperature	**Endothermic Reaction**	
	Increasing the temperature	Products (forward reaction)
	Decreasing the temperature	Reactants (reverse reaction)
	Exothermic Reaction	
	Increasing the temperature	Reactants (reverse reaction)
	Decreasing the temperature	Products (forward reaction)
Catalyst	Increasing the rates equally	No effect

SAMPLE PROBLEM 13.7 Using Le Châtelier's Principle

TRY IT FIRST

Methanol, CH_4O, is finding use as a fuel additive. Describe the effect of each of the following changes on the equilibrium mixture for the combustion of methanol:

$$2CH_4O(g) + 3O_2(g) \rightleftharpoons 2CO_2(g) + 4H_2O(g) + 1450 \text{ kJ}$$

a. adding more $CO_2(g)$
b. adding more $O_2(g)$
c. increasing the volume of the container
d. increasing the temperature
e. adding a catalyst

SOLUTION

a. When the concentration of the product CO_2 increases, the equilibrium shifts in the direction of the reactants.
b. When the concentration of the reactant O_2 increases, the equilibrium shifts in the direction of the products.
c. When the volume increases, the equilibrium shifts in the direction of the greater number of moles of gas, which is the products.
d. When the temperature is increased for an exothermic reaction, the equilibrium shifts in the direction of the reactants.
e. When a catalyst is added, there is no change in the equilibrium mixture.

SELF TEST 13.7

Describe the effect of each of the following changes on the equilibrium mixture for the following reaction:

$$2HF(g) + Cl_2(g) + 357 \text{ kJ} \rightleftharpoons 2HCl(g) + F_2(g)$$

a. adding more $Cl_2(g)$
b. decreasing the volume of the container
c. decreasing the temperature

ANSWER

a. When the concentration of the reactant Cl_2 increases, the equilibrium shifts in the direction of the products.
b. There is no change in equilibrium mixture because the moles of reactants are equal to the moles of products.
c. When the temperature for an endothermic reaction decreases, the equilibrium shifts in the direction of the reactants.

PRACTICE PROBLEMS

Try Practice Problems 13.35 to 13.42

Chemistry Link to Health

Homeostasis: Regulation of Body Temperature

In a physiological system of equilibrium called *homeostasis*, changes in our environment are balanced by changes in our bodies. It is crucial to our survival that we balance heat gain with heat loss. If we do not lose enough heat, our body temperature rises. At high temperatures, the body can no longer regulate our metabolic reactions. If we lose too much heat, body temperature drops. At low temperatures, essential functions proceed too slowly.

The skin plays an important role in the maintenance of body temperature. When the outside temperature rises, receptors in the skin send signals to the brain. The temperature-regulating part of the brain stimulates the sweat glands to produce perspiration. As perspiration evaporates from the skin, heat is removed and the body temperature is decreased.

In cold temperatures, epinephrine is released, causing an increase in metabolic rate, which increases the production of heat. Receptors on the skin signal the brain to constrict the blood vessels. Less blood flows through the skin, and heat is conserved. The production of perspiration stops, thereby lessening the heat lost by evaporation.

Blood vessels dilate
• sweat production increases
• sweat evaporates
• skin cools

Blood vessels constrict and epinephrine is released
• metabolic activity increases
• muscular activity increases
• shivering occurs
• sweat production stops

PRACTICE PROBLEMS

13.5 Changing Equilibrium Conditions: Le Châtelier's Principle

13.35 In the lower atmosphere, oxygen is converted to ozone (O_3) by the energy provided from lightning.

$$3O_2(g) + heat \rightleftharpoons 2O_3(g)$$

For each of the following changes at equilibrium, indicate whether the equilibrium shifts in the direction of product, reactants, or does not change:
a. adding more $O_2(g)$
b. adding more $O_3(g)$
c. increasing the temperature
d. increasing the volume of the container
e. adding a catalyst

13.36 Ammonia is produced by reacting nitrogen gas and hydrogen gas.

$$N_2(g) + 3H_2(g) \rightleftharpoons 2NH_3(g) + 92 \text{ kJ}$$

For each of the following changes at equilibrium, indicate whether the equilibrium shifts in the direction of product, reactants, or does not change:
a. removing some $N_2(g)$
b. decreasing the temperature
c. adding more $NH_3(g)$
d. adding more $H_2(g)$
e. increasing the volume of the container

13.37 Hydrogen chloride can be made by reacting hydrogen gas and chlorine gas.

$$H_2(g) + Cl_2(g) + heat \rightleftharpoons 2HCl(g)$$

For each of the following changes at equilibrium, indicate whether the equilibrium shifts in the direction of product, reactants, or does not change:
a. adding more $H_2(g)$
b. increasing the temperature
c. removing some $HCl(g)$
d. adding a catalyst
e. removing some $Cl_2(g)$

13.38 When heated, carbon monoxide reacts with water to produce carbon dioxide and hydrogen.

$$CO(g) + H_2O(g) \rightleftharpoons CO_2(g) + H_2(g) + heat$$

For each of the following changes at equilibrium, indicate whether the equilibrium shifts in the direction of products, reactants, or does not change:

a. decreasing the temperature
b. adding more $H_2(g)$
c. removing $CO_2(g)$ as its forms
d. adding more $H_2O(g)$
e. decreasing the volume of the container

Applications

Use the following equation for the equilibrium of hemoglobin in the blood to answer problems 13.39 to 13.42:

$$Hb(aq) + O_2(g) \rightleftharpoons HbO_2(aq)$$

13.39 For the hemoglobin equilibrium, indicate if each of the following changes shifts the equilibrium in the direction of the product, the reactants, or does not change:
a. increasing $[O_2]$
b. increasing $[HbO_2]$
c. decreasing $[Hb]$

13.40 For the hemoglobin equilibrium, indicate if each of the following changes shifts the equilibrium in the direction of the product, the reactants, or does not change:
a. decreasing $[O_2]$
b. increasing $[Hb]$
c. decreasing $[HbO_2]$

13.41 Athletes who train at high altitudes initially experience hypoxia, which causes their body to produce more hemoglobin. When an athlete first arrives at high altitude,
a. What is the stress on the hemoglobin equilibrium?
b. In what direction does the hemoglobin equilibrium shift?

13.42 A person who has been a smoker and has a low oxygen blood saturation uses an oxygen tank for supplemental oxygen. When oxygen is first supplied,
a. What is the stress on the hemoglobin equilibrium?
b. In what direction does the hemoglobin equilibrium shift?

13.6 Equilibrium in Saturated Solutions

LEARNING GOAL Write the solubility product expression for a slightly soluble ionic compound and calculate the K_{sp}; use K_{sp} to determine the solubility.

CORE CHEMISTRY SKILL

Writing the Solubility Product Expression

Until now, we have looked primarily at equilibrium systems that involve gases. However, there are also equilibrium systems that involve aqueous saturated solutions between solid solutes of slightly soluble ionic compounds and their ions. Everyday examples of solubility equilibrium in solution are found in the slightly soluble ionic compounds that are found in bone and kidney stones. Bone is composed of calcium phosphate, $Ca_3(PO_4)_2$, which produces ions during bone loss. Kidney stones are composed of compounds such as calcium oxalate, CaC_2O_4.

Solubility Product Expression

In a saturated solution, a solid slightly soluble ionic compound is in equilibrium with its ions. As long as the temperature remains constant, the concentration of the ions in the saturated solution is constant. Let us look at the solubility equilibrium equation for CaC_2O_4, which is written with the solid solute on the left and the ions in solution on the right.

$$CaC_2O_4(s) \rightleftharpoons Ca^{2+}(aq) + C_2O_4^{2-}(aq)$$

The *solubility* of a substance is the quantity that dissolves to form a saturated solution. We represent the solubility in a saturated aqueous solution of solid CaC_2O_4 by the **solubility product expression**, which is the product of the ion concentrations. As in other heterogeneous equilibria, the concentration of the solid CaC_2O_4 is constant and is not included in the solubility product expression.

$$K_{sp} = [Ca^{2+}][C_2O_4^{2-}]$$

In another example, we look at the equilibrium of solid calcium phosphate and its ions Ca^{2+} and PO_4^{3-}.

$$Ca_3(PO_4)_2(s) \rightleftharpoons 3Ca^{2+}(aq) + 2PO_4^{3-}(aq)$$

As with other equilibrium expressions, the molar concentration of each product ion is raised to a power that is equal to its coefficient in the balanced equilibrium equation. For this equilibrium equation, the solubility product expression consists of $[Ca^{2+}]$ raised to the power of 3 and $[PO_4^{3-}]$ raised to the power of 2.

$$K_{sp} = [Ca^{2+}]^3[PO_4^{3-}]^2$$

Solubility Product Constant

The numerical value of the solubility product expression is the **solubility product constant, K_{sp}**. In this text, solubility will be expressed as the molar solubility, which is the moles of solute that dissolve in 1 liter of saturated solution. The calculation of a solubility product constant is shown in Sample Problem 13.8.

Calcium oxalate produces small amounts of Ca^{2+} and $C_2O_4^{2-}$ in aqueous solution.

ENGAGE 13.13

Why is the solubility product expression for Cu_2CO_3 written $K_{sp} = [Cu^+]^2[CO_3^{2-}]$?

PRACTICE PROBLEMS

Try Practice Problems 13.43 and 13.44

CORE CHEMISTRY SKILL

Calculating a Solubility Product Constant

SAMPLE PROBLEM 13.8 Calculating the Solubility Product Constant

TRY IT FIRST

We can make a saturated solution of $BaCO_3$ by adding solid $BaCO_3$ to water and stirring until equilibrium is reached. What is the numerical value of K_{sp} for $BaCO_3$ if the equilibrium mixture contains 5.1×10^{-5} M Ba^{2+} and 5.1×10^{-5} M CO_3^{2-}?

SOLUTION

STEP 1 State the given and needed quantities.

ANALYZE THE PROBLEM	Given	Need	Connect
	$[Ba^{2+}] = 5.1 \times 10^{-5}$ M, $[CO_3^{2-}] = 5.1 \times 10^{-5}$ M	numerical value of K_{sp}	solubility product expression

STEP 2 Write the equilibrium equation for the dissociation of the slightly soluble ionic compound.

$$BaCO_3(s) \rightleftharpoons Ba^{2+}(aq) + CO_3^{2-}(aq)$$

STEP 3 Write the solubility product expression.

$$K_{sp} = [Ba^{2+}][CO_3^{2-}]$$

Terraces in Pamukkale, Turkey, are composed of calcium carbonate, which is slightly soluble.

TABLE 13.6 Examples of Solubility Product Constants (K_{sp})

Formula	K_{sp}
AgCl	1.8×10^{-10}
Ag_2SO_4	1.2×10^{-5}
$BaSO_4$	1.1×10^{-10}
$CaCO_3$	5.0×10^{-9}
CaF_2	3.2×10^{-11}
$Ca(OH)_2$	6.5×10^{-6}
$CaSO_4$	2.4×10^{-5}
$PbCl_2$	1.5×10^{-6}
$PbCO_3$	7.4×10^{-14}

STEP 4 Substitute the molar concentration of each ion into the solubility product expression and calculate.

$$K_{sp} = [5.1 \times 10^{-5}][5.1 \times 10^{-5}] = 2.6 \times 10^{-9}$$

SELF TEST 13.8

a. A saturated solution of AgBr contains 7.3×10^{-7} M Ag^+ and 7.3×10^{-7} M Br^-. What is the numerical value of K_{sp} for AgBr?
b. A saturated solution of CoS contains 6.3×10^{-11} M Co^{2+} and 6.3×10^{-11} M S^{2-}. What is the numerical value of K_{sp} for CoS?

ANSWER

a. $K_{sp} = 5.3 \times 10^{-13}$ **b.** $K_{sp} = 4.0 \times 10^{-21}$

TABLE 13.6 gives values of K_{sp} for a selected group of slightly soluble ionic compounds at 25 °C.

SAMPLE PROBLEM 13.9 Calculating the Solubility Product Constant Using Coefficients

TRY IT FIRST

A saturated solution of strontium fluoride, SrF_2, contains 8.7×10^{-4} M Sr^{2+} and 1.7×10^{-3} M F^-. What is the numerical value of K_{sp} for SrF_2?

SOLUTION

STEP 1 State the given and needed quantities.

ANALYZE THE PROBLEM	Given	Need	Connect
	$[Sr^{2+}] = 8.7 \times 10^{-4}$ M, $[F^-] = 1.7 \times 10^{-3}$ M	numerical value of K_{sp}	solubility product expression

STEP 2 Write the equilibrium equation for the dissociation of the slightly soluble ionic compound.

$$SrF_2(s) \rightleftharpoons Sr^{2+}(aq) + 2F^-(aq)$$

STEP 3 Write the solubility product expression.

$$K_{sp} = [Sr^{2+}][F^-]^2$$

STEP 4 Substitute the molar concentration of each ion into the solubility product expression and calculate.

$$K_{sp} = [8.7 \times 10^{-4}][1.7 \times 10^{-3}]^2 = 2.5 \times 10^{-9}$$

SELF TEST 13.9

a. What is the numerical value of K_{sp} for silver oxalate, $Ag_2C_2O_4$, if a saturated solution contains 2.2×10^{-4} M Ag^+ and 1.1×10^{-4} M $C_2O_4^{2-}$?
b. A saturated solution of PbI_2 contains 1.2×10^{-3} M Pb^{2+} and 2.4×10^{-3} M I^-. What is the numerical value of K_{sp} for PbI_2?

ANSWER

a. $K_{sp} = 5.3 \times 10^{-12}$ **b.** $K_{sp} = 6.9 \times 10^{-9}$

PRACTICE PROBLEMS

Try Practice Problems 13.45 to 13.48

Molar Solubility, S

The molar solubility, S, of a slightly soluble ionic compound is the number of moles of solute that dissolves in 1 liter of solution. For example, the molar solubility of cadmium sulfide, CdS, is found experimentally to be 1×10^{-12} M.

$$CdS(s) \rightleftharpoons Cd^{2+}(aq) + S^{2-}(aq)$$

Because CdS dissociates into Cd^{2+} and S^{2-} ions, they each have a concentration equal to the solubility S.

$$S = [Cd^{2+}] = [S^{2-}] = 1 \times 10^{-12} \text{ M}$$

If we know the K_{sp} of a slightly soluble ionic compound, we can determine its molar solubility, as shown in Sample Problem 13.10.

Cadmium sulfide is slightly soluble.

SAMPLE PROBLEM 13.10 Calculating the Molar Solubility from K_{sp}

TRY IT FIRST

Calculate the molar solubility, S, of $PbSO_4$ if it has a $K_{sp} = 1.6 \times 10^{-8}$.

SOLUTION

STEP 1 State the given and needed quantities.

	Given	Need	Connect
ANALYZE THE PROBLEM	$K_{sp} = 1.6 \times 10^{-8}$	molar solubility (S) of $PbSO_4(s)$	solubility product expression

STEP 2 Write the equilibrium equation for the dissociation of the slightly soluble ionic compound.

$$PbSO_4(s) \rightleftharpoons Pb^{2+}(aq) + SO_4{}^{2-}(aq)$$

STEP 3 Write the solubility product expression using **S**.

$$K_{sp} = [Pb^{2+}][SO_4{}^{2-}] = S \times S = S^2 = 1.6 \times 10^{-8}$$

STEP 4 Calculate the molar solubility, **S**.

$$S^2 = 1.6 \times 10^{-8}$$
$$S = \sqrt{1.6 \times 10^{-8}} = 1.3 \times 10^{-4} \text{ M}$$

Thus, 1.3×10^{-4} mol of $PbSO_4$ will dissolve in 1 L of solution.

SELF TEST 13.10

a. Calculate the molar solubility, S, of MnS if it has a $K_{sp} = 2.5 \times 10^{-10}$.
b. Calculate the molar solubility, S, of AgOH if it has a $K_{sp} = 2.0 \times 10^{-8}$.

ANSWER

a. $S = 1.6 \times 10^{-5}$ M   b. $S = 1.4 \times 10^{-4}$ M

CORE CHEMISTRY SKILL

Calculating the Molar Solubility

PRACTICE PROBLEMS

Try Practice Problems 13.49 and 13.50

PRACTICE PROBLEMS

13.6 Equilibrium in Saturated Solutions

13.43 For each of the following slightly soluble ionic compounds, write the equilibrium equation for dissociation and the solubility product expression:
   a. $MgCO_3$ b. CaF_2 c. Ag_3PO_4

13.44 For each of the following slightly soluble ionic compounds, write the equilibrium equation for dissociation and the solubility product expression:
   a. Ag_2S b. $Al(OH)_3$ c. BaF_2

13.45 A saturated solution of barium sulfate, $BaSO_4$, has $[Ba^{2+}] = 1 \times 10^{-5}$ M and $[SO_4^{2-}] = 1 \times 10^{-5}$ M. What is the numerical value of K_{sp} for $BaSO_4$?

13.46 A saturated solution of copper(II) sulfide, CuS, has $[Cu^{2+}] = 1.1 \times 10^{-18}$ M and $[S^{2-}] = 1.1 \times 10^{-18}$ M. What is the numerical value of K_{sp} for CuS?

13.47 A saturated solution of silver carbonate, Ag_2CO_3, has $[Ag^+] = 2.6 \times 10^{-4}$ M and $[CO_3^{2-}] = 1.3 \times 10^{-4}$ M. What is the numerical value of K_{sp} for Ag_2CO_3?

13.48 A saturated solution of barium fluoride, BaF_2, has $[Ba^{2+}] = 3.6 \times 10^{-3}$ M and $[F^-] = 7.2 \times 10^{-3}$ M. What is the numerical value of K_{sp} for BaF_2?

13.49 Calculate the molar solubility, S, of CuI if it has a K_{sp} of 1×10^{-12}.

13.50 Calculate the molar solubility, S, of SnS if it has a K_{sp} of 1×10^{-26}.

UPDATE   Equilibrium of CO_2 in the Ocean

The CO_2 level in the atmosphere has increased over the last few decades due to burning fossil fuels and deforestation. About 35% of the CO_2 released into the atmosphere dissolves in the oceans, where it forms $H_2CO_3(aq)$, which breaks up into $H^+(aq)$ and $HCO_3^-(aq)$. The increase in H^+

makes the ocean water more acidic. One of Peter's research projects is to study the influence of a change in ocean water acidity on coral species, which have calcium carbonate skeletons.

Applications

13.51 **a.** Write the equation for the dissociation of the slightly soluble ionic compound calcium carbonate, $CaCO_3$.
   b. Write the solubility product expression for $CaCO_3$.

13.52 **a.** What is the numerical value of K_{sp} for $CaCO_3$ if the equilibrium mixture has $[Ca^{2+}]$ and $[CO_3^{2-}]$ equal to 7.1×10^{-5} M?
   b. If increasing the acidity of the ocean has the effect of decreasing the concentration of CO_3^{2-}, will calcium carbonate be more or less soluble if the acidity increases?

Coral reefs are affected by the changing acidity of the ocean.

CONCEPT MAP

REACTION RATES AND CHEMICAL EQUILIBRIUM

involves

in

Reaction Rates

are affected by

Concentrations of Reactants

Temperature

Catalyst

Reversible Reactions

when equal in rate give

Equilibrium Expression

is written as

$$K_c = \frac{[Products]}{[Reactants]}$$

Small K_c has Mostly Reactants

Large K_c has Mostly Products

Le Châtelier's Principle

indicates that equilibrium shifts for changes in

Concentration, Temperature, and Volume

Saturated Solutions

is called

Solubility Product Expression

is written as

$$K_{sp} = [cation][anion]$$

CHAPTER REVIEW

13.1 Rates of Reactions

LEARNING GOAL
Describe how temperature, concentration, and catalysts affect the rate of a reaction.

- The rate of a reaction is the speed at which the reactants are converted to products.
- Increasing the concentrations of reactants, increasing the temperature, or adding a catalyst can increase the rate of a reaction.

13.2 Chemical Equilibrium

LEARNING GOAL Use the concept of reversible reactions to explain chemical equilibrium.

- Chemical equilibrium occurs in a reversible reaction when the rate of the forward reaction becomes equal to the rate of the reverse reaction.
- At equilibrium, no further change occurs in the concentrations of the reactants and products as the forward and reverse reactions continue.

13.3 Equilibrium Constants

LEARNING GOAL Calculate the equilibrium constant for a reversible reaction given the concentrations of reactants and products at equilibrium.

- An equilibrium constant, K_c, is the ratio of the concentrations of the products to the concentrations of the reactants, with each concentration raised to a power equal to its coefficient in the balanced chemical equation.
- For heterogeneous reactions, only the molar concentrations of gases are placed in the equilibrium expression.

13.4 Using Equilibrium Constants

LEARNING GOAL
Use an equilibrium constant to predict the extent of reaction and to calculate equilibrium concentrations.

- A large value of K_c indicates that an

equilibrium mixture contains mostly products, whereas a small value of K_c indicates that the equilibrium mixture contains mostly reactants.
- Equilibrium constants can be used to calculate the concentration of a component in the equilibrium mixture.

13.5 Changing Equilibrium Conditions: Le Châtelier's Principle

LEARNING GOAL Use Le Châtelier's principle to describe the changes made in equilibrium concentrations when reaction conditions change.

- When reactants are removed or products are added to an equilibrium mixture, the system shifts in the direction of the reactants.
- When reactants are added or products are removed from an equilibrium mixture, the system shifts in the direction of the products.
- A decrease in the volume of a reaction container causes a shift in the direction of the smaller number of moles of gas.
- An increase in the volume of a reaction container causes a shift in the direction of the greater number of moles of gas.
- Increasing the temperature of an endothermic reaction or decreasing the temperature of an exothermic reaction will cause the system to shift in the direction of the products.
- Decreasing the temperature of an endothermic reaction or increasing the temperature of an exothermic reaction will cause the system to shift in the direction of the reactants.

13.6 Equilibrium in Saturated Solutions

LEARNING GOAL Write the solubility product expression for a slightly soluble ionic compound and calculate K_{sp}; use K_{sp} to determine the solubility.

- In a saturated solution, a slightly soluble ionic compound is in equilibrium with its ions.
- In a saturated solution, the concentrations of the ions from the slightly soluble ionic compound are constant and can be used to calculate the solubility product constant, K_{sp}.
- If K_{sp} for a slightly soluble ionic compound is known, its solubility can be calculated.

KEY TERMS

activation energy The energy that must be provided by a collision to break apart the bonds of the reacting molecules.

catalyst A substance that increases the rate of reaction by lowering the activation energy.

chemical equilibrium The point at which the rate of forward and reverse reactions are equal so that no further change in concentrations of reactants and products takes place.

collision theory A model for a chemical reaction stating that molecules must collide with sufficient energy and proper orientation to form products.

equilibrium constant, K_c The numerical value obtained by substituting the equilibrium concentrations of the components into the equilibrium expression.

equilibrium expression The ratio of the concentrations of products to the concentrations of reactants, with each component raised to an exponent equal to the coefficient of that compound in the balanced chemical equation.

heterogeneous equilibrium An equilibrium system in which the components are in different states.

homogeneous equilibrium An equilibrium system in which all components are in the same state.

Le Châtelier's principle When a stress is placed on a system at equilibrium, the equilibrium shifts to relieve that stress.

rate of reaction The speed at which reactants form products.

reversible reaction A reaction in which a forward reaction occurs from reactants to products, and a reverse reaction occurs from products back to reactants.

solubility product constant, K_{sp} The product of the concentrations of the ions in a saturated solution of a slightly soluble ionic compound, with each concentration raised to a power equal to its coefficient in the balanced equilibrium equation.

solubility product expression The product of the ion concentrations, with each concentration raised to an exponent equal to the coefficient in the balanced chemical equation.

CORE CHEMISTRY SKILLS

The chapter section containing each Core Chemistry Skill is shown in parentheses at the end of each heading.

Writing the Equilibrium Expression (13.3)

- An equilibrium expression for a reversible reaction is written by multiplying the concentrations of the products in the numerator and dividing by the product of the concentrations of the reactants in the denominator.
- Each concentration is raised to a power equal to its coefficient in the balanced chemical equation:

$$K_c = \frac{[\text{Products}]}{[\text{Reactants}]} = \frac{[\text{C}]^c [\text{D}]^d}{[\text{A}]^a [\text{B}]^b} \quad \text{Coefficients}$$

Example: Write the equilibrium expression for the following chemical reaction:

$$2NO_2(g) \rightleftharpoons N_2O_4(g)$$

Answer: $K_c = \dfrac{[N_2O_4]}{[NO_2]^2}$

Calculating an Equilibrium Constant (13.3)

- The equilibrium constant, K_c, is the numerical value obtained by substituting experimentally measured molar concentrations at equilibrium into the equilibrium expression.

Example: Calculate the numerical value of K_c for the following reaction when the equilibrium mixture contains 0.025 M NO_2 and 0.087 M N_2O_4:

$$2NO_2(g) \rightleftharpoons N_2O_4(g)$$

Answer: Write the equilibrium expression, substitute the molar concentrations, and calculate.

$$K_c = \frac{[N_2O_4]}{[NO_2]^2} = \frac{[0.087]}{[0.025]^2} = 140$$

Calculating Equilibrium Concentrations (13.4)

- To determine the concentration of a product or a reactant at equilibrium, we use the equilibrium expression to solve for the unknown concentration.

Example: Calculate the equilibrium concentration for CF_4 if $K_c = 2.0$, and the equilibrium mixture contains 0.10 M COF_2 and 0.050 M CO_2.

$$2COF_2(g) \rightleftharpoons CO_2(g) + CF_4(g)$$

Answer: Write the equilibrium expression.

$$K_c = \frac{[CO_2][CF_4]}{[COF_2]^2}$$

Solve the equation for the unknown concentration, substitute the molar concentrations, and calculate.

$$[CF_4] = K_c \times \frac{[COF_2]^2}{[CO_2]} = 2.0 \times \frac{[0.10]^2}{[0.050]} = 0.40 \text{ M}$$

Using Le Châtelier's Principle (13.5)

- Le Châtelier's principle states that when a system at equilibrium is disturbed by changes in concentration, volume, or temperature, the system will shift in the direction that will reduce that stress.

Example: Nitrogen and oxygen form dinitrogen pentoxide in an exothermic reaction.

$$2N_2(g) + 5O_2(g) \rightleftharpoons 2N_2O_5(g) + \text{heat}$$

For each of the following changes at equilibrium, indicate whether the equilibrium shifts in the direction of the product, the reactants, or does not change:

  a. removing some $N_2(g)$
  b. decreasing the temperature
  c. increasing the volume of the container

Answer: **a.** Removing a reactant shifts the equilibrium in the direction of the reactants.
  b. Decreasing the temperature shifts the equilibrium of an exothermic reaction in the direction of the product.
  c. Increasing the volume of the container shifts the equilibrium in the direction of the greater number of moles of gas, which is in the direction of the reactants.

Writing the Solubility Product Expression (13.6)

- In a saturated solution, a solid slightly soluble ionic compound is in equilibrium with its ions.

$$FeF_2(s) \rightleftharpoons Fe^{2+}(aq) + 2F^-(aq)$$

- We represent the solubility of solid FeF_2 in a saturated aqueous solution by the solubility product expression, which is the product of the ion concentrations raised to the powers equal to the coefficients.

$$K_{sp} = [Fe^{2+}][F^-]^2$$

Example: Write the equilibrium equation for the dissociation of the slightly soluble ionic compound PbI_2, and its solubility product expression.

Answer: $PbI_2(s) \rightleftharpoons Pb^{2+}(aq) + 2I^-(aq)$ $K_{sp} = [Pb^{2+}][I^-]^2$

Calculating a Solubility Product Constant (13.6)

- The solubility product constant, K_{sp}, is calculated by substituting the molar concentrations of the ions into the solubility product expression.

Example: What is the numerical value of K_{sp} for a saturated solution of PbI_2 that contains 1.3×10^{-3} M Pb^{2+} and 2.6×10^{-3} M I^-?

Answer: The K_{sp} is calculated by substituting the molar concentrations into the solubility product expression.

$$K_{sp} = [Pb^{2+}][I^-]^2 = [1.3 \times 10^{-3}][2.6 \times 10^{-3}]^2$$

$$= 8.8 \times 10^{-9}$$

Calculating the Molar Solubility (13.6)

- The molar solubility, S, of a slightly soluble ionic compound is the number of moles of solute that dissolves in 1 liter of solution.
- If we know the K_{sp} of a slightly soluble ionic compound, we can calculate the molar solubility, S, of the slightly soluble ionic compound.

Example: What is the solubility product expression and the molar solubility, S, of NiS, which has a K_{sp} of 4.0×10^{-20}?

Answer: $K_{sp} = [Ni^{2+}][S^{2-}] = S \times S = S^2 = 4.0 \times 10^{-20}$

$$S = \sqrt{4.0 \times 10^{-20}} = 2.0 \times 10^{-10} \text{ M}$$

Therefore, 2.0×10^{-10} mol of NiS will dissolve in 1 L of solution.

UNDERSTANDING THE CONCEPTS

The chapter sections to review are shown in parentheses at the end of each problem.

13.53 Write the equilibrium expression for each of the following reactions: (13.3)
    a. $CH_4(g) + 2O_2(g) \rightleftharpoons CO_2(g) + 2H_2O(g)$
    b. $4NH_3(g) + 3O_2(g) \rightleftharpoons 2N_2(g) + 6H_2O(g)$
    c. $C(s) + 2H_2(g) \rightleftharpoons CH_4(g)$

13.54 Write the equilibrium expression for each of the following reactions: (13.3)
    a. $2C_2H_6(g) + 7O_2(g) \rightleftharpoons 4CO_2(g) + 6H_2O(g)$
    b. $2KHCO_3(s) \rightleftharpoons K_2CO_3(s) + CO_2(g) + H_2O(g)$
    c. $4NH_3(g) + 5O_2(g) \rightleftharpoons 4NO(g) + 6H_2O(g)$

13.55 Would the equilibrium constant, K_c, for the reaction in the diagrams have a large or small value? (13.4)

Initially   At equilibrium

13.56 Would the equilibrium constant, K_c, for the reaction in the diagrams have a large or small value? (13.4)

Initially   At equilibrium

13.57 Would T_2 be higher or lower than T_1 for the reaction shown in the diagrams? (13.5)

$T_1 = 300\ °C$ $T_2 = ?$

13.58 Would the reaction shown in the diagrams be exothermic or endothermic? (13.5)

$T_1 = 100\ °C$ $T_2 = 200\ °C$

ADDITIONAL PRACTICE PROBLEMS

13.59 For each of the following changes at equilibrium, indicate whether the equilibrium shifts in the direction of products, reactants, or does not change: (13.5)

$$C_2H_4(g) + Cl_2(g) \rightleftharpoons C_2H_4Cl_2(g) + heat$$

a. increasing the temperature
b. decreasing the volume of the container
c. adding a catalyst
d. adding more $Cl_2(g)$

13.60 For each of the following changes at equilibrium, indicate whether the equilibrium shifts in the direction of products, reactants, or does not change: (13.5)

$$N_2(g) + O_2(g) + heat \rightleftharpoons 2NO(g)$$

a. increasing the temperature
b. decreasing the volume of the container
c. adding a catalyst
d. adding more $N_2(g)$

13.61 For each of the following reactions, indicate if the equilibrium mixture contains mostly products, mostly reactants, or both reactants and products: (13.4)

a. $H_2(g) + Cl_2(g) \rightleftharpoons 2HCl(g)$ $K_c = 1.3 \times 10^{34}$
b. $2NOBr(g) \rightleftharpoons 2NO(g) + Br_2(g)$ $K_c = 2.0$
c. $2H_2S(g) + CH_4(g) \rightleftharpoons CS_2(g) + 4H_2(g)$
$$K_c = 5.3 \times 10^{-8}$$
d. $C(s) + H_2O(g) \rightleftharpoons CO(g) + H_2(g)$ $K_c = 6.3 \times 10^{-1}$

13.62 For each of the following reactions, indicate if the equilibrium mixture contains mostly products, mostly reactants, or both reactants and products: (13.4)

a. $2H_2O(g) \rightleftharpoons 2H_2(g) + O_2(g)$ $K_c = 4 \times 10^{-48}$
b. $N_2(g) + 3H_2(g) \rightleftharpoons 2NH_3(g)$ $K_c = 0.30$
c. $2SO_2(g) + O_2(g) \rightleftharpoons 2SO_3(g)$ $K_c = 1.2 \times 10^9$
d. $H_2(g) + S(s) \rightleftharpoons 2H_2S(g)$ $K_c = 7.8 \times 10^5$

13.63 Consider the reaction: (13.3)

$$2NH_3(g) \rightleftharpoons N_2(g) + 3H_2(g)$$

a. Write the equilibrium expression.
b. What is the numerical value of K_c for the reaction if the concentrations at equilibrium are 0.20 M NH_3, 3.0 M N_2, and 0.50 M H_2?

13.64 Consider the reaction: (13.3)

$$2SO_2(g) + O_2(g) \rightleftharpoons 2SO_3(g)$$

a. Write the equilibrium expression.
b. What is the numerical value of K_c for the reaction if the concentrations at equilibrium are 0.10 M SO_2, 0.12 M O_2, and 0.60 M SO_3?

13.65 The K_c for the following reaction is 5.0 at 100 °C. If an equilibrium mixture contains 0.50 M NO_2, what is the molar concentration of N_2O_4? (13.3, 13.4)

$$2NO_2(g) \rightleftharpoons N_2O_4(g)$$

13.66 The K_c for the following reaction is 15 at 220 °C. If an equilibrium mixture contains 0.40 M CO and 0.20 M H_2, what is the molar concentration of CH_4O? (13.3, 13.4)

$$CO(g) + 2H_2(g) \rightleftharpoons CH_4O(g)$$

13.67 According to Le Châtelier's principle, does the equilibrium shift in the direction of products or reactants when O_2 is added to the equilibrium mixture of each of the following reactions? (13.5)

a. $3O_2(g) \rightleftharpoons 2O_3(g)$
b. $2CO_2(g) \rightleftharpoons 2CO(g) + O_2(g)$
c. $2SO_2(g) + O_2(g) \rightleftharpoons 2SO_3(g)$
d. $2SO_2(g) + 2H_2O(g) \rightleftharpoons 2H_2S(g) + 3O_2(g)$

13.68 According to Le Châtelier's principle, does the equilibrium shift in the direction of products or reactants when N_2 is added to the equilibrium mixture of each of the following reactions? (13.5)

a. $2NH_3(g) \rightleftharpoons 3H_2(g) + N_2(g)$
b. $N_2(g) + O_2(g) \rightleftharpoons 2NO(g)$
c. $2NO_2(g) \rightleftharpoons N_2(g) + 2O_2(g)$
d. $4NH_3(g) + 3O_2(g) \rightleftharpoons 2N_2(g) + 6H_2O(g)$

13.69 Would decreasing the volume of the container for each of the following reactions cause the equilibrium to shift in the direction of products, reactants, or not change? (13.5)

a. $3O_2(g) \rightleftharpoons 2O_3(g)$
b. $2CO_2(g) \rightleftharpoons 2CO(g) + O_2(g)$
c. $P_4(g) + 5O_2(g) \rightleftharpoons P_4O_{10}(s)$
d. $2SO_2(g) + 2H_2O(g) \rightleftharpoons 2H_2S(g) + 3O_2(g)$

13.70 Would increasing the volume of the container for each of the following reactions cause the equilibrium to shift in the direction of products, reactants, or not change? (13.5)

a. $2NH_3(g) \rightleftharpoons 3H_2(g) + N_2(g)$
b. $N_2(g) + O_2(g) \rightleftharpoons 2NO(g)$
c. $N_2(g) + 2O_2(g) \rightleftharpoons 2NO_2(g)$
d. $4NH_3(g) + 3O_2(g) \rightleftharpoons 2N_2(g) + 6H_2O(g)$

13.71 The numerical value of the equilibrium constant, K_c, for the decomposition of $COCl_2$ to CO and Cl_2 is 0.68. If an equilibrium mixture contains 0.40 M CO and 0.74 M Cl_2, what is the molar concentration of $COCl_2$? (13.3, 13.4)

$$COCl_2(g) \rightleftharpoons CO(g) + Cl_2(g)$$

13.72 The numerical value of the equilibrium constant, K_c, for the reaction of carbon and water to form carbon monoxide and hydrogen is 0.20 at 1000 °C. If an equilibrium mixture contains solid carbon, 0.40 M H_2O, and 0.40 M CO, what is the molar concentration of H_2? (13.3, 13.4)

$$C(s) + H_2O(g) \rightleftharpoons CO(g) + H_2(g)$$

13.73 For each of the following slightly soluble ionic compounds, write the equilibrium equation for dissociation and the solubility product expression: (13.6)

a. $CuCO_3$ **b.** PbF_2 **c.** $Fe(OH)_3$

13.74 For each of the following slightly soluble ionic compounds, write the equilibrium equation for dissociation and the solubility product expression: (13.6)

a. CuS   **b.** Ag_2SO_4 **c.** $Zn(OH)_2$

13.75 A saturated solution of iron(II) sulfide, FeS, has $[Fe^{2+}] = 7.7 \times 10^{-10}$ M and $[S^{2-}] = 7.7 \times 10^{-10}$ M. What is the numerical value of K_{sp} for FeS? (13.6)

13.76 A saturated solution of copper(I) chloride, CuCl, has $[Cu^+] = 1.1 \times 10^{-3}$ M and $[Cl^-] = 1.1 \times 10^{-3}$ M. What is the numerical value of K_{sp} for CuCl? (13.6)

13.77 A saturated solution of manganese(II) hydroxide, $Mn(OH)_2$, has $[Mn^{2+}] = 3.7 \times 10^{-5}$ M and $[OH^-] = 7.4 \times 10^{-5}$ M. What is the numerical value of K_{sp} for $Mn(OH)_2$? (13.6)

13.78 A saturated solution of silver chromate, Ag_2CrO_4, has $[Ag^+] = 1.3 \times 10^{-4}$ M and $[CrO_4^{2-}] = 6.5 \times 10^{-5}$ M. What is the numerical value of K_{sp} for Ag_2CrO_4? (13.6)

13.79 What is the molar solubility, S, of CdS if it has a K_{sp} of 1.0×10^{-24}? (13.6)

13.80 What is the molar solubility, S, of $CuCO_3$ if it has K_{sp} of 1×10^{-26}? (13.6)

CHALLENGE PROBLEMS

The following problems are related to the topics in this chapter. However, they do not all follow the chapter order, and they require you to combine concepts and skills from several sections. These problems will help you increase your critical thinking skills and prepare for your next exam.

13.81 The K_c at 100 °C is 2.0 for the decomposition reaction of NOBr. (13.3, 13.4, 13.5)

$$2NOBr(g) \rightleftharpoons 2NO(g) + Br_2(g)$$

In an experiment, 1.0 mol of NOBr, 1.0 mol of NO, and 1.0 mol of Br_2 were placed in a 1.0-L container.

a. Write the equilibrium expression for the reaction.
b. Is the system at equilibrium?
c. If not, will the rate of the forward or reverse reaction initially speed up?
d. At equilibrium, which concentration(s) will be greater than 1.0 mol/L, and which will be less than 1.0 mol/L?

13.82 Consider the following reaction: (13.3, 13.4, 13.5)

$$PCl_5(g) \rightleftharpoons PCl_3(g) + Cl_2(g)$$

a. Write the equilibrium expression for the reaction.
b. Initially, 0.60 mol of PCl_5 is placed in a 1.0-L flask. At equilibrium, there is 0.16 mol of PCl_3 in the flask. What are the equilibrium concentrations of PCl_5 and Cl_2?
c. What is the numerical value of the equilibrium constant, K_c, for the reaction?
d. If 0.20 mol of Cl_2 is added to the equilibrium mixture, will the concentration of PCl_5 increase or decrease?

13.83 Indicate how each of the following will affect the equilibrium concentration of CO in the following reaction: (13.3, 13.5)

$$C(s) + H_2O(g) + 31\ kcal \rightleftharpoons CO(g) + H_2(g)$$

a. adding more $H_2(g)$
b. increasing the temperature
c. increasing the volume of the container
d. decreasing the volume of the container
e. adding a catalyst

13.84 Indicate how each of the following will affect the equilibrium concentration of NH_3 in the following reaction: (13.3, 13.5)

$$4NH_3(g) + 5O_2(g) \rightleftharpoons 4NO(g) + 6H_2O(g) + 906\ kJ$$

a. adding more $O_2(g)$
b. increasing the temperature
c. increasing the volume of the container
d. adding more $NO(g)$
e. removing some $H_2O(g)$

13.85 Indicate if you would increase or decrease the volume of the container to *increase* the yield of the products in each of the following: (13.5)

a. $2C(s) + O_2(g) \rightleftharpoons 2CO(g)$
b. $2CH_4(g) \rightleftharpoons C_2H_2(g) + 3H_2(g)$
c. $2H_2(g) + O_2(g) \rightleftharpoons 2H_2O(g)$

13.86 Indicate if you would increase or decrease the volume of the container to *increase* the yield of the products in each of the following: (13.5)

a. $Cl_2(g) + 2NO(g) \rightleftharpoons 2NOCl(g)$
b. $N_2(g) + 2H_2(g) \rightleftharpoons N_2H_4(g)$
c. $N_2O_4(g) \rightleftharpoons 2NO_2(g)$

13.87 The antacid milk of magnesia, which contains $Mg(OH)_2$, is used to neutralize excess stomach acid. If the solubility of $Mg(OH)_2$ in water is 9.7×10^{-3} g/L, what is the numerical value of K_{sp} for $Mg(OH)_2$? (13.6)

13.88 The slightly soluble ionic compound PbF_2 has a solubility in water of 0.74 g/L. What is the numerical value of K_{sp} for PbF_2? (13.6)

ANSWERS TO ENGAGE QUESTIONS

13.1 If molecules collide without the minimum activation energy, they do not react.

13.2 An increase in temperature will increase the rate of reaction because the molecules move faster with more energy, which means there are more collisions with the necessary activation energy.

13.3 A decrease in the concentration of a reactant decreases the rate of reaction because there are fewer molecules to collide.

13.4 Before equilibrium is reached, the forward reaction is faster, decreasing the concentrations of reactants.

13.5 Once a chemical reaction reaches equilibrium, the rates of the forward and reverse reactions are equal.

13.6 In the equilibrium expression, any coefficient in the chemical equation is indicated as an exponent.

13.7 The concentrations of solids or liquids are not included in the equilibrium expression for K_c because they are constant.

13.8 The equilibrium expression has the concentrations of products over the concentrations of reactants, thus an equilibrium mixture containing mostly products will have a large numerical value for K_c.

13.9 An equilibrium reaction with a $K_c = 1.2$, which is close to 1, contains almost equal amounts of both reactants and products.

13.10 The equilibrium expression has the concentrations of products over the concentrations of reactants. An equilibrium mixture with a K_c of 2×10^{-9} means the numerator, which is reactants, is much smaller than the denominator.

13.11 Because HI is a product, adding more HI will shift the equilibrium in the direction of the reactants.

13.12 In an exothermic reaction, heat is a product. To decrease the amount of product, the temperature should be increased.

13.13 In the equation for the reaction when the slightly soluble ionic compound Cu_2CO_3 dissolves, Cu^+ has a coefficient of 2 and CO_3^{2-} has a coefficient of 1. Thus, the K_{sp} expression is written as $K_{sp} = [Cu^+]^2[CO_3^{2-}]$.

ANSWERS TO SELECTED PROBLEMS

13.1 **a.** The rate of the reaction indicates how fast the products form or how fast the reactants are used up.
b. At room temperature, the reactions involved in the growth of bread mold will proceed at a faster rate than at the lower temperature of the refrigerator.

13.3 The number of collisions will increase when the number of Br_2 molecules is increased.

13.5 **a.** increase **b.** increase **c.** increase **d.** decrease

13.7 A reversible reaction is one in which a forward reaction converts reactants to products, whereas a reverse reaction converts products to reactants.

13.9 **a.** not at equilibrium **b.** at equilibrium
c. at equilibrium

13.11 The reaction has reached equilibrium because the number of reactants and products does not change.

13.13 **a.** $K_c = \dfrac{[CS_2][H_2]^4}{[CH_4][H_2S]^2}$ **b.** $K_c = \dfrac{[N_2][O_2]}{[NO]^2}$

c. $K_c = \dfrac{[CS_2][O_2]^4}{[SO_3]^2[CO_2]}$ **d.** $K_c = \dfrac{[H_2]^3[CO]}{[CH_4][H_2O]}$

13.15 $K_c = \dfrac{[XY]^2}{[X_2][Y_2]} = 36$

13.17 $K_c = 1.5$

13.19 $K_c = 260$

13.21 **a.** homogeneous equilibrium **b.** heterogeneous equilibrium
c. homogeneous equilibrium **d.** heterogeneous equilibrium

13.23 **a.** $K_c = \dfrac{[O_2]^3}{[O_3]^2}$ **b.** $K_c = [CO_2][H_2O]$

c. $K_c = \dfrac{[C_6H_{12}]}{[C_6H_6][H_2]^3}$ **d.** $K_c = \dfrac{[H_2]^2}{[HCl]^4}$

13.25 $K_c = 0.26$

13.27 Diagram **B** represents the equilibrium mixture.

13.29 **a.** mostly products **b.** mostly products
c. mostly reactants

13.31 $[H_2] = 1.1 \times 10^{-3}$ M

13.33 $[NOBr] = 1.4$ M

13.35 **a.** Equilibrium shifts in the direction of the product.
b. Equilibrium shifts in the direction of the reactants.
c. Equilibrium shifts in the direction of the product.
d. Equilibrium shifts in the direction of the reactants.
e. No shift in equilibrium occurs.

13.37 **a.** Equilibrium shifts in the direction of the product.
b. Equilibrium shifts in the direction of the product.
c. Equilibrium shifts in the direction of the product.
d. No shift in equilibrium occurs.
e. Equilibrium shifts in the direction of the reactants.

13.39 **a.** Equilibrium shifts in the direction of the product.
b. Equilibrium shifts in the direction of the reactants.
c. Equilibrium shifts in the direction of the reactants.

13.41 **a.** The oxygen concentration is lowered.
b. Equilibrium shifts in the direction of the reactants.

13.43 **a.** $MgCO_3(s) \rightleftharpoons Mg^{2+}(aq) + CO_3^{2-}(aq)$;
$K_{sp} = [Mg^{2+}][CO_3^{2-}]$
b. $CaF_2(s) \rightleftharpoons Ca^{2+}(aq) + 2F^-(aq)$; $K_{sp} = [Ca^{2+}][F^-]^2$
c. $Ag_3PO_4(s) \rightleftharpoons 3Ag^+(aq) + PO_4^{3-}(aq)$;
$K_{sp} = [Ag^+]^3[PO_4^{3-}]$

13.45 $K_{sp} = 1 \times 10^{-10}$

13.47 $K_{sp} = 8.8 \times 10^{-12}$

13.49 $S = 1 \times 10^{-6}$ M

13.51 **a.** $CaCO_3(s) \rightleftharpoons Ca^{2+}(aq) + CO_3^{2-}(aq)$
b. $K_{sp} = [Ca^{2+}][CO_3^{2-}]$

13.53 **a.** $K_c = \dfrac{[CO_2][H_2O]^2}{[CH_4][O_2]^2}$ **b.** $K_c = \dfrac{[N_2]^2[H_2O]^6}{[NH_3]^4[O_2]^3}$

c. $K_c = \dfrac{[CH_4]}{[H_2]^2}$

13.55 The equilibrium constant for the reaction would have a small value.

13.57 T_2 is lower than T_1.

13.59 **a.** Equilibrium shifts in the direction of the reactants.
b. Equilibrium shifts in the direction of the product.
c. There is no change in equilibrium.
d. Equilibrium shifts in the direction of the product.

13.61 **a.** mostly products
b. both reactants and products
c. mostly reactants
d. both reactants and products

13.63 **a.** $K_c = \dfrac{[N_2][H_2]^3}{[NH_3]^2}$ **b.** $K_c = 9.4$

13.65 $[N_2O_4] = 1.3$ M

13.67 **a.** Equilibrium shifts in the direction of the product.
b. Equilibrium shifts in the direction of the reactant.
c. Equilibrium shifts in the direction of the product.
d. Equilibrium shifts in the direction of the reactants.

13.69 **a.** Equilibrium shifts in the direction of the product.
b. Equilibrium shifts in the direction of the reactant.
c. Equilibrium shifts in the direction of the product.
d. Equilibrium shifts in the direction of the reactants.

13.71 $[COCl_2] = 0.44$ M

13.73 **a.** $CuCO_3(s) \rightleftharpoons Cu^{2+}(aq) + CO_3^{2-}(aq)$;
$K_{sp} = [Cu^{2+}][CO_3^{2-}]$
b. $PbF_2(s) \rightleftharpoons Pb^{2+}(aq) + 2F^-(aq)$;
$K_{sp} = [Pb^{2+}][F^-]^2$
c. $Fe(OH)_3(s) \rightleftharpoons Fe^{3+}(aq) + 3OH^-(aq)$;
$K_{sp} = [Fe^{3+}][OH^-]^3$

13.75 $K_{sp} = 5.9 \times 10^{-19}$

13.77 $K_{sp} = 2.0 \times 10^{-13}$

13.79 $S = 1.0 \times 10^{-12}$ M

13.81 **a.** $K_c = \dfrac{[NO]^2[Br_2]}{[NOBr]^2}$

b. When the concentrations are placed in the equilibrium expression, the result is 1.0, which is not equal to K_c. The system is not at equilibrium.
c. The rate of the forward reaction will increase.
d. The $[Br_2]$ and $[NO]$ will increase and $[NOBr]$ will decrease.

13.83 **a.** decrease **b.** increase **c.** increase
d. decrease **e.** no change

13.85 **a.** increase **b.** increase **c.** decrease

13.87 $K_{sp} = 2.0 \times 10^{-11}$

Acids and Bases

After an automobile accident, Larry, a 30-year-old man, is brought to the emergency room, where he is unresponsive. One of the emergency room nurses takes a blood sample, which is then sent to Brianna, a medical laboratory technologist, who begins the process of analyzing the pH, the partial pressures of O_2 and CO_2, and the concentrations of glucose and electrolytes.

Brianna determines that Larry's blood pH is 7.30, and the partial pressure of CO_2 gas is above the desired level. Blood pH is typically in the range of 7.35 to 7.45, and a value less than 7.35 indicates a state of acidosis. Respiratory acidosis occurs because an increase in the partial pressure of CO_2 gas in the bloodstream increases the concentration of H_2CO_3, which leads to an increase in the concentration of H_3O^+.

Brianna recognizes these signs and immediately contacts the emergency room to inform them that Larry's airway may be blocked. In the emergency room, they provide Larry with an IV containing bicarbonate to increase the blood pH and begin the process of unblocking his airway. Shortly afterward, Larry's airway is cleared, and his blood pH and partial pressure of CO_2 gas return to normal.

CAREER

Medical Laboratory Technologist

Medical laboratory technologists perform a wide variety of tests on body fluids and cells that help in the diagnosis and treatment of patients. These tests range from determining blood concentrations of glucose and cholesterol to determining drug levels in the blood for transplant patients or a patient undergoing treatment. Medical laboratory technologists also prepare specimens in the detection of cancerous tumors and type blood samples for transfusions. Medical laboratory technologists must also interpret and analyze the test results, which are then passed on to the physician.

UPDATE Acid Reflux Disease

After Larry was discharged from the hospital, he complained of a sore throat and dry cough, which his doctor diagnosed as acid reflux. You can view the symptoms of acid reflux disease (GERD) in the **UPDATE Acid Reflux Disease**, pages 463–464, and learn about the pH changes in the stomach and how the condition is treated.

REVIEW
Writing Ionic Formulas (6.2)

Citrus fruits are sour because of the presence of acids.

Sulfuric acid dissolves in water to produce one or two H⁺ and an anion.

14.1 Acids and Bases

LEARNING GOAL Describe and name acids and bases.

Acids and bases are important substances in health, industry, and the environment. One of the most common characteristics of acids is their sour taste. Lemons and grapefruits taste sour because they contain acids such as citric and ascorbic acid (vitamin C). Vinegar tastes sour because it contains acetic acid. We produce lactic acid in our muscles when we exercise. Acid from bacteria turns milk sour in the production of yogurt and cottage cheese. We have hydrochloric acid in our stomachs that helps us digest food. Sometimes we take antacids, which are bases such as sodium bicarbonate or milk of magnesia, to neutralize the effects of too much stomach acid.

The term *acid* comes from the Latin word *acidus*, which means "sour." You are probably already familiar with the sour tastes of vinegar and lemons.

In 1887, Swedish chemist Svante Arrhenius was the first to describe **acids** as substances that produce hydrogen ions (H^+) when they dissolve in water. Because acids produce ions in water, they are electrolytes. For example, hydrogen chloride dissociates in water to give hydrogen ions, H^+, and chloride ions, Cl^-. The hydrogen ions give acids a sour taste, change the blue litmus indicator to red, and corrode some metals.

$$HCl(g) \xrightarrow{H_2O} H^+(aq) + Cl^-(aq)$$

Polar molecular compound   Dissociation   Hydrogen ion

Naming Acids

Acids dissolve in water to produce hydrogen ions, along with a negative ion that may be a simple nonmetal anion or a polyatomic ion. When an acid dissolves in water to produce a hydrogen ion and a simple nonmetal anion, the prefix *hydro* is used before the name of the nonmetal, and its *ide* ending is changed to *ic acid*. For example, hydrogen chloride (HCl) dissolves in water to form HCl(*aq*), which is named hydrochloric acid. An exception is hydrogen cyanide (HCN), which as an acid is named hydrocyanic acid.

When an acid contains oxygen, it dissolves in water to produce a hydrogen ion and an oxygen-containing polyatomic anion. The most common form of an oxygen-containing acid has a name that ends with *ic acid*. The name of its polyatomic anion ends in *ate*. If the acid contains a polyatomic ion with an *ite* ending, its name ends in *ous acid*. When the acid has one oxygen atom less than the common form, the suffix *ous* is used and the polyatomic ion is named with an *ite* ending. The names of some common acids and their anions are listed in **TABLE 14.1**.

TABLE 14.1 Names of Common Acids and Their Anions			
Acid	Name of Acid	Anion	Name of Anion
HF	**Hydro**fluoric acid	F^-	Fluoride
HCl	**Hydro**chloric acid	Cl^-	Chloride
HBr	**Hydro**bromic acid	Br^-	Bromide
HI	**Hydro**iodic acid	I^-	Iodide
HCN	**Hydro**cyanic acid	CN^-	Cyanide
HNO_3	Nitr**ic acid**	NO_3^-	Nitrate
HNO_2	Nitr**ous acid**	NO_2^-	Nitrite
H_2SO_4	Sulfur**ic acid**	SO_4^{2-}	Sulfate
H_2SO_3	Sulfur**ous acid**	SO_3^{2-}	Sulfite
H_2CO_3	Carbon**ic acid**	CO_3^{2-}	Carbonate
$HC_2H_3O_2$	Acet**ic acid**	$C_2H_3O_2^-$	Acetate
H_3PO_4	Phosphor**ic acid**	PO_4^{3-}	Phosphate
H_3PO_3	Phosphor**ous acid**	PO_3^{3-}	Phosphite

The halogens in Group 7A (17) can form more than two oxygen-containing acids. For chlorine, the common form is chloric acid, $HClO_3$, which contains the polyatomic ion chlorate, ClO_3^-. For the acid that contains one more oxygen atom than the common form, the prefix *per* is used; $HClO_4$ is *perchloric acid*. When the polyatomic ion in the acid has one oxygen atom less than the common form, the suffix *ous* is used. Thus, $HClO_2$ is *chlorous acid*; it contains the chlorite ion, ClO_2^-. The prefix *hypo* is used for the acid that has two oxygen atoms less than the common form; $HClO$ is *hypochlorous acid*.

ENGAGE 14.1

Why is HBr named hydrobromic acid?

Names of Common Acids and Their Anions

Acid	Name of Acid	Anion	Name of Anion
$HClO_4$	**Per**chloric acid	ClO_4^-	Perchlorate
$HClO_3$	Chlor**ic** acid	ClO_3^-	Chlorate
$HClO_2$	Chlor**ous** acid	ClO_2^-	Chlor**ite**
$HClO$	**Hypo**chlor**ous** acid	ClO^-	Hypochlor**ite**

Bases

You may be familiar with household bases such as antacids, drain openers, and oven cleaners. According to the Arrhenius theory, **bases** are ionic compounds that dissociate into cations and hydroxide ions (OH^-) when they dissolve in water. They are strong electrolytes. For example, sodium hydroxide is an Arrhenius base that dissociates completely in water to give sodium ions Na^+, and hydroxide ions OH^-.

Most Arrhenius bases are formed from Groups 1A (1) and 2A (2) metals, such as NaOH, KOH, LiOH, and $Ca(OH)_2$. The hydroxide ions (OH^-) give Arrhenius bases common characteristics, such as a bitter taste and a slippery feel. A base turns litmus indicator blue and phenolphthalein indicator pink. **TABLE 14.2** compares some characteristics of acids and bases.

$NaOH(s) \xrightarrow{H_2O} Na^+(aq) + OH^-(aq)$

Ionic compound / Dissociation / Hydroxide ion

An Arrhenius base produces cations and OH^- anions in an aqueous solution.

TABLE 14.2 Some Characteristics of Acids and Bases

Characteristic	Acids	Bases
Arrhenius	Produce H^+	Produce OH^-
Electrolytes	Yes	Yes
Taste	Sour	Bitter, chalky
Feel	May sting	Soapy, slippery
Litmus	Red	Blue
Phenolphthalein	Colorless	Pink
Neutralization	Neutralize bases	Neutralize acids

Naming Bases

Typical Arrhenius bases are named as *hydroxides*.

Base	Name
LiOH	Lithium **hydroxide**
NaOH	Sodium **hydroxide**
KOH	Potassium **hydroxide**
$Ca(OH)_2$	Calcium **hydroxide**
$Al(OH)_3$	Aluminum **hydroxide**

Calcium hydroxide, $Ca(OH)_2$, is used in the food industry to produce beverages, and in dentistry as a filler for root canals.

SAMPLE PROBLEM 14.1　Names and Formulas of Acids and Bases

TRY IT FIRST

a. Identify each of the following as an acid or a base, and give its name:
　1. H_3PO_4, ingredient in soft drinks　　**2.** NaOH, ingredient in oven cleaner

b. Write the formula for each of the following:
　1. magnesium hydroxide, ingredient in antacids
　2. hydrobromic acid, used industrially to prepare bromide compounds

A soft drink contains H_3PO_4 and H_2CO_3.

SOLUTION

a. 1. acid, phosphoric acid
b. 1. $Mg(OH)_2$

2. base, sodium hydroxide
2. HBr

SELF TEST 14.1

a. Determine whether H_2CO_3 is an acid or base, and give its name.
b. Determine whether iron(III) hydroxide is an acid or base, and write its formula.

PRACTICE PROBLEMS
Try Practice Problems 14.1 to 14.6

ANSWER

a. acid, carbonic acid

b. base, $Fe(OH)_3$

PRACTICE PROBLEMS

14.1 Acids and Bases

14.1 Indicate whether each of the following statements is characteristic of an acid, a base, or both:
a. has a sour taste
b. neutralizes bases
c. produces H^+ ions in water
d. is named barium hydroxide
e. is an electrolyte

14.2 Indicate whether each of the following statements is characteristic of an acid, a base, or both:
a. neutralizes acids
b. produces OH^- ions in water
c. has a slippery feel
d. conducts an electrical current in solution
e. turns litmus red

14.3 Name each of the following acids or bases:
a. HCl   b. $Ca(OH)_2$ c. $HClO_4$
d. $Sr(OH)_2$ e. H_2SO_3 f. $HBrO_2$

14.4 Name each of the following acids or bases:
a. $Al(OH)_3$ b. HBr   c. H_2SO_4
d. KOH   e. HNO_2 f. $HClO_2$

14.5 Write formulas for each of the following acids and bases:
a. rubidium hydroxide   b. hydrofluoric acid
c. phosphoric acid   d. lithium hydroxide
e. ammonium hydroxide   f. periodic acid

14.6 Write formulas for each of the following acids and bases:
a. barium hydroxide   b. hydroiodic acid
c. nitric acid   d. strontium hydroxide
e. acetic acid   f. hypochlorous acid

14.2 Brønsted–Lowry Acids and Bases

LEARNING GOAL Identify conjugate acid–base pairs for Brønsted–Lowry acids and bases.

In 1923, J. N. Brønsted in Denmark and T. M. Lowry in Great Britain expanded the definition of acids and bases to include bases that do not contain OH^- ions. A **Brønsted–Lowry acid** can donate a hydrogen ion, H^+, and a **Brønsted–Lowry base** can accept a hydrogen ion.

A Brønsted–Lowry acid is a substance that donates H^+.

A Brønsted–Lowry base is a substance that accepts H^+.

A free hydrogen ion, H^+, does not actually exist in water. Its attraction to polar water molecules is so strong that the H^+ bonds to a water molecule and forms a **hydronium ion, H_3O^+**.

Water   Hydrogen ion   Hydronium ion

We can write the formation of a hydrochloric acid solution as a transfer of H^+ from hydrogen chloride to water. By accepting an H^+ in the reaction, water is acting as a base according to the Brønsted–Lowry concept.

$$HCl \ + \ H_2O \longrightarrow H_3O^+ \ + \ Cl^-$$

Hydrogen   Water   Hydronium   Chloride
chloride   ion   ion

Acid   Base
(H^+ donor) (H^+ acceptor)   Acidic solution

In another reaction, ammonia, NH_3, acts as a base by accepting H^+ when it reacts with water. Because the nitrogen atom of NH_3 has a stronger attraction for H^+ than oxygen, water acts as an acid by donating H^+.

$$NH_3 \ + \ H_2O \ \rightleftharpoons \ NH_4^+ \ + \ OH^-$$

Ammonia   Water   Ammonium   Hydroxide
                             ion   ion

Base   Acid
(H^+ acceptor) (H^+ donor)   Basic solution

SAMPLE PROBLEM 14.2   Acids and Bases

TRY IT FIRST

In each of the following equations, identify the reactant that is a Brønsted–Lowry acid and the reactant that is a Brønsted–Lowry base:

a. $HBr(aq) + H_2O(l) \longrightarrow H_3O^+(aq) + Br^-(aq)$
b. $CN^-(aq) + H_2O(l) \rightleftharpoons HCN(aq) + OH^-(aq)$

SOLUTION

a. HBr, Brønsted–Lowry acid; H_2O, Brønsted–Lowry base
b. H_2O, Brønsted–Lowry acid; CN^-, Brønsted–Lowry base

SELF TEST 14.2

a. When HNO_3 reacts with water, water acts as a Brønsted–Lowry base. Write the equation for the reaction.
b. When hypochlorite ion, ClO^-, reacts with water, water acts as a Brønsted–Lowry acid. Write the equation for the reaction.

ANSWER

a. $HNO_3(aq) + H_2O(l) \longrightarrow H_3O^+(aq) + NO_3^-(aq)$
b. $ClO^-(aq) + H_2O(l) \rightleftharpoons HClO(aq) + OH^-(aq)$

PRACTICE PROBLEMS
Try Practice Problems 14.7 and 14.8

Conjugate Acid–Base Pairs

According to the Brønsted–Lowry theory, a **conjugate acid–base pair** consists of molecules or ions related by the loss of one H^+ by an acid, and the gain of one H^+ by a base. Every acid–base reaction contains two conjugate acid–base pairs because an H^+ is transferred in both the forward and reverse directions. When an acid such as HF loses one H^+, the conjugate base F^- is formed. When the base H_2O gains an H^+, its conjugate acid, H_3O^+, is formed.

Because the overall reaction of HF is *reversible*, the conjugate acid H_3O^+ can donate H^+ to the conjugate base F^- and re-form the acid HF and the base H_2O. Using the relationship of loss and gain of one H^+, we can now identify the conjugate acid–base pairs as HF/F^- along with H_3O^+/H_2O.

Conjugate acid–base pair

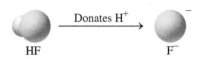

HF   Donates H^+ F^-

Conjugate acid–base pair

H_2O Accepts H^+ H_3O^+

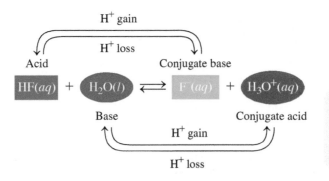

HF, an acid, loses one H^+ to form its conjugate base F^-. Water acts as a base by gaining one H^+ to form its conjugate acid H_3O^+.

ENGAGE 14.2

Why is $HBrO_2$ the conjugate acid of BrO_2^-?

In another reaction, ammonia (NH_3) accepts H^+ from H_2O to form the conjugate acid NH_4^+ and conjugate base OH^-. Each of these conjugate acid–base pairs, NH_4^+/NH_3 and H_2O/OH^-, is related by the loss and gain of one H^+.

Ammonia, NH_3, acts as a base when it gains one H^+ to form its conjugate acid NH_4^+. Water acts as an acid by losing one H^+ to form its conjugate base OH^-.

ENGAGE 14.3

Why can H_2O be both the conjugate base of H_3O^+ and the conjugate acid of OH^-?

In these two examples, we see that water can act as an acid when it donates H^+ or as a base when it accepts H^+. Substances that can act as both acids and bases are **amphoteric** or *amphiprotic*. For water, the most common amphoteric substance, the acidic or basic behavior depends on the other reactant. Water donates H^+ when it reacts with a stronger base, and it accepts H^+ when it reacts with a stronger acid. Another example of an amphoteric substance is bicarbonate (HCO_3^-). With a base, HCO_3^- acts as an acid and donates one H^+ to give CO_3^{2-}. However, when HCO_3^- reacts with an acid, it acts as a base and accepts one H^+ to form H_2CO_3.

Amphoteric substances act as both acids and bases.

SAMPLE PROBLEM 14.3   Identifying Conjugate Acid–Base Pairs

TRY IT FIRST

Identify the conjugate acid–base pairs in the following reaction:

$$HBr(aq) + NH_3(aq) \longrightarrow Br^-(aq) + NH_4^+(aq)$$

SOLUTION

ANALYZE THE PROBLEM	Given		Need	Connect
	HBr	Br^-	conjugate	lose/gain
	NH_3	NH_4^+	acid–base pairs	one H^+

STEP 1   **Identify the reactant that loses H^+ as the acid.** In the reaction, HBr loses H^+ to form the product Br^-. Thus HBr is the acid and Br^- is its conjugate base.

STEP 2   **Identify the reactant that gains H^+ as the base.** In the reaction, NH_3 gains H^+ to form the product NH_4^+. Thus, NH_3 is the base and NH_4^+ is its conjugate acid.

STEP 3   **Write the conjugate acid–base pairs.**

HBr/Br^- and NH_4^+/NH_3

SELF TEST 14.3

Identify the conjugate acid–base pairs in each of the following reactions:

a. $HCN(aq) + SO_4^{2-}(aq) \rightleftharpoons CN^-(aq) + HSO_4^-(aq)$
b. $H_2O(l) + S^{2-}(aq) \rightleftharpoons OH^-(aq) + HS^-(aq)$

ANSWER

a. The conjugate acid–base pairs are HCN/CN^- and HSO_4^-/SO_4^{2-}.
b. The conjugate acid–base pairs are H_2O/OH^- and HS^-/S^{2-}.

PRACTICE PROBLEMS

Try Practice Problems 14.9 to 14.16

PRACTICE PROBLEMS

14.2 Brønsted–Lowry Acids and Bases

14.7 Identify the reactant that is a Brønsted–Lowry acid and the reactant that is a Brønsted–Lowry base in each of the following:
a. $HI(aq) + H_2O(l) \longrightarrow I^-(aq) + H_3O^+(aq)$
b. $F^-(aq) + H_2O(l) \rightleftharpoons HF(aq) + OH^-(aq)$
c. $H_2S(aq) + C_2H_5—NH_2(aq) \rightleftharpoons$
$HS^-(aq) + C_2H_5—NH_3^+(aq)$

14.8 Identify the reactant that is a Brønsted–Lowry acid and the reactant that is a Brønsted–Lowry base in each of the following:
a. $CO_3^{2-}(aq) + H_2O(l) \rightleftharpoons HCO_3^-(aq) + OH^-(aq)$
b. $H_2SO_4(aq) + H_2O(l) \longrightarrow HSO_4^-(aq) + H_3O^+(aq)$
c. $C_2H_3O_2^-(aq) + H_3O^+(aq) \rightleftharpoons HC_2H_3O_2(aq) + H_2O(l)$

14.9 Write the formula for the conjugate base for each of the following acids:
a. HF   **b.** H_2O **c.** $H_2PO_3^-$
d. HSO_4^- **e.** $HClO_2$

14.10 Write the formula for the conjugate base for each of the following acids:
a. HCO_3^- **b.** $CH_3—NH_3^+$ **c.** HPO_4^{2-}
d. HNO_2 **e.** HBrO

14.11 Write the formula for the conjugate acid for each of the following bases:
a. CO_3^{2-} **b.** H_2O **c.** $H_2PO_4^-$
d. Br^- **e.** ClO_4^-

14.12 Write the formula for the conjugate acid for each of the following bases:
a. SO_4^{2-} **b.** CN^- **c.** NH_3
d. ClO_2^- **e.** HS^-

14.13 Identify the Brønsted–Lowry acid–base pairs in each of the following equations:
a. $H_2CO_3(aq) + H_2O(l) \rightleftharpoons HCO_3^-(aq) + H_3O^+(aq)$
b. $HCN(aq) + NO_2^-(aq) \rightleftharpoons CN^-(aq) + HNO_2(aq)$
c. $CHO_2^-(aq) + HF(aq) \rightleftharpoons HCHO_2(aq) + F^-(aq)$

14.14 Identify the Brønsted–Lowry acid–base pairs in each of the following equations:
a. $H_3PO_4(aq) + H_2O(l) \rightleftharpoons H_2PO_4^-(aq) + H_3O^+(aq)$
b. $H_3PO_4(aq) + NH_3(aq) \rightleftharpoons H_2PO_4^-(aq) + NH_4^+(aq)$
c. $HNO_2(aq) + C_2H_5—NH_2(aq) \rightleftharpoons$
$C_2H_5—NH_3^+(aq) + NO_2^-(aq)$

14.15 When ammonium chloride dissolves in water, the ammonium ion, NH_4^+, acts as an acid. Write a balanced equation for the reaction of the ammonium ion with water.

14.16 When sodium carbonate dissolves in water, the carbonate ion, CO_3^{2-}, acts as a base. Write a balanced equation for the reaction of the carbonate ion with water.

14.3 Strengths of Acids and Bases

LEARNING GOAL Write equations for the dissociation of strong and weak acids; identify the direction of reaction.

In the process called **dissociation**, an acid or a base separates into ions in water. The *strength* of an acid is determined by the moles of H_3O^+ that are produced for each mole of acid that dissolves. The *strength* of a base is determined by the moles of OH^- that are produced for each mole of base that dissolves. Strong acids and strong bases dissociate completely in water, whereas weak acids and weak bases dissociate only slightly, leaving most of the initial acid or base undissociated.

Strong and Weak Acids

Strong acids are examples of strong electrolytes because they donate H^+ so easily that their dissociation in water is essentially complete. For example, when HCl, a strong acid, dissociates in water, H^+ is transferred to H_2O; the resulting solution contains essentially only the ions H_3O^+ and Cl^-. We consider the reaction of HCl in H_2O as going 100% to products. Thus, one mole of a strong acid dissociates in water to yield one mole of H_3O^+ and one mole of its conjugate base. We write the equation for a strong acid such as HCl with a single arrow.

$$HCl(g) + H_2O(l) \longrightarrow H_3O^+(aq) + Cl^-(aq)$$

There are only six common strong acids, which are stronger acids than H_3O^+. All other acids are weak. **TABLE 14.3** lists the relative strengths of acids and bases. **Weak acids** are weak electrolytes because they dissociate slightly in water, forming only a small amount of H_3O^+ ions. A weak acid has a strong conjugate base, which is why the reverse reaction is more prevalent. Even at high concentrations, weak acids produce low concentrations of H_3O^+ ions (see **FIGURE 14.1**).

Many of the products you use at home contain weak acids. Citric acid is a weak acid found in fruits and fruit juices such as lemons, oranges, and grapefruit. The vinegar used in salad dressings is typically a 5% (m/v) acetic acid, $HC_2H_3O_2$, solution. In water, a few $HC_2H_3O_2$ molecules donate H^+ to H_2O to form H_3O^+ ions and acetate ions $C_2H_3O_2^-$. The reverse reaction also takes place, which converts the H_3O^+ ions and acetate ions $C_2H_3O_2^-$ back to reactants. The formation of hydronium ions from vinegar is the reason we notice the sour taste of vinegar. We write the equation for a weak acid in an aqueous solution with a double arrow to indicate that the forward and reverse reactions are at equilibrium.

$$\underset{\text{Acetic acid}}{HC_2H_3O_2(aq)} + H_2O(l) \rightleftharpoons \underset{\text{Acetate ion}}{C_2H_3O_2^-(aq)} + H_3O^+(aq)$$

ENGAGE 14.4

What is the difference between a strong acid and a weak acid?

Weak acids are found in foods and household products.

ENGAGE 14.5

Which is the weaker acid: H_2SO_4 or H_2S?

TABLE 14.3 Relative Strengths of Acids and Bases

Acid		Conjugate Base	
Strong Acids		**Weak Bases**	
Hydroiodic acid	HI	I^-	Iodide ion
Hydrobromic acid	HBr	Br^-	Bromide ion
Perchloric acid	$HClO_4$	ClO_4^-	Perchlorate ion
Hydrochloric acid	HCl	Cl^-	Chloride ion
Sulfuric acid	H_2SO_4	HSO_4^-	Hydrogen sulfate ion
Nitric acid	HNO_3	NO_3^-	Nitrate ion
Hydronium ion	H_3O^+	H_2O	Water
Weak Acids		**Strong Bases**	
Hydrogen sulfate ion	HSO_4^-	SO_4^{2-}	Sulfate ion
Phosphoric acid	H_3PO_4	$H_2PO_4^-$	Dihydrogen phosphate ion
Nitrous acid	HNO_2	NO_2^-	Nitrite ion
Hydrofluoric acid	HF	F^-	Fluoride ion
Acetic acid	$HC_2H_3O_2$	$C_2H_3O_2^-$	Acetate ion
Carbonic acid	H_2CO_3	HCO_3^-	Bicarbonate ion
Hydrosulfuric acid	H_2S	HS^-	Hydrogen sulfide ion
Dihydrogen phosphate ion	$H_2PO_4^-$	HPO_4^{2-}	Hydrogen phosphate ion
Ammonium ion	NH_4^+	NH_3	Ammonia
Hydrocyanic acid	HCN	CN^-	Cyanide ion
Bicarbonate ion	HCO_3^-	CO_3^{2-}	Carbonate ion
Methylammonium ion	$CH_3-NH_3^+$	CH_3-NH_2	Methylamine
Hydrogen phosphate ion	HPO_4^{2-}	PO_4^{3-}	Phosphate ion
Hydrogen sulfide ion	HS^-	S^{2-}	Sulfide ion
Water	H_2O	OH^-	Hydroxide ion

Acid Strength Increases ↑

Base Strength Increases ↓

PRACTICE PROBLEMS

Try Practice Problems 14.17 to 14.22

HCl: Completely
dissociated

HC$_2$H$_3$O$_2$: Mostly
molecules and a
few ions

FIGURE 14.1 A strong acid such as HCl is completely dissociated ($\approx$100%), whereas a weak acid such as HC$_2$H$_3$O$_2$ contains mostly molecules and a few ions.

Diprotic Acids

Some weak acids, such as carbonic acid, are *diprotic acids* that have two H$^+$, which dissociate one at a time. For example, carbonated soft drinks are prepared by dissolving CO$_2$ in water to form carbonic acid, H$_2$CO$_3$. A weak acid such as H$_2$CO$_3$ reaches equilibrium between the mostly undissociated H$_2$CO$_3$ molecules and the ions H$_3$O$^+$ and HCO$_3^-$.

$$H_2CO_3(aq) + H_2O(l) \rightleftharpoons H_3O^+(aq) + HCO_3^-(aq)$$
Carbonic acid   Bicarbonate ion
                                       (hydrogen carbonate)

Because HCO$_3^-$ is also a weak acid, a second dissociation can take place to produce another hydronium ion and the carbonate ion, CO$_3^{2-}$.

$$HCO_3^-(aq) + H_2O(l) \rightleftharpoons H_3O^+(aq) + CO_3^{2-}(aq)$$
Bicarbonate ion   Carbonate ion
(hydrogen carbonate)

H$_2$CO$_3$ HCO$_3^-$ CO$_3^{2-}$

Carbonic acid, a weak acid, loses one H$^+$ to form hydrogen carbonate ion, which loses a second H$^+$ to form carbonate ion.

Sulfuric acid, H$_2$SO$_4$, is also a diprotic acid. However, its first dissociation is complete (100%), which means H$_2$SO$_4$ is a strong acid. The product, hydrogen sulfate HSO$_4^-$, can dissociate again but only slightly, which means that the hydrogen sulfate ion is a weak acid.

$$H_2SO_4(aq) + H_2O(l) \longrightarrow H_3O^+(aq) + HSO_4^-(aq)$$
Sulfuric acid   Bisulfate ion
                                      (hydrogen sulfate)

$$HSO_4^-(aq) + H_2O(l) \rightleftharpoons H_3O^+(aq) + SO_4^{2-}(aq)$$
Bisulfate ion   Sulfate ion
(hydrogen sulfate)

In summary, a strong acid such as HI in water dissociates completely to form an aqueous solution of the ions H_3O^+ and I^-. A weak acid such as HF dissociates only slightly in water to form an aqueous solution that consists mostly of HF molecules and a few H_3O^+ and F^- ions (see **FIGURE 14.2**).

Strong acid: $HI(aq) + H_2O(l) \longrightarrow H_3O^+(aq) + I^-(aq)$ Completely dissociated

Weak acid: $HF(aq) + H_2O(l) \rightleftharpoons H_3O^+(aq) + F^-(aq)$ Slightly dissociated

FIGURE 14.2 Strong and weak acids in water

Strong and Weak Bases

As strong electrolytes, **strong bases** dissociate completely in water. Because these strong bases are ionic compounds, they dissociate in water to give an aqueous solution of metal ions and hydroxide ions. The Group 1A (1) hydroxides are very soluble in water, which can give high concentrations of OH^- ions. A few strong bases are less soluble in water, but what does dissolve dissociates completely as ions. For example, when KOH forms a KOH solution, it contains only the ions K^+ and OH^-.

$$KOH(s) \xrightarrow{H_2O} K^+(aq) + OH^-(aq)$$

Bases in household products are used to remove grease.

Bases in Household Products

Weak Bases

Window cleaner, ammonia, NH_3

Bleach, NaOCl

Laundry detergent, Na_2CO_3, Na_3PO_4

Toothpaste and baking soda, $NaHCO_3$

Baking powder, scouring powder, Na_2CO_3

Lime for lawns and agriculture, $CaCO_3$

Laxatives, antacids, $Mg(OH)_2$, $Al(OH)_3$

Strong Bases

Drain cleaner, oven cleaner, NaOH

Strong Bases	
Lithium hydroxide	LiOH
Sodium hydroxide	NaOH
Potassium hydroxide	KOH
Rubidium hydroxide	RbOH
Cesium hydroxide	CsOH
Calcium hydroxide	$Ca(OH)_2$*
Strontium hydroxide	$Sr(OH)_2$*
Barium hydroxide	$Ba(OH)_2$*
*Low solubility, but dissociates completely	

Weak bases are weak electrolytes that are poor acceptors of hydrogen ions and produce very few ions in solution. A typical weak base, ammonia, NH_3, is found in window cleaners. In an aqueous solution, only a few ammonia molecules accept hydrogen ions to form NH_4^+ and OH^-.

$$NH_3(g) + H_2O(l) \rightleftharpoons NH_4^+(aq) + OH^-(aq)$$
Ammonia   Ammonium hydroxide

Direction of Reaction

There is a relationship between the components in each conjugate acid–base pair. Strong acids have weak conjugate bases that do not readily accept H^+. As the strength of the acid decreases, the strength of its conjugate base increases.

In any acid–base reaction, there are two acids and two bases. However, one acid is stronger than the other acid, and one base is stronger than the other base. By comparing their relative strengths, we can determine the direction of the reaction. For example, the strong acid H_2SO_4 readily gives up H^+ to water. The hydronium ion H_3O^+ produced is a weaker acid than H_2SO_4, and the conjugate base HSO_4^- is a weaker base than water.

$$H_2SO_4(aq) + H_2O(l) \longrightarrow H_3O^+(aq) + HSO_4^-(aq) \quad \text{Mostly products}$$
Stronger acid | Stronger base | Weaker acid | Weaker base

Let's look at another reaction in which water donates one H^+ to carbonate, CO_3^{2-}, to form HCO_3^- and OH^-. From **TABLE 14.3**, we see that HCO_3^- is a stronger acid than H_2O. We also see that OH^- is a stronger base than CO_3^{2-}. To reach equilibrium, the stronger acid and stronger base react in the direction of the weaker acid and weaker base.

$$CO_3^{2-}(aq) + H_2O(l) \longleftarrow HCO_3^-(aq) + OH^-(aq) \quad \text{Mostly reactants}$$
Weaker base | Weaker acid | Stronger acid | Stronger base

SAMPLE PROBLEM 14.4 Direction of Reaction

TRY IT FIRST

Does the mixture of the following reaction contain mostly reactants or products?

$$HF(aq) + H_2O(l) \rightleftharpoons H_3O^+(aq) + F^-(aq)$$

SOLUTION

From **TABLE 14.3**, we see that HF is a weaker acid than H_3O^+ and that H_2O is a weaker base than F^-. Thus, the mixture contains mostly reactants.

$$HF(aq) + H_2O(l) \rightleftharpoons H_3O^+(aq) + F^-(aq)$$
Weaker acid | Weaker base | Stronger acid | Stronger base

SELF TEST 14.4

Does the mixture from the reaction of nitric acid and water contain mostly reactants or products?

ANSWER

$$HNO_3(aq) + H_2O(l) \longrightarrow H_3O^+(aq) + NO_3^-(aq)$$

The mixture contains mostly products because HNO_3 is a stronger acid than H_3O^+, and H_2O is a stronger base than NO_3^-.

Hydrofluoric acid is the only halogen acid that is a weak acid.

PRACTICE PROBLEMS

Try Practice Problems 14.23 to 14.26

PRACTICE PROBLEMS

14.3 Strengths of Acids and Bases

14.17 What is meant by the phrase "A strong acid has a weak conjugate base"?

14.18 What is meant by the phrase "A weak acid has a strong conjugate base"?

14.19 Identify the stronger acid in each of the following pairs:
a. HBr or HNO_2 b. H_3PO_4 or HSO_4^- c. HCN or H_2CO_3

14.20 Identify the stronger acid in each of the following pairs:
a. NH_4^+ or H_3O^+ b. H_2SO_4 or HCl c. H_2O or H_2CO_3

14.21 Identify the weaker acid in each of the following pairs:
a. HCl or HSO_4^- b. HNO_2 or HF c. HCO_3^- or NH_4^+

14.22 Identify the weaker acid in each of the following pairs:
a. HNO_3 or HCO_3^- b. HSO_4^- or H_2O c. H_2SO_4 or H_2CO_3

14.23 Predict whether each of the following reactions contains mostly reactants or products at equilibrium:

a. $H_2CO_3(aq) + H_2O(l) \rightleftharpoons HCO_3^-(aq) + H_3O^+(aq)$
b. $NH_4^+(aq) + H_2O(l) \rightleftharpoons NH_3(aq) + H_3O^+(aq)$
c. $HNO_2(aq) + NH_3(aq) \rightleftharpoons NO_2^-(aq) + NH_4^+(aq)$

14.24 Predict whether each of the following reactions contains mostly reactants or products at equilibrium:

a. $H_3PO_4(aq) + H_2O(l) \rightleftharpoons H_3O^+(aq) + H_2PO_4^-(aq)$
b. $CO_3^{2-}(aq) + H_2O(l) \rightleftharpoons OH^-(aq) + HCO_3^-(aq)$
c. $HS^-(aq) + F^-(aq) \rightleftharpoons HF(aq) + S^{2-}(aq)$

14.25 Write an equation for the acid–base reaction between ammonium ion and sulfate ion. Why does the equilibrium mixture contain mostly reactants?

14.26 Write an equation for the acid–base reaction between nitrous acid and hydroxide ion. Why does the equilibrium mixture contain mostly products?

REVIEW

Writing the Equilibrium Expression (13.3)

Using Le Châtelier's Principle (13.5)

14.4 Dissociation of Weak Acids and Bases

LEARNING GOAL Write the expression for the dissociation of a weak acid or weak base.

As we have seen, acids have different strengths depending on how much they dissociate in water. Because the dissociation of strong acids in water is essentially complete, the reaction is not considered to be an equilibrium situation. However, because weak acids in water dissociate only slightly, the ion products reach equilibrium with the undissociated weak acid molecules. For example, formic acid $HCHO_2$, the acid found in bee and ant stings, is a weak acid. Formic acid is a weak acid that dissociates in water to form hydronium ion, H_3O^+, and formate ion, CHO_2^-.

$$HCHO_2(aq) + H_2O(l) \rightleftharpoons H_3O^+(aq) + CHO_2^-(aq)$$

HCHO₂

CHO₂⁻

Formic acid, a weak acid, loses one H⁺ to form formate ion.

Writing Dissociation Expressions

An acid dissociation expression can be written for weak acids that gives the ratio of the concentrations of products to the reactants. As with other equilibrium expressions, the molar concentration of the products is divided by the molar concentration of the reactants. (Recall that the brackets in an equilibrium expression represent the molar concentrations of the reactants and products.) Because water is a pure liquid with a constant concentration, it is omitted.

$$K_a = \frac{[H_3O^+][CHO_2^-]}{[HCHO_2]} \quad \text{Acid dissociation expression}$$

The numerical value of the acid dissociation expression is the **acid dissociation constant, K_a**. The value of the K_a for formic acid at 25 °C is determined by experiment to be 1.8×10^{-4}. Thus, for the weak acid $HCHO_2$, the K_a is written

$$K_a = \frac{[H_3O^+][CHO_2^-]}{[HCHO_2]} = 1.8 \times 10^{-4} \quad \text{Acid dissociation constant}$$

The K_a for formic acid is small, which means that the equilibrium mixture of formic acid in water contains mostly reactants and only small amounts of the products. Weak acids have small K_a values. However, strong acids, which are essentially 100% dissociated, have very large K_a values, but these values are not usually given. **TABLE 14.4** gives K_a and K_b values for selected weak acids and bases.

Similar to what we saw for weak acids, weak bases have small K_b values. Strong bases, which are essentially 100% dissociated, have very large K_b values, but these values are not usually given.

Let us now consider the dissociation of the weak base methylamine:

$$CH_3-NH_2(aq) + H_2O(l) \rightleftharpoons CH_3-NH_3^+(aq) + OH^-(aq)$$

As we did with the acid dissociation expression, the concentration of water is omitted from the base dissociation expression. The **base dissociation constant, K_b,** for methylamine is written

$$K_b = \frac{[CH_3{-}NH_3{}^+][OH^-]}{[CH_3{-}NH_2]} = 4.4 \times 10^{-4}$$

TABLE 14.4 gives K_a and K_b values for selected weak acids and bases.

TABLE 14.4 K_a and K_b Values for Selected Weak Acids and Bases

Acids		K_a
Phosphoric acid	H_3PO_4	7.5×10^{-3}
Nitrous acid	HNO_2	4.5×10^{-4}
Hydrofluoric acid	HF	3.5×10^{-4}
Formic acid	$HCHO_2$	1.8×10^{-4}
Acetic acid	$HC_2H_3O_2$	1.8×10^{-5}
Carbonic acid	H_2CO_3	4.3×10^{-7}
Hydrosulfuric acid	H_2S	9.1×10^{-8}
Dihydrogen phosphate	$H_2PO_4{}^-$	6.2×10^{-8}
Hypochlorous acid	$HClO$	3.0×10^{-8}
Hydrocyanic acid	HCN	4.9×10^{-10}
Hydrogen carbonate	$HCO_3{}^-$	5.6×10^{-11}
Hydrogen phosphate	$HPO_4{}^{2-}$	2.2×10^{-13}
Bases		K_b
Methylamine	$CH_3{-}NH_2$	4.4×10^{-4}
Carbonate	$CO_3{}^{2-}$	2.2×10^{-4}
Ammonia	NH_3	1.8×10^{-5}

TABLE 14.5 summarizes the characteristics of acids and bases in terms of strength and equilibrium position.

TABLE 14.5 Characteristics of Acids and Bases

Characteristic	Strong Acids	Weak Acids
Equilibrium Position	Toward products	Toward reactants
K_a	Large	Small
$[H_3O^+]$ and $[A^-]$	100% of [HA] dissociates	Small percent of [HA] dissociates
Conjugate Base	Weak	Strong
Characteristic	**Strong Bases**	**Weak Bases**
Equilibrium Position	Toward products	Toward reactants
K_b	Large	Small
$[BH^+]$ and $[OH^-]$	100% of [B] reacts	Small percent of [B] reacts
Conjugate Acid	Weak	Strong

SAMPLE PROBLEM 14.5 Writing an Acid Dissociation Expression

TRY IT FIRST

Write the acid dissociation expression for the weak acid, nitrous acid.

SOLUTION

The equation for the dissociation of nitrous acid is written

$$HNO_2(aq) + H_2O(l) \rightleftharpoons H_3O^+(aq) + NO_2{}^-(aq)$$

The acid dissociation expression is written as the concentration of the products divided by the concentration of the undissociated weak acid.

$$K_a = \frac{[H_3O^+][NO_2^-]}{[HNO_2]}$$

SELF TEST 14.5

Write the acid dissociation expression for each of the following:

a. hydrogen phosphate
b. hypochlorous acid

ANSWER

a. $K_a = \dfrac{[H_3O^+][PO_4^{3-}]}{[HPO_4^{2-}]}$

b. $K_a = \dfrac{[H_3O^+][ClO^-]}{[HClO]}$

PRACTICE PROBLEMS

Try Practice Problems 14.27 to 14.32

PRACTICE PROBLEMS

14.4 Dissociation of Weak Acids and Bases

14.27 Answer *True* or *False* for each of the following: A strong acid
  a. is completely dissociated in aqueous solution
  b. has a small value of K_a
  c. has a strong conjugate base
  d. has a weak conjugate base
  e. is slightly dissociated in aqueous solution

14.28 Answer *True* or *False* for each of the following: A weak acid
  a. is completely dissociated in aqueous solution
  b. has a small value of K_a
  c. has a strong conjugate base
  d. has a weak conjugate base
  e. is slightly dissociated in aqueous solution

14.29 Consider the following acids and their dissociation constants:

$$H_2SO_3(aq) + H_2O(l) \rightleftarrows H_3O^+(aq) + HSO_3^-(aq)$$
$$K_a = 1.2 \times 10^{-2}$$

$$HS^-(aq) + H_2O(l) \rightleftarrows H_3O^+(aq) + S^{2-}(aq)$$
$$K_a = 1.3 \times 10^{-19}$$

  a. Which is the stronger acid, H_2SO_3 or HS^-?
  b. What is the conjugate base of H_2SO_3?
  c. Which acid has the weaker conjugate base?

  d. Which acid has the stronger conjugate base?
  e. Which acid produces more ions?

14.30 Consider the following acids and their dissociation constants:

$$HPO_4^{2-}(aq) + H_2O(l) \rightleftarrows H_3O^+(aq) + PO_4^{3-}(aq)$$
$$K_a = 2.2 \times 10^{-13}$$

$$HCHO_2(aq) + H_2O(l) \rightleftarrows H_3O^+(aq) + CHO_2^-(aq)$$
$$K_a = 1.8 \times 10^{-4}$$

  a. Which is the weaker acid, HPO_4^{2-} or $HCHO_2$?
  b. What is the conjugate base of HPO_4^{2-}?
  c. Which acid has the weaker conjugate base?
  d. Which acid has the stronger conjugate base?
  e. Which acid produces more ions?

14.31 Phosphoric acid dissociates to form hydronium ion and dihydrogen phosphate. Phosphoric acid has a K_a of 7.5×10^{-3}. Write the equation for the reaction and the acid dissociation expression for phosphoric acid.

14.32 Aniline, $C_6H_5-NH_2$, a weak base with a K_b of 4.0×10^{-10}, reacts with water to form $C_6H_5-NH_3^+$ and hydroxide ion. Write the equation for the reaction and the base dissociation expression for aniline.

14.5 Dissociation of Water

LEARNING GOAL Use the water dissociation expression to calculate the $[H_3O^+]$ and $[OH^-]$ in an aqueous solution.

In many acid–base reactions, water is *amphoteric*, which means that it can act either as an acid or as a base. In pure water, there is a forward reaction between two water molecules that transfers H^+ from one water molecule to the other. One molecule acts as an acid by losing H^+, and the water molecule that gains H^+ acts as a base. Every time H^+ is transferred between two water molecules, the products are one H_3O^+ and one OH^-, which react in the reverse direction to re-form two water molecules. Thus, equilibrium is reached between the conjugate acid–base pairs of water molecules.

ENGAGE 14.6

Why is the $[H_3O^+]$ equal to the $[OH^-]$ in pure water?

Conjugate acid–base pair

$$:\ddot{O}-H \ + \ :\ddot{O}-H \ \rightleftharpoons \ H-\ddot{O}-H^+ \ + \ \bar{:}\ddot{O}-H$$

Conjugate acid–base pair

Acid	**Base**	**Acid**	**Base**
(H^+ donor)	(H^+ acceptor)	(H^+ donor)	(H^+ acceptor)

Writing the Water Dissociation Expression

Using the equation for the dissociation of water, we can write the equilibrium expression that shows the concentrations of the products divided by the concentrations of the reactants.

$$H_2O(l) + H_2O(l) \rightleftharpoons H_3O^+(aq) + OH^-(aq)$$

$$K = \frac{[H_3O^+][OH^-]}{[H_2O][H_2O]}$$

By omitting the constant concentration of pure water, we can write the **water dissociation expression**.

$$K_w = [H_3O^+][OH^-]$$

Experiments have determined that, in pure water, the concentration of H_3O^+ and OH^- at 25 °C are each 1.0×10^{-7} M.

Pure water $[H_3O^+] = [OH^-] = 1.0 \times 10^{-7}$ M

When we place the $[H_3O^+]$ and $[OH^-]$ into the water dissociation expression, we obtain the numerical value of the **water dissociation constant, K_w**, which is 1.0×10^{-14} at 25 °C. As before, the concentration units are omitted in the K_w value.

$$K_w = [H_3O^+][OH^-]$$
$$= [1.0 \times 10^{-7}][1.0 \times 10^{-7}] = 1.0 \times 10^{-14}$$

PRACTICE PROBLEMS

Try Practice Problems 14.33 and 14.34

Neutral, Acidic, and Basic Solutions

The K_w value (1.0×10^{-14}) applies to any aqueous solution at 25 °C because all aqueous solutions contain both H_3O^+ and OH^- (see **FIGURE 14.3**). When the $[H_3O^+]$ and $[OH^-]$ in a solution are equal, the solution is **neutral**. However, most solutions are not neutral; they have different concentrations of H_3O^+ and OH^-. If acid is added to water, there is an increase

FIGURE 14.3 In a neutral solution, $[H_3O^+]$ and $[OH^-]$ are equal. In acidic solutions, the $[H_3O^+]$ is greater than the $[OH^-]$. In basic solutions, the $[OH^-]$ is greater than the $[H_3O^+]$.

ENGAGE 14.7

Is a solution that has a $[H_3O^+]$ of 1.0×10^{-3} M acidic, basic, or neutral?

PRACTICE PROBLEMS

Try Practice Problems 14.35 and 14.36

CORE CHEMISTRY SKILL

Calculating $[H_3O^+]$ and $[OH^-]$ in Solutions

ENGAGE 14.8

If you know the $[H_3O^+]$ of a solution, how do you use the K_w to calculate the $[OH^-]$?

ENGAGE 14.9

Why does the $[H_3O^+]$ of an aqueous solution increase if the $[OH^-]$ decreases?

in $[H_3O^+]$ and a decrease in $[OH^-]$, which makes an acidic solution. If base is added, $[OH^-]$ increases and $[H_3O^+]$ decreases, which gives a basic solution. However, for any aqueous solution, whether it is neutral, acidic, or basic, the product $[H_3O^+][OH^-]$ is equal to K_w (1.0×10^{-14} at 25 °C) (see **TABLE 14.6**).

TABLE 14.6 Examples of $[H_3O^+]$ and $[OH^-]$ in Neutral, Acidic, and Basic Solutions

Type of Solution	$[H_3O^+]$	$[OH^-]$	K_w (25 °C)
Neutral	1.0×10^{-7} M	1.0×10^{-7} M	1.0×10^{-14}
Acidic	1.0×10^{-2} M	1.0×10^{-12} M	1.0×10^{-14}
Acidic	2.5×10^{-5} M	4.0×10^{-10} M	1.0×10^{-14}
Basic	1.0×10^{-8} M	1.0×10^{-6} M	1.0×10^{-14}
Basic	5.0×10^{-11} M	2.0×10^{-4} M	1.0×10^{-14}

Using the K_w to Calculate $[H_3O^+]$ and $[OH^-]$ in a Solution

If we know the $[H_3O^+]$ of a solution, we can use the K_w to calculate $[OH^-]$. If we know the $[OH^-]$ of a solution, we can calculate $[H_3O^+]$ from their relationship in the K_w, as shown in Sample Problem 14.6.

$$K_w = [H_3O^+][OH^-]$$

$$[OH^-] = \frac{K_w}{[H_3O^+]} \qquad [H_3O^+] = \frac{K_w}{[OH^-]}$$

SAMPLE PROBLEM 14.6 Calculating the $[H_3O^+]$ of a Solution

TRY IT FIRST

A vinegar solution has a $[OH^-] = 5.0 \times 10^{-12}$ M at 25 °C. What is the $[H_3O^+]$ of the vinegar solution? Is the solution acidic, basic, or neutral?

SOLUTION

STEP 1 State the given and needed quantities.

ANALYZE THE PROBLEM	Given	Need	Connect
	$[OH^-] = 5.0 \times 10^{-12}$ M	$[H_3O^+]$	$K_w = [H_3O^+][OH^-]$

STEP 2 Write the K_w for water and solve for the unknown $[H_3O^+]$.

$$K_w = [H_3O^+][OH^-] = 1.0 \times 10^{-14}$$

Solve for $[H_3O^+]$ by dividing both sides by $[OH^-]$.

$$\frac{K_w}{[OH^-]} = \frac{[H_3O^+][OH^-]}{[OH^-]}$$

$$[H_3O^+] = \frac{1.0 \times 10^{-14}}{[OH^-]}$$

STEP 3 Substitute the known $[OH^-]$ into the equation and calculate.

$$[H_3O^+] = \frac{1.0 \times 10^{-14}}{[5.0 \times 10^{-12}]} = 2.0 \times 10^{-3} \text{ M}$$

Because the $[H_3O^+]$ of 2.0×10^{-3} M is larger than the $[OH^-]$ of 5.0×10^{-12} M, the solution is acidic.

a. What is the $[H_3O^+]$ of an ammonia cleaning solution with $[OH^-] = 4.0 \times 10^{-4}$ M? Is the solution acidic, basic, or neutral?

b. The $[H_3O^+]$ of tomato juice is 6.3×10^{-5} M. What is the $[OH^-]$ of the juice? Is the tomato juice acidic, basic, or neutral?

ANSWER

a. $[H_3O^+] = 2.5 \times 10^{-11}$ M, basic   **b.** $[OH^-] = 1.6 \times 10^{-10}$ M, acidic

PRACTICE PROBLEMS
Try Practice Problems 14.37 to 14.42

PRACTICE PROBLEMS

14.5 Dissociation of Water

14.33 Why are the concentrations of H_3O^+ and OH^- equal in pure water?

14.34 What is the meaning and value of K_w at 25 °C?

14.35 In an acidic solution, how does the concentration of H_3O^+ compare to the concentration of OH^-?

14.36 If a base is added to pure water, why does the $[H_3O^+]$ decrease?

14.37 Indicate whether each of the following solutions is acidic, basic, or neutral:
  a. $[H_3O^+] = 2.0 \times 10^{-5}$ M
  b. $[H_3O^+] = 1.4 \times 10^{-9}$ M
  c. $[OH^-] = 8.0 \times 10^{-3}$ M
  d. $[OH^-] = 3.5 \times 10^{-10}$ M

14.38 Indicate whether each of the following solutions is acidic, basic, or neutral:
  a. $[H_3O^+] = 6.0 \times 10^{-12}$ M
  b. $[H_3O^+] = 1.4 \times 10^{-4}$ M
  c. $[OH^-] = 5.0 \times 10^{-12}$ M
  d. $[OH^-] = 4.5 \times 10^{-2}$ M

14.39 Calculate the $[H_3O^+]$ of each aqueous solution with the following $[OH^-]$:
  a. coffee, 1.0×10^{-9} M   **b.** soap, 1.0×10^{-6} M
  c. cleanser, 2.0×10^{-5} M   **d.** lemon juice, 4.0×10^{-13} M

14.40 Calculate the $[H_3O^+]$ of each aqueous solution with the following $[OH^-]$:
  a. NaOH solution, 1.0×10^{-2} M
  b. milk of magnesia, 1.0×10^{-5} M
  c. aspirin, 1.8×10^{-11} M
  d. seawater, 2.5×10^{-6} M

Applications

14.41 Calculate the $[OH^-]$ of each aqueous solution with the following $[H_3O^+]$:
  a. stomach acid, 4.0×10^{-2} M
  b. urine, 5.0×10^{-6} M
  c. orange juice, 2.0×10^{-4} M
  d. bile, 7.9×10^{-9} M

14.42 Calculate the $[OH^-]$ of each aqueous solution with the following $[H_3O^+]$:
  a. baking soda, 1.0×10^{-8} M
  b. blood, 4.2×10^{-8} M
  c. milk, 5.0×10^{-7} M
  d. pancreatic juice, 4.0×10^{-9} M

14.6 The pH Scale

LEARNING GOAL Calculate pH from $[H_3O^+]$; given the pH, calculate the $[H_3O^+]$ and $[OH^-]$ of a solution.

In the environment, the acidity, or pH, of rain can have significant effects. When rain becomes too acidic, it can dissolve marble statues and accelerate the corrosion of metals. In lakes and ponds, the acidity of water can affect the ability of plants and fish to survive. The acidity of soil around plants affects their growth. If the soil pH is too acidic or too basic, the roots of the plant cannot take up some nutrients. Most plants thrive in soil with a nearly neutral pH, although certain plants, such as orchids, camellias, and blueberries, require a more acidic soil.

Although we have expressed H_3O^+ and OH^- as molar concentrations, it is more convenient to describe the acidity of solutions using the *pH scale*. On this scale, a number between 0 and 14 represents the H_3O^+ concentration for common solutions. A neutral solution has a pH of 7.0 at 25 °C. An acidic solution has a pH less than 7.0; a basic solution has a pH greater than 7.0 (see **FIGURE 14.4**).

When we relate acidity and pH, we are using an inverse relationship, which is when one component increases while the other component decreases. When an acid is added to pure water, the $[H_3O^+]$ (acidity) of the solution increases but its pH decreases. When a base is added to pure water, it becomes more basic, which means its acidity decreases and the pH increases.

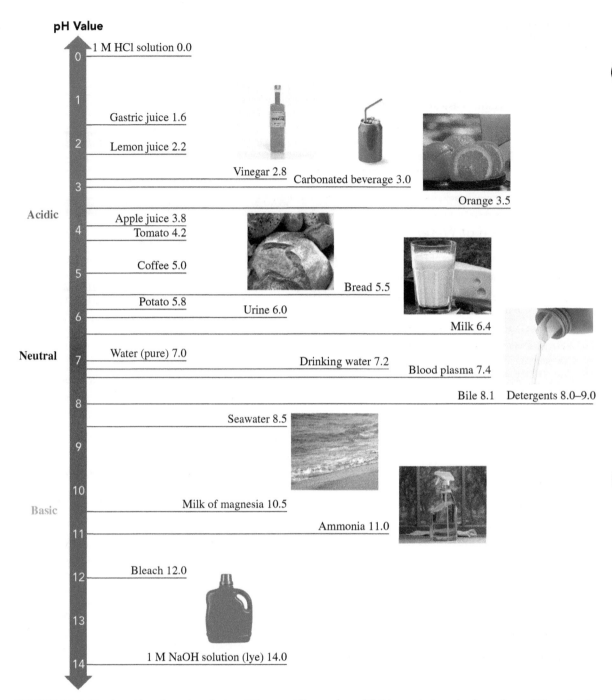

pH Value

0 — 1 M HCl solution 0.0

1 —

Gastric juice 1.6

2 —
Lemon juice 2.2

Vinegar 2.8   Carbonated beverage 3.0

3 —
Orange 3.5

Acidic

Apple juice 3.8
4 — Tomato 4.2

Coffee 5.0

5 —
Bread 5.5

Potato 5.8
Urine 6.0
6 —
Milk 6.4

Neutral 7 — Water (pure) 7.0

Drinking water 7.2
Blood plasma 7.4

Bile 8.1   Detergents 8.0–9.0
8 —

Seawater 8.5

9 —

10 —

Basic
Milk of magnesia 10.5

Ammonia 11.0
11 —

12 — Bleach 12.0

13 —

1 M NaOH solution (lye) 14.0
14 —

FIGURE 14.4 On the pH scale, values below 7.0 are acidic, a value of 7.0 is neutral, and values above 7.0 are basic.

ENGAGE 14.10

Is apple juice an acidic, a basic, or a neutral solution?

In the laboratory, a pH meter is commonly used to determine the pH of a solution. There are also various indicators and pH papers that turn specific colors when placed in solutions of different pH values. The pH is found by comparing the color on the test paper or the color of the solution to a color chart (see **FIGURE 14.5**).

FIGURE 14.5 The pH of a solution can be determined using different methods.

pH meter   pH paper   pH indicator

SAMPLE PROBLEM 14.7 pH of Solutions

TRY IT FIRST

Consider the pH of the following body fluids:

Body Fluid	pH
Stomach acid	1.4
Pancreatic juice	8.4
Sweat	4.8
Urine	5.3
Cerebrospinal fluid	7.3

a. Place the pH values of the body fluids on the list in order of most acidic to most basic.
b. Which body fluid has the highest $[H_3O^+]$?

SOLUTION

a. The most acidic body fluid is the one with the lowest pH, and the most basic is the body fluid with the highest pH: stomach acid (1.4), sweat (4.8), urine (5.3), cerebrospinal fluid (7.3), pancreatic juice (8.4).
b. The body fluid with the highest $[H_3O^+]$ would have the lowest pH value, which is stomach acid.

SELF TEST 14.7

a. Which body fluid has the highest $[OH^-]$?
b. Which body fluid is more acidic, urine or cerebrospinal fluid?

ANSWER

a. pancreatic juice
b. urine

Calculating the pH of Solutions

The pH scale is a *logarithmic scale* that corresponds to the $[H_3O^+]$ of aqueous solutions. Mathematically, **pH** is the negative logarithm (base 10) of the $[H_3O^+]$.

$$pH = -\log[H_3O^+]$$

Essentially, the negative powers of 10 in the molar concentrations are converted to positive numbers. For example, a lemon juice solution with $[H_3O^+] = 1.0 \times 10^{-2}$ M has a pH of 2.00. This can be calculated using the pH equation:

$$pH = -\log[1.0 \times 10^{-2}]$$
$$pH = -(-2.00)$$
$$= 2.00$$

The number of *decimal places* in the pH value is the same as the number of significant figures in the $[H_3O^+]$. The number to the left of the decimal point in the pH value is the power of 10.

$$[H_3O^+] = 1.0 \times 10^{-2} \qquad pH = 2.00$$

Two SFs   Two SFs

Because pH is a log scale, a change of one pH unit corresponds to a tenfold change in $[H_3O^+]$. It is important to recall that the pH decreases as the $[H_3O^+]$ increases. For example, a solution with a pH of 2.00 has a $[H_3O^+]$ that is ten times greater than a solution with a pH of 3.00 and 100 times greater than a solution with a pH of 4.00. The pH of a solution is calculated from the $[H_3O^+]$ by using the *log* key and changing the sign as shown in Sample Problem 14.8.

ENGAGE 14.11

If a pH meter reads 4.00, why is the solution acidic?

A dipstick is used to measure the pH of a urine sample.

PRACTICE PROBLEMS

Try Practice Problems 14.43 to 14.46

KEY MATH SKILL

Calculating pH from $[H_3O^+]$

If soil is too acidic, nutrients are not absorbed by crops. Then lime ($CaCO_3$), which acts as a base, may be added to increase the soil pH.

ENGAGE 14.12

Explain why 6.00, but not 6.0, is the correct pH for $[H_3O^+] = 1.0 \times 10^{-6}$ M.

SAMPLE PROBLEM 14.8 Calculating pH from [H₃O⁺]

TRY IT FIRST

Aspirin, which is acetylsalicylic acid, was the first nonsteroidal anti-inflammatory drug (NSAID) used to alleviate pain and fever. If a solution of aspirin has a $[H_3O^+] = 1.7 \times 10^{-3}$ M, what is the pH of the solution?

Acidic H
that dissociates in
aqueous solution

Aspirin, acetylsalicylic acid, is a weak acid.

SOLUTION

STEP 1 State the given and needed quantities.

ANALYZE THE PROBLEM	Given	Need	Connect
	$[H_3O^+] = 1.7 \times 10^{-3}$ M	pH	pH equation

STEP 2 Enter the [H₃O⁺] into the pH equation and calculate.

$$pH = -\log[H_3O^+] = -\log[1.7 \times 10^{-3}]$$

Calculator Procedure   **Calculator Display**

1.7 [EE or EXP] [+/−] 3 [log] [+/−] [=] or [+/−] [log] 1.7 [EE or EXP] [+/−] 3 [=]   2.769551079

Be sure to check the instructions for your calculator.

STEP 3 Adjust the number of SFs on the *right* of the decimal point. In a pH value, the number to the *left* of the decimal point is an *exact* number derived from the power of 10. Thus, the two SFs in the coefficient determine that there are two SFs after the decimal point in the pH value.

Coefficient   **Power of ten**

   1.7   × 10^{-3} M $pH = -\log[1.7 \times 10^{-3}] = 2.77$

Two SFs   Exact   Exact   Two SFs

SELF TEST 14.8

a. What is the pH of bleach with $[H_3O^+] = 4.2 \times 10^{-12}$ M?
b. What is the pH of a borax solution with $[H_3O^+] = 6.1 \times 10^{-10}$ M?

ANSWER

a. pH = 11.38   **b.** pH = 9.21

PRACTICE PROBLEMS

Try Practice Problems 14.47 and 14.48

pOH

The **pOH** scale is similar to the pH scale except that pOH is associated with the [OH⁻] of an aqueous solution.

$$pOH = -\log[OH^-]$$

Solutions with high [OH⁻] have low pOH values; solutions with low [OH⁻] have high pOH values. In any aqueous solution, the sum of the pH and pOH is equal to 14.00, which is the negative logarithm of the K_w.

$$pH + pOH = 14.00$$

For example, if the pH of a solution is 3.50, the pOH can be calculated as follows:

$$pH + pOH = 14.00$$
$$pOH = 14.00 - pH = 14.00 - 3.50 = 10.50$$

A comparison of [H₃O⁺], [OH⁻], and their corresponding pH and pOH values is given in **TABLE 14.7**.

TABLE 14.7 A Comparison of pH and pOH Values at 25 °C, [H₃O⁺], and [OH⁻]

pH	[H₃O⁺]	[OH⁻]	pOH	
0	10^0	10^{-14}	14	
1	10^{-1}	10^{-13}	13	
2	10^{-2}	10^{-12}	12	
3	10^{-3}	10^{-11}	11	Acidic
4	10^{-4}	10^{-10}	10	
5	10^{-5}	10^{-9}	9	
6	10^{-6}	10^{-8}	8	
7	10^{-7}	10^{-7}	7	Neutral
8	10^{-8}	10^{-6}	6	
9	10^{-9}	10^{-5}	5	
10	10^{-10}	10^{-4}	4	Basic
11	10^{-11}	10^{-3}	3	
12	10^{-12}	10^{-2}	2	
13	10^{-13}	10^{-1}	1	
14	10^{-14}	10^0	0	

Acids produce the sour taste of the fruits we eat.

When we need to calculate the pH from [OH⁻], we use the K_w to calculate [H₃O⁺], place it in the pH equation, and calculate the pH of the solution as shown in Sample Problem 14.9.

SAMPLE PROBLEM 14.9   Calculating pH from [OH⁻]

TRY IT FIRST

What is the pH of an ammonia solution with [OH⁻] = 3.7 × 10⁻³ M?

SOLUTION

STEP 1   State the given and needed quantities.

	Given	Need	Connect
ANALYZE THE PROBLEM	[OH⁻] = 3.7 × 10⁻³ M	[H₃O⁺], pH	K_w = [H₃O⁺][OH⁻], pH equation

STEP 2 **Enter the $[H_3O^+]$ into the pH equation and calculate.** Because $[OH^-]$ is given for the ammonia solution, we have to calculate $[H_3O^+]$. Using the water dissociation expression, we divide both sides by $[OH^-]$ to obtain $[H_3O^+]$.

$$K_w = [H_3O^+][OH^-] = 1.0 \times 10^{-14}$$

$$\frac{K_w}{[OH^-]} = \frac{[H_3O^+][\cancel{OH^-}]}{[\cancel{OH^-}]}$$

$$[H_3O^+] = \frac{1.0 \times 10^{-14}}{[3.7 \times 10^{-3}]} = 2.7 \times 10^{-12} \text{ M}$$

Now, we enter the $[H_3O^+]$ into the pH equation.

$$pH = -\log[H_3O^+] = -\log[2.7 \times 10^{-12}]$$

Calculator Procedure **Calculator Display**

2.7 (EE or EXP) (+/−) 12 (log) (+/−) (=) or (+/−) (log) 2.7 (EE or EXP) (+/−) 12 (=)   *11.56863624*

STEP 3 **Adjust the number of SFs on the *right* of the decimal point.**

2.7×10^{-12} M   pH = 11.57
Two SFs   Two SFs to the *right* of the decimal point

SELF TEST 14.9

a. Calculate the pH of a sample of bile that has $[OH^-] = 1.3 \times 10^{-6}$ M.
b. What is the pH of egg whites with $[OH^-] = 1.2 \times 10^{-5}$ M?

ANSWER

a. pH = 8.11   **b.** pH = 9.08

PRACTICE PROBLEMS

Try Practice Problems 14.49 and 14.50

KEY MATH SKILL

Calculating $[H_3O^+]$ from pH

Calculating $[H_3O^+]$ from pH

If we are given the pH of the solution and asked to determine the $[H_3O^+]$, we need to reverse the calculation of pH.

$$[H_3O^+] = 10^{-pH}$$

For example, if the pH of a solution is 3.0, we can substitute it into this equation. The number of significant figures in $[H_3O^+]$ is equal to the number of decimal places in the pH value.

$$[H_3O^+] = 10^{-pH} = 10^{-3.0} = 1 \times 10^{-3} \text{ M}$$

For pH values that are not whole numbers, the calculation requires the use of the 10^x key, which is usually a *2nd function* key. On some calculators, this operation is done using the inverse log equation as shown in Sample Problem 14.10.

SAMPLE PROBLEM 14.10   Calculating $[H_3O^+]$ from pH

TRY IT FIRST

Calculate $[H_3O^+]$ for a urine sample, which has a pH of 7.5.

SOLUTION

STEP 1 **State the given and needed quantities.**

ANALYZE THE PROBLEM	Given	Need	Connect
	pH = 7.5	$[H_3O^+]$	$[H_3O^+] = 10^{-pH}$

STEP 2 **Enter the pH value into the inverse log equation and calculate.**

$$[H_3O^+] = 10^{-pH} = 10^{-7.5}$$

Calculator Procedure **Calculator Display**

(2nd) (log) (+/−) 7.5 (=) or 7.5 (+/−) (2nd) (log) (=)   *3.16227766E−08*

STEP 3   **Adjust the SFs for the coefficient.** Because the pH value 7.5 has one digit to the *right* of the decimal point, the coefficient for $[H_3O^+]$ is written with one SF.

$$[H_3O^+] = 3 \times 10^{-8} \text{ M}$$
One SF

SELF TEST 14.10

a. What are the $[H_3O^+]$ and $[OH^-]$ of cola that has a pH of 3.17?
b. What are the $[H_3O^+]$ and $[OH^-]$ of pancreatic juice that has a pH of 8.3?

ANSWER

a. $[H_3O^+] = 6.8 \times 10^{-4} \text{ M}$, $[OH^-] = 1.5 \times 10^{-11} \text{ M}$
b. $[H_3O^+] = 5 \times 10^{-9} \text{ M}$, $[OH] = 2 \times 10^{-6} \text{ M}$

The pH of a liquid is measured using a pH meter.

PRACTICE PROBLEMS

Try Practice Problems 14.51 to 14.54

Chemistry Link to Health

Stomach Acid, HCl

Gastric acid, which contains HCl, is produced by parietal cells that line the stomach. When the stomach expands with the intake of food, the gastric glands begin to secrete a strongly acidic solution of HCl. In a single day, a person may secrete 2000 mL of gastric juice, which contains hydrochloric acid, mucins, and the enzymes pepsin and lipase.

The HCl in the gastric juice activates a digestive enzyme from the chief cells called *pepsinogen* to form *pepsin*, which breaks down proteins in food entering the stomach. The secretion of HCl continues until the stomach has a pH of about 2, which is the optimum for activating the digestive enzymes without ulcerating the stomach lining. In addition, the low pH destroys bacteria that reach the stomach. Normally, large quantities of viscous mucus are secreted within the stomach to protect its lining from acid and enzyme damage. Gastric acid may also form under conditions of stress when the nervous system activates the production of HCl. As the contents of the stomach move into the small intestine, cells produce bicarbonate that neutralizes the gastric acid until the pH is about 5.

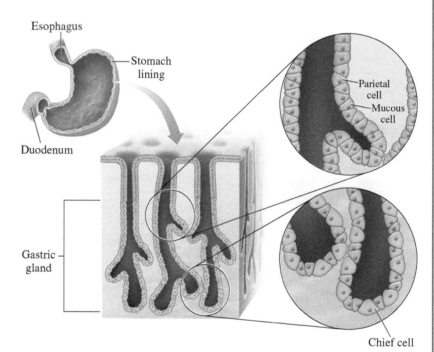

Parietal cells in the lining of the stomach secrete gastric acid HCl.

PRACTICE PROBLEMS

14.6 The pH Scale

14.43 Why does a neutral solution have a pH of 7.0?

14.44 If you know the $[OH^-]$, how can you determine the pH of a solution?

14.45 State whether each of the following solutions is acidic, basic, or neutral:
    a. blood plasma, pH 7.38   **b.** vinegar, pH 2.8
    c. drain cleaner, pOH 2.8   **d.** coffee, pH 5.52
    e. tomatoes, pH 4.2   **f.** chocolate cake, pH 7.6

14.46 State whether each of the following solutions is acidic, basic, or neutral:
    a. soda, pH 3.22
    b. shampoo, pOH 8.3
    c. laundry detergent, pOH 4.56
    d. rain, pH 5.8
    e. honey, pH 3.9
    f. cheese, pH 4.9

14.47 A solution with a pH of 3 is 10 times more acidic than a solution with pH 4. Explain.

14.48 A solution with a pH of 10 is 100 times more basic than a solution with pH 8. Explain.

14.49 Calculate the pH of each solution given the following:
a. $[H_3O^+] = 1 \times 10^{-4}$ M   b. $[H_3O^+] = 3 \times 10^{-9}$ M
c. $[OH^-] = 1 \times 10^{-5}$ M   d. $[OH^-] = 2.5 \times 10^{-11}$ M
e. $[H_3O^+] = 6.7 \times 10^{-8}$ M   f. $[OH^-] = 8.2 \times 10^{-4}$ M

14.50 Calculate the pOH of each solution given the following:
a. $[H_3O^+] = 1 \times 10^{-8}$ M   b. $[H_3O^+] = 5 \times 10^{-6}$ M
c. $[OH^-] = 1 \times 10^{-2}$ M   d. $[OH^-] = 8.0 \times 10^{-3}$ M
e. $[H_3O^+] = 4.7 \times 10^{-2}$ M   f. $[OH^-] = 3.9 \times 10^{-6}$ M

Applications

14.51 Complete the following table:

$[H_3O^+]$	$[OH^-]$	pH	pOH	Acidic, Basic, or Neutral?
	1.0×10^{-6} M			
		3.49		
2.8×10^{-5} M				
			2.00	

14.52 Complete the following table:

$[H_3O^+]$	$[OH^-]$	pH	pOH	Acidic, Basic, or Neutral?
		10.00		
				Neutral
			5.66	
6.4×10^{-12} M				

14.53 A patient with severe metabolic acidosis has a blood plasma pH of 6.92. What is the $[H_3O^+]$ of the blood plasma?

14.54 A patient with respiratory alkalosis has a blood plasma pH of 7.58. What is the $[H_3O^+]$ of the blood plasma?

REVIEW
Balancing a Chemical Equation (8.2)

Magnesium reacts rapidly with acid and forms H_2 gas and a salt of magnesium.

Sodium bicarbonate (baking soda) reacts with an acid (vinegar) to form carbon dioxide gas, water, and a salt.

14.7 Reactions of Acids and Bases

LEARNING GOAL Write balanced equations for reactions of acids with metals, carbonates or bicarbonates, and bases.

Typical reactions of acids and bases include the reactions of acids with metals, carbonates or bicarbonates, and bases. For example, when you drop an antacid tablet in water, the bicarbonate ion and citric acid in the tablet react to produce carbon dioxide bubbles, water, and a salt. A **salt** is an ionic compound that does not have H^+ as the cation or OH^- as the anion.

Acids and Metals

Acids react with certain metals to produce hydrogen gas (H_2) and a salt. Active metals include potassium, sodium, calcium, magnesium, aluminum, zinc, iron, and tin. In these single replacement reactions, the metal ion replaces the hydrogen in the acid.

$$2HCl(aq) + Mg(s) \longrightarrow H_2(g) + MgCl_2(aq)$$
$\quad$ Acid $\qquad$ Metal $\qquad$ Hydrogen $\quad$ Salt

$$2HNO_3(aq) + Zn(s) \longrightarrow H_2(g) + Zn(NO_3)_2(aq)$$
$\quad$ Acid $\qquad$ Metal $\qquad$ Hydrogen $\quad$ Salt

Acids React with Carbonates or Bicarbonates

When an acid is added to a carbonate or bicarbonate, the products are carbon dioxide gas, water, and a salt. The acid reacts with CO_3^{2-} or HCO_3^- to produce carbonic acid, H_2CO_3, which breaks down rapidly to CO_2 and H_2O.

$$2HCl(aq) + Na_2CO_3(aq) \longrightarrow CO_2(g) + H_2O(l) + 2NaCl(aq)$$
$\quad$ Acid $\qquad$ Carbonate $\qquad$ Carbon $\quad$ Water $\qquad$ Salt
$\qquad\qquad\qquad\qquad\qquad$ dioxide

$$HBr(aq) + NaHCO_3(aq) \longrightarrow CO_2(g) + H_2O(l) + NaBr(aq)$$
$\quad$ Acid $\qquad$ Bicarbonate $\qquad$ Carbon $\quad$ Water $\qquad$ Salt
$\qquad\qquad\qquad\qquad\qquad$ dioxide

Acids and Hydroxides: Neutralization

Neutralization is a reaction between a strong or weak acid with a strong base to produce a water and a salt. The H^+ of the acid and the OH^- of the base combine to form water. The salt is the combination of the cation from the base and the anion from the acid. We can write the following equation for the neutralization reaction between HCl and NaOH:

$$HCl(aq) + NaOH(aq) \longrightarrow H_2O(l) + NaCl(aq)$$
Acid   Base   Water   Salt

If we write the strong acid HCl and the strong base NaOH as ions, we see that H^+ combines with OH^- to form water, leaving the ions Na^+ and Cl^- in solution.

$$H^+(aq) + Cl^-(aq) + Na^+(aq) + OH^-(aq) \longrightarrow H_2O(l) + Na^+(aq) + Cl^-(aq)$$

When we omit the ions that do not change during the reaction (*spectator ions*), we obtain the *net ionic equation*.

$$H^+(aq) + \cancel{Cl^-(aq)} + \cancel{Na^+(aq)} + OH^-(aq) \longrightarrow H_2O(l) + \cancel{Na^+(aq)} + \cancel{Cl^-(aq)}$$

The net ionic equation for the neutralization of H^+ and OH^- to form H_2O is

$$H^+(aq) + OH^-(aq) \longrightarrow H_2O(l) \quad \text{Net ionic equation}$$

Balancing Neutralization Equations

In a neutralization reaction, one H^+ always reacts with one OH^-. Therefore, a neutralization equation may need coefficients to balance the H^+ from the acid with the OH^- from the base as shown in Sample Problem 14.11.

SAMPLE PROBLEM 14.11   Balancing Equations for Acids

TRY IT FIRST

Write the balanced equation for the neutralization of $HCl(aq)$ and $Ba(OH)_2(s)$.

SOLUTION

STEP 1   **Write the reactants and products.**

$$HCl(aq) + Ba(OH)_2(s) \longrightarrow H_2O(l) + \text{salt}$$

STEP 2   **Balance the H^+ in the acid with the OH^- in the base.** Placing a coefficient of 2 in front of the HCl provides $2H^+$ for the $2OH^-$ from $Ba(OH)_2$.

$$2HCl(aq) + Ba(OH)_2(s) \longrightarrow H_2O(l) + \text{salt}$$

STEP 3   **Balance the H_2O with the H^+ and the OH^-.** Use a coefficient of 2 in front of H_2O to balance $2H^+$ and $2OH^-$.

$$2HCl(aq) + Ba(OH)_2(s) \longrightarrow 2H_2O(l) + \text{salt}$$

STEP 4   **Write the salt from the remaining ions.** Use the ions Ba^{2+} and $2Cl^-$ and write the formula for the salt as $BaCl_2$.

$$2HCl(aq) + Ba(OH)_2(s) \longrightarrow 2H_2O(l) + BaCl_2(aq)$$

SELF TEST 14.11

Write the balanced equation for the reaction between each of the following:

a. $H_2SO_4(aq)$ and $K_2CO_3(s)$
b. $H_3PO_4(aq)$ and $LiOH(aq)$

ANSWER

a. $H_2SO_4(aq) + K_2CO_3(s) \longrightarrow CO_2(g) + H_2O(l) + K_2SO_4(aq)$
b. $H_3PO_4(aq) + 3LiOH(aq) \longrightarrow Li_3PO_4(aq) + 3H_2O(l)$

Chemistry Link to Health

Antacids

Antacids are substances used to neutralize excess stomach acid (HCl). Some antacids are mixtures of aluminum hydroxide and magnesium hydroxide. These hydroxides are not very soluble in water, so the levels of available OH⁻ are not damaging to the intestinal tract. However, aluminum hydroxide has the side effects of producing constipation and binding phosphate in the intestinal tract, which may cause weakness and loss of appetite. Magnesium hydroxide has a laxative effect. These side effects are less likely when a combination of the antacids is used.

$$3HCl(aq) + Al(OH)_3(s) \longrightarrow 3H_2O(l) + AlCl_3(aq)$$

$$2HCl(aq) + Mg(OH)_2(s) \longrightarrow 2H_2O(l) + MgCl_2(aq)$$

Antacids neutralize excess stomach acid.

Some antacids use calcium carbonate to neutralize excess stomach acid. About 10% of the calcium is absorbed into the bloodstream, where it elevates the level of serum calcium. Calcium carbonate is not recommended for patients who have peptic ulcers or a tendency to form kidney stones, which typically consist of an insoluble calcium salt.

$$2HCl(aq) + CaCO_3(s) \longrightarrow CO_2(g) + H_2O(l) + CaCl_2(aq)$$

Still other antacids contain sodium bicarbonate. This type of antacid neutralizes excess gastric acid, increases blood pH, but also elevates sodium levels in the body fluids. It also is not recommended in the treatment of peptic ulcers.

$$HCl(aq) + NaHCO_3(s) \longrightarrow CO_2(g) + H_2O(l) + NaCl(aq)$$

The neutralizing substances in some antacid preparations are given in **TABLE 14.8**.

TABLE 14.8 Basic Compounds in Some Antacids

Antacid	Base(s)
Amphojel	$Al(OH)_3$
Milk of magnesia	$Mg(OH)_2$
Mylanta, Maalox, Di-Gel, Gelusil, Riopan	$Mg(OH)_2$, $Al(OH)_3$
Bisodol, Rolaids	$CaCO_3$, $Mg(OH)_2$
Titralac, Tums, Pepto-Bismol	$CaCO_3$
Alka-Seltzer	$NaHCO_3$, $KHCO_3$

PRACTICE PROBLEMS

14.7 Reactions of Acids and Bases

14.55 Complete and balance the equation for each of the following reactions:
  a. $ZnCO_3(s) + HBr(aq) \longrightarrow$
  b. $Zn(s) + HCl(aq) \longrightarrow$
  c. $HCl(aq) + NaHCO_3(s) \longrightarrow$
  d. $H_2SO_4(aq) + Mg(OH)_2(s) \longrightarrow$

14.56 Complete and balance the equation for each of the following reactions:
  a. $KHCO_3(s) + HBr(aq) \longrightarrow$
  b. $Ca(s) + H_2SO_4(aq) \longrightarrow$
  c. $H_2SO_4(aq) + Ca(OH)_2(s) \longrightarrow$
  d. $Na_2CO_3(s) + H_2SO_4(aq) \longrightarrow$

14.57 Balance each of the following neutralization reactions:
  a. $HCl(aq) + Mg(OH)_2(s) \longrightarrow H_2O(l) + MgCl_2(aq)$
  b. $H_3PO_4(aq) + LiOH(aq) \longrightarrow H_2O(l) + Li_3PO_4(aq)$

14.58 Balance each of the following neutralization reactions:
  a. $HNO_3(aq) + Ba(OH)_2(s) \longrightarrow H_2O(l) + Ba(NO_3)_2(aq)$
  b. $H_2SO_4(aq) + Al(OH)_3(s) \longrightarrow H_2O(l) + Al_2(SO_4)_3(aq)$

14.59 Write a balanced equation for the neutralization of each of the following:
  a. $H_2SO_4(aq)$ and $NaOH(aq)$
  b. $HCl(aq)$ and $Fe(OH)_3(s)$
  c. $H_2CO_3(aq)$ and $Mg(OH)_2(s)$

14.60 Write a balanced equation for the neutralization of each of the following:
  a. $H_3PO_4(aq)$ and $NaOH(aq)$
  b. $HI(aq)$ and $LiOH(aq)$
  c. $HNO_3(aq)$ and $Ca(OH)_2(s)$

14.8 Acid–Base Titration

LEARNING GOAL Calculate the molarity or volume of an acid or base solution from titration information.

Suppose we need to find the molarity of a solution of HCl, which has an unknown concentration. We can do this by a laboratory procedure called **titration** in which we neutralize an acid sample with a known amount of base. In a titration, we place a measured volume of the acid in a flask and add a few drops of an **indicator**, such as phenolphthalein. An indicator is a compound that dramatically changes color when pH of the solution changes. In an acidic solution, phenolphthalein is colorless. Then we fill a buret with a NaOH solution of known molarity and carefully add NaOH solution to neutralize the acid in the flask (see **FIGURE 14.6**). We know that neutralization has taken place when the phenolphthalein in the solution changes from colorless to pink. This is called the neutralization **endpoint**. From the measured volume of the NaOH solution and its molarity, we calculate the number of moles of NaOH, the moles of acid, and use the measured volume of acid to calculate its concentration.

REVIEW

Using Concentration as a Conversion Factor (12.4)

ENGAGE 14.14

What data is needed to determine the molarity of the acid in the flask?

FIGURE 14.6 The titration of an acid. A known volume of an acid is placed in a flask with an indicator and titrated with a measured volume of a base solution, such as NaOH, to the neutralization endpoint.

SAMPLE PROBLEM 14.12   Titration of an Acid

TRY IT FIRST

If 16.3 mL of a 0.185 M Sr(OH)$_2$ solution is used to titrate the HCl in 0.0250 L of gastric juice, what is the molarity of the HCl solution?

$$2HCl(aq) + Sr(OH)_2(aq) \longrightarrow 2H_2O(l) + SrCl_2(aq)$$

CORE CHEMISTRY SKILL

Calculating Molarity or Volume of an Acid or Base in a Titration

SOLUTION

STEP 1   State the given and needed quantities and concentrations.

	Given	Need	Connect
ANALYZE THE PROBLEM	16.3 mL of 0.185 M Sr(OH)$_2$ solution, 0.0250 L of HCl solution	molarity of the HCl solution	molarity, mole–mole factor
	Neutralization Equation		
	2HCl(aq) + Sr(OH)$_2$(aq) $\longrightarrow$ 2H$_2$O(l) + SrCl$_2$(aq)		

STEP 2   Write a plan to calculate the molarity.

mL of Sr(OH)$_2$ solution → Metric factor → L of Sr(OH)$_2$ solution → Molarity → moles of Sr(OH)$_2$ → Mole–mole factor → moles of HCl → Divide by liters → molarity of HCl solution

STEP 3   State equalities and conversion factors, including concentrations.

$$\dfrac{1\ \text{L of Sr(OH)}_2\ \text{solution}}{1000\ \text{mL Sr(OH)}_2\ \text{solution}} = \dfrac{1000\ \text{mL of Sr(OH)}_2\ \text{solution}}{1\ \text{L Sr(OH)}_2\ \text{solution}}$$

1 L of Sr(OH)$_2$ solution = 1000 mL of Sr(OH)$_2$ solution

$$\dfrac{1000\ \text{mL Sr(OH)}_2\ \text{solution}}{1\ \text{L Sr(OH)}_2\ \text{solution}}\ \text{and}\ \dfrac{1\ \text{L Sr(OH)}_2\ \text{solution}}{1000\ \text{mL Sr(OH)}_2\ \text{solution}}$$

1 L of Sr(OH)$_2$ solution = 0.185 mol of Sr(OH)$_2$

$$\dfrac{0.185\ \text{mol Sr(OH)}_2}{1\ \text{L Sr(OH)}_2\ \text{solution}}\ \text{and}\ \dfrac{1\ \text{L Sr(OH)}_2\ \text{solution}}{0.185\ \text{mol Sr(OH)}_2}$$

2 mol of HCl = 1 mol of Sr(OH)$_2$

$$\dfrac{2\ \text{mol HCl}}{1\ \text{mol Sr(OH)}_2}\ \text{and}\ \dfrac{1\ \text{mol Sr(OH)}_2}{2\ \text{mol HCl}}$$

STEP 4   Set up the problem to calculate the needed quantity.

$$16.3\ \text{mL Sr(OH)}_2\ \text{solution} \times \dfrac{1\ \text{L Sr(OH)}_2\ \text{solution}}{1000\ \text{mL Sr(OH)}_2\ \text{solution}} \times \dfrac{0.185\ \text{mol Sr(OH)}_2}{1\ \text{L Sr(OH)}_2\ \text{solution}} \times \dfrac{2\ \text{mol HCl}}{1\ \text{mol Sr(OH)}_2}$$

$$= 0.006\ 03\ \text{mol of HCl}$$

$$\text{molarity of HCl solution} = \dfrac{0.006\ 03\ \text{mol HCl}}{0.0250\ \text{L HCl solution}} = 0.241\ \text{M HCl solution}$$

INTERACTIVE VIDEO

Acid–Base Titration

PRACTICE PROBLEMS

Try Practice Problems 14.61 to 14.68

SELF TEST 14.12

a. What is the molarity of an HCl solution if 28.6 mL of a 0.175 M NaOH solution is needed to titrate a 25.0-mL sample of the HCl solution?
b. What volume, in milliliters, of a 0.220 M KOH solution is needed to titrate 8.00 mL of 0.425 M H$_2$SO$_4$ solution?

ANSWER

a. 0.200 M HCl solution   b. 30.9 mL of KOH solution

PRACTICE PROBLEMS

14.8 Acid–Base Titration

14.61 If you need to determine the molarity of a formic acid solution, HCHO$_2$, how would you proceed?

14.62 If you need to determine the molarity of an acetic acid solution, HC$_2$H$_3$O$_2$, how would you proceed?

14.63 What is the molarity of a solution of HCl if 5.00 mL of the HCl solution is titrated with 28.6 mL of a 0.145 M NaOH solution?

$$\text{HCl}(aq) + \text{NaOH}(aq) \longrightarrow \text{H}_2\text{O}(l) + \text{NaCl}(aq)$$

14.64 What is the molarity of an acetic acid solution if 25.0 mL of the HC$_2$H$_3$O$_2$ solution is titrated with 29.7 mL of a 0.205 M KOH solution?

$$\text{HC}_2\text{H}_3\text{O}_2(aq) + \text{KOH}(aq) \longrightarrow \text{H}_2\text{O}(l) + \text{KC}_2\text{H}_3\text{O}_2(aq)$$

14.65 If 38.2 mL of a 0.163 M KOH solution is required to neutralize completely 25.0 mL of a solution of H$_2$SO$_4$, what is the molarity of the H$_2$SO$_4$ solution?

$$\text{H}_2\text{SO}_4(aq) + 2\text{KOH}(aq) \longrightarrow 2\text{H}_2\text{O}(l) + \text{K}_2\text{SO}_4(aq)$$

14.66 A solution of 0.162 M NaOH is used to titrate 25.0 mL of a solution of H$_2$SO$_4$. If 32.8 mL of the NaOH solution is required to reach the endpoint, what is the molarity of the H$_2$SO$_4$ solution?

$$\text{H}_2\text{SO}_4(aq) + 2\text{NaOH}(aq) \longrightarrow 2\text{H}_2\text{O}(l) + \text{Na}_2\text{SO}_4(aq)$$

14.67 A solution of 0.204 M NaOH is used to titrate 50.0 mL of a 0.0224 M H$_3$PO$_4$ solution. What volume, in milliliters, of the NaOH solution is required?

$$\text{H}_3\text{PO}_4(aq) + 3\text{NaOH}(aq) \longrightarrow 3\text{H}_2\text{O}(l) + \text{Na}_3\text{PO}_4(aq)$$

14.68 A solution of 0.312 M KOH is used to titrate 15.0 mL of a 0.186 M H$_3$PO$_4$ solution. What volume, in milliliters, of the KOH solution is required?

$$\text{H}_3\text{PO}_4(aq) + 3\text{KOH}(aq) \longrightarrow 3\text{H}_2\text{O}(l) + \text{K}_3\text{PO}_4(aq)$$

14.9 Buffers

LEARNING GOAL Describe the role of buffers in maintaining the pH of a solution; calculate the pH of a buffer.

The lungs and the kidneys are the primary organs that regulate the pH of body fluids, including blood and urine. Major changes in the pH of the body fluids can severely affect biological activities within the cells. *Buffers* are present to prevent large fluctuations in pH.

The pH of water and most solutions changes drastically when a small amount of acid or base is added. However, when an acid or a base is added to a buffer solution, there is little change in pH. A **buffer solution** maintains the pH of a solution by neutralizing small amounts of added acid or base. In the human body, whole blood contains plasma, white blood cells and platelets, and red blood cells. Blood plasma contains buffers that maintain a consistent pH of about 7.4. If the pH of the blood plasma goes slightly above or below 7.4, changes in our oxygen levels and our metabolic processes can be drastic enough to cause death. Even though we obtain acids and bases from foods and cellular reactions, the buffers in the body neutralize those compounds so effectively that the pH of our blood plasma remains essentially unchanged (see **FIGURE 14.7**).

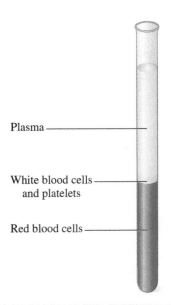

Whole blood consists of plasma, white blood cells and platelets, and red blood cells.

FIGURE 14.7 Adding an acid or a base to water changes the pH drastically, but a buffer resists pH change when small amounts of acid or base are added.

In a buffer, an acid must be present to react with any OH^- that is added, and a base must be available to react with any added H_3O^+. However, that acid and base must not neutralize each other. Therefore, a combination of an acid–base conjugate pair is used in buffers. Most buffer solutions consist of nearly equal concentrations of a weak acid and a salt containing its conjugate base. Buffers may also contain a weak base and the salt of the weak base, which contains its conjugate acid.

For example, a typical buffer can be made from the weak acid acetic acid ($HC_2H_3O_2$) and its salt, sodium acetate ($NaC_2H_3O_2$). As a weak acid, acetic acid dissociates slightly in water to form H_3O^+ and a very small amount of $C_2H_3O_2^-$. The addition of its salt, sodium acetate, provides a much larger concentration of acetate ion ($C_2H_3O_2^-$), which is necessary for its buffering capability.

$$HC_2H_3O_2(aq) + H_2O(l) \rightleftharpoons H_3O^+(aq) + C_2H_3O_2^-(aq)$$

Large amount   Large amount

ENGAGE 14.15

Why does a buffer require the presence of a weak acid or weak base and the salt of that weak acid or weak base?

We can now describe how this buffer solution maintains the $[H_3O^+]$. When a small amount of acid is added, the additional H_3O^+ combines with the acetate ion, $C_2H_3O_2^-$, causing the equilibrium to shift in the direction of the reactants, acetic acid and water. There will be a slight decrease in the $[C_2H_3O_2^-]$ and a slight increase in the $[HC_2H_3O_2]$, but both the $[H_3O^+]$ and pH are maintained.

$$HC_2H_3O_2(aq) + H_2O(l) \longleftarrow H_3O^+(aq) + C_2H_3O_2^-(aq)$$

Equilibrium shifts in the
direction of the reactants

If a small amount of base is added to this same buffer solution, it is neutralized by the acetic acid, $HC_2H_3O_2$, which shifts the equilibrium in the direction of the products, acetate ion and water. The $[HC_2H_3O_2]$ decreases slightly and the $[C_2H_3O_2^-]$ increases slightly, but again the $[H_3O^+]$ and thus the pH of the solution are maintained (see **FIGURE 14.8**).

$$HC_2H_3O_2(aq) + OH^-(aq) \longrightarrow H_2O(l) + C_2H_3O_2^-(aq)$$

Equilibrium shifts in the
direction of the products

ENGAGE 14.16

Which part of a buffer neutralizes any H_3O^+ that is added?

PRACTICE PROBLEMS

Try Practice Problems 14.69 to 14.72

FIGURE 14.8 The buffer described here consists of about equal concentrations of acetic acid ($HC_2H_3O_2$) and its conjugate base acetate ion ($C_2H_3O_2^-$). The pH of the solution is maintained as long as the added amount of acid or base is small compared to the concentrations of the buffer components.

Calculating the pH of a Buffer

CORE CHEMISTRY SKILL

Calculating the pH of a Buffer

By rearranging the K_a expression to give $[H_3O^+]$, we can obtain the ratio of the acetic acid/acetate buffer.

$$K_a = \frac{[H_3O^+][C_2H_3O_2^-]}{[HC_2H_3O_2]}$$

Solving for $[H_3O^+]$ gives:

INTERACTIVE VIDEO

Calculating the pH of a Buffer

$$[H_3O^+] = K_a \times \frac{[HC_2H_3O_2]}{[C_2H_3O_2^-]} \quad \longleftarrow \text{Weak acid} \\ \quad \longleftarrow \text{Conjugate base}$$

In this rearrangement of K_a, the weak acid is in the numerator and the conjugate base in the denominator. We can now calculate the $[H_3O^+]$ and pH for an acetic acid buffer as shown in Sample Problem 14.13.

SAMPLE PROBLEM 14.13   Calculating the pH of a Buffer

TRY IT FIRST

The K_a for acetic acid, $HC_2H_3O_2$, is 1.8×10^{-5}. What is the pH of a buffer prepared with 1.0 M $HC_2H_3O_2$ and 1.0 M $C_2H_3O_2^-$?

$$HC_2H_3O_2(aq) + H_2O(l) \rightleftharpoons H_3O^+(aq) + C_2H_3O_2^-(aq)$$

SOLUTION

STEP 1 **State the given and needed quantities.**

	Given	Need	Connect
ANALYZE THE PROBLEM	1.0 M $HC_2H_3O_2$, 1.0 M $C_2H_3O_2^-$	pH	K_a expression
	Equation		
	$HC_2H_3O_2(aq) + H_2O(l) \rightleftharpoons H_3O^+(aq) + C_2H_3O_2^-(aq)$		

STEP 2 Write the K_a expression and rearrange for $[H_3O^+]$.

$$K_a = \frac{[H_3O^+][C_2H_3O_2^-]}{[HC_2H_3O_2]}$$

$$[H_3O^+] = K_a \times \frac{[HC_2H_3O_2]}{[C_2H_3O_2^-]}$$

STEP 3 Substitute [HA] and [A⁻] into the K_a expression.

$$[H_3O^+] = 1.8 \times 10^{-5} \times \frac{[1.0]}{[1.0]}$$

$$[H_3O^+] = 1.8 \times 10^{-5}\,M$$

STEP 4 Use $[H_3O^+]$ to calculate pH. Placing the $[H_3O^+]$ into the pH equation gives the pH of the buffer.

$$pH = -\log[1.8 \times 10^{-5}] = 4.74$$

SELF TEST 14.13

a. One of the conjugate acid–base pairs that buffers the blood plasma is $H_2PO_4^-/HPO_4^{2-}$. The K_a for $H_2PO_4^-$ is 6.2×10^{-8}. What is the pH of a buffer that is prepared from 0.10 M $H_2PO_4^-$ and 0.50 M HPO_4^{2-}?
b. What is the pH of a buffer made from 0.10 M formic acid ($HCHO_2$) and 0.010 M formate (CHO_2^-)? The K_a for formic acid is 1.8×10^{-4}.

ANSWER

a. pH = 7.91   b. pH = 2.74

Because K_a is a constant at a given temperature, the $[H_3O^+]$ is determined by the $[HC_2H_3O_2]/[C_2H_3O_2^-]$ ratio. As long as the addition of small amounts of either acid or base changes the ratio of $[HC_2H_3O_2]/[C_2H_3O_2^-]$ only slightly, the changes in $[H_3O^+]$ will be small, and the pH will be maintained. If a large amount of acid or base is added, the *buffering capacity* of the system may be exceeded. Buffers can be prepared from conjugate acid–base pairs such as $H_2PO_4^-/HPO_4^{2-}$, HPO_4^{2-}/PO_4^{3-}, HCO_3^-/CO_3^{2-}, or NH_4^+/NH_3. The pH of the buffer solution will depend on the conjugate acid–base pair chosen.

Using a common phosphate buffer for biological specimens, we can look at the effect of using different ratios of $[H_2PO_4^-]/[HPO_4^{2-}]$ on the $[H_3O^+]$ and pH. The K_a of $H_2PO_4^-$ is 6.2×10^{-8}. The equation and the $[H_3O^+]$ are written as follows:

$$H_2PO_4^-(aq) + H_2O(l) \rightleftharpoons H_3O^+(aq) + HPO_4^{2-}(aq)$$

$$[H_3O^+] = K_a \times \frac{[H_2PO_4^-]}{[HPO_4^{2-}]}$$

ENGAGE 14.17

How would a solution composed of $H_2PO_4^-$ and HPO_4^{2-} act as a buffer?

K_a	$\dfrac{[H_2PO_4^-]}{[HPO_4^{2-}]}$	Ratio	$[H_3O^+]$	pH
6.2×10^{-8}	$\dfrac{1.0\,M}{0.10\,M}$	$\dfrac{10}{1}$	6.2×10^{-7}	6.21
6.2×10^{-8}	$\dfrac{1.0\,M}{1.0\,M}$	$\dfrac{1}{1}$	6.2×10^{-8}	7.21
6.2×10^{-8}	$\dfrac{0.10\,M}{1.0\,M}$	$\dfrac{1}{10}$	6.2×10^{-9}	8.21

To prepare a phosphate buffer with a pH close to the pH of a biological sample, 7.4, we would choose concentrations that are about equal, such as 1.0 M $H_2PO_4^-$ and 1.0 M HPO_4^{2-}.

PRACTICE PROBLEMS

Try Practice Problems 14.73 to 14.78

Chemistry Link to Health

Buffers in the Blood Plasma

The arterial blood plasma has a normal pH of 7.35 to 7.45. If changes in H_3O^+ lower the pH below 6.8 or raise it above 8.0, cells cannot function properly, and death may result. In our cells, CO_2 is continually produced as an end product of cellular metabolism. Some CO_2 is carried to the lungs for elimination, and the rest dissolves in body fluids such as plasma and saliva, forming carbonic acid, H_2CO_3. As a weak acid, carbonic acid dissociates to give bicarbonate, HCO_3^-, and H_3O^+. More of the anion HCO_3^- is supplied by the kidneys to give an important buffer system in the body fluid—the H_2CO_3/HCO_3^- buffer.

$$CO_2(g) + H_2O(l) \rightleftharpoons H_2CO_3(aq) \rightleftharpoons$$
$$H_3O^+(aq) + HCO_3^-(aq)$$

Excess H_3O^+ entering the body fluids reacts with the HCO_3^-, and excess OH^- reacts with the carbonic acid.

$$H_2CO_3(aq) + H_2O(l) \longleftarrow H_3O^+(aq) + HCO_3^-(aq)$$
Equilibrium shifts in the direction of the reactants

$$H_2CO_3(aq) + OH^-(aq) \longrightarrow H_2O(l) + HCO_3^-(aq)$$
Equilibrium shifts in the direction of the products

For carbonic acid, we can write the equilibrium expression as

$$K_a = \frac{[H_3O^+][HCO_3^-]}{[H_2CO_3]}$$

To maintain the normal blood plasma pH (7.35 to 7.45), the ratio of $[H_2CO_3]/[HCO_3^-]$ needs to be about 1 to 10, which is obtained by the concentrations in the blood plasma of 0.0024 M H_2CO_3 and 0.024 M HCO_3^-.

$$[H_3O^+] = K_a \times \frac{[H_2CO_3]}{[HCO_3^-]}$$
$$[H_3O^+] = 4.3 \times 10^{-7} \times \frac{[0.0024]}{[0.024]}$$
$$= 4.3 \times 10^{-7} \times 0.10 = 4.3 \times 10^{-8} \text{ M}$$
$$pH = -\log[4.3 \times 10^{-8}] = 7.37$$

In the body, the concentration of carbonic acid is closely associated with the partial pressure of CO_2, P_{CO_2}. **TABLE 14.9** lists the normal values for arterial blood. If the CO_2 level rises, increasing $[H_2CO_3]$, the equilibrium shifts to produce more H_3O^+, which lowers the pH. This condition is called *acidosis*. Difficulty with ventilation or gas diffusion can lead to respiratory acidosis, which can happen in emphysema or when an accident or depressive drugs affect the medulla of the brain.

A lowering of the CO_2 level leads to a high blood pH, a condition called *alkalosis*. Excitement, trauma, or a high temperature may cause a person to hyperventilate, which expels large amounts of CO_2. As the partial pressure of CO_2 in the blood falls below normal, the equilibrium shifts from H_2CO_3 to CO_2 and H_2O. This shift decreases the $[H_3O^+]$ and raises the pH. The kidneys also regulate H_3O^+ and HCO_3^-, but they do so more slowly than the adjustment made by the lungs during ventilation.

TABLE 14.10 lists some of the conditions that lead to changes in the blood pH and some possible treatments.

TABLE 14.9 Normal Values for Blood Buffer in Arterial Blood

P_{CO_2}	40 mmHg
H_2CO_3	2.4 mmol/L of plasma
HCO_3^-	24 mmol/L of plasma
pH	7.35 to 7.45

TABLE 14.10 Acidosis and Alkalosis: Symptoms, Causes, and Treatments

Respiratory Acidosis: $CO_2\uparrow$ pH$\downarrow$	
Symptoms	Failure to ventilate, suppression of breathing, disorientation, weakness, coma
Causes	Lung disease blocking gas diffusion (e.g., emphysema, pneumonia, bronchitis, asthma); depression of respiratory center by drugs, cardiopulmonary arrest, stroke, poliomyelitis, or nervous system disorders
Treatment	Correction of disorder, infusion of bicarbonate
Metabolic Acidosis: $H^+\uparrow$ pH$\downarrow$	
Symptoms	Increased ventilation, fatigue, confusion
Causes	Renal disease, including hepatitis and cirrhosis; increased acid production in diabetes mellitus, hyperthyroidism, alcoholism, and starvation; loss of alkali in diarrhea; acid retention in renal failure
Treatment	Sodium bicarbonate given orally, dialysis for renal failure, insulin treatment for diabetic ketosis
Respiratory Alkalosis: $CO_2\downarrow$ pH$\uparrow$	
Symptoms	Increased rate and depth of breathing, numbness, light-headedness, tetany
Causes	Hyperventilation because of anxiety, hysteria, fever, exercise; reaction to drugs such as salicylate, quinine, and antihistamines; conditions causing hypoxia (e.g., pneumonia, pulmonary edema, heart disease)
Treatment	Elimination of anxiety-producing state, rebreathing into a paper bag
Metabolic Alkalosis: $H^+\downarrow$ pH$\uparrow$	
Symptoms	Depressed breathing, apathy, confusion
Causes	Vomiting, diseases of the adrenal glands, ingestion of excess alkali
Treatment	Infusion of saline solution, treatment of underlying diseases

PRACTICE PROBLEMS

14.9 Buffers

14.69 Which of the following make a buffer system when added to water? Explain.
  a. NaOH and NaCl   **b.** H_2CO_3 and $NaHCO_3$
  c. HF and KF   **d.** KCl and NaCl

14.70 Which of the following make a buffer system when added to water? Explain.
  a. $HClO_2$ **b.** $NaNO_3$
  c. $HC_2H_3O_2$ and $NaC_2H_3O_2$ **d.** HCl and NaOH

14.71 Consider the buffer system of hydrofluoric acid, HF, and its salt, NaF.

$$HF(aq) + H_2O(l) \rightleftarrows H_3O^+(aq) + F^-(aq)$$

  a. The purpose of this buffer system is to:
    1. maintain [HF]   **2.** maintain [F$^-$]
    3. maintain pH
  b. The salt of the weak acid is needed to:
    1. provide the conjugate base   **2.** neutralize added H_3O^+
    3. provide the conjugate acid
  c. If OH$^-$ is added, it is neutralized by:
    1. the salt   **2.** H_2O **3.** H_3O^+
  d. When H_3O^+ is added, the equilibrium shifts in the direction of the:
    1. reactants   **2.** products   **3.** does not change

14.72 Consider the buffer system of nitrous acid, HNO_2, and its salt, $NaNO_2$.

$$HNO_2(aq) + H_2O(l) \rightleftarrows H_3O^+(aq) + NO_2^-(aq)$$

  a. The purpose of this buffer system is to:
    1. maintain [HNO_2]   **2.** maintain [NO_2^-]
    3. maintain pH

  b. The weak acid is needed to:
    1. provide the conjugate base   **2.** neutralize added OH$^-$
    3. provide the conjugate acid
  c. If H_3O^+ is added, it is neutralized by:
    1. the salt   **2.** H_2O **3.** OH$^-$
  d. When OH$^-$ is added, the equilibrium shifts in the direction of the:
    1. reactants   **2.** products   **3.** does not change

14.73 Nitrous acid has a K_a of 4.5×10^{-4}. What is the pH of a buffer solution containing 0.10 M HNO_2 and 0.10 M NO_2^-?

14.74 Acetic acid has a K_a of 1.8×10^{-5}. What is the pH of a buffer solution containing 0.15 M $HC_2H_3O_2$ and 0.15 M $C_2H_3O_2^-$?

14.75 Using Table 14.4 for K_a values, compare the pH of a HF buffer that contains 0.10 M HF and 0.10 M NaF with another HF buffer that contains 0.060 M HF and 0.120 M NaF.

14.76 Using Table 14.4 for K_a values, compare the pH of a H_2CO_3 buffer that contains 0.10 M H_2CO_3 and 0.10 M $NaHCO_3$ with another H_2CO_3 buffer that contains 0.15 M H_2CO_3 and 0.050 M $NaHCO_3$.

Applications

14.77 Someone with kidney failure excretes urine with large amounts of HCO_3^-. How would this loss of HCO_3^- affect the pH of the blood plasma?

14.78 Someone with severe diabetes obtains energy by the breakdown of fats, which produce large amounts of acidic substances. How would this affect the pH of the blood plasma?

UPDATE   Acid Reflux Disease

Larry has not been feeling well lately. He tells his doctor that he has discomfort and a burning feeling in his chest, and a sour taste in his throat and mouth. At times, Larry says he feels bloated after a big meal, has a dry cough, is hoarse, and sometimes has a sore throat. He has tried antacids, but they do not bring any relief.

The doctor tells Larry that he thinks he has acid reflux. At the top of the stomach there is a valve, the lower esophageal sphincter, that normally closes after food passes through it. However, if the valve does not close completely, acid produced in the stomach to digest food can move up into the esophagus, a condition called *acid reflux*. The acid, which is hydrochloric acid, HCl, is produced in the stomach to kill microorganisms, and activate the enzymes we need to break down food.

If acid reflux occurs, the strong acid HCl comes in contact with the lining of the esophagus, where it causes irritation and produces a burning feeling in the chest. Sometimes the pain in the chest is called *heartburn*. If the HCl reflux goes high enough to reach the throat, a sour taste may be noticed in the mouth. If Larry's symptoms occur three or more times a week, he may have a chronic condition known as *acid reflux disease* or *gastroesophageal reflux disease* (GERD).

Larry's doctor orders an *esophageal pH test* in which the amount of acid entering the esophagus from the stomach is measured over 24 h. A probe that measures the pH is inserted into the lower esophagus above the esophageal sphincter. The pH measurements indicate a reflux episode each time the pH drops to 4 or less.

In the 24-h period, Larry has several reflux episodes, and his doctor determines that he has chronic GERD. He and Larry discuss treatment for GERD, which includes eating smaller meals, not lying down for 3 h after eating, making dietary changes, and losing weight. Antacids may be used

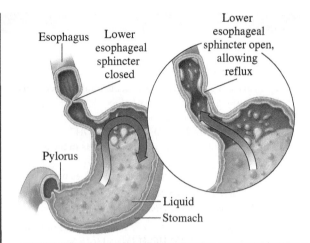

In acid reflux disease, the lower esophageal sphincter opens, allowing acidic fluid from the stomach to enter the esophagus.

to neutralize the acid coming up from the stomach. Other medications known as *proton pump inhibitors* (PPIs), such as Prilosec and Nexium, may be used to suppress the production of HCl in the stomach (gastric parietal cells), which raises the pH in the stomach to between 4 and 5, and gives the esophagus time to heal. Nexium may be given in oral doses of 40 mg once a day for 4 weeks. In severe GERD cases, an artificial valve may be created at the top of the stomach to strengthen the lower esophageal sphincter.

Applications

14.79 At rest, the $[H_3O^+]$ of the stomach fluid is 2.0×10^{-4} M. What is the pH of the stomach fluid?

14.80 When food enters the stomach, HCl is released and the $[H_3O^+]$ of the stomach fluid rises to 4.2×10^{-2} M. What is the pH of the stomach fluid while eating?

14.81 In Larry's esophageal pH test, a pH value of 3.60 was recorded in the esophagus. What is the $[H_3O^+]$ in his esophagus?

14.82 After Larry had taken Nexium for 4 weeks, the pH in his stomach was raised to 4.52. What is the $[H_3O^+]$ in his stomach?

14.83 Write the balanced chemical equation for the neutralization reaction of stomach acid HCl with $CaCO_3$, an ingredient in some antacids.

14.84 Write the balanced chemical equation for the neutralization reaction of stomach acid HCl with $Al(OH)_3$, an ingredient in some antacids.

14.85 How many grams of $CaCO_3$ are required to neutralize 100. mL of stomach acid HCl, which is 0.0400 M HCl?

14.86 How many grams of $Al(OH)_3$ are required to neutralize 150. mL of stomach acid HCl with a pH of 1.50?

CONCEPT MAP

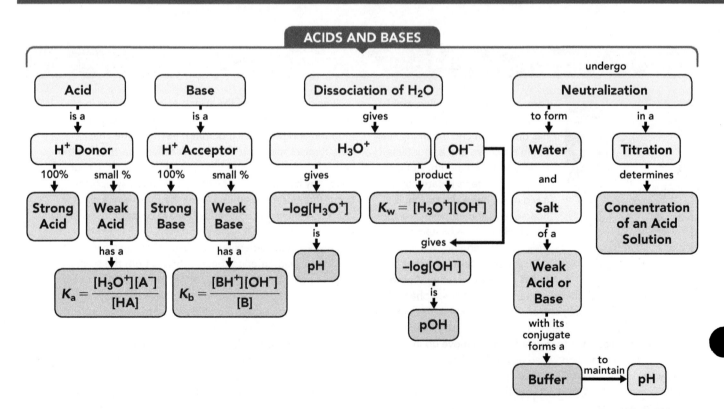

CHAPTER REVIEW

14.1 Acids and Bases

LEARNING GOAL Describe and name acids and bases.

NaOH(s)

- An Arrhenius acid produces H⁺ and an Arrhenius base produces OH⁻ in aqueous solutions.
- Acids taste sour, may sting, and neutralize bases.
- Bases taste bitter, feel slippery, and neutralize acids.
- Acids containing a simple anion use a *hydro* prefix, whereas acids with oxygen-containing polyatomic anions are named as *ic* or *ous acids*.

$$\text{NaOH}(s) \xrightarrow{\text{H}_2\text{O}} \text{Na}^+(aq) + \text{OH}^-(aq)$$

Ionic   Dissociation   Hydroxide
compound   ion

14.2 Brønsted–Lowry Acids and Bases

LEARNING GOAL Identify conjugate acid–base pairs for Brønsted–Lowry acids and bases.

$$\text{HCl} + \text{H}_2\text{O} \longrightarrow \text{H}_3\text{O}^+ + \text{Cl}^-$$

Hydrogen   Water   Hydronium   Chloride
chloride   ion   ion

Acid   Base
(H⁺ donor) (H⁺ acceptor)   Acidic solution

- According to the Brønsted–Lowry theory, acids are H⁺ donors and bases are H⁺ acceptors.
- A conjugate acid–base pair is related by the loss or gain of one H⁺.
- For example, when the acid HF donates H⁺, the F⁻ is its conjugate base. The other acid–base pair would be $\text{H}_3\text{O}^+/\text{H}_2\text{O}$.

$$\text{HF}(aq) + \text{H}_2\text{O}(l) \rightleftharpoons \text{H}_3\text{O}^+(aq) + \text{F}^-(aq)$$

14.3 Strengths of Acids and Bases

LEARNING GOAL Write equations for the dissociation of strong and weak acids; identify the direction of reaction.

- Strong acids dissociate completely in water, and the H⁺ is accepted by H_2O acting as a base.
- A weak acid dissociates slightly in water, producing only a small percentage of H_3O^+.
- Strong bases are hydroxides of Groups 1A (1) and 2A (2) that dissociate completely in water.
- An important weak base is ammonia, NH_3.

14.4 Dissociation of Weak Acids and Bases

LEARNING GOAL Write the expression for the dissociation of a weak acid or weak base.

HCHO₂   CHO₂⁻

- In water, weak acids and weak bases produce only a few ions when equilibrium is reached.
- Weak acids have small K_a values whereas strong acids, which are essentially 100% dissociated, have very large K_a values.
- The reaction for a weak acid can be written:

$$\text{HA} + \text{H}_2\text{O} \rightleftharpoons \text{H}_3\text{O}^+ + \text{A}^-$$

- The acid dissociation expression is written:

$$K_a = \frac{[\text{H}_3\text{O}^+][\text{A}^-]}{[\text{HA}]}$$

- For a weak base, $\text{B} + \text{H}_2\text{O} \rightleftharpoons \text{BH}^+ + \text{OH}^-$, the base dissociation expression is written:

$$K_b = \frac{[\text{BH}^+][\text{OH}^-]}{[\text{B}]}$$

14.5 Dissociation of Water

LEARNING GOAL Use the water dissociation expression to calculate the $[\text{H}_3\text{O}^+]$ and $[\text{OH}^-]$ in an aqueous solution.

$[\text{H}_3\text{O}^+] = [\text{OH}^-]$

- In pure water, a few water molecules transfer H⁺ to other water molecules, producing small, but equal, $[\text{H}_3\text{O}^+]$ and $[\text{OH}^-]$.
- In pure water, the molar concentrations of H_3O^+ and OH^- are each 1.0×10^{-7} mol/L.
- The water dissociation expression, is written as $K_w = [\text{H}_3\text{O}^+][\text{OH}^-]$.
- At 25 °C, $K_w = 1.0 \times 10^{-14}$.
- In acidic solutions, the $[\text{H}_3\text{O}^+]$ is greater than the $[\text{OH}^-]$.
- In basic solutions, the $[\text{OH}^-]$ is greater than the $[\text{H}_3\text{O}^+]$.

14.6 The pH Scale

LEARNING GOAL Calculate pH from $[\text{H}_3\text{O}^+]$; given the pH, calculate the $[\text{H}_3\text{O}^+]$ and $[\text{OH}^-]$ of a solution.

- The pH scale is a range of numbers typically from 0 to 14, which represents the $[\text{H}_3\text{O}^+]$ of the solution.
- A neutral solution has a pH of 7.0. In acidic solutions, the pH is below 7.0; in basic solutions, the pH is above 7.0.
- Mathematically, pH is the negative logarithm of the hydronium ion concentration, $\text{pH} = -\log[\text{H}_3\text{O}^+]$.
- The pOH is the negative log of the hydroxide ion concentration, $\text{pOH} = -\log[\text{OH}^-]$.
- The sum of the pH + pOH is 14.00.

14.7 Reactions of Acids and Bases

LEARNING GOAL Write balanced equations for reactions of acids with metals, carbonates or bicarbonates, and bases.

- An acid reacts with a metal to produce hydrogen gas and a salt.
- The reaction of an acid with a carbonate or bicarbonate produces carbon dioxide, water, and a salt.
- In neutralization, an acid reacts with a base to produce water and a salt.

14.8 Acid–Base Titration

LEARNING GOAL Calculate the molarity or volume of an acid or base solution from titration information.

- In a titration, an acid sample is neutralized with a known amount of a base.
- From the volume and molarity of the base, the concentration of the acid is calculated.

14.9 Buffers

LEARNING GOAL
Describe the role of buffers in maintaining the pH of a solution; calculate the pH of a buffer.

Buffer

- A buffer solution resists changes in pH when small amounts of an acid or a base are added.

- A buffer contains either a weak acid and its salt or a weak base and its salt.
- In a buffer, the weak acid reacts with added OH^-, and the anion of the salt reacts with added H_3O^+.
- Most buffer solutions consist of nearly equal concentrations of a weak acid and a salt containing its conjugate base.
- The pH of a buffer is calculated by solving the K_a expression for $[H_3O^+]$.

KEY TERMS

acid A substance that dissolves in water and produces hydrogen ions (H^+), according to the Arrhenius theory. All acids are hydrogen ion donors, according to the Brønsted–Lowry theory.

acid dissociation constant, K_a The numerical value of the product of the ions from the dissociation of a weak acid divided by the concentration of the weak acid.

amphoteric Substances that can act as either an acid or a base in water.

base A substance that dissolves in water and produces hydroxide ions (OH^-), according to the Arrhenius theory. All bases are hydrogen ion acceptors, according to the Brønsted–Lowry theory.

base dissociation constant, K_b The numerical value of the product of the ions from the dissociation of a weak base divided by the concentration of the weak base.

Brønsted–Lowry acids and bases An acid is a hydrogen ion donor; a base is a hydrogen ion acceptor.

buffer solution A solution of a weak acid and its conjugate base or a weak base and its conjugate acid that maintains the pH by neutralizing added acid or base.

conjugate acid–base pair An acid and a base that differ by one H^+. When an acid donates a hydrogen ion, the product is its conjugate base, which is capable of accepting a hydrogen ion in the reverse reaction.

dissociation The separation of an acid or a base into ions in water.

endpoint The point at which an indicator changes color. For the indicator phenolphthalein, the color change occurs when the number of moles of OH^- is equal to the number of moles of H_3O^+ in the sample.

hydronium ion, H_3O^+ The ion formed by the attraction of a hydrogen ion, H^+, to a water molecule.

indicator A substance added to a titration sample that changes color when the pH of the solution changes.

neutral The term that describes a solution with equal concentrations of H_3O^+ and OH^-.

neutralization A reaction between an acid and a base to form water and a salt.

pH A measure of the $[H_3O^+]$ in a solution; $pH = -\log[H_3O^+]$.

pOH A measure of the $[OH^-]$ in a solution; $pOH = -\log[OH^-]$.

salt An ionic compound that contains a metal ion or NH_4^+ and a nonmetal or polyatomic ion other than OH^-.

strong acid An acid that completely dissociates in water.

strong base A base that completely dissociates in water.

titration The addition of base to an acid sample to determine the concentration of the acid.

water dissociation constant, K_w The numerical value of the product of $[H_3O^+]$ and $[OH^-]$ in solution; $K_w = 1.0 \times 10^{-14}$.

water dissociation expression The product of $[H_3O^+]$ and $[OH^-]$ in solution; $K_w = [H_3O^+][OH^-]$.

weak acid An acid that is a poor donor of H^+ and dissociates only slightly in water.

weak base A base that is a poor acceptor of H^+ and produces only a small number of ions in water.

KEY MATH SKILLS

The chapter section containing each Key Math Skill is shown in parentheses at the end of each heading.

Calculating pH from $[H_3O^+]$ (14.6)

- The pH of a solution is calculated from the negative log of the $[H_3O^+]$.

$$pH = -\log[H_3O^+]$$

Example: What is the pH of a solution that has

$$[H_3O^+] = 2.4 \times 10^{-11}\ M?$$

Answer: We substitute the given $[H_3O^+]$ into the pH equation and calculate the pH.

$$pH = -\log[H_3O^+]$$

$$= -\log[2.4 \times 10^{-11}]$$ **Calculator Display**

$\boxed{+/-}$ $\boxed{\log}$ 2.4 $\boxed{\text{EE or EXP}}$ $\boxed{+/-}$ 11 $\boxed{=}$ *10.61978876*

$$= 10.62$$

Two decimal places in the pH equal the two SFs in the $[H_3O^+]$ coefficient.

Calculating $[H_3O^+]$ from pH (14.6)

- The calculation of $[H_3O^+]$ from the pH is done by reversing the pH calculation using the negative pH.

$$[H_3O^+] = 10^{-pH}$$

Example: What is the $[H_3O^+]$ of a solution with a pH of 4.80?

Answer: $[H_3O^+] = 10^{-pH}$

$$= 10^{-4.80}$$ **Calculator Display**

$\boxed{2^{nd}}$ $\boxed{\log}$ $\boxed{+/-}$ 4.80 $\boxed{=}$ *1.584893192−05*

$$= 1.6 \times 10^{-5}\ M$$

Two SFs in the $[H_3O^+]$ equal the two decimal places in the pH.

CORE CHEMISTRY SKILLS

The chapter section containing each Core Chemistry Skill is shown in parentheses at the end of each heading.

Identifying Conjugate Acid–Base Pairs (14.2)

- According to the Brønsted–Lowry theory, a conjugate acid–base pair consists of molecules or ions related by the loss of one H^+ by an acid, and the gain of one H^+ by a base.
- Every acid–base reaction contains two conjugate acid–base pairs because an H^+ is transferred in both the forward and reverse directions.
- When an acid such as HF loses one H^+, the conjugate base F^- is formed. When H_2O acts as a base, it gains one H^+, which forms its conjugate acid, H_3O^+.

Example: Identify the conjugate acid–base pairs in the following reaction:
$$H_2SO_4(aq) + H_2O(l) \rightleftharpoons HSO_4^-(aq) + H_3O^+(aq)$$

Answer: $H_2SO_4(aq) + H_2O(l) \rightleftharpoons HSO_4^-(aq) + H_3O^+(aq)$
        Acid   Base   Conjugate base   Conjugate acid

Conjugate acid–base pairs: H_2SO_4/HSO_4^- and H_3O^+/H_2O

Calculating [H₃O⁺] and [OH⁻] in Solutions (14.5)

- For all aqueous solutions, the product of $[H_3O^+]$ and $[OH^-]$ is equal to the water dissociation constant, K_w.

$$K_w = [H_3O^+][OH^-]$$

- Because pure water contains equal numbers of OH^- ions and H_3O^+ ions, each with molar concentrations of 1.0×10^{-7} M, the numerical value of K_w is 1.0×10^{-14} at 25 °C.

$$K_w = [H_3O^+][OH^-] = [1.0 \times 10^{-7}][1.0 \times 10^{-7}]$$
$$= 1.0 \times 10^{-14}$$

- If we know the $[H_3O^+]$ of a solution, we can use the K_w expression to calculate the $[OH^-]$. If we know the $[OH^-]$ of a solution, we can calculate the $[H_3O^+]$ using the K_w expression.

$$[OH^-] = \frac{K_w}{[H_3O^+]} \quad [H_3O^+] = \frac{K_w}{[OH^-]}$$

Example: What is the $[OH^-]$ in a solution that has $[H_3O^+] = 2.4 \times 10^{-11}$ M? Is the solution acidic or basic?
Answer: We solve the K_w expression for $[OH^-]$ and substitute in the known values of K_w and $[H_3O^+]$.

$$[OH^-] = \frac{K_w}{[H_3O^+]} = \frac{1.0 \times 10^{-14}}{[2.4 \times 10^{-11}]} = 4.2 \times 10^{-4} \text{ M}$$

Because the $[OH^-]$ is greater than the $[H_3O^+]$, this is a basic solution.

Writing Equations for Reactions of Acids and Bases (14.7)

- Acids react with certain metals to produce hydrogen gas (H_2) and a salt.

$$2HCl(aq) + Mg(s) \longrightarrow H_2(g) + MgCl_2(aq)$$
  Acid   Metal   Hydrogen   Salt

- When an acid is added to a carbonate or bicarbonate, the products are carbon dioxide gas, water, and a salt.

$$2HCl(aq) + Na_2CO_3(aq) \longrightarrow CO_2(g) + H_2O(l) + 2NaCl(aq)$$
  Acid   Carbonate   Carbon   Water   Salt
                                   dioxide

- Neutralization is a reaction between a strong or weak acid with a strong base to produce water and a salt.

$$HCl(aq) + NaOH(aq) \longrightarrow H_2O(l) + NaCl(aq)$$
  Acid   Base   Water   Salt

Example: Write the balanced chemical equation for the reaction of hydrobromic acid $HBr(aq)$ and $ZnCO_3(s)$.
Answer:
$$2HBr(aq) + ZnCO_3(s) \longrightarrow CO_2(g) + H_2O(l) + ZnBr_2(aq)$$

Calculating Molarity or Volume of an Acid or Base in a Titration (14.7)

- In a titration, a measured volume of acid is neutralized by a strong base solution of known molarity.
- From the measured volume of the strong base solution required for titration and its molarity, the number of moles of the strong base, the moles of acid, and the concentration of the acid are calculated.

Example: A 15.0-mL sample of a H_2SO_4 solution is titrated with 24.0 mL of a 0.245 M NaOH solution. What is the molarity of the H_2SO_4 solution?

$$H_2SO_4(aq) + 2NaOH(aq) \longrightarrow 2H_2O(l) + Na_2SO_4(aq)$$

Answer:
$$24.0 \text{ mL NaOH solution} \times \frac{1 \text{ L NaOH solution}}{1000 \text{ mL NaOH solution}}$$
$$\times \frac{0.245 \text{ mol NaOH}}{1 \text{ L NaOH solution}} \times \frac{1 \text{ mol } H_2SO_4}{2 \text{ mol NaOH}} = 0.002\ 94 \text{ mol of } H_2SO_4$$

$$\text{Molarity (M)} = \frac{0.002\ 94 \text{ mol } H_2SO_4}{0.0150 \text{ L } H_2SO_4 \text{ solution}} = 0.196 \text{ M } H_2SO_4 \text{ solution}$$

Calculating the pH of a Buffer (14.9)

- A buffer solution maintains pH by neutralizing small amounts of added acid or base.
- Most buffer solutions consist of nearly equal concentrations of a weak acid and a salt containing its conjugate base such as acetic acid, $HC_2H_3O_2$, and its salt, $NaC_2H_3O_2$.
- The $[H_3O^+]$ is calculated by solving the K_a expression for $[H_3O^+]$, then substituting the values of $[H_3O^+]$, $[HA]$, and K_a into the equation.

$$K_a = \frac{[H_3O^+][C_2H_3O_2^-]}{[HC_2H_3O_2]}$$

Solving for $[H_3O^+]$ gives:

$$[H_3O^+] = K_a \times \frac{[HC_2H_3O_2]}{[C_2H_3O_2^-]} \quad \longleftarrow \text{ Weak acid} \\ \longleftarrow \text{ Conjugate base}$$

- The pH of the buffer is calculated from the $[H_3O^+]$.

$$pH = -\log[H_3O^+]$$

Example: What is the pH of a buffer prepared with 0.40 M $HC_2H_3O_2$ and 0.20 M $C_2H_3O_2^-$, if the K_a of acetic acid is 1.8×10^{-5}?
Answer: $[H_3O^+] = K_a \times \frac{[HC_2H_3O_2]}{[C_2H_3O_2^-]} = 1.8 \times 10^{-5} \times \frac{[0.40]}{[0.20]}$
$$= 3.6 \times 10^{-5} \text{ M}$$
$$pH = -\log[3.6 \times 10^{-5}] = 4.44$$

UNDERSTANDING THE CONCEPTS

The chapter sections to review are shown in parentheses at the end of each problem.

14.87 Determine if each of the following diagrams represents a strong acid or a weak acid. The acid has the formula HX. (14.3)

14.88 Adding a few drops of a strong acid to water will lower the pH appreciably. However, adding the same number of drops to a buffer does not appreciably alter the pH. Why? (14.9)

14.89 Identify each of the following as an acid or a base: (14.1)
a. H_2SO_4
b. RbOH
c. $Ca(OH)_2$
d. HI

14.90 Identify each of the following as an acid or a base: (14.1)
a. $Sr(OH)_2$
b. H_2SO_3
c. $HC_2H_3O_2$
d. CsOH

14.91 Complete the following table: (14.2)

Acid	Conjugate Base
H_2O	
	CN^-
HNO_2	
	$H_2PO_4^-$

14.92 Complete the following table: (14.2)

Base	Conjugate Acid
	HS^-
	H_3O^+
NH_3	
HCO_3^-	

Applications

14.93 State whether each of the following solutions is acidic, basic, or neutral: (14.6)
a. sweat, pH 5.2
b. tears, pH 7.5
c. bile, pH 8.1
d. stomach acid, pH 2.5

14.94 State whether each of the following solutions is acidic, basic, or neutral: (14.6)
a. saliva, pH 6.8
b. urine, pH 5.9
c. pancreatic juice, pH 8.0
d. blood, pH 7.45

14.95 Sometimes, during stress or trauma, a person can start to hyperventilate. Then the person might breathe into a paper bag to avoid fainting. (14.9)
a. What changes occur in the blood pH during hyperventilation?
b. How does breathing into a paper bag help return blood pH to normal?

Breathing into a paper bag can help a person who is hyperventilating.

14.96 In the blood plasma, pH is maintained by the carbonic acid–bicarbonate buffer system. (14.9)
a. How is pH maintained when acid is added to the buffer system?
b. How is pH maintained when base is added to the buffer system?

ADDITIONAL PRACTICE PROBLEMS

14.97 Identify each of the following as an acid, base, or salt, and give its name: (14.1)
a. $HBrO_2$
b. CsOH
c. $Mg(NO_3)_2$
d. $HClO_4$

14.98 Identify each of the following as an acid, base, or salt, and give its name: (14.1)
a. HNO_2
b. $MgBr_2$
c. NH_3
d. Li_2SO_3

14.99 Complete the following table: (14.2)

Acid	Conjugate Base
HI	
	Cl^-
NH_4^+	
	HS^-

14.100 Complete the following table: (14.2)

Base	Conjugate Acid
F^-	
	$HC_2H_3O_2$
	HSO_3^-
ClO^-	

14.101 Using Table 14.3, identify the stronger acid in each of the following pairs: (14.3)
 a. HF or HCN
 b. H_3O^+ or H_2S
 c. HNO_2 or $HC_2H_3O_2$
 d. H_2O or HCO_3^-

14.102 Using Table 14.3, identify the stronger base in each of the following pairs: (14.3)
 a. H_2O or Cl^-
 b. OH^- or NH_3
 c. SO_4^{2-} or NO_2^-
 d. CO_3^{2-} or H_2O

14.103 Determine the pH and pOH for each of the following solutions: (14.6)
 a. $[H_3O^+] = 2.0 \times 10^{-8}$ M
 b. $[H_3O^+] = 5.0 \times 10^{-2}$ M
 c. $[OH^-] = 3.5 \times 10^{-4}$ M
 d. $[OH^-] = 0.0054$ M

14.104 Determine the pH and pOH for each of the following solutions: (14.6)
 a. $[OH^-] = 1.0 \times 10^{-7}$ M
 b. $[H_3O^+] = 4.2 \times 10^{-3}$ M
 c. $[H_3O^+] = 0.0001$ M
 d. $[OH^-] = 8.5 \times 10^{-9}$ M

14.105 Are the solutions in problem 14.103 acidic, basic, or neutral? (14.6)

14.106 Are the solutions in problem 14.104 acidic, basic, or neutral? (14.6)

14.107 Calculate the $[H_3O^+]$ and $[OH^-]$ for a solution with each of the following pH values: (14.6)
 a. 3.00
 b. 6.2
 c. 8.85
 d. 11.00

14.108 Calculate the $[H_3O^+]$ and $[OH^-]$ for a solution with each of the following pH values: (14.6)
 a. 10.00
 b. 5.0
 c. 6.54
 d. 1.82

14.109 Solution A has a pH of 4.5, and solution B has a pH of 6.7. (14.6)
 a. Which solution is more acidic?
 b. What is the $[H_3O^+]$ in each?
 c. What is the $[OH^-]$ in each?

14.110 Solution X has a pH of 9.5, and solution Y has a pH of 7.5. (14.6)
 a. Which solution is more acidic?
 b. What is the $[H_3O^+]$ in each?
 c. What is the $[OH^-]$ in each?

14.111 What is the $[OH^-]$ in a solution that contains 0.225 g of NaOH in 0.250 L of solution? (14.7)

14.112 What is the $[H_3O^+]$ in a solution that contains 1.54 g of HNO_3 in 0.500 L of solution? (14.7)

14.113 What are the pH and pOH values of a solution prepared by dissolving 2.5 g of HCl in water to make 425 mL of solution? (14.6)

14.114 What are the pH and pOH values of a solution prepared by dissolving 1.0 g of $Ca(OH)_2$ in water to make 875 mL of solution? (14.6)

CHALLENGE PROBLEMS

The following problems are related to the topics in this chapter. However, they do not all follow the chapter order, and they require you to combine concepts and skills from several sections. These problems will help you increase your critical thinking skills and prepare for your next exam.

14.115 For each of the following: (14.2, 14.3)
 1. H_2S
 2. H_3PO_4
 a. Write the formula for the conjugate base.
 b. Write the K_a expression.
 c. Which is the weaker acid?

14.116 For each of the following: (14.2, 14.3)
 1. HCO_3^-
 2. $HC_2H_3O_2$
 a. Write the formula for the conjugate base.
 b. Write the K_a expression.
 c. Which is the stronger acid?

14.117 Using Table 14.3, identify the conjugate acid–base pairs in each of the following equations and whether the equilibrium mixture contains mostly products or mostly reactants: (14.2, 14.3)
 a. $NH_3(aq) + HNO_3(aq) \rightleftharpoons NH_4^+(aq) + NO_3^-(aq)$
 b. $HBr(aq) + H_2O(l) \rightleftharpoons H_3O^+(aq) + Br^-(aq)$

14.118 Using Table 14.3, identify the conjugate acid–base pairs in each of the following equations and whether the equilibrium mixture contains mostly products or mostly reactants: (14.2, 14.3)
 a. $HNO_2(aq) + HS^-(aq) \rightleftharpoons H_2S(g) + NO_2^-(aq)$
 b. $Cl^-(aq) + H_2O(l) \rightleftharpoons OH^-(aq) + HCl(aq)$

14.119 Complete and balance each of the following: (14.7)
 a. $H_2SO_4(aq) + ZnCO_3(s) \longrightarrow$
 b. $HNO_3(aq) + Al(s) \longrightarrow$

14.120 Complete and balance each of the following: (14.7)
 a. $H_3PO_4(aq) + Ca(OH)_2(s) \longrightarrow$
 b. $HNO_3(aq) + KHCO_3(s) \longrightarrow$

14.121 Determine each of the following for a 0.050 M KOH solution: (14.6, 14.7)
 a. $[H_3O^+]$
 b. pH
 c. pOH
 d. the balanced equation for the reaction with H_2SO_4
 e. milliliters of KOH solution required to neutralize 40.0 mL of a 0.035 M H_2SO_4 solution

14.122 Determine each of the following for a 0.10 M HBr solution: (14.6, 14.7)
 a. $[H_3O^+]$
 b. pH
 c. pOH
 d. the balanced equation for the reaction with LiOH
 e. milliliters of HBr solution required to neutralize 36.0 mL of a 0.25 M LiOH solution

14.123 Calculate the volume, in milliliters, of a 0.150 M NaOH solution that will completely neutralize each of the following: (14.8)
 a. 25.0 mL of a 0.288 M HCl solution
 b. 10.0 mL of a 0.560 M H_2SO_4 solution

14.124 Calculate the volume, in milliliters, of a 0.215 M NaOH solution that will completely neutralize each of the following: (14.8)
 a. 3.80 mL of a 1.25 M HNO_3 solution
 b. 8.50 mL of a 0.825 M H_3PO_4 solution

14.125 A solution of 0.205 M NaOH is used to titrate 20.0 mL of a H_2SO_4 solution. If 45.6 mL of the NaOH solution is required to reach the endpoint, what is the molarity of the H_2SO_4 solution? (14.8)

$$H_2SO_4(aq) + 2NaOH(aq) \longrightarrow 2H_2O(l) + Na_2SO_4(aq)$$

14.126 A 10.0-mL sample of vinegar, which is an aqueous solution of acetic acid, $HC_2H_3O_2$, requires 16.5 mL of a 0.500 M NaOH solution to reach the endpoint in a titration. What is the molarity of the acetic acid solution? (14.8)

$$HC_2H_3O_2(aq) + NaOH(aq) \longrightarrow H_2O(l) + NaC_2H_3O_2(aq)$$

14.127 A buffer solution is made by dissolving H_3PO_4 and NaH_2PO_4 in water. (14.9)
 a. Write an equation that shows how this buffer neutralizes added acid.
 b. Write an equation that shows how this buffer neutralizes added base.
 c. Calculate the pH of this buffer if it contains 0.50 M H_3PO_4 and 0.20 M $H_2PO_4^-$. The K_a for H_3PO_4 is 7.5×10^{-3}.

14.128 A buffer solution is made by dissolving $HC_2H_3O_2$ and $NaC_2H_3O_2$ in water. (14.9)
 a. Write an equation that shows how this buffer neutralizes added acid.
 b. Write an equation that shows how this buffer neutralizes added base.
 c. Calculate the pH of this buffer if it contains 0.20 M $HC_2H_3O_2$ and 0.40 M $C_2H_3O_2^-$. The K_a for $HC_2H_3O_2$ is 1.8×10^{-5}.

Applications

14.129 One of the most acidic lakes in the United States is Little Echo Pond in the Adirondacks in New York. Recently, this lake had a pH of 4.2, well below the recommended pH of 6.5. (14.6, 14.8)
 a. What are the $[H_3O^+]$ and $[OH^-]$ of Little Echo Pond?
 b. What are the $[H_3O^+]$ and $[OH^-]$ of a lake that has a pH of 6.5?

A helicopter drops calcium carbonate on an acidic lake to increase its pH.

 c. One way to raise the pH of an acidic lake (and restore aquatic life) is to add limestone ($CaCO_3$). How many grams of $CaCO_3$ are needed to neutralize 1.0 kL of the acidic water from the lake if the acid is sulfuric acid?

$$H_2SO_4(aq) + CaCO_3(s) \longrightarrow CO_2(g) + H_2O(l) + CaSO_4(aq)$$

14.130 The daily output of stomach acid (gastric juice) is 1000 mL to 2000 mL. Prior to a meal, stomach acid (HCl) typically has a pH of 1.42. (14.6, 14.7, 14.8)
 a. What is the $[H_3O^+]$ of the stomach acid?
 b. One chewable tablet of the antacid Maalox contains 600. mg of $CaCO_3$. Write the neutralization equation, and calculate the milliliters of stomach acid neutralized by two tablets of Maalox.
 c. The antacid milk of magnesia contains 400. mg of $Mg(OH)_2$ per teaspoon. Write the neutralization equation, and calculate the number of milliliters of stomach acid that are neutralized by 1 tablespoon of milk of magnesia (1 tablespoon = 3 teaspoons).

ANSWERS TO ENGAGE QUESTIONS

14.1 An acid composed of a hydrogen ion and a simple nonmetal anion is named with the prefix *hydro* before the name of the nonmetal, and the *ide* ending of the most common form is changed to *ic acid*.

14.2 The base BrO_2^- accepts H^+ to form its conjugate acid $HBrO_2$.

14.3 H_2O loses one H^+ to form OH^- but gains one H^+ to form H_3O^+.

14.4 A strong acid dissociates nearly 100% to form H^+ and a negative ion, whereas a weak acid dissociates slightly in water, forming a small amount of H^+ and a negative ion.

14.5 According to Table 14.3, H_2SO_4 is a stronger acid than H_2S.

14.6 In pure water, when one H_2O molecule donates H^+ to another water molecule, forming equal amounts of H_3O^+ and OH^-.

14.7 A solution that has a $[H_3O^+]$ of 1.0×10^{-3} M would have a $[OH^-]$ of 1.0×10^{-11} M. Because $[H_3O^+] > [OH^-]$, the solution is acidic.

14.8 You obtain the $[OH^-]$ of a solution by dividing the K_w by the $[H_3O^+]$ of the solution.

14.9 Because there is an inverse relationship between $[H_3O^+]$ and $[OH^-]$, if $[OH^-]$ decreases, $[H_3O^+]$ must increase.

14.10 According to Figure 14.4, apple juice is acidic.

14.11 A pH of 4.00, which is less than 7.0, is acidic.

14.12 A solution with $[H_3O^+]$ of 1.0×10^{-6} M has a pH of 6.00. There are two zeros after the decimal point because there are two significant figures in the coefficient 1.0.

14.13 When KOH completely neutralizes H_2SO_4, the salt produced is K_2SO_4.

14.14 To determine the molarity of an acid solution, the volume of the acid and the volume and molarity of the base used to neutralize the acid are needed.

14.15 A weak acid or base and the salt of that weak acid or base are necessary in a buffer to react with added acid or base.

14.16 When H_3O^+ is added to a buffer, it is neutralized by the weak base or the salt of the weak acid.

14.17 A solution of $H_2PO_4^-$ and HPO_4^{2-} acts as a buffer; $H_2PO_4^-$ will neutralize added OH^-, and HPO_4^{2-} will neutralize added H_3O^+.

ANSWERS TO SELECTED PROBLEMS

14.1 a. acid b. acid c. acid
d. base e. both

14.3 a. hydrochloric acid b. calcium hydroxide
c. perchloric acid d. strontium hydroxide
e. sulfurous acid f. bromous acid

14.5 a. RbOH b. HF c. H_3PO_4
d. LiOH e. NH_4OH f. HIO_4

14.7 a. HI is the acid (hydrogen ion donor), and H_2O is the base (hydrogen ion acceptor).
b. H_2O is the acid (hydrogen ion donor), and F^- is the base (hydrogen ion acceptor).
c. H_2S is the acid (hydrogen ion donor), and $C_2H_5-NH_2$ is the base (hydrogen ion acceptor).

14.9 a. F^- b. OH^- c. HPO_3^{2-}
d. SO_4^{2-} e. ClO_2^-

14.11 a. HCO_3^- b. H_3O^+ c. H_3PO_4
d. HBr e. $HClO_4$

14.13 a. The conjugate acid–base pairs are H_2CO_3/HCO_3^- and H_3O^+/H_2O.
b. The conjugate acid–base pairs are HCN/CN^- and HNO_2/NO_2^-.
c. The conjugate acid–base pairs are HF/F^- and $HCHO_2/CHO_2^-$.

14.15 $NH_4^+(aq) + H_2O(l) \rightleftharpoons NH_3(aq) + H_3O^+(aq)$

14.17 A strong acid is a good hydrogen ion donor, whereas its conjugate base is a poor hydrogen ion acceptor.

14.19 a. HBr b. HSO_4^- c. H_2CO_3

14.21 a. HSO_4^- b. HF c. HCO_3^-

14.23 a. reactants b. reactants c. products

14.25 $NH_4^+(aq) + SO_4^{2-}(aq) \rightleftharpoons NH_3(aq) + HSO_4^-(aq)$

The equilibrium mixture contains mostly reactants because NH_4^+ is a weaker acid than HSO_4^-, and SO_4^{2-} is a weaker base than NH_3.

14.27 a. True b. False c. False
d. True e. False

14.29 a. H_2SO_3 b. HSO_3^- c. H_2SO_3
d. HS^- e. H_2SO_3

14.31 $H_3PO_4(aq) + H_2O(l) \rightleftharpoons H_3O^+(aq) + H_2PO_4^-(aq)$

$$K_a = \frac{[H_3O^+][H_2PO_4^-]}{[H_3PO_4]}$$

14.33 In pure water, $[H_3O^+] = [OH^-]$ because one of each is produced every time a hydrogen ion is transferred from one water molecule to another.

14.35 In an acidic solution, the $[H_3O^+]$ is greater than the $[OH^-]$.

14.37 a. acidic b. basic
c. basic d. acidic

14.39 a. 1.0×10^{-5} M b. 1.0×10^{-8} M
c. 5.0×10^{-10} M d. 2.5×10^{-2} M

14.41 a. 2.5×10^{-13} M b. 2.0×10^{-9} M
c. 5.0×10^{-11} M d. 1.3×10^{-6} M

14.43 In a neutral solution, the $[H_3O^+]$ is 1.0×10^{-7} M and the pH is 7.00, which is the negative value of the power of 10.

14.45 a. basic b. acidic c. acidic
d. acidic e. basic

14.47 An increase or decrease of one pH unit changes the $[H_3O^+]$ by a factor of 10. Thus a pH of 3 is 10 times more acidic than a pH of 4.

14.49 a. 4.0 b. 8.5 c. 9.0
d. 3.40 e. 7.17 f. 10.92

14.51

$[H_3O^+]$	$[OH^-]$	pH	pOH	Acidic, Basic, or Neutral?
1.0×10^{-8} M	1.0×10^{-6} M	8.00	6.00	Basic
3.2×10^{-4} M	3.1×10^{-11} M	3.49	10.51	Acidic
2.8×10^{-5} M	3.6×10^{-10} M	4.55	9.45	Acidic
1.0×10^{-12} M	1.0×10^{-2} M	12.00	2.00	Basic

14.53 1.2×10^{-7} M

14.55 a. $2HBr(aq) + ZnCO_3(s) \longrightarrow CO_2(g) + H_2O(l) + ZnBr_2(aq)$
b. $2HCl(aq) + Zn(s) \longrightarrow H_2(g) + ZnCl_2(aq)$
c. $HCl(aq) + NaHCO_3(s) \longrightarrow CO_2(g) + H_2O(l) + NaCl(aq)$
d. $H_2SO_4(aq) + Mg(OH)_2(s) \longrightarrow 2H_2O(l) + MgSO_4(aq)$

14.57 a. $2HCl(aq) + Mg(OH)_2(s) \longrightarrow 2H_2O(l) + MgCl_2(aq)$
b. $H_3PO_4(aq) + 3LiOH(aq) \longrightarrow 3H_2O(l) + Li_3PO_4(aq)$

14.59 a. $H_2SO_4(aq) + 2NaOH(aq) \longrightarrow 2H_2O(l) + Na_2SO_4(aq)$
b. $3HCl(aq) + Fe(OH)_3(s) \longrightarrow 3H_2O(l) + FeCl_3(aq)$
c. $H_2CO_3(aq) + Mg(OH)_2(s) \longrightarrow 2H_2O(l) + MgCO_3(s)$

14.61 To a known volume of a formic acid solution, add a few drops of indicator. Place a solution of NaOH of known molarity in a buret. Add base to acid until one drop changes the color of the solution. Use the volume and molarity of NaOH and the volume of a formic acid solution to calculate the concentration of the formic acid in the sample.

14.63 0.829 M HCl solution

14.65 0.124 M H_2SO_4 solution

14.67 16.5 mL

14.69 b and c are buffer systems. b contains the weak acid H_2CO_3 and its salt $NaHCO_3$. c contains HF, a weak acid, and its salt, KF.

14.71 a. 3 b. 1 and 2
c. 3 d. 1

14.73 pH = 3.35

14.75 The pH of the 0.10 M HF/0.10 M NaF buffer is 3.46.
The pH of the 0.060 M HF/0.120 M NaF buffer is 3.76.

14.77 If large amounts of HCO_3^- are lost, equilibrium shifts to higher H_3O^+, which lowers the pH.

14.79 pH = 3.70

14.81 2.5×10^{-4} M

14.83 $2HCl(aq) + CaCO_3(s) \longrightarrow CO_2(g) + H_2O(l) + CaCl_2(aq)$

14.85 0.200 g of $CaCO_3$

14.87 a. This diagram represents a weak acid; only a few HX molecules separate into H_3O^+ and X^- ions.
b. This diagram represents a strong acid; all the HX molecules separate into H_3O^+ and X^- ions.

14.89 a. acid b. base
c. base d. acid

14.91

Acid	Conjugate Base
H_2O	OH^-
HCN	CN^-
HNO_2	NO_2^-
H_3PO_4	$H_2PO_4^-$

14.93 a. acidic   **b.** basic
     c. basic   **d.** acidic

14.95 a. During hyperventilation, a person will lose CO_2 and the blood pH will rise.
     b. Breathing into a paper bag will increase the CO_2 concentration and lower the blood pH.

14.97 a. acid, bromous acid   **b.** base, cesium hydroxide
     c. salt, magnesium nitrate   **d.** acid, perchloric acid

14.99

Acid	Conjugate Base
HI	I^-
HCl	Cl^-
NH_4^+	NH_3
H_2S	HS^-

14.101 a. HF **b.** H_3O^+
     c. HNO_2 **d.** HCO_3^-

14.103 a. pH = 7.70; pOH = 6.30   **b.** pH = 1.30; pOH = 12.70
     c. pH = 10.54; pOH = 3.46   **d.** pH = 11.73; pOH = 2.27

14.105 a. basic   **b.** acidic
     c. basic   **d.** basic

14.107 a. $[H_3O^+] = 1.0 \times 10^{-3}$ M; $[OH^-] = 1.0 \times 10^{-11}$ M
     b. $[H_3O^+] = 6 \times 10^{-7}$ M; $[OH^-] = 2 \times 10^{-8}$ M
     c. $[H_3O^+] = 1.4 \times 10^{-9}$ M; $[OH^-] = 7.1 \times 10^{-6}$ M
     d. $[H_3O^+] = 1.0 \times 10^{-11}$ M; $[OH^-] = 1.0 \times 10^{-3}$ M

14.109 a. Solution A
     b. Solution A $[H_3O^+] = 3 \times 10^{-5}$ M;
      Solution B $[H_3O^+] = 2 \times 10^{-7}$ M
     c. Solution A $[OH^-] = 3 \times 10^{-10}$ M;
      Solution B $[OH^-] = 5 \times 10^{-8}$ M

14.111 $[OH^-] = 0.0225$ M

14.113 pH = 0.80; pOH = 13.20

14.115 a. 1. HS^-
       2. $H_2PO_4^-$

    b. 1. $\dfrac{[H_3O^+][HS^-]}{[H_2S]}$

      2. $\dfrac{[H_3O^+][H_2PO_4^-]}{[H_3PO_4]}$

    c. H_2S

14.117 a. HNO_3/NO_3^- and NH_4^+/NH_3; equilibrium mixture contains mostly products
     b. HBr/Br^- and H_3O^+/H_2O; equilibrium mixture contains mostly products

14.119 a. $H_2SO_4(aq) + ZnCO_3(s) \longrightarrow$
$$CO_2(g) + H_2O(l) + ZnSO_4(aq)$$
     b. $6HNO_3(aq) + 2Al(s) \longrightarrow 3H_2(g) + 2Al(NO_3)_3(aq)$

14.121 a. $[H_3O^+] = 2.0 \times 10^{-13}$ M
     b. pH = 12.70
     c. pOH = 1.30
     d. $H_2SO_4(aq) + 2KOH(aq) \longrightarrow 2H_2O(l) + K_2SO_4(aq)$
     e. 56 mL of the KOH solution

14.123 a. 48.0 mL of NaOH solution
     b. 74.7 mL of NaOH solution

14.125 0.234 M H_2SO_4

14.127 a. acid: $H_2PO_4^-(aq) + H_3O^+(aq) \longrightarrow H_3PO_4(aq) + H_2O(l)$
     b. base: $H_3PO_4(aq) + OH^-(aq) \longrightarrow H_2PO_4^-(aq) + H_2O(l)$
     c. pH = 1.72

14.129 a. $[H_3O^+] = 6 \times 10^{-5}$ M; $[OH^-] = 2 \times 10^{-10}$ M
     b. $[H_3O^+] = 3 \times 10^{-7}$ M; $[OH^-] = 3 \times 10^{-8}$ M
     c. 3 g of $CaCO_3$

CI.21 Methane is a major component of purified natural gas used for heating and cooking. When 1.0 mol of methane gas burns with oxygen to produce carbon dioxide and water vapor, 883 kJ of heat is produced. At STP, methane gas has a density of 0.715 g/L. For transport, the natural gas is cooled to $-163\,°C$ to form liquefied natural gas (LNG) with a density of 0.45 g/mL. A tank on a ship can hold 7.0 million gallons of LNG. (2.7, 7.2, 7.3, 8.2, 8.3, 9.5, 10.1, 11.6)

An LNG carrier transports liquefied natural gas.

a. Draw the Lewis structures for methane, which has the formula CH_4.
b. What is the mass, in kilograms, of LNG (assume that LNG is all methane) transported in one tank on a ship?
c. What is the volume, in liters, of LNG (methane) from one tank when the LNG (methane) from one tank is converted to methane gas at STP?
d. Write the balanced chemical equation for the combustion of methane and oxygen in a gas burner, including the heat of reaction.

Methane is the fuel burned in a gas cooktop.

e. How many kilograms of oxygen are needed to react with all of the methane in one tank of LNG?
f. How much heat, in kilojoules, is released after burning all of the methane from one tank of LNG?

CI.22 Automobile exhaust is a major cause of air pollution. One pollutant is nitrogen oxide, which forms from nitrogen and oxygen gases in the air at the high temperatures in an automobile engine. Once emitted into the air, nitrogen oxide reacts with oxygen to produce nitrogen dioxide, a reddish brown gas with a sharp, pungent odor that makes up smog. One component of gasoline is octane, C_8H_{18}, which has a density of 0.803 g/mL. In one year, a

Two gases found in automobile exhaust are carbon dioxide and nitrogen oxide.

typical automobile uses 550 gal of gasoline and produces 41 lb of nitrogen oxide. (2.7, 7.2, 7.3, 8.2, 8.3, 9.2, 11.6)
a. Write balanced chemical equations for the production of nitrogen oxide and nitrogen dioxide.
b. If all the nitrogen oxide emitted by one automobile is converted to nitrogen dioxide in the atmosphere, how many kilograms of nitrogen dioxide are produced in one year by a single automobile?
c. Write a balanced chemical equation for the combustion of octane.
d. How many moles of C_8H_{18} are present in 15.2 gal of octane?
e. How many liters of CO_2 at STP are produced in one year from the gasoline used by the typical automobile?

CI.23 A mixture of 25.0 g of CS_2 gas and 30.0 g of O_2 gas is placed in a 10.0-L container and heated to 125 °C. The products of the reaction are carbon dioxide gas and sulfur dioxide gas. (7.2, 7.3, 8.2, 9.2, 9.3, 11.1, 11.7)
a. Write a balanced chemical equation for the reaction.
b. How many grams of CO_2 are produced?
c. What is the partial pressure, in millimeters of mercury, of the remaining reactant?
d. What is the final pressure, in millimeters of mercury, in the container?

CI.24 In wine-making, glucose ($C_6H_{12}O_6$) from grapes undergoes fermentation in the absence of oxygen to produce ethanol and carbon dioxide. A bottle of vintage port wine has a volume of 750 mL and contains 135 mL of ethanol (C_2H_6O). Ethanol has a density of 0.789 g/mL. In 1.5 lb of grapes, there are 26 g of glucose. (2.7, 7.2, 7.3, 8.2, 9.2, 12.4)

Port is a type of fortified wine that is produced in Portugal.

When the glucose in grapes is fermented, ethanol is produced.

a. Calculate the volume percent (v/v) of ethanol in the port wine.
b. What is the molarity (M) of ethanol in the port wine?
c. Write the balanced chemical equation for the fermentation reaction of sugar in grapes.
d. How many grams of sugar from grapes are required to produce one bottle of port wine?
e. How many bottles of port wine can be produced from 1.0 ton of grapes (1 ton = 2000 lb)?

CI.25 Consider the following reaction at equilibrium:

$$2H_2(g) + S_2(g) \rightleftharpoons 2H_2S(g) + \text{heat}$$

In a 10.0-L container, an equilibrium mixture contains 2.02 g of H_2, 10.3 g of S_2, and 68.2 g of H_2S. (7.2, 7.3, 13.2, 13.3, 13.4, 13.5)
a. What is the numerical value of K_c for this equilibrium mixture?
b. If more H_2 is added to the equilibrium mixture, how will the equilibrium shift?

473

c. How will the equilibrium shift if the mixture is placed in a 5.00-L container with no change in temperature?

d. If a 5.00-L container has an equilibrium mixture of 0.300 mol of H_2 and 2.50 mol of H_2S, what is the $[S_2]$ if temperature does not change?

CI.26 A saturated solution of silver hydroxide has a pH of 10.15. (7.2, 7.3, 13.2, 13.6)

a. Write the solubility product expression for silver hydroxide.

b. Calculate the numerical value of K_{sp} for silver hydroxide.

c. How many grams of silver hydroxide will dissolve in 2.0 L of water?

CI.27 A metal M with a mass of 0.420 g completely reacts with 34.8 mL of a 0.520 M HCl solution to form H_2 gas and aqueous MCl_3. (7.2, 7.3, 8.2, 9.2, 11.7)

When a metal reacts with a strong acid, bubbles of hydrogen gas form.

a. Write a balanced chemical equation for the reaction of the metal M(s) and HCl(aq).

b. What volume, in milliliters, of H_2 at 720. mmHg and 24 °C is produced?

c. How many moles of metal M reacted?

d. Using your results from part **c**, determine the molar mass and name of metal M.

e. Write the balanced chemical equation for the reaction.

CI.28 In a teaspoon (5.0 mL) of a liquid antacid, there are 400. mg of $Mg(OH)_2$ and 400. mg of $Al(OH)_3$. A 0.080 M HCl solution, which is similar to stomach acid, is used to neutralize 5.0 mL of the liquid antacid. (12.4, 14.6, 14.7, 14.8)

An antacid neutralizes stomach acid and raises the pH.

a. Write the equation for the neutralization of HCl and $Mg(OH)_2$.

b. Write the equation for the neutralization of HCl and $Al(OH)_3$.

c. What is the pH of the HCl solution?

d. How many milliliters of the HCl solution are needed to neutralize the $Mg(OH)_2$?

e. How many milliliters of the HCl solution are needed to neutralize the $Al(OH)_3$?

CI.29 A KOH solution is prepared by dissolving 8.57 g of KOH in enough water to make 850. mL of KOH solution. (12.4, 14.6, 14.7, 14.8)

a. What is the molarity of the KOH solution?

b. What is the $[H_3O^+]$ and pH of the KOH solution?

c. Write the balanced chemical equation for the neutralization of KOH by H_2SO_4.

d. How many milliliters of a 0.250 M H_2SO_4 solution is required to neutralize 10.0 mL of the KOH solution?

CI.30 A solution of HCl is prepared by diluting 15.0 mL of a 12.0 M HCl solution with enough water to make 750. mL of HCl solution. (12.4, 14.6, 14.7, 14.8)

a. What is the molarity of the HCl solution?

b. What is the $[H_3O^+]$ and pH of the HCl solution?

c. Write the balanced chemical equation for the reaction of HCl and $MgCO_3$.

d. How many milliliters of the diluted HCl solution is required to completely react with 350. mg of $MgCO_3$?

CI.31 A volume of 200.0 mL of a carbonic acid buffer for blood plasma is prepared that contains 0.403 g of $NaHCO_3$ and 0.0149 g of H_2CO_3. At body temperature (37 °C), the K_a of carbonic acid is 7.9×10^{-7}. (7.2, 7.3, 8.4, 14.9)

$$H_2CO_3(aq) + H_2O(l) \rightleftharpoons HCO_3^-(aq) + H_3O^+(aq)$$

a. What is the $[H_2CO_3]$?

b. What is the $[HCO_3^-]$?

c. What is the $[H_3O^+]$?

d. What is the pH of the buffer?

e. Write a balanced chemical equation that shows how this buffer neutralizes added acid.

f. Write a balanced chemical equation that shows how this buffer neutralizes added base.

CI.32 In the kidneys, the ammonia buffer system buffers high H_3O^+. Ammonia, which is produced in renal tubules from amino acids, combines with H^+ to be excreted as NH_4Cl. At body temperature (37 °C), the $K_a = 5.6 \times 10^{-10}$. A buffer solution with a volume of 125 mL contains 3.34 g of NH_4Cl and 0.0151 g of NH_3. (7.2, 7.3, 8.4, 14.9)

$$NH_4^+(aq) + H_2O(l) \rightleftharpoons NH_3(aq) + H_3O^+(aq)$$

a. What is the $[NH_4^+]$?

b. What is the $[NH_3]$?

c. What is the $[H_3O^+]$?

d. What is the pH of the buffer?

e. Write a balanced chemical equation that shows how this buffer neutralizes added acid.

f. Write a balanced chemical equation that shows how this buffer neutralizes added base.

ANSWERS

CI.21 **a.**

$$H:\overset{\displaystyle H}{\underset{\displaystyle H}{\overset{..}{\underset{..}{C}}}}:H \quad \text{or} \quad H-\overset{\displaystyle H}{\underset{\displaystyle H}{\overset{|}{\underset{|}{C}}}}-H$$

b. 1.2×10^7 kg of LNG (methane)

c. 1.7×10^{10} L of LNG (methane)

d. $CH_4(g) + 2O_2(g) \xrightarrow{\Delta} CO_2(g) + 2H_2O(g) + 883$ kJ

e. 4.8×10^7 kg of O_2

f. 6.6×10^{11} kJ

CI.23 **a.** $CS_2(g) + 3O_2(g) \xrightarrow{\Delta} CO_2(g) + 2SO_2(g)$

b. 13.8 g of CO_2

c. 37 mmHg

d. 2370 mmHg

CI.25 **a.** $K_c = 248$

b. If H_2 is added, the equilibrium will shift in the direction of the products.

c. If the volume decreases, the equilibrium will shift in the direction of the products.

d. $[S_2] = 0.280$ M

CI.27 **a.** $6HCl(aq) + 2M(s) \longrightarrow 3H_2(g) + 2MCl_3(aq)$

b. 233 mL of H_2

c. 6.03×10^{-3} mol of M

d. 69.7 g/mol; gallium

e. $6HCl(aq) + 2Ga(s) \longrightarrow 3H_2(g) + 2GaCl_3(aq)$

CI.29 **a.** 0.180 M

b. $[H_3O^+] = 5.56 \times 10^{-14}$ M; pH = 13.255

c. $H_2SO_4(aq) + 2KOH(aq) \longrightarrow 2H_2O(l) + K_2SO_4(aq)$

d. 3.60 mL

CI.31 **a.** 1.20×10^{-3} M

b. 0.0240 M

c. 4.0×10^{-8} M

d. 7.40

e. $HCO_3^-(aq) + H_3O^+(aq) \longrightarrow H_2CO_3(aq) + H_2O(l)$

f. $H_2CO_3(aq) + OH^-(aq) \longrightarrow HCO_3^-(aq) + H_2O(l)$

Oxidation and Reduction

15

Kimberly's teeth are badly stained by food, coffee, and drinks that form a layer on top of the enamel of her teeth. Kimberly's dentist, Jane, removes some of this layer by scraping her teeth, then brushes Kimberly's teeth with an abrasive toothpaste. After that, Jane begins the process of whitening Kimberly's teeth. She explains to Kimberly that she uses a gel containing 35% hydrogen peroxide (H_2O_2) that penetrates into the enamel of the tooth, where it undergoes a chemical reaction that whitens the teeth. The chemical reaction is an oxidation–reduction reaction where one chemical (hydrogen peroxide) is reduced and the other chemical (the coffee stain) is oxidized. During the oxidation, the coffee stains on the teeth become lighter or colorless, and therefore, the teeth are whiter.

CAREER

Dentist

A dentist assesses and maintains the health of the teeth, gums, mouth, and bones of the jaw. A dentist examines X-rays to determine if there is disease or decay of the teeth and gums. Dentists remove decay and fill cavities, or they may extract teeth if necessary. Local anesthetics are often used for dental procedures. Dentists also educate patients on proper care of teeth and gums. Many dentists operate their own businesses with a staff that includes dental hygienists and a dental technician. A dentist may specialize in certain areas of dentistry. An orthodontist fits a patient's teeth with wires to straighten teeth or correct a bite. A periodontist treats problems with soft tissues of the gums. Pediatric dentists treat children's teeth.

UPDATE   Whitening Kimberly's Teeth

After Kimberly's teeth are cleaned and whitened, her dentist recommends she use a toothpaste to strengthen her teeth. You can see more details about the whitening process in the **UPDATE Whitening Kimberly's Teeth**, page 499, and learn about compounds in teeth and toothpaste.

15.1 Oxidation and Reduction

LEARNING GOAL Identify a reaction as an oxidation or a reduction. Assign and use oxidation numbers to identify elements that are oxidized or reduced.

Oxidation and *reduction* are types of chemical reactions that are prevalent in our everyday lives. When we use natural gas to heat our home, turn on a computer, start a car, or eat food, we are using energy from the processes of oxidation and reduction.

In all of these processes, electrons are transferred during oxidation and reduction. In an **oxidation** reaction, a reactant loses one or more electrons. We say the substance is *oxidized*. In a **reduction** reaction, a reactant gains one or more electrons. We say the substance is *reduced*. When a substance undergoes oxidation, the electrons it loses must be gained by another substance, which is reduced. In an **oxidation–reduction reaction**, processes of oxidation and reduction must occur together. Sometimes we shorten this to *redox reactions*.

Writing Oxidation and Reduction Reactions

Perhaps you have seen glasses with photochromic lenses that darken when exposed to UV light from the Sun. Once the wearer returns indoors, the lenses become clear again. The change from clear to dark and back to clear is the result of oxidation and reduction reactions. We can use the compounds in the photochromic lenses to illustrate some oxidation and reduction reactions.

To make the lenses photosensitive, silver chloride (AgCl) and copper(I) chloride (CuCl) are embedded in the lens material. Away from sunlight, visible light goes through the clear lens. However, when exposed to sunlight, there is an oxidation–reduction reaction when the UV light is absorbed by AgCl, and Cl atoms and Ag atoms are produced. It is the silver atoms that are black and darken the lenses. This redox reaction is written:

$$AgCl \longrightarrow Ag + Cl \quad \text{or}$$

$$Ag^+ + Cl^- \longrightarrow Ag + Cl \qquad \text{Oxidation–reduction reaction}$$

The chloride (Cl^-) ions lose electrons; they are oxidized. The silver (Ag^+) ions gain electrons; they are reduced.

Half-Reactions

To identify the oxidation and the reduction reactions, we write each of the reactants with its product as a *half-reaction*. The half-reaction with Cl^- is an oxidation because Cl^- loses an electron when it forms Cl. The half-reaction with Ag^+ is a reduction because the Ag^+ gains an electron when it forms Ag.

$$Cl^- \longrightarrow Cl + e^- \qquad \text{Oxidation} \quad \text{Loss of electron}$$

$$Ag^+ + e^- \longrightarrow Ag \qquad \text{Reduction} \quad \text{Gain of electron} \quad \text{Lenses darken}$$

To prevent the lenses from clearing immediately, some Cu^+ ions are added to the lenses. These Cu^+ ions convert the Cl atoms produced from the darkening process to Cl^- ions.

$$Cu^+ + Cl \longrightarrow Cu^{2+} + Cl^-$$

$$Cu^+ \longrightarrow Cu^{2+} + e^- \qquad \text{Oxidation}$$

$$Cl + e^- \longrightarrow Cl^- \qquad \text{Reduction}$$

When the wearer of the lenses moves out of sunlight, the reactions take place in the opposite direction. The Cu^{2+} ions combine with the silver atoms to re-form Cu^+ and Ag^+, which causes the lens to become clear again.

$$Cu^{2+} + Ag \longrightarrow Cu^+ + Ag^+$$

$$Ag \longrightarrow Ag^+ + e^- \qquad \text{Oxidation} \quad \text{Lenses clear}$$

$$Cu^{2+} + e^- \longrightarrow Cu^+ \qquad \text{Reduction}$$

REVIEW

Using Positive and Negative Numbers in Calculations (1.4)

Writing Positive and Negative Ions (6.1)

Writing Ionic Formulas (6.2)

Balancing a Chemical Equation (8.2)

Identifying Oxidized and Reduced Substances (8.4)

Glasses with photochromic lenses darken when exposed to sunlight.

ENGAGE 15.1

How can you identify a half-reaction that is a reduction?

SAMPLE PROBLEM 15.1 Identifying Oxidation and Reduction Reactions

TRY IT FIRST

For each of the following, identify the reaction as an oxidation or a reduction:

a. $Cu^{2+}(aq) + 2\,e^- \longrightarrow Cu(s)$
b. $Fe(s) \longrightarrow Fe^{3+}(aq) + 3\,e^-$
c. $Cr^{2+}(aq) \longrightarrow Cr^{3+}(aq) + e^-$
d. $Li^+(aq) + e^- \longrightarrow Li(s)$

SOLUTION

a. Since Cu^{2+} gained electrons, this is a reduction.
b. Since Fe lost electrons, this is an oxidation.
c. Since Cr^{2+} lost an electron, this is an oxidation.
d. Since Li^+ gained an electron, this is a reduction.

SELF TEST 15.1

Identify each of the following reactions as an oxidation or a reduction:

a. $Na(s) \longrightarrow Na^+(aq) + e^-$ **b.** $Zn^{2+}(aq) + 2\,e^- \longrightarrow Zn(s)$

ANSWER

a. oxidation   **b.** reduction

PRACTICE PROBLEMS

Try Practice Problems 15.1 and 15.2

Oxidation Numbers

In more complex oxidation–reduction reactions, the identification of the substances oxidized and reduced is not always obvious. To help identify the atoms or ions that are oxidized or reduced, we assign values called **oxidation numbers** (sometimes called *oxidation states*) to the elements of the reactants and products. It is important to recognize that oxidation numbers do not always represent actual charges, but they help us identify loss or gain of electrons.

Rules for Assigning Oxidation Numbers

CORE CHEMISTRY SKILL

Assigning Oxidation Numbers

The rules for assigning oxidation numbers to the atoms or ions in the reactants and products are given in **TABLE 15.1**.

ENGAGE 15.2

Why is the oxidation number 0 assigned to Mg and to Ba, whereas the oxidation number +2 is assigned to Mg^{2+} and to Ba^{2+}?

TABLE 15.1 Rules for Assigning Oxidation Numbers

1. The sum of the oxidation numbers in a molecule is zero (0), or for a polyatomic ion is equal to its charge.
2. The oxidation number of an element (monatomic or diatomic) is zero (0).
3. The oxidation number of a monatomic ion is equal to its charge.
4. In compounds, the oxidation number of Group 1A (1) metals is +1, and that of Group 2A (2) metals is +2.
5. In compounds, the oxidation number of fluorine is −1. Other nonmetals in Group 7A (17) are −1 except when combined with oxygen or fluorine.
6. In compounds, the oxidation number of oxygen is −2 except in OF_2, where O is +1; in H_2O_2 and other peroxides, O is −1.
7. In compounds with nonmetals, the oxidation number of hydrogen is +1; in compounds with metals, the oxidation number of hydrogen is −1.

We can now look at how these rules are used to assign oxidation numbers. For each formula, the oxidation numbers are written *below* the symbols of the elements (see **TABLE 15.2**).

TABLE 15.2 Examples of Using Rules to Assign Oxidation Numbers

Formula	Oxidation Numbers	Explanation
Br_2	Br_2 0	Each Br atom in diatomic bromine has an oxidation number of 0 (Rule 2).
Ba^{2+}	Ba^{2+} +2	The oxidation number of a monatomic ion is equal to its charge (Rule 3).
CO_2	CO_2 +4 −2	In compounds, O has an oxidation number of −2 (Rule 6). Because CO_2 is neutral, the oxidation number of C is calculated as +4 (Rule 1). $C + 2O = 0$ $C + 2(-2) = 0$ $C = +4$
Al_2O_3	Al_2O_3 +3 −2	In compounds, the oxidation number of O is −2 (Rule 6). For Al_2O_3 (neutral), the oxidation number of Al is calculated as +3 (Rule 1). $2Al + 3O = 0$ $2Al + 3(-2) = 0$ $2Al = +6$ $Al = +3$
$HClO_3$	$HClO_3$ +1 +5 −2	In compounds, the oxidation number of H is +1 (Rule 7), and O is −2 (Rule 6). For $HClO_3$ (neutral), the oxidation number of Cl is calculated as +5 (Rule 1). $H + Cl + 3O = 0$ $(+1) + Cl + 3(-2) = 0$ $Cl - 5 = 0$ $Cl = +5$
SO_4^{2-}	SO_4^{2-} +6 −2	The oxidation number of O is −2 (Rule 6). For SO_4^{2-} (−2 charge), the oxidation number of S is calculated as +6 (Rule 1). $S + 4O = -2$ $S + 4(-2) = -2$ $S = +6$
CH_2O	CH_2O 0 +1 −2	In compounds, the oxidation number of H is +1 (Rule 7), and O is −2 (Rule 6). For CH_2O (neutral), the oxidation number of C is calculated as 0 (Rule 1). $C + 2H + O = 0$ $C + 2(+1) + (-2) = 0$ $C = 0$

ENGAGE 15.3

Why is the S in the formula H_2SO_3 assigned an oxidation number of +4?

SAMPLE PROBLEM 15.2   Assigning Oxidation Numbers

TRY IT FIRST

Assign oxidation numbers to the elements in each of the following:

a. NCl_3

b. CO_3^{2-}

SOLUTION

a. NCl_3: The oxidation number of Cl is −1 (Rule 5). For NCl_3 (neutral), the sum of the oxidation numbers of N and 3Cl must be equal to zero (Rule 1). Thus, the oxidation number of N is calculated as +3.

$N + 3Cl = 0$
$N + 3(-1) = 0$
$N = +3$

The oxidation numbers are written as

NCl_3
$+3-1$

b. CO_3^{2-}: The oxidation number of O is -2 (Rule 6). For CO_3^{2-}, the sum of the oxidation numbers is equal to -2 (Rule 1). The oxidation number of C is calculated as $+4$.

$C + 3O = -2$
$C + 3(-2) = -2$
$C = +4$

The oxidation numbers are written as

CO_3^{2-}
$+4-2$

SELF TEST 15.2

Assign oxidation numbers to the elements in each of the following:

a. H_3PO_4 **b.** MnO_4^-

ANSWER

a. Because H has an oxidation number of $+1$, and O is -2, P has an oxidation number of $+5$ to maintain a neutral charge.

H_3PO_4
$+1+5-2$

b. Because O has an oxidation number of -2, Mn has an oxidation number of $+7$ to give an overall charge of -1.

MnO_4^-
$+7-2$

PRACTICE PROBLEMS

Try Practice Problems 15.3 to 15.14

Using Oxidation Numbers to Identify Oxidation–Reduction

CORE CHEMISTRY SKILL

Using Oxidation Numbers

Oxidation numbers can be used to identify the elements that are oxidized and the elements that are reduced in a reaction. In oxidation, the loss of electrons increases the oxidation number so that it is higher (more positive) in the product than in the reactant. In reduction, the gain of electrons decreases the oxidation number so that it is lower (more negative) in the product than in the reactant.

Reduction: oxidation number decreases

$-7\ -6\ -5\ -4\ -3\ -2\ -1\ \ 0\ \ +1\ +2\ +3\ +4\ +5\ +6\ +7$

Oxidation: oxidation number increases

ENGAGE 15.4

Why is the reaction of the hydrogen atoms in H_2 to hydrogen ions ($2H^+$) called an oxidation reaction?

An oxidation reaction occurs when the charge becomes more positive. A reduction reaction occurs when the charge becomes more negative.

SAMPLE PROBLEM 15.3 **Using Oxidation Numbers to Determine Oxidation and Reduction**

TRY IT FIRST

Identify the element that is oxidized and the element that is reduced in the following equation:

$$CO_2(g) + H_2(g) \longrightarrow CO(g) + H_2O(g)$$

SOLUTION

STEP 1 **Assign oxidation numbers to each element.** In H_2, the oxidation number of H is 0. In H_2O, the oxidation number of H is +1. In CO_2, CO, and H_2O, the oxidation number of O is −2. Using the oxidation number −2 for O, the oxidation number of C would be +4 in CO_2, and +2 in CO.

$$CO_2(g) + H_2(g) \longrightarrow CO(g) + H_2O(g)$$
+4 −2   0   +2 −2   +1 −2   Oxidation numbers

STEP 2 **Identify the increase in oxidation number as oxidation; the decrease as reduction.** H is oxidized because its oxidation number increases from 0 to +1. C is reduced because its oxidation number decreases from +4 to +2.

SELF TEST 15.3

a. In the following equation, which reactant is oxidized?

$$2Al(s) + 3Sn^{2+}(aq) \longrightarrow 2Al^{3+}(aq) + 3Sn(s)$$

b. In the following equation, which reactant is reduced?

$$Al(s) + 3Ag^+(aq) \longrightarrow Al^{3+}(aq) + 3Ag(s)$$

ANSWER

a. The oxidation number of Al increases from 0 to +3; Al is oxidized.
b. The oxidation number of Ag^+ decreases from +1 to 0; Ag^+ is reduced.

Oxidizing and Reducing Agents

We have seen that an oxidation reaction must always be accompanied by a reduction reaction. The substance that loses electrons is oxidized, and the substance that gains electrons is reduced. For example, Zn is oxidized to Zn^{2+} by losing 2 electrons and Cl_2 reduced to $2Cl^-$ by gaining 2 electrons.

$$Zn(s) + Cl_2(g) \longrightarrow ZnCl_2(s)$$

In an oxidation–reduction reaction, the substance that is oxidized is called the **reducing agent** because it provides electrons for reduction. The substance that is reduced is called the **oxidizing agent** because it accepts electrons from oxidation. Because Zn is oxidized in this reaction, it is the *reducing agent*. In the same reaction Cl_2 is reduced, which makes it the *oxidizing agent*.

$$Zn(s) \longrightarrow Zn^{2+}(aq) + 2\ e^-$$ Zn is oxidized; Zn is the *reducing agent*.
$$Cl_2(g) + 2\ e^- \longrightarrow 2Cl^-(aq)$$ Cl (in Cl_2) is reduced; Cl_2 is the *oxidizing agent*.

The terms we use to describe oxidation and reduction are listed in the margin.

CORE CHEMISTRY SKILL
Identifying Oxidizing and Reducing Agents

ENGAGE 15.5
In the following oxidation–reduction reaction, why is Zn the reducing agent, and Co^{2+} the oxidizing agent?

$$Zn(s) + CoBr_2(aq) \longrightarrow$$
$$ZnBr_2(aq) + Co(s)$$

Oxidation–Reduction Terminology

Oxidation	**Reduction**
Loss of electrons	Gain of electrons
Oxidation number increases	Oxidation number decreases
Reducing agent	Oxidizing agent

SAMPLE PROBLEM 15.4   Identifying Oxidizing and Reducing Agents

TRY IT FIRST

Identify the oxidizing agent and the reducing agent for the reaction of lead(II) oxide and carbon monoxide.

$$PbO(s) + CO(g) \longrightarrow Pb(s) + CO_2(g)$$

Lead oxides, once used in paint, are now banned due to the toxicity of lead.

SOLUTION

ANALYZE THE PROBLEM	Given	Need	Connect
	equation	oxidizing agent, reducing agent	oxidation numbers

STEP 1 **Assign oxidation numbers to each element.** Using the oxidation number -2 for O, the oxidation number of Pb in PbO is $+2$, the oxidation number of C in CO is $+2$, and the C in CO_2 would have an oxidation number of $+4$. The element Pb has an oxidation number of 0.

$$PbO(s) + CO(g) \longrightarrow Pb(s) + CO_2(g)$$
$$\quad +2\ -2 \qquad +2\ -2 \qquad\qquad 0 \qquad +4\ -2 \quad \text{Oxidation numbers}$$

STEP 2 **Identify the increase in oxidation number as oxidation; the decrease as reduction.** The C in CO is oxidized because the oxidation number of C increases from $+2$ to $+4$. The Pb in PbO is reduced because the oxidation number of Pb decreases from $+2$ to 0.

Oxidation

$$PbO(s) + CO(g) \longrightarrow Pb(s) + CO_2(g)$$
$$\quad +2\ -2 \qquad +2\ -2 \qquad\qquad 0 \qquad +4\ -2 \quad \text{Oxidation numbers}$$

Reduction

STEP 3 **Identify the oxidized substance as the reducing agent and the reduced substance as the oxidizing agent.** The compound CO is oxidized; it is the reducing agent. The compound PbO is reduced; it is the oxidizing agent.

SELF TEST 15.4

Identify the oxidizing agent and the reducing agent in each of the following reactions:

a. $2Al(s) + 3CuO(s) \longrightarrow Al_2O_3(s) + 3Cu(s)$
b. $2HCl(aq) + Mg(s) \longrightarrow MgCl_2(aq) + H_2(g)$

ANSWER

a. CuO is the oxidizing agent; Al is the reducing agent.
b. HCl is the oxidizing agent; Mg is the reducing agent.

PRACTICE PROBLEMS

Try Practice Problems 15.15 to 15.20

PRACTICE PROBLEMS

15.1 Oxidation and Reduction

15.1 Identify each of the following as an oxidation or a reduction reaction:
a. $Al^{3+}(aq) + 3\,e^- \longrightarrow Al(s)$
b. $Ca(s) \longrightarrow Ca^{2+}(aq) + 2\,e^-$
c. $Fe^{3+}(aq) + e^- \longrightarrow Fe^{2+}(aq)$

15.2 Identify each of the following as an oxidation or a reduction reaction:
a. $Ni^{2+}(aq) \longrightarrow Ni^{3+}(aq) + e^-$
b. $K^+(aq) + e^- \longrightarrow K(s)$
c. $Br_2(l) + 2\,e^- \longrightarrow 2Br^-(aq)$

15.3 In each of the following reactions, identify the reactant that is oxidized and the reactant that is reduced:
a. $4Al(s) + 3O_2(g) \longrightarrow 2Al_2O_3(s)$
b. $Zn(s) + 2H^+(aq) \longrightarrow Zn^{2+}(aq) + H_2(g)$
c. $F_2(g) + 2Br^-(aq) \longrightarrow 2F^-(aq) + Br_2(l)$

15.4 In each of the following reactions, identify the reactant that is oxidized and the reactant that is reduced:
a. $2Ag^+(aq) + Zn(s) \longrightarrow 2Ag(s) + Zn^{2+}(aq)$
b. $Ca(s) + S(s) \longrightarrow CaS(s)$
c. $Sn^{2+}(aq) + 2Cr^{3+}(aq) \longrightarrow Sn^{4+}(aq) + 2Cr^{2+}(aq)$

15.5 Assign oxidation numbers to each of the following:
a. Cu b. F_2 c. Fe^{2+} d. Cl^-

15.6 Assign oxidation numbers to each of the following:
a. Al b. Al^{3+} c. F^- d. N_2

15.7 Assign oxidation numbers to all the elements in each of the following:
a. KCl b. MnO_2 c. NO d. Mn_2O_3

15.8 Assign oxidation numbers to all the elements in each of the following:
a. H_2S b. NO_2 c. CCl_4 d. PCl_3

15.9 Assign oxidation numbers to all the elements in each of the following:
a. Li_3PO_4 b. SO_3^{2-} c. Cr_2S_3 d. NO_3^-

15.10 Assign oxidation numbers to all the elements in each of the following:
a. $C_2H_3O_2^-$ b. $AlCl_3$ c. NH_4^+ d. $HBrO_4$

15.11 Assign oxidation numbers to all the elements in each of the following:
a. HSO_4^- b. H_3PO_3 c. $Cr_2O_7^{2-}$ d. Na_2CO_3

15.12 Assign oxidation numbers to all the elements in each of the following:
a. N_2O b. LiOH c. SbO_2^- d. IO_4^-

15.13 What is the oxidation number of the specified element in each of the following?
a. N in HNO_3 b. C in C_3H_6
c. P in K_3PO_4 d. Cr in CrO_4^{2-}

15.14 What is the oxidation number of the specified element in each of the following?
a. C in $BaCO_3$ b. Fe in $FeBr_3$
c. Cl in ClF_4^- d. S in $S_2O_3^{2-}$

15.15 Indicate whether each of the following describes the oxidizing agent or the reducing agent in an oxidation–reduction reaction:
a. the substance that is oxidized
b. the substance that gains electrons

15.16 Indicate whether each of the following describes the oxidizing agent or the reducing agent in an oxidation–reduction reaction:
a. the substance that is reduced
b. the substance that loses electrons

15.17 For each of the following reactions, identify the substance that is oxidized, the substance that is reduced, the oxidizing agent, and the reducing agent:
a. $2Li(s) + Cl_2(g) \longrightarrow 2LiCl(s)$
b. $Cl_2(g) + 2NaBr(aq) \longrightarrow 2NaCl(aq) + Br_2(l)$
c. $2Pb(s) + O_2(g) \longrightarrow 2PbO(s)$
d. $Al(s) + 3Ag^+(aq) \longrightarrow Al^{3+}(aq) + 3Ag(s)$

15.18 For each of the following reactions, identify the substance that is oxidized, the substance that is reduced, the oxidizing agent, and the reducing agent:
a. $2Li(s) + F_2(g) \longrightarrow 2LiF(s)$
b. $Cl_2(g) + 2KI(aq) \longrightarrow 2KCl(aq) + I_2(s)$
c. $Zn(s) + Ni^{2+}(aq) \longrightarrow Zn^{2+}(aq) + Ni(s)$
d. $Fe(s) + CuSO_4(aq) \longrightarrow FeSO_4(aq) + Cu(s)$

15.19 For each of the following reactions, identify the substance that is oxidized, the substance that is reduced, the oxidizing agent, and the reducing agent:
a. $2NiS(s) + 3O_2(g) \longrightarrow 2NiO(s) + 2SO_2(g)$
b. $Sn^{2+}(aq) + 2Fe^{3+}(aq) \longrightarrow Sn^{4+}(aq) + 2Fe^{2+}(aq)$
c. $CH_4(g) + 2O_2(g) \longrightarrow CO_2(g) + 2H_2O(g)$
d. $2Cr_2O_3(s) + 3Si(s) \longrightarrow 4Cr(s) + 3SiO_2(s)$

15.20 For each of the following reactions, identify the substance that is oxidized, the substance that is reduced, the oxidizing agent, and the reducing agent:
a. $2HgO(s) \longrightarrow 2Hg(l) + O_2(g)$
b. $Zn(s) + 2HCl(aq) \longrightarrow ZnCl_2(aq) + H_2(g)$
c. $2Na(s) + 2H_2O(l) \longrightarrow 2Na^+(aq) + 2OH^-(aq) + H_2(g)$
d. $14H^+(aq) + 6Fe^{2+}(aq) + Cr_2O_7^{2-}(aq) \longrightarrow$
$6Fe^{3+}(aq) + 2Cr^{3+}(aq) + 7H_2O(l)$

15.2 Balancing Oxidation–Reduction Equations Using Half-Reactions

LEARNING GOAL Balance oxidation–reduction equations using the half-reaction method.

In the **half-reaction method** for balancing equations, an oxidation–reduction reaction is written as two *half-reactions*. The half-reaction method uses ionic charges and electrons to balance each half-reaction. Oxidation numbers are not used. The loss of electrons by one reactant is used to identify the oxidized substance, and the gain of electrons by another reactant is used to identify the reduced substance. Once the loss and gain of electrons are equalized for the half-reactions, they are combined to obtain the overall balanced equation. The half-reaction method is typically used to balance equations that are written as ionic equations. Let us consider the reaction between aluminum metal and a solution of Cu^{2+} as shown in Sample Problem 15.5.

CORE CHEMISTRY SKILL

Using Half-Reactions to Balance Redox Equations

SAMPLE PROBLEM 15.5   Using Half-Reactions to Balance Equations

TRY IT FIRST

Use half-reactions to balance the following equation:

$Al(s) + Cu^{2+}(aq) \longrightarrow Al^{3+}(aq) + Cu(s)$

SOLUTION

	Given	Need	Connect
ANALYZE THE PROBLEM	Equation: $Al(s) + Cu^{2+}(aq) \longrightarrow Al^{3+}(aq) + Cu(s)$	balanced equation	half-reactions, electron balance

STEP 1 Write two half-reactions for the equation.

$$Al(s) \longrightarrow Al^{3+}(aq)$$
$$Cu^{2+}(aq) \longrightarrow Cu(s)$$

STEP 2 For each half-reaction, balance the elements other than H and O. In these half-reactions, the Al and Cu are already balanced.

$$Al(s) \longrightarrow Al^{3+}(aq)$$
$$Cu^{2+}(aq) \longrightarrow Cu(s)$$

STEP 3 Balance each half-reaction for charge by adding electrons. For the aluminum half-reaction, we need to add three electrons on the product side to balance the charge. With a loss of electrons, this is an oxidation.

$$Al(s) \longrightarrow \underbrace{Al^{3+}(aq) + 3\,e^-}_{} \quad \text{Oxidation}$$
$$\underset{0 \text{ charge}}{} = \underset{0 \text{ charge}}{}$$

For the Cu^{2+} half-reaction, we need to add two electrons on the reactant side to balance the charge. With a gain of electrons, this is a reduction.

$$\underbrace{Cu^{2+}(aq) + 2\,e^-}_{0 \text{ charge}} \longrightarrow Cu(s) \quad \text{Reduction}$$
$$= 0 \text{ charge}$$

STEP 4 Multiply each half-reaction by factors that equalize the loss and gain of electrons. To obtain the same number of electrons in each half-reaction, we need to multiply the oxidation half-reaction by 2 and the reduction half-reaction by 3.

$$2 \times [Al(s) \longrightarrow Al^{3+}(aq) + 3\,e^-]$$
$$2Al(s) \longrightarrow 2Al^{3+}(aq) + 6\,e^- \quad 6\,e^- \text{ lost}$$
$$3 \times [Cu^{2+}(aq) + 2\,e^- \longrightarrow Cu(s)]$$
$$3Cu^{2+}(aq) + 6\,e^- \longrightarrow 3Cu(s) \qquad 6\,e^- \text{ gained}$$

STEP 5 Add half-reactions and cancel electrons, and any identical ions or molecules. Check the balance of atoms and charge.

$$2Al(s) \longrightarrow 2Al^{3+}(aq) + 6\,e^-$$
$$3Cu^{2+}(aq) + 6\,e^- \longrightarrow 3Cu(s)$$
$$\overline{2Al(s) + 3Cu^{2+}(aq) + \cancel{6\,e^-} \longrightarrow 2Al^{3+}(aq) + \cancel{6\,e^-} + 3Cu(s)}$$

Final balanced equation:
$$2Al(s) + 3Cu^{2+}(aq) \longrightarrow 2Al^{3+}(aq) + 3Cu(s)$$

Check the balance of atoms and charge.

Reactants		Products
2Al	=	2Al
3Cu	=	3Cu
Charge: 6+	=	6+

ENGAGE 15.6

How do we know that the final oxidation–reduction equation is balanced?

SELF TEST 15.5

Use the half-reaction method to balance each of the following equations:

a. $Zn(s) + Fe^{3+}(aq) \longrightarrow Zn^{2+}(aq) + Fe^{2+}(aq)$
b. $Ag^+(aq) + Sn(s) \longrightarrow Ag(s) + Sn^{4+}(aq)$

ANSWER

a. $Zn(s) + 2Fe^{3+}(aq) \longrightarrow Zn^{2+}(aq) + 2Fe^{2+}(aq)$
b. $4Ag^+(aq) + Sn(s) \longrightarrow 4Ag(s) + Sn^{4+}(aq)$

Balancing Oxidation–Reduction Equations in Acidic Solution

When we use the half-reaction method for balancing equations for reactions in acidic solution, we include balancing O by adding H_2O and balancing H by adding H^+ as shown in Sample Problem 15.6.

SAMPLE PROBLEM 15.6   Using Half-Reactions to Balance Equations in Acidic Solution

TRY IT FIRST

Use half-reactions to balance the following equation for a reaction that takes place in acidic solution:

$$I^-(aq) + Cr_2O_7{}^{2-}(aq) \longrightarrow I_2(s) + Cr^{3+}(aq)$$

A dichromate solution (yellow) and an iodide solution (colorless) form a brown solution of Cr^{3+} and iodine (I_2).

SOLUTION

STEP 1   Write two half-reactions for the equation.

$$I^-(aq) \longrightarrow I_2(s)$$
$$Cr_2O_7{}^{2-}(aq) \longrightarrow Cr^{3+}(aq)$$

STEP 2   For each half-reaction, balance the elements other than H and O. Balance O by adding H_2O, and H by adding H^+. The two I atoms in I_2 are balanced with a coefficient of 2 for I^-.

$$2I^-(aq) \longrightarrow I_2(s)$$

The two Cr atoms are balanced with a coefficient of 2 for Cr^{3+}.

$$Cr_2O_7{}^{2-}(aq) \longrightarrow 2Cr^{3+}(aq)$$

Now balance O by adding H_2O to the product side.

$$Cr_2O_7{}^{2-}(aq) \longrightarrow 2Cr^{3+}(aq) + \mathbf{7H_2O}(l) \quad \text{\small H_2O balances O}$$

Balance H by adding H^+ to the reactant side.

$$\mathbf{14H^+}(aq) + Cr_2O_7{}^{2-}(aq) \longrightarrow 2Cr^{3+}(aq) + 7H_2O(l) \quad \text{\small H^+ balances H}$$

STEP 3   Balance each half-reaction for charge by adding electrons. A charge of -2 is balanced with two electrons on the product side.

$$2I^-(aq) \longrightarrow I_2(s) + \mathbf{2\,e^-} \quad \text{\small Oxidation}$$
$$\underset{-2}{} \quad = \quad \underset{-2}{}$$

A charge of $+6$ is obtained on the reactant side by adding six electrons to the reactant side.

$$\mathbf{6\,e^-} + 14H^+(aq) + Cr_2O_7{}^{2-}(aq) \longrightarrow 2Cr^{3+}(aq) + 7H_2O(l) \quad \text{\small Reduction}$$
$$\underbrace{}_{+6} \quad = \quad +6$$

STEP 4   Multiply each half-reaction by factors that equalize the loss and gain of electrons. The half-reaction with I is multiplied by 3 to equal the gain of $6\,e^-$ by the Cr half-reaction.

$$\mathbf{3} \times [2I^-(aq) \longrightarrow I_2(s) + \mathbf{2\,e^-}]$$
$$6I^-(aq) \longrightarrow 3I_2(s) + \mathbf{6\,e^-} \qquad \text{\small $6\,e^-$ lost}$$
$$\mathbf{6\,e^-} + 14H^+(aq) + Cr_2O_7{}^{2-}(aq) \longrightarrow 2Cr^{3+}(aq) + 7H_2O(l) \quad \text{\small $6\,e^-$ gained}$$

STEP 5   Add half-reactions and cancel electrons, and any identical ions or molecules. Check the balance of atoms and charge.

$$6I^-(aq) \longrightarrow 3I_2(s) + \mathbf{6\,e^-}$$
$$\mathbf{6\,e^-} + 14H^+(aq) + Cr_2O_7{}^{2-}(aq) \longrightarrow 2Cr^{3+}(aq) + 7H_2O(l)$$

$$\cancel{6e^-} + 14H^+(aq) + Cr_2O_7{}^{2-}(aq) + 6I^-(aq) \longrightarrow 2Cr^{3+}(aq) + 3I_2(s) + 7H_2O(l) + \cancel{6e^-}$$

ENGAGE 15.7

Why are $7H_2O$ added to the product side of this half-reaction?

ENGAGE 15.8

Why are $6\,e^-$ added to the reactant side of this half-reaction?

Final balanced equation:

$$14H^+(aq) + Cr_2O_7^{2-}(aq) + 6I^-(aq) \longrightarrow 2Cr^{3+}(aq) + 3I_2(s) + 7H_2O(l)$$

Check the balance of atoms and charge.

Reactants		Products
6I	=	6I
2Cr	=	2Cr
14H	=	14H
7O	=	7O
Charge: 6+	=	6+

SELF TEST 15.6

Use half-reactions to balance each of the following equations in acidic solution:

a. $Fe^{2+}(aq) + IO_3^-(aq) \longrightarrow Fe^{3+}(aq) + I_2(s)$
b. $Mn^{2+}(aq) + BiO_3^-(aq) \longrightarrow MnO_4^-(aq) + Bi^{2+}(aq)$

ANSWER

a. $12H^+(aq) + 10Fe^{2+}(aq) + 2IO_3^-(aq) \longrightarrow 10Fe^{3+}(aq) + I_2(s) + 6H_2O(l)$
b. $6H^+(aq) + 3Mn^{2+}(aq) + 5BiO_3^-(aq) \longrightarrow 3MnO_4^-(aq) + 5Bi^{2+}(aq) + 3H_2O(l)$

PRACTICE PROBLEMS

Try Practice Problems 15.21 to 15.24

Balancing Oxidation–Reduction Equations in Basic Solution

An oxidation–reduction reaction can also take place in basic solution. In that case, we use the same half-reaction method, but once we have the balanced equation, we will neutralize the H^+ with OH^- to form water. The H^+ is neutralized by adding OH^- to both sides of the equation to form H_2O as shown in Sample Problem 15.7.

SAMPLE PROBLEM 15.7   Using Half-Reactions to Balance Equations in Basic Solution

TRY IT FIRST

Use half-reactions to balance the following equation that takes place in basic solution:

$$Fe^{2+}(aq) + MnO_4^-(aq) \longrightarrow Fe^{3+}(aq) + MnO_2(s)$$

SOLUTION

STEP 1 Write two half-reactions for the equation.

$$Fe^{2+}(aq) \longrightarrow Fe^{3+}(aq)$$
$$MnO_4^-(aq) \longrightarrow MnO_2(s)$$

STEP 2 For each half-reaction, balance the elements other than H and O. Balance O by adding H_2O, and H by adding H^+.

$$Fe^{2+}(aq) \longrightarrow Fe^{3+}(aq)$$
$$MnO_4^-(aq) \longrightarrow MnO_2(s) + 2H_2O(l) \quad \text{H}_2\text{O balances O}$$
$$4H^+(aq) + MnO_4^-(aq) \longrightarrow MnO_2(s) + 2H_2O(l) \quad \text{H}^+ \text{ balances H}$$

STEP 3 Balance each half-reaction for charge by adding electrons.

$$Fe^{2+}(aq) \longrightarrow \underbrace{Fe^{3+}(aq) + e^-} \quad \text{Oxidation}$$
$$\phantom{Fe^{2+}}+2 \quad = \quad +2$$

$$\underbrace{3\,e^- + 4H^+(aq) + MnO_4^-(aq)} \longrightarrow \underbrace{MnO_2(s) + 2H_2O(l)} \quad \text{Reduction}$$
$$0 \text{ charge} \quad = \quad 0 \text{ charge}$$

STEP 4 **Multiply each half-reaction by factors that equalize the loss and gain of electrons.** The half-reaction with Fe is multiplied by 3 to equal the gain of 3 e^- by Mn.

$$3 \times [Fe^{2+}(aq) \longrightarrow Fe^{3+}(aq) + e^-]$$

$$3Fe^{2+}(aq) \longrightarrow 3Fe^{3+}(aq) + 3\,e^- \qquad 3\,e^- \text{ lost}$$

$$3\,e^- + 4H^+(aq) + MnO_4^-(aq) \longrightarrow MnO_2(s) + 2H_2O(l) \quad 3\,e^- \text{ gained}$$

STEP 5 **Add half-reactions and cancel electrons, and any identical ions or molecules. In base, add OH$^-$ to neutralize H$^+$. Check the balance of atoms and charge.**

$$3Fe^{2+}(aq) \longrightarrow 3Fe^{3+}(aq) + 3\,e^-$$

$$3\,e^- + 4H^+(aq) + MnO_4^-(aq) \longrightarrow MnO_2(s) + 2H_2O(l)$$

$$\overline{3e^- + 4H^+(aq) + 3Fe^{2+}(aq) + MnO_4^-(aq) \longrightarrow 3Fe^{3+}(aq) + MnO_2(s) + 2H_2O(l) + 3e^-}$$

Final balanced equation:

$$4H^+(aq) + 3Fe^{2+}(aq) + MnO_4^-(aq) \longrightarrow 3Fe^{3+}(aq) + MnO_2(s) + 2H_2O(l)$$

ENGAGE 15.9

Why is 4OH$^-$ added to the product side of this equation?

To convert the equation to an oxidation–reduction reaction in basic solution, we neutralize H$^+$ with OH$^-$ to form H$_2$O. For this equation, we add 4OH$^-$(aq) to both sides of the equation.

$$4OH^-(aq) + 4H^+(aq) + 3Fe^{2+}(aq) + MnO_4^-(aq) \longrightarrow 3Fe^{3+}(aq) + MnO_2(s) + 2H_2O(l) + 4OH^-(aq)$$

Combining 4H$^+$ and 4OH$^-$ gives 4H$_2$O on the reactant side.

$$4H_2O(l) + 3Fe^{2+}(aq) + MnO_4^-(aq) \longrightarrow 3Fe^{3+}(aq) + MnO_2(s) + 2H_2O(l) + 4OH^-(aq)$$

Canceling 2H$_2$O on both the reactant and the product side gives the balanced equation in basic solution.

Final balanced equation in basic solution:

$$2H_2O(l) + 3Fe^{2+}(aq) + MnO_4^-(aq) \longrightarrow 3Fe^{3+}(aq) + MnO_2(s) + 4OH^-(aq)$$

Check the balance of atoms and charge.

Reactants		Products
3Fe	=	3Fe
1Mn	=	1Mn
4H	=	4H
6O	=	6O
Charge: 5+	=	5+

SELF TEST 15.7

For the following equation in basic solution:

$$N_2O(g) + ClO^-(aq) \longrightarrow NO_2^-(aq) + Cl^-(aq)$$

a. write the balanced half-reactions
b. write the balanced equation

ANSWER

a. $3H_2O(l) + N_2O(g) \longrightarrow 2NO_2^-(aq) + 6H^+(aq) + 4\,e^-$

$2\,e^- + 2H^+(aq) + ClO^-(aq) \longrightarrow Cl^-(aq) + H_2O(l)$

b. $N_2O(g) + 2ClO^-(aq) + 2OH^-(aq) \longrightarrow 2Cl^-(aq) + 2NO_2^-(aq) + H_2O(l)$

PRACTICE PROBLEMS

Try Practice Problems 15.25 and 15.26

15.2 Balancing Oxidation–Reduction Equations Using Half-Reactions

15.21 Balance each of the following half-reactions in acidic solution:
a. $Sn^{2+}(aq) \longrightarrow Sn^{4+}(aq)$
b. $Mn^{2+}(aq) \longrightarrow MnO_4^-(aq)$
c. $NO_2^-(aq) \longrightarrow NO_3^-(aq)$
d. $ClO_3^-(aq) \longrightarrow ClO_2(aq)$

15.22 Balance each of the following half-reactions in acidic solution:
a. $Cu(s) \longrightarrow Cu^{2+}(aq)$
b. $SO_4^{2-}(aq) \longrightarrow SO_3^{2-}(aq)$
c. $BrO_3^-(aq) \longrightarrow Br^-(aq)$
d. $IO_3^-(aq) \longrightarrow I_2(s)$

15.23 Use the half-reaction method to balance each of the following in acidic solution:
a. $Ag(s) + NO_3^-(aq) \longrightarrow Ag^+(aq) + NO_2(g)$
b. $NO_3^-(aq) + S(s) \longrightarrow NO(g) + SO_2(g)$
c. $S_2O_3^{2-}(aq) + Cu^{2+}(aq) \longrightarrow S_4O_6^{2-}(aq) + Cu(s)$

15.24 Use the half-reaction method to balance each of the following in acidic solution:
a. $Mn(s) + NO_3^-(aq) \longrightarrow Mn^{2+}(aq) + NO_2(g)$
b. $C_2O_4^{2-}(aq) + MnO_4^-(aq) \longrightarrow CO_2(g) + Mn^{2+}(aq)$
c. $ClO_3^-(aq) + SO_3^{2-}(aq) \longrightarrow Cl^-(aq) + SO_4^{2-}(aq)$

15.25 Use the half-reaction method to balance each of the following in basic solution:
a. $Fe(s) + CrO_4^{2-}(aq) \longrightarrow Fe_2O_3(s) + Cr_2O_3(s)$
b. $CN^-(aq) + MnO_4^-(aq) \longrightarrow CNO^-(aq) + MnO_2(s)$

15.26 Use the half-reaction method to balance each of the following in basic solution:
a. $Al(s) + ClO^-(aq) \longrightarrow AlO_2^-(aq) + Cl^-(aq)$
b. $Sn^{2+}(aq) + IO_4^-(aq) \longrightarrow Sn^{4+}(aq) + I^-(aq)$

15.3 Electrical Energy from Oxidation–Reduction Reactions

LEARNING GOAL Use the activity series to determine if an oxidation–reduction reaction is spontaneous. Write the half-reactions that occur in a voltaic cell and the cell notation.

Because Zn is above Cu on the activity series, a spontaneous oxidation–reduction reaction occurs.

When we place a zinc metal strip in a solution of Cu^{2+}, reddish-brown Cu metal accumulates on the Zn strip according to the following spontaneous reaction:

$$Zn(s) + Cu^{2+}(aq) \longrightarrow Zn^{2+}(aq) + Cu(s) \quad \text{Spontaneous}$$

However, if we place a Cu metal strip in a Zn^{2+} solution, nothing will happen. The reaction does not occur spontaneously in the reverse direction because Cu does not lose electrons as easily as Zn.

We can determine the direction of a spontaneous reaction from the *activity series*, which ranks the metals and H_2 in terms of how easily they lose electrons.

In the **activity series**, the metals that lose electrons most easily are placed at the top, and the metals that do not lose electrons easily are at the bottom. Thus the metals that are more easily oxidized are above the metals whose ions are more easily reduced (see **TABLE 15.3**). Active metals include K, Na, Ca, Mg, Al, Zn, Fe, and Sn. In single replacement reactions, the metal ion replaces the H in the acid. Metals listed below $H_2(g)$ will not react with H^+ from acids.

According to the activity series, a metal will oxidize spontaneously when it is combined with the reverse of the half-reaction for any metal below it on the list. We use the activity series to predict the direction of the spontaneous reaction. Suppose we have two beakers. In one, we place a Mg strip in a solution containing Ni^{2+} ions. In the other, we place a Ni strip in a solution containing Mg^{2+} ions. Looking at the activity series we see that the half-reaction for the oxidation of Mg is listed above that for Ni, which means that Mg is the more active metal and loses electrons more easily than Ni. Using the activity series table, we write these two half-reactions as follows:

$$Mg(s) \longrightarrow Mg^{2+}(aq) + 2\ e^-$$
$$Ni(s) \longrightarrow Ni^{2+}(aq) + 2\ e^-$$

The reaction that will be spontaneous is the oxidation of Mg combined with the reverse (reduction) of Ni^{2+}.

$$Mg(s) \longrightarrow Mg^{2+}(aq) + 2\ e^-$$
$$Ni^{2+}(aq) + 2\ e^- \longrightarrow Ni(s)$$

TABLE 15.3 Activity Series for Some Metals and Hydrogen

		Metal		Ion
Most Reactive		Li(s)	$\longrightarrow$	$Li^+(aq) + e^-$
		K(s)	$\longrightarrow$	$K^+(aq) + e^-$
		Ca(s)	$\longrightarrow$	$Ca^{2+}(aq) + 2\,e^-$
		Na(s)	$\longrightarrow$	$Na^+(aq) + e^-$
		Mg(s)	$\longrightarrow$	$Mg^{2+}(aq) + 2\,e^-$
		Al(s)	$\longrightarrow$	$Al^{3+}(aq) + 3\,e^-$
		Zn(s)	$\longrightarrow$	$Zn^{2+}(aq) + 2\,e^-$
		Cr(s)	$\longrightarrow$	$Cr^{3+}(aq) + 3\,e^-$
		Fe(s)	$\longrightarrow$	$Fe^{2+}(aq) + 2\,e^-$
		Ni(s)	$\longrightarrow$	$Ni^{2+}(aq) + 2\,e^-$
		Sn(s)	$\longrightarrow$	$Sn^{2+}(aq) + 2\,e^-$
		Pb(s)	$\longrightarrow$	$Pb^{2+}(aq) + 2\,e^-$
		$H_2(g)$	$\longrightarrow$	$2H^+(aq) + 2\,e^-$
		Cu(s)	$\longrightarrow$	$Cu^{2+}(aq) + 2\,e^-$
		Ag(s)	$\longrightarrow$	$Ag^+(aq) + e^-$
Least Reactive		Au(s)	$\longrightarrow$	$Au^{3+}(aq) + 3\,e^-$

Ease of Oxidation Decreases

ENGAGE 15.10

Why would the following oxidation–reduction reaction be spontaneous?
Fe(s) + Pb²⁺(aq) $\longrightarrow$ Fe²⁺(aq) + Pb(s)

Therefore, we combine the half-reactions, which gives the following overall reaction that occurs spontaneously:

$Mg(s) + Ni^{2+}(aq) + 2e^- \longrightarrow Mg^{2+}(aq) + Ni(s) + 2e^-$
$Mg(s) + Ni^{2+}(aq) \longrightarrow Mg^{2+}(aq) + Ni(s)$ Spontaneous

However, a reaction between a Mg strip and a solution containing K^+ ions will not occur spontaneously. We determine this by looking at the two half-reactions needed.

$Mg(s) \longrightarrow Mg^{2+}(aq) + 2\,e^-$
$K^+(aq) + e^- \longrightarrow K(s)$

Because the half-reaction for the oxidation of K is above that for Mg, there will not be a spontaneous reaction between Mg and K^+.

$Mg(s) + 2K^+(aq) \not\longrightarrow Mg^{2+}(aq) + 2K(s)$ Not spontaneous

SAMPLE PROBLEM 15.8 Predicting Spontaneous Reactions

TRY IT FIRST

Determine if the reaction for each of the following metals with an HCl (H^+) solution is spontaneous:

a. $Zn(s) + 2H^+(aq) \longrightarrow Zn^{2+}(aq) + H_2(g)$
b. $Cu(s) + 2H^+(aq) \longrightarrow Cu^{2+}(aq) + H_2(g)$

SOLUTION

a. Using the activity series (see Table 15.3), we see that Zn oxidizes more easily than H_2. Thus the oxidation half-reaction for Zn, combined with the reverse half-reaction for H_2, is spontaneous.

$Zn(s) \longrightarrow Zn^{2+}(aq) + 2\,e^-$
$2H^+(aq) + 2\,e^- \longrightarrow H_2(g)$
$Zn(s) + 2H^+(aq) \longrightarrow Zn^{2+}(aq) + H_2(g)$ Spontaneous

b. Using the activity series (see Table 15.3), we see that H_2 oxidizes more easily than Cu. In this reaction, we would be combining an oxidation half-reaction with one that is above it in the activity series.

$$2H^+(aq) + 2\,e^- \longrightarrow H_2(g)$$
$$Cu(s) \longrightarrow Cu^{2+}(aq) + 2\,e^-$$
$$Cu(s) + 2H^+(aq) \not\longrightarrow Cu^{2+}(aq) + H_2(g) \quad \text{Not spontaneous}$$

SELF TEST 15.8

Determine if each of the following reactions is spontaneous:

a. $2Al(s) + 3Cu^{2+}(aq) \longrightarrow 2Al^{3+}(aq) + 3Cu(s)$
b. $Fe(s) + Mg^{2+}(aq) \longrightarrow Fe^{2+}(aq) + Mg(s)$

ANSWER

a. Yes, this reaction is spontaneous.   **b.** No, this reaction is not spontaneous.

PRACTICE PROBLEMS

Try Practice Problems 15.27 and 15.28

Voltaic Cells

We can generate electrical energy from a spontaneous oxidation–reduction reaction by using a *voltaic cell*. A piece of zinc metal placed in a Cu^{2+} solution becomes coated with a rusty-brown coating of Cu, while the blue color (Cu^{2+}) of the solution fades. The oxidation of the zinc metal provides electrons for the reduction of the Cu^{2+} ions. We can write the two half-reactions as

$$Zn(s) \longrightarrow Zn^{2+}(aq) + 2\,e^- \quad \text{Oxidation}$$
$$Cu^{2+}(aq) + 2\,e^- \longrightarrow Cu(s) \quad \text{Reduction}$$

The overall reaction is

$$Zn(s) + Cu^{2+}(aq) \longrightarrow Zn^{2+}(aq) + Cu(s)$$

As long as the Zn metal and Cu^{2+} ions are in the same container, the electrons are transferred directly from Zn to Cu^{2+}. In a **voltaic cell**, the components of the two half-reactions are separated into *half-cells*. The electrons can only flow from one half-cell to the other through an external circuit, producing an electrical current. In each half-cell, there is a strip of metal, called an *electrode*, in contact with the ionic solution. The electrode where oxidation takes place is called the **anode**; the electrode where reduction takes place is called the **cathode**. In this example, the anode is a zinc metal strip placed in a Zn^{2+}($ZnSO_4$) solution. The cathode is a copper metal strip placed in a Cu^{2+}($CuSO_4$) solution. In this voltaic cell, the Zn anode and Cu cathode are connected by a wire that allows electrons to move from the oxidation half-cell to the reduction half-cell.

Anode is where

oxidation takes place
electrons are produced $\Big\}$ $Zn(s) \longrightarrow Zn^{2+}(aq) + 2\,e^-$

Cathode is where

reduction takes place
electrons are used up $\Big\}$ $Cu^{2+}(aq) + 2\,e^- \longrightarrow Cu(s)$

ENGAGE 15.11

What is the purpose of a salt bridge in a voltaic cell?

The circuit is completed by a *salt bridge* containing positive and negative ions that are placed in the half-cell solutions. The purpose of the salt bridge is to provide ions, such as Na^+ and SO_4^{2-} ions, to maintain an electrical balance in each half-cell solution. As oxidation occurs, there is an increase in Zn^{2+} ions, which is balanced by SO_4^{2-} anions from the salt bridge. At the cathode, there is a loss of Cu^{2+}, which is balanced by SO_4^{2-} moving into the salt bridge. The complete circuit involves the flow of electrons from the anode to the cathode and the flow of anions from the cathode solution to the anode solution (see **FIGURE 15.1**).

FIGURE 15.1 In this voltaic cell, the Zn anode is in a Zn^{2+} solution, and the Cu cathode is in a Cu^{2+} solution. Electrons produced by the oxidation of Zn flow from the anode through the wire to the cathode where they reduce Cu^{2+} to Cu. The circuit is completed by the flow of SO_4^{2-} through the salt bridge.

ENGAGE 15.12

Which electrode will be heavier when the reaction in Figure 15.1 ends?

Oxidation Half-Reaction   **Reduction Half-Reaction**

$$Zn(s) \longrightarrow Zn^{2+}(aq) + 2\,e^- \qquad Cu^{2+}(aq) + 2\,e^- \longrightarrow Cu(s)$$

Overall Oxidation–Reduction Reaction

$$Zn(s) + Cu^{2+}(aq) \longrightarrow Zn^{2+}(aq) + Cu(s)$$

We can diagram the oxidation and reduction reactions that take place in the cell using a *shorthand notation* as follows:

$$Zn(s)\,|\,Zn^{2+}(aq)\,\|\,Cu^{2+}(aq)\,|\,Cu(s)$$

The components of the oxidation half-cell (anode) are written on the left side, and the components of the reduction half-cell (cathode) are written on the right. A single vertical line separates the solid Zn anode from the Zn^{2+} solution, and another vertical line separates the Cu^{2+} solution from the Cu cathode. A double vertical line separates the two half-cells.

Electrons →

Oxidation half-cell   Reduction half-cell

Anode   Cathode

$$Zn(s) \mid Zn^{2+}(aq) \parallel Cu^{2+}(aq) \mid Cu(s)$$

Salt bridge

In some voltaic cells, there is no component in the half-reactions that can be used as an electrode. When this is the case, inert electrodes made of graphite or platinum are used for the transfer of electrons. If there are two ionic components in a cell, their symbols are separated by a comma. For example, suppose a voltaic cell consists of a platinum anode placed in a Sn^{2+} solution, and a silver cathode placed in a Ag^+ solution. The notation for the cell would be written as

$$Pt(s)\,|\,Sn^{2+}(aq),\,Sn^{4+}(aq)\,\|\,Ag^+(aq)\,|\,Ag(s)$$

The oxidation reaction at the anode is

$$Sn^{2+}(aq) \longrightarrow Sn^{4+}(aq) + 2\,e^-$$

The reduction reaction at the cathode is

$$Ag^+(aq) + e^- \longrightarrow Ag(s)$$

To balance the overall cell reaction, we multiply the cathode reduction by 2 and combine the two half-reactions.

$$2Ag^+(aq) + 2\,e^- \longrightarrow 2Ag(s) \qquad \text{Reduction}$$
$$Sn^{2+}(aq) \longrightarrow Sn^{4+}(aq) + 2\,e^- \qquad \text{Oxidation}$$
$$\overline{Sn^{2+}(aq) + 2Ag^+(aq) \longrightarrow Sn^{4+}(aq) + 2Ag(s)} \qquad \text{Oxidation–reduction reaction}$$

SAMPLE PROBLEM 15.9   Diagramming a Voltaic Cell

TRY IT FIRST

A voltaic cell consists of an iron (Fe) anode in a Fe^{2+} solution and a tin (Sn) cathode placed in a Sn^{2+} solution. Write the cell notation, the oxidation and reduction half-reactions, and the overall cell reaction.

SOLUTION

The notation for the cell would be written as

$$Fe(s)\,|\,Fe^{2+}(aq)\,\|\,Sn^{2+}(aq)\,|\,Sn(s)$$

The oxidation reaction at the anode is

$$Fe(s) \longrightarrow Fe^{2+}(aq) + 2\,e^-$$

The reduction reaction at the cathode is

$$Sn^{2+}(aq) + 2\,e^- \longrightarrow Sn(s)$$

To write the overall cell reaction, we combine the two half-reactions.

$$Fe(s) \longrightarrow Fe^{2+}(aq) + 2\,e^- \qquad \text{Oxidation}$$
$$Sn^{2+}(aq) + 2\,e^- \longrightarrow Sn(s) \qquad \text{Reduction}$$
$$\overline{Fe(s) + Sn^{2+}(aq) \longrightarrow Fe^{2+}(aq) + Sn(s)} \qquad \text{Oxidation–reduction reaction}$$

SELF TEST 15.9

Using the following notation for a voltaic cell, write the following:

$$Co(s)\,|\,Co^{2+}(aq)\,\|\,Cu^{2+}(aq)\,|\,Cu(s)$$

a. the half-reactions at the anode and cathode
b. the overall cell reaction

ANSWER

a. Anode reaction: $\qquad\qquad\qquad\qquad Co(s) \longrightarrow Co^{2+}(aq) + 2\,e^-$
Cathode reaction: $\qquad\qquad Cu^{2+}(aq) + 2\,e^- \longrightarrow Cu(s)$
b. Overall cell reaction: $\quad Co(s) + Cu^{2+}(aq) \longrightarrow Co^{2+}(aq) + Cu(s)$

PRACTICE PROBLEMS

Try Practice Problems 15.29 to 15.32

Batteries

Batteries are needed to power your cell phone, watch, and calculator. Batteries also are needed to make cars start, and flashlights produce light. Within each of these batteries are voltaic cells that produce electrical energy. Let's look at some examples of commonly used batteries.

Lead Storage Battery

A lead storage battery is used to operate the electrical system in a car. We need a car battery to start the engine, turn on the lights, or operate the radio. If the battery runs down, the car won't start and the lights won't turn on. A car battery or a lead storage battery is a type of voltaic cell. In a typical 12-V battery, there are six voltaic cells

Batteries come in many shapes and sizes.

linked together. Each of the cells consists of a lead (Pb) plate that acts as the anode and a lead(IV) oxide (PbO_2) plate that acts as the cathode. Both half-cells contain a sulfuric acid (H_2SO_4) solution. When the car battery is producing electrical energy (discharging), the following half-reactions take place:

Anode (oxidation): $Pb(s) + SO_4^{2-}(aq) \longrightarrow PbSO_4(s) + 2\ e^-$

Cathode (reduction): $2\ e^- + 4H^+(aq) + PbO_2(s) + SO_4^{2-}(aq) \longrightarrow PbSO_4(s) + 2H_2O(l)$

Overall cell reaction: $4H^+(aq) + Pb(s) + PbO_2(s) + 2SO_4^{2-}(aq) \longrightarrow 2PbSO_4(s) + 2H_2O(l)$

In both half-reactions, Pb^{2+} is produced, which combines with SO_4^{2-} to form an insoluble ionic compound $PbSO_4$. As a car battery is used, there is a buildup of $PbSO_4$ on the electrodes. At the same time, there is a decrease in the concentrations of the sulfuric acid components, H^+ and SO_4^{2-}. As a car runs, the battery is continuously recharged by an alternator, which is powered by the engine. The recharging reactions restore the Pb and PbO_2 electrodes as well as H_2SO_4. Without recharging, the car battery cannot continue to produce electrical energy.

Dry-Cell Batteries

Dry-cell batteries are used in calculators, watches, flashlights, and battery-operated toys. The term *dry cell* describes a battery that uses a paste rather than an aqueous solution. Dry cells can be acidic or alkaline. In an acidic dry cell, the anode is a zinc metal case that contains a paste of MnO_2, NH_4Cl, $ZnCl_2$, H_2O, and starch. Within this MnO_2 electrolyte mixture is a graphite cathode.

e^- Flow

Anode: Pb metal grids

Cathode: PbO_2 grids

$H_2SO_4(aq)$ solution

A 12-volt car battery is also known as a lead storage battery.

Acidic dry-cell battery

Graphite core

MnO_2 paste around graphite

Zinc metal can

NH_4Cl and $ZnCl_2$ paste

In an acidic dry-cell battery, the cathode is graphite, and the anode is a zinc case.

Anode (oxidation): $Zn(s) \longrightarrow Zn^{2+}(aq) + 2\ e^-$

Cathode (reduction): $2\ e^- + 2MnO_2(s) + 2NH_4^+(aq) \longrightarrow Mn_2O_3(s) + 2NH_3(aq) + H_2O(l)$

Overall cell reaction: $Zn(s) + 2MnO_2(s) + 2NH_4^+(aq) \longrightarrow Zn^{2+}(aq) + Mn_2O_3(s) + 2NH_3(aq) + H_2O(l)$

An alkaline battery has similar components except that NaOH or KOH replaces the NH_4Cl electrolyte. Under basic conditions, the product of oxidation is zinc oxide (ZnO). Alkaline batteries tend to be more expensive, but they last longer and produce more power than acidic dry-cell batteries.

Anode (oxidation): $Zn(s) + 2OH^-(aq) \longrightarrow ZnO(s) + H_2O(l) + 2\ e^-$

Cathode (reduction): $2\ e^- + 2MnO_2(s) + H_2O(l) \longrightarrow Mn_2O_3(s) + 2OH^-(aq)$

Overall cell reaction: $Zn(s) + 2MnO_2(s) \longrightarrow ZnO(s) + Mn_2O_3(s)$

PRACTICE PROBLEMS

Try Practice Problems 15.33 to 15.36

Nickel–Cadmium (NiCad) Batteries

Nickel–cadmium (NiCad) batteries, used in cordless power tools, can be recharged. They consist of a cadmium anode and a cathode of solid nickel oxide $NiO(OH)(s)$.

Anode (oxidation): $\qquad Cd(s) + 2OH^-(aq) \longrightarrow Cd(OH)_2(s) + 2\,e^-$

Cathode (reduction): $\quad 2\,e^- + 2NiO(OH)(s) + 2H_2O(l) \longrightarrow$
$$2Ni(OH)_2(s) + 2OH^-(aq)$$

Overall cell reaction: $\quad Cd(s) + 2NiO(OH)(s) + 2H_2O(l) \longrightarrow$
$$Cd(OH)_2(s) + 2Ni(OH)_2(s)$$

NiCad batteries are expensive, but they can be recharged many times. A charger provides an electrical current that converts the solid $Cd(OH)_2$ and $Ni(OH)_2$ products in the NiCad battery back to the reactants. Because NiCad batteries contain cadmium, which is toxic, they need to be disposed of properly.

A battery in a cell phone can be recharged many times.

Lithium-Ion Batteries

Lithium-ion batteries are now used in many of our electronic devices including cell phones, laptops, digital cameras, and tablets. Lithium-ion batteries charge quickly, last longer, are rechargeable, and are light due to the low density of lithium. When the battery is in use and discharging, lithium ions, which are part of a lithium cobalt oxide compound, $LiCoO_2(s)$, move from the anode to the cathode. The electrons move through a circuit that provides the energy for the device. When the lithium ion battery is charging, the lithium ions move in the opposite direction from cathode to anode to build up the charge. A simplified reaction is:

$$Li(s) + CoO_2(s) \underset{\text{discharging}}{\overset{\text{charging}}{\rightleftarrows}} LiCoO_2(s)$$

Chemistry Link to the Environment

Corrosion: Oxidation of Metals

Metals used in building materials, such as iron, eventually oxidize, which causes deterioration of the metal. This oxidation process, known as *corrosion*, produces rust on cars, bridges, ships, and underground pipes.

$$4Fe(s) + 3O_2(g) \longrightarrow 2Fe_2O_3(s)$$
$$\text{Rust}$$

The formation of rust requires both oxygen and water. The process of rusting requires an anode and cathode in different places on the surface of a piece of iron. In one area of the iron surface, called the *anode region*, the oxidation half-reaction takes place (see **FIGURE 15.2**).

Anode (oxidation): $Fe(s) \longrightarrow Fe^{2+}(aq) + 2\,e^-$
$$\text{or}$$
$$2Fe(s) \longrightarrow 2Fe^{2+}(aq) + 4\,e^-$$

The electrons move through the iron metal from the anode to an area called the *cathode region* where oxygen dissolved in water is reduced to water.

Cathode (reduction): $4\,e^- + 4H^+(aq) + O_2(g) \longrightarrow 2H_2O(l)$

By combining the half-reactions that occur in the anode and cathode regions, we can write the overall oxidation–reduction equation.

$$2Fe(s) + 4H^+(aq) + O_2(g) \longrightarrow 2Fe^{2+}(aq) + 2H_2O(l)$$

The formation of rust occurs as Fe^{2+} ions move out of the anode region and come in contact with dissolved oxygen (O_2). The Fe^{2+} oxidizes to Fe^{3+}, which reacts with oxygen to form rust.

$$4H_2O(l) + 4Fe^{2+}(aq) + O_2(g) \longrightarrow 2Fe_2O_3(s) + 8H^+(aq)$$
$$\text{Rust}$$

FIGURE 15.2 Rust forms when electrons from the oxidation of Fe flow from the anode region to the cathode region where oxygen is reduced. As Fe^{2+} ions come in contact with O_2 and H_2O, rust forms.

We can write the formation of rust starting with solid Fe reacting with O_2 as follows. There is no H^+ in the overall equation because H^+ is used and produced in equal quantities.

Corrosion of iron

$$4Fe(s) + 3O_2(g) \longrightarrow 2Fe_2O_3(s)$$
$$\text{Rust}$$

Other metals such as aluminum, copper, and silver also undergo corrosion, but at a slower rate than iron. The oxidation of Al on the surface of an aluminum object produces Al^{3+}, which reacts with oxygen in the air to form a protective coating of Al_2O_3. This Al_2O_3 coating prevents further oxidation of the aluminum underneath it.

$$Al(s) \longrightarrow Al^{3+}(aq) + 3\,e^-$$

When copper is used on a roof, dome, or a steeple, it oxidizes to Cu^{2+}, which is converted to a green patina of $Cu_2(OH)_2CO_3$.

$$Cu(s) \longrightarrow Cu^{2+}(aq) + 2\,e^-$$

When we use silver dishes and utensils, the Ag^+ ion from oxidation reacts with sulfides in food to form Ag_2S, which we call "tarnish."

$$Ag(s) \longrightarrow Ag^+(aq) + e^-$$

Prevention of Corrosion

Billions of dollars are spent each year to prevent corrosion and repair building materials made of iron. One way to prevent corrosion is to paint the bridges, cars, and ships with paints containing materials that seal the iron surface from H_2O and O_2. But it is necessary to repaint often; a scratch in the paint exposes the iron, which then begins to rust.

A more effective way to prevent corrosion is to place the iron in contact with a metal that substitutes for the anode region of iron. Metals such as Zn, Mg, or Al lose electrons more easily than iron. When one of these metals is in contact with iron, the metal acts as the anode instead of iron. For example, in a process called *galvanization*, an iron object is coated with zinc. The zinc becomes the anode because zinc is more easily oxidized than Fe. As long as Fe does not act as an anode, rust cannot form.

In a method called *cathodic protection*, structures such as iron pipes and underground storage containers are placed in contact with a piece of metal such as Mg, Al, or Zn, which is called the *sacrificial anode*. Again, because these metals lose electrons more easily than Fe, they become the anode, thereby preventing the rusting of the iron. A magnesium plate that is welded or bolted to a ship's hull loses electrons more easily than iron or steel and protects the hull from rusting. Occasionally, a new magnesium plate is added to replace the magnesium as it is used up. Magnesium stakes placed in the ground are connected to underground pipelines and storage containers to prevent corrosion damage.

Oxidation (corrosion) is prevented by painting a metal surface.

Rusting of an iron pipe is prevented by attaching a piece of magnesium, which oxidizes more easily than iron.

PRACTICE PROBLEMS

15.3 Electrical Energy from Oxidation–Reduction Reactions

15.27 Use the activity series in Table 15.3 to predict whether each of the following reactions will occur spontaneously:
a. $2Au(s) + 6H^+(aq) \longrightarrow 2Au^{3+}(aq) + 3H_2(g)$
b. $Ni^{2+}(aq) + Fe(s) \longrightarrow Ni(s) + Fe^{2+}(aq)$
c. $2Ag(s) + Cu^{2+}(aq) \longrightarrow 2Ag^+(aq) + Cu(s)$

15.28 Use the activity series in Table 15.3 to predict whether each of the following reactions will occur spontaneously:
a. $2Ag(s) + 2H^+(aq) \longrightarrow 2Ag^+(aq) + H_2(g)$
b. $Mg(s) + Cu^{2+}(aq) \longrightarrow Mg^{2+}(aq) + Cu(s)$
c. $2Al(s) + 3Cu^{2+}(aq) \longrightarrow 2Al^{3+}(aq) + 3Cu(s)$

15.29 Write the half-reactions and the overall cell reaction for each of the following voltaic cells:
a. $Pb(s)\,|\,Pb^{2+}(aq)\,\|\,Cu^{2+}(aq)\,|\,Cu(s)$
b. $Cr(s)\,|\,Cr^{2+}(aq)\,\|\,Ag^+(aq)\,|\,Ag(s)$

15.30 Write the half-reactions and the overall cell reaction for each of the following voltaic cells:
a. $Al(s)\,|\,Al^{3+}(aq)\,\|\,Cd^{2+}(aq)\,|\,Cd(s)$
b. $Sn(s)\,|\,Sn^{2+}(aq)\,\|\,Fe^{3+}(aq),\,Fe^{2+}(aq)\,|\,C$ (electrode)

15.31 Describe the voltaic cell and half-cell components, and write the shorthand notation for the following oxidation–reduction reactions:
a. $Cd(s) + Sn^{2+}(aq) \longrightarrow Cd^{2+}(aq) + Sn(s)$
b. $Zn(s) + Cl_2(g) \longrightarrow Zn^{2+}(aq) + 2Cl^-(aq)\,|\,C$ (electrode)

15.32 Describe the voltaic cell and half-cell components, and write the shorthand notation for the following oxidation–reduction reactions:
a. $Mn(s) + Sn^{2+}(aq) \longrightarrow Mn^{2+}(aq) + Sn(s)$
b. $Ni(s) + 2Ag^+(aq) \longrightarrow Ni^{2+}(aq) + 2Ag(s)$

Applications

15.33 The following half-reaction takes place in a nickel–cadmium battery used in a cordless drill:

$$Cd(s) + 2OH^-(aq) \longrightarrow Cd(OH)_2(s) + 2\,e^-$$

a. Is the half-reaction an oxidation or a reduction?
b. What substance is oxidized or reduced?
c. At which electrode would this half-reaction occur?

15.34 The following half-reaction takes place in a mercury battery used in hearing aids:

$$HgO(s) + H_2O(l) + 2\,e^- \longrightarrow Hg(l) + 2OH^-(aq)$$

a. Is the half-reaction an oxidation or a reduction?
b. What substance is oxidized or reduced?
c. At which electrode would this half-reaction occur?

15.35 The following half-reaction takes place in a mercury battery used in pacemakers and watches:

$$Zn(s) + 2OH^-(aq) \longrightarrow ZnO(s) + H_2O(l) + 2\,e^-$$

a. Is the half-reaction an oxidation or a reduction?
b. What substance is oxidized or reduced?
c. At which electrode would this half-reaction occur?

15.36 The following half-reaction takes place in a lead storage battery used in automobiles:

$$Pb(s) + SO_4^{2-}(aq) \longrightarrow PbSO_4(s) + 2\,e^-$$

a. Is the half-reaction an oxidation or a reduction?
b. What substance is oxidized or reduced?
c. At which electrode would this half-reaction occur?

Chemistry Link to the Environment

Fuel Cells: Clean Energy for the Future

Fuel cells are of interest to scientists because they provide an alternative source of electrical energy that is more efficient, does not use up oil reserves, and generates products that do not pollute the atmosphere. Fuel cells are considered to be a clean way to produce energy.

Like other electrochemical cells, a fuel cell consists of an anode and a cathode connected by a wire. But unlike other cells, the reactants must continuously enter the fuel cell to produce energy; electrical current is generated only as long as the fuels are supplied. One type of hydrogen–oxygen fuel cell has been used in automobile prototypes. In this hydrogen cell, gas enters the fuel cell and comes in contact with a platinum catalyst embedded in a plastic membrane. The catalyst assists in the oxidation of hydrogen atoms to hydrogen ions and electrons (see **FIGURE 15.3**).

The electrons produce an electric current as they travel through the wire from the anode to the cathode. The hydrogen ions flow through the plastic membrane to the cathode. At the cathode, oxygen molecules are reduced to oxide ions that combine with the hydrogen ions to form water. The overall hydrogen–oxygen fuel cell reaction can be written as

$$2H_2(g) + O_2(g) \longrightarrow 2H_2O(l)$$

Fuel cells have already been used for power on the space station and are now used to produce energy for cars and buses. An advantage of fuel cell cars is that the hydrogen fuel produces water and heat, which gives zero pollution.

In homes, fuel cells may one day replace the batteries currently used to provide electrical power for cell phones, DVD players, and laptop computers. Fuel cell design is still in the prototype phase, although there is much interest in their development. We already know they can work, but modifications must still be made before they become reasonably priced and part of our everyday lives.

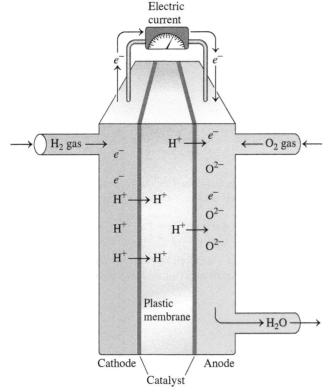

Oxidation
$$2H_2(g) \longrightarrow 4\,e^- + 4H^+(aq)$$

Reduction
$$4\,e^- + 4H^+(aq) + O_2(g) \longrightarrow 2H_2O(l)$$

FIGURE 15.3 With a supply of hydrogen and oxygen, a fuel cell can generate electricity continuously.

Fuel cells in cars use hydrogen as the fuel and produce water and heat.

15.4 Oxidation–Reduction Reactions That Require Electrical Energy

LEARNING GOAL Describe the half-cell reactions and the overall reactions that occur in electrolysis.

When we look at the activity series in Table 15.3, we see that the oxidation of Cu is below Zn. This means that the following oxidation–reduction reaction is not spontaneous:

$$Cu(s) + Zn^{2+}(aq) \longrightarrow Cu^{2+}(aq) + Zn(s) \quad \text{Not spontaneous}$$
Less active   More active

To make this reaction take place, we need an **electrolytic cell** that uses an electrical current to drive a nonspontaneous oxidation–reduction reaction. This process is called **electrolysis** (see **FIGURE 15.4**).

ENGAGE 15.13

Why does the reaction in an electrolytic cell require an electrical current?

$$Cu(s) \longrightarrow Cu^{2+}(aq) + 2\ e^- \qquad Zn^{2+}(aq) + 2\ e^- \longrightarrow Zn(s)$$
$$Cu(s) + Zn^{2+}(aq) \longrightarrow Cu^{2+}(aq) + Zn(s)$$

FIGURE 15.4 In this electrolytic cell, the Cu anode is in a Cu^{2+} solution, and the Zn cathode is in a Zn^{2+} solution. Electrons provided by a battery reduce Zn^{2+} to Zn and drive the oxidation of Cu to Cu^{2+} at the Cu anode.

Electrolysis of Sodium Chloride

When molten sodium chloride is electrolyzed, the products are sodium metal and chlorine gas. In this electrolytic cell, electrodes are placed in the mixture of Na^+ and Cl^- and connected to a battery. The products are separated to prevent them from reacting spontaneously with each other. As electrons flow to the cathode, Na^+ is reduced to sodium metal. At the same time, electrons leave the anode as Cl^- is oxidized to Cl_2. The half-reactions and the overall reactions are

Anode (oxidation): $\qquad\qquad 2Cl^-(l) \longrightarrow Cl_2(g) + 2\ e^-$

Cathode (reduction): $\quad 2Na^+(l) + 2\ e^- \longrightarrow 2Na(l)$

Overall cell reaction: $2Na^+(l) + 2Cl^-(l) \longrightarrow 2Na(l) + Cl_2(g)$ $\quad$ Not spontaneous

$$\boxed{\text{Electrical energy} \rangle}$$

Electroplating

In industry, the process of electroplating uses electrolysis to coat an object with a thin layer of a metal such as silver, platinum, or gold. Steel objects are electroplated with chromium to prevent rusting. Silver-plated utensils, bowls, and platters are made by electroplating objects with a layer of silver.

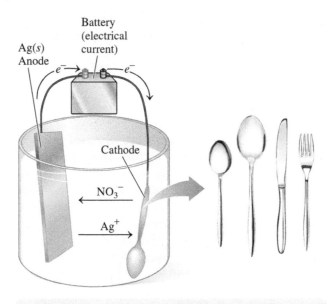

Utensils are silver-plated by electrolysis.

Using electrolysis, a thin layer of chromium is plated onto a hubcap.

SAMPLE PROBLEM 15.10   Electrolysis

TRY IT FIRST

Electrolysis is used to chrome plate an iron hubcap by placing the hubcap in a Cr^{3+} solution.

a. What half-reaction takes place to plate the hubcap with metallic chromium?
b. Is the iron hubcap the anode or the cathode?

SOLUTION

a. The Cr^{3+} ions in solution would gain electrons (reduction).

$$Cr^{3+}(aq) + 3\ e^- \longrightarrow Cr(s)$$

b. The iron hubcap is the cathode where reduction takes place.

SELF TEST 15.10

a. Why is energy needed to chrome plate the iron in Sample Problem 15.10?
b. Why is a chromium bar used as the anode in the reaction?

ANSWER

a. Since Fe is below Cr in the activity series, the plating of Cr^{3+} onto Fe is not spontaneous. Energy is needed to make the reaction proceed.
b. A chromium bar is used as the anode in the reaction in order to supply Cr^{3+} for the plating reaction. At the anode, the reaction is $Cr(s) \longrightarrow Cr^{3+}(aq) + 3\ e^-$.

PRACTICE PROBLEMS

Try Practice Problems 15.37 to 15.40

PRACTICE PROBLEMS

15.4 Oxidation–Reduction Reactions That Require Electrical Energy

Applications

15.37 What we call "tin cans" are really iron cans coated with a thin layer of tin. The anode is a bar of tin, and the cathode is the iron can. An electrical current is used to oxidize the Sn to Sn^{2+} in solution, where the Sn^{2+} is reduced to produce a thin coating of Sn on the can.
  a. What half-reaction takes place to tin plate an iron can?
  b. Why is the iron can the cathode?
  c. Why is the tin bar the anode?

An iron can is coated with tin to prevent rusting.

15.38 Electrolysis is used to gold plate jewelry made of stainless steel.
  a. What half-reaction takes place when a Au^{3+} solution is used to gold plate a stainless steel ring?
  b. Is the ring the anode or the cathode?
  c. Why is energy needed to gold plate the ring?

15.39 When the tin coating on an iron can is scratched, rust will form. Use the activity series in Table 15.3 to explain why this happens.

15.40 When the zinc coating on an iron can is scratched, rust does not form. Use the activity series in Table 15.3 to explain why this happens.

UPDATE Whitening Kimberly's Teeth

The reaction of hydrogen peroxide, $H_2O_2(aq)$, placed on Kimberly's teeth produces water, $H_2O(l)$, and oxygen gas. After Kimberly had her teeth cleaned and whitened, Jane recommended that Kimberly use a toothpaste containing tin(II) fluoride. Tooth enamel, which is composed of hydroxyapatite, $Ca_5(PO_4)_3OH$, is strengthened when it reacts with fluoride ions to form fluoroapatite, $Ca_5(PO_4)_3F$.

15.41 The simplified reaction for the production of aqueous H_2O_2 involves the combination of H_2 gas and O_2 gas.
  a. What reactant is oxidized and what reactant is reduced in this reaction?
  b. What is the oxidizing agent?
  c. What is the reducing agent?
  d. Write a balanced chemical equation for the reaction.

15.42 The reaction of solid Sn and F_2 gas forms solid SnF_2.
  a. What reactant is oxidized and what reactant is reduced in this reaction?
  b. What is the oxidizing agent?
  c. What is the reducing agent?
  d. Write a balanced chemical equation for the reaction.

CONCEPT MAP

OXIDATION AND REDUCTION

involve

| Oxidation | and | Reduction |

when — **Electrons Are Lost** — to make — **The Oxidation Numbers Increase**

when — **Electrons Are Gained** — to make — **The Oxidation Numbers Decrease**

are balanced using

Oxidation Numbers   or   **Half-Reactions**

involve

| Voltaic Cell | Electrolytic Cell |

that produces — **Electrical Energy** — using a — **Spontaneous Reaction**

that requires — **Electrical Energy** — using a — **Nonspontaneous Reaction**

as determined by the

Activity Series

CHAPTER REVIEW

15.1 Oxidation and Reduction

LEARNING GOAL Identify a reaction as an oxidation or a reduction. Assign and use oxidation numbers to identify elements that are oxidized or reduced.

- In an oxidation–reduction reaction, electrons are transferred from one reactant to another.
- The reactant that loses electrons is oxidized, and the reactant that gains electrons is reduced.
- Oxidation numbers assigned to elements keep track of the changes in the loss and gain of electrons.
- Oxidation is an increase in oxidation number; reduction is a decrease in oxidation number.
- The reducing agent is the substance that provides electrons for reduction.
- The oxidizing agent is the substance that accepts the electrons from oxidation.
- In molecular compounds and polyatomic ions, oxidation numbers are assigned using a set of rules.
- The oxidation number of an element is zero, and the oxidation number of a monatomic ion is the same as the ionic charge of the ion.
- The sum of the oxidation numbers for a compound is equal to zero and for a polyatomic ion is equal to the overall charge.
- Balancing oxidation–reduction equations using oxidation numbers involves the following:

  1. assigning oxidation numbers
  2. determining the loss and gain of electrons
  3. equalizing the loss and gain of electrons
  4. balancing the remaining substances by inspection

15.2 Balancing Oxidation–Reduction Equations Using Half-Reactions

LEARNING GOAL Balance oxidation–reduction equations using the half-reaction method.

- Balancing oxidation–reduction equations using half-reactions involves the following:

  1. separate the equation into half-reactions
  2. balance elements other than H and O
  3. balance O with H_2O and H with H^+
  4. balance charge with electrons
  5. equalize the loss and gain of electrons
  6. combine half-reactions, cancele electrons, and combine H_2O and H^+
  7. use OH^- to neutralize H^+ to H_2O for an oxidation–reduction reaction in basic solution

15.3 Electrical Energy from Oxidation–Reduction Reactions

LEARNING GOAL Use the activity series to determine if an oxidation–reduction reaction is spontaneous. Write the half-reactions that occur in a voltaic cell and the cell notation.

- The activity series, which lists metals with the most easily oxidized metal at the top, is used to predict the direction of a spontaneous reaction.
- In a voltaic cell, the components of the two half-reactions of a spontaneous oxidation–reduction reaction are placed in separate containers called half-cells.
- With a wire connecting the half-cells, an electrical current is generated as electrons move from the anode where oxidation takes place to the cathode where reduction takes place.

15.4 Oxidation–Reduction Reactions That Require Electrical Energy

LEARNING GOAL Describe the half-cell reactions and the overall reactions that occur in electrolysis.

- In an electrolytic cell, electrical energy from an external source is used to make reactions take place that are not spontaneous.

- A method called electrolysis is used to plate chrome on hubcaps, zinc on iron, or gold on stainless steel jewelry.

KEY TERMS

activity series A table of half-reactions with the metals that oxidize most easily at the top, and the metals that do not oxidize easily at the bottom.

anode The electrode where oxidation takes place.

cathode The electrode where reduction takes place.

electrolysis The use of electrical energy to run a nonspontaneous oxidation–reduction reaction in an electrolytic cell.

electrolytic cell A cell in which electrical energy is used to make a nonspontaneous oxidation–reduction reaction happen.

half-reaction method A method of balancing oxidation–reduction reactions in which the half-reactions are balanced separately and then combined to give the complete reaction.

oxidation The loss of electrons by a substance.

oxidation number A number equal to zero in an element or the charge of a monatomic ion; in molecular compounds and polyatomic ions, oxidation numbers are assigned using a set of rules.

oxidation–reduction reaction A reaction in which electrons are transferred from one reactant to another.

oxidizing agent The reactant that gains electrons and is reduced.

reducing agent The reactant that loses electrons and is oxidized.

reduction The gain of electrons by a substance.

voltaic cell A type of cell with two compartments that uses spontaneous oxidation–reduction reactions to produce electrical energy.

CORE CHEMISTRY SKILLS

The chapter section containing each Core Chemistry Skill is shown in parentheses at the end of each heading.

Assigning Oxidation Numbers (15.1)

- Oxidation numbers are assigned to atoms or ions in a reaction to determine the substance that is oxidized and the substance that is reduced.
  1. The sum of the oxidation numbers for a molecule is 0; for a polyatomic ion the sum is its charge

  The oxidation number for:
  2. an element is 0
  3. a monatomic ion is its charge
  4. Group 1A (1) metals is $+1$, Group 2A (2) metals is $+2$
  5. fluorine is -1; for other Group 7A (17) metals is -1, except when combined with O or F
  6. oxygen is -2, except in OF_2 or peroxides
  7. H in nonmetal compounds is $+1$; for H with a metal is -1

Example: Assign oxidation numbers to each of the elements in the following:

     a. Sn^{4+} **b.** Mn   **c.** MnO_2 **d.** $MgCO_3$

Answer: **a.** Sn^{4+} **b.** Mn   **c.** MnO_2 **d.** $MgCO_3$
          $+4$ 0 $+4-2$ $+2+4-2$

Using Oxidation Numbers (15.1)

- An atom is oxidized when its oxidation number increases from reactant to product.
- An atom is reduced when its oxidation number decreases from reactant to product.

Example: Assign oxidation numbers to each element and identify which is oxidized and which is reduced.

$$Fe_2O_3(s) + 2Al(s) \longrightarrow Al_2O_3(s) + 2Fe(l)$$

Answer: $Fe_2O_3(s) + 2Al(s) \longrightarrow Al_2O_3(s) + 2Fe(l)$
        $+3-2$ 0 $+3-2$ 0
       $Fe(+3) \longrightarrow Fe(0)$ Fe is reduced.
       $Al(0) \longrightarrow Al(+3)$ Al is oxidized.

Identifying Oxidizing and Reducing Agents (15.1)

- In an oxidation–reduction reaction, the substance that is reduced is the oxidizing agent.
- In an oxidation–reduction reaction, the substance that is oxidized is the reducing agent.

Example: Identify the oxidizing agent and the reducing agent in the following:

$$SnO_2(s) + 2H_2(g) \longrightarrow Sn(s) + 2H_2O(l)$$

Answer: Oxidation numbers are assigned to identify the substance that is oxidized as the reducing agent and the substance that is reduced as the oxidizing agent.

     $SnO_2(s) + 2H_2(g) \longrightarrow Sn(s) + 2H_2O(l)$
     $+4-2$ 0   0 $+1-2$

     $Sn(+4) \longrightarrow Sn(0)$
     Sn is reduced; SnO_2 is the oxidizing agent.
     $H(0) \longrightarrow H(+1)$
     H is oxidized; H_2 is the reducing agent.

Using Half-Reactions to Balance Redox Equations (15.2)

- The half-reaction method for balancing equations involves:
  1. separate the oxidation–reduction reaction into two *half-reactions*
  2. balance the elements in each except H and O; then if the reaction takes place in acid or base balancing O with H_2O and H with H^+
  3. balance charge by adding electrons

4. multiply each half-reaction by factors that equalize the loss and gain of electrons

5. combine half-reactions and cancel electrons, identical ions or molecules, and use OH^- in base to neutralize H^+

Example: Use the half-reaction method to balance the following oxidation–reduction equation in acidic solution:

$$NO_3^-(aq) + Sn^{2+}(aq) \longrightarrow NO(g) + Sn^{4+}(aq)$$

Answer:

1. $NO_3^-(aq) \longrightarrow NO(g)$
   $Sn^{2+}(aq) \longrightarrow Sn^{4+}(aq)$

2. $4H^+(aq) + NO_3^-(aq) \longrightarrow NO(g) + 2H_2O(l)$
   $Sn^{2+}(aq) \longrightarrow Sn^{4+}(aq)$

3. $3\,e^- + 4H^+(aq) + NO_3^-(aq) \longrightarrow NO(g) + 2H_2O(l)$
   $Sn^{2+}(aq) \longrightarrow Sn^{4+}(aq) + 2\,e^-$

4. $2 \times [3\,e^- + 4H^+(aq) + NO_3^-(aq) \longrightarrow NO(g) + 2H_2O(l)]$
   $\overline{6\,e^- + 8H^+(aq) + 2NO_3^-(aq) \longrightarrow 2NO(g) + 4H_2O(l)}$

   $3 \times [Sn^{2+}(aq) \longrightarrow Sn^{4+}(aq) + 2\,e^-]$
   $\overline{3Sn^{2+}(aq) \longrightarrow 3Sn^{4+}(aq) + 6\,e^-}$

5. $8H^+(aq) + 2NO_3^-(aq) + 3Sn^{2+}(aq) \longrightarrow$
   $\qquad\qquad 2NO(g) + 3Sn^{4+}(aq) + 4H_2O(l)$

Identifying Spontaneous Reactions (15.3)

- The activity series in Table 15.3 places the metals that are easily oxidized at the top, and the metals that do not oxidize easily at the bottom.
- A metal will oxidize spontaneously when it is combined with the reverse of the half-reaction of any metal below it on the activity series in Table 15.3.

Example: Write a balanced redox equation for the spontaneous reaction for the following half-reactions on the activity series:

$$Ca(s) \longrightarrow Ca^{2+}(aq) + 2\,e^-$$
$$Ni(s) \longrightarrow Ni^{2+}(aq) + 2\,e^-$$

Answer: The spontaneous reaction is a combination of the oxidation half-reaction higher on the activity series with the reverse of the half-reaction below it.

$Ca(s) \longrightarrow Ca^{2+}(aq) + 2\,e^-$ Oxidation
$Ni^{2+}(aq) + 2\,e^- \longrightarrow Ni(s)$ Reduction
$\overline{Ca(s) + Ni^{2+}(aq) \longrightarrow Ca^{2+}(aq) + Ni(s)}$ Spontaneous reaction

---

UNDERSTANDING THE CONCEPTS

The chapter sections to review are shown in parentheses at the end of each problem.

15.43 Classify each of the following as oxidation or reduction: (15.1)

  a. Electrons are lost.
  b. Reaction of an oxidizing agent.
  c. $O_2(g) \longrightarrow OH^-(aq)$
  d. $Br_2(l) \longrightarrow 2Br^-(aq)$
  e. $Sn^{2+}(aq) \longrightarrow Sn^{4+}(aq)$

15.44 Classify each of the following as oxidation or reduction: (15.1)

  a. Electrons are gained.
  b. Reaction of a reducing agent.
  c. $Ni(s) \longrightarrow Ni^{2+}(aq)$
  d. $MnO_4^-(aq) \longrightarrow MnO_2(s)$
  e. $Sn^{4+}(aq) \longrightarrow Sn^{2+}(aq)$

15.45 Assign oxidation numbers to the elements in each of the following: (15.1)

  a. VO_2 **b.** Ag_2CrO_4
  c. $S_2O_8^{2-}$ **d.** $FeSO_4$

15.46 Assign oxidation numbers to the elements in each of the following: (15.1)

  a. $NbCl_3$ **b.** NbO
  c. NbO_2 **d.** Nb_2O_5

15.47 Which of the following are oxidation–reduction reactions? (15.1)

  a. $Ca(s) + 2H_2O(l) \longrightarrow Ca(OH)_2(aq) + H_2(g)$
  b. $CaCO_3(s) \longrightarrow CaO(s) + CO_2(g)$
  c. $4Al(s) + 3O_2(g) \longrightarrow 2Al_2O_3(s)$

15.48 Which of the following are oxidation–reduction reactions? (15.1)

  a. $BaCl_2(aq) + Na_2SO_4(aq) \longrightarrow BaSO_4(s) + 2NaCl(aq)$
  b. $Cl_2(g) + 2NaBr(aq) \longrightarrow Br_2(l) + 2NaCl(aq)$
  c. $2KClO_3(s) \longrightarrow 2KCl(s) + 3O_2(g)$

15.49 For this reaction, identify each of the following: (15.1)

$$2Cr_2O_3(s) + 3Si(s) \longrightarrow 4Cr(s) + 3SiO_2(s)$$

  a. the substance reduced   **b.** the substance oxidized
  c. the oxidizing agent   **d.** the reducing agent

Chromium(III) oxide and silicon undergo an oxidation–reduction reaction.

15.50 For this reaction, identify each of the following: (15.1)

$$C_2H_4(g) + 3O_2(g) \xrightarrow{\Delta} 2CO_2(g) + 2H_2O(g)$$

  a. the substance reduced   **b.** the substance oxidized
  c. the oxidizing agent   **d.** the reducing agent

15.51 Balance each of the following half-reactions in acidic solution: (15.2)

  a. $Zn(s) \longrightarrow Zn^{2+}(aq)$
  b. $SnO_2^{2-}(aq) \longrightarrow SnO_3^{2-}(aq)$
  c. $SO_3^{2-}(aq) \longrightarrow SO_4^{2-}(aq)$
  d. $NO_3^-(aq) \longrightarrow NO(g)$

15.52 Balance each of the following half-reactions in acidic solution: (15.2)

  a. $I_2(s) \longrightarrow I^-(aq)$
  b. $MnO_4^-(aq) \longrightarrow Mn^{2+}(aq)$
  c. $Br_2(l) \longrightarrow BrO_3^-(aq)$
  d. $ClO_3^-(aq) \longrightarrow ClO_4^-(aq)$

15.53 Consider the following voltaic cell: (15.3)

a. What is the oxidation half-reaction?
b. What is the reduction half-reaction?
c. What metal is the anode?
d. What metal is the cathode?
e. What is the direction of electron flow?
f. What is the overall reaction that takes place?
g. Write the shorthand cell notation.

15.54 Consider the following voltaic cell: (15.3)

a. What is the oxidation half-reaction?
b. What is the reduction half-reaction?
c. What metal is the anode?
d. What metal is the cathode?
e. What is the direction of electron flow?
f. What is the overall reaction that takes place?
g. Write the shorthand cell notation.

ADDITIONAL PRACTICE PROBLEMS

15.55 Which of the following are oxidation–reduction reactions? (15.1)

a. $AgNO_3(aq) + NaCl(aq) \longrightarrow AgCl(s) + NaNO_3(aq)$
b. $6Li(s) + N_2(g) \longrightarrow 2Li_3N(s)$
c. $Ni(s) + Pb(NO_3)_2(aq) \longrightarrow Ni(NO_3)_2(aq) + Pb(s)$
d. $2K(s) + 2H_2O(l) \longrightarrow 2KOH(aq) + H_2(g)$

15.56 Which of the following are oxidation–reduction reactions? (15.1)

a. $Ca(s) + F_2(g) \longrightarrow CaF_2(s)$
b. $Fe(s) + 2HCl(aq) \longrightarrow FeCl_2(aq) + H_2(g)$
c. $2NaCl(aq) + Pb(NO_3)_2(aq) \longrightarrow$
$PbCl_2(s) + 2NaNO_3(aq)$
d. $2CuCl(aq) \longrightarrow Cu(s) + CuCl_2(aq)$

15.57 In the mitochondria of human cells, energy is provided by the oxidation and reduction of the iron ions in the cytochromes. Identify each of the following reactions as an oxidation or reduction: (15.1)

a. $Fe^{3+}(aq) + e^- \longrightarrow Fe^{2+}(aq)$
b. $Fe^{2+}(aq) \longrightarrow Fe^{3+}(aq) + e^-$

15.58 Chlorine (Cl_2) is used as a germicide to kill microbes in swimming pools. If the product is Cl^-, was the elemental chlorine oxidized or reduced? (15.1)

15.59 Assign oxidation numbers to all the elements in each of the following: (15.1)

a. Co_2O_3 b. $KMnO_4$ c. $SbCl_5$
d. ClO_3^- e. PO_4^{3-}

15.60 Assign oxidation numbers to all the elements in each of the following: (15.1)

a. PO_3^{2-} b. NH_4^+ c. $Fe(OH)_2$
d. HNO_3 e. H_2CO

15.61 Assign oxidation numbers to all the elements in each of the following reactions, and identify the reactant that is oxidized, the reactant that is reduced, the oxidizing agent, and the reducing agent: (15.1)

a. $2FeCl_2(aq) + Cl_2(g) \longrightarrow 2FeCl_3(aq)$
b. $2H_2S(g) + 3O_2(g) \longrightarrow 2H_2O(l) + 2SO_2(g)$
c. $P_2O_5(s) + 5C(s) \longrightarrow 2P(s) + 5CO(g)$

15.62 Assign oxidation numbers to all the elements in each of the following reactions, and identify the reactant that is oxidized, the reactant that is reduced, the oxidizing agent, and the reducing agent: (15.1)

a. $4Al(s) + 3O_2(g) \longrightarrow 2Al_2O_3(s)$
b. $I_2O_5(s) + 5CO(g) \longrightarrow I_2(g) + 5CO_2(g)$
c. $2Cr_2O_3(s) + 3C(s) \longrightarrow 4Cr(s) + 3CO_2(g)$

15.63 Write the balanced half-reactions and a balanced redox equation for each of the following reactions in acidic solution: (15.2)

a. $Zn(s) + NO_3^-(aq) \longrightarrow Zn^{2+}(aq) + NO_2(g)$
b. $MnO_4^-(aq) + SO_3^{2-}(aq) \longrightarrow Mn^{2+}(aq) + SO_4^{2-}(aq)$
c. $ClO_3^-(aq) + I^-(aq) \longrightarrow Cl^-(aq) + I_2(s)$
d. $Cr_2O_7^{2-}(aq) + C_2O_4^{2-}(aq) \longrightarrow Cr^{3+}(aq) + CO_2(g)$

15.64 Write the balanced half-reactions and a balanced redox equation for each of the following reactions in acidic solution: (15.2)

a. $Sn^{2+}(aq) + IO_4^-(aq) \longrightarrow Sn^{4+}(aq) + I^-(aq)$
b. $S_2O_3^{2-}(aq) + I_2(s) \longrightarrow S_4O_6^{2-}(aq) + I^-(aq)$
c. $Mg(s) + VO_4^{3-}(aq) \longrightarrow Mg^{2+}(aq) + V^{2+}(aq)$
d. $Al(s) + Cr_2O_7^{2-}(aq) \longrightarrow Al^{3+}(aq) + Cr^{3+}(aq)$

15.65 Use the activity series in Table 15.3 to predict whether each of the following reactions will occur spontaneously: (15.3)

a. $2Cr(s) + 3Ni^{2+}(aq) \longrightarrow 2Cr^{3+}(aq) + 3Ni(s)$
b. $Cu(s) + Zn^{2+}(aq) \longrightarrow Cu^{2+}(aq) + Zn(s)$
c. $Zn(s) + Pb^{2+}(aq) \longrightarrow Zn^{2+}(aq) + Pb(s)$

15.66 Use the activity series in Table 15.3 to predict whether each of the following reactions will occur spontaneously: (15.3)

a. $Zn(s) + Mg^{2+}(aq) \longrightarrow Zn^{2+}(aq) + Mg(s)$
b. $3Na(s) + Al^{3+}(aq) \longrightarrow 3Na^+(aq) + Al(s)$
c. $Mg(s) + Ni^{2+}(aq) \longrightarrow Mg^{2+}(aq) + Ni(s)$

15.67 In a voltaic cell, one half-cell consists of nickel metal in a Ni^{2+} solution, and the other half-cell consists of magnesium metal in a Mg^{2+} solution. Give each of the following: (15.3)

a. the anode
b. the cathode
c. the half-reaction at the anode
d. the half-reaction at the cathode
e. the overall reaction
f. the shorthand cell notation

15.68 In a voltaic cell, one half-cell consists of zinc metal in a Zn^{2+} solution, and the other half-cell consists of copper metal in a Cu^{2+} solution. Give each of the following: (15.3)

a. the anode
b. the cathode
c. the half-reaction at the anode
d. the half-reaction at the cathode
e. the overall reaction
f. the shorthand cell notation

15.69 Use the activity series in Table 15.3 to determine which of the following ions will be reduced when an iron strip is placed in an aqueous solution of that ion: (15.3)

a. $Ca^{2+}(aq)$ b. $Ag^+(aq)$ c. $Ni^{2+}(aq)$
d. $Al^{3+}(aq)$ e. $Pb^{2+}(aq)$

15.70 Use the activity series in Table 15.3 to determine which of the following ions will be reduced when an aluminum strip is placed in an aqueous solution of that ion: (15.3)

a. $Fe^{2+}(aq)$ b. $Au^{3+}(aq)$ c. $Mg^{2+}(aq)$
d. $H^+(aq)$ e. $Pb^{2+}(aq)$

15.71 In a lead storage battery, the following unbalanced half-reaction takes place: (15.3)

$$Pb(s) + SO_4^{2-}(aq) \longrightarrow PbSO_4(s)$$

a. Balance the half-reaction.
b. Is $Pb(s)$ oxidized or reduced?
c. Indicate whether the half-reaction takes place at the anode or cathode.

15.72 In an acidic dry-cell battery, the following unbalanced half-reaction takes place in acidic solution: (15.3)

$$MnO_2(s) \longrightarrow Mn_2O_3(s)$$

a. Balance the half-reaction.
b. Is $MnO_2(s)$ oxidized or reduced?
c. Indicate whether the half-reaction takes place at the anode or cathode.

15.73 Steel bolts made for sailboats are coated with zinc. Add the necessary components (electrodes, wires, batteries) to this diagram of an electrolytic cell with a zinc nitrate solution to show how it could be used to zinc plate a steel bolt. (15.3, 15.4)

$Zn^{2+}(aq)$
$2NO_3^-(aq)$

a. What is the anode?
b. What is the cathode?
c. What is the half-reaction that takes place at the anode?
d. What is the half-reaction that takes place at the cathode?
e. If steel is mostly iron, what is the purpose of the zinc coating?

15.74 Copper cooking pans are stainless steel pans plated with a layer of copper. Add the necessary components (electrodes, wires, batteries) to this diagram of an electrolytic cell with a copper(II) nitrate solution to show how it could be used to copper plate a stainless steel (iron) pan. (15.3, 15.4)

$Cu^{2+}(aq)$
$2NO_3^-(aq)$

a. What is the anode?
b. What is the cathode?
c. What is the half-reaction that takes place at the anode?
d. What is the half-reaction that takes place at the cathode?

CHALLENGE PROBLEMS

The following problems are related to the topics in this chapter. However, they do not all follow the chapter order, and they require you to combine concepts and skills from several sections. These problems will help you increase your critical thinking skills and prepare for your next exam.

15.75 Determine the oxidation number of Br in each of the following: (15.1)

a. Br_2 b. $HBrO_2$
c. BrO_3^- d. $NaBrO_4$

15.76 Determine the oxidation number of Cr in each of the following: (15.1)

a. CrO b. $HCrO_4^-$
c. CrO_3^- d. CrF_4

15.77 Use half-reactions to balance the following equation in acidic solution: (15.2)

$$Cr_2O_7^{2-}(aq) + NO_2^-(aq) \longrightarrow Cr^{3+}(aq) + NO_3^-(aq)$$

15.78 Use half-reactions to balance the following equation in acidic solution: (15.2)

$$Mn(s) + Cr^{3+}(aq) \longrightarrow Mn^{2+}(aq) + Cr(s)$$

15.79 The following unbalanced reaction takes place in acidic solution: (11.6, 15.2)

$$Ag(s) + NO_3^-(aq) \longrightarrow Ag^+(aq) + NO(g)$$

a. Write the balanced equation.
b. How many liters of $NO(g)$ are produced at STP when 15.0 g of silver reacts with excess nitric acid?

15.80 The following unbalanced reaction takes place in acidic solution: (12.6, 15.2)

$$MnO_4^-(aq) + Fe^{2+}(aq) \longrightarrow Mn^{2+}(aq) + Fe^{3+}(aq)$$

  a. Write the balanced equation.
  b. How many milliliters of a 0.150 M $KMnO_4$ solution are needed to react with 25.0 mL of a 0.400 M $FeSO_4$ solution?

15.81 The following unbalanced reaction takes place in acidic solution: (12.6, 15.2)

$$S^{2-}(aq) + NO_3^-(aq) \longrightarrow SO_4^{2-}(aq) + NO(g)$$

  a. Write the balanced equation.
  b. How many milliliters of a 16.0 M HNO_3 solution are needed to react with 24.8 g of CuS?

15.82 The following unbalanced reaction takes place in acidic solution: (12.6, 15.2)

$$Cr_2O_7^{2-}(aq) + Fe^{2+}(aq) \longrightarrow Cr^{3+}(aq) + Fe^{3+}(aq)$$

  a. Write the balanced equation.
  b. How many milliliters of a 0.211 M $K_2Cr_2O_7$ solution are needed to react with 5.00 g of $FeSO_4$?

15.83 Use half-reactions to balance the following equation in basic solution: (15.2)

$$Br^-(aq) + MnO_4^-(aq) \longrightarrow BrO_3^-(aq) + MnO_2(s)$$

15.84 Use half-reactions to balance the following equation in basic solution: (15.2)

$$CN^-(aq) + IO_3^-(aq) \longrightarrow CNO^-(aq) + I^-(aq)$$

15.85 Using the activity series in Table 15.3, indicate whether each of the following reactions is spontaneous: (15.3)
  a. $Zn(s) + Ca^{2+}(aq) \longrightarrow Zn^{2+}(aq) + Ca(s)$
  b. $2Al(s) + 3Sn^{2+}(aq) \longrightarrow 2Al^{3+}(aq) + 3Sn(s)$
  c. $Mg(s) + 2H^+(aq) \longrightarrow Mg^{2+}(aq) + H_2(g)$

15.86 Using the activity series in Table 15.3, indicate whether each of the following reactions is spontaneous: (15.3)
  a. $Cu(s) + Ni^{2+}(aq) \longrightarrow Cu^{2+}(aq) + Ni(s)$
  b. $2Cr(s) + 3Fe^{2+}(aq) \longrightarrow 2Cr^{3+}(aq) + 3Fe(s)$
  c. $Fe(s) + Mg^{2+}(aq) \longrightarrow Fe^{2+}(aq) + Mg(s)$

15.87 Draw a diagram of a voltaic cell for $Ni(s)\,|\,Ni^{2+}(aq)\,\|\,Ag^+(aq)\,|\,Ag(s)$. (15.3)
  a. What is the anode?
  b. What is the cathode?
  c. What is the half-reaction that takes place at the anode?
  d. What is the half-reaction that takes place at the cathode?
  e. What is the overall reaction for the cell?

15.88 Draw a diagram of a voltaic cell for $Mg(s)\,|\,Mg^{2+}(aq)\,\|\,Al^{3+}(aq)\,|\,Al(s)$. (15.3)
  a. What is the anode?
  b. What is the cathode?
  c. What is the half-reaction that takes place at the anode?
  d. What is the half-reaction that takes place at the cathode?
  e. What is the overall reaction for the cell?

ANSWERS TO ENGAGE QUESTIONS

15.1 In a reduction half-reaction, there is a gain of electrons by one of the reactants.

15.2 Mg and Ba are elements, which have oxidation numbers of 0. The ions of Mg and Ba have the oxidation number of the monoatomic ions Mg^{2+}, Ba^{2+}.

15.3 The oxidation number of S in H_2SO_3 is calculated by adding +2 from 2 H and −6 from 3 O, which gives −4. Because H_2SO_3 has an overall 0 charge, S must have an oxidation number of +4.

15.4 The reaction of H_2 to $2H^+$ is an oxidation because there is an increase in the oxidation number of H, from 0 to +1.

15.5 Zn is oxidized when it loses electrons to form Zn^{2+}; it is the reducing agent. Co^{2+} is reduced when it gains electrons to form Co; it is the oxidizing agent.

15.6 To be sure the final oxidation–reduction reaction is balanced, check the balance of all the atoms and the overall charge.

15.7 The 7 H_2O are added to the product side of the half-reaction to balance the 7 O in $Cr_2O_7^{2-}$ on the reactant side.

15.8 On the product side of the half-reaction, the charge is +6. On the reactant side, the charge is +14 + (−2) = +12. To balance the two sides, −6 (6 e^-) must be added to the reactant side.

15.9 The $4OH^-$ added to the product side of the equation to neutralize the $4H^+$.

15.10 Because Fe is above Pb in the activity series, the reaction $Fe(s) + Pb^{2+}(aq) \longrightarrow Fe^{2+}(aq) + Pb(s)$ is spontaneous.

15.11 A salt bridge allows ions to move from one side of the cell to the other to complete the electrical circuit.

15.12 The copper electrode will be heavier because solid copper is deposited on it.

15.13 An electrolytic cell uses a nonspontaneous reaction, and it needs an outside source of energy, an electrical current, in order to proceed.

ANSWERS TO SELECTED PROBLEMS

15.1 **a.** reduction
  b. oxidation
  c. reduction

15.3 **a.** Al is oxidized and O_2 is reduced.
  b. Zn is oxidized and H^+ is reduced.
  c. Br^- is oxidized and F_2 is reduced.

15.5 **a.** 0   **b.** 0   **c.** +2   **d.** −1

15.7 **a.** K is +1, Cl is −1.   **b.** Mn is +4, O is −2.
  c. N is +2, O is −2.   **d.** Mn is +3, O is −2.

15.9 **a.** Li is +1, P is +5, and O is −2. **b.** S is +4, O is −2.
  c. Cr is +3, S is −2.   **d.** N is +5, O is −2.

15.11 **a.** H is +1, S is +6, and O is −2.
  b. H is +1, P is +3, and O is −2.
  c. Cr is +6, O is −2.
  d. Na is +1, C is +4, and O is −2.

15.13 **a.** +5   **b.** −2   **c.** +5   **d.** +6

15.15 **a.** The substance that is oxidized is the reducing agent.
  b. The substance that gains electrons is reduced and is the oxidizing agent.

15.17 a. Li is oxidized; Li is the reducing agent. Cl (in Cl_2) is reduced; Cl_2 is the oxidizing agent.

b. Br^- (in NaBr) is oxidized; NaBr is the reducing agent. Cl (in Cl_2) is reduced; Cl_2 is the oxidizing agent.

c. Pb is oxidized; Pb is the reducing agent. O (in O_2) is reduced; O_2 is the oxidizing agent.

d. Al is oxidized; Al is the reducing agent. Ag^+ is reduced; Ag^+ is the oxidizing agent.

15.19 a. S^{2-} (in NiS) is oxidized; NiS is the reducing agent. O (in O_2) is reduced; O_2 is the oxidizing agent.

b. Sn^{2+} is oxidized; Sn^{2+} is the reducing agent. Fe^{3+} is reduced; Fe^{3+} is the oxidizing agent.

c. C (in CH_4) is oxidized; CH_4 is the reducing agent. O (in O_2) is reduced; O_2 is the oxidizing agent.

d. Si is oxidized; Si is the reducing agent. Cr^{3+} (in Cr_2O_3) is reduced; Cr_2O_3 is the oxidizing agent.

15.21 a. $Sn^{2+}(aq) \longrightarrow Sn^{4+}(aq) + 2\,e^-$

b. $Mn^{2+}(aq) + 4H_2O(l) \longrightarrow MnO_4^-(aq) + 8H^+(aq) + 5\,e^-$

c. $NO_2^-(aq) + H_2O(l) \longrightarrow NO_3^-(aq) + 2H^+(aq) + 2\,e^-$

d. $e^- + 2H^+(aq) + ClO_3^-(aq) \longrightarrow ClO_2(aq) + H_2O(l)$

15.23 a. $2H^+(aq) + Ag(s) + NO_3^-(aq) \longrightarrow$
$Ag^+(aq) + NO_2(g) + H_2O(l)$

b. $4H^+(aq) + 4NO_3^-(aq) + 3S(s) \longrightarrow$
$4NO(g) + 3SO_2(g) + 2H_2O(l)$

c. $2S_2O_3^{2-}(aq) + Cu^{2+}(aq) \longrightarrow S_4O_6^{2-}(aq) + Cu(s)$

15.25 a. $2Fe(s) + 2CrO_4^{2-}(aq) + 2H_2O(l) \longrightarrow$
$Fe_2O_3(s) + Cr_2O_3(s) + 4OH^-(aq)$

b. $3CN^-(aq) + 2MnO_4^-(aq) + H_2O(l) \longrightarrow$
$3CNO^-(aq) + 2MnO_2(s) + 2OH^-(aq)$

15.27 a. Since Au is below H_2 in the activity series, the reaction will not be spontaneous.

b. Since Fe is above Ni in the activity series, the reaction will be spontaneous.

c. Since Ag is below Cu in the activity series, the reaction will not be spontaneous.

15.29 a. Anode reaction: $Pb(s) \longrightarrow Pb^{2+}(aq) + 2\,e^-$

Cathode reaction: $Cu^{2+}(aq) + 2\,e^- \longrightarrow Cu(s)$

Overall cell reaction: $Pb(s) + Cu^{2+}(aq) \longrightarrow$
$Pb^{2+}(aq) + Cu(s)$

b. Anode reaction: $Cr(s) \longrightarrow Cr^{2+}(aq) + 2\,e^-$

Cathode reaction: $Ag^+(aq) + e^- \longrightarrow Ag(s)$

Overall cell reaction: $Cr(s) + 2Ag^+(aq) \longrightarrow$
$Cr^{2+}(aq) + 2Ag(s)$

15.31 a. The anode is a Cd metal electrode in a Cd^{2+} solution. The anode reaction is

$Cd(s) \longrightarrow Cd^{2+}(aq) + 2\,e^-$
The cathode is a Sn metal electrode in a Sn^{2+} solution. The cathode reaction is

$Sn^{2+}(aq) + 2\,e^- \longrightarrow Sn(s)$
The shorthand notation for this cell is

$Cd(s)\,|\,Cd^{2+}(aq)\,\|\,Sn^{2+}(aq)\,|\,Sn(s)$

b. The anode is a Zn metal electrode in a Zn^{2+} solution. The anode reaction is

$Zn(s) \longrightarrow Zn^{2+}(aq) + 2\,e^-$
The cathode is a C (graphite) electrode, where Cl_2 gas is reduced to Cl^-. The cathode reaction is

$Cl_2(g) + 2\,e^- \longrightarrow 2Cl^-(aq)$
The shorthand notation for this cell is

$Zn(s)\,|\,Zn^{2+}(aq)\,\|\,Cl_2(g)\,|\,Cl^-(aq)\,|\,C$

15.33 a. The half-reaction is an oxidation.

b. Cd metal is oxidized.

c. Oxidation takes place at the anode.

15.35 a. The half-reaction is an oxidation.

b. Zn metal is oxidized.

c. Oxidation takes place at the anode.

15.37 a. $Sn^{2+}(aq) + 2\,e^- \longrightarrow Sn(s)$

b. The reduction of Sn^{2+} to Sn occurs at the cathode, which is the iron can.

c. The oxidation of Sn to Sn^{2+} occurs at the anode, which is the tin bar.

15.39 Since Fe is above Sn in the activity series, if the Fe is exposed to air and water, Fe will be oxidized and rust will form. To protect iron, Sn would have to be more active than Fe and it is not.

15.41 a. H_2 is oxidized and O_2 is reduced.

b. O_2 is the oxidizing agent.

c. H_2 is the reducing agent.

d. $H_2(g) + O_2(g) \longrightarrow H_2O_2(aq)$

15.43 a. oxidation   **b.** reduction   **c.** reduction

d. reduction   **e.** oxidation

15.45 a. $V = +4, O = -2$

b. $Ag = +1, Cr = +6, O = -2$

c. $S = +7, O = -2$

d. $Fe = +2, S = +6, O = -2$

15.47 Reactions **a** and **c** involve loss and gain of electrons; **a** and **c** are oxidation–reduction reactions.

15.49 a. Cr in Cr_2O_3 is reduced.

b. Si is oxidized.

c. Cr_2O_3 is the oxidizing agent.

d. Si is the reducing agent.

15.51 a. $Zn(s) \longrightarrow Zn^{2+}(aq) + 2\,e^-$

b. $SnO_2^{2-}(aq) + H_2O(l) \longrightarrow SnO_3^{2-}(aq) + 2H^+(aq) + 2\,e^-$

c. $SO_3^{2-}(aq) + H_2O(l) \longrightarrow SO_4^{2-}(aq) + 2H^+(aq) + 2\,e^-$

d. $3\,e^- + 4H^+(aq) + NO_3^-(aq) \longrightarrow NO(g) + 2H_2O(l)$

15.53 a. $Fe(s) \longrightarrow Fe^{2+}(aq) + 2\,e^-$

b. $Ni^{2+}(aq) + 2\,e^- \longrightarrow Ni(s)$

c. Fe is the anode.

d. Ni is the cathode.

e. The electrons flow from Fe to Ni.

f. $Fe(s) + Ni^{2+}(aq) \longrightarrow Fe^{2+}(aq) + Ni(s)$

g. $Fe(s)\,|\,Fe^{2+}(aq)\,\|\,Ni^{2+}(aq)\,|\,Ni(s)$

15.55 Reactions **b**, **c**, and **d** all involve loss and gain of electrons; **b**, **c**, and **d** are oxidation–reduction reactions.

15.57 a. Fe^{3+} is gaining electrons; this is a reduction.

b. Fe^{2+} is losing electrons; this is an oxidation.

15.59 a. $Co = +3, O = -2$ **b.** $K = +1, Mn = +7, O = -2$

c. $Sb = +5, Cl = -1$ **d.** $Cl = +5, O = -2$

e. $P = +5, O = -2$

15.61 a. $2FeCl_2(aq) + Cl_2(g) \longrightarrow 2FeCl_3(aq)$
${+2-1}{0}{+3-1}$
Fe in $FeCl_2$ is oxidized; $FeCl_2$ is the reducing agent.
Cl in Cl_2 is reduced; Cl_2 is the oxidizing agent.

b. $2H_2S(g) + 3O_2(g) \longrightarrow 2H_2O(l) + 2SO_2(g)$
${+1-2}{0}{+1-2}{+4-2}$
S in H_2S is oxidized; H_2S is the reducing agent.
O in O_2 is reduced; O_2 is the oxidizing agent.

c. $P_2O_5(s) + 5C(s) \longrightarrow 2P(s) + 5CO(g)$
${+5-2}{0}{0}{+2-2}$
C is oxidized; C is the reducing agent.
P in P_2O_5 is reduced; P_2O_5 is the oxidizing agent.

15.63 a. $Zn(s) \longrightarrow Zn^{2+}(aq) + 2\,e^-$;

$e^- + 2H^+(aq) + NO_3^-(aq) \longrightarrow NO_2(g) + H_2O(l)$

Overall:

$4H^+(aq) + Zn(s) + 2NO_3^-(aq) \longrightarrow$
$\qquad Zn^{2+}(aq) + 2NO_2(g) + 2H_2O(l)$

b. $5\,e^- + 8H^+(aq) + MnO_4^-(aq) \longrightarrow$
$\qquad Mn^{2+}(aq) + 4H_2O(l)$;

$SO_3^{2-}(aq) + H_2O(l) \longrightarrow SO_4^{2-}(aq) + 2H^+(aq) + 2\,e^-$

Overall:

$6H^+(aq) + 2MnO_4^-(aq) + 5SO_3^{2-}(aq) \longrightarrow$
$\qquad 2Mn^{2+}(aq) + 5SO_4^{2-}(aq) + 3H_2O(l)$

c. $2I^-(aq) \longrightarrow I_2(s) + 2\,e^-$;

$6\,e^- + 6H^+(aq) + ClO_3^-(aq) \longrightarrow Cl^-(aq) + 3H_2O(l)$

Overall:

$6H^+(aq) + ClO_3^-(aq) + 6I^-(aq) \longrightarrow$
$\qquad Cl^-(aq) + 3I_2(s) + 3H_2O(l)$

d. $C_2O_4^{2-}(aq) \longrightarrow 2CO_2(g) + 2\,e^-$;

$6\,e^- + 14H^+(aq) + Cr_2O_7^{2-}(aq) \longrightarrow$
$\qquad 2Cr^{3+}(aq) + 7H_2O(l)$

Overall:

$14H^+(aq) + Cr_2O_7^{2-}(aq) + 3C_2O_4^{2-}(aq) \longrightarrow$
$\qquad 2Cr^{3+}(aq) + 6CO_2(g) + 7H_2O(l)$

15.65 a. Since Cr is above Ni in the activity series, the reaction will be spontaneous.

b. Since Cu is below Zn in the activity series, the reaction will not be spontaneous.

c. Since Zn is above Pb in the activity series, the reaction will be spontaneous.

15.67 a. The anode is Mg.

b. The cathode is Ni.

c. The half-reaction at the anode is
$Mg(s) \longrightarrow Mg^{2+}(aq) + 2\,e^-$

d. The half-reaction at the cathode is
$Ni^{2+}(aq) + 2\,e^- \longrightarrow Ni(s)$

e. The overall reaction is
$Mg(s) + Ni^{2+}(aq) \longrightarrow Mg^{2+}(aq) + Ni(s)$

f. The shorthand cell notation is
$Mg(s)\,|\,Mg^{2+}(aq)\,\|\,Ni^{2+}(aq)\,|\,Ni(s)$

15.69 a. $Ca^{2+}(aq)$ will not be reduced by an iron strip.

b. $Ag^+(aq)$ will be reduced by an iron strip.

c. $Ni^{2+}(aq)$ will be reduced by an iron strip.

d. $Al^{3+}(aq)$ will not be reduced by an iron strip.

e. $Pb^{2+}(aq)$ will be reduced by an iron strip.

15.71 a. $Pb(s) + SO_4^{2-}(aq) \longrightarrow PbSO_4(s) + 2\,e^-$

b. $Pb(s)$ is oxidized.

c. The half-reaction takes place at the anode.

15.73

a. The anode is a bar of zinc.

b. The cathode is the steel bolt.

c. The half-reaction at the anode is
$Zn(s) \longrightarrow Zn^{2+}(aq) + 2\,e^-$

d. The half-reaction at the cathode is
$Zn^{2+}(aq) + 2\,e^- \longrightarrow Zn(s)$

e. The purpose of the zinc coating is to prevent rusting of the bolt by H_2O and O_2.

15.75 a. 0   **b.** +3   **c.** +5   **d.** +7

15.77 $8H^+(aq) + Cr_2O_7^{2-}(aq) + 3NO_2^-(aq) \longrightarrow$
$\qquad 2Cr^{3+}(aq) + 3NO_3^-(aq) + 4H_2O(l)$

15.79 a. $4H^+(aq) + 3Ag(s) + NO_3^-(aq) \longrightarrow$
$\qquad 3Ag^+(aq) + NO(g) + 2H_2O(l)$

b. 1.04 L of $NO(g)$ are produced at STP.

15.81 a. $8H^+(aq) + 3S^{2-}(aq) + 8NO_3^-(aq) \longrightarrow$
$\qquad 3SO_4^{2-}(aq) + 8NO(g) + 4H_2O(l)$

b. 43.2 mL of HNO_3 solution

15.83 $H_2O(l) + Br^-(aq) + 2MnO_4^-(aq) \longrightarrow$
$\qquad BrO_3^-(aq) + 2MnO_2(s) + 2OH^-(aq)$

15.85 a. Since Zn is below Ca in the activity series, the reaction will not be spontaneous.

b. Since Al is above Sn in the activity series, the reaction will be spontaneous.

c. Since Mg is above H_2 in the activity series, the reaction will be spontaneous.

15.87

a. $Ni(s)$ is the anode.

b. $Ag(s)$ is the cathode.

c. The half-reaction at the anode is
$Ni(s) \longrightarrow Ni^{2+}(aq) + 2\,e^-$

d. The half-reaction at the cathode is
$Ag^+(aq) + e^- \longrightarrow Ag(s)$

e. The overall cell reaction is
$Ni(s) + 2Ag^+(aq) \longrightarrow Ni^{2+}(aq) + 2Ag(s)$

Nuclear Chemistry

Simone's doctor is concerned about her elevated cholesterol, which could lead to coronary heart disease and a heart attack. He sends her to a nuclear medicine center to undergo a cardiac stress test. Pauline, the radiation technologist, explains to Simone that a stress test measures the blood flow to her heart muscle at rest and then during stress. The test is performed in a similar way as a routine exercise stress test, but images are produced that show areas of blood flow through the heart. It involves taking two sets of images of her heart, one when the heart is at rest, and one when she is walking on a treadmill.

Pauline tells Simone that she will inject thallium-201 into her bloodstream. She explains that Tl-201 is a radioactive isotope that has a half-life of 3.0 days. Simone is curious about the term "half-life." Pauline explains that a half-life is the amount of time it takes for one-half of a radioactive sample to break down. She assures Simone that after four half-lives, the radiation emitted will be almost zero. Pauline tells Simone that Tl-201 decays to Hg-201 and emits energy similar to X-rays. When the Tl-201 reaches any areas within her heart with restricted blood supply, smaller amounts of the radioisotope will accumulate.

CAREER

Radiation Technologist

A radiation technologist works in a hospital or imaging center where nuclear medicine is used to diagnose and treat a variety of medical conditions. In a diagnostic test, a radiation technologist uses a scanner, which converts radiation into images. The images are evaluated to determine any abnormalities in the body. A radiation technologist performs tests such as computed tomography (CT), magnetic resonance imaging (MRI), and positron emission tomography (PET) and operates the instrumentation and computers associated with these tests. In addition, they must physically and mentally prepare patients for imaging. A radiation technologist must know how to handle radioisotopes safely, use the necessary type of shielding, and give radioactive isotopes to patients. A patient may be given radioactive tracers such as technetium-99m, iodine-131, gallium-67, and thallium-201 that emit gamma radiation, which is detected and used to develop an image of the kidneys or thyroid or to follow the blood flow in the heart muscle.

UPDATE Cardiac Imaging Using a Radioisotope

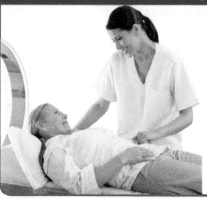

When Simone arrives at the clinic for her nuclear stress test, a radioactive isotope is injected that helps the technologist make images of her heart muscle. You can read about Simone's scans and the results of her nuclear stress test in the **UPDATE Cardiac Imaging Using a Radioisotope**, page 532.

16.1 Natural Radioactivity

LEARNING GOAL Describe alpha, beta, positron, and gamma radiation.

Most naturally occurring isotopes of elements up to atomic number 19 have stable nuclei. Elements with atomic numbers 20 and higher usually have one or more isotopes that have unstable nuclei in which the nuclear forces cannot offset the repulsions between the protons. An unstable nucleus is *radioactive*, which means that it spontaneously emits small amounts of energy called **radiation** to become more stable. Radiation may take the form of alpha (α) and beta (β) particles, positrons (β^+), or pure energy such as gamma (γ) rays. An isotope of an element that emits radiation is called a **radioisotope**. For most types of radiation, there is a change in the number of protons in the nucleus, which means that an atom is converted into an atom of a different element. Elements with atomic numbers of 93 and higher are produced artificially in nuclear laboratories and consist only of radioactive isotopes.

Symbols for Radioisotopes

The atomic symbols for the different isotopes are written with the mass number in the upper left corner and the atomic number in the lower left corner. The mass number is the sum of the number of protons and neutrons in the nucleus, and the atomic number is equal to the number of protons. For example, a radioactive isotope of carbon used for archaeological dating has a mass number of 14 and an atomic number of 6.

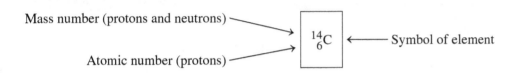

Mass number (protons and neutrons) ⟶ $^{14}_{6}\text{C}$ ⟵ Symbol of element

Atomic number (protons) ⟶

$^{14}_{6}\text{C}$ carbon-14 C-14

6 protons (red)

8 neutrons (white)

Radioactive isotopes are identified by writing the mass number after the element's name or symbol. Thus, in this example, the isotope is called carbon-14 or C-14. **TABLE 16.1** compares some stable, nonradioactive isotopes with some radioactive isotopes.

TABLE 16.1 Stable and Radioactive Isotopes of Some Elements

Magnesium	Iodine	Uranium
Stable Isotopes		
$^{24}_{12}\text{Mg}$	$^{127}_{53}\text{I}$	None
Magnesium-24	Iodine-127	
Radioactive Isotopes		
$^{23}_{12}\text{Mg}$	$^{125}_{53}\text{I}$	$^{235}_{92}\text{U}$
Magnesium-23	Iodine-125	Uranium-235
$^{27}_{12}\text{Mg}$	$^{131}_{53}\text{I}$	$^{238}_{92}\text{U}$
Magnesium-27	Iodine-131	Uranium-238

Types of Radiation

By emitting radiation, an unstable nucleus forms a more stable, lower-energy nucleus. One type of radiation consists of *alpha particles*. An **alpha particle** is identical to a helium (He) nucleus, which has two protons and two neutrons. An alpha particle has a mass number of 4, an atomic number of 2, and a charge of 2+. The symbol for an alpha particle is the Greek letter alpha (α) or the symbol of a helium nucleus except that the 2+ charge is omitted.

Another type of radiation is a **beta particle**, which is a high-energy electron, with a charge of 1− and a mass number of 0. It is represented by the Greek letter beta (β) or by the symbol for the electron including the mass number and the charge ($^{0}_{-1}e$).

Alpha particle $^{4}_{2}\text{He}$ or α

Beta particle $^{0}_{-1}e$ or β

| Positron | $_{+1}^{0}e$ or β^+ |

| Gamma ray | $_{0}^{0}\gamma$ or γ |

A **positron**, similar to a beta particle, has a positive (1+) charge with a mass number of 0. It is represented by the Greek letter beta with a 1+ charge, β^+, or by the symbol, $_{+1}^{0}e$. A positron is an example of *antimatter*, a term physicists use to describe a particle that is the opposite of another particle, in this case, an electron.

Gamma rays are high-energy radiation, released when an unstable nucleus undergoes a rearrangement of its particles to give a more stable, lower-energy nucleus. Gamma rays are often emitted along with other types of radiation. A gamma ray is written as the Greek letter gamma, γ. Because gamma rays are energy only, zeros are used to show that a gamma ray has no mass or charge ($_{0}^{0}\gamma$).

TABLE 16.2 summarizes the types of radiation we use in nuclear equations.

ENGAGE 16.1

What is the charge and the mass number of an alpha particle emitted by a radioactive atom?

TABLE 16.2 Some Forms of Radiation

Type of Radiation	Symbol		Mass Number	Charge
Alpha Particle	$_{2}^{4}He$	α	4	2+
Beta Particle	$_{-1}^{0}e$	β	0	1−
Positron	$_{+1}^{0}e$	β^+	0	1+
Gamma Ray	$_{0}^{0}\gamma$	γ	0	0
Proton	$_{1}^{1}H$	p	1	1+
Neutron	$_{0}^{1}n$	n	1	0

SAMPLE PROBLEM 16.1   Radiation Particles

TRY IT FIRST

Identify and write the symbol for each of the following types of radiation:

a. contains two protons and two neutrons
b. has a mass number of 0 and a 1− charge

SOLUTION

a. An alpha particle, $_{2}^{4}He$ or α, has two protons and two neutrons.
b. A beta particle, $_{-1}^{0}e$ or β, has a mass number of 0 and a 1− charge.

PRACTICE PROBLEMS

Try Practice Problems 16.1 to 16.10

SELF TEST 16.1

a. Identify and write the symbol for the type of radiation that has a mass number of zero and a 1+ charge.
b. Identify and write the symbol for the type of radiation that has a mass number of 0 and a 0 charge.

ANSWER

a. A positron, $_{+1}^{0}e$, has a mass number of 0 and a 1+ charge.
b. Gamma radiation, $_{0}^{0}\gamma$, has a mass number of 0 and a 0 charge.

Biological Effects of Radiation

When radiation strikes molecules in its path, electrons may be knocked away, forming unstable ions. If this *ionizing radiation* passes through the human body, it may interact with water molecules, removing electrons and producing H_2O^+, which can cause undesirable chemical reactions.

The cells most sensitive to radiation are the ones undergoing rapid division—those of the bone marrow, skin, reproductive organs, and intestinal lining, as well as all cells of growing children. Damaged cells may lose their ability to produce necessary materials. For example, if radiation damages cells of the bone marrow, red blood cells may no longer be produced. If sperm cells, ova, or the cells of a fetus are damaged, birth defects may result. In contrast, cells of the nerves, muscles, liver, and adult bones are much less sensitive to radiation because they undergo little or no cellular division.

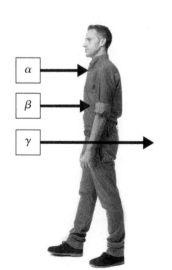

Different types of radiation penetrate the body to different depths.

Cancer cells are another example of rapidly dividing cells. Because cancer cells are highly sensitive to radiation, large doses of radiation are used to destroy them. The normal tissue that surrounds cancer cells divides at a slower rate and suffers less damage from radiation. However, radiation may cause malignant tumors, leukemia, anemia, and genetic mutations.

Radiation Protection

Radiation technologists, chemists, doctors, and nurses who work with radioactive isotopes must use proper radiation protection. Proper *shielding* is necessary to prevent exposure. Alpha particles, which have the largest mass and charge of the radiation particles, travel only a few centimeters in the air before they collide with air molecules, acquire electrons, and become helium atoms. A piece of paper, clothing, and our skin are protection against alpha particles. Lab coats and gloves will also provide sufficient shielding. However, if alpha emitters are ingested or inhaled, the alpha particles they give off can cause serious internal damage.

Beta particles have a very small mass and move much faster and farther than alpha particles, traveling as much as several meters through air. They can pass through paper and penetrate as far as 4 to 5 mm into body tissue. External exposure to beta particles can burn the surface of the skin, but they do not travel far enough to reach the internal organs. Heavy clothing such as lab coats and gloves are needed to protect the skin from beta particles.

Gamma rays travel great distances through the air and pass through many materials, including body tissues. Because gamma rays penetrate so deeply, exposure to gamma rays can be extremely hazardous. Only very dense shielding, such as lead or concrete, will stop them. Syringes used for injections of radioactive materials use shielding made of lead or heavyweight materials such as tungsten and plastic composites.

When working with radioactive materials, medical personnel wear protective clothing and gloves and stand behind a shield (see **FIGURE 16.1**). Long tongs may be used to pick up vials of radioactive material, keeping them away from the hands and body. **TABLE 16.3** summarizes the shielding materials required for the various types of radiation.

FIGURE 16.1 In a nuclear pharmacy, a person working with radioisotopes wears protective clothing and gloves and uses a lead glass shield on a syringe.

TABLE 16.3 Properties of Radiation and Shielding Required			
Property	Alpha (α) Particle	Beta (β) Particle	Gamma (γ) Ray
Travel Distance in Air	2 to 4 cm	200 to 300 cm	500 m
Tissue Depth	0.05 mm	4 to 5 mm	50 cm or more
Shielding	Paper, clothing	Heavy clothing, lab coats, gloves	Lead, thick concrete
Typical Source	Radium-226	Carbon-14	Technetium-99m

ENGAGE 16.2

What types of radiation does a lead shield block?

People who work in nuclear medicine minimize the time they spend close to radioactive materials to reduce exposure. Remaining in a radioactive area twice as long increases exposure of a person to twice the amount of radiation. The greater the distance from the radioactive source, the lower the intensity of radiation received. By doubling the distance from the radiation source, the intensity of radiation drops to $\left(\frac{1}{2}\right)^2$ or one-fourth of its previous value.

PRACTICE PROBLEMS

Try Practice Problems 16.11 and 16.12

PRACTICE PROBLEMS

16.1 Natural Radioactivity

16.1 Identify the type of particle or radiation for each of the following:
   a. ^4_2He **b.** $^0_{+1}e$ **c.** $^0_0\gamma$

16.2 Identify the type of particle or radiation for each of the following:
   a. $^0_{-1}e$ **b.** ^1_1H **c.** ^1_0n

16.3 Naturally occurring potassium consists of three isotopes: potassium-39, potassium-40, and potassium-41.
   a. Write the atomic symbol for each isotope.
   b. In what ways are the isotopes similar, and in what ways do they differ?

16.4 Naturally occurring iodine is iodine-127. Medically, radioactive isotopes of iodine-125 and iodine-131 are used.
    a. Write the atomic symbol for each isotope.
    b. In what ways are the isotopes similar, and in what ways do they differ?

16.5 Identify each of the following:
    a. $_{-1}^{0}X$ **b.** $_{2}^{4}X$ **c.** $_{0}^{1}X$
    d. $_{18}^{38}X$ **e.** $_{6}^{14}X$

16.6 Identify each of the following:
    a. $_{1}^{1}X$ **b.** $_{35}^{81}X$ **c.** $_{0}^{0}X$
    d. $_{26}^{59}X$ **e.** $_{+1}^{0}X$

Applications

16.7 Write the atomic symbol for each of the following isotopes used in nuclear medicine:
    a. copper-64   **b.** selenium-75
    c. sodium-24   **d.** nitrogen-15

16.8 Write the atomic symbol for each of the following isotopes used in nuclear medicine:
    a. indium-111   **b.** palladium-103
    c. barium-131   **d.** rubidium-82

16.9 Supply the missing information in the following table:

Medical Use	Atomic Symbol	Mass Number	Number of Protons	Number of Neutrons
Heart imaging	$_{81}^{201}Tl$			
Radiation therapy		60	27	
Abdominal scan			31	36
Hyperthyroidism	$_{53}^{131}I$			
Leukemia treatment		32		17

16.10 Supply the missing information in the following table:

Medical Use	Atomic Symbol	Mass Number	Number of Protons	Number of Neutrons
Cancer treatment	$_{55}^{131}Cs$			
Brain scan			43	56
Blood flow		141	58	
Bone scan			85	47
Lung function	$_{54}^{133}Xe$			

16.11 Match the type of radiation (**1** to **3**) with each of the following statements:
    1. alpha particle
    2. beta particle
    3. gamma radiation

    a. does not penetrate skin
    b. shielding protection includes lead or thick concrete
    c. can be very harmful if ingested

16.12 Match the type of radiation (**1** to **3**) with each of the following statements:
    1. alpha particle
    2. beta particle
    3. gamma radiation

    a. penetrates farthest into skin and body tissues
    b. shielding protection includes lab coats and gloves
    c. travels only a short distance in air

REVIEW

Using Positive and Negative Numbers in Calculations (1.4)

Solving Equations (1.4)

Counting Protons and Neutrons (4.4)

CORE CHEMISTRY SKILL

Writing Nuclear Equations

16.2 Nuclear Reactions

LEARNING GOAL Write a balanced nuclear equation for radioactive decay, showing mass numbers and atomic numbers.

In a process called **radioactive decay**, a nucleus spontaneously breaks down by emitting radiation. This process is shown by writing a *nuclear equation* with the atomic symbols of the original radioactive nucleus on the left, an arrow, and the new nucleus and the type of radiation emitted on the right.

$$\text{Radioactive nucleus} \longrightarrow \text{new nucleus} + \text{radiation } (\alpha, \beta, \beta^+, \gamma)$$

In a nuclear equation, the sum of the mass numbers and the sum of the atomic numbers on one side of the arrow must equal the sum of the mass numbers and the sum of the atomic numbers on the other side.

Alpha Decay

An unstable nucleus may emit an alpha particle, which consists of two protons and two neutrons. Thus, the mass number of the radioactive nucleus decreases by 4, and its atomic number decreases by 2. For example, when uranium-238 emits an alpha particle, the new nucleus that forms has a mass number of 234 and an atomic number of 90. Uranium is transformed into a different element, thorium, an example of *transmutation*.

ENGAGE 16.3

What happens to the U-238 nucleus when an alpha particle is emitted?

In the nuclear equation for alpha decay, the mass number of the new nucleus decreases by 4 and its atomic number decreases by 2.

We can look at writing a balanced nuclear equation for americium-241, which undergoes alpha decay as shown in Sample Problem 16.2.

SAMPLE PROBLEM 16.2   Writing a Nuclear Equation for Alpha Decay

TRY IT FIRST

Smoke detectors that are used in homes contain americium-241, which undergoes alpha decay. When alpha particles collide with air molecules, charged particles are produced that generate an electrical current. If smoke particles enter the detector, they interfere with the formation of charged particles in the air, and the electrical current is interrupted. This causes the alarm to sound and warns the occupants of the danger of fire. Write the balanced nuclear equation for the alpha decay of americium-241.

A smoke detector contains the radioactive isotope americium-241, which undergoes alpha decay.

SOLUTION

ANALYZE THE PROBLEM	Given	Need	Connect
	Am-241, alpha decay	balanced nuclear equation	mass number, atomic number of new nucleus

STEP 1   Write the incomplete nuclear equation.

$$^{241}_{95}\text{Am} \longrightarrow ? + ^{4}_{2}\text{He}$$

STEP 2   Determine the missing mass number. In the equation, the mass number, 241, is equal to the sum of the mass numbers of the new nucleus and the alpha particle.

$$241 = ? + 4$$
$$? = 241 - 4 = 237 \text{ (mass number of new nucleus)}$$

STEP 3   Determine the missing atomic number. The atomic number, 95, must equal the sum of the atomic numbers of the new nucleus and the alpha particle.

$$95 = ? + 2$$
$$? = 95 - 2 = 93 \text{ (atomic number of new nucleus)}$$

STEP 4   Determine the symbol of the new nucleus and complete the nuclear equation. On the periodic table, the element that has atomic number 93 is neptunium, Np. The atomic symbol for this isotope of Np is written $^{237}_{93}\text{Np}$.

$$^{241}_{95}\text{Am} \longrightarrow ^{237}_{93}\text{Np} + ^{4}_{2}\text{He}$$

SELF TEST 16.2

a. Write the balanced nuclear equation for the alpha decay of Po-214.
b. Bismuth-213, used to treat leukemia, emits alpha particles that destroy cancer cells. Write the balanced nuclear equation for the alpha decay of Bi-213.

PRACTICE PROBLEMS

Try Practice Problems 16.13 and 16.14

ANSWER

a. $^{214}_{84}\text{Po} \longrightarrow {}^{210}_{82}\text{Pb} + {}^{4}_{2}\text{He}$ **b.** $^{213}_{83}\text{Bi} \longrightarrow {}^{209}_{81}\text{Tl} + {}^{4}_{2}\text{He}$

Chemistry Link to Health

Radon in Our Homes

The presence of radon gas has become an environmental and health issue because of the radiation danger it poses. Radioactive isotopes such as radium-226 are naturally present in many types of rocks and soils. Radium-226 emits an alpha particle and is converted into radon gas, which diffuses out of the rocks and soil.

$$^{226}_{88}\text{Ra} \longrightarrow {}^{222}_{86}\text{Rn} + {}^{4}_{2}\text{He}$$

Outdoors, radon gas poses little danger because it disperses in the air. However, if the radioactive source is under a house or building, the radon gas can enter the house through cracks in the foundation or other openings. Those who live or work there may inhale the radon. In the lungs, radon-222 emits alpha particles to form polonium-218, which is known to cause lung cancer.

$$^{222}_{86}\text{Rn} \longrightarrow {}^{218}_{84}\text{Po} + {}^{4}_{2}\text{He}$$

The U.S. Environmental Protection Agency (EPA) estimates that radon causes about 20 000 lung cancer deaths in one year. The EPA recommends that the maximum level of radon not exceed 4 pico-curies (pCi) per liter of air in a home. One picocurie (pCi) is equal to 10^{-12} curies (Ci); curies are described in Section 16.3. The EPA estimates that about one out of every 15 homes in the United States have radon levels that exceed this maximum.

A radon test kit is used to determine radon levels in buildings.

Beta Decay

A beta particle forms as the result of the breakdown of a neutron into a proton and an electron (beta particle). Because the proton remains in the nucleus, the number of protons increases by one, whereas the number of neutrons decreases by one. Thus, in a nuclear equation for beta decay, the mass number of the radioactive nucleus and the mass number of the new nucleus are the same. However, the atomic number of the new nucleus increases by one. For example, the beta decay of a carbon-14 nucleus produces a nitrogen-14 nucleus.

ENGAGE 16.4

What happens to a C-14 nucleus when a beta particle is emitted?

In the nuclear equation for beta decay, the mass number of the new nucleus remains the same, and its atomic number increases by 1.

8 neutrons
6 protons

7 neutrons
7 protons

0 neutrons
0 protons
1− charge

$$^{14}_{6}\text{C} \longrightarrow {}^{14}_{7}\text{N} + {}^{0}_{-1}e$$

Radioactive nucleus → New nucleus + Beta particle

SAMPLE PROBLEM 16.3 Writing a Nuclear Equation for Beta Decay

TRY IT FIRST

The radioactive isotope yttrium-90, a beta emitter, is used in cancer treatment and as a colloidal injection into joints to relieve arthritis pain. Write the balanced nuclear equation for the beta decay of yttrium-90.

A radioisotope is injected into the joint to relieve the pain caused by arthritis.

SOLUTION

ANALYZE THE PROBLEM	Given	Need	Connect
	Y-90, beta decay	balanced nuclear equation	mass number, atomic number of new nucleus

STEP 1 Write the incomplete nuclear equation.

$$^{90}_{39}Y \longrightarrow \; ? + \; ^{0}_{-1}e$$

STEP 2 **Determine the missing mass number.** In the equation, the mass number, 90, is equal to the sum of the mass numbers of the new nucleus and the beta particle.

$$90 = ? + 0$$
$$? = 90 - 0 = 90 \text{ (mass number of new nucleus)}$$

STEP 3 **Determine the missing atomic number.** The atomic number, 39, must equal the sum of the atomic numbers of the new nucleus and the beta particle.

$$39 = ? - 1$$
$$? = 39 + 1 = 40 \text{ (atomic number of new nucleus)}$$

STEP 4 **Determine the symbol of the new nucleus and complete the nuclear equation.** On the periodic table, the element that has atomic number 40 is zirconium, Zr. The atomic symbol for this isotope of Zr is written $^{90}_{40}Zr$.

$$^{90}_{39}Y \longrightarrow \; ^{90}_{40}Zr + \; ^{0}_{-1}e$$

SELF TEST 16.3

a. Write the balanced nuclear equation for the beta decay of chromium-51.
b. Write the balanced nuclear equation for the beta decay of Cu-64, used in cancer therapy.

ANSWER

a. $^{51}_{24}Cr \longrightarrow \; ^{51}_{25}Mn + \; ^{0}_{-1}e$

b. $^{64}_{29}Cu \longrightarrow \; ^{64}_{30}Zn + \; ^{0}_{-1}e$

PRACTICE PROBLEMS

Try Practice Problems 16.15 and 16.16

Positron Emission

In positron emission, a proton in an unstable nucleus is converted to a neutron and a positron. The neutron remains in the nucleus, but the positron is emitted from the nucleus. In a nuclear equation for positron emission, the mass number of the radioactive nucleus and the mass number of the new nucleus are the same. However, the atomic number of the new nucleus decreases by one, indicating a change of one element into another. For example, an aluminum-24 nucleus undergoes positron emission to produce a magnesium-24 nucleus. The atomic number of magnesium (12) and the charge of the positron (1+) give the atomic number of aluminum (13).

Proton in the nucleus	New neutron remains in the nucleus	Positron emitted

$$^{24}_{13}Al \longrightarrow \; ^{24}_{12}Mg + \; ^{0}_{+1}e$$
Positron

SAMPLE PROBLEM 16.4 Writing a Nuclear Equation for Positron Emission

TRY IT FIRST

Write the balanced nuclear equation for the positron emission of manganese-49.

SOLUTION

	Given	Need	Connect
ANALYZE THE PROBLEM	Mn-49, positron emission	balanced nuclear equation	mass number, atomic number of new nucleus

STEP 1 Write the incomplete nuclear equation.

$$^{49}_{25}\text{Mn} \longrightarrow \text{?} + {}^{0}_{+1}e$$

STEP 2 Determine the missing mass number. In the equation, the mass number, 49, is equal to the sum of the mass numbers of the new nucleus and the positron.

$$49 = \text{?} + 0$$

$$\text{?} = 49 - 0 = 49 \text{ (mass number of new nucleus)}$$

STEP 3 Determine the missing atomic number. The atomic number, 25, must equal the sum of the atomic numbers of the new nucleus and the positron.

$$25 = \text{?} + 1$$

$$\text{?} = 25 - 1 = 24 \text{ (atomic number of new nucleus)}$$

STEP 4 Determine the symbol of the new nucleus and complete the nuclear equation. On the periodic table, the element that has atomic number 24 is chromium, Cr. The atomic symbol for this isotope of Cr is written $^{49}_{24}\text{Cr}$.

$$^{49}_{25}\text{Mn} \longrightarrow {}^{49}_{24}\text{Cr} + {}^{0}_{+1}e$$

SELF TEST 16.4

a. Write the balanced nuclear equation for the positron emission of xenon-118.
b. Write the balanced nuclear equation for the positron emission of arsenic-74, used in cancer imaging.

ANSWER

a. $^{118}_{54}\text{Xe} \longrightarrow {}^{118}_{53}\text{I} + {}^{0}_{+1}e$
b. $^{74}_{33}\text{As} \longrightarrow {}^{74}_{32}\text{Ge} + {}^{0}_{+1}e$

ENGAGE 16.5

Why does the atomic number but not the mass number change in positron emission?

$^{0}_{+1}e$

$^{49}_{24}\text{Cr}$ $^{49}_{25}\text{Mn}$

PRACTICE PROBLEMS

Try Practice Problems 16.17 and 16.18

Gamma Emission

Pure gamma emitters are rare, although gamma radiation accompanies most alpha and beta radiation. In radiology, one of the most commonly used gamma emitters is technetium (Tc). The unstable isotope of technetium is written as the *metastable* (symbol m) isotope technetium-99m, Tc-99m, or $^{99m}_{43}\text{Tc}$. By emitting energy in the form of gamma rays, the nucleus becomes more stable.

$$^{99m}_{43}\text{Tc} \longrightarrow {}^{99}_{43}\text{Tc} + {}^{0}_{0}\gamma$$

FIGURE 16.2 summarizes the changes in the nucleus for alpha, beta, positron, and gamma radiation.

Radiation Source	Radiation	New Nucleus
Alpha emitter	^4_2He +	New element

Mass number − 4
Atomic number − 2

Beta emitter	$^0_{-1}e$ +	New element

Mass number same
Atomic number + 1

Positron emitter	$^0_{+1}e$ +	New element

Mass number same
Atomic number − 1

Gamma emitter	$^0_0\gamma$ +	Stable nucleus of the same element

Mass number same
Atomic number same

FIGURE 16.2 When the nuclei of alpha, beta, positron, and gamma emitters emit radiation, new, more stable nuclei are produced.

PRACTICE PROBLEMS
Try Practice Problems 16.19 and 16.20

Producing Radioactive Isotopes

Today, many radioisotopes are produced in small amounts by bombarding stable, nonradioactive isotopes with high-speed particles such as alpha particles, protons, neutrons, and small nuclei. When one of these particles is absorbed, the stable nucleus is converted to a radioactive isotope and usually some type of radiation particle.

$$^4_2He \quad + \quad ^{10}_5B \longrightarrow ^{13}_7N \quad + \quad ^1_0n$$

Bombarding particle | Stable nucleus | New radioactive nucleus | Neutron

When nonradioactive B-10 is bombarded by an alpha particle, the products are radioactive N-13 and a neutron.

All elements that have an atomic number greater than 92 have been produced by bombardment. Most have been produced in only small amounts and exist for only a short time, making it difficult to study their properties. For example, when californium-249 is bombarded with nitrogen-15, the radioactive element dubnium-260 and four neutrons are produced.

$$^{15}_7N + ^{249}_{98}Cf \longrightarrow ^{260}_{105}Db + 4^1_0n$$

Technetium-99m is a radioisotope used in nuclear medicine for several diagnostic procedures, including brain tumor detection and liver and spleen examinations. The source of technetium-99m is molybdenum-99, which is produced in a nuclear reactor by neutron bombardment of molybdenum-98.

$$^1_0n + ^{98}_{42}Mo \longrightarrow ^{99}_{42}Mo$$

Many radiology laboratories have small generators containing molybdenum-99, which decays to the technetium-99m radioisotope.

$$^{99}_{42}Mo \longrightarrow ^{99m}_{43}Tc + ^0_{-1}e$$

INTERACTIVE VIDEO

Pearson eText Writing Equations for an Isotope Produced by Bombardment

A generator is used to prepare technetium-99m.

The technetium-99m radioisotope decays by emitting gamma rays. Gamma emission is desirable for diagnostic work because the gamma rays pass through the body to the detection equipment.

$$^{99m}_{43}\text{Tc} \longrightarrow ^{99}_{43}\text{Tc} + ^{0}_{0}\gamma$$

SAMPLE PROBLEM 16.5 Writing a Nuclear Equation for an Isotope Produced by Bombardment

TRY IT FIRST

Write the balanced nuclear equation for the bombardment of nickel-58 by a proton, $^{1}_{1}\text{H}$, which produces a radioactive isotope and an alpha particle.

SOLUTION

	Given	Need	Connect
ANALYZE THE PROBLEM	Ni-58, proton bombardment, to produce alpha particle	new isotope, balanced nuclear equation	mass number, atomic number of new nucleus

STEP 1 Write the incomplete nuclear equation.

$$^{1}_{1}\text{H} + ^{58}_{28}\text{Ni} \longrightarrow ? + ^{4}_{2}\text{He}$$

STEP 2 **Determine the missing mass number.** In the equation, the sum of the mass numbers of the proton, 1, and the nickel, 58, must equal the sum of the mass numbers of the new nucleus and the alpha particle.

$$1 + 58 = ? + 4$$
$$? = 59 - 4 = 55 \text{ (mass number of new nucleus)}$$

STEP 3 **Determine the missing atomic number.** The sum of the atomic numbers of the proton, 1, and nickel, 28, must equal the sum of the atomic numbers of the new nucleus and the alpha particle.

$$1 + 28 = ? + 2$$
$$? = 29 - 2 = 27 \text{ (atomic number of new nucleus)}$$

STEP 4 **Determine the symbol of the new nucleus, and complete the nuclear equation.** On the periodic table, the element that has atomic number 27 is cobalt, Co. The atomic symbol for this isotope of Co is written $^{55}_{27}\text{Co}$.

$$^{1}_{1}\text{H} + ^{58}_{28}\text{Ni} \longrightarrow ^{55}_{27}\text{Co} + ^{4}_{2}\text{He}$$

$^{4}_{2}\text{He}$

$^{1}_{1}\text{H}$

$^{55}_{27}\text{Co}$ $^{58}_{28}\text{Ni}$

SELF TEST 16.5

a. The first radioactive isotope was produced in 1934 by the bombardment of aluminum-27 by an alpha particle to produce a radioactive isotope and one neutron. Write the balanced nuclear equation for this bombardment.

b. When iron-54 is bombarded by an alpha particle, the products are a new isotope and two protons. Write the balanced nuclear equation for this bombardment.

ANSWER

a. $^{4}_{2}\text{He} + ^{27}_{13}\text{Al} \longrightarrow ^{30}_{15}\text{P} + ^{1}_{0}n$

b. $^{4}_{2}\text{He} + ^{54}_{26}\text{Fe} \longrightarrow ^{56}_{26}\text{Fe} + 2^{1}_{1}\text{H}$

PRACTICE PROBLEMS

Try Practice Problems 16.21 and 16.22

PRACTICE PROBLEMS

16.2 Nuclear Reactions

16.13 Write a balanced nuclear equation for the alpha decay of each of the following radioactive isotopes:
  a. $^{208}_{84}\text{Po}$ **b.** $^{232}_{90}\text{Th}$ **c.** $^{251}_{102}\text{No}$ **d.** radon-220

16.14 Write a balanced nuclear equation for the alpha decay of each of the following radioactive isotopes:
  a. curium-243   **b.** $^{252}_{99}\text{Es}$ **c.** $^{251}_{98}\text{Cf}$ **d.** $^{261}_{107}\text{Bh}$

16.15 Write a balanced nuclear equation for the beta decay of each of the following radioactive isotopes:
    a. $^{25}_{11}Na$ **b.** $^{20}_{8}O$ **c.** strontium-92 **d.** iron-60

16.16 Write a balanced nuclear equation for the beta decay of each of the following radioactive isotopes:
    a. $^{44}_{19}K$ **b.** iron-59   **c.** potassium-42 **d.** $^{141}_{56}Ba$

16.17 Write a balanced nuclear equation for the positron emission of each of the following radioactive isotopes:
    a. silicon-26 **b.** cobalt-54 **c.** $^{77}_{37}Rb$ **d.** $^{93}_{45}Rh$

16.18 Write a balanced nuclear equation for the positron emission of each of the following radioactive isotopes:
    a. boron-8 **b.** $^{15}_{8}O$ **c.** $^{40}_{19}K$ **d.** nitrogen-13

16.19 Complete each of the following nuclear equations, and describe the type of radiation:
    a. $^{28}_{13}Al \longrightarrow ? + ^{0}_{-1}e$ **b.** $^{180m}_{73}Ta \longrightarrow ^{180}_{73}Ta + ?$
    c. $^{66}_{29}Cu \longrightarrow ^{66}_{30}Zn + ?$ **d.** $? \longrightarrow ^{234}_{90}Th + ^{4}_{2}He$
    e. $^{188}_{80}Hg \longrightarrow ? + ^{0}_{+1}e$

16.20 Complete each of the following nuclear equations, and describe the type of radiation:
    a. $^{11}_{6}C \longrightarrow ^{11}_{5}B + ?$ **b.** $^{35}_{16}S \longrightarrow ? + ^{0}_{-1}e$
    c. $? \longrightarrow ^{90}_{39}Y + ^{0}_{-1}e$ **d.** $^{210}_{83}Bi \longrightarrow ? + ^{4}_{2}He$
    e. $? \longrightarrow ^{89}_{39}Y + ^{0}_{+1}e$

16.21 Complete each of the following bombardment reactions:
    a. $^{1}_{0}n + ^{9}_{4}Be \longrightarrow ?$
    b. $^{1}_{0}n + ^{131}_{52}Te \longrightarrow ? + ^{0}_{-1}e$
    c. $^{1}_{0}n + ? \longrightarrow ^{24}_{11}Na + ^{4}_{2}He$
    d. $^{4}_{2}He + ^{14}_{7}N \longrightarrow ? + ^{1}_{1}H$

16.22 Complete each of the following bombardment reactions:
    a. $? + ^{40}_{18}Ar \longrightarrow ^{43}_{19}K + ^{1}_{1}H$
    b. $^{1}_{0}n + ^{238}_{92}U \longrightarrow ?$
    c. $^{1}_{0}n + ? \longrightarrow ^{14}_{6}C + ^{1}_{1}H$
    d. $? + ^{64}_{28}Ni \longrightarrow ^{272}_{111}Rg + ^{1}_{0}n$

16.3 Radiation Measurement

LEARNING GOAL Describe how radiation is detected and measured.

One of the most common instruments for detecting beta and gamma radiation is the Geiger counter. It consists of a metal tube filled with a gas such as argon. When radiation enters a window on the end of the tube, it forms charged particles in the gas, which produce an electrical current. Each burst of current is amplified to give a click and a reading on a meter.

$$Ar + radiation \longrightarrow Ar^+ + e^-$$

Measuring Radiation

Radiation is measured in several different ways. When a radiology laboratory obtains a radioisotope, the *activity* of the sample is measured in terms of the number of nuclear disintegrations per second. The **curie (Ci)**, the original unit of activity, was defined as the number of disintegrations that occur in 1 s for 1 g of radium, which is equal to 3.7×10^{10} disintegrations/s. The unit was named for Polish scientist Marie Curie, who along with her husband, Pierre, discovered the radioactive elements radium and polonium. The SI unit of radiation activity is the **becquerel (Bq)**, which is 1 disintegration/s.

Radiation levels in workers at the Fukushima Daiichi nuclear power plant are measured.

A Geiger counter detects alpha particles, beta particles, and gamma rays using the ionization effect produced in a Geiger–Müller tube.

    The **rad (radiation absorbed dose)** is a unit that measures the amount of radiation absorbed by a gram of material such as body tissue. The SI unit for absorbed dose is the **gray (Gy)**, which is defined as the joules of energy absorbed by 1 kg of body tissue. The gray is equal to 100 rad.
    The **rem (radiation equivalent in humans)** is a unit that measures the biological effects of different kinds of radiation. Although alpha particles do not penetrate the skin,

if they should enter the body by some other route, they can cause extensive damage within a short distance in tissue. High-energy radiation, such as beta particles, high-energy protons, and neutrons that travel into tissue, causes more damage. Gamma rays are damaging because they travel a long way through body tissue.

To determine the **equivalent dose** or rem dose, the absorbed dose (rad) is multiplied by a factor that adjusts for biological damage caused by a particular form of radiation. For beta and gamma radiation the factor is 1, so the biological damage in rems is the same as the absorbed radiation (rad). For high-energy protons and neutrons, the factor is about 10, and for alpha particles it is 20.

Biological damage (rem) = Absorbed dose (rad) × Factor

Often, the measurement for an equivalent dose will be in units of millirems (mrem). One rem is equal to 1000 mrem. The SI unit is the **sievert (Sv)**. One sievert is equal to 100 rem. **TABLE 16.4** summarizes the units used to measure radiation.

TABLE 16.4 Units of Radiation Measurement

Measurement	Common Unit	SI Unit	Relationship
Activity	curie (Ci) 1 Ci = 3.7 × 10¹⁰ disintegrations/s	becquerel (Bq) 1 Bq = 1 disintegration/s	1 Ci = 3.7 × 10¹⁰ Bq
Absorbed Dose	rad	gray (Gy) 1 Gy = 1 J/kg of tissue	1 Gy = 100 rad
Biological Damage	rem	sievert (Sv)	1 Sv = 100 rem

Chemistry Link to Health

Radiation and Food

Foodborne illnesses caused by pathogenic bacteria such as *Salmonella*, *Listeria*, and *Escherichia coli* (*E. coli*) have become a major health concern in the United States. *E. coli* has been responsible for outbreaks of illness from contaminated ground beef, fruit juices, lettuce, and alfalfa sprouts.

The U.S. Food and Drug Administration (FDA) has approved the use of 0.3 to 1 kGy of radiation produced by cobalt-60 or cesium-137 for the treatment of foods. The irradiation technology is much like that used to sterilize medical supplies. Radioactive cobalt pellets are placed in stainless steel tubes, which are arranged in racks. When food moves through the series of racks, the gamma rays pass through the food and kill the bacteria.

It is important for consumers to understand that when food is irradiated, it never comes in contact with the radioactive source. The gamma rays pass through the food to kill bacteria, but that does not make the food radioactive. The radiation kills bacteria because it stops their ability to divide and grow. We cook or heat food thoroughly for the same purpose. Radiation, as well as heat, has little effect on the food itself because its cells are no longer dividing or growing. Thus irradiated food is not harmed, although a small amount of vitamins B_1 and C may be lost.

Currently, tomatoes, blueberries, strawberries, and mushrooms are being irradiated to allow them to be harvested when completely ripe and extend their shelf life (see **FIGURE 16.3**). The FDA has also approved the irradiation of pork, poultry, and beef to decrease potential infections and to extend shelf life. Currently, irradiated

vegetable and meat products are available in retail markets in more than 40 countries. In the United States, irradiated foods such as tropical fruits, spinach, and ground meats are found in some stores. *Apollo 17* astronauts ate irradiated foods on the Moon, and some U.S. hospitals and nursing homes now use irradiated poultry to reduce the possibility of salmonella infections among residents. The extended shelf life of irradiated food also makes it useful for campers and military personnel. Soon, consumers concerned about food safety will have a choice of irradiated meats, fruits, and vegetables at the market.

The FDA requires this symbol to appear on irradiated retail foods.

After two weeks, the irradiated strawberries on the right show no spoilage. Mold is growing on the nonirradiated ones on the left.

FIGURE 16.3 Irradiation of food

People who work in radiology laboratories wear dosimeters attached to their clothing to determine any exposure to radiation such as X-rays, gamma rays, or beta particles. A dosimeter can be thermoluminescent (TLD), optically stimulated luminescence (OSL), or electronic personal (EPD). Dosimeters provide real-time radiation levels measured by monitors in the work area.

A dosimeter measures radiation exposure.

SAMPLE PROBLEM 16.6 Radiation Measurement

TRY IT FIRST

One treatment for bone pain involves intravenous administration of the radioisotope phosphorus-32, which is incorporated into bone. A typical dose of 7.0 mCi can produce up to 450 rad in the bone. What is the difference between the units of mCi and rad?

SOLUTION

The millicuries (mCi) indicate the activity of the P-32 in terms of the number of nuclei that break down in 1 s. The radiation absorbed dose (rad) is a measure of the amount of radiation absorbed by the bone.

SELF TEST 16.6

a. What is the absorbed dose of 450 rad in grays (Gy)?
b. What is the activity of 7.0 mCi in becquerels (Bq)?

ANSWER

a. 4.5 Gy   **b.** 2.6×10^8 Bq

PRACTICE PROBLEMS

Try Practice Problems 16.23 to 16.28

Exposure to Radiation

Every day, we are exposed to low levels of radiation from naturally occurring radioactive isotopes in the buildings where we live and work, in our food and water, and in the air we breathe. For example, potassium-40, a naturally occurring radioactive isotope, is present in any potassium-containing food. Other naturally occurring radioisotopes in air and food are carbon-14, radon-222, strontium-90, and iodine-131. The average person in the United States is exposed to about 3.6 mSv of radiation annually. Medical sources of radiation, including dental, hip, spine, and chest X-rays and mammograms, add to our radiation exposure. **TABLE 16.5** lists some common sources of radiation.

Another source of background radiation is cosmic radiation produced in space by the Sun. People who live at high altitudes or travel by airplane receive a greater amount of cosmic radiation because there are fewer molecules in the atmosphere to absorb the radiation. For example, a person living in Denver receives about twice the cosmic radiation as a person living in Los Angeles. A person living close to a nuclear power plant normally does not receive much additional radiation, perhaps 0.001 mSv in 1 yr. However, in the accident at the Chernobyl nuclear power plant in 1986 in Ukraine, it is estimated that people in a nearby town received as much as 10 mSv/h.

Radiation Sickness

The larger the dose of radiation received at one time, the greater the effect on the body. Exposure to radiation of less than 0.25 Sv usually cannot be detected. Whole-body exposure of 1 Sv produces a temporary decrease in the number of white blood cells. If the exposure to radiation is greater than 1 Sv, a person may suffer the symptoms of radiation sickness: nausea, vomiting, fatigue, and a reduction in white-cell count. A whole-body dosage greater than 3 Sv can decrease the white-cell count to zero. The person suffers diarrhea, hair loss, and infection. Exposure to radiation of 5 Sv is expected to cause death in 50% of the people receiving that dose. This amount of radiation to the whole body is called the *lethal dose for one-half the population*, or the LD_{50}. The LD_{50} varies for different life forms, as **TABLE 16.6** shows. Whole-body radiation of 6 Sv or greater would be fatal to all humans within a few weeks.

TABLE 16.5 Average Annual Radiation Received by a Person in the United States

Source	Dose (mSv)
Natural	
Ground	0.2
Air, water, food	0.3
Cosmic rays	0.4
Wood, concrete, brick	0.5
Medical	
Chest X-ray	0.2
Dental X-ray	0.2
Mammogram	0.4
Hip X-ray	0.6
Lumbar spine X-ray	0.7
Upper gastrointestinal tract X-ray	2
Other	
Nuclear power plants	0.001
Television	0.2
Air travel	0.1
Radon	2*
*Varies widely.	

TABLE 16.6 Lethal Doses of Radiation for Some Life Forms

Life Form	LD_{50} (Sv)
Insect	1000
Bacterium	500
Rat	8
Human	5
Dog	3

PRACTICE PROBLEMS

16.3 Radiation Measurement

16.23 Match each property (**1** to **3**) with its unit of measurement.
1. activity
2. absorbed dose
3. biological damage

   a. rad   **b.** mrem   **c.** mCi   **d.** Gy

16.24 Match each property (**1** to **3**) with its unit of measurement.
1. activity
2. absorbed dose
3. biological damage

   a. mrad   **b.** gray   **c.** becquerel   **d.** Sv

Applications

16.25 Two technicians in a nuclear laboratory were accidentally exposed to radiation. If one was exposed to 8 mGy and the other to 5 rad, which technician received more radiation?

16.26 Two samples of a radioisotope were spilled in a nuclear laboratory. The activity of one sample was 8 kBq and the other 15 mCi. Which sample produced the higher amount of radiation?

16.27 **a.** The recommended dosage of iodine-131 is 4.20 μCi/kg of body mass. How many microcuries of iodine-131 are needed for a 70.0-kg person with hyperthyroidism?
    b. A person receives 50 rad of gamma radiation. What is that amount in grays?

16.28 **a.** The dosage of technetium-99m for a lung scan is 20. μCi/kg of body mass. How many millicuries of technetium-99m should be given to a 50.0-kg person (1 mCi = 1000 μCi)?
    b. Suppose a person absorbed 50 mrad of alpha radiation. What would be the equivalent dose in millisieverts?

REVIEW

Interpreting Graphs (1.4)
Using Conversion Factors (2.6)

16.4 Half-Life of a Radioisotope

LEARNING GOAL Given the half-life of a radioisotope, calculate the amount of radioisotope remaining after one or more half-lives.

The **half-life** of a radioisotope is the amount of time it takes for one-half of a sample to decay. For example, $^{131}_{53}\text{I}$ has a half-life of 8.0 days. As $^{131}_{53}\text{I}$ decays, it produces the nonradioactive isotope $^{131}_{54}\text{Xe}$ and a beta particle.

$$^{131}_{53}\text{I} \longrightarrow {}^{131}_{54}\text{Xe} + {}^{0}_{-1}e$$

ENGAGE 16.7

If a 24-mg sample of Tc-99m has a half-life of 6.0 h, why are only 3.0 mg of Tc-99m radioactive after 18 h?

Suppose we have a sample that initially contains 20. mg of $^{131}_{53}\text{I}$. In 8.0 days, one-half (10. mg) of all the I-131 nuclei in the sample will decay, which leaves 10. mg of I-131. After 16 days (two half-lives), 5.0 mg of the remaining I-131 decays, which leaves 5.0 mg of I-131. After 24 days (three half-lives), 2.5 mg of the remaining I-131 decays, which leaves 2.5 mg of I-131 nuclei still capable of producing radiation.

In one half-life, the activity of an isotope decreases by half.

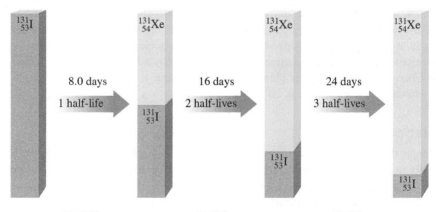

A **decay curve** is a diagram of the decay of a radioactive isotope. **FIGURE 16.4** shows such a curve for the $^{131}_{53}\text{I}$ we have discussed.

FIGURE 16.4 The decay curve for iodine-131 shows that one-half of the radioactive sample decays and one-half remains radioactive after each half-life of 8.0 days.

SAMPLE PROBLEM 16.7 Using Half-Lives of a Radioisotope in Medicine

TRY IT FIRST

Phosphorus-32, a radioisotope used in the treatment of leukemia, has a half-life of 14.3 days. If a sample contains 8.0 mg of phosphorus-32, how many milligrams of phosphorus-32 remain after 42.9 days?

SOLUTION

STEP 1 **State the given and needed quantities.**

ANALYZE THE PROBLEM	Given	Need	Connect
	8.0 mg of P-32, 42.9 days elapsed, half-life = 14.3 days	milligrams of P-32 remaining	number of half-lives

STEP 2 **Write a plan to calculate the unknown quantity.**

days $\xrightarrow{\text{Half-life}}$ number of half-lives

milligrams of $^{32}_{15}P$ $\xrightarrow{\text{Number of half-lives}}$ milligrams of $^{32}_{15}P$ remaining

STEP 3 **Write the half-life equality and conversion factors.**

1 half-life = 14.3 days

$$\frac{14.3 \text{ days}}{1 \text{ half-life}} \quad \text{and} \quad \frac{1 \text{ half-life}}{14.3 \text{ days}}$$

STEP 4 **Set up the problem to calculate the needed quantity.** First, we determine the number of half-lives in the amount of time that has elapsed.

$$\text{number of half-lives} = 42.9 \text{ days} \times \frac{1 \text{ half-life}}{14.3 \text{ days}} = 3.00 \text{ half-lives}$$

Now we can determine how much of the sample decays in three half-lives and how many milligrams of the phosphorus remain.

8.0 mg of $^{32}_{15}P$ $\xrightarrow[14.3 \text{ days}]{1 \text{ half-life}}$ 4.0 mg of $^{32}_{15}P$ $\xrightarrow[28.6 \text{ days}]{2 \text{ half-lives}}$ 2.0 mg of $^{32}_{15}P$ $\xrightarrow[42.9 \text{ days}]{3 \text{ half-lives}}$ 1.0 mg of $^{32}_{15}P$

SELF TEST 16.7

a. Iron-59 has a half-life of 44 days. If a nuclear laboratory receives a sample of 8.0 μg of iron-59, how many micrograms of iron-59 are still active after 176 days?

b. Rubidium-82, used to diagnose myocardial disease, has a half-life of 1.27 min. If the initial dose is 1480 MBq, how many half-lives have elapsed, and what is the activity of Rb-82 after 5.08 min?

ANSWER

a. 0.50 μg of iron-59   **b.** 4 half-lives; 92.5 MBq

Half-Lives

ENGAGE 16.8

Why do radioisotopes used in nuclear medicine have short half-lives?

Naturally occurring isotopes of the elements usually have long half-lives, as shown in TABLE 16.7. They disintegrate slowly and produce radiation over a long period of time, even hundreds or millions of years. In contrast, the radioisotopes used in nuclear medicine have much shorter half-lives. They disintegrate rapidly and produce almost all their radiation in a short period of time. For example, technetium-99m emits half of its radiation in the first 6 h. This means that a small amount of the radioisotope given to a patient is essentially gone within two days. The decay products of technetium-99m are totally eliminated by the body.

TABLE 16.7 Half-Lives of Some Radioisotopes

Element	Radioisotope	Half-Life	Type of Radiation
Naturally Occurring Radioisotopes			
Carbon-14	$^{14}_{6}C$	5730 yr	Beta
Potassium-40	$^{40}_{19}K$	1.3×10^9 yr	Beta, gamma
Radium-226	$^{226}_{88}Ra$	1600 yr	Alpha
Strontium-90	$^{90}_{38}Sr$	38.1 yr	Alpha
Uranium-238	$^{238}_{92}U$	4.5×10^9 yr	Alpha
Some Medical Radioisotopes			
Carbon-11	$^{11}_{6}C$	20. min	Positron
Chromium-51	$^{51}_{24}Cr$	28 days	Gamma
Iodine-131	$^{131}_{53}I$	8.0 days	Gamma
Oxygen-15	$^{15}_{8}O$	2.0 min	Positron
Iron-59	$^{59}_{26}Fe$	44 days	Beta, gamma
Radon-222	$^{222}_{86}Rn$	3.8 days	Alpha
Technetium-99m	$^{99m}_{43}Tc$	6.0 h	Gamma

PRACTICE PROBLEMS

Try Practice Problems 16.29 to 16.34

Chemistry Link to the Environment

Dating Ancient Objects

Radiological dating is a technique used by geologists, archaeologists, and historians to determine the age of ancient objects. The age of an object derived from plants or animals (such as wood, fiber, natural pigments, bone, and cotton or woolen clothing) is determined by measuring the amount of carbon-14, a naturally occurring radioactive form of carbon. In 1960, Willard Libby received the Nobel Prize for the work he did developing carbon-14 dating techniques during the 1940s. Carbon-14 is produced in the upper atmosphere by the bombardment of $^{14}_{7}N$ by high-energy neutrons from cosmic rays.

$$^{1}_{0}n + ^{14}_{7}N \longrightarrow ^{14}_{6}C + ^{1}_{1}H$$

Neutron from cosmic rays | Nitrogen in atmosphere | Radioactive carbon-14 | Proton

The carbon-14 reacts with oxygen to form radioactive carbon dioxide, $^{14}_{6}CO_2$. Living plants continuously absorb carbon dioxide, which incorporates carbon-14 into the plant material. The uptake of carbon-14 stops when the plant dies. As the carbon-14 undergoes beta decay, the amount of radioactive carbon-14 in the plant material steadily decreases.

$$^{14}_{6}C \longrightarrow ^{14}_{7}N + ^{0}_{-1}e$$

In a process called *carbon dating*, scientists use the half-life of carbon-14 (5730 yr) to calculate the length of time since the plant died. For example, a wooden beam found in an ancient dwelling might have one-half of the carbon-14 found in living plants today.

Because one half-life of carbon-14 is 5730 yr, the dwelling was constructed about 5730 yr ago. Carbon-14 dating was used to determine that the Dead Sea Scrolls are about 2000 yr old.

A radiological dating method used for determining the age of much older items is based on the radioisotope uranium-238, which decays through a series of reactions to lead-206. The uranium-238 isotope has an incredibly long half-life, about 4×10^9 yr. Measurements of the amounts of uranium-238 and lead-206 enable geologists to determine the age of rock samples. The older rocks will have a higher percentage of lead-206 because more of the uranium-238 has decayed. The age of rocks brought back from the Moon by the *Apollo* missions, for example, was determined using uranium-238. They were found to be about 4×10^9 yr old, approximately the same age calculated for Earth.

The age of the Dead Sea Scrolls was determined using carbon-14.

SAMPLE PROBLEM 16.8 Dating Using Half-Lives

TRY IT FIRST

The bones of humans and animals assimilate carbon until death. Using radiocarbon dating, the number of half-lives of carbon-14 from a bone sample determines the age of the bone. Suppose a sample is obtained from a prehistoric animal and used for radiocarbon dating. We can calculate the age of the bone or the years elapsed since the animal died by using the half-life of carbon-14, which is 5730 yr. A bone sample from the skeleton of a prehistoric animal has 25% of the activity of C-14 found in a living animal. How many years ago did the prehistoric animal die?

The age of a bone sample from a skeleton can be determined by carbon dating.

SOLUTION

STEP 1 State the given and needed quantities.

ANALYZE THE PROBLEM	Given	Need	Connect
	half-life = 5730 yr, 25% of initial C-14 activity	years elapsed	number of half-lives

STEP 2 Write a plan to calculate the unknown quantity.

Activity: 100% $\xrightarrow{\text{1.0 half-life}}$ 50% $\xrightarrow{\text{2.0 half-lives}}$ 25%
(initial)

STEP 3 Write the half-life equality and conversion factors.

$$1 \text{ half-life} = 5730 \text{ yr}$$

$$\frac{5730 \text{ yr}}{1 \text{ half-life}} \quad \text{and} \quad \frac{1 \text{ half-life}}{5730 \text{ yr}}$$

STEP 4 Set up the problem to calculate the needed quantity.

$$\text{Years elapsed} = 2.0 \text{ half-lives} \times \frac{5730 \text{ yr}}{1 \text{ half-life}} = 11\,000 \text{ yr}$$

We would estimate that the animal died 11 000 yr ago.

SELF TEST 16.8

a. Suppose that a piece of wood found in a cave had one-eighth of its original carbon-14 activity. About how many years ago was the wood part of a living tree?

b. How many micrograms of C-14 were initially in an organism if 4.0 mcg remain after 11 460 yr?

ANSWER

a. 17 000 yr   **b.** 16 mcg

PRACTICE PROBLEMS

16.4 Half-Life of a Radioisotope

16.29 For each of the following, indicate if the number of half-lives elapsed is:
  1. one half-life
  2. two half-lives
  3. three half-lives

  a. a sample of Pd-103 with a half-life of 17 days after 34 days
  b. a sample of C-11 with a half-life of 20 min after 20 min
  c. a sample of At-211 with a half-life of 7 h after 21 h

16.30 For each of the following, indicate if the number of half-lives elapsed is:
  1. one half-life
  2. two half-lives
  3. three half-lives

  a. a sample of Ce-141 with a half-life of 32.5 days after 32.5 days
  b. a sample of F-18 with a half-life of 110 min after 330 min
  c. a sample of Au-198 with a half-life of 2.7 days after 5.4 days

Applications

16.31 Technetium-99m is an ideal radioisotope for scanning organs because it has a half-life of 6.0 h and is a pure gamma emitter. Suppose that 80.0 mg were prepared in the technetium generator this morning. How many milligrams of technetium-99m would remain after each of the following intervals?
    a. one half-life   **b.** two half-lives
    c. 18 h   **d.** 1.5 days

16.32 A sample of sodium-24 with an activity of 12 mCi is used to study the rate of blood flow in the circulatory system. If sodium-24 has a half-life of 15 h, what is the activity after each of the following intervals?
    a. one half-life   **b.** 30 h
    c. three half-lives   **d.** 2.5 days

16.33 Strontium-85, used for bone scans, has a half-life of 65 days.
    a. How long will it take for the radiation level of strontium-85 to drop to one-fourth of its original level?
    b. How long will it take for the radiation level of strontium-85 to drop to one-eighth of its original level?

16.34 Fluorine-18, which has a half-life of 110 min, is used in PET scans.
    a. If 100. mg of fluorine-18 is shipped at 8:00 A.M., how many milligrams of the radioisotope are still active after 110 min?
    b. If 100. mg of fluorine-18 is shipped at 8:00 A.M., how many milligrams of the radioisotope are still active when the sample arrives at the radiology laboratory at 1:30 P.M.?

16.5 Medical Applications Using Radioactivity

LEARNING GOAL Describe the use of radioisotopes in medicine.

The first medical radioactive isotope was used to treat a person with leukemia at the University of California at Berkeley. In 1946, radioactive iodine was successfully used to diagnose thyroid function and to treat hyperthyroidism and thyroid cancer. Radioactive isotopes are now used to produce images of organs including the liver, spleen, thyroid, kidneys, brain, and heart.

To determine the condition of an organ in the body, a radiation technologist may use a radioisotope that concentrates in that organ. The cells in the body do not differentiate between a nonradioactive atom and a radioactive one, so these radioisotopes are easily incorporated. Then the radioactive atoms are detected because they emit radiation. Some radioisotopes used in nuclear medicine are listed in **TABLE 16.8**.

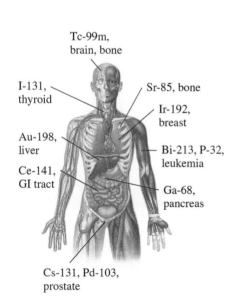

Tc-99m, brain, bone

I-131, thyroid

Sr-85, bone

Ir-192, breast

Au-198, liver

Bi-213, P-32, leukemia

Ce-141, GI tract

Ga-68, pancreas

Cs-131, Pd-103, prostate

Different radioisotopes are used to diagnose and treat a number of diseases.

TABLE 16.8 Medical Applications of Radioisotopes			
Isotope	Half-Life	Radiation	Medical Application
Au-198	2.7 days	Beta	Liver imaging; treatment of abdominal carcinoma
Bi-213	46 min	Alpha	Treatment of leukemia
Ce-141	32.5 days	Beta	Gastrointestinal tract diagnosis; measuring blood flow to the heart
Cs-131	9.7 days	Gamma	Prostate brachytherapy
F-18	110 min	Positron	Positron emission tomography (PET)
Ga-67	78 h	Gamma	Abdominal imaging; tumor detection
Ga-68	68 min	Positron	Detection of pancreatic cancer
I-123	13.2 h	Gamma	Treatment of thyroid, brain, and prostate cancer
I-131	8.0 days	Beta	Treatment of Graves' disease, goiter, hyperthyroidism, thyroid and prostate cancer
Ir-192	74 days	Gamma	Treatment of breast and prostate cancer
P-32	14.3 days	Beta	Treatment of leukemia, excess red blood cells, and pancreatic cancer
Pd-103	17 days	Gamma	Prostate brachytherapy
Sm-153	46 h	Beta	Treatment of bone cancer
Sr-85	65 days	Gamma	Detection of bone lesions; brain scans
Tc-99m	6.0 h	Gamma	Imaging of skeleton and heart muscle, brain, liver, heart, lungs, bone, spleen, kidney, and thyroid; most widely used radioisotope in nuclear medicine
Xe-133	5.2 days	Beta	Pulmonary function diagnosis
Y-90	2.7 days	Beta	Treatment of liver cancer

Scans with Radioisotopes

After a person receives a radioisotope, the radiation technologist determines the level and location of radioactivity emitted by the radioisotope. An apparatus called a *scanner* is used to produce an image of the organ. The scanner moves slowly across the body above the region where the organ containing the radioisotope is located. The gamma rays emitted from the radioisotope in the organ can be used to expose a photographic plate, producing a *scan* of the organ. On a scan, an area of decreased or increased radiation can indicate conditions such as a disease of the organ, a tumor, a blood clot, or edema.

A common method of determining thyroid function is the use of *radioactive iodine uptake*. Taken orally, the radioisotope iodine-131 mixes with the iodine already present in the thyroid. Twenty-four hours later, the amount of iodine taken up by the thyroid is determined. A detection tube held up to the area of the thyroid gland detects the radiation coming from the iodine-131 that has located there (see **FIGURE 16.5**).

A person with a hyperactive thyroid will have a higher than normal level of radioactive iodine, whereas a person with a hypoactive thyroid will have lower values. If a person has hyperthyroidism, treatment is begun to lower the activity of the thyroid. One treatment involves giving a therapeutic dosage of radioactive iodine, which has a higher radiation level than the diagnostic dose. The radioactive iodine goes to the thyroid where its radiation destroys some of the thyroid cells. The thyroid produces less thyroid hormone, bringing the hyperthyroid condition under control.

Positron Emission Tomography

Positron emitters with short half-lives such as carbon-11, oxygen-15, nitrogen-13, and fluorine-18 are used in an imaging method called *positron emission tomography* (PET). A positron-emitting isotope such as fluorine-18 combined with substances in the body such as glucose is used to study brain function, metabolism, and blood flow.

$$^{18}_{9}F \longrightarrow \, ^{18}_{8}O + \, ^{0}_{+1}e$$

As positrons are emitted, they combine with electrons to produce gamma rays that are detected by computerized equipment to create a three-dimensional image of the organ (see **FIGURE 16.6**).

Nonradioactive Imaging

Computed Tomography

Another imaging method used to scan organs such as the brain, lungs, and heart is *computed tomography* (CT). A computer monitors the absorption of 30 000 X-ray beams directed at successive layers of the target organ. Based on the densities of the tissues and fluids in the organ, the differences in absorption of the X-rays provide a series of images of the organ. This technique is successful in the identification of hemorrhages, tumors, and atrophy.

Magnetic Resonance Imaging

Magnetic resonance imaging (MRI) is a powerful imaging technique that does not involve nuclear radiation. It is the least invasive imaging method available. MRI is based on the absorption of energy when the protons in hydrogen atoms are excited by a strong magnetic field. The difference in energy between the two states is released, which produces the

A scanner detects radiation from a radioisotope in an organ.

A thyroid scan uses radioactive iodine-131.

FIGURE 16.5 Detecting radiation from an organ

FIGURE 16.6 These PET scans of the brain show a normal brain on the left and a brain affected by Alzheimer's disease on the right.

A CT scan shows a tumor (purple) in the brain.

An MRI scan provides images of the heart and lungs.

electromagnetic signal that the scanner detects. These signals are sent to a computer system, where a color image of the body is generated. MRI is particularly useful in obtaining images of soft tissues, which contain large amounts of hydrogen atoms.

Titanium "seeds" filled with a radioactive isotope are implanted in the body to treat cancer.

SAMPLE PROBLEM 16.9 Medical Applications of Radioactivity

TRY IT FIRST

In the treatment of abdominal carcinoma, a person is treated with "seeds" of gold-198, which is a beta emitter. Write the balanced nuclear equation for the beta decay of gold-198.

SOLUTION

ANALYZE THE PROBLEM	Given	Need	Connect
	gold-198, beta decay	balanced nuclear equation	mass number, atomic number of new nucleus

STEP 1 **Write the incomplete nuclear equation.**

$$^{198}_{79}\text{Au} \longrightarrow ? + _{-1}^{0}e$$

STEP 2 **Determine the missing mass number.** In beta decay, the mass number, 198, does not change.

$$^{198}_{79}\text{Au} \longrightarrow ^{198}? + _{-1}^{0}e$$

STEP 3 **Determine the missing atomic number.** The atomic number of the new isotope increases by one.

$$^{198}_{79}\text{Au} \longrightarrow ^{198}_{80}? + _{-1}^{0}e$$

STEP 4 **Determine the symbol of the new nucleus.** On the periodic table, mercury, Hg, has the atomic number of 80.

STEP 5 **Complete the nuclear equation.**

$$^{198}_{79}\text{Au} \longrightarrow ^{198}_{80}\text{Hg} + _{-1}^{0}e$$

SELF TEST 16.9

a. In an experimental treatment, a person is given boron-10, which is taken up by malignant tumors. When bombarded with neutrons, boron-10 decays by emitting alpha particles that destroy the surrounding tumor cells. Write the balanced nuclear equation for the reaction for this experimental procedure.

b. Fluorine-18, a source of positrons for PET scans, is synthesized by bombarding oxygen-18 with protons. Write the balanced nuclear equation for this reaction.

ANSWER

a. $^{1}_{0}n + ^{10}_{5}\text{B} \longrightarrow ^{7}_{3}\text{Li} + ^{4}_{2}\text{He}$ **b.** $^{1}_{1}\text{H} + ^{18}_{8}\text{O} \longrightarrow ^{18}_{9}\text{F} + ^{1}_{0}n$

PRACTICE PROBLEMS

Try Practice Problems 16.35 to 16.40

Chemistry Link to Health

Brachytherapy

The process called *brachytherapy*, or seed implantation, is an internal form of radiation therapy. The prefix *brachy* is from the Greek word for short distance. With internal radiation, a high dose of radiation is delivered to a cancerous area, whereas normal tissue sustains minimal damage. Because higher doses are used, fewer treatments of shorter duration are needed. Conventional external treatment delivers a lower dose per treatment, but requires six to eight weeks of treatment.

Permanent Brachytherapy

One of the most common forms of cancer in males is prostate cancer. In addition to surgery and chemotherapy, one treatment option is to place 40 or more titanium capsules, or "seeds," in the malignant area. Each seed, which is the size of a grain of rice, contains radioactive iodine-125, palladium-103, or cesium-131, which decay by gamma emission. The radiation from the seeds destroys the cancer by interfering with the reproduction of cancer cells with minimal damage to adjacent normal tissues. Ninety percent (90%) of the radioisotopes decay within a few months because they have short half-lives.

Isotope	I-125	Pd-103	Cs-131
Radiation	Gamma	Gamma	Gamma
Half-Life	60 days	17 days	10 days
Time Required to Deliver 90% of Radiation	7 months	2 months	1 month

Almost no radiation passes out of the patient's body. The amount of radiation received by a family member is no greater than that received on a long plane flight. Because the radioisotopes decay to products that are not radioactive, the inert titanium capsules can be left in the body.

Temporary Brachytherapy
In another type of treatment for prostate cancer, long needles containing iridium-192 are placed in the tumor. However, the needles are removed after 5 to 10 min, depending on the activity of the iridium isotope. Compared to permanent brachytherapy, temporary brachytherapy can deliver a higher dose of radiation over a shorter time. The procedure may be repeated in a few days.

Brachytherapy is also used following breast cancer lumpectomy. An iridium-192 isotope is inserted into the catheter implanted in the space left by the removal of the tumor. Radiation is delivered primarily to the tissue surrounding the cavity that contained the tumor and where the cancer is most likely to recur. The procedure is repeated twice a day for five days to give an absorbed dose of 34 Gy (3400 rad). The catheter is removed, and no radioactive material remains in the body.

In conventional external beam therapy for breast cancer, a patient is given 2 Gy once a day for six to seven weeks, which gives a total absorbed dose of about 80 Gy or 8000 rad. The external beam therapy irradiates the entire breast, including the tumor cavity.

A catheter is placed temporarily in the breast for radiation from Ir-192.

PRACTICE PROBLEMS

16.5 Medical Applications Using Radioactivity

Applications

16.35 Bone and bony structures contain calcium and phosphorus.
 a. Why would the radioisotopes calcium-47 and phosphorus-32 be used in the diagnosis and treatment of bone diseases?
 b. During nuclear tests, scientists were concerned that strontium-85, a radioactive product, would be harmful to the growth of bone in children. Explain.

16.36 a. Technetium-99m emits only gamma radiation. Why would this type of radiation be used in diagnostic imaging rather than an isotope that also emits beta or alpha radiation?
 b. A person with *polycythemia vera* (excess production of red blood cells) receives radioactive phosphorus-32. Why would this treatment reduce the production of red blood cells in the bone marrow of the patient?

16.37 In a diagnostic test for leukemia, a person receives 4.0 mL of a solution containing selenium-75. If the activity of the selenium-75 is 45 μCi/mL, what dose, in microcuries, does the patient receive?

16.38 A vial contains radioactive iodine-131 with an activity of 2.0 mCi/mL. If a thyroid test requires 3.0 mCi in an "atomic cocktail," how many milliliters are used to prepare the iodine-131 solution?

16.39 Gallium-68 is taken up by tumors; the emission of positrons allows the tumors to be located.
 a. Write an equation for the positron emission of Ga-68.
 b. If the half-life is 68 min, how much of a 64-mcg sample is active after 136 min?

16.40 Xenon-133 is used to test lung function; it decays by emitting a beta particle.
 a. Write an equation for the beta decay of Xe-133.
 b. If the half-life of Xe-133 is 5.2 h, how much of a 20.-mCi sample is still active after 15.6 h?

16.6 Nuclear Fission and Fusion

LEARNING GOAL Describe the processes of nuclear fission and fusion.

During the 1930s, scientists bombarding uranium-235 with neutrons discovered that the U-235 nucleus splits into two smaller nuclei and produces a great amount of energy. This was the discovery of nuclear **fission**. The energy generated by splitting the atom was called *atomic energy*.

A typical equation for nuclear fission is

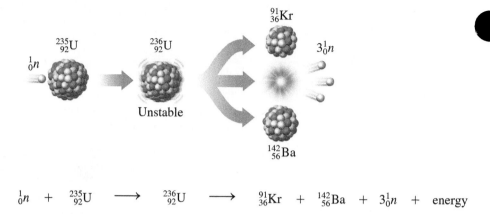

$$\,^1_0n \;+\; \,^{235}_{92}U \;\longrightarrow\; \,^{236}_{92}U \;\longrightarrow\; \,^{91}_{36}Kr \;+\; \,^{142}_{56}Ba \;+\; 3\,^1_0n \;+\; energy$$

If we could determine the mass of the products krypton, barium, and three neutrons with great accuracy, we would find that their total mass is slightly less than the mass of the starting materials. The missing mass has been converted into an enormous amount of energy, consistent with the famous equation derived by Albert Einstein:

$$E = mc^2$$

where E is the energy released, m is the mass lost, and c is the speed of light, 3×10^8 m/s. Even though the mass loss is very small, when it is multiplied by the speed of light squared, the result is a large value for the energy released. The fission of 1 g of uranium-235 produces about as much energy as the burning of 3 tons of coal.

Chain Reaction

Fission begins when a neutron collides with the nucleus of a uranium atom. The resulting nucleus is unstable and splits into smaller nuclei. This fission process also releases several neutrons and large amounts of gamma radiation and energy. The neutrons emitted have high energies and bombard other uranium-235 nuclei. In a **chain reaction**, there is a rapid increase in the number of high-energy neutrons available to react with more uranium. To sustain a nuclear chain reaction, sufficient quantities of uranium-235 must be brought together to provide a *critical mass* in which almost all the neutrons immediately collide with more uranium-235 nuclei. So much heat and energy build up that an atomic explosion can occur (see **FIGURE 16.7**).

Nuclear Fusion

In **fusion**, two small nuclei combine to form a larger nucleus. Mass is lost, and a tremendous amount of energy is released, even more than the energy released from nuclear fission. However, a fusion reaction requires a temperature of 100 000 000 °C to overcome the repulsion of the hydrogen nuclei and cause them to undergo fusion. Fusion reactions occur continuously in the Sun and other stars, providing us with heat and light. The huge amounts of energy produced by our sun come from the fusion of 6×10^{11} kg of hydrogen every second. In a fusion reaction, isotopes of hydrogen combine to form helium and large amounts of energy.

$$\,^3_1H \;+\; \,^2_1H \;\longrightarrow\; \,^4_2He \;+\; \,^1_0n \;+\; energy$$

Hydrogen isotopes combine in a fusion reaction to produce helium, a neutron, and energy.

ENGAGE 16.9

Why is a critical mass of uranium-235 necessary to sustain a nuclear chain reaction?

PRACTICE PROBLEMS

Try Practice Problems 16.41 to 16.44

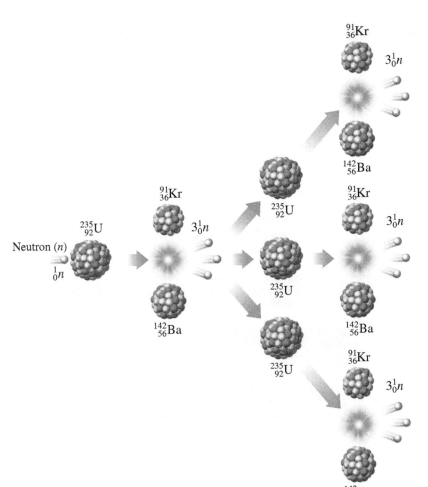

FIGURE 16.7 In a nuclear chain reaction, the fission of each uranium-235 atom produces three neutrons that cause the nuclear fission of more and more uranium-235 atoms.

Scientists expect less radioactive waste with shorter half-lives from fusion reactors. However, fusion is still in the experimental stage because the extremely high temperatures needed have been difficult to reach and even more difficult to maintain.

SAMPLE PROBLEM 16.10   Identifying Fission and Fusion

TRY IT FIRST

Indicate whether each of the following is characteristic of the fission or fusion process, or both:
a. A large nucleus breaks apart to produce smaller nuclei.
b. Large amounts of energy are released.
c. Extremely high temperatures are needed for reaction.

SOLUTION

a. When a large nucleus breaks apart to produce smaller nuclei, the process is fission.
b. Large amounts of energy are released in both the fusion and fission processes.
c. An extremely high temperature is required for fusion.

SELF TEST 16.10

Classify each of the following nuclear equations as fission or fusion:

a. $^{3}_{2}He + ^{3}_{2}He \longrightarrow ^{4}_{2}He + 2^{1}_{1}H$ **b.** $^{1}_{0}n + ^{235}_{92}U \longrightarrow ^{140}_{54}Xe + ^{94}_{38}Sr + 2^{1}_{0}n$

ANSWER

a. fusion   **b.** fission

PRACTICE PROBLEMS

Try Practice Problems 16.45 and 16.46

PRACTICE PROBLEMS

16.6 Nuclear Fission and Fusion

16.41 What is nuclear fission?

16.42 How does a chain reaction occur in nuclear fission?

16.43 Complete the following fission reaction:

$$_{0}^{1}n + _{92}^{235}U \longrightarrow _{50}^{131}Sn + ? + 2_{0}^{1}n + energy$$

16.44 In another fission reaction, uranium-235 bombarded with a neutron produces strontium-94, another smaller nucleus, and three neutrons. Write the balanced nuclear equation for the fission reaction.

16.45 Indicate whether each of the following is characteristic of the fission or fusion process, or both:

a. Neutrons bombard a nucleus.
b. The nuclear process occurs in the Sun.
c. A large nucleus splits into smaller nuclei.
d. Small nuclei combine to form larger nuclei.

16.46 Indicate whether each of the following is characteristic of the fission or fusion process, or both:

a. Very high temperatures are required to initiate the reaction.
b. Less radioactive waste is produced.
c. Hydrogen nuclei are the reactants.
d. Large amounts of energy are released when the nuclear reaction occurs.

UPDATE   Cardiac Imaging Using a Radioisotope

As part of her nuclear stress test, Simone starts to walk on the treadmill. When she reaches the maximum level, Pauline injects a radioactive isotope containing Tl-201 with an activity of 74 MBq. The radiation emitted from areas of the heart is detected by a scanner and produces images of her heart muscle. A thallium stress test can determine how effectively coronary arteries provide blood to the heart. If Simone has any damage to her coronary arteries, reduced blood flow during stress would show a narrowing of an artery or a blockage. After Simone rests for 3 h, Pauline injects more Tl-201, and she is placed under the scanner again. A second set of images of her heart muscle at rest are taken. When Simone's doctor reviews her scans, he assures her that she had normal blood flow to her heart muscle, both at rest and under stress.

A thallium stress test can show narrowing of an artery during stress.

Applications

16.47 What is the activity of the radioactive isotope injection for Simone
a. in curies?   **b.** in millicuries?

16.48 If the half-life of Tl-201 is 3.0 days, what is its activity, in megabecquerels
a. after 3.0 days?   **b.** after 6.0 days?

16.49 How many days will it take until the activity of the Tl-201 in Simone's body is one-eighth of the initial activity?

16.50 Radiation from Tl-201 to Simone's kidneys can be 24 mGy. What is this amount of radiation in rads?

CONCEPT MAP

CHAPTER REVIEW

16.1 Natural Radioactivity

LEARNING GOAL Describe alpha, beta, positron, and gamma radiation.

$^4_2$He or α

Alpha particle

- Radioactive isotopes have unstable nuclei that break down (decay), spontaneously emitting alpha (α), beta (β), positron (β^+), and gamma (γ) radiation.
- Because radiation can damage the cells in the body, shielding is used to protect from radiation.

16.2 Nuclear Reactions

LEARNING GOAL Write a balanced nuclear equation for radioactive decay, showing mass numbers and atomic numbers.

- A balanced nuclear equation is used to represent the changes that take place in the nuclei of the reactants and products.
- The new isotopes and the type of radiation emitted can be determined from the symbols that show the mass numbers and atomic numbers of the isotopes in the nuclear equation.
- A radioisotope is produced artificially when a nonradioactive isotope is bombarded by a small particle.

16.3 Radiation Measurement

LEARNING GOAL Describe how radiation is detected and measured.

- In a Geiger counter, radiation produces charged particles in the gas contained in a tube, which generates an electrical current.
- The curie (Ci) and the becquerel (Bq) measure the activity, which is the number of nuclear disintegrations per second.
- The amount of radiation absorbed by a substance is measured in the rad or the gray (Gy).
- The rem and the sievert (Sv) are units used to measure the biological damage from the different types of radiation.

16.4 Half-Life of a Radioisotope

LEARNING GOAL Given the half-life of a radioisotope, calculate the amount of radioisotope remaining after one or more half-lives.

- Every radioisotope has its own rate of emitting radiation.
- The time it takes for one-half of a radioactive sample to decay is called its half-life.
- For many medical radioisotopes, such as Tc-99m and I-131, half-lives are short.
- For other isotopes, usually naturally occurring ones such as C-14, Ra-226, and U-238, half-lives are extremely long.

16.5 Medical Applications Using Radioactivity

LEARNING GOAL Describe the use of radioisotopes in medicine.

- In nuclear medicine, radioisotopes that go to specific sites in the body are given to the patient.
- By detecting the radiation they emit, an evaluation can be made about the location and extent of an injury, disease, tumor, or the level of function of a particular organ.
- Higher levels of radiation are used to treat or destroy tumors.

16.6 Nuclear Fission and Fusion

LEARNING GOAL Describe the processes of nuclear fission and fusion.

- In fission, the bombardment of a large nucleus breaks it apart into smaller nuclei, releasing one or more types of radiation and a great amount of energy.
- In fusion, small nuclei combine to form larger nuclei while great amounts of energy are released.

KEY TERMS

alpha particle A nuclear particle identical to a helium nucleus, symbol α or $^4_2$He.

becquerel (Bq) A unit of activity of a radioactive sample equal to one disintegration per second.

beta particle A particle identical to an electron, symbol $^0_{-1}e$ or β, that forms in the nucleus when a neutron changes to a proton and an electron.

chain reaction A fission reaction that will continue once it has been initiated by a high-energy neutron bombarding a heavy nucleus such as uranium-235.

curie (Ci) A unit of activity of a radioactive sample equal to 3.7×10^{10} disintegrations/s.

decay curve A diagram of the decay of a radioactive element.

equivalent dose The measure of biological damage from an absorbed dose that has been adjusted for the type of radiation.

fission A process in which large nuclei are split into smaller nuclei, releasing large amounts of energy.

fusion A reaction in which large amounts of energy are released when small nuclei combine to form larger nuclei.

gamma ray High-energy radiation, symbol $^0_0\gamma$, emitted by an unstable nucleus.

gray (Gy) A unit of absorbed dose equal to 100 rad.

half-life The length of time it takes for one-half of a radioactive sample to decay.

positron A particle of radiation with no mass and a positive charge, symbol β^+ or $^0_{+1}e$, produced when a proton is transformed into a neutron and a positron.

rad (radiation absorbed dose) A measure of an amount of radiation absorbed by the body.

radiation Energy or particles released by radioactive atoms.

radioactive decay The process by which an unstable nucleus breaks down with the release of high-energy radiation.

radioisotope A radioactive atom of an element.

rem (radiation equivalent in humans) A measure of the biological damage caused by radiation (rad × radiation biological factor).

sievert (Sv) A unit of biological damage (equivalent dose) equal to 100 rem.

CORE CHEMISTRY SKILLS

The chapter section containing each Core Chemistry Skill is shown in parentheses at the end of each heading.

Writing Nuclear Equations (16.2)

- A nuclear equation is written with the atomic symbols of the original radioactive nucleus on the left, an arrow, and the new nucleus and the type of radiation emitted on the right.
- The sum of the mass numbers and the sum of the atomic numbers on one side of the arrow must equal the sum of the mass numbers and the sum of the atomic numbers on the other side.
- When an alpha particle is emitted, the mass number of the new nucleus decreases by 4, and its atomic number decreases by 2.
- When a beta particle is emitted, there is no change in the mass number of the new nucleus, but its atomic number increases by one.
- When a positron is emitted, there is no change in the mass number of the new nucleus, but its atomic number decreases by one.
- In gamma emission, there is no change in the mass number or the atomic number of the new nucleus.

Example: **a.** Write a balanced nuclear equation for the alpha decay of Po-210.

b. Write a balanced nuclear equation for the beta decay of Co-60.

Answer: **a.** When an alpha particle is emitted, we calculate the decrease of 4 in the mass number (210) of the polonium, and a decrease of 2 in its atomic number.

$$^{210}_{84}\text{Po} \longrightarrow ^{206}_{82}? + ^{4}_{2}\text{He}$$

Because lead has atomic number 82, the new nucleus must be an isotope of lead.

$$^{210}_{84}\text{Po} \longrightarrow ^{206}_{82}\text{Pb} + ^{4}_{2}\text{He}$$

b. When a beta particle is emitted, there is no change in the mass number (60) of the cobalt, but there is an increase of 1 in its atomic number.

$$^{60}_{27}\text{Co} \longrightarrow ^{60}_{28}? + ^{0}_{-1}e$$

Because nickel has atomic number 28, the new nucleus must be an isotope of nickel.

$$^{60}_{27}\text{Co} \longrightarrow ^{60}_{28}\text{Ni} + ^{0}_{-1}e$$

Using Half-Lives (16.4)

- The half-life of a radioisotope is the amount of time it takes for one-half of a sample to decay.
- The remaining amount of a radioisotope is calculated by dividing its quantity or activity by one-half for each half-life that has elapsed.

Example: Co-60 has a half-life of 5.3 yr. If the initial sample of Co-60 has an activity of 1200 mCi, what is its activity after 15.9 yr?

Answer:

$$\text{number of half-lives} = 15.9 \, \cancel{yr} \times \frac{1 \text{ half-life}}{5.3 \, \cancel{yr}} = 3.0 \text{ half-lives}$$

$$1200 \text{ mCi} \xrightarrow{1 \text{ half-life}} 600 \text{ mCi} \xrightarrow{2 \text{ half-lives}} 300 \text{ mCi} \xrightarrow{3 \text{ half-lives}} 150 \text{ mCi}$$

In 15.9 yr, three half-lives have passed. The activity was reduced from 1200 mCi to 150 mCi.

UNDERSTANDING THE CONCEPTS

The chapter sections to review are shown in parentheses at the end of each problem.

In problems 16.51 to 16.54, a nucleus is shown with protons and neutrons.

● proton   ○ neutron

16.51 Draw the new nucleus when this isotope emits a positron to complete the following: (16.2)

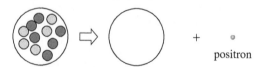

positron

16.52 Draw the nucleus that emits a beta particle to complete the following: (16.2)

beta particle

16.53 Draw the nucleus of the isotope that is bombarded in the following: (16.2)

16.54 Complete the bombardment reaction by drawing the nucleus of the new isotope that is produced in the following: (16.2)

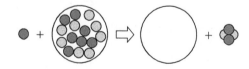

16.55 Carbon dating of small bits of charcoal used in cave paintings has determined that some of the paintings are from 10 000 to 30 000 yr old. Carbon-14 has a half-life of 5730 yr. In a 1-μg sample of carbon from a live tree, the activity of carbon-14 is 6.4 μCi. If researchers determine that 1 μg of charcoal from a prehistoric cave painting in France has an activity of 0.80 μCi, what is the age of the painting? (16.4)

The technique of carbon dating is used to determine the age of ancient cave paintings.

16.56 Use the following decay curve for iodine-131 to answer problems **a** to **c**: (16.4)

    a. Complete the values for the mass of radioactive iodine-131 on the vertical axis.

    b. Complete the number of days on the horizontal axis.

    c. What is the half-life, in days, of iodine-131?

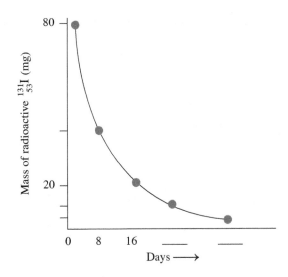

ADDITIONAL PRACTICE PROBLEMS

16.57 Determine the number of protons and number of neutrons in the nucleus of each of the following: (16.1)
    a. sodium-25   **b.** nickel-61
    c. rubidium-84   **d.** silver-110

16.58 Determine the number of protons and number of neutrons in the nucleus of each of the following: (16.1)
    a. boron-10   **b.** zinc-72
    c. iron-59   **d.** gold-198

16.59 Identify each of the following as alpha decay, beta decay, positron emission, or gamma emission: (16.2)
    a. $^{27m}_{13}Al \longrightarrow ^{27}_{13}Al + ^{0}_{0}\gamma$ **b.** $^{8}_{5}B \longrightarrow ^{8}_{4}Be + ^{0}_{+1}e$
    c. $^{220}_{86}Rn \longrightarrow ^{216}_{84}Po + ^{4}_{2}He$

16.60 Identify each of the following as alpha decay, beta decay, positron emission, or gamma emission: (16.2)
    a. $^{127}_{55}Cs \longrightarrow ^{127}_{54}Xe + ^{0}_{+1}e$ **b.** $^{90}_{38}Sr \longrightarrow ^{90}_{39}Y + ^{0}_{-1}e$
    c. $^{218}_{85}At \longrightarrow ^{214}_{83}Bi + ^{4}_{2}He$

16.61 Write the balanced nuclear equation for each of the following: (16.2)
    a. Th-225 (α decay)   **b.** Bi-210 (α decay)
    c. cesium-137 (β decay)   **d.** tin-126 (β decay)
    e. F-18 (β^+ emission)

16.62 Write the balanced nuclear equation for each of the following: (16.2)
    a. potassium-40 (β decay)   **b.** sulfur-35 (β decay)
    c. platinum-190 (α decay)   **d.** Ra-210 (α decay)
    e. In-113m (γ emission)

16.63 Complete each of the following nuclear equations: (16.2)
    a. $^{4}_{2}He + ^{14}_{7}N \longrightarrow ? + ^{1}_{1}H$
    b. $^{4}_{2}He + ^{27}_{13}Al \longrightarrow ^{30}_{14}Si + ?$
    c. $^{1}_{0}n + ^{235}_{92}U \longrightarrow ^{90}_{38}Sr + 3^{1}_{0}n + ?$
    d. $^{23m}_{12}Mg \longrightarrow ? + ^{0}_{0}\gamma$

16.64 Complete each of the following nuclear equations: (16.2)
    a. $? + ^{59}_{27}Co \longrightarrow ^{56}_{25}Mn + ^{4}_{2}He$
    b. $? \longrightarrow ^{14}_{7}N + ^{0}_{-1}e$
    c. $^{0}_{-1}e + ^{76}_{36}Kr \longrightarrow ?$
    d. $^{4}_{2}He + ^{241}_{95}Am \longrightarrow ? + 2^{1}_{0}n$

16.65 Write the balanced nuclear equation for each of the following: (16.2)
    a. When two oxygen-16 atoms collide, one of the products is an alpha particle.
    b. When californium-249 is bombarded by oxygen-18, a new element, seaborgium-263, and four neutrons are produced.
    c. Radon-222 undergoes alpha decay.
    d. An atom of strontium-80 emits a positron.

16.66 Write the balanced nuclear equation for each of the following: (16.2)
    a. Actinium-225 decays to give francium-221.
    b. Bismuth-211 emits an alpha particle.
    c. A radioisotope emits a positron to form titanium-48.
    d. An atom of germanium-69 emits a positron.

16.67 A 120-mg sample of technetium-99m is used for a diagnostic test. If technetium-99m has a half-life of 6.0 h, how many milligrams of the technetium-99m sample remains active 24 h after the test? (16.4)

16.68 The half-life of oxygen-15 is 124 s. If a sample of oxygen-15 has an activity of 4000 Bq, how many minutes will elapse before it has an activity of 500 Bq? (16.4)

16.69 What is the difference between fission and fusion? (16.6)

16.70 What are the products in the fission of uranium-235 that make possible a nuclear chain reaction? (16.6)

16.71 Where does fusion occur naturally? (16.6)

16.72 Why are scientists continuing to try to build a fusion reactor even though the very high temperatures it requires have been difficult to reach and maintain? (16.6)

Applications

16.73 The activity of K-40 in a 70.-kg human body is estimated to be 120 nCi. What is this activity in becquerels? (16.3)

16.74 The activity of C-14 in a 70.-kg human body is estimated to be 3.7 kBq. What is this activity in microcuries? (16.3)

16.75 If the amount of radioactive phosphorus-32, used to treat leukemia, in a sample decreases from 1.2 mg to 0.30 mg in 28.6 days, what is the half-life of phosphorus-32? (16.4)

16.76 If the amount of radioactive iodine-123, used to treat thyroid cancer, in a sample decreases from 0.4 mg to 0.1 mg in 26.4 h, what is the half-life of iodine-123? (16.4)

16.77 Calcium-47, used to evaluate disorders in calcium metabolism, has a half-life of 4.5 days. (16.2, 16.4)
   a. Write the balanced nuclear equation for the beta decay of calcium-47.
   b. How many milligrams of a 16-mg sample of calcium-47 remain after 18 days?
   c. How many days have passed if 4.8 mg of calcium-47 decayed to 1.2 mg of calcium-47?

16.78 Cesium-137, used in cancer treatment, has a half-life of 30 yr. (16.2, 16.4)
   a. Write the balanced nuclear equation for the beta decay of cesium-137.
   b. How many milligrams of a 16-mg sample of cesium-137 remain after 90 yr?
   c. How many years are required for 28 mg of cesium-137 to decay to 3.5 mg of cesium-137?

CHALLENGE PROBLEMS

The following problems are related to the topics in this chapter. However, they do not all follow the chapter order, and they require you to combine concepts and skills from several sections. These problems will help you increase your critical thinking skills and prepare for your next exam.

16.79 Write the balanced nuclear equation for each of the following radioactive emissions: (16.2)
   a. an alpha particle from Hg-180
   b. a beta particle from Au-198
   c. a positron from Rb-82

16.80 Write the balanced nuclear equation for each of the following radioactive emissions: (16.2)
   a. an alpha particle from Gd-148
   b. a beta particle from Sr-90
   c. a positron from Al-25

16.81 All the elements beyond uranium, the transuranium elements, have been prepared by bombardment and are not naturally occurring elements. The first transuranium element neptunium, Np, was prepared by bombarding U-238 with neutrons to form a neptunium atom and a beta particle. Complete the following equation: (16.2)

$$ {}^{1}_{0}n + {}^{238}_{92}U \longrightarrow \text{?} + \text{?} $$

16.82 One of the most recently produced transuranium elements, oganesson-294 (Og-294), atomic number 118, was prepared by bombarding californium-249 with another isotope. Complete the following equation for the preparation of this new element: (16.2)

$$ \text{?} + {}^{249}_{98}Cf \longrightarrow {}^{294}_{118}Og + 3{}^{1}_{0}n $$

16.83 A 64-μCi sample of Tl-201 decays to 4.0 μCi in 12 days. What is the half-life, in days, of Tl-201? (16.3, 16.4)

16.84 A wooden object from the site of an ancient temple has a carbon-14 activity of 10 counts/min compared with a reference piece of wood cut today that has an activity of 40 counts/min. If the half-life for carbon-14 is 5730 yr, what is the age of the ancient wood object? (16.3, 16.4)

16.85 Element 114 was recently named flerovium, symbol Fl. The reaction for its synthesis involves bombarding Pu-244 with Ca-48. Write the balanced nuclear equation for the synthesis of flerovium. (16.2)

16.86 Element 116 was recently named livermorium, symbol Lv. The reaction for its synthesis involves bombarding Cm-248 with Ca-48. Write the balanced nuclear equation for the synthesis of livermorium. (16.2)

Applications

16.87 The half-life for the radioactive decay of calcium-47 is 4.5 days. If a sample has an activity of 1.0 μCi after 27 days, what was the initial activity, in microcuries, of the sample? (16.3, 16.4)

16.88 The half-life for the radioactive decay of Ce-141 is 32.5 days. If a sample has an activity of 4.0 μCi after 130. days have elapsed, what was the initial activity, in microcuries, of the sample? (16.3, 16.4)

16.89 A nuclear technician was accidentally exposed to potassium-42 while doing brain scans for possible tumors. The error was not discovered until 36 h later when the activity of the potassium-42 sample was 2.0 μCi. If potassium-42 has a half-life of 12 h, what was the activity of the sample at the time the technician was exposed? (16.3, 16.4)

16.90 The radioisotope sodium-24 is used to determine the levels of electrolytes in the body. A 16-μg sample of sodium-24 decays to 2.0 μg in 45 h. What is the half-life, in hours, of sodium-24? (16.4)

ANSWERS TO ENGAGE QUESTIONS

16.1 An alpha particle has a charge of 2+ and a mass number of 4.

16.2 Lead shielding will block alpha, beta, and gamma radiation.

16.3 When an alpha particle is emitted by a U-238 nucleus, a thorium-234 nucleus is formed.

16.4 When a beta particle is emitted by a C-14 nucleus, a nitrogen-14 nucleus is formed.

16.5 When a positron is emitted, a proton in the nucleus is changed to a neutron. The atomic number decreases but the mass number, the total number of protons plus neutrons, stays the same.

16.6 The becquerel measures the number of disintegrations/s, whereas the rem measures the biological effect of radiation.

16.7 After 18 h, the Tc-99m sample has had 3.0 half-lives. The amount remaining is 1/8 ($\frac{1}{2} \times \frac{1}{2} \times \frac{1}{2}$) of the original 24 mg, or 3.0 mg.

16.8 Radioisotopes used in nuclear medicine have short half-lives, which allows them to be eliminated quickly from the body.

16.9 If there is less than a critical mass of U-235 available, too many neutrons escape, and a chain reaction cannot be sustained.

ANSWERS TO SELECTED PROBLEMS

16.1   **a.** alpha particle   **b.** positron   **c.** gamma radiation

16.3   **a.** $^{39}_{19}K$ $^{40}_{19}K$ $^{41}_{19}K$

       b. They all have 19 protons and 19 electrons, but they differ in the number of neutrons.

16.5   **a.** $^{0}_{-1}e$ or β **b.** $^{4}_{2}He$ or α **c.** $^{1}_{0}n$ or n

       d. $^{38}_{18}Ar$ **e.** $^{14}_{6}C$

16.7   **a.** $^{64}_{29}Cu$ **b.** $^{75}_{34}Se$ **c.** $^{24}_{11}Na$ **d.** $^{15}_{7}N$

16.9

Medical Use	Atomic Symbol	Mass Number	Number of Protons	Number of Neutrons
Heart imaging	$^{201}_{81}Tl$	201	81	120
Radiation therapy	$^{60}_{27}Co$	60	27	33
Abdominal scan	$^{67}_{31}Ga$	67	31	36
Hyperthyroidism	$^{131}_{53}I$	131	53	78
Leukemia treatment	$^{32}_{15}P$	32	15	17

16.11   **a.** 1, alpha particle   **b.** 3, gamma radiation

        c. 1, alpha particle

16.13   **a.** $^{208}_{84}Po \longrightarrow ^{204}_{82}Pb + ^{4}_{2}He$ **b.** $^{232}_{90}Th \longrightarrow ^{228}_{88}Ra + ^{4}_{2}He$

        c. $^{251}_{102}No \longrightarrow ^{247}_{100}Fm + ^{4}_{2}He$ **d.** $^{220}_{86}Rn \longrightarrow ^{216}_{84}Po + ^{4}_{2}He$

16.15   **a.** $^{25}_{11}Na \longrightarrow ^{25}_{12}Mg + ^{0}_{-1}e$ **b.** $^{20}_{8}O \longrightarrow ^{20}_{9}F + ^{0}_{-1}e$

        c. $^{92}_{38}Sr \longrightarrow ^{92}_{39}Y + ^{0}_{-1}e$ **d.** $^{60}_{26}Fe \longrightarrow ^{60}_{27}Co + ^{0}_{-1}e$

16.17   **a.** $^{26}_{14}Si \longrightarrow ^{26}_{13}Al + ^{0}_{+1}e$ **b.** $^{54}_{27}Co \longrightarrow ^{54}_{26}Fe + ^{0}_{+1}e$

        c. $^{77}_{37}Rb \longrightarrow ^{77}_{36}Kr + ^{0}_{+1}e$ **d.** $^{93}_{45}Rh \longrightarrow ^{93}_{44}Ru + ^{0}_{+1}e$

16.19   **a.** $^{28}_{14}Si$, beta decay   **b.** $^{0}_{0}\gamma$, gamma emission

        c. $^{0}_{-1}e$, beta decay   **d.** $^{238}_{92}U$, alpha decay

        e. $^{188}_{79}Au$, positron emission

16.21   **a.** $^{10}_{4}Be$ **b.** $^{132}_{53}I$ **c.** $^{27}_{13}Al$ **d.** $^{17}_{8}O$

16.23   **a.** 2, absorbed dose   **b.** 3, biological damage

        c. 1, activity   **d.** 2, absorbed dose

16.25   The technician exposed to 5 rad received more radiation.

16.27   **a.** 294 μCi   **b.** 0.5 Gy

16.29   **a.** two half-lives   **b.** one half-life   **c.** three half-lives

16.31   **a.** 40.0 mg   **b.** 20.0 mg   **c.** 10.0 mg   **d.** 1.25 mg

16.33   **a.** 130 days   **b.** 195 days

16.35   **a.** The radioactive isotopes of Ca and P will become part of the bony structures of the body, where their radiation can be used to diagnose or treat bone diseases.

        b. The body will accumulate radioactive strontium in bones in the same way that it incorporates calcium. Radioactive strontium is harmful to children because its radiation causes more damage in cells that are dividing rapidly.

16.37   180 μCi

16.39   **a.** $^{68}_{31}Ga \longrightarrow ^{68}_{30}Mn + ^{0}_{+1}e$ **b.** 16 mcg

16.41   Nuclear fission is the splitting of a large atom into smaller fragments with the release of a large amount of energy.

16.43 $^{103}_{42}Mo$

16.45   **a.** fission   **b.** fusion   **c.** fission   **d.** fusion

16.47   **a.** 2.0×10^{-3} Ci   **b.** 2.0 mCi

16.49   9.0 days

16.51

16.53

16.55   17 000 yr old

16.57   **a.** 11 protons and 14 neutrons

        b. 28 protons and 33 neutrons

        c. 37 protons and 47 neutrons

        d. 47 protons and 63 neutrons

16.59   **a.** gamma emission   **b.** positron emission

        c. alpha decay

16.61   **a.** $^{225}_{90}Th \longrightarrow ^{221}_{88}Ra + ^{4}_{2}He$ **b.** $^{210}_{83}Bi \longrightarrow ^{206}_{81}Tl + ^{4}_{2}He$

        c. $^{137}_{55}Cs \longrightarrow ^{137}_{56}Ba + ^{0}_{-1}e$ **d.** $^{126}_{50}Sn \longrightarrow ^{126}_{51}Sb + ^{0}_{-1}e$

        e. $^{18}_{9}F \longrightarrow ^{18}_{8}O + ^{0}_{+1}e$

16.63   **a.** $^{17}_{8}O$ **b.** $^{1}_{1}H$ **c.** $^{143}_{54}Xe$ **d.** $^{23}_{12}Mg$

16.65   **a.** $^{16}_{8}O + ^{16}_{8}O \longrightarrow ^{28}_{14}Si + ^{4}_{2}He$

        b. $^{18}_{8}O + ^{249}_{98}Cf \longrightarrow ^{263}_{106}Sg + 4^{1}_{0}n$

        c. $^{222}_{86}Rn \longrightarrow ^{218}_{84}PO + ^{4}_{2}He$

        d. $^{80}_{38}Sr \longrightarrow ^{80}_{37}Rb + ^{0}_{+1}e$

16.67   7.5 mg of Tc-99m

16.69   In the fission process, an atom splits into smaller nuclei. In fusion, small nuclei combine (fuse) to form a larger nucleus.

16.71   Fusion occurs naturally in the Sun and other stars.

16.73 4.4×10^{3} Bq

16.75   14.3 days

16.77   **a.** $^{47}_{20}Ca \longrightarrow ^{47}_{21}Sc + ^{0}_{-1}e$ **b.** 1.0 mg of Ca-47

        c. 9.0 days

16.79   **a.** $^{180}_{80}Hg \longrightarrow ^{176}_{78}Pt + ^{4}_{2}He$ **b.** $^{198}_{79}Au \longrightarrow ^{198}_{80}Hg + ^{0}_{-1}e$

        c. $^{82}_{37}Rb \longrightarrow ^{82}_{36}Kr + ^{0}_{+1}e$

16.81 $^{1}_{0}n + ^{238}_{92}U \longrightarrow ^{239}_{93}Np + ^{0}_{-1}e$

16.83   3.0 days

16.85 $^{48}_{20}Ca + ^{244}_{94}Pu \longrightarrow ^{292}_{114}Fl$

16.87   64 μCi

16.89   16 μCi

CI.33 Consider the reaction of sodium oxalate ($Na_2C_2O_4$) and potassium permanganate ($KMnO_4$) in acidic solution. The unbalanced equation is the following: (9.2, 9.3, 12.4, 12.5, 15.2)

$$MnO_4^-(aq) + C_2O_4^{2-}(aq) \longrightarrow Mn^{2+}(aq) + CO_2(g)$$

In an oxidation–reduction titration, the $KMnO_4$ from the buret reacts with $Na_2C_2O_4$.

a. What is the balanced oxidation half-reaction?
b. What is the balanced reduction half-reaction?
c. What is the balanced ionic equation for the reaction?
d. If 24.6 mL of a $KMnO_4$ solution is needed to titrate a solution containing 0.758 g of sodium oxalate ($Na_2C_2O_4$), what is the molarity of the $KMnO_4$ solution?

CI.34 A strip of magnesium metal dissolves rapidly in 6.00 mL of a 0.150 M HCl solution, producing hydrogen gas and magnesium chloride. (9.2, 9.3, 12.4, 12.6, 14.6, 15.1, 15.2)

$$HCl(aq) + Mg(s) \longrightarrow H_2(g) + MgCl_2(aq) \quad \text{Unbalanced}$$

Magnesium metal reacts vigorously with hydrochloric acid.

a. Assign oxidation numbers to all of the elements in the reactants and products.
b. What is the balanced chemical equation for the reaction?
c. What is the oxidizing agent?
d. What is the reducing agent?
e. What is the pH of the 0.150 M HCl solution?
f. How many grams of magnesium can dissolve in the HCl solution?

CI.35 A piece of magnesium with a mass of 0.121 g is added to 50.0 mL of a 1.00 M HCl solution at a temperature of 22 °C. When the magnesium dissolves, the solution reaches a temperature of 33 °C. For the equation, see your answer to problem CI.34b. (9.2, 9.3, 9.4, 9.6, 12.4, 12.6)

a. What is the limiting reactant?
b. What volume, in milliliters, of hydrogen gas would be produced at 33 °C when the pressure is 750. mmHg?

c. How many joules were released by the reaction of the magnesium? Assume the density of the HCl solution is 1.00 g/mL and the specific heat of the HCl solution is the same as that of water.
d. What is the heat of reaction for magnesium in joules/gram? in kilojoules/mol?

CI.36 The iceman known as Ötzi was discovered in a high mountain pass on the Austrian–Italian border. Samples of his hair and bones had carbon-14 activity that was 50% of that present in new hair or bone. Carbon-14 undergoes beta decay and has a half-life of 5730 yr. (16.2, 16.4)

The mummified remains of Ötzi were discovered in 1991.

a. How long ago did Ötzi live?
b. Write a balanced nuclear equation for the decay of carbon-14.

CI.37 Some of the isotopes of silicon are listed in the following table: (4.4, 5.4, 10.1, 10.3, 16.2, 16.4)

Isotope	% Natural Abundance	Atomic Mass	Half-Life (radioactive)	Radiation
$^{27}_{14}Si$		26.987	4.9 s	Positron
$^{28}_{14}Si$	92.230	27.977	Stable	None
$^{29}_{14}Si$	4.683	28.976	Stable	None
$^{30}_{14}Si$	3.087	29.974	Stable	None
$^{31}_{14}Si$		30.975	2.6 h	Beta

a. In the following table, indicate the number of protons, neutrons, and electrons for each isotope listed:

Isotope	Number of Protons	Number of Neutrons	Number of Electrons
$^{27}_{14}Si$			
$^{28}_{14}Si$			
$^{29}_{14}Si$			
$^{30}_{14}Si$			
$^{31}_{14}Si$			

b. What are the electron configuration and the abbreviated electron configuration of silicon?
c. Calculate the atomic mass for silicon using the isotopes that have a natural abundance.
d. Write the balanced nuclear equations for the positron emission of Si-27 and the beta decay of Si-31.
e. How many hours are needed for a sample of Si-31 with an activity of 16 μCi to decay to 2.0 μCi?
f. Draw the Lewis structures and predict the shape of $SiCl_4$.

CI.38 K^+ is an electrolyte required by the human body and found in many foods as well as salt substitutes. One of the isotopes of potassium is potassium-40, which has a natural abundance of 0.012% and a half-life of 1.3×10^9 yr. The isotope potassium-40 decays to calcium-40 or to argon-40. A typical activity for potassium-40 is 7.0 μCi per gram of potassium. (16.2, 16.3, 16.4)

Potassium chloride is used as a salt substitute.

a. Write a balanced nuclear equation for each type of decay, and identify the particle emitted.
b. A shaker of salt substitute contains 1.6 oz of K. What is the activity, in millicuries and becquerels, of the potassium-40 in the shaker?

CI.39 Uranium-238 decays in a series of nuclear changes until stable lead-206 is produced. Complete the following nuclear equations that are part of the uranium-238 decay series: (16.2, 16.3, 16.4)

a. $^{238}_{92}U \longrightarrow ^{234}_{90}Th + ?$ **b.** $^{234}_{90}Th \longrightarrow ? + ^{0}_{-1}e$

c. $? \longrightarrow ^{222}_{86}Rn + ^{4}_{2}He$

CI.40 Radon levels in a home can be measured with a home radon-detection kit. Environmental agencies have set the maximum level of radon-222 in a home at 4 picocuries per liter (pCi/L) of air. (11.7, 16.2, 16.3, 16.4)

A home detection kit is used to measure the level of radon-222.

a. Write the balanced nuclear equation for the decay of Ra-226.
b. Write the balanced nuclear equation for the decay of Rn-222.
c. If a room contains 24 000 atoms of radon-222, how many atoms of radon-222 remain after 15.2 days?
d. Suppose a room has a volume of 72 000 L (7.2×10^4 L). If the radon level is the maximum allowed (4 pCi/L), how many alpha particles are emitted from Rn-222 in one day? ($1 Ci = 3.7 \times 10^{10}$ disintegrations per second)

ANSWERS

CI.33 a. $C_2O_4^{2-}(aq) \longrightarrow 2CO_2(g) + 2 e^-$

b. $5 e^- + 8H^+(aq) + MnO_4^-(aq) \longrightarrow$
$$Mn^{2+}(aq) + 4H_2O(l)$$

c. $16H^+(aq) + 2MnO_4^-(aq) + 5C_2O_4^{2-}(aq) \longrightarrow$
$$10CO_2(g) + 2Mn^{2+}(aq) + 8H_2O(l)$$

d. 0.0920 M $KMnO_4$ solution

CI.35 a. Mg is the limiting reactant.
b. 127 mL of $H_2(g)$
c. 2.3×10^3 J
d. 1.9×10^4 J/g; 460 kJ/mol

CI.37 a.

Isotope	Number of Protons	Number of Neutrons	Number of Electrons
$^{27}_{14}Si$	14	13	14
$^{28}_{14}Si$	14	14	14
$^{29}_{14}Si$	14	15	14
$^{30}_{14}Si$	14	16	14
$^{31}_{14}Si$	14	17	14

b. $1s^22s^22p^63s^23p^2$; $[Ne]3s^23p^2$
c. 28.085 amu
d. $^{27}_{14}Si \longrightarrow ^{27}_{13}Al + ^{0}_{+1}e$
$^{31}_{14}Si \longrightarrow ^{31}_{15}P + ^{0}_{-1}e$
e. 7.8 h

f. tetrahedral

CI.39 a. $^{238}_{92}U \longrightarrow ^{234}_{90}Th + ^{4}_{2}He$
b. $^{234}_{90}Th \longrightarrow ^{234}_{91}Pa + ^{0}_{-1}e$
c. $^{226}_{88}Ra \longrightarrow ^{222}_{86}Rn + ^{4}_{2}He$

Organic Chemistry

At 4:35 A.M., a rescue crew responds to a call about a house fire. At the scene, Jack, a firefighter/emergency medical technician (EMT), finds Diane lying in the front yard of her house. In his assessment, Jack reports that Diane has second- and third-degree burns over 40% of her body as well as a broken leg. He places an oxygen re-breather mask on Diane to provide a high concentration of oxygen. Another firefighter/EMT, Nancy, begins dressing the burns with sterile water and cling film, a first aid material made of polyvinyl chloride, which does not stick to the skin and is protective. Jack and his crew transport Diane to the burn center for further treatment.

At the scene of the fire, arson investigators use trained dogs to search for traces of accelerants and fuel. Gasoline, which is often found at arson scenes, is a mixture of organic molecules called alkanes. Alkanes or hydrocarbons are chains of carbon and hydrogen atoms. The alkanes present in gasoline consist of a mixture of compounds with five to eight carbon atoms in a chain. Alkanes are extremely combustible; they react with oxygen to form carbon dioxide, water, and large amounts of heat. Because alkanes undergo combustion reactions, they can be used to start arson fires.

CAREER

Firefighter/Emergency Medical Technician

Firefighters/emergency medical technicians are first responders to fires, accidents, and other emergency situations. They are required to have an emergency medical technician certification in order to be able to treat seriously injured people. By combining the skills of a firefighter and an emergency medical technician, they increase the survival rates of the injured. The physical demands of firefighters are extremely high as they fight, extinguish, and prevent fires while wearing heavy protective clothing. They also train for and participate in firefighting drills, and maintain fire equipment so that it is always working and ready. Firefighters must also be knowledgeable about fire codes, arson, and the handling and disposal of hazardous materials. Because firefighters also provide emergency care for sick and injured people, they need to be aware of emergency medical and rescue procedures, as well as the proper methods for controlling the spread of infectious disease.

UPDATE   Diane's Treatment in the Burn Unit

When Diane arrives at the hospital, she is diagnosed with second- and third-degree burns. You can see Diane's treatment in the **UPDATE Diane's Treatment in the Burn Unit**, page 580, and see the results of the arson investigation into the house fire.

17.1 Alkanes

LEARNING GOAL Identify the properties of organic or inorganic compounds. Write the IUPAC names and draw the condensed structural and line-angle formulas for alkanes.

Organic chemistry is the study of carbon compounds. The element carbon has a special role because many carbon atoms can bond together to give a vast array of molecular compounds. Organic compounds always contain carbon and hydrogen, and sometimes other nonmetals such as oxygen, sulfur, nitrogen, phosphorus, or a halogen. We find organic compounds in many common products we use every day, such as gasoline, medicines, shampoos, plastics, and perfumes. The food we eat is composed of organic compounds such as carbohydrates, fats, and proteins that supply us with fuel for energy and the carbon atoms needed to build and repair the cells of our bodies.

The formulas of organic compounds are written with carbon first, followed by hydrogen, and then any other elements. Organic compounds typically have low melting and boiling points, are not soluble in water, and are less dense than water. For example, vegetable oil, which is a mixture of organic compounds, does not dissolve in water but floats on top. Many organic compounds undergo combustion and burn vigorously in air. By contrast, many inorganic compounds have high melting and boiling points. Inorganic compounds that are ionic are usually soluble in water, and most do not burn in air. **TABLE 17.1** contrasts some of the properties associated with organic and inorganic compounds, such as propane, C_3H_8, and sodium chloride, NaCl (see **FIGURE 17.1**).

REVIEW

Balancing a Chemical Equation (8.2)

Drawing Lewis Structures (10.1)

Predicting Shape (10.3)

TABLE 17.1 Some Properties of Organic and Inorganic Compounds

Property	Organic	Example: C_3H_8	Inorganic	Example: NaCl
Elements Present	C and H, sometimes O, S, N, P, or Cl (F, Br, I)	C and H	Most metals and nonmetals	Na and Cl
Particles	Molecules	C_3H_8	Mostly ions	Na^+ and Cl^-
Bonding	Mostly covalent	Covalent	Many are ionic, some covalent	Ionic
Polarity of Bonds	Nonpolar unless a strongly electronegative atom is present	Nonpolar	Most are ionic or polar covalent, a few are nonpolar covalent	Ionic
Melting Point	Usually low	$-188\,°C$	Usually high	$801\,°C$
Boiling Point	Usually low	$-42\,°C$	Usually high	$1413\,°C$
Flammability	High	Burns in air	Low	Does not burn
Solubility in Water	Not soluble unless a polar group is present	No	Most are soluble unless nonpolar	Yes

Vegetable oil, a mixture of organic compounds, is not soluble in water.

$CH_3 - CH_2 - CH_3$

Na^+

Cl^-

FIGURE 17.1 Propane, C_3H_8, is an organic compound, whereas sodium chloride, NaCl, is an inorganic compound.

SAMPLE PROBLEM 17.1 Properties of Organic Compounds

TRY IT FIRST

Indicate whether the following properties are more typical of organic or inorganic compounds:

a. is not soluble in water **b.** has a high melting point **c.** burns in air

SOLUTION

a. Many organic compounds are not soluble in water.
b. Inorganic compounds are more likely to have high melting points.
c. Organic compounds are more likely to burn in air.

SELF TEST 17.1

a. What elements are always found in organic compounds?
b. If butane in a lighter is used to start a fire, is butane an inorganic or organic compound?

ANSWER

a. C and H   **b.** organic compound

PRACTICE PROBLEMS

Try Practice Problems 17.1 to 17.4

Representations of Carbon Compounds

Hydrocarbons are organic compounds that consist of only carbon and hydrogen. In organic molecules, every carbon atom has four bonds. In the simplest hydrocarbon, methane (CH_4), the carbon atom forms an octet by sharing its four valence electrons with four hydrogen atoms.

$$\cdot \overset{\cdot\cdot}{\underset{\cdot\cdot}{C}} \cdot \ + \ 4H\cdot \ \longrightarrow \ H:\overset{H}{\underset{H}{\overset{\cdot\cdot}{C}}}:H \ = \ H-\overset{\overset{\textstyle H}{|}}{\underset{\underset{\textstyle H}{|}}{C}}-H$$

Methane

The most accurate representation of methane is the three-dimensional *space-filling model* (**a**) in which spheres show the relative size and shape of all the atoms. Another type of three-dimensional representation is the *ball-and-stick model* (**b**), where the atoms are shown as balls and the bonds between them are shown as sticks. In the ball-and-stick model of methane, CH_4, the covalent bonds from the carbon atom to each hydrogen atom are directed to the corners of a tetrahedron with bond angles of 109°. In the *wedge–dash model* (**c**), the three-dimensional shape is represented by symbols of the atoms with lines for bonds in the plane of the page, wedges for bonds that project out from the page, and dashes for bonds that are behind the page.

ENGAGE 17.1

Why does methane have a tetrahedral shape?

Three-Dimensional and Two-Dimensional Representations of Methane

(a) Space-filling model   (b) Ball-and-stick model   (c) Wedge–dash model   (d) Expanded structural formula   (e) Condensed structural formula

However, the three-dimensional models are awkward to draw and view for more complex molecules. Therefore, it is more practical to use their corresponding two-dimensional formulas. The **expanded structural formula (d)** shows all of the atoms and the bonds connected to each atom. A **condensed structural formula (e)** shows the carbon atoms each grouped with the attached number of hydrogen atoms.

The hydrocarbon ethane with two carbon atoms and six hydrogen atoms can be represented by a similar set of three- and two-dimensional models and formulas in which each carbon atom is bonded to another carbon and three hydrogen atoms. As in methane, each carbon atom in ethane retains a tetrahedral shape. A hydrocarbon is referred to as a *saturated hydrocarbon* when all the bonds in the molecule are single bonds.

Three-Dimensional and Two-Dimensional Representations of Ethane

| (a) Space-filling model | (b) Ball-and-stick model | (c) Wedge–dash model | (d) Expanded structural formula | (e) Condensed structural formula |

Naming Alkanes

More than 90% of the compounds in the world are organic compounds. The large number of carbon compounds is possible because the covalent bond between carbon atoms ($C-C$) is very strong, allowing carbon atoms to form long, stable chains.

The **alkanes** are a type of hydrocarbon in which the carbon atoms are connected only by single bonds. One of the most common uses of alkanes is as fuels. Methane, used in gas heaters and gas cooktops, is an alkane with one carbon atom. The alkanes ethane, propane, and butane contain two, three, and four carbon atoms, respectively, connected in a row or a *continuous chain*. As we can see, the names for alkanes end in *ane*. Such names are part of the **IUPAC system** (International Union of Pure and Applied Chemistry) used by chemists to name organic compounds. Alkanes with five or more carbon atoms in a chain are named using Greek prefixes: *pent* (5), *hex* (6), *hept* (7), *oct* (8), *non* (9), and *dec* (10) (see **TABLE 17.2**).

CORE CHEMISTRY SKILL

Naming and Drawing Alkanes

PRACTICE PROBLEMS

Try Practice Problems 17.5 and 17.6

TABLE 17.2 IUPAC Names and Formulas of the First 10 Alkanes

Number of Carbon Atoms	IUPAC Name	Molecular Formula	Condensed Structural Formula	Line-Angle Formula
1	Methane	CH_4	CH_4	
2	Ethane	C_2H_6	CH_3-CH_3	
3	Propane	C_3H_8	$CH_3-CH_2-CH_3$	
4	Butane	C_4H_{10}	$CH_3-CH_2-CH_2-CH_3$	
5	Pentane	C_5H_{12}	$CH_3-CH_2-CH_2-CH_2-CH_3$	
6	Hexane	C_6H_{14}	$CH_3-CH_2-CH_2-CH_2-CH_2-CH_3$	
7	Heptane	C_7H_{16}	$CH_3-CH_2-CH_2-CH_2-CH_2-CH_2-CH_3$	
8	Octane	C_8H_{18}	$CH_3-CH_2-CH_2-CH_2-CH_2-CH_2-CH_2-CH_3$	
9	Nonane	C_9H_{20}	$CH_3-CH_2-CH_2-CH_2-CH_2-CH_2-CH_2-CH_2-CH_3$	
10	Decane	$C_{10}H_{22}$	$CH_3-CH_2-CH_2-CH_2-CH_2-CH_2-CH_2-CH_2-CH_2-CH_3$	

Condensed Structural and Line-Angle Formulas

In a condensed structural formula, each carbon atom and its attached hydrogen atoms are written as a group. A subscript indicates the number of hydrogen atoms bonded to each carbon atom.

$$
\begin{array}{c}
H \\
| \\
H-C- \\
| \\
H
\end{array}
= CH_3-
\qquad
\begin{array}{c}
H \\
| \\
-C- \\
| \\
H
\end{array}
= -CH_2-
$$

Expanded   Condensed   Expanded   Condensed

When an organic molecule consists of a chain of three or more carbon atoms, the carbon atoms do not lie in a straight line. Rather, they are arranged in a zigzag pattern.

A simplified formula called the **line-angle formula** shows a zigzag line in which carbon atoms are represented as the ends of each line and as corners. For example, in the line-angle formula of pentane, each line in the zigzag drawing represents a single bond. The carbon atoms on the ends are bonded to three hydrogen atoms. However, the carbon atoms in the middle of the carbon chain are each bonded to two carbons and two hydrogen atoms as shown in Sample Problem 17.2.

ENGAGE 17.2

How does a line-angle formula represent an organic compound of carbon and hydrogen with single bonds?

SAMPLE PROBLEM 17.2 Drawing Formulas for an Alkane

TRY IT FIRST

Draw the expanded, condensed structural, and line-angle formulas for pentane.

SOLUTION

	Given	Need	Connect
ANALYZE THE PROBLEM	pentane	expanded, condensed structural, and line-angle formulas	carbon chain, zigzag line

STEP 1 **Draw the carbon chain.** A molecule of pentane has five carbon atoms in a continuous chain.

$$C-C-C-C-C$$

STEP 2 **Draw the expanded structural formula by adding the hydrogen atoms using single bonds to each of the carbon atoms.**

$$
\begin{array}{ccccc}
H & H & H & H & H \\
| & | & | & | & | \\
H-C-&C-&C-&C-&C-H \\
| & | & | & | & | \\
H & H & H & H & H
\end{array}
$$

STEP 3 **Draw the condensed structural formula by combining the H atoms with each C atom.**

$$
\begin{array}{ccccc}
H & H & H & H & H \\
| & | & | & | & | \\
H-C-&C-&C-&C-&C-H \\
| & | & | & | & | \\
H & H & H & H & H
\end{array}
$$
Expanded structural formula

$$CH_3-CH_2-CH_2-CH_2-CH_3$$ Condensed structural formula

STEP 4 Draw the line-angle formula as a zigzag line in which the ends and corners represent C atoms.

$CH_3 — CH_2 — CH_2 — CH_2 — CH_3$ Condensed structural formula

Line-angle formula

SELF TEST 17.2

Draw the condensed structural formula and write the name for each of the following line-angle formulas:

a.

b.

ANSWER

a. $CH_3 — CH_2 — CH_2 — CH_2 — CH_2 — CH_2 — CH_3$ heptane

b. $CH_3 — CH_2 — CH_2 — CH_3$ butane

Structural Isomers

When an alkane has four or more carbon atoms, the atoms can be arranged so that a side group called a *branch* or **substituent** is attached to a carbon chain. For example, FIGURE 17.2 shows different models and formulas for two compounds that have the same molecular formula, C_4H_{10}. One model has a chain of four carbon atoms. In the other model, a carbon atom is attached as a branch or substituent to a carbon in a chain of three atoms. An alkane with at least one branch is called a *branched alkane*. When the two compounds have the same molecular formula but different arrangements of atoms, they are **structural isomers**.

Structural Isomers of C_4H_{10}

$CH_3 — CH_2 — CH_2 — CH_3$

$$CH_3 — \overset{\displaystyle CH_3}{\overset{|}{CH}} — CH_3$$

ENGAGE 17.3

How can two or more structural isomers have the same molecular formula?

FIGURE 17.2 The structural isomers of C_4H_{10} have the same number and type of atoms but are bonded in a different order.

In another example, we can draw the condensed structural and line-angle formulas for three different structural isomers with the molecular formula C_5H_{12} as follows:

	Structural Isomers of C_5H_{12}		
Condensed Structural Formula	$CH_3 — CH_2 — CH_2 — CH_2 — CH_3$	$CH_3 — \overset{\displaystyle CH_3}{\overset{|}{CH}} — CH_2 — CH_3$	$CH_3 — \overset{\displaystyle CH_3}{\underset{\displaystyle CH_3}{\overset{|}{\underset{|}{C}}}} — CH_3$
Line-Angle Formula			

SAMPLE PROBLEM 17.3 Structural Isomers

TRY IT FIRST

Identify each pair of formulas as structural isomers or the same molecule.

a.
$$
\begin{array}{cc}
CH_3 & CH_3 \\
| & | \\
CH_2{-}CH_2
\end{array}
\quad \text{and} \quad
\begin{array}{c}
CH_2{-}CH_2{-}CH_3 \\
| \\
CH_3
\end{array}
$$

b. [line-angle structure] and [line-angle structure]

SOLUTION

a. Both have the same molecular formula C_4H_{10}. Both have continuous four-carbon chains even though the $-CH_3$ ends are drawn above or below the chain. Thus, both formulas represent the same molecule.

b. Both have the same molecular formula C_6H_{14}. The formula on the left has a five-carbon chain with a $-CH_3$ substituent on the second carbon of the chain. The formula on the right has a four-carbon chain with two $-CH_3$ substituents. Thus, there is a different order of bonding of atoms, which represents structural isomers.

SELF TEST 17.3

Why do each of the following line-angle formulas represent a different structural isomer of the molecules in Sample Problem 17.3, part **b**?

a. [line-angle structure]   b. [line-angle structure]

ANSWER

a. This isomer with the molecular formula C_6H_{14} has a different arrangement of carbon atoms with a $-CH_3$ group on the third carbon of a five-carbon chain.

b. This isomer with the molecular formula C_6H_{14} has a different arrangement of carbon atoms with two $-CH_3$ groups on the second carbon of a four-carbon chain.

PRACTICE PROBLEMS

Try Practice Problems 17.7 and 17.8

Substituents in Alkanes

In the IUPAC names for alkanes, a carbon branch is named as an **alkyl group**, which is an alkane that is missing one hydrogen atom. The alkyl group is named by replacing the *ane* ending of the corresponding alkane name with *yl*. Alkyl groups cannot exist on their own: They must be attached to a carbon chain. When a halogen atom is attached to a carbon chain, it is named as a *halo* group: *fluoro*, *chloro*, *bromo*, or *iodo*. Some of the common groups attached to carbon chains are illustrated in **TABLE 17.3**.

TABLE 17.3 Formulas and Names of Some Common Substituents

Formula	$CH_3{-}$		$CH_3{-}CH_2{-}$	
Name	methyl		ethyl	
Formula	$CH_3{-}CH_2{-}CH_2{-}$		$CH_3{-}\overset{\textstyle |}{CH}{-}CH_3$	
Name	propyl		isopropyl	
Formula	$CH_3{-}CH_2{-}CH_2{-}CH_2{-}$		$CH_3{-}\overset{\textstyle CH_3}{\overset{\textstyle |}{CH}}{-}CH_2{-}$	
Name	butyl		isobutyl	
Formula	$F{-}$	$Cl{-}$	$Br{-}$	$I{-}$
Name	fluoro	chloro	bromo	iodo

Naming Alkanes with Substituents

In the IUPAC system of naming, a carbon chain is numbered to give the location of the substituents. Let's take a look at how we use the IUPAC system to name the alkane shown in Sample Problem 17.4.

INTERACTIVE VIDEO

Pearson eText

Naming Alkanes

SAMPLE PROBLEM 17.4 Writing IUPAC Names for Alkanes with Substituents

TRY IT FIRST

Write the IUPAC name for the following alkane:

$$CH_3-CH-CH_2-CH_2-C-CH_3$$

with CH_3 above the second carbon, and Br above and CH_3 below the fifth carbon.

SOLUTION

	Given	Need	Connect
ANALYZE THE PROBLEM	six-carbon chain, two methyl groups, one bromo group	IUPAC name	position of substituents on the carbon chain

STEP 1 **Write the alkane name for the longest chain of carbon atoms.** The longest chain contains six carbon atoms; it is hexane.

$$CH_3-CH-CH_2-C-CH_2-CH_3 \quad \text{hexane}$$

with CH_3 above the second carbon, Br above and CH_3 below the fourth carbon.

STEP 2 **Number the carbon atoms from the end nearer a substituent.** The first substituent is attached to carbon 2 counting from the left, and two more substituents are on carbon 4.

$$CH_3-CH-CH_2-C-CH_2-CH_3 \quad \text{hexane}$$

with CH_3 above the second carbon, Br above and CH_3 below the fourth carbon; numbered 1 2 3 4 5 6.

STEP 3 **Give the location and name for each substituent (alphabetical order) as a prefix to the name of the main chain.** The substituents are listed in alphabetical order (bromo first, then methyl). A hyphen is placed between the number and the substituent name. When there are two or more of the same substituent, a prefix (*di, tri, tetra*) is used, and commas separate the numbers. However, prefixes are not used to determine the alphabetical order of the substituents.

$$CH_3-CH-CH_2-C-CH_2-CH_3 \quad \text{4-bromo-2,4-dimethylhexane}$$

with CH_3 above the second carbon, Br above and CH_3 below the fourth carbon; numbered 1 2 3 4 5 6.

SELF TEST 17.4

Write the IUPAC name for each of the following compounds:

a.

b. $CH_3-CH-CH_2-C-CH_2-CH_3$ with CH_3 groups at the CH and C positions (two CH_3 on the C)

PRACTICE PROBLEMS
Try Practice Problems 17.9 and 17.10

ANSWER

a. 4-isopropylheptane   b. 2,4,4-trimethylhexane

Drawing Structural Formulas for Alkanes with Substituents

The IUPAC name gives all the information needed to draw the condensed structural formula for an alkane. Suppose you are asked to draw the condensed structural formula for 2,3-dimethylbutane. The alkane name gives the number of carbon atoms in the longest chain. The other names indicate the substituents and where they are attached. We can break down the name in the following way:

2,3-Dimethylbutane				
2,3-	Di	methyl	but	ane
Substituents on carbons 2 and 3	two identical groups	—CH_3 alkyl groups	four C atoms in the main chain	single (C—C) bonds

SAMPLE PROBLEM 17.5   Drawing Condensed Structural and Line-Angle Formulas from IUPAC Names

TRY IT FIRST

Draw the condensed structural and line-angle formulas for 2,3-dimethylbutane.

SOLUTION

	Given	Need	Connect
ANALYZE THE PROBLEM	2,3-dimethylbutane	condensed structural and line-angle formulas	four-carbon chain, two methyl groups

STEP 1   Draw the main chain of carbon atoms. For butane, we draw a chain and a zigzag line of four carbon atoms.

C—C—C—C

STEP 2   Number the chain and place the substituents on the carbons indicated by the numbers. The first part of the name indicates two methyl groups —CH_3: one on carbon 2 and one on carbon 3.

STEP 3   For the condensed structural formula, add the correct number of hydrogen atoms to give four bonds to each C atom.

SELF TEST 17.5

Draw the condensed structural and line-angle formulas for each of the following:

a. 2-bromo-4-methylpentane

b. 2-chloro-2-methylpropane

ANSWER

a. $CH_3-CH(Br)-CH_2-CH(CH_3)-CH_3$

b. $CH_3-C(CH_3)(Cl)-CH_3$

PRACTICE PROBLEMS
Try Practice Problems 17.11 to 17.14

Uses of Alkanes

The first four alkanes—methane, ethane, propane, and butane—are gases at room temperature and are widely used as heating fuels.

Alkanes having five to eight carbon atoms (pentane, hexane, heptane, and octane) are liquids at room temperature. They are highly volatile, which makes them useful in fuels such as gasoline.

Liquid alkanes with 9 to 17 carbon atoms have higher boiling points and are found in kerosene, diesel, and jet fuels. Motor oil is a mixture of high-molecular-weight liquid hydrocarbons and is used to lubricate the internal components of engines. Mineral oil is a mixture of liquid hydrocarbons and is used as a laxative and a lubricant. Alkanes with 18 or more carbon atoms are waxy solids at room temperature. Known as paraffins, they are used in waxy coatings added to fruits and vegetables to retain moisture, inhibit mold, and enhance appearance. Petrolatum jelly, or Vaseline, is a semisolid mixture of hydrocarbons with more than 25 carbon atoms used in ointments and cosmetics and as a lubricant.

The solid alkanes that make up waxy coatings on fruits and vegetables help retain moisture, inhibit mold, and enhance appearance.

Combustion of Alkanes

The carbon–carbon single bonds in alkanes are difficult to break, which makes them the least reactive family of organic compounds. However, alkanes burn readily in oxygen to produce carbon dioxide, water, and energy.

Propane is the gas used in portable heaters and gas barbecues (see **FIGURE 17.3**). The equation for the combustion of propane (C_3H_8) is written:

$$C_3H_8(g) + 5O_2(g) \xrightarrow{\Delta} 3CO_2(g) + 4H_2O(g) + energy$$

Propane

Solubility and Density

Alkanes are nonpolar, which makes them insoluble in water. However, they are soluble in nonpolar solvents. Alkanes have densities from 0.62 g/mL to about 0.79 g/mL, which is less than the density of water (1.0 g/mL).

If there is an oil spill in the ocean, the alkanes in the oil, which do not mix with water, form a thin layer on the surface that spreads over a large area (see **FIGURE 17.4**).

FIGURE 17.3 The propane fuel in the tank undergoes combustion, which provides energy.

FIGURE 17.4 In oil spills, large quantities of oil spread out to form a thin layer on top of the ocean surface.

In April 2010, an explosion on an oil-drilling rig in the Gulf of Mexico caused the largest oil spill in U.S. history. At its maximum, an estimated 10 million liters of oil was leaked every day. If the crude oil reaches land, there can be considerable damage to beaches, shellfish, birds, and wildlife habitats. When animals such as birds are covered with oil, they must be cleaned quickly because ingestion of the hydrocarbons when they try to clean themselves is fatal.

PRACTICE PROBLEMS

17.1 Alkanes

17.1 Identify each of the following as an organic or inorganic compound:
- **a.** KCl
- **b.** C_3H_7Cl
- **c.** is soluble in water
- **d.** has a low boiling point
- **e.** contains carbon and hydrogen
- **f.** contains ionic bonds

17.2 Identify each of the following as an organic or inorganic compound:
- **a.** contains Li and F
- **b.** K_3PO_4
- **c.** contains covalent bonds
- **d.** produces ions in water
- **e.** is a gas at room temperature
- **f.** C_4H_9Br

17.3 Match each of the following physical and chemical properties with ethane, C_2H_6, or sodium bromide, NaBr:
- **a.** boils at $-89\,°C$
- **b.** burns vigorously in air
- **c.** is a solid at $250\,°C$
- **d.** dissolves in water

17.4 Match each of the following physical and chemical properties with hexane, C_6H_{14} or calcium nitrate, $Ca(NO_3)_2$:
- **a.** melts at $500\,°C$
- **b.** is insoluble in water
- **c.** does not burn in air
- **d.** is a liquid at room temperature

17.5 Write the IUPAC name for each of the following alkanes:

a. $CH_3 — CH_2 — CH_2 — CH_2 — CH_2 — CH_3$

b.

c.

17.6 Write the IUPAC name for each of the following alkanes:
a. $CH_3 — CH_2 — CH_2 — CH_3$

b. (zigzag structure)

c. (ball-and-stick model)

17.7 Indicate whether each of the following pairs of formulas represents structural isomers or the same molecule:

a.
$$CH_3 — \overset{\overset{CH_3}{|}}{CH} — CH_3 \quad \text{and} \quad \overset{\overset{CH_3}{|}}{CH} — CH_3 \ \overset{|}{\underset{CH_3}{}}$$

b.
$$\overset{\overset{CH_3}{|}}{CH_2} — \overset{\overset{CH_3}{|}}{CH} — CH_2 — CH_3 \quad \text{and}$$
$$CH_3 — \overset{\overset{CH_3}{|}}{CH} — \overset{\overset{CH_3}{|}}{CH} — CH_3$$

c.
and

17.8 Indicate whether each of the following pairs of formulas represents structural isomers or the same molecule:

a.
$$CH_3 — \overset{\overset{CH_3}{|}}{\underset{\underset{CH_3}{|}}{C}} — CH_3 \quad \text{and} \quad \overset{\overset{CH_3}{|}}{\underset{\underset{CH_3}{|}}{CH}} — CH_2 — CH_3$$

b.
$$CH_3 — \overset{\overset{CH_3}{|}}{CH} — \overset{\overset{CH_3}{|}}{CH} — \overset{\overset{CH_3}{|}}{CH_2} \quad \text{and}$$
$$CH_3 — \overset{\overset{CH_3}{|}}{CH} — CH_2 — \overset{\overset{CH_3}{|}}{CH} — CH_3$$

c. 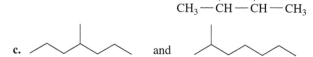 and

17.9 Write the IUPAC name for each of the following:

a. $CH_3 — \overset{\overset{F}{|}}{CH} — CH_2 — CH_3$

b. $CH_3 — \overset{\overset{CH_3}{|}}{\underset{\underset{CH_3}{|}}{C}} — CH_3$

c. (structure with Cl)

17.10 Write the IUPAC name for each of the following:

a. $CH_3 — \overset{\overset{CH_3}{|}}{CH} — CH_2 — CH_2 — CH_2 — Br$

b. $CH_3 — \overset{\overset{CH_3}{|}}{CH} — \overset{\overset{CH_3}{|}}{CH} — CH_2 — CH_2 — CH_3$

c. (structure with Br and Cl)

17.11 Draw the condensed structural formula for each of the following alkanes:
  a. 3,3-dimethylpentane
  b. 2,3,5-trimethylhexane
  c. 1-bromo-2-chloroethane

17.12 Draw the condensed structural formula for each of the following alkanes:
  a. 3-ethylpentane
  b. 4-isopropyl-3-methylheptane
  c. 2-bromopropane

17.13 Draw the line-angle formula for each of the following:
  a. 3-methylheptane
  b. 1-chloro-3-methylpentane
  c. 2,3-dichlorohexane

17.14 Draw the line-angle formula for each of the following:
  a. 1-bromo-2-methylpentane
  b. 2,2,3-trimethylbutane
  c. 3-ethylhexane

17.15 Heptane, used as a solvent for rubber cement, has a density of 0.68 g/mL and boils at 98 °C.
  a. Draw the condensed structural and line-angle formulas for heptane.
  b. Is heptane a solid, liquid, or gas at room temperature?

  c. Is heptane soluble in water?
  d. Will heptane float on water or sink?
  e. Write the balanced chemical equation for the complete combustion of heptane.

17.16 Nonane, found in kerosene, has a density of 0.79 g/mL and boils at 151 °C.
  a. Draw the condensed structural and line-angle formulas for nonane.
  b. Is nonane a solid, liquid, or gas at room temperature?
  c. Is nonane soluble in water?
  d. Will nonane float on water or sink?
  e. Write the balanced chemical equation for the complete combustion of nonane.

Applications

17.17 Write the balanced chemical equation for the complete combustion of each of the following:
  a. methane
  b. 3-methylpentane

17.18 Write the balanced chemical equation for the complete combustion of each of the following:
  a. pentane
  b. methylpropane

17.2 Alkenes, Alkynes, and Polymers

LEARNING GOAL Write the IUPAC names and draw the condensed structural and line-angle formulas for alkenes and alkynes.

We organize organic compounds into *classes* or *families* by their **functional groups**, which are groups of specific atoms. Compounds that contain the same functional group have similar physical and chemical properties. Identifying functional groups allows us to classify organic compounds according to their structure, to name compounds within each family, predict their chemical reactions, and draw the structures for their products. A list of the common functional groups in organic compounds is shown in **TABLE 17.4**.

  Alkenes and alkynes are classes of unsaturated hydrocarbons that contain *double* and *triple* bonds, respectively. The functional group of an **alkene** contains at least one double bond between carbons. The double bond forms when two adjacent carbon atoms share two pairs of valence electrons. The simplest alkene is ethene, C_2H_4, which is often called by its common name, ethylene. In ethene, each carbon atom is attached to two H atoms and the other carbon atom in the double bond. The resulting molecule has a flat geometry because the carbon and hydrogen atoms all lie in the same plane. The functional group of an **alkyne** contains a triple bond, which occurs when two carbon atoms share three pairs of valence electrons.

Ethene   Ethyne

TABLE 17.4 Classes of Organic Compounds

Class	Functional Group	Example/Name	Occurrence/Use
Alkene	$>C=C<$	$H_2C=CH_2$ Ethene (ethylene)	Ripening of fruit, used to make polyethylene
Alkyne	$-C\equiv C-$	$H-C\equiv C-H$ Ethyne (acetylene)	Welding fuel
Alcohol	$-OH$	$CH_3-CH_2-\mathbf{OH}$ Ethanol (ethyl alcohol)	Solvent
Ether	$-O-$	$CH_3-CH_2-\mathbf{O}-CH_2-CH_3$ Diethyl ether	Solvent
Aldehyde	$\overset{\displaystyle O}{\overset{\|}{-C-H}}$	$CH_3-\overset{\displaystyle O}{\overset{\|}{C}}-H$ Ethanal (acetaldehyde)	Preparation of acetic acid
Ketone	$\overset{\displaystyle O}{\overset{\|}{-C-}}$	$CH_3-\overset{\displaystyle O}{\overset{\|}{C}}-CH_3$ Propanone (acetone)	Solvent, paint and fingernail polish remover
Carboxylic acid	$\overset{\displaystyle O}{\overset{\|}{-C-OH}}$	$CH_3-\overset{\displaystyle O}{\overset{\|}{C}}-\mathbf{OH}$ Ethanoic acid (acetic acid)	Component of vinegar
Ester	$\overset{\displaystyle O}{\overset{\|}{-C-O-}}$	$CH_3-\overset{\displaystyle O}{\overset{\|}{C}}-\mathbf{O}-CH_3$ Methyl ethanoate (methyl acetate)	Rum flavoring
Amine	$-\overset{\|}{N}-$	$CH_3-\mathbf{NH_2}$ Methylamine	Odor of fish
Amide	$\overset{\displaystyle O}{\overset{\|}{-C}}-\overset{\|}{N}-$	$CH_3-\overset{\displaystyle O}{\overset{\|}{C}}-\mathbf{NH_2}$ Ethanamide (acetamide)	Odor of mice

ENGAGE 17.5

How does the functional group in an alcohol differ from the functional group in a ketone?

Naming Alkenes and Alkynes

The IUPAC names for alkenes and alkynes are similar to those of alkanes. Using the alkane name with the same number of carbon atoms, the *ane* ending is replaced with *ene* for an alkene and *yne* for an alkyne (see **TABLE 17.5**).

TABLE 17.5 Comparison of Names for Alkanes, Alkenes, and Alkynes

Alkane	Alkene	Alkyne
CH_3-CH_3	$H_2C=CH_2$	$HC\equiv CH$
Ethane	Ethene (ethylene)	Ethyne (acetylene)
$CH_3-CH_2-CH_2$	$CH_3-CH=CH_2$	$CH_3-C\equiv CH$
Propane	Propene	Propyne

Fruit is ripened with ethene, a plant hormone.

PRACTICE PROBLEMS

Try Practice Problems 17.19 and 17.20

SAMPLE PROBLEM 17.6 Naming Alkenes and Alkynes

TRY IT FIRST

Write the IUPAC name for the following:

$$CH_3-\underset{\underset{\displaystyle CH_3}{|}}{CH}-CH{=}CH-CH_3$$

SOLUTION

ANALYZE THE PROBLEM	Given	Need	Connect
	five-carbon chain, double bond, methyl group	IUPAC name	replace the *ane* of the alkane name with *ene*

STEP 1 **Name the longest carbon chain that contains the double bond.** There are five carbon atoms in the longest carbon chain containing the double bond. Replace the *ane* in the corresponding alkane name with *ene* to give pentene.

$$CH_3-\underset{\underset{\displaystyle CH_3}{|}}{CH}-CH{=}CH-CH_3 \qquad \text{pentene}$$

STEP 2 **Number the carbon chain starting from the end nearer the double bond.** Place the number of the first carbon in the double bond in front of the alkene name.

$$\underset{5}{CH_3}-\underset{4}{\underset{\underset{\displaystyle CH_3}{|}}{CH}}-\underset{3}{CH}{=}\underset{2}{CH}-\underset{1}{CH_3} \qquad \text{2-pentene}$$

Alkenes with two or three carbons do not need numbers.

STEP 3 **Give the location and name for each substituent (alphabetical order) as a prefix to the alkene name.**

$$\underset{5}{CH_3}-\underset{4}{\underset{\underset{\displaystyle CH_3}{|}}{CH}}-\underset{3}{CH}{=}\underset{2}{CH}-\underset{1}{CH_3} \qquad \text{4-methyl-2-pentene}$$

SELF TEST 17.6

Draw the condensed structural formula for each of the following:
a. 1-chloro-3-hexyne
b. 1-bromo-1-pentene

ANSWER

a. $Cl-CH_2-CH_2-C{\equiv}C-CH_2-CH_3$ **b.** $Br-CH{=}CH-CH_2-CH_2-CH_3$

PRACTICE PROBLEMS
Try Practice Problems 17.21 to 17.24

Hydrogenation

In a reaction called *hydrogenation*, H atoms add to each of the carbon atoms in a double bond of an alkene. During hydrogenation, the double bonds are converted to single bonds in alkanes. A catalyst such as finely divided platinum (Pt), nickel (Ni), or palladium (Pd) is used to speed up the reaction. The equation for the hydrogenation of 2-butene is written as follows:

CORE CHEMISTRY SKILL
Writing Equations for
Hydrogenation and Polymerization

$$\underset{\text{2-Butene}}{CH_3-CH{=}CH-CH_3} + H_2 \xrightarrow{\text{Pt}} \underset{\text{Butane}}{CH_3-CH_2-CH_2-CH_3}$$

SAMPLE PROBLEM 17.7 Writing an Equation for Hydrogenation

TRY IT FIRST

Draw the condensed structural formula for the product of the following hydrogenation reaction:

$$CH_3-CH=CH_2 + H_2 \xrightarrow{Pt}$$

SOLUTION

ANALYZE THE PROBLEM	Given	Need	Connect
	alkene + H_2	structural formula of product	add 2H to double bond

$$CH_3-CH_2-CH_3$$

SELF TEST 17.7

a. Draw the condensed structural formula for the product of the hydrogenation of 2-methyl-1-butene, using a platinum catalyst.
b. Draw the line-angle formula for the product of the hydrogenation of 2-pentene using a palladium catalyst.

ANSWER

PRACTICE PROBLEMS

Try Practice Problems 17.25 and 17.26

$$\text{a.} \quad CH_3-\overset{\overset{\displaystyle CH_3}{|}}{CH}-CH_2-CH_3 \qquad\qquad \text{b.} \quad \text{/\textbackslash/\textbackslash/}$$

Chemistry Link to Health

Hydrogenation of Unsaturated Fats

Vegetable oils such as corn oil or safflower oil are unsaturated fats composed of fatty acids that contain double bonds. The process of hydrogenation is used commercially to convert the double bonds in the unsaturated fats in vegetable oils to saturated fats such as margarine, which are more solid. Adjusting the amount of added hydrogen produces partially hydrogenated fats such as soft margarine, solid margarine and shortenings, which are used in cooking. For example, oleic acid is a typical unsaturated fatty acid in olive oil and has a double bond at carbon 9. When oleic acid is hydrogenated, it is converted to stearic acid, a saturated fatty acid.

The unsaturated fats in vegetable oils are converted to saturated fats to make a more solid product.

Double bond

O

OH

Oleic acid (found in olive oil and other unsaturated fats)

Single bond

O

OH

Stearic acid (found in saturated fats)

Polymerization

A **polymer** is a large molecule that consists of small repeating units called **monomers**. In the past hundred years, the plastics industry has made synthetic polymers that are in many of the materials we use every day, such as carpeting, plastic wrap, nonstick pans, plastic cups, and rain gear. In medicine, synthetic polymers are used to replace diseased or damaged body parts such as hip joints, teeth, heart valves, and blood vessels. There are about 100 billion kg of plastics produced every year, which is about 15 kg for every person on Earth.

Many of the synthetic polymers are made by addition reactions of small alkene monomers. In an addition reaction, a polymer grows longer as each monomer is added at the end of the chain. A polymer may contain as many as 1000 monomers. Many polymerization reactions require high temperature, a catalyst, and high pressure (over 1000 atm). Polyethylene, a polymer made from ethylene monomers, is used in plastic bottles, film, and plastic dinnerware. More polyethylene is produced worldwide than any other polymer. Low-density polyethylene (LDPE) is flexible, breakable, less dense, and more branched than high-density polyethylene (HDPE). High-density polyethylene is stronger, is more dense, and melts at a higher temperature than LDPE.

Ethene (ethylene) monomers → Polyethylene section

TABLE 17.6 shows the recycling code, the alkene monomer, a polymer section, and common uses for some polymers. The alkane-like nature of these plastic synthetic polymers makes them unreactive. Thus, they do not decompose easily (they are not biodegradable). As a result, they have become significant contributors to pollution, on land and in the oceans. Efforts are being made to make them more degradable.

TABLE 17.6 Some Alkenes and Their Polymers

Recycling Code	Monomer	Polymer Section		Common Uses
2 HDPE 4 LDPE	$H_2C=CH_2$ Ethene (ethylene)	Polyethylene (PE)		Plastic bottles, film, insulation materials, shopping bags, plastic furniture
3 V	$H_2C=CH$–Cl Chloroethene (vinyl chloride)	Polyvinyl chloride (PVC)		Plastic pipes and tubing, garden hoses, garbage bags, shower curtains, credit cards, medical containers, film for packaging foods, drink bottles, blister packs for pills
5 PP	$H_2C=CH$–CH_3 Propene (propylene)	Polypropylene (PP)		Ski and hiking clothing, carpets, artificial joints, plastic bottles, food containers, medical face masks, tubing, plastic bags
	$F-C=C-F$ (F,F) Tetrafluoroethene	Polytetrafluoroethylene		Nonstick coatings, Teflon, Gore-Tex rainwear
6 PS	$H_2C=CH$–(phenyl) Phenylethene (styrene)	Polystyrene (PS)		Plastic coffee cups and cartons, insulation, plastic tableware

You can identify the type of polymer used to manufacture a plastic item by looking for the recycling code (arrows in a triangle) found on the label or on the bottom of the plastic container. For example, the number 5 or the letters PP inside the triangle is the code for a polypropylene plastic.

SAMPLE PROBLEM 17.8   Polymers

TRY IT FIRST

A firefighter/EMT arrives at a home where a premature baby has been delivered. To prevent hypothermia during transport to the neonatal facility, she wraps the baby in clear wrap, which is polyvinyl chloride.

a. Draw the monomer unit and give its IUPAC name.
b. Draw a portion of the polymer formed from three monomer units.

SOLUTION

a.
$$\underset{\text{Chloroethene}}{H_2C=\overset{\overset{\displaystyle Cl}{\displaystyle |}}{CH}}$$

b.
$$-\overset{\overset{\displaystyle H}{|}}{\underset{\underset{\displaystyle H}{|}}{C}}-\overset{\overset{\displaystyle Cl}{|}}{\underset{\underset{\displaystyle H}{|}}{C}}-\overset{\overset{\displaystyle H}{|}}{\underset{\underset{\displaystyle H}{|}}{C}}-\overset{\overset{\displaystyle Cl}{|}}{\underset{\underset{\displaystyle H}{|}}{C}}-\overset{\overset{\displaystyle H}{|}}{\underset{\underset{\displaystyle H}{|}}{C}}-\overset{\overset{\displaystyle Cl}{|}}{\underset{\underset{\displaystyle H}{|}}{C}}-$$

SELF TEST 17.8

a. Draw the condensed structural formula for the monomer used in the manufacturing of polypropylene (PP), used for plastic dishes.
b. What is the name of the polymer that contains monomers of phenylethene?

ANSWER

a.
$$H_2C=\overset{\overset{\displaystyle CH_3}{|}}{CH}$$

b. polystyrene

PRACTICE PROBLEMS

Try Practice Problems 17.27 to 17.32

PRACTICE PROBLEMS

17.2 Alkenes, Alkynes, and Polymers

17.19 Identify each of the following as an alkane, alkene, or alkyne:

a. $CH_3-CH=CH_2$

b. $CH_3-CH_2-C\equiv CH$

c.

17.20 Identify each of the following as an alkane, alkene, or alkyne:

a.

b.
$$CH_3-\overset{\overset{\displaystyle CH_3}{|}}{\underset{\underset{\displaystyle CH_3}{|}}{C}}=C-CH_3$$

c.
$$CH_3-CH_2-\overset{\overset{\displaystyle CH_3}{|}}{CH}-CH_3$$

17.21 Write the IUPAC name for each of the following:

a. $H_2C=CH_2$

b.
$$CH_3-\overset{\overset{\displaystyle CH_3}{|}}{C}=CH_2$$

c. $CH_3-CH_2-C\equiv C-CH_3$

17.22 Write the IUPAC name for each of the following:

a. $H_2C=CH-CH_2-CH_3$

b.
$$CH_3-C\equiv C-CH_2-\overset{\overset{\displaystyle CH_3}{|}}{CH}-CH_3$$

c. $CH_3-CH_2-CH=CH-CH_3$

17.23 Draw the condensed structural formula for each of the following compounds:

a. propene   **b.** 1-hexyne
c. 2-methyl-1-butene

17.24 Draw the condensed structural formula for each of the following compounds:
   a. 1-pentene
   b. 3-methyl-1-butyne
   c. 3,4-dimethyl-1-pentene

17.25 Draw the condensed structural formula for the product in each of the following:

   a. $CH_3-CH_2-CH_2-CH=CH_2 + H_2 \xrightarrow{Pt}$

   b. $CH_3-CH=CH-CH_3 + H_2 \xrightarrow{Ni}$

   c. $+ H_2 \xrightarrow{Pt}$

17.26 Draw the condensed structural formula for the product in each of the following:

   a. $CH_3-CH_2-CH=CH_2 + H_2 \xrightarrow{Pt}$

   b. $CH_3-\overset{\overset{\displaystyle CH_3}{|}}{C}=CH-CH_2-CH_3 + H_2 \xrightarrow{Pt}$

   c. $+ H_2 \xrightarrow{Pt}$

17.27 What is a polymer?

17.28 What is a monomer?

17.29 Write an equation that represents the formation of a portion of polypropylene from three of its monomers.

17.30 Write an equation that represents the formation of a portion of polyvinyl fluoride (PVF) from three monomers of fluoroethene.

Applications

17.31 The plastic polyvinylidene difluoride, PVDF, is made from monomers of 1,1-difluoroethene. Draw the expanded structural formula for a portion of the polymer formed from three monomers of 1,1-difluoroethene.

17.32 The polymer polyacrylonitrile, PAN, used in the fabric material Orlon is made from monomers of acrylonitrile. Draw the expanded structural formula for a portion of the polymer formed from three monomers of acrylonitrile.

Acrylonitrile

17.3 Aromatic Compounds

LEARNING GOAL Describe the bonding in benzene; name aromatic compounds and draw their line-angle formulas.

In 1825, Michael Faraday isolated a hydrocarbon called **benzene**, which consists of a ring of six carbon atoms with one hydrogen atom attached to each carbon. In benzene, each carbon atom uses three valence electrons to bond to the hydrogen atom and two adjacent carbons. That leaves one valence electron, which scientists first thought was shared in a double bond with an adjacent carbon. In 1865, August Kekulé proposed that the carbon atoms in benzene were arranged in a flat ring with alternating single and double bonds between the adjacent carbon atoms. There are two possible structural representations of benzene in which the double bonds can form between two different carbon atoms. If there were double bonds as in alkenes, then benzene should be much more reactive than it is.

However, unlike the alkenes and alkynes, aromatic hydrocarbons do not easily undergo addition reactions. If their reaction behavior is quite different, they must also differ in how the atoms are bonded in their structures. Today, we know that the six electrons are shared equally among the six carbon atoms. This unique feature of benzene makes it especially stable. Benzene is most often represented as a line-angle formula, which shows a hexagon with a circle in the center. Some of the ways to represent benzene are shown as follows:

Equivalent structures for benzene   Structural formulas for benzene

Because many compounds containing benzene had fragrant odors, the family of benzene compounds became known as **aromatic compounds**. Some common examples of aromatic compounds that we use for flavor are anisole from anise, estragole from tarragon, and thymol from thyme.

Anisole (anise)   Estragole (tarragon)   Thymol (thyme)

The aroma and flavor of the herbs anise, tarragon, and thyme are because of aromatic compounds.

Naming Aromatic Compounds

Many compounds containing benzene have been important in chemistry for years and still use their common names. Toluene is benzene with a methyl group ($-CH_3$), aniline is benzene with an amino group ($-NH_2$), and phenol is benzene with a hydroxyl group ($-OH$). The names toluene, aniline, and phenol are allowed by IUPAC rules.

ENGAGE 17.6

How is the formula of toluene similar to and different from that of phenol?

CH_3

Toluene

Line-angle formula for toluene

NH_2

Aniline

OH

Phenol

When benzene has only one substituent, the ring is not numbered. When there are two substituents, the benzene ring is numbered to give the lowest numbers to the substituents. In common names, when the benzene ring is a substituent, it is named as a phenyl group. When a name such as toluene, phenol, or aniline is used, the carbon atom attached to the methyl, hydroxyl, or amine group is numbered as carbon 1.

Cl
Cl

1,2-Dichlorobenzene

Cl
Cl

1,3-Dichlorobenzene

Cl
Cl

1,4-Dichlorobenzene

NH_2
Br

3-Bromoaniline

OH
Br

4-Bromophenol

Br

2-Bromotoluene

SAMPLE PROBLEM 17.9 Naming Aromatic Compounds

TRY IT FIRST

Write the IUPAC name for the following:

SOLUTION

ANALYZE THE PROBLEM	Given	Need	Connect
	structural formula	IUPAC name	name of aromatic compound, number substituent

STEP 1 **Write the name for the aromatic compound.** A benzene ring with a methyl group is named toluene.

STEP 2 **Number and name a substituent as a prefix.** The methyl group of toluene is attached to carbon 1, and the ring is numbered to give the lower number to the chlorine substituent.

3-chlorotoluene

SELF TEST 17.9

Write the IUPAC name for each of the following:

a.

b.

ANSWER

a. 1,3-diethylbenzene

b. 2-chlorophenol

PRACTICE PROBLEMS

Try Practice Problems 17.33 to 17.36

Chemistry Link to Health

Some Common Aromatic Compounds

Aromatic compounds are common in nature and in medicine. Toluene is used as a reactant to make drugs, dyes, and explosives such as TNT (trinitrotoluene). The benzene ring is found in pain relievers such as aspirin and acetaminophen, and in flavorings such as vanillin.

TNT (2,4,6-trinitrotoluene)

Aspirin

Acetaminophen

Vanillin

Chemistry Link to Health

Polycyclic Aromatic Hydrocarbons (PAHs)

Large aromatic compounds known as *polycyclic aromatic hydrocarbons* (PAHs) are formed by fusing together two or more benzene rings edge to edge. In a fused-ring compound, neighboring benzene rings share two carbon atoms. Naphthalene, with two benzene rings, is well known for its use in mothballs; anthracene, with three rings, is used in the manufacture of dyes.

Naphthalene   Anthracene   Phenanthrene   Benzo[*a*]pyrene

When a polycyclic compound contains phenanthrene, it may act as a carcinogen, a substance known to cause cancer.

Compounds containing five or more fused benzene rings such as benzo[*a*]pyrene are potent carcinogens. The molecules interact with the DNA in the cells, causing abnormal cell growth and cancer. Increased exposure to carcinogens increases the chance of DNA

alterations in the cells. Benzo[*a*]pyrene, a product of combustion, has been identified in coal tar, tobacco smoke, barbecued meats, and automobile exhaust.

Aromatic compounds such as benzo[a]pyrene are strongly associated with lung cancers.

PRACTICE PROBLEMS

17.3 Aromatic Compounds

17.33 Write the IUPAC name for each of the following:

a. [structure with Cl] b. [structure with ethyl] c. [structure with OH]

17.34 Write the IUPAC name for each of the following:

a. [structure with Br] b. [structure with methyl] c. [structure with Cl, Cl]

17.35 Draw the line-angle formula for each of the following compounds:
a. aniline
b. 1-chloro-4-fluorobenzene
c. 2-ethyltoluene

17.36 Draw the line-angle formula for each of the following compounds:
a. 1-bromo-2-chlorobenzene
b. 4-chlorotoluene
c. propylbenzene

17.4 Alcohols and Ethers

LEARNING GOAL Write the IUPAC and common names for alcohols and the common names for ethers; draw the condensed structural and line-angle formulas when given their names.

ENGAGE 17.7

How does the functional group of an ether differ from that of an alcohol?

In the functional group of an **alcohol**, a hydroxyl group (—OH) replaces a hydrogen atom in a hydrocarbon. In a *phenol*, the hydroxyl group replaces a hydrogen atom attached to a benzene ring. Molecules of alcohols and phenols have a bent shape around the oxygen atom. In the functional group of an **ether**, an oxygen atom is attached by single bonds to two carbon atoms (see **FIGURE 17.5**).

Ball-and-Stick Model			
Condensed Structural Formula	CH₃—O—H	⬡—O—H	CH₃—O—CH₃
Line-Angle Formula	—OH	⬡—OH	∕O∖
Name	Methanol	Phenol	Dimethyl ether

FIGURE 17.5 An alcohol or a phenol has a hydrogen atom replaced by a hydroxyl group (—OH); an ether contains an oxygen atom (—O—) bonded to two carbon groups.

Naming Alcohols

In the IUPAC system, an alcohol is named by replacing the final *e* in the corresponding alkane name with *ol*. The common name of a simple alcohol uses the name of the alkyl group followed by *alcohol*.

$$CH_3—H \qquad CH_3—OH \qquad\qquad CH_3—CH_2—H \qquad CH_3—CH_2—OH$$
Methane   Methanol   Ethane   Ethanol
             (methyl alcohol)   (ethyl alcohol)

Alcohols with one or two carbon atoms do not require a number for the hydroxyl group. When an alcohol consists of a chain with three or more carbon atoms, the chain is numbered to give the position of the —OH group and any substituents on the chain.

$$\underset{3}{CH_3}—\underset{2}{CH_2}—\underset{1}{CH_2}—OH \qquad\qquad \underset{1}{CH_3}—\underset{2}{\overset{\overset{\displaystyle OH}{|}}{CH}}—\underset{3}{CH_3}$$

          1-Propanol   2-Propanol
       (propyl alcohol)   (isopropyl alcohol)

We can also draw the line-angle formulas for alcohols as shown for 2-propanol and 2-butanol.

  OH   OH

2-Propanol   2-Butanol

SAMPLE PROBLEM 17.10 Naming Alcohols

TRY IT FIRST

Write the IUPAC name for the following:

$$CH_3—\overset{\overset{\displaystyle CH_3}{|}}{CH}—CH_2—\overset{\overset{\displaystyle OH}{|}}{CH}—CH_3$$

SOLUTION

	Given	Need	Connect
ANALYZE THE PROBLEM	five carbon chain, hydroxyl group, methyl group	IUPAC name	position of methyl and hydroxyl groups, replace *e* in alkane name with *ol*

STEP 1 **Name the longest carbon chain attached to the —OH group by replacing the e in the corresponding alkane name with ol.** To name the alcohol, the *e* in the alkane name pentane is replaced by *ol*.

$$CH_3-CH-CH_2-CH-CH_3 \qquad \text{pentanol}$$

with CH₃ above the second carbon and OH above the fourth carbon

STEP 2 **Number the chain starting at the end nearer to the —OH group.** This carbon chain is numbered from right to left to give the position of the —OH group as carbon 2, which is shown as a prefix in the name 2-pentanol.

$$CH_3-CH-CH_2-CH-CH_3 \qquad \text{2-pentanol}$$
$$\;\;\;\;5\;\;\;\;\;4\;\;\;\;\;\;3\;\;\;\;\;\;\;2\;\;\;\;\;1$$

STEP 3 **Give the location and name of each substituent.**

$$CH_3-CH-CH_2-CH-CH_3 \qquad \text{4-methyl-2-pentanol}$$
$$\;\;\;\;5\;\;\;\;\;4\;\;\;\;\;\;3\;\;\;\;\;\;\;2\;\;\;\;\;1$$

SELF TEST 17.10

Write the IUPAC name for each of the following:

a.

b.

ANSWER

a. 3-chloro-1-butanol

b. 4,6-dimethyl-2-heptanol

PRACTICE PROBLEMS

Try Practice Problems 17.37 and 17.38

Naming Ethers

An *ether* consists of an oxygen atom that is attached by single bonds to two carbon groups that are alkyl or aromatic groups. In the common name of an ether, the names of the alkyl or aromatic groups attached to the oxygen atom are written in alphabetical order, followed by the word *ether*.

Methyl group → $\boxed{CH_3}$—O—$\boxed{CH_2-CH_2-CH_3}$ ← Propyl group

Common name: methyl propyl ether

PRACTICE PROBLEMS

Try Practice Problems 17.39 to 17.42

Chemistry Link to Health

Some Important Alcohols, Phenols, and Ethers

Methanol (*methyl alcohol*), the simplest alcohol, is found in many solvents and paint removers. If ingested, methanol is oxidized to formaldehyde, which can cause headaches, blindness, and death. Methanol is used to make plastics, medicines, and fuels. In car racing, it is used as a fuel because it is less flammable and has a higher octane rating than does gasoline.

Ethanol (*ethyl alcohol*) has been known since prehistoric times as an intoxicating product formed by the fermentation of grains, sugars, and starches. Recent interest in biofuels has led to increased production of ethanol by fermentation. "Gasohol" is a mixture of ethanol and gasoline used as a fuel.

$$C_6H_{12}O_6 \xrightarrow{\text{Fermentation}} 2CH_3-CH_2-OH + 2CO_2$$

Glucose   Ethanol

Ethanol is also used in hand sanitizers. In an alcohol-containing sanitizer, the amount of ethanol can be as high as 85% (v/v). This amount of ethanol can make hand sanitizers a fire hazard in the home because ethanol is highly flammable. It is recommended that sanitizers containing ethanol be placed in storage areas that are away from heat sources in the home.

Today, ethanol for commercial use is produced by reacting ethene and water at high temperatures and pressures.

$$H_2C=CH_2 + H_2O \xrightarrow{\text{300 °C, 200 atm, } H^+} CH_3-CH_2-OH$$
Ethene   Ethanol

1,2-Ethanediol (*ethylene glycol*) is used as antifreeze in heating and cooling systems. It is also a solvent for paints, inks, and plastics, and it is used in the production of synthetic fibers such as Dacron. If ingested, it is extremely toxic. In the body, it is oxidized to oxalic acid, which forms insoluble salts in the kidneys that cause renal damage, convulsions, and death. Because its sweet taste is attractive to pets and children, ethylene glycol solutions must be carefully stored.

$$HO-CH_2-CH_2-OH \xrightarrow{[O]} HO-\overset{\overset{\displaystyle O}{\|}}{C}-\overset{\overset{\displaystyle O}{\|}}{C}-OH$$
1,2-Ethanediol   Oxalic acid
(ethylene glycol)

Antifreeze (ethylene glycol) raises the boiling point and decreases the freezing point of water in a radiator.

1,2,3-Propanetriol (*glycerol* or *glycerin*), a trihydroxy alcohol, is a viscous liquid obtained from oils and fats during the production of soaps. The presence of several polar —OH groups makes it strongly attracted to water, a feature that makes glycerol useful as a skin softener in products such as skin lotions, cosmetics, shaving creams, and liquid soaps.

$$HO-CH_2-\overset{\overset{\displaystyle OH}{|}}{CH}-CH_2-OH$$
1,2,3-Propanetriol
(glycerol)

Phenols found in essential oils of plants produce the odor or flavor of the plant.

Ethers as Anesthetics

Anesthesia is the loss of sensation and consciousness. The term *ether* has been associated with anesthesia because diethyl ether was the most widely used anesthetic for more than a hundred years. Although it is easy to administer, ether is very volatile and highly flammable. A small spark in the operating room could cause an explosion. Since the 1950s, anesthetics such as Forane (isoflurane), Ethrane (enflurane), Suprane (desflurane), and Sevoflurane have been developed that are not as flammable. Most of these anesthetics retain the ether group, but the addition of halogen atoms reduces the volatility and flammability of the ethers.

Isoflurane (Forane) is an inhaled anesthetic.

Forane
(isoflurane)

Ethrane
(enflurane)

Suprane
(desflurane)

Sevoflurane

PRACTICE PROBLEMS

17.4 Alcohols and Ethers

17.37 Write the IUPAC name for each of the following:
 a. $CH_3—CH_2—OH$

 b. $CH_3—CH_2—\overset{\displaystyle OH}{\underset{\displaystyle \vert}{CH}}—CH_3$

 c.

17.38 Write the IUPAC name for each of the following:

 a.

 b. $CH_3—CH_2—\overset{\displaystyle CH_3}{\underset{\displaystyle \vert}{CH}}—\overset{\displaystyle CH_3}{\underset{\displaystyle \vert}{CH}}—CH_2—OH$

 c.

17.39 Write the common name for each of the following:
 a. $CH_3—O—CH_2—CH_3$
 b. $CH_3—CH_2—CH_2—O—CH_2—CH_2—CH_3$

 c.

17.40 Write the common name for each of the following:
 a. $CH_3—CH_2—O—CH_2—CH_2—CH_3$

 b.

 c. $CH_3—O—CH_3$

17.41 Draw the condensed structural and line-angle formulas for each of the following:
 a. 1-propanol
 b. ethyl propyl ether
 c. diethyl ether
 d. 2-methyl-2-butanol

17.42 Draw the condensed structural and line-angle formulas for each of the following:
 a. ethyl methyl ether
 b. 3-methyl-1-butanol
 c. 2,4-dichloro-3-hexanol
 d. butyl ethyl ether

17.5 Aldehydes and Ketones

LEARNING GOAL Write the IUPAC and common names for aldehydes and ketones; draw the condensed structural and line-angle formulas when given their names.

ENGAGE 17.8

If aldehydes and ketones both contain a carbonyl group, how can you differentiate between compounds from each family?

Aldehydes and ketones contain a *carbonyl group* (C═O) that has a carbon–oxygen double bond with two groups of atoms attached to the carbon atom at angles of 120°. Because the oxygen atom in the carbonyl group is much more electronegative than the carbon atom, the carbonyl group has a dipole with a partial negative charge (δ^-) on the oxygen and a partial positive charge (δ^+) on the carbon. The polarity of the carbonyl group strongly influences the physical and chemical properties of aldehydes and ketones.

In an **aldehyde,** the carbon of the carbonyl group is bonded to at least one hydrogen atom. That carbon may also be bonded to another hydrogen atom, a carbon of an alkyl group, or an aromatic ring (see **FIGURE 17.6**). In a **ketone,** the carbonyl group is bonded to two alkyl groups or aromatic rings.

Naming Aldehydes

In the IUPAC system, an aldehyde is named by replacing the *e* in the corresponding alkane name with *al*. No number is needed for the aldehyde group because it always appears at the end of the chain. The aldehydes with carbon chains of one to four carbon atoms are often referred to by their common names, which end in *aldehyde* (see **FIGURE 17.7**). The roots (*form, acet, propion,* and *butyr*) of these common names are derived from Latin or Greek words.

FIGURE 17.6 The carbonyl group is found in aldehydes and ketones.

CORE CHEMISTRY SKILL

Naming Aldehydes and Ketones

ENGAGE 17.9

Why is the carbon in the carbonyl group in aldehydes always at the end of the chain?

The carbonyl carbon is at the end of the chain

| IUPAC | Methanal | Ethanal | Propanal | Butanal |
| Common | (**form**aldehyde) | (**acet**aldehyde) | (**propion**aldehyde) | (**butyr**aldehyde) |

FIGURE 17.7 In the structures of aldehydes, the carbonyl group is always the end carbon.

The IUPAC system names the aldehyde of benzene as benzaldehyde. Any substituents are numbered from the carbonyl group, which is attached to carbon 1.

Benzaldehyde

SAMPLE PROBLEM 17.11   Naming Aldehydes

TRY IT FIRST

Write the IUPAC name for the following:

$$CH_3-CH_2-\overset{\overset{\displaystyle CH_3}{|}}{C}H-CH_2-\overset{\overset{\displaystyle O}{||}}{C}-H$$

SOLUTION

	Given	Need	Connect
ANALYZE THE PROBLEM	five-carbon chain, methyl substituent	IUPAC name	position of methyl group, replace *e* in alkane name with *al*

STEP 1   **Name the longest carbon chain by replacing the *e* in the alkane name with *al*.** The longest carbon chain has five carbon atoms.

$$CH_3-CH_2-\overset{\overset{\displaystyle CH_3}{|}}{C}H-CH_2-\overset{\overset{\displaystyle O}{||}}{C}-H \qquad \text{pentanal}$$

STEP 2 **Name and number any substituents by counting the carbonyl group as carbon 1.** Counting from the right, the methyl substituent is on carbon 3.

$$CH_3-CH_2-\underset{3}{CH}-CH_2-\overset{O}{\overset{\|}{C}}-H \qquad \text{3-methylpentanal}$$

with CH₃ on carbon 3, carbons numbered 5 4 3 2 1

SELF TEST 17.11

a. What is the IUPAC name of the following compound?

b. Draw the line-angle formula for 4-chlorobenzaldehyde.

ANSWER

a. 5-methylhexanal

b.

Naming Ketones

In the IUPAC system, the name of a ketone is obtained by replacing the *e* in the corresponding alkane name with *one*. Carbon chains with five carbon atoms or more are numbered from the end nearer the carbonyl group.

Because ketones play a major role in organic chemistry, they are still often referred to by their common names. In the common names for unbranched ketones, the alkyl groups bonded to the carbonyl group are named as substituents and are listed alphabetically, followed by *ketone*. Acetone, which is another name for propanone, has been retained by the IUPAC system.

ENGAGE 17.10

Why is ethyl propyl ketone the same compound as 3-hexanone?

$$CH_3-\overset{O}{\overset{\|}{C}}-CH_3 \qquad CH_3-CH_2-\overset{O}{\overset{\|}{C}}-CH_3 \qquad CH_3-CH_2-\overset{O}{\overset{\|}{C}}-CH_2-CH_3$$

Propanone (dimethyl ketone; acetone)   Butanone (ethyl methyl ketone)   3-Pentanone (diethyl ketone)

SAMPLE PROBLEM 17.12 Naming Ketones

TRY IT FIRST

Write the IUPAC name for the following ketone:

SOLUTION

	Given	Need	Connect
ANALYZE THE PROBLEM	five-carbon chain, methyl substituent	IUPAC name	position of methyl and carbonyl groups, replace *e* in alkane name with *one*

STEP 1 Name the longest carbon chain by replacing the *e* in the alkane name with *one.* The longest chain has five carbon atoms, which is named pentanone.

pentanone

STEP 2 Number the carbon chain starting from the end nearer the carbonyl group and indicate its location. Counting from the right, the carbonyl group is on carbon 2.

2-pentanone

5 4 3 2 1

STEP 3 Name and number any substituents on the carbon chain. Counting from the right, the methyl group is on carbon 4. The IUPAC name is 4-methyl-2-pentanone.

4-methyl-2-pentanone

5 4 3 2 1

SELF TEST 17.12

a. What is the IUPAC name of the following compound?

b. Draw the line-angle formula for 3-chlorobutanone.

ANSWER

a. 3-hexanone

b.

Cl

PRACTICE PROBLEMS

Try Practice Problems 17.43 to 17.48

Chemistry Link to Health

Some Important Aldehydes and Ketones

Formaldehyde, the simplest aldehyde, is a colorless gas with a pungent odor. An aqueous solution called *formalin,* which contains 40% formaldehyde, is used as a germicide and to preserve biological specimens. Industrially, it is a reactant in the synthesis of polymers used to make fabrics, insulation materials, carpeting, pressed wood products such as plywood, and plastics for kitchen counters. Exposure to formaldehyde fumes can irritate the eyes, nose, and upper respiratory tract and cause skin rashes, headaches, dizziness, and general fatigue.

Several naturally occurring aromatic aldehydes are used to flavor food and as fragrances in perfumes. Benzaldehyde is found in almonds, vanillin in vanilla beans, and cinnamaldehyde in cinnamon.

Benzaldehyde (almond)

Vanillin (vanilla)

Cinnamaldehyde (cinnamon)

$$H-\overset{\overset{\textstyle O}{\|}}{C}-H$$

Methanal (formaldehyde)

The simplest ketone, known as *acetone* or propanone (dimethyl ketone), is a colorless liquid with a mild odor that has wide use as a solvent in cleaning fluids, paint and nail polish removers, and rubber cement.

Acetone is extremely flammable, and care must be taken when using it. In the body, acetone may be produced in uncontrolled diabetes, fasting, and high-protein diets when large amounts of fats are metabolized for energy.

Muscone is a ketone used to make musk perfumes, and oil of spearmint contains carvone.

Muscone
(musk)

Carvone
(spearmint oil)

Acetone is used as a solvent in paint and nail polish removers.

PRACTICE PROBLEMS

17.5 Aldehydes and Ketones

17.43 Write the common name for each of the following:

a. $CH_3-\overset{\displaystyle O}{\overset{\displaystyle \|}{C}}-H$

b.

c. $H-\overset{\displaystyle O}{\overset{\displaystyle \|}{C}}-H$

17.44 Write the common name for each of the following:

a. $CH_3-CH_2-\overset{\displaystyle O}{\overset{\displaystyle \|}{C}}-CH_3$

b.

c. $CH_3-CH_2-\overset{\displaystyle O}{\overset{\displaystyle \|}{C}}-H$

17.45 Write the IUPAC name for each of the following:

a. $CH_3-CH_2-\overset{\displaystyle O}{\overset{\displaystyle \|}{C}}-H$

b.

c.

17.46 Write the IUPAC name for each of the following:

a. $CH_3-CH_2-CH_2-\overset{\displaystyle O}{\overset{\displaystyle \|}{C}}-H$

b.

c.

17.47 Draw the condensed structural formula for **a** and **b** and the line-angle formula for **c**.
   a. acetaldehyde   b. 2-pentanone
   c. butyl methyl ketone

17.48 Draw the condensed structural formula for **a** and **b** and the line-angle formula for **c**.
   a. propionaldehyde   b. 3,4-dimethylhexanal
   c. 4-bromobutanone

17.6 Carboxylic Acids and Esters

LEARNING GOAL Write the IUPAC and common names for carboxylic acids and esters; draw the condensed structural and line-angle formulas when given their names.

In the functional group of a **carboxylic acid**, the carbon atom of a carbonyl group is attached to a hydroxyl group (—OH), which forms a *carboxyl group*. Some ways to represent the carboxyl group in carboxylic acids are shown for propanoic acid.

Carbonyl group

$-\overset{\displaystyle O}{\overset{\displaystyle \|}{C}}-\boxed{OH}$ *Hydroxyl group*

Carboxyl group

$CH_3-CH_2-\overset{\displaystyle O}{\overset{\displaystyle \|}{C}}-OH$

$\overset{\displaystyle O}{\overset{\displaystyle \|}{\diagup}}OH$

Propanoic acid
(propionic acid)

CH_3-CH_2-COOH

IUPAC Names of Carboxylic Acids

The IUPAC name of a carboxylic acid replaces the *e* in the corresponding alkane name with *oic acid*. If there are substituents, the carbon chain is numbered beginning with the carboxyl carbon.

$$\underset{\text{Methanoic acid}}{H-\overset{\displaystyle O}{\overset{\|}{C}}-OH} \qquad \underset{\text{2-Methylpropanoic acid}}{CH_3-\overset{\displaystyle CH_3}{\overset{|}{CH}}-\overset{\displaystyle O}{\overset{\|}{C}}-OH} \qquad \underset{\text{3-Chlorobutanoic acid}}{CH_3-\overset{\displaystyle Cl}{\overset{|}{CH}}-CH_2-\overset{\displaystyle O}{\overset{\|}{C}}-OH}$$

As with the aldehydes, carboxylic acids with one to four carbon atoms have common names, which are derived from their natural sources. The common names use the prefixes of *form*, *acet*, *propion*, and *butyr*; these common names are derived from Latin or Greek words (see **TABLE 17.7**).

The sour taste of vinegar is due to ethanoic acid (acetic acid).

TABLE 17.7 IUPAC and Common Names of Selected Carboxylic Acids

Condensed Structural Formula	Line-Angle Formula	IUPAC Name	Common Name	Ball-and-Stick Model
$H-\overset{O}{\overset{\|}{C}}-OH$		Methanoic acid	Formic acid	
$CH_3-\overset{O}{\overset{\|}{C}}-OH$		Ethanoic acid	Acetic acid	
$CH_3-CH_2-\overset{O}{\overset{\|}{C}}-OH$		Propanoic acid	Propionic acid	
$CH_3-CH_2-CH_2-\overset{O}{\overset{\|}{C}}-OH$		Butanoic acid	Butyric acid	

The simplest aromatic carboxylic acid is benzoic acid. The carbon of the carboxyl group is bonded to carbon 1 in the ring, and the ring is numbered to give the lowest numbers for any substituent.

Benzoic acid   3,4-Dichlorobenzoic acid   2-Bromobenzoic acid

TRY IT FIRST

Write the IUPAC name for the following:

SOLUTION

	Given	Need	Connect
ANALYZE THE PROBLEM	four-carbon chain, methyl substituent	IUPAC name	position of methyl group, replace e in alkane name with *oic acid*

STEP 1 **Identify the longest carbon chain and replace the e in the alkane name with *oic acid*.** The IUPAC name of a carboxylic acid with four carbons is butanoic acid.

butanoic acid

STEP 2 **Name and number any substituents by counting the carboxyl carbon as 1.** Counting from the right, the methyl substituent is on carbon 2. The IUPAC name is 2-methylbutanoic acid.

2-methylbutanoic acid

4 3 2 1

SELF TEST 17.13

a. Write the IUPAC name for the following:

$$CH_3 - CH_2 - CH_2 - CH_2 - \overset{\displaystyle O}{\overset{\|}{C}} - OH$$

b. Draw the line-angle formula for 3-chlorobenzoic acid.

ANSWER

a. pentanoic acid

b.

PRACTICE PROBLEMS
Try Practice Problems 17.49 to 17.52

Chemistry Link to Health

Carboxylic Acids in Metabolism

Several carboxylic acids are part of the metabolic processes within our cells. For example, during glycolysis, a molecule of glucose is broken down into two molecules of pyruvic acid, or actually, its carboxylate ion, pyruvate. During strenuous exercise when oxygen levels are low (anaerobic), pyruvic acid is reduced to give lactic acid or the lactate ion.

$$CH_3 - \overset{\displaystyle O}{\overset{\|}{C}} - \overset{\displaystyle O}{\overset{\|}{C}} - OH + 2H \xrightarrow{\text{Reduction}} CH_3 - \overset{\displaystyle OH}{\overset{|}{CH}} - \overset{\displaystyle O}{\overset{\|}{C}} - OH$$

Pyruvic acid   Lactic acid

In the *citric acid cycle*, also called the Krebs cycle, di- and tricarboxylic acids are oxidized and decarboxylated (loss of CO_2) to produce energy for the cells of the body. These carboxylic acids

During exercise, pyruvic acid is converted to lactic acid in the muscles.

are normally referred to by their common names. At the start of the citric acid cycle, citric acid with six carbons is converted to five-carbon α-ketoglutaric acid. Citric acid is also the acid that gives the sour taste to citrus fruits such as lemons and grapefruits.

At the pH of the aqueous environment in the cells, the carboxylic acids are ionized, which means it is actually the carboxylate ions that take part in the reactions of the citric acid cycle. For example, in water, succinic acid is in equilibrium with its carboxylate ion, succinate.

$$
\begin{array}{ccc}
\text{COOH} & & \text{COOH} \\
| & & | \\
\text{CH}_2 & & \text{CH}_2 \\
| & [O] & | \\
\text{HO}-\text{C}-\text{COOH} & \xrightarrow{} & \text{CH}_2 + \text{CO}_2 \\
| & & | \\
\text{CH}_2 & & \text{C}=\text{O} \\
| & & | \\
\text{COOH} & & \text{COOH} \\
\text{Citric acid} & & \alpha\text{-Ketoglutaric acid}
\end{array}
$$

$$
\begin{array}{ccc}
\text{COOH} & & \text{COO}^- \\
| & & | \\
\text{CH}_2 & & \text{CH}_2 \\
| & +\,2\text{H}_2\text{O} \rightleftharpoons & | + 2\text{H}_3\text{O}^+ \\
\text{CH}_2 & & \text{CH}_2 \\
| & & | \\
\text{COOH} & & \text{COO}^- \\
\text{Succinic acid} & & \text{Succinate ion}
\end{array}
$$

The citric acid cycle continues as α-ketoglutaric acid loses CO_2 to give a four-carbon succinic acid. Succinic acid is then oxidized to fumaric acid. We see that some of the functional groups we have studied along with reactions such as oxidation are part of the metabolic processes that take place in our cells.

$$
\begin{array}{ccc}
\text{COOH} & & \text{COOH} \\
| & & | \\
\text{CH}_2 & [O] & \text{C}-\text{H} \\
| & \xrightarrow{} & || \\
\text{CH}_2 & & \text{H}-\text{C} \\
| & & | \\
\text{COOH} & & \text{COOH} \\
\text{Succinic acid} & & \text{Fumaric acid}
\end{array}
$$

Citric acid gives the sour taste to citrus fruits.

Esters

In the functional group of an **ester**, the —H of the carboxylic acid is replaced by an alkyl group. Fats and oils in our diets contain esters of glycerol and fatty acids, which are long-chain carboxylic acids. The pleasant aromas and flavors of many fruits including bananas, oranges, and strawberries are due to esters.

Carboxylic Acid　　　**Ester**

$$
\begin{array}{cc}
\overset{\displaystyle O}{\overset{||}{\text{CH}_3-\text{C}-\text{OH}}} & \overset{\displaystyle O}{\overset{||}{\text{CH}_3-\text{C}-\text{O}-\text{CH}_3}}
\end{array}
$$

Ethanoic acid　　　Methyl ethanoate
(acetic acid)　　　(methyl acetate)

Esterification

In a reaction called *esterification*, an ester is produced when a carboxylic acid and an alcohol react in the presence of an acid catalyst (usually H_2SO_4) and heat. An excess of the alcohol reactant is used to shift the equilibrium in the direction of the formation of the ester product. In this esterification reaction, the —OH removed from the carboxylic acid and the —H removed from the alcohol combine to form water.

CORE CHEMISTRY SKILL

Forming Esters

$$
\overset{\displaystyle O}{\overset{||}{\text{CH}_3-\text{C}-\text{OH}}} + \text{H}-\text{O}-\text{CH}_3 \underset{}{\overset{\text{H}^+,\ \text{heat}}{\rightleftharpoons}} \overset{\displaystyle O}{\overset{||}{\text{CH}_3-\text{C}-\text{O}-\text{CH}_3}} + \text{H}-\text{OH}
$$

Ethanoic acid　　　Methanol　　　　Methyl ethanoate
(acetic acid)　　(methyl alcohol)　　(methyl acetate)

SAMPLE PROBLEM 17.14 Writing Esterification Equations

TRY IT FIRST

An ester that has the smell of pineapples can be synthesized from butanoic acid and methanol. Write the balanced chemical equation for the formation of this ester.

SOLUTION

ANALYZE THE PROBLEM	Given	Need	Connect
	butanoic acid, methanol	esterification equation	ester and H_2O

$$CH_3-CH_2-CH_2-\overset{\overset{\displaystyle O}{\|}}{C}-\textbf{OH} + \textbf{H}-O-CH_3 \underset{}{\overset{H^+,\ heat}{\rightleftharpoons}} CH_3-CH_2-CH_2-\overset{\overset{\displaystyle O}{\|}}{C}-O-CH_3 + \textbf{H}_2\textbf{O}$$

Butanoic acid (butyric acid)   Methanol (methyl alcohol)   Methyl butanoate (methyl butyrate)

SELF TEST 17.14

a. What are the IUPAC names of the carboxylic acid and alcohol that are needed to form the following ester, which has the odor of apples?

$$CH_3-CH_2-\overset{\overset{\displaystyle O}{\|}}{C}-O-CH_2-CH_2-CH_2-CH_2-CH_3$$

b. The ester that smells like plums can be synthesized from methanoic acid and 1-butanol. Write the balanced chemical equation, using line-angle formulas, for the formation of this ester.

ANSWER

a. propanoic acid and 1-pentanol

PRACTICE PROBLEMS
Try Practice Problems 17.53 to 17.56

b.

ENGAGE 17.11

What is the IUPAC name of the ester formed from propanoic acid and ethanol?

Methyl ethanoate (methyl acetate)

FIGURE 17.8 The ester methyl ethanoate (methyl acetate) is made from methyl alcohol and ethanoic acid (acetic acid).

Naming Esters

The name of an ester consists of two words that are derived from the names of the alcohol and the acid in that ester. The first word indicates the *alkyl* part from the alcohol. The second word is the *carboxylate* part from the carboxylic acid. The IUPAC names of esters use the IUPAC names for the carbon chain of the acid, while the common names of esters use the common names of the acids. Let's take a look at the following ester, which has a fruity odor. We start by separating the ester into two parts, which gives us the alkyl of the alcohol and the carboxylate of the acid. Then we name the ester as an alkyl carboxylate (see **FIGURE 17.8**).

SAMPLE PROBLEM 17.15 Naming Esters

TRY IT FIRST

Write the IUPAC and common names for the following:

$$CH_3-CH_2-\overset{\overset{\displaystyle O}{\|}}{C}-O-CH_2-CH_2-CH_3$$

SOLUTION

	Given	Need	Connect
ANALYZE THE PROBLEM	ester	IUPAC name, common name	write the alkyl name for the alcohol, change *ic acid* to *ate*

STEP 1 Write the name for the carbon chain from the alcohol as an *alkyl* group. The alcohol that is used for the ester is propanol, which is named as the alkyl group propyl.

$$CH_3-CH_2-\overset{\overset{\text{O}}{\|}}{C}-O-CH_2-CH_2-CH_3 \qquad \text{propyl}$$

STEP 2 Change the *ic acid* of the acid name to *ate*. The carboxylic acid with three carbon atoms is propanoic acid. Replacing the *ic acid* with *ate* gives the IUPAC name propyl propanoate. The common name of propionic acid gives the common name for the ester of propyl propionate.

$$CH_3-CH_2-\overset{\overset{\text{O}}{\|}}{C}-O-CH_2-CH_2-CH_3 \qquad \text{propyl propanoate}$$

(propyl propionate)

PRACTICE PROBLEMS

Try Practice Problems 17.57 to 17.60

SELF TEST 17.15

a. Write the IUPAC name for the following ester, which gives the odor and flavor to grapes.

b. Draw the line-angle formula for ethyl methanoate, which is present in bee stings.

ANSWER

a. ethyl heptanoate

b.

The odor of grapes is due to an ester.

Esters in Plants

Many of the fragrances of perfumes and flowers and the flavors of fruits are due to esters. Small esters are volatile, so we can smell them, and soluble in water, so we can taste them. Several esters and their flavors and odors are listed in **TABLE 17.8**.

TABLE 17.8 Some Esters in Fruits and Flavorings

Condensed Structural Formula and Name	Flavor/Odor
$CH_3-\overset{\displaystyle O}{\overset{\|}{C}}-O-CH_2-CH_2-CH_3$ Propyl ethanoate (propyl acetate)	Pears
$CH_3-\overset{\displaystyle O}{\overset{\|}{C}}-O-CH_2-CH_2-CH_2-CH_2-CH_3$ Pentyl ethanoate (pentyl acetate)	Bananas
$CH_3-\overset{\displaystyle O}{\overset{\|}{C}}-O-CH_2-CH_2-CH_2-CH_2-CH_2-CH_2-CH_2-CH_3$ Octyl ethanoate (octyl acetate)	Oranges
$CH_3-CH_2-CH_2-\overset{\displaystyle O}{\overset{\|}{C}}-O-CH_2-CH_3$ Ethyl butanoate (ethyl butyrate)	Pineapples
$CH_3-CH_2-CH_2-\overset{\displaystyle O}{\overset{\|}{C}}-O-CH_2-CH_2-CH_2-CH_2-CH_3$ Pentyl butanoate (pentyl butyrate)	Apricots

Esters such as ethyl butanoate give the odor and flavor to many fruits.

PRACTICE PROBLEMS

17.6 Carboxylic Acids and Esters

17.49 Write the IUPAC and common name (if any) for each of the following carboxylic acids:

a. $CH_3-\overset{\displaystyle O}{\overset{\|}{C}}-OH$ b. $CH_3-CH_2-\overset{\displaystyle O}{\overset{\|}{C}}-OH$

c.   d.

17.50 Write the IUPAC and common name (if any) for each of the following carboxylic acids:

a. $H-\overset{\displaystyle O}{\overset{\|}{C}}-OH$ b.

c. (structure with benzene ring, COOH, and Cl)   d. $CH_3-CH_2-\overset{CH_3}{\overset{\|}{C}}H-\overset{\displaystyle O}{\overset{\|}{C}}-OH$

17.51 Draw the condensed structural formulas for **a** and **b** and the line-angle formulas for **c** and **d**.
   a. butyric acid   b. 2-chloroethanoic acid
   c. benzoic acid   d. 3-bromopropanoic acid

17.52 Draw the condensed structural formulas for **a** and **b** and the line-angle formulas for **c** and **d**.
   a. 2-methylhexanoic acid   b. 3-ethylbenzoic acid
   c. 4-chloropentanoic acid   d. 2,4-dibromobutanoic acid

17.53 Draw the condensed structural formula for the ester formed when each of the following reacts with ethyl alcohol:
   a. acetic acid   b. butyric acid

17.54 Draw the condensed structural formula for the ester formed when each of the following reacts with methyl alcohol:
   a. formic acid   b. propionic acid

17.55 Draw the line-angle formula for the ester formed in each of the following reactions.

a. (structure) OH + HO (structure) $\xrightarrow{H^+, heat}$

b. (structure) OH + HO (structure) $\xrightarrow{H^+, heat}$

17.56 Draw the line-angle formula for the ester formed in each of the following reactions.

a. (structure) OH + HO (structure) $\xrightarrow{H^+, heat}$

b.

17.57 Write the IUPAC and common names, if any, for each of the following:

a.

b.

c. $CH_3-CH_2-\overset{\displaystyle O}{\overset{\|}{C}}-O-CH_2-CH_3$

17.58 Write the IUPAC and common names, if any, for each of the following:

a.

b. $CH_3-CH_2-CH_2-CH_2-CH_2-\overset{\displaystyle O}{\overset{\|}{C}}-O-CH_3$

c. $CH_3-CH_2-CH_2-\overset{\displaystyle O}{\overset{\|}{C}}-O-CH_3$

17.59 Draw the condensed structural formulas for **a** and **b** and the line-angle formulas for **c** and **d**.
  a. methyl acetate
  b. butyl formate
  c. ethyl pentanoate
  d. propyl propanoate

17.60 Draw the condensed structural formulas for **a** and **b** and the line-angle formulas for **c** and **d**.
  a. hexyl acetate
  b. ethyl formate
  c. ethyl hexanoate
  d. methyl benzoate

17.7 Amines and Amides

LEARNING GOAL Write the common names for amines and the IUPAC and common names for amides; draw the condensed structural and line-angle formulas when given their names.

In the functional group of an **amine**, a nitrogen atom is attached to one or more carbon atoms. In methylamine, a methyl group replaces one hydrogen atom in ammonia. The bonding of two methyl groups gives dimethylamine. In trimethylamine, methyl groups replace all three hydrogen atoms attached to the nitrogen atom (see **FIGURE 17.9**).

$H-\overset{\displaystyle \cdot\cdot}{\underset{\displaystyle |}{N}}-H$ $CH_3-\overset{\displaystyle \cdot\cdot}{\underset{\displaystyle |}{N}}-H$ $CH_3-\overset{\displaystyle \cdot\cdot}{\underset{\displaystyle |}{N}}-CH_3$ $CH_3-\overset{\displaystyle \cdot\cdot}{\underset{\displaystyle |}{N}}-CH_3$
$\qquad H \qquad\qquad H \qquad\qquad\qquad H \qquad\qquad\qquad CH_3$

Ammonia   Methylamine   Dimethylamine   Trimethylamine

FIGURE 17.9 Amines have one or more carbon atoms bonded to the N atom.

Naming Amines

Several systems are used for naming amines. For simple amines, the common names are often used. In the common name, the alkyl groups bonded to the nitrogen atom are listed in alphabetical order. The prefixes *di* and *tri* are used to indicate two and three identical substituents.

ENGAGE 17.12

What is the difference between diethylamine and triethylamine?

Aromatic Amines

The aromatic amines use the name *aniline*, which is approved by IUPAC.

Aniline   4-Bromoaniline

Aniline is used to make many dyes, which give color to wool, cotton, and silk fibers, as well as blue jeans. It is also used to make the polymer polyurethane and in the synthesis of the pain reliever acetaminophen.

Indigo

Indigo used in blue dyes can be obtained from tropical plants such as *Indigofera tinctoria*.

SAMPLE PROBLEM 17.16 Naming Amines

TRY IT FIRST

Write the common name for each of the following amines:

a. $CH_3-CH_2-NH_2$

b. $CH_3-\overset{\overset{\displaystyle CH_3}{|}}{N}-CH_3$

SOLUTION

a. This amine has one ethyl group attached to the nitrogen atom; its name is ethylamine.

b. The common name for an amine with three methyl groups attached to the nitrogen atom is trimethylamine.

SELF TEST 17.16

What is the name of each of the following amines?

a. (benzene ring with NH_2)

b. $CH_3-\overset{\overset{\displaystyle H}{|}}{N}-CH_2-CH_3$

ANSWER

a. aniline

b. ethylmethylamine

PRACTICE PROBLEMS
Try Practice Problems 17.61 and 17.62

Chemistry Link to the Environment

Alkaloids: Amines in Plants

Alkaloids are physiologically active nitrogen-containing compounds produced by plants. The term *alkaloid* refers to the "alkali-like" or basic characteristics of amines. Certain alkaloids are used in anesthetics, in antidepressants, and as stimulants, and many are habit forming.

As a stimulant, nicotine increases the level of adrenaline in the blood, which increases the heart rate and blood pressure. Nicotine is addictive because it activates pleasure centers in the brain. Coniine, which is obtained from hemlock, is extremely toxic.

Nicotine   Coniine

Caffeine is a central nervous system stimulant. Present in coffee, tea, soft drinks, energy drinks, chocolate, and cocoa, caffeine increases alertness, but it may cause nervousness and insomnia. Caffeine is also used in certain pain relievers to counteract the drowsiness caused by an antihistamine.

Caffeine

Caffeine is a stimulant found in coffee, tea, energy drinks, and chocolate.

For many centuries, morphine and codeine, alkaloids found in the oriental poppy plant, have been used as effective painkillers. Codeine, which is structurally similar to morphine, is used in some prescription painkillers and cough syrups. Heroin, obtained by a chemical modification of morphine, is strongly addicting and is not used medically. The structure of the prescription drug OxyContin (oxycodone) used to relieve severe pain is similar to heroin. Today, there are an increasing number of deaths from OxyContin abuse because its physiological effects are also similar to those of heroin.

The green, unripe poppy seed capsule contains a milky sap (opium) that is the source of the alkaloids morphine and codeine.

Morphine

Codeine

Heroin

OxyContin

Amides

In the functional group of an **amide**, the hydroxyl group of a carboxylic acid is replaced by a nitrogen atom (see **FIGURE 17.10**).

Carboxylic Acid

Amide

Ethanoic acid
(acetic acid)

Ethanamide
(acetamide)

FIGURE 17.10 Amides are derivatives of carboxylic acids in which a nitrogen atom replaces the hydroxyl group (—OH).

Preparation of Amides

An amide is produced in a reaction called *amidation*, in which a carboxylic acid reacts with ammonia or an amine. A molecule of water is eliminated, and the fragments of the carboxylic acid and amine molecules join to form the amide, much like the formation of esters.

CORE CHEMISTRY SKILL

Forming Amides

$$CH_3-CH_2-\overset{\overset{\displaystyle O}{\|}}{C}-OH \; + \; H-\overset{\overset{\displaystyle H}{|}}{N}-CH_3 \xrightarrow{\text{Heat}} CH_3-CH_2-\overset{\overset{\displaystyle O}{\|}}{C}-\overset{\overset{\displaystyle H}{|}}{N}-CH_3 \; + \; \mathbf{H_2O}$$

Propanoic acid   Methylamine   N-Methylpropanamide
(propionic acid)   (N-methylpropionamide)

SAMPLE PROBLEM 17.17   Formation of Amides

TRY IT FIRST

Draw the condensed structural formula for the amide product in the following reaction:

$$CH_3-\overset{\overset{\displaystyle O}{\|}}{C}-OH \; + \; H_2N-CH_2-CH_3 \xrightarrow{\text{Heat}}$$

SOLUTION

The condensed structural formula for the amide product can be drawn by attaching the carbonyl group from the acid to the nitrogen atom of the amine. The —OH group removed from the acid and —H from the amine combine to form water.

$$CH_3-\overset{\overset{\displaystyle O}{\|}}{C}-\overset{\overset{\displaystyle H}{|}}{N}-CH_2-CH_3$$

SELF TEST 17.17

Draw **a.** the condensed structural formula and **b.** the line-angle formula for the amide formed from the reaction of propanoic acid and ethylamine.

ANSWER

a. $CH_3-CH_2-\overset{\overset{\displaystyle O}{\|}}{C}-\overset{\overset{\displaystyle H}{|}}{N}-CH_2-CH_3$

b.

PRACTICE PROBLEMS

Try Practice Problems 17.63 and 17.64

Naming Amides

In the IUPAC and common names for amides, the *oic acid* or *ic acid* from the carboxylic acid name is replaced with *amide*. The amide of benzoic acid is benzamide. We can diagram the name of an amide in the following way:

From butanoic acid
(butyric acid)

From ammonia

$$CH_3-CH_2-CH_2-\overset{\overset{\displaystyle O}{\|}}{C}-NH_2$$

IUPAC   Butanamide
Common   (butyramide)

ENGAGE 17.13

Why is the product formed by the amidation of pentanoic acid with ammonia named pentanamide?

$$H-\overset{\overset{\displaystyle O}{\|}}{C}-NH_2$$

Methanamide
(formamide)

$$CH_3-\overset{\overset{\displaystyle O}{\|}}{C}-NH_2$$

Ethanamide
(acetamide)

Benzamide

SAMPLE PROBLEM 17.18 Naming Amides

TRY IT FIRST

Write the IUPAC and common names for the following amide:

$$CH_3-CH_2-\overset{\displaystyle O}{\overset{\displaystyle \|}{C}}-NH_2$$

SOLUTION

The IUPAC name of the carboxylic acid is propanoic acid; the common name is propionic acid. Replacing the *oic acid* or *ic acid* ending with *amide* gives the IUPAC name of propanamide and common name of propionamide.

SELF TEST 17.18

Write the IUPAC name for each of the following:

a.

b.

ANSWER

a. hexanamide

b. 2-bromobenzamide

PRACTICE PROBLEMS

Try Practice Problems 17.65 to 17.68

PRACTICE PROBLEMS

17.7 Amines and Amides

17.61 Write the common name for each of the following:

a. CH_3-NH_2

b.

c. $CH_3-CH_2-\overset{\displaystyle CH_3}{\overset{\displaystyle |}{N}}-CH_2-CH_3$

17.62 Write the common name for each of the following:

a. $CH_3-CH_2-CH_2-NH_2$ b. $CH_3-\overset{\displaystyle CH_3}{\overset{\displaystyle |}{N}}-CH_2-CH_3$

c.

17.63 Draw the condensed structural or line-angle formula for the amide formed in each of the following reactions:

a. $+ NH_3 \xrightarrow{\text{Heat}}$

b. $CH_3-\overset{\displaystyle O}{\overset{\displaystyle \|}{C}}-OH + H_2N-CH_2-CH_3 \xrightarrow{\text{Heat}}$

c. $+ H_2N\diagup\diagdown \xrightarrow{\text{Heat}}$

17.64 Draw the condensed structural or line-angle formula for the amide formed in each of the following reactions:

a. $CH_3-CH_2-CH_2-CH_2-\overset{\displaystyle O}{\overset{\displaystyle \|}{C}}-OH + NH_3 \xrightarrow{\text{Heat}}$

b. $CH_3-\overset{\displaystyle CH_3}{\overset{\displaystyle |}{CH}}-\overset{\displaystyle O}{\overset{\displaystyle \|}{C}}-OH + H_2N-CH_2-CH_3 \xrightarrow{\text{Heat}}$

c. $+ H_2N$ $\xrightarrow{\text{Heat}}$

17.65 Write the IUPAC and common name (if any) for each of the following amides:

a. $CH_3-\overset{\displaystyle O}{\overset{\displaystyle \|}{C}}-NH_2$

b.

c. $H-\overset{\displaystyle O}{\overset{\displaystyle \|}{C}}-NH_2$

17.66 Write the IUPAC and common name (if any) for each of the following amides:

a. $CH_3-\overset{\displaystyle Br}{\overset{\displaystyle |}{CH}}-\overset{\displaystyle O}{\overset{\displaystyle \|}{C}}-NH_2$

b.

c.

17.67 Draw the condensed structural formulas for **a** and **b** and the line-angle formulas for **c** and **d**.
  a. propionamide
  b. 2-methylpentanamide
  c. ethanamide
  d. 4-chlorobenzamide

17.68 Draw the condensed structural formulas for **a** and **b** and the line-angle formulas for **c** and **d**.
  a. heptanamide
  b. 3-methylbutanamide
  c. benzamide
  d. 3-bromopentanamide

UPDATE Diane's Treatment in the Burn Unit

When Diane arrives at the hospital, she is transferred to the ICU burn unit. She has second-degree burns that cause damage to the underlying layers of skin and third-degree burns that cause damage to all the layers of the skin. The most common complications of burns are related to infection. To prevent infection, her skin is covered with a topical antibiotic. Over a period of 3 months, grafts of Diane's unburned skin are used to cover burned areas. She remains in the burn unit for 3 months and then is discharged. However, Diane returns to the hospital for more skin grafts and plastic surgery.

The arson investigators determine that gasoline was the primary accelerant used to start the fire at Diane's house. Because there was a lot of paper and dry wood in the area, the fire spread quickly. Some of the hydrocarbons found in gasoline include hexane, heptane, octane, nonane, decane, and toluene.

Applications

17.69 The medications for Diane are contained in blister packs made of polychlorotrifluoroethylene (PCTFE). Draw the expanded structural formula for a portion of PCTFE polymer from three monomers of chlorotrifluoroethylene:

Chlorotrifluoroethylene

Tablets are contained in blister packs made from polychlorotrifluoroethylene (PCTFE).

17.70 New polymers have been synthesized to replace PVC in IV bags and cling film. One of these is poly (ethylene-vinyl acetate) (PEVA). Draw the expanded structural formula for a portion of PEVA using an alternating sequence that has two of each monomer shown below:

Ethylene   Vinyl acetate

17.71 Write the balanced chemical equation for the complete combustion of each of the following hydrocarbons found in gasoline:
  a. hexane
  b. toluene

17.72 Write the balanced chemical equation for the complete combustion of each of the following hydrocarbons found in gasoline:
  a. nonane
  b. octane

CONCEPT MAP

CHAPTER REVIEW

17.1 Alkanes

LEARNING GOAL Identify the properties of organic or inorganic compounds. Write the IUPAC names and draw the condensed structural and line-angle formulas for alkanes.

- Organic compounds contain C and H, have covalent bonds, most form nonpolar molecules, have low melting points and low boiling points, are not very soluble in water, dissolve as molecules in solutions, and burn vigorously in air.
- Inorganic compounds are often ionic or contain polar covalent bonds and form polar molecules, have high melting and boiling points, are usually soluble in water, produce ions in water, and do not burn in air.
- Alkanes are hydrocarbons that have only C—C single bonds.
- In the expanded structural formula, a separate line is drawn for every bonded atom.
- A condensed structural formula depicts groups composed of each carbon atom and its attached hydrogen atoms.
- A line-angle formula represents the carbon skeleton as ends and corners of a zigzag line.
- Substituents such as alkyl groups and halogen atoms (named as fluoro, chloro, bromo, or iodo) can replace hydrogen atoms on the main chain.
- The IUPAC system is used to name organic compounds by indicating the number of carbon atoms and the position of any substituents.

17.2 Alkenes, Alkynes, and Polymers

LEARNING GOAL Write the IUPAC names and draw the condensed structural and line-angle formulas for alkenes and alkynes.

- Alkenes are unsaturated hydrocarbons that contain carbon–carbon double bonds (C=C).
- Alkynes contain a carbon–carbon triple bond (C≡C).

- The IUPAC names of alkenes end with *ene*; alkyne names end with *yne*.
- The main chain is numbered from the end nearer the double or triple bond.
- Alkenes react with hydrogen using a metal catalyst to form alkanes.
- Polymers are long-chain molecules that consist of many repeating units of smaller carbon molecules called monomers.

17.3 Aromatic Compounds

LEARNING GOAL Describe the bonding in benzene; name aromatic compounds, and draw their line-angle formulas.

- Most aromatic compounds contain benzene, C_6H_6, a cyclic structure containing six carbon atoms and six hydrogen atoms.
- The structure of benzene is represented as a hexagon with a circle in the center.
- The IUPAC system uses the names of benzene, toluene, aniline, and phenol.
- In the IUPAC name, two substituents are numbered and listed in alphabetical order.

17.4 Alcohols and Ethers

LEARNING GOAL Write the IUPAC and common names for alcohols and the common names for ethers; draw the condensed structural and line-angle formulas when given their names.

- The functional group of an alcohol is the hydroxyl group (—OH) bonded to a carbon chain.

$$CH_3 \diagup \!\! ^O \!\! \diagdown H$$

Methanol

- In a phenol, the hydroxyl group is bonded to an aromatic ring.
- In the IUPAC system, the names of alcohols have *ol* endings, and the location of the —OH group is given by numbering the carbon chain.
- In an ether, an oxygen atom (—O—) is connected by single bonds to two alkyl or aromatic groups.
- In the common names of ethers, the alkyl groups are listed alphabetically followed by the word *ether*.

17.5 Aldehydes and Ketones

LEARNING GOAL Write the IUPAC and common names for aldehydes and ketones; draw the condensed structural and line-angle formulas when given their names.

- Aldehydes and ketones contain a carbonyl group (C=O), which is strongly polar.
- In aldehydes, the carbonyl group appears at the end of the carbon chain attached to at least one hydrogen atom.
- In ketones, the carbonyl group occurs between two carbon atoms.
- In the IUPAC system, the *e* in the alkane name is replaced with *al* for aldehydes and *one* for ketones.
- For ketones with more than four carbon atoms in the main chain, the carbonyl group is numbered to show its location.
- Many of the simple aldehydes and ketones use common names.

17.6 Carboxylic Acids and Esters

LEARNING GOAL Write the IUPAC and common names for carboxylic acids and esters; draw the condensed structural and line-angle formulas when given their names.

- A carboxylic acid contains the carboxyl functional group, which is a hydroxyl group connected to a carbonyl group.
- The IUPAC name of a carboxylic acid is obtained by replacing the *e* in the alkane name with *oic acid*.
- The common names of carboxylic acids with one to four carbon atoms are formic acid, acetic acid, propionic acid, and butyric acid.
- In an ester, a carbon atom replaces the H of the hydroxyl group of a carboxylic acid.
- When heated in the presence of a strong acid, a carboxylic acid reacts with an alcohol to produce an ester. A molecule of water is removed: —OH from the carboxylic acid and —H from the alcohol molecule.
- The names of esters consist of two words, the alkyl group from the alcohol and the name of the carboxylate obtained by replacing the *ic acid* ending with *ate*.

17.7 Amines and Amides

LEARNING GOAL Write the common names for amines and the IUPAC and common names for amides; draw the condensed structural and line-angle formulas when given their names.

Dimethylamine

- A nitrogen atom attached to one or more carbon atoms forms an amine.
- In the common names of simple amines, the alkyl groups are listed alphabetically followed by *amine*.
- Amides are derivatives of carboxylic acids in which the hydroxyl group is replaced by a nitrogen atom. Amides are named by replacing the *ic acid* or *oic acid* ending with *amide*.
- Amides are formed by the reaction of a carboxylic acid with ammonia or an amine.

SUMMARY OF NAMING

Family	Condensed Structural Formula	IUPAC Name	Common Name
Alkane	$CH_3-CH_2-CH_3$	Propane	
Alkane with a substituent	$CH_3-CH-CH_3$ with CH_3 above the CH	Methylpropane	
Alkene	$CH_3-CH=CH_2$	Propene	Propylene
Alkyne	$CH_3-C\equiv CH$	Propyne	
Aromatic	(benzene ring)	Benzene	
Alcohol	CH_3-OH	Methanol	Methyl alcohol
Ether	CH_3-O-CH_3		Dimethyl ether
Aldehyde	$H-\overset{\overset{O}{\|\|}}{C}-H$	Methanal	Formaldehyde
Ketone	$CH_3-\overset{\overset{O}{\|\|}}{C}-CH_3$	Propanone	Acetone; dimethyl ketone

Family	Condensed Structural Formula	IUPAC Name	Common Name
Carboxylic acid	$\begin{matrix} & O \\ & \| \\ CH_3 - & C - OH \end{matrix}$	Ethanoic acid	Acetic acid
Ester	$\begin{matrix} & O \\ & \| \\ CH_3 - & C - O - CH_3 \end{matrix}$	Methyl ethanoate	Methyl acetate
Amine	$CH_3 - CH_2 - NH_2$		Ethylamine
Amide	$\begin{matrix} & O \\ & \| \\ CH_3 - & C - NH_2 \end{matrix}$	Ethanamide	Acetamide

SUMMARY OF REACTIONS

The chapter sections to review are shown after the name of each reaction.

Combustion (17.1)

$$CH_3 - CH_2 - CH_3(g) + 5O_2(g) \xrightarrow{\Delta} 3CO_2(g) + 4H_2O(g) + energy$$
Propane

Hydrogenation (17.2)

$$H_2C = CH - CH_3 + H_2 \xrightarrow{Pt} CH_3 - CH_2 - CH_3$$
Propene   Propane

Polymerization (17.2)

$$H_2C = CH_2 + H_2C = CH_2 + H_2C = CH_2 \xrightarrow[\text{catalyst}]{\text{Heat, pressure,}} -CH_2 - CH_2 - CH_2 - CH_2 - CH_2 - CH_2 -$$
Ethene monomers   Polyethylene

Esterification (17.6)

$$CH_3 - \overset{\overset{\displaystyle O}{\|}}{C} - OH + HO - CH_3 \underset{}{\overset{H^+, heat}{\rightleftharpoons}} CH_3 - \overset{\overset{\displaystyle O}{\|}}{C} - O - CH_3 + H_2O$$
Ethanoic acid   Methanol   Methyl ethanoate
(acetic acid)   (methyl alcohol)   (methyl acetate)

Amidation (17.7)

$$CH_3 - CH_2 - \overset{\overset{\displaystyle O}{\|}}{C} - OH + H - \overset{\overset{\displaystyle H}{|}}{N} - H \xrightarrow{Heat} CH_3 - CH_2 - \overset{\overset{\displaystyle O}{\|}}{C} - \overset{\overset{\displaystyle H}{|}}{N} - H + H_2O$$
Propanoic acid   Ammonia   Propanamide
(propionic acid)   (propionamide)

KEY TERMS

alcohol An organic compound that contains a hydroxyl group (—OH) attached to a carbon chain.

aldehyde An organic compound that contains a carbonyl group (C=O) bonded to at least one hydrogen atom.

alkane A type of hydrocarbon in which the carbon atoms are connected only by single bonds.

alkene A type of hydrocarbon that contains a carbon–carbon double bond (C=C).

alkyl group An alkane minus one hydrogen atom. Alkyl groups are named like the corresponding alkanes except a *yl* ending replaces *ane*.

alkyne A type of hydrocarbon that contains a carbon–carbon triple bond (C≡C).

amide An organic compound in which the hydroxyl group of a carboxylic acid is replaced by a nitrogen atom.

amine An organic compound that contains a nitrogen atom bonded to one or more carbon atoms.

aromatic compound A compound that contains the ring structure of benzene.

benzene A ring of six carbon atoms each of which is attached to one hydrogen atom, C_6H_6.

carboxylic acid An organic compound that contains the carboxyl group (—COOH).

condensed structural formula A formula that shows the carbon atoms grouped with the attached number of hydrogen atoms.

ester An organic compound in which the —H of a carboxyl group is replaced by a carbon atom.

ether An organic compound in which an oxygen atom is bonded to two carbon groups that are alkyl or aromatic.

expanded structural formula A formula that shows all of the atoms and the bonds connected to each atom.

functional group A group of atoms that determines the physical and chemical properties and naming of a class of organic compounds.

hydrocarbon A type of organic compound that contains only carbon and hydrogen.

IUPAC system A naming system used for organic compounds.

ketone An organic compound in which a carbonyl group ($C=O$) is bonded to two alkyl or aromatic groups.

line-angle formula A simplified structure that shows a zigzag line in which carbon atoms are represented as the ends of each line and as corners.

monomer The small organic molecule that is repeated many times in a polymer.

polymer A very large molecule that is composed of many small, repeating structural units that are identical.

structural isomers Two compounds that have the same molecular formula but different arrangements of atoms.

substituent A group of atoms such as an alkyl group or a halogen bonded to the main chain of carbon atoms.

CORE CHEMISTRY SKILLS

The chapter section containing each Core Chemistry Skill is shown in parentheses at the end of each heading.

Naming and Drawing Alkanes (17.1)

* The alkanes ethane, propane, and butane contain two, three, and four carbon atoms, respectively, connected in a row or a *continuous* chain.
* Alkanes with five or more carbon atoms in a chain are named using the prefixes *pent* (5), *hex* (6), *hept* (7), *oct* (8), *non* (9), and *dec* (10).
* In the condensed structural formula, the carbon and hydrogen atoms on the ends are written as $-CH_3$ and the carbon and hydrogen atoms in the middle are written as $-CH_2-$.

Example: a. What is the name of $CH_3-CH_2-CH_2-CH_3$?
b. Draw the condensed structural formula for pentane.

Answer: a. An alkane with a four-carbon chain is named butane.
b. Pentane is an alkane with a five-carbon chain. The carbon atoms on the ends are attached to three H atoms each, and the carbon atoms in the middle are attached to two H each. $CH_3-CH_2-CH_2-CH_2-CH_3$

Writing Equations for Hydrogenation and Polymerization (17.2)

* Hydrogenation adds hydrogen atoms to the double bond of an alkene, using a metal catalyst, to form an alkane.
* In polymerization, many small molecules (monomers) join together to form a polymer.

Example: a. Draw the condensed structural formula for 2-methyl-2-butene.
b. Draw the condensed structural formula for the product of the hydrogenation of 2-methyl-2-butene.
c. Draw the condensed structural formula of a portion of the polymer polybutene (polybutylene), used in synthetic rubber, formed from three monomers of 1-butene.

Answer: a.
$$\underset{\displaystyle CH_3-C=CH-CH_3}{CH_3}$$

b. When H_2 adds to a double bond, the product is an alkane with the same number of carbon atoms.
$$\underset{\displaystyle CH_3-CH-CH_2-CH_3}{CH_3}$$

c.
$$\underset{\displaystyle -CH_2-\underset{CH_2}{\underset{|}{\overset{|}{\overset{CH_3}{CH}}}}-CH_2-\underset{CH_2}{\underset{|}{\overset{|}{\overset{CH_3}{CH}}}}-CH_2-\underset{CH_2}{\underset{|}{\overset{|}{\overset{CH_3}{CH}}}}-}{}$$

Naming Aldehydes and Ketones (17.5)

* In the IUPAC system, an aldehyde is named by replacing the *e* in the alkane name with *al* and a ketone by replacing the *e* with *one*.
* The position of a substituent on an aldehyde is indicated by numbering the carbon chain from the carbonyl group and given in front of the name.
* For a ketone, the carbon chain is numbered from the end nearer the carbonyl group.

Example: Give the IUPAC name for the following:

Answer: 5-methyl-3-hexanone

Naming Carboxylic Acids (17.6)

* The IUPAC name of a carboxylic acid is obtained by replacing the *e* in the corresponding alkane name with *oic acid*.
* The common names of carboxylic acids with one to four carbon atoms are formic acid, acetic acid, propionic acid, and butyric acid.

Example: Write the IUPAC name for the following:

$$CH_3-\underset{\underset{\displaystyle}{|}}{\overset{\overset{\displaystyle Br}{|}}{CH}}-CH_2-\overset{\overset{\displaystyle O}{\|}}{C}-OH$$

Answer: 3-bromobutanoic acid

Forming Esters (17.6)

* Esters are formed when carboxylic acids react with alcohols in the presence of a strong acid and heat.

Example: Draw the condensed structural formula for the ester product of the reaction of propanoic acid and methanol.

Answer: $CH_3-CH_2-\overset{\overset{\displaystyle O}{\|}}{C}-O-CH_3$

Forming Amides (17.7)

* Amides are formed when carboxylic acids react with ammonia or amines in the presence of heat.

Example: Draw the condensed structural formula for the amide produced by the reaction of methanoic acid and ethylamine.

Answer: $H-\overset{\overset{\displaystyle O}{\|}}{C}-\overset{\overset{\displaystyle H}{|}}{N}-CH_2-CH_3$

UNDERSTANDING THE CONCEPTS

The chapter sections to review are shown in parentheses at the end of each problem.

17.73 Draw a portion of the polymer Teflon, which is made from 1,1,2,2-tetrafluoroethene (use three monomers). (17.2)

Teflon is used as a nonstick coating on cooking pans.

17.74 A garden hose is made of polyvinyl chloride (PVC), a polymer of chloroethene (vinyl chloride). Draw a portion of the polymer (use three monomers) for PVC. (17.2)

A plastic garden hose is made of PVC.

17.75 Citronellal, found in oil of citronella, lemon, and lemon grass, is used in perfumes and as an insect repellent. (17.2, 17.5)

An insect-repelling candle contains citronellal, which is found in natural sources.

Citronellal has the following structure:

$$CH_3-\overset{\overset{\displaystyle CH_3}{|}}{C}=CH-CH_2-CH_2-\overset{\overset{\displaystyle CH_3}{|}}{CH}-CH_2-\overset{\overset{\displaystyle O}{||}}{C}-H$$

a. Complete the IUPAC name for citronellal:

_____, _____-di _____-_____-octenal.

b. What does the *en* in octenal signify?

c. What does the *al* in octenal signify?

17.76 Draw the condensed structural formula for each of the following: (17.5)

a. 2-heptanone, an alarm pheromone of bees

b. 2,6-dimethyl-3-heptanone, a communication pheromone of bees

Bees emit chemicals called pheromones to communicate.

ADDITIONAL PRACTICE PROBLEMS

17.77 Draw the condensed structural formula for each of the following: (17.1, 17.2)

a. 3-ethylhexane

b. 2-pentene

c. 2-hexyne

17.78 Draw the condensed structural formula for each of the following: (17.1, 17.2)

a. 1-chloro-2-butyne

b. 2,3-dimethylpentane

c. 3-hexene

17.79 Write the IUPAC name for each of the following: (17.1, 17.2)

a.

b. $CH_3-CH_2-C{\equiv}CH$

c. $CH_3-CH=CH-CH_2-CH_3$

17.80 Write the IUPAC name for each of the following: (17.1, 17.2)

a. $H_2C=\overset{\overset{\displaystyle CH_3}{|}}{C}-CH_2-CH_2-CH_3$

b. CH_3-CH_2-Cl

c. $CH_3-CH_2-\overset{\overset{\displaystyle CH_3-CH_2}{|}}{CH}-CH_2-\overset{\overset{\displaystyle Br}{|}}{CH}-CH_3$

17.81 Classify each of the following according to its functional group: (17.2)

a. $CH_3-CH_2-CH_2-\overset{\overset{\displaystyle O}{||}}{C}-CH_3$

b. $CH_3-CH=CH_2$

c.

d. $CH_3-CH_2-CH_2-NH_2$

17.82 Classify each of the following according to its functional group: (17.2)

a.

b. $CH_3-CH_2-CH_2-\overset{\overset{\displaystyle O}{\|}}{C}-OH$

c. $CH_3-\overset{\overset{\displaystyle CH_3}{|}}{CH}-CH_2-OH$

d. $CH_3-\overset{\overset{\displaystyle CH_3}{|}}{CH}-\overset{\overset{\displaystyle O}{\|}}{C}-H$

17.83 Select the class of organic compound that matches each of the following definitions (**a** to **d**): alkane, alkene, alkyne, alcohol, ether, aldehyde, ketone, carboxylic acid, ester, amine, or amide. (17.2)
a. contains a hydroxyl group
b. contains one or more carbon–carbon double bonds
c. contains a carbonyl group bonded to a hydrogen
d. contains only carbon–carbon single bonds

17.84 Select the class of organic compound that matches each of the following definitions (**a** to **d**): alkane, alkene, alkyne, alcohol, ether, aldehyde, ketone, carboxylic acid, ester, amine, or amide. (17.2)
a. contains a carboxyl group in which the hydrogen atom is replaced by a carbon atom
b. contains a carbonyl group bonded to a hydroxyl group
c. contains a nitrogen atom bonded to one or more carbon atoms
d. contains a carbonyl group bonded to two carbon atoms

17.85 Classify each of the following according to its functional group(s): (17.2)

a.

Almonds

b.

Cinnamon sticks

17.86 Classify each of the following according to its functional group(s): (17.2)
a. BHA is an antioxidant used as a preservative in foods such as baked goods, butter, meats, and snack foods.

Baked goods contain
BHA as a preservative.

b. $CH_3-\overset{\overset{\displaystyle O}{\|}}{C}-\overset{\overset{\displaystyle O}{\|}}{C}-CH_3$

Butter

17.87 Name each of the following aromatic compounds: (17.3)

a.   b.   c.

17.88 Name each of the following aromatic compounds: (17.3)

a.   b.   c.

17.89 Draw the structural formula for each of the following: (17.3)
a. ethylbenzene
b. 4-chlorotoluene
c. 1,4-dibromobenzene

17.90 Draw the structural formula for each of the following: (17.3)
a. 3-fluorotoluene
b. 2-methylaniline
c. 1,2-dimethylbenzene

17.91 Write the IUPAC name for **a** and **b** and the common name for **c**. (17.4)

a. $CH_3-CH_2-\overset{\overset{\displaystyle CH_3}{|}}{CH}-OH$ b.

c. $CH_3-CH_2-O-CH_2-CH_2-CH_2-CH_3$

17.92 Write the IUPAC name for **a** and **b** and the common name for **c**. (17.4)

a.

b. $CH_3-\overset{\overset{\displaystyle CH_3}{|}}{CH}-CH_2-CH_2-OH$

c.

17.93 Draw the condensed structural formula for each of the following: (17.4)
a. dibutyl ether
b. 2-methyl-3-pentanol
c. 2-methyl-2-propanol

17.94 Draw the condensed structural formula for each of the following: (17.4)
a. 3-hexanol
b. 2-pentanol
c. ethyl propyl ether

17.95 Write the IUPAC name for each of the following: (17.5)

a.

b. $Cl-CH_2-CH_2-\overset{\overset{\displaystyle O}{\|}}{C}-H$

c. $CH_3-\overset{\overset{\displaystyle Cl}{|}}{CH}-\overset{\overset{\displaystyle O}{\|}}{C}-CH_2-CH_3$

17.96 Write the IUPAC name for each of the following: (17.5)

a. $CH_3-\overset{\overset{\displaystyle CH_3}{|}}{CH}-\overset{\overset{\displaystyle O}{\|}}{C}-CH_3$

b. [structure: 3-chlorobenzaldehyde]

c. $CH_3-\overset{\overset{\displaystyle CH_3}{|}}{CH}-CH_2-CH_2-\overset{\overset{\displaystyle O}{\|}}{C}-OH$

17.97 Draw the condensed structural formulas for **a** and **b** and the line-angle formula for **c**. (17.5)
a. ethyl methyl ketone
b. 3-chloropropionaldehyde
c. 2-bromobenzaldehyde

17.98 Draw the condensed structural formulas for **a** and **b** and the line-angle formula for **c**. (17.5)
a. butyraldehyde
b. 2-bromobutanal
c. 3,5-dimethyl-2-hexanone

17.99 Write the IUPAC name for each of the following: (17.6)

a. $CH_3-\overset{\overset{\displaystyle CH_3}{|}}{CH}-CH_2-\overset{\overset{\displaystyle O}{\|}}{C}-OH$

b. [structure: methyl benzoate]

c. $CH_3-CH_2-\overset{\overset{\displaystyle O}{\|}}{C}-O-CH_2-CH_3$

17.100 Write the IUPAC name for each of the following: (17.6)

a. $CH_3-\overset{\overset{\displaystyle CH_3}{|}}{CH}-CH_2-CH_2-\overset{\overset{\displaystyle O}{\|}}{C}-OH$

b. [structure: 3-chlorobenzoic acid]

c. [structure: ethyl benzoate]

17.101 Draw the condensed structural formula for each of the following: (17.7)
a. ethylamine
b. hexanamide
c. triethylamine

17.102 Draw the condensed structural formula for each of the following: (17.7)
a. formamide
b. ethylpropylamine
c. diethylmethylamine

17.103 Write the name for each of the following: (17.7)

a. [structure: line-angle amine]

b. $CH_3-CH_2-\overset{\overset{\displaystyle O}{\|}}{C}-NH_2$

c. $CH_3-CH_2-CH_2-CH_2-CH_2-NH_2$

17.104 Write the name for each of the following: (17.7)

a. $Cl-CH_2-CH_2-CH_2-\overset{\overset{\displaystyle O}{\|}}{C}-NH_2$

b. $CH_3-CH_2-CH_2-CH_2-\overset{\overset{\displaystyle O}{\|}}{C}-NH_2$

c. $CH_3-CH_2-CH_2-\overset{\overset{\displaystyle CH_2-CH_3}{|}}{N}-CH_2-CH_3$

CHALLENGE PROBLEMS

The following problems are related to the topics in this chapter. However, they do not all follow the chapter order, and they require you to combine concepts and skills from several sections. These problems will help you increase your critical thinking skills and prepare for your next exam.

17.105 Draw the condensed structural formulas and write the IUPAC names for all eight alcohols that have the molecular formula $C_5H_{12}O$. (17.4)

17.106 There are four amine isomers with the molecular formula C_3H_9N. Draw their condensed structural formulas. (17.7)

17.107 The insect repellent DEET can be made from an amidation reaction of 3-methylbenzoic acid and diethylamine. Draw the line-angle formula for DEET. (17.7)

Many insect repellents contain DEET, which is an amide.

17.108 One of the compounds that give blackberries their odor and flavor can be made by heating hexanoic acid and 1-propanol with an acid catalyst. Draw the condensed structural formula for this compound. (17.6)

17.109 Complete and balance each of the following reactions: (17.1, 17.2)

a. $CH_3 — CH_2 — C \equiv CH + O_2 \xrightarrow{\Delta}$

b. $CH_3 — CH_2 — CH = CH — CH_2 — CH_3 + H_2 \xrightarrow{Pt}$

17.110 Complete and balance each of the following reactions: (17.1, 17.2)

a. $H_2C = CH_2 + H_2 \xrightarrow{Pt}$

b. $CH_3 — CH_2 — CH_2 — CH_2 — CH = CH_2 + O_2 \xrightarrow{\Delta}$

ANSWERS TO ENGAGE QUESTIONS

17.1 The central carbon atom in methane has four electron groups. To minimize repulsion, the groups are arranged in a tetrahedral shape.

17.2 In a line-angle formula, each line is a single bond, and carbon atoms are represented as the corners and ends of each line.

17.3 Structural isomers have the same molecular formula, but the atoms in the formulas are bonded in different ways.

17.4 Oil, which is a mixture of hydrocarbons, is not soluble in water and is less dense than water. Thus, oil will remain on the surface when added to water.

17.5 The functional group in an alcohol is an —OH group bonded to carbon. The functional group in a ketone is an oxygen atom bonded to a carbon by a double bond. This carbon is bonded to two other carbon atoms.

17.6 In toluene, a methyl group is bonded to benzene. In phenol, an —OH group is bonded to benzene.

17.7 The functional group in an ether is an oxygen atom bonded by single bonds to two carbon atoms. The functional group in an alcohol is an —OH group bonded to carbon.

17.8 Both aldehydes and ketones contain carbonyl functional groups. In an aldehyde, the carbonyl group is bonded to at least one hydrogen atom. In a ketone, the carbonyl group is bonded to two carbon atoms.

17.9 In an aldehyde, the carbonyl group is bonded to at least one hydrogen atom, which means that the carbonyl group must be at the end of the chain.

17.10 Ethyl propyl ketone and 3-hexanone each have six carbons with a two-carbon group and a three-carbon group attached to a carbonyl group on the third carbon. Thus, they are the same compound.

17.11 The ester formed from propanoic acid and ethanol is ethyl propanoate.

17.12 Diethylamine has two ethyl groups attached to the nitrogen atom, whereas triethylamine has three ethyl groups attached to the nitrogen atom.

17.13 When an amide is formed, the product is named by changing the *oic acid* in the name of the carboxylic acid to *amide*.

ANSWERS TO SELECTED PROBLEMS

17.1 **a.** inorganic **b.** organic
c. inorganic **d.** organic
e. organic **f.** inorganic

17.3 **a.** ethane **b.** ethane
c. NaBr **d.** NaBr

17.5 **a.** hexane
b. heptane
c. pentane

17.7 **a.** same molecule
b. structural isomers
c. structural isomers

17.9 **a.** 2-fluorobutane
b. dimethylpropane
c. 2-chloro-3-methylpentane

17.11 **a.**
$$CH_3 — CH_2 — \overset{\overset{\displaystyle CH_3}{|}}{\underset{\underset{\displaystyle CH_3}{|}}{C}} — CH_2 — CH_3$$

b.
$$CH_3 — \overset{\overset{\displaystyle CH_3}{|}}{CH} — \overset{\overset{\displaystyle CH_3}{|}}{CH} — CH_2 — \overset{\overset{\displaystyle CH_3}{|}}{CH} — CH_3$$

c. $Br — CH_2 — CH_2 — Cl$

17.13 a.

b.

c.

17.15 **a.** $CH_3 — CH_2 — CH_2 — CH_2 — CH_2 — CH_2 — CH_3$

b. liquid
c. no
d. float
e. $C_7H_{16}(l) + 11O_2(g) \xrightarrow{\Delta} 7CO_2(g) + 8H_2O(g) + energy$

17.17 **a.** $CH_4(g) + 2O_2(g) \xrightarrow{\Delta} CO_2(g) + 2H_2O(g) + energy$

b. $2C_6H_{14}(l) + 19O_2(g) \xrightarrow{\Delta} 12CO_2(g) + 14H_2O(g) + energy$

17.19 **a.** alkene **b.** alkyne **c.** alkene

17.21 **a.** ethene **b.** 2-methylpropene **c.** 2-pentyne

17.23 **a.** $CH_3 — CH = CH_2$

b. $HC \equiv C — CH_2 — CH_2 — CH_2 — CH_3$

c. $H_2C = \overset{\overset{\displaystyle CH_3}{|}}{C} — CH_2 — CH_3$

17.25 a. $CH_3-CH_2-CH_2-CH_2-CH_3$

b. $CH_3-CH_2-CH_2-CH_3$

c.

17.27 A polymer is a large molecule composed of small units that are repeated many times.

17.29 $3H_2C=CH$ (CH_3) $\longrightarrow$

17.31

F H F H F H
| | | | | |
—C—C—C—C—C—C—
| | | | | |
F H F H F H

17.33 a. 2-chlorotoluene
b. ethylbenzene
c. phenol

17.35 a. (benzene ring with NH_2)
b. (benzene ring with Cl and F)
c. (benzene ring with methyl and ethyl)

17.37 a. ethanol
b. 2-butanol
c. 2-methyl-2-pentanol

17.39 a. ethyl methyl ether
b. dipropyl ether
c. butyl propyl ether

17.41 a. $CH_3-CH_2-CH_2-OH$

b. $CH_3-CH_2-O-CH_2-CH_2-CH_3$

c. $CH_3-CH_2-O-CH_2-CH_3$

d. $CH_3-\overset{\overset{OH}{|}}{\underset{\underset{CH_3}{|}}{C}}-CH_2-CH_3$

17.43 a. acetaldehyde
b. methyl propyl ketone
c. formaldehyde

17.45 a. propanal
b. 2-methyl-3-pentanone
c. 2-chlorobenzaldehyde

17.47 a. $CH_3-\overset{\overset{O}{\|}}{C}-H$

b. $CH_3-\overset{\overset{O}{\|}}{C}-CH_2-CH_2-CH_3$

c.

17.49 a. ethanoic acid (acetic acid)
b. propanoic acid (propionic acid)
c. 3-methylhexanoic acid
d. 3-bromobenzoic acid

17.51 a. $CH_3-CH_2-CH_2-\overset{\overset{O}{\|}}{C}-OH$

b. $Cl-CH_2-\overset{\overset{O}{\|}}{C}-OH$

c. (benzene ring with $\overset{\overset{O}{\|}}{C}-OH$)

d. Br~~~~~$\overset{\overset{O}{\|}}{C}$~~OH

17.53 a. $CH_3-\overset{\overset{O}{\|}}{C}-O-CH_2-CH_3$

b. $CH_3-CH_2-CH_2-\overset{\overset{O}{\|}}{C}-O-CH_2-CH_3$

17.55 a.

b.

17.57 a. methyl methanoate (methyl formate)
b. methyl butanoate (methyl butyrate)
c. ethyl propanoate (ethyl propionate)

17.59 a. $CH_3-\overset{\overset{O}{\|}}{C}-O-CH_3$

b. $H-\overset{\overset{O}{\|}}{C}-O-CH_2-CH_2-CH_2-CH_3$

c.

d.

17.61 a. methylamine
b. methylpropylamine
c. diethylmethylamine

17.63 a.

b. $CH_3 - \overset{\overset{\displaystyle O}{\|}}{C} - \overset{\overset{\displaystyle H}{|}}{N} - CH_2 - CH_3$

c.

17.65 a. ethanamide (acetamide)
b. 2-chlorobutanamide
c. methanamide (formamide)

17.67 a. $CH_3 - CH_2 - \overset{\overset{\displaystyle O}{\|}}{C} - NH_2$

b. $CH_3 - CH_2 - CH_2 - \overset{\overset{\displaystyle CH_3}{|}}{CH} - \overset{\overset{\displaystyle O}{\|}}{C} - NH_2$

c.

d.

17.69

17.71 a. $2C_6H_{14}(l) + 19O_2(g) \xrightarrow{\Delta}$

$12CO_2(g) + 14H_2O(g) + energy$

b. $C_7H_8(l) + 9O_2(g) \xrightarrow{\Delta} 7CO_2(g) + 4H_2O(g) + energy$

17.73

17.75 a. 3,7-dimethyl-6-octenal
b. The *en* signifies that a double bond is present.
c. The *al* signifies that an aldehyde is present.

17.77 a. $CH_3 - CH_2 - \overset{\overset{\displaystyle CH_2 - CH_3}{|}}{CH} - CH_2 - CH_2 - CH_3$

b. $CH_3 - CH = CH - CH_2 - CH_3$

c. $CH_3 - C \equiv C - CH_2 - CH_2 - CH_3$

17.79 a. 2,2-dimethylbutane **b.** 1-butyne
c. 2-pentene

17.81 a. ketone **b.** alkene
c. ester **d.** amine

17.83 a. alcohol **b.** alkene
c. aldehyde **d.** alkane

17.85 a. aromatic, aldehyde
b. aromatic, aldehyde, alkene

17.87 a. 3-bromotoluene
b. 1,2-dichlorobenzene
c. 3-ethyltoluene

17.89 a. **b.** **c.**

17.91 a. 2-butanol
b. 3-methyl-2-pentanol
c. butyl ethyl ether

17.93 a. $CH_3 - CH_2 - CH_2 - CH_2 - O - CH_2 - CH_2 - CH_2 - CH_3$

b. $CH_3 - \overset{\overset{\displaystyle CH_3}{|}}{CH} - \overset{\overset{\displaystyle OH}{|}}{CH} - CH_2 - CH_3$

c. $CH_3 - \overset{\overset{\displaystyle OH}{|}}{\underset{\underset{\displaystyle CH_3}{|}}{C}} - CH_3$

17.95 a. 4-chlorobenzaldehyde
b. 3-chloropropanal
c. 2-chloro-3-pentanone

17.97 a. $CH_3 - CH_2 - \overset{\overset{\displaystyle O}{\|}}{C} - CH_3$

b. $Cl - CH_2 - CH_2 - \overset{\overset{\displaystyle O}{\|}}{C} - H$

c.

17.99 a. 3-methylbutanoic acid
b. methyl benzoate
c. ethyl propanoate

17.101 a. $CH_3 - CH_2 - NH_2$

b. $CH_3 - CH_2 - CH_2 - CH_2 - CH_2 - \overset{\overset{\displaystyle O}{\|}}{C} - NH_2$

c. $CH_3 - CH_2 - \overset{\overset{\displaystyle CH_2 - CH_3}{|}}{N} - CH_2 - CH_3$

17.103 a. butyldimethylamine
    c. pentylamine

b. propanamide

17.105

$$CH_3-CH_2-CH_2-CH_2-CH_2-OH$$ 1-Pentanol

$$CH_3-\overset{\overset{\displaystyle OH}{|}}{CH}-CH_2-CH_2-CH_3$$ 2-Pentanol

$$CH_3-CH_2-\overset{\overset{\displaystyle OH}{|}}{CH}-CH_2-CH_3$$ 3-Pentanol

$$HO-CH_2-\overset{\overset{\displaystyle CH_3}{|}}{CH}-CH_2-CH_3$$ 2-Methyl-1-butanol

$$HO-CH_2-CH_2-\overset{\overset{\displaystyle CH_3}{|}}{CH}-CH_3$$ 3-Methyl-1-butanol

$$CH_3-\overset{\overset{\displaystyle CH_3}{|}}{\underset{\underset{\displaystyle OH}{|}}{C}}-CH_2-CH_3$$ 2-Methyl-2-butanol

$$CH_3-\overset{\overset{\displaystyle OH}{|}}{CH}-\overset{\overset{\displaystyle CH_3}{|}}{CH}-CH_3$$ 3-Methyl-2-butanol

$$CH_3-\overset{\overset{\displaystyle CH_3}{|}}{\underset{\underset{\displaystyle CH_3}{|}}{C}}-CH_2-OH$$ 2,2-Dimethyl-1-propanol

17.107

17.109 a. $2CH_3-CH_2-C\equiv CH + 11O_2 \xrightarrow{\Delta}$

$$8CO_2 + 6H_2O + energy$$

    b. $CH_3-CH_2-CH=CH-CH_2-CH_3 + H_2 \xrightarrow{Pt}$

$$CH_3-CH_2-CH_2-CH_2-CH_2-CH_3$$

Biochemistry

During her annual physical examination, Kate, a 64-year-old woman, complains that her vision is blurry, she feels a frequent need to urinate, and she has gained 22 lb over the past year. She tried to lose weight and increase her exercise for the past six months without success. Her diet is high in carbohydrates. For dinner, Kate typically eats two cups of pasta and three to four slices of bread with butter or olive oil. She also eats eight to ten pieces of fresh fruit per day at meals and as snacks. A lab test shows that her fasting blood glucose level is 178 mg/dL, indicating type 2 diabetes. She is referred to the diabetes specialty clinic, where she meets Paula, a diabetes nurse. Paula explains to Kate that foods like pasta, bread, and fruit can raise blood glucose levels because they contain large amounts of molecules called carbohydrates, which are broken down in the body into glucose molecules.

CAREER

Diabetes Nurse

Diabetes nurses teach patients about diabetes so they can self-manage and control their condition. This includes education on proper diets and nutrition for both diabetic and pre-diabetic patients. Diabetes nurses help patients learn to monitor their medication and blood sugar levels and look for symptoms like diabetic nerve damage and vision loss. Diabetes nurses may also work with patients who have been hospitalized because of complications from their disease. Working with patients with diabetes requires a thorough knowledge of the endocrine system, the system that regulates metabolism, so there is some overlap between diabetes and endocrinology nursing.

UPDATE   Kate's Program for Type 2 Diabetes

Now Kate uses a glucose meter to check her blood glucose level. You can read about changes she makes to her lifestyle in the **UPDATE Kate's Program for Type 2 Diabetes**, pages 633–634, and see how a change in eating habits and an increase in exercise help Kate to decrease her blood sugar and lose weight.

18.1 Carbohydrates

LEARNING GOAL Classify a monosaccharide as an aldose or a ketose; draw the open chain and Haworth structures for monosaccharides.

We have many carbohydrates in our food. There are polysaccharides called starches in bread and pasta. The table sugar used to sweeten cereal, tea, or coffee is sucrose, a disaccharide that consists of two simple sugars, glucose and fructose. **Carbohydrates** such as table sugar, lactose in milk, and cellulose are all made of carbon, hydrogen, and oxygen. Simple sugars, which have formulas of $C_n(H_2O)_n$, were once thought to be hydrates of carbon, thus the name *carbohydrate*.

Glucose ($C_6H_{12}O_6$) is the most important simple carbohydrate in metabolism. In a series of reactions called *photosynthesis*, energy from the Sun is used to combine the carbon atoms from carbon dioxide (CO_2) and the hydrogen and oxygen atoms of water (H_2O) to form glucose.

$$6CO_2 + 6H_2O + \text{ energy} \underset{\text{Respiration}}{\overset{\text{Photosynthesis}}{\rightleftharpoons}} \underset{\text{Glucose}}{C_6H_{12}O_6} + 6O_2$$

In the body, glucose is oxidized in a series of metabolic reactions known as *respiration* to release chemical energy to do work in the cells. Carbon dioxide and water are produced and returned to the atmosphere. The combination of photosynthesis and respiration is called the *carbon cycle*, in which energy from the Sun is stored in plants by photosynthesis and made available to us when the carbohydrates in our diets are metabolized (see **FIGURE 18.1**).

Monosaccharides

The simplest carbohydrates are the **monosaccharides**. A monosaccharide cannot be split into smaller carbohydrates. A monosaccharide has a chain of three to seven carbon atoms, one in a carbonyl group and all the others are attached to hydroxyl groups (—OH). In an *aldose*, the carbonyl group is on the first carbon (—CHO); a *ketose* contains the carbonyl group on the second carbon atom as a ketone (C=O).

Monosaccharides are also classified by the number of carbon atoms. A monosaccharide with three carbon atoms is a *triose*, one with four carbon atoms is a *tetrose*; a *pentose* has five carbons, and a *hexose* contains six carbons. We can use both classification systems to indicate the aldehyde or ketone group and the number of carbon atoms. An aldopentose is a five-carbon monosaccharide that is an aldehyde; a ketohexose is a six-carbon monosaccharide that is a ketone.

Carbohydrates contained in foods such as pasta and bread provide energy for the body.

FIGURE 18.1 During photosynthesis, energy from the Sun combines CO_2 and H_2O to form glucose ($C_6H_{12}O_6$) and O_2. During respiration in the body, carbohydrates are oxidized to CO_2 and H_2O, while energy is released.

ENGAGE 18.1

What are the reactants and products of respiration?

CH$_2$OH
|
C=O
|
H—C—OH
|
H—C—OH
|
CH$_2$OH

Ribulose

H$\quad$O
\\C//
|
H—C—OH
|
HO—C—H
|
H—C—OH
|
H—C—OH
|
CH$_2$OH

Glucose

H$\quad$O
\\C//
|
H—C—OH
|
CH$_2$OH

Glyceraldehyde
(aldotriose)

H$\quad$O
\\C//
|
HO—C—H
|
H—C—OH
|
CH$_2$OH

Threose
(aldotetrose)

H$\quad$O
\\C//
|
H—C—OH
|
H—C—OH
|
H—C—OH
|
CH$_2$OH

Ribose
(aldopentose)

CH$_2$OH
|
C=O
|
HO—C—H
|
H—C—OH
|
H—C—OH
|
CH$_2$OH

Fructose
(ketohexose)

Open-Chain Structures of Some Important Monosaccharides

Glucose, galactose, and fructose are the most important monosaccharides. They are all hexoses with the molecular formula $C_6H_{12}O_6$. In nature and the cells of the body, their most common form is the D isomer, which has the —OH group attached to carbon 5 on the right side of the chain.

H$\quad$O
\\C//
| 1
H—C—OH
| 2
HO—C—H
| 3
H—C—OH
| 4
H—C—OH
| 5
CH$_2$OH
6

D-Glucose

H$\quad$O
\\C//
| 1
H—C—OH
| 2
HO—C—H
| 3
HO—C—H
| 4
H—C—OH
| 5
CH$_2$OH
6

D-Galactose

CH$_2$OH
| 1
C=O
| 2
HO—C—H
| 3
H—C—OH
| 4
H—C—OH
| 5
CH$_2$OH
6

D-Fructose

The sweet taste of honey is due to the monosaccharides D-glucose and D-fructose.

The most common hexose, **D-glucose**, $C_6H_{12}O_6$, an aldohexose also known as dextrose and blood sugar, is found in fruits, vegetables, corn syrup, and honey. It is a building block of the disaccharides sucrose, lactose, and maltose, and polysaccharides such as starch, cellulose, and glycogen.

D-Galactose is an aldohexose that does not occur in the free form in nature. It is obtained from the disaccharide lactose, a sugar found in milk and milk products. D-Galactose is important in the cellular membranes of the brain and nervous system. The only difference in the structures of D-glucose and D-galactose is the arrangement of the —OH group on carbon 4.

In contrast to glucose and galactose, **D-fructose** is a ketohexose. The structure of D-fructose differs from D-glucose at carbons 1 and 2 by the location of the carbonyl group. D-Fructose is the sweetest of the carbohydrates, twice as sweet as sucrose (table sugar). This makes D-fructose popular with dieters because less D-fructose and, therefore, fewer calories are needed to provide a pleasant taste. D-Fructose is found in fruit juices and honey.

Chemistry Link to Health

Hyperglycemia and Hypoglycemia

Kate's doctor ordered an oral glucose tolerance test (OGTT) to evaluate her body's ability to return to normal glucose concentrations (70 to 99 mg/dL) in response to the ingestion of a specified amount of glucose (dextrose). After Kate fasts for 12 h, she drinks a solution containing 75 g of glucose. A blood sample is taken immediately, followed by more blood samples each half-hour for 2 h, and then every hour for a total of 5 h. After her test, Kate is told that her blood glucose was 178 mg/dL, which indicates hyperglycemia. The prefix *hyper* means above or over; *hypo* means below or under. The term *glyc* or *gluco* refers to "sugar." Thus, the blood sugar level in *hyperglycemia* is above normal and in *hypoglycemia* it is below normal.

A glucose solution is given to determine blood glucose levels.

A disease that can cause hyperglycemia is *type 2 diabetes*, which occurs when the pancreas is unable to produce sufficient quantities of insulin. As a result, glucose levels in the body fluids can rise as high as 350 mg/dL. Kate's symptoms of type 2 diabetes include thirst, excessive urination, and increased appetite. In older adults, type 2 diabetes is sometimes a consequence of excessive weight gain.

When a person is hypoglycemic, the blood glucose level rises and then decreases rapidly to levels as low as 40 mg/dL. In some cases, hypoglycemia is caused by overproduction of insulin by the pancreas. Low blood glucose can cause dizziness, general weakness, and muscle tremors. A diet may be prescribed that consists of several small meals high in protein and low in carbohydrate. Some hypoglycemic patients are finding success with diets that include more complex carbohydrates rather than simple sugars.

Haworth Structures of Monosaccharides

Up until now, we have drawn the structures for monosaccharides such as D-glucose as open chains. However, the most stable form of hexoses is a six-atom ring, known as a **Haworth structure**. We can draw the Haworth structure for D-glucose as shown in Sample Problem 18.1.

CORE CHEMISTRY SKILL

Drawing Haworth Structures

SAMPLE PROBLEM 18.1 Drawing the Haworth Structure for D-Glucose

TRY IT FIRST

D-Glucose has the following open-chain structure. Draw the Haworth structures for D-glucose.

$$
\begin{array}{c}
H\diagdown \;\; \diagup O \\
C \\
| \\
H-C-OH \\
| \\
HO-C-H \\
| \\
H-C-OH \\
| \\
H-C-OH \\
| \\
CH_2OH
\end{array}
$$

D-Glucose

SOLUTION

STEP 1 Turn the open-chain structure clockwise by 90°. The —H and —OH groups on the right of the vertical carbon chain are below the horizontal carbon chain. Those on the left of the open chain are above the horizontal carbon chain.

D-Glucose (open chain)

STEP 2   **Fold the horizontal carbon chain into a hexagon, rotate the groups on carbon 5, and bond the O on carbon 5 to carbon 1.** With carbons 2 and 3 as the base of a hexagon, move the remaining carbons upward. Rotate the groups on carbon 5 so that the —OH group is close to carbon 1. To complete the Haworth structure, draw a bond between the oxygen of the —OH group on carbon 5 to carbon 1. By convention, the carbon atoms in the ring are drawn as corners.

Rotation of groups on carbon 5   Carbon-5 oxygen bonds to carbon 1   Cyclic structure

New hydroxyl group on carbon 1

STEP 3   **Draw the new —OH group on carbon 1 below the ring to give the α form or above the ring to give the β form.** Because the new —OH group can form above or below the plane of the Haworth structure, there are two forms of D-glucose, which differ only by the position of the —OH group at carbon 1.

β Form

α Form

α-D-Glucose   β-D-Glucose

SELF TEST 18.1

a. Draw the Haworth structure for α-D-galactose.

D-Galactose

b. Draw the Haworth structure for β-D-mannose.

D-Mannose

ANSWER

a.   b.

In contrast to D-glucose and D-galactose, D-fructose is a ketohexose. It forms the five-atom ring when the hydroxyl group on carbon 5 reacts with the carbon of the ketone group. The new hydroxyl group is on carbon 2.

α-D-Fructose   β-D-Fructose

PRACTICE PROBLEMS

Try Practice Problems 18.5 to 18.8

PRACTICE PROBLEMS

18.1 Carbohydrates

18.1 What functional groups are found in all monosaccharides?

18.2 What is the difference between an aldose and a ketose?

18.3 What are the functional groups and number of carbons in a ketopentose?

18.4 What are the functional groups and number of carbons in an aldohexose?

18.5 What are the kind and number of atoms in the ring portion of the Haworth structure of D-glucose?

18.6 What are the kind and number of atoms in the ring portion of the Haworth structure of D-fructose?

18.7 Identify each of the following Haworth structures as the α or β form:

a.   b.

18.8 Identify each of the following Haworth structures as the α or β form:

a.   b.

Applications

18.9 Classify each of the following monosaccharides as an aldopentose, aldohexose, ketopentose, or ketohexose:

a. Psicose is present in low levels in foods.

b. Lyxose is a component of bacterial lipids.

$$CH_2OH$$
$$|$$
$$C{=}O$$
$$|$$
$$H{-}C{-}OH$$
$$|$$
$$H{-}C{-}OH$$
$$|$$
$$H{-}C{-}OH$$
$$|$$
$$CH_2OH$$
Psicose

$$H{\diagdown}$$
$$C{=}O$$
$$|$$
$$HO{-}C{-}H$$
$$|$$
$$HO{-}C{-}H$$
$$|$$
$$H{-}C{-}OH$$
$$|$$
$$CH_2OH$$
Lyxose

18.10 Classify each of the following monosaccharides as an aldopentose, aldohexose, ketopentose, or ketohexose:

a. A solution of xylose is given to test its absorption by the intestines.

b. Tagatose, found in fruit, is similar in sweetness to sugar.

Xylose   Tagatose

18.11 An infant with galactosemia can utilize D-glucose in milk but not D-galactose. How does the open-chain structure of D-galactose differ from that of D-glucose?

18.12 D-Fructose is the sweetest monosaccharide. How does the open-chain structure of D-fructose differ from that of D-glucose?

18.2 Disaccharides and Polysaccharides

LEARNING GOAL Describe the monosaccharide units and linkages in disaccharides and polysaccharides.

A **disaccharide** is composed of two monosaccharides linked together. The most common disaccharides are maltose, lactose, and sucrose.

Maltose, or malt sugar, is a dissacharide obtained from starch and is found in germinating grains. Maltose is used in cereals, candies, and the brewing of beverages. Maltose has a *glycosidic bond* between two glucose molecules. To form maltose, the —OH group on carbon 1 in the first glucose forms a bond with the —OH group on carbon 4 in a second glucose molecule, which is designated as an $\alpha(1\rightarrow4)$ linkage. Because the second glucose molecule in maltose has a free —OH group on carbon 1, there are α and β forms of maltose.

ENGAGE 18.4

How does α-maltose differ from β-maltose?

CH$_2$OH   CH$_2$OH

α-D-Glucose α-D-Glucose

$\alpha(1\rightarrow4)$-Glycosidic bond

+ H$_2$O

α Form

α-Maltose, a disaccharide

Lactose, milk sugar, is a disaccharide found in milk and milk products (see **FIGURE 18.2**). The bond in lactose is a $\beta(1\rightarrow4)$-glycosidic bond because the β form of galactose links to the hydroxyl group on carbon 4 of glucose. It makes up 6 to 8% of human milk and about 4 to 5% of cow's milk. Some people do not produce sufficient quantities of the enzyme needed to break down lactose, and the sugar remains undigested and undergoes fermentation, producing gas that causes abdominal cramps and diarrhea. In some commercial milk products, an enzyme

ENGAGE 18.5

What type of glycosidic bond links galactose and glucose in lactose?

β-D-Galactose α-D-Glucose

$\beta(1\rightarrow4)$-Glycosidic bond

+ H$_2$O

α Form

α-Lactose, a disaccharide

FIGURE 18.2 Lactose is a disaccharide found in milk and milk products.

called lactase is added to break down lactose. Because the glucose molecule in lactose has a free —OH group on carbon 1, there are α and β forms of lactose.

Sucrose, ordinary table sugar, is the most abundant disaccharide in the world. Sucrose consists of an α-D-glucose and a β-D-fructose molecule joined by an $\alpha,\beta(1\rightarrow2)$-glycosidic bond. The glycosidic bond in sucrose is between carbon 1 of glucose and carbon 2 of fructose. Most of the sucrose for table sugar comes from sugar cane (20% by mass) or sugar beets (15% by mass) (see **FIGURE 18.3**). Both the raw and refined forms of sugar are sucrose. Some estimates indicate that each person in the United States consumes an average of 68 kg (150 lb) of sucrose every year either by itself or in a variety of food products.

ENGAGE 18.6

What monosaccharides form sucrose?

FIGURE 18.3 Sucrose, a disaccharide obtained from sugar beets and sugar cane, contains glucose and fructose.

SAMPLE PROBLEM 18.2 Glycosidic Bonds in Disaccharides

TRY IT FIRST

Melibiose is a disaccharide that is 30 times sweeter than sucrose.

Melibiose

a. What are the monosaccharide units in melibiose?
b. What type of glycosidic bond links the monosaccharides?
c. Identify the structure as α- or β-melibiose.

SOLUTION

a. First monosaccharide (left)	When the —OH group on carbon 4 is above the plane, it is D-galactose. When the —OH group on carbon 1 is below the plane, it is α-D-galactose.
Second monosaccharide (right)	When the —OH group on carbon 4 is below the plane, it is α-D-glucose.
b. Type of glycosidic bond	When the —OH group at carbon 1 of α-D-galactose bonds with the —OH group on carbon 6 of glucose, it is an $\alpha(1\rightarrow6)$-glycosidic bond.
c. Name of disaccharide	The —OH group on carbon 1 of glucose is below the plane, which is α-melibiose.

a. Isomaltulose is a sugar substitute that is digested more slowly than sucrose. What are the monosaccharide units in isomaltulose, and what is the glycosidic link in isomaltulose?

Isomaltulose

b. Cellobiose is a disaccharide composed of two D-glucose molecules connected by a $\beta(1\rightarrow4)$-glycosidic linkage. Draw the Haworth structure for β-cellobiose.

ANSWER

a. The two monosaccharides in isomaltulose are glucose and fructose. They are linked by an $\alpha(1\rightarrow6)$-glycosidic bond.

b.

Chemistry Link to Health

How Sweet is My Sweetener?

Although many of the monosaccharides and disaccharides taste sweet, they differ considerably in their degree of sweetness. Dietetic foods contain sweeteners that are noncarbohydrate or carbohydrates that are sweeter than sucrose. Some examples of sweeteners compared with sucrose are shown in **TABLE 18.1**.

Sucralose, which is known as Splenda, is made from sucrose by replacing some of the hydroxyl groups with chlorine atoms.

Sucralose (Splenda)

Aspartame, which is marketed as NutraSweet and Equal, is used in a large number of sugar-free products. It is a noncarbohydrate sweetener made of aspartate and a methyl ester of phenylalanine. It does have some caloric value, but it is so sweet that only a very small quantity is needed. However, phenylalanine, one of the breakdown products, poses a danger to anyone who cannot metabolize it properly, a condition called *phenylketonuria* (PKU).

From aspartate   From phenylalanine
Aspartame (NutraSweet)

Another artificial sweetener, Neotame, is a modification of the aspartame structure. The addition of a large alkyl group to the amine group prevents enzymes from breaking the amide bond between aspartate and phenylalanine. Thus, phenylalanine is not produced when Neotame is used as a sweetener. Very small amounts of Neotame are needed because it is about 10 000 times sweeter than sucrose.

Stevia is a sugar substitute obtained from the leaves of a plant *Stevia rebaudiana*. It is composed of steviol glycosides that are about 150 times sweeter than sucrose. Stevia has been used to sweeten tea and medicines in South America for 1500 years.

Steviol

Large alkyl group to modify Aspartame   From aspartate   From phenylalanine

Neotame

Advantame is another modification of aspartame with a complex organic group bonded to the amine. Advantame is 20 000 times sweeter than sucrose.

Advantame

Saccharin, which is marketed as Sweet'N Low, has been used as a noncarbohydrate artificial sweetener for about 100 years.

Saccharin (Sweet'N Low)

Artificial sweeteners are used as sugar substitutes.

TABLE 18.1 Relative Sweetness of Sugars and Artificial Sweeteners

	Sweetness Relative to Sucrose ($=100$)
Monosaccharides	
Galactose	30
Glucose	75
Fructose	175
Disaccharides	
Lactose	16
Maltose	33
Sucrose	100
Sugar Alcohols	
Sorbitol	60
Maltitol	80
Xylitol	100
Noncarbohydrate Sweeteners	
Stevia	15 000
Aspartame	18 000
Saccharin	45 000
Sucralose	60 000
Neotame	1 000 000
Advantame	2 000 000

Polysaccharides

A **polysaccharide** is a polymer of many monosaccharides joined together. Four biologically important polysaccharides—*amylose, amylopectin, glycogen, and cellulose*—are polymers of D-glucose that differ only in the type of glycosidic bonds and the amount of branching in the molecule.

Starch, a storage form of glucose in plants, is found as insoluble granules in rice, wheat, potatoes, beans, and cereals. Starch is composed of two kinds of polysaccharides, amylose and amylopectin. **Amylose**, which makes up about 20% of starch, consists of 250 to 4000 α-D-glucose molecules connected by $\alpha(1\rightarrow4)$-glycosidic bonds in a continuous chain. Sometimes called a straight-chain polymer, polymers of amylose are actually coiled in helical fashion.

Amylopectin, which makes up as much as 80% of plant starch, is a branched-chain polysaccharide. Like amylose, the glucose molecules are connected by $\alpha(1\rightarrow4)$-glycosidic bonds. However, at about every 25 glucose units, there is a branch of glucose molecules attached by an $\alpha(1\rightarrow6)$-glycosidic bond between carbon 1 of the branch and carbon 6 in the main chain (see **FIGURE 18.4**).

ENGAGE 18.7

What types of glycosidic bonds link glucose molecules in amylose and amylopectin?

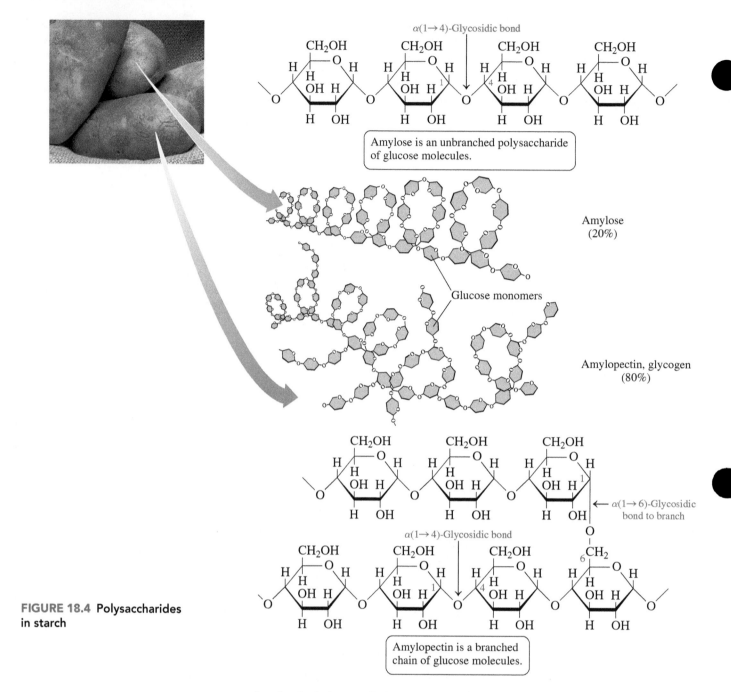

α(1→4)-Glycosidic bond

Amylose is an unbranched polysaccharide of glucose molecules.

Amylose (20%)

Glucose monomers

Amylopectin, glycogen (80%)

α(1→6)-Glycosidic bond to branch

α(1→4)-Glycosidic bond

Amylopectin is a branched chain of glucose molecules.

FIGURE 18.4 Polysaccharides in starch

Starches hydrolyze easily in water and acid to give smaller saccharides called *dextrins*, which then hydrolyze to maltose and finally glucose. In our bodies, these complex carbohydrates are digested by the enzymes amylase (in saliva) and maltase (in the intestine). The glucose obtained provides about 50% of our nutritional calories.

$$\text{Amylose, amylopectin} \xrightarrow{\text{H}^+ \text{ or amylase}} \text{dextrins} \xrightarrow{\text{H}^+ \text{ or amylase}} \text{maltose} \xrightarrow{\text{H}^+ \text{ or maltase}} \text{many D-glucose units}$$

Glycogen, or animal starch, is a polymer of glucose that is stored in the liver and muscle of animals. It is used in our cells at a rate that maintains the blood level of glucose and provides energy between meals. The structure of glycogen is very similar to that of amylopectin, found in plants, except that glycogen is more highly branched.

Cellulose is the major structural material of wood and plants. Cotton is almost pure cellulose. In cellulose, glucose molecules form a long unbranched chain similar to that of amylose. However, the glucose units in cellulose are linked by β(1→4)-glycosidic bonds. The cellulose chains do not form coils like amylose but are aligned in parallel rows held in place by hydrogen bonds between hydroxyl groups in adjacent chains, making cellulose insoluble in water. This gives a rigid structure to the cell wells in wood and fiber that is more resistant to hydrolysis than are the starches (see **FIGURE 18.5**).

FIGURE 18.5 The polysaccharide cellulose is composed of glucose units linked by β(1→4)-glycosidic bonds.

Humans have enzymes in saliva and pancreatic juices that hydrolyze the α(1→4)-glycosidic bonds of starches, but not the β(1→4)-glycosidic bonds of cellulose. Thus, humans cannot digest cellulose. Animals such as horses, cows, and goats can obtain glucose from cellulose because their digestive systems contain bacteria that provide enzymes to hydrolyze β(1→4)-glycosidic bonds.

ENGAGE 18.8

Why are humans unable to digest cellulose?

SAMPLE PROBLEM 18.3   Structures of Polysaccharides

TRY IT FIRST

Give the name of one or more polysaccharides described by each of the following:

a. a polysaccharide that is stored in the liver and muscle tissues
b. an unbranched polysaccharide containing β(1→4)-glycosidic bonds
c. a polysaccharide containing α(1→4)- and α(1→6)-glycosidic bonds

SOLUTION

a. glycogen   **b.** cellulose   **c.** amylopectin, glycogen

SELF TEST 18.3

a. Cellulose and amylose are both unbranched glucose polymers. How do they differ?
b. Amylose and glycogen are both glucose polymers. How do they differ?

ANSWER

a. Cellulose contains glucose units connected by β(1→4)-glycosidic bonds, whereas the glucose units in amylose are connected by α(1→4)-glycosidic bonds.
b. Amylose is an unbranched polymer of glucose units, whereas glycogen is a branched polymer of glucose.

PRACTICE PROBLEMS

Try Practice Problems 18.13 to 18.20

PRACTICE PROBLEMS

18.2 Disaccharides and Polysaccharides

18.13 For each of the following, state the monosaccharide units, the type of glycosidic bond, and the name of the disaccharide, including the α or β form:

a.

b.

18.14 For each of the following, state the monosaccharide units, the type of glycosidic bond, and the name of the disaccharide, including the α or β form:

a.

b.

Applications

18.15 Isomaltose, obtained from the breakdown of starch, has the following Haworth structure:

Isomaltose

a. Is isomaltose a mono-, di-, or polysaccharide?
b. What are the monosaccharides in isomaltose?
c. What is the glycosidic link in isomaltose?
d. Is this the α or β form of isomaltose?

18.16 Sophorose, found in certain types of beans, has the following Haworth structure:

Sophorose

a. Is sophorose a mono-, di-, or polysaccharide?
b. What are the monosaccharides in sophorose?
c. What is the glycosidic link in sophorose?
d. Is this the α or β form of sophorose?

18.17 Identify the disaccharide that fits each of the following descriptions:
a. ordinary table sugar
b. found in milk and milk products
c. also called malt sugar
d. contains galactose and glucose

18.18 Identify the disaccharide that fits each of the following descriptions:
a. used in brewing
b. composed of two glucose units
c. also called milk sugar
d. contains glucose and fructose

18.19 Give the name of one or more polysaccharides that matches each of the following descriptions:
a. not digestible by humans
b. the storage form of carbohydrates in plants
c. contains only α(1→4)-glycosidic bonds
d. the most highly branched polysaccharide

18.20 Give the name of one or more polysaccharides that matches each of the following descriptions:
a. the storage form of carbohydrates in animals
b. contains only β(1→4)-glycosidic bonds
c. contains both α(1→4)- and α(1→6)-glycosidic bonds
d. produces only maltose during digestion

18.3 Lipids

LEARNING GOAL Draw the condensed structural or line-angle formula for a fatty acid, a triacylglycerol, and the products of hydrogenation or saponification. Identify the steroid nucleus.

REVIEW

Naming Carboxylic Acids (17.6)

Forming Esters (17.6)

Lipids are a family of biomolecules that have the common property of being soluble in organic solvents but not very soluble in water. The word *lipid* comes from the Greek word *lipos*, meaning "fat" or "lard." Within the lipid family, there are certain structures that distinguish the different types of lipids. Lipids that contain *fatty acids* include waxes and triacylglycerols, commonly known as fats and oils, which are esters of glycerol and fatty acids. Lipids that are *steroids* do not contain fatty acids but are characterized by the steroid nucleus of four fused carbon rings.

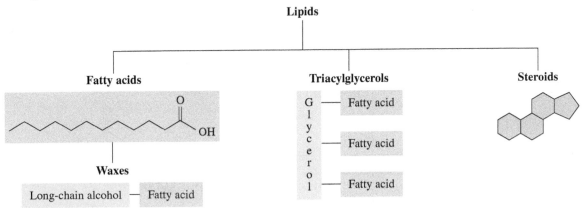

Lipids are naturally occurring biomolecules in cells and tissues, which are soluble in organic solvents but not in water.

Fatty Acids

A **fatty acid** contains a long unbranched chain of carbon atoms, usually 12 to 20, with a carboxylic acid group at one end. The long carbon chain makes fatty acids insoluble in water. An example is lauric acid, a 12-carbon acid found in coconut oil, which has a structure that can be drawn in several forms as follows:

CORE CHEMISTRY SKILL

Identifying Fatty Acids

Ball-and-stick model of lauric acid

$CH_3-(CH_2)_{10}-\overset{\displaystyle O}{\overset{\|}{C}}-OH$ $\qquad$ $CH_3-(CH_2)_{10}-COOH$

$CH_3-CH_2-CH_2-CH_2-CH_2-CH_2-CH_2-CH_2-CH_2-CH_2-CH_2-C\overset{\displaystyle O}{\underset{\displaystyle OH}{}}$

Condensed structural formulas of lauric acid

Line-angle formula of lauric acid

Saturated fatty acids, such as lauric acid, contain only single bonds between carbons. *Monounsaturated fatty acids* have one double bond in the carbon chain, and *polyunsaturated fatty acids* have two or more double bonds. **TABLE 18.2** lists some of the typical fatty acids in lipids.

ENGAGE 18.9

Why is stearic acid classified as a saturated fatty acid and oleic acid as a monounsaturated fatty acid?

TABLE 18.2 Structures and Melting Points of Common Fatty Acids

Name	Carbon Atoms: Double Bonds	Present in	Melting Point (°C)	Structures
Saturated Fatty Acids				
Lauric acid	12:0	Coconut	44	$CH_3-(CH_2)_{10}-COOH$
Myristic acid	14:0	Nutmeg	55	$CH_3-(CH_2)_{12}-COOH$
Palmitic acid	16:0	Palm	63	$CH_3-(CH_2)_{14}-COOH$
Stearic acid	18:0	Animal fat	69	$CH_3-(CH_2)_{16}-COOH$
Monounsaturated Fatty Acids				
Palmitoleic acid	16:1	Butter	0	$CH_3-(CH_2)_5-CH=CH-(CH_2)_7-COOH$
Oleic acid	18:1	Olives, pecan, grapeseed	14	$CH_3-(CH_2)_7-CH=CH-(CH_2)_7-COOH$
Polyunsaturated Fatty Acids				
Linoleic acid	18:2	Soybean, safflower, sunflower	−5	$CH_3-(CH_2)_4-CH=CH-CH_2-CH=CH-(CH_2)_7-COOH$
Linolenic acid	18:3	Corn	−11	$CH_3-CH_2-CH=CH-CH_2-CH=CH-CH_2-CH=CH-(CH_2)_7-COOH$

SAMPLE PROBLEM 18.4 Structures and Properties of Fatty Acids

TRY IT FIRST

The line-angle formula for vaccenic acid, a fatty acid found in dairy products and human milk, is shown below.

a. Why is this substance an acid?
b. How many carbon atoms are in vaccenic acid?
c. Is the fatty acid saturated, monounsaturated, or polyunsaturated?
d. Would it be soluble in water?

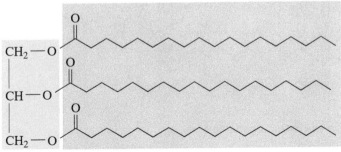

SOLUTION

a. Vaccenic acid contains a carboxylic acid group.
b. It contains 18 carbon atoms.
c. It is a monounsaturated fatty acid.
d. No; its long hydrocarbon chain makes it insoluble in water.

SELF TEST 18.4

Palmitoleic acid is a fatty acid with the following condensed structural formula:

$$CH_3-(CH_2)_5-CH=CH-(CH_2)_7-\overset{\overset{\displaystyle O}{\|}}{C}-OH$$

a. How many carbon atoms are in palmitoleic acid?
b. Is the fatty acid saturated, monounsaturated, or polyunsaturated?
c. Would it be soluble in water?

ANSWER

a. 16   **b.** monounsaturated   **c.** no

Waxes

Waxes are found in many plants and animals. A wax is an ester of a long-chain fatty acid and a long-chain alcohol. The formulas of some common waxes are given in **TABLE 18.3**. Beeswax obtained from honeycombs and carnauba wax from palm trees are used to give a protective coating to furniture, cars, and floors. Jojoba wax is used in making candles and cosmetics such as lipstick. Lanolin, a mixture of waxes obtained from wool, is used in hand and facial lotions to aid retention of water, which softens the skin.

TABLE 18.3 Some Typical Waxes

Type	Condensed Structural Formula	Source	Uses
Beeswax	$CH_3-(CH_2)_{14}-\overset{\overset{\displaystyle O}{\|}}{C}-O-(CH_2)_{29}-CH_3$	Honeycomb	Candles, shoe polish
Carnauba wax	$CH_3-(CH_2)_{24}-\overset{\overset{\displaystyle O}{\|}}{C}-O-(CH_2)_{29}-CH_3$	Brazilian palm tree	Waxes for furniture, cars, floors, shoes
Jojoba wax	$CH_3-(CH_2)_{18}-\overset{\overset{\displaystyle O}{\|}}{C}-O-(CH_2)_{19}-CH_3$	Jojoba bush	Candles, soaps, cosmetics

Honeycomb (beeswax) is an ester of a saturated fatty acid and a long-chain alcohol.

Triacylglycerols

In the body, fatty acids are stored as **triacylglycerols**, also called *triglycerides*, which are triesters of glycerol (a trihydroxy alcohol) and fatty acids. The general structure of a triacylglycerol follows:

CORE CHEMISTRY SKILL

Drawing Structures for Triacylglycerols

Triacylglycerol

Tristearin

In a triacylglycerol, three hydroxyl groups of glycerol form ester bonds with the carboxyl groups of three fatty acids. For example, glycerol and three molecules of stearic acid form glyceryl tristearate (tristearin).

$$
\begin{array}{l}
CH_2-O-H \;+\; HO-\overset{\displaystyle O}{\overset{\|}{C}}-(CH_2)_{16}-CH_3 \\[4pt]
CH-O-H \;+\; HO-\overset{\displaystyle O}{\overset{\|}{C}}-(CH_2)_{16}-CH_3 \longrightarrow \\[4pt]
CH_2-O-H \;+\; HO-\overset{\displaystyle O}{\overset{\|}{C}}-(CH_2)_{16}-CH_3
\end{array}
$$

Glycerol   3 Stearic acid molecules

Ester bond

$$
\begin{array}{l}
CH_2-O-\overset{\displaystyle O}{\overset{\|}{C}}-(CH_2)_{16}-CH_3 \\[4pt]
CH-O-\overset{\displaystyle O}{\overset{\|}{C}}-(CH_2)_{16}-CH_3 \;+\; 3H_2O \\[4pt]
CH_2-O-\overset{\displaystyle O}{\overset{\|}{C}}-(CH_2)_{16}-CH_3
\end{array}
$$

Glyceryl tristearate (tristearin)

Prior to hibernation, a polar bear eats food with a high caloric content.

Triacylglycerols are the major form of energy storage for animals. Animals that hibernate eat large quantities of plants, seeds, and nuts that are high in calories. Prior to hibernation, these animals, such as polar bears, gain as much as 14 kg a week. As the external temperature drops, the animal goes into hibernation. The body temperature drops to nearly freezing, and cellular activity, respiration, and heart rate are drastically reduced. Animals that live in extremely cold climates will hibernate for 4 to 7 months. During this time, stored fat is the only source of energy.

Melting Points of Fats and Oils

A **fat** is a triacylglycerol that is solid at room temperature, and comes from animal sources such as meat, whole milk, butter, and cheese. An **oil** is a triacylglycerol commonly containing one or more double bonds that is usually liquid at room temperature, and is obtained from a plant source (see **FIGURE 18.6**).

Glyceryl trioleate (triolein)

FIGURE 18.6 Vegetable oils such as olive oil contain unsaturated fats.

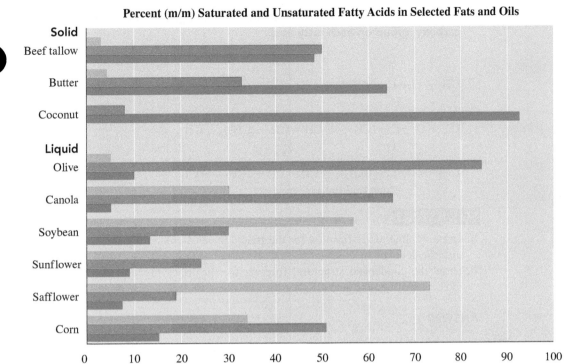

Percent (m/m) Saturated and Unsaturated Fatty Acids in Selected Fats and Oils

Solid
Beef tallow

Butter

Coconut

Liquid
Olive

Canola

Soybean

Sunflower

Safflower

Corn

0 10 20 30 40 50 60 70 80 90 100

■ Polyunsaturated ■ Monounsaturated ■ Saturated

FIGURE 18.7 Vegetable oils are liquids at room temperature because they have a higher percentage of unsaturated fatty acids than do animal fats.

The amounts of saturated, monounsaturated, and polyunsaturated fatty acids in some typical fats and oils are shown in **FIGURE 18.7**. Saturated fatty acids have higher melting points than unsaturated fatty acids because they pack together more tightly. Animal fats usually contain more saturated fatty acids than do vegetable oils. Therefore, the melting points of animal fats are higher than those of vegetable oils.

ENGAGE 18.10

Why is butter a solid at room temperature, whereas canola oil is a liquid?

SAMPLE PROBLEM 18.5 Drawing the Structure for a Triacylglycerol

TRY IT FIRST

Draw the condensed structural formula for glyceryl tripalmitoleate (tripalmitolein), which is used in cosmetic creams and lotions.

SOLUTION

	Given	Need	Connect
ANALYZE THE PROBLEM	glyceryl tripalmitoleate (tripalmitolein)	condensed structural formula	ester of glycerol and three fatty acids

STEP 1 **Draw the condensed structural formulas for glycerol and the fatty acids.**

$$CH_2-OH \ + \ HO-\overset{\overset{\displaystyle O}{\|}}{C}-(CH_2)_7-CH=CH-(CH_2)_5-CH_3 \quad \text{Palmitoleic acid}$$

$$CH-OH \ + \ HO-\overset{\overset{\displaystyle O}{\|}}{C}-(CH_2)_7-CH=CH-(CH_2)_5-CH_3 \quad \text{Palmitoleic acid}$$

$$CH_2-OH \ + \ HO-\overset{\overset{\displaystyle O}{\|}}{C}-(CH_2)_7-CH=CH-(CH_2)_5-CH_3 \quad \text{Palmitoleic acid}$$

Glycerol

Triacylglycerols are used to thicken creams and lotions.

STEP 2 **Form ester bonds between the hydroxyl groups on glycerol and the carboxyl group on each fatty acid.**

$$CH_2-O-\overset{\overset{\displaystyle O}{\|}}{C}-(CH_2)_7-CH=CH-(CH_2)_5-CH_3$$
$$CH-O-\overset{\overset{\displaystyle O}{\|}}{C}-(CH_2)_7-CH=CH-(CH_2)_5-CH_3$$
$$CH_2-O-\overset{\overset{\displaystyle O}{\|}}{C}-(CH_2)_7-CH=CH-(CH_2)_5-CH_3$$

Glyceryl tripalmitoleate (tripalmitolein)

SELF TEST 18.5

a. Draw the line-angle formula for the triacylglycerol containing three molecules of myristic acid (14:0).
b. Draw the condensed structural formula for the triacylglycerol containing three molecules of linoleic acid (18:2).

ANSWER

a.

b.

$$CH_2-O-\overset{\overset{\displaystyle O}{\|}}{C}-(CH_2)_7-CH=CH-CH_2-CH=CH-(CH_2)_4-CH_3$$
$$CH-O-\overset{\overset{\displaystyle O}{\|}}{C}-(CH_2)_7-CH=CH-CH_2-CH=CH-(CH_2)_4-CH_3$$
$$CH_2-O-\overset{\overset{\displaystyle O}{\|}}{C}-(CH_2)_7-CH=CH-CH_2-CH=CH-(CH_2)_4-CH_3$$

Reactions of Triacylglycerols

The double bonds in unsaturated fatty acids react with hydrogen to produce saturated fatty acids. For example, when hydrogen is added to glyceryl trioleate (triolein) using a nickel catalyst, the product is the saturated fat glyceryl tristearate (tristearin).

Double bonds

$$CH_2-O-\overset{\overset{\displaystyle O}{\|}}{C}-(CH_2)_7-CH=CH-(CH_2)_7-CH_3$$
$$CH-O-\overset{\overset{\displaystyle O}{\|}}{C}-(CH_2)_7-CH=CH-(CH_2)_7-CH_3 + 3H_2 \xrightarrow{Ni}$$
$$CH_2-O-\overset{\overset{\displaystyle O}{\|}}{C}-(CH_2)_7-CH=CH-(CH_2)_7-CH_3$$

Glyceryl trioleate
(triolein)

Single bonds

$$CH_2-O-\overset{\overset{\displaystyle O}{\|}}{C}-(CH_2)_7-CH_2-CH_2-(CH_2)_7-CH_3$$
$$CH-O-\overset{\overset{\displaystyle O}{\|}}{C}-(CH_2)_7-CH_2-CH_2-(CH_2)_7-CH_3$$
$$CH_2-O-\overset{\overset{\displaystyle O}{\|}}{C}-(CH_2)_7-CH_2-CH_2-(CH_2)_7-CH_3$$

Glyceryl tristearate
(tristearin)

Saponification occurs when a fat is heated with a strong base such as sodium hydroxide to give glycerol and the sodium salts of the fatty acids, which is soap. When NaOH is used, a solid soap is produced that can be molded into a desired shape; KOH produces a softer, liquid soap. Oils that are polyunsaturated produce softer soaps. Names like "coconut soap" or "avocado shampoo" tell you the sources of the oil used in the reaction.

PRACTICE PROBLEMS

Try Practice Problems 18.21 to 18.30

Fat or oil   +   strong base $\xrightarrow{\text{Heat}}$ glycerol   +   salts of fatty acids (soap)

$$CH_2-O-\overset{\displaystyle O}{\overset{\|}{C}}-(CH_2)_{14}-CH_3$$
$$CH-O-\overset{\displaystyle O}{\overset{\|}{C}}-(CH_2)_{14}-CH_3 + 3NaOH \xrightarrow{\text{Heat}}$$
$$CH_2-O-\overset{\displaystyle O}{\overset{\|}{C}}-(CH_2)_{14}-CH_3$$

Glyceryl tripalmitate
(tripalmitin)

$$CH_2-OH \quad Na^+ \; {}^-O-\overset{\displaystyle O}{\overset{\|}{C}}-(CH_2)_{14}-CH_3$$
$$CH-OH + Na^+ \; {}^-O-\overset{\displaystyle O}{\overset{\|}{C}}-(CH_2)_{14}-CH_3$$
$$CH_2-OH \quad Na^+ \; {}^-O-\overset{\displaystyle O}{\overset{\|}{C}}-(CH_2)_{14}-CH_3$$

Glycerol   3 Sodium palmitate
(soap)

Steroids: Cholesterol

Steroids are compounds containing the steroid nucleus, which consists of four carbon rings fused together. Although they are lipids, steroids do not contain fatty acids.

A cross section of a normal, open artery shows no buildup of plaque.

Steroid nucleus   HO   Cholesterol

Attaching other atoms and groups of atoms to the steroid nucleus forms a wide variety of steroid compounds. *Cholesterol*, which is one of the most important and abundant steroids in the body, is a *sterol* because it contains an oxygen atom as a hydroxyl group (—OH). Like many steroids, cholesterol has methyl groups, a double bond, and a carbon side chain. In other steroids, the hydroxyl group is replaced by a carbonyl group (C=O). Cholesterol is obtained from eating meats, milk, and eggs, and it is also synthesized by the liver. There is no cholesterol in vegetable and plant products.

A cross section of an artery that is almost completely clogged by atherosclerotic plaque.

Cholesterol in the Body

If a diet is high in cholesterol, the liver produces less cholesterol. A typical daily American diet includes 400 to 500 mg of cholesterol, one of the highest in the world. The American Heart Association has recommended that we consume no more than 300 mg of cholesterol a day. Researchers suggest that saturated fats and cholesterol are associated with diseases such as diabetes; cancers of the breast, pancreas, and colon; and atherosclerosis. In *atherosclerosis*, deposits of a protein–lipid complex (plaque) accumulate in the coronary blood vessels, restricting the flow of blood to the tissue and causing necrosis (death) of the tissue (see **FIGURE 18.8**). In the heart, plaque accumulation could result in a *myocardial infarction* (heart attack). Other factors that may also increase the risk of heart disease are family history, lack of exercise, smoking, obesity, diabetes, gender, and age.

FIGURE 18.8 Excess cholesterol forms plaque that can block an artery.

PRACTICE PROBLEMS

Try Practice Problems 18.31 and 18.32

PRACTICE PROBLEMS

18.3 Lipids

18.21 Classify each of the following fatty acids as saturated, monoun-saturated, or polyunsaturated (see Table 18.2):
   a. lauric acid   **b.** linolenic acid   **c.** stearic acid

18.22 Classify each of the following fatty acids as saturated, monoun-saturated, or polyunsaturated (see Table 18.2):
   a. linoleic acid   **b.** palmitoleic acid   **c.** myristic acid

18.23 Caprylic acid is an 8-carbon saturated fatty acid that occurs in coconut oil (10%) and palm kernel oil (4%). Draw the line-angle formula for glyceryl tricaprylate (tricaprylin).

18.24 Draw the condensed structural formula for glyceryl trilaurate (trilaurin).

Applications

18.25 Safflower oil is polyunsaturated, whereas olive oil is monoun-saturated. Why would safflower oil have a lower melting point than olive oil?

18.26 Olive oil is monounsaturated, whereas butter fat is saturated. Why does olive oil have a lower melting point than butter fat?

18.27 Use condensed structural formulas to write the balanced chemical equation for the hydrogenation of glyceryl tripalmitoleate, a fat containing glycerol and three palmitoleic acid molecules.

18.28 Use condensed structural formulas to write the balanced chemical equation for the hydrogenation of glyceryl trilinolenate, a fat containing glycerol and three linolenic acid molecules.

18.29 Use condensed structural formulas to write the balanced chemical equation for the NaOH saponification of glyceryl trimyristate, a fat containing glycerol and three myristic acid molecules.

18.30 Use condensed structural formulas to write the balanced chemical equation for the NaOH saponification of glyceryl trioleate, a fat containing glycerol and three oleic acid molecules.

18.31 Draw the structure for the steroid nucleus.

18.32 What are the functional groups on the cholesterol molecule?

REVIEW

Forming Amides (17.7)

18.4 Amino Acids and Proteins

LEARNING GOAL Describe protein functions and draw structures for amino acids and peptides.

Proteins perform different functions in the body. Some proteins form structural components such as cartilage, muscles, hair, and nails. Wool, silk, feathers, and horns in animals are made of proteins. Proteins that function as enzymes regulate biological reactions such as digestion and cellular metabolism. Other proteins, such as hemoglobin and myoglobin, transport oxygen in the blood and muscle. **TABLE 18.4** gives examples of proteins that are classified by their functions in biological systems.

The horns of animals are made of proteins.

TABLE 18.4 Classification of Some Proteins and Their Functions

Class of Protein	Function	Examples
Structural	Provide structural components	*Collagen* is in tendons and cartilage. *Keratin* is in hair, skin, wool, horns, and nails.
Contractile	Make muscles move	*Myosin* and *actin* contract muscle fibers.
Transport	Carry essential substances throughout the body	*Hemoglobin* transports oxygen. *Lipoproteins* transport lipids.
Storage	Store nutrients	*Casein* stores protein in milk. *Ferritin* stores iron in the spleen and liver.
Hormone	Regulate body metabolism and the nervous system	*Insulin* regulates blood glucose level. *Growth hormone* regulates body growth.
Enzyme	Catalyze biochemical reactions in the cells	*Sucrase* catalyzes the hydrolysis of sucrose. *Trypsin* catalyzes the hydrolysis of proteins.
Protection	Recognize and destroy foreign substances	*Immunoglobulins* stimulate immune responses.

Amino Acids

Proteins are composed of molecular building blocks called *amino acids*. Every **amino acid**, which is ionized in biological environments, has a central alpha carbon atom (α carbon) bonded to an ammonium group ($-NH_3^+$), a carboxylate group ($-COO^-$), a hydrogen atom, and an R group. The differences in the 20 α-amino acids present in human proteins are due to the unique characteristics of the R groups.

Ionized form of alanine

Ball-and-stick model of alanine

Classification of Amino Acids

We classify amino acids using their specific R groups, which determine their properties in aqueous solution. The *nonpolar amino acids* have hydrogen, alkyl, or aromatic R groups, which make them *hydrophobic* ("water fearing"). The *polar amino acids* have R groups that interact with water, which makes them *hydrophilic* ("water loving"). There are three groups of polar amino acids. The R groups of polar *neutral amino acids* contain hydroxyl ($-$OH), thiol ($-$SH), or amide ($-$CONH$_2$) R groups. The R group of a polar *acidic amino acid* contains a carboxylate group ($-$COO$^-$). The R group of a polar *basic amino acid* contains an ammonium group ($-$NH$_3{}^+$). The names and structures of the 20 alpha amino acids commonly found in proteins, along with their three-letter and one-letter abbreviations, are shown at physiological pH (7.4) in **TABLE 18.5**.

CORE CHEMISTRY SKILL

Drawing the Structure for an
Amino Acid at Physiological pH

ENGAGE 18.11

Why is valine classified as a nonpolar amino acid whereas histidine is classified as a polar amino acid?

TABLE 18.5 Structures, Names, and Abbreviations of 20 Common Amino Acids at Physiological pH (7.4)

Nonpolar Amino Acids (Hydrophobic)

(continued)

Polar Neutral Amino Acids (Hydrophilic)

Serine (Ser, S) Threonine (Thr, T) Tyrosine (Tyr, Y) Cysteine (Cys, C) Asparagine (Asn, N) Glutamine (Gln, Q)

Polar Acidic Amino Acids (Hydrophilic)

Aspartate (Asp, D) Glutamate (Glu, E)

Polar Basic Amino Acids (Hydrophilic)

Histidine (His, H)[b] Lysine (Lys, K) Arginine (Arg, R)

[a]Proline is considered to be a cyclic amino acid because its R group bonds to the nitrogen atom attached to the α carbon.
[b]At physiological pH, some histidine molecules have a positively charged R group, while other histidine molecules have a neutral R group. The molecule with the positively charged R group is shown here.

SAMPLE PROBLEM 18.6   Structural Formula and Polarity of an Amino Acid

TRY IT FIRST

Draw the structure for the amino acid serine at physiological pH, and write the three-letter and one-letter abbreviations.

SOLUTION

The structure of a specific amino acid is drawn by attaching the R group to the central carbon atom of the general structure of an amino acid.

The abbreviations for serine are Ser and S.

SELF TEST 18.6

a. Classify serine as polar or nonpolar and hydrophobic or hydrophilic.
b. Classify valine as polar or nonpolar and hydrophobic or hydrophilic.

ANSWER

a. polar, hydrophilic   b. nonpolar, hydrophobic

PRACTICE PROBLEMS

Try Practice Problems 18.33 to 18.38

A summary of the classification of amino acids follows:

Type of Amino Acid	Number	Type of R Groups	Interaction with Water
Nonpolar	9	Nonpolar	Hydrophobic
Polar, neutral	6	Contain O and S atoms, but no charge	Hydrophilic
Polar, acidic	2	Contain carboxylate groups, negative charge	Hydrophilic
Polar, basic	3	Contain ammonium groups, positive charge	Hydrophilic

Chemistry Link to Health

Essential Amino Acids and Complete Proteins

Of the 20 common amino acids used to build the proteins in the body, only 11 can be synthesized in the body. The other 9 amino acids, listed in **TABLE 18.6**, are *essential amino acids* that must be obtained from the proteins in the diet.

TABLE 18.6 Essential Amino Acids

Histidine (His, H)	Phenylalanine (Phe, F)
Isoleucine (Ile, I)	Threonine (Thr, T)
Leucine (Leu, L)	Tryptophan (Trp, W)
Lysine (Lys, K)	Valine (Val, V)
Methionine (Met, M)	

Complete proteins such as eggs, milk, meat, and fish contain all of the essential amino acids. Incomplete proteins from plants such as grains, beans, and nuts are deficient in one or more essential amino acids.

Complete proteins, which contain all of the essential amino acids, are found in most animal products such as eggs, milk, meat, fish, and poultry. However, gelatin and plant proteins such as grains, beans, and nuts are *incomplete proteins* because they are deficient in one or more of the essential amino acids. Diets that rely on plant foods for protein must contain a variety of protein sources to obtain all the essential amino acids. Some examples of *complementary protein* sources include rice and beans, or a peanut butter sandwich on whole grain bread. Rice contains the amino acid methionine that is deficient in beans, whereas peanuts are rich in lysine that is lacking in grains (see **TABLE 18.7**).

TABLE 18.7 Amino Acid Deficiencies in Selected Vegetables and Grains

Food Source	Amino Acid Deficiency
Eggs, milk, meat, fish, poultry	None
Wheat, rice, oats	Lysine
Corn	Lysine, tryptophan
Beans	Methionine, tryptophan
Peas, peanuts	Methionine
Almonds, walnuts	Lysine, tryptophan
Soy	Methionine

Peptides

A *peptide bond* is an amide bond that forms when the $-COO^-$ group of one amino acid reacts with the $-NH_3^+$ group of the next amino acid. We can write the formation of the dipeptide glycylalanine (Gly–Ala, GA) between glycine and alanine. The amino acid written on the left, glycine has a free (unbonded) NH_3^+ group. Therefore, it is the amino acid at the *N-terminus* of the peptide. The amino acid written on the right, alanine, has a free (unbonded) $-COO^-$ group. Therefore, it is the amino acid at the *C-terminus* of the peptide. In the name of a peptide, each amino acid beginning from the N-terminus has the *ine, an,* or *ate* replaced by *yl*. The last amino acid at the C-terminus uses its full name.

PRACTICE PROBLEMS

Try Practice Problems 18.41 and 18.42

ENGAGE 18.12

In the dipeptide Val–Thr, how do you know that valine is at the N-terminus and threonine is at the C-terminus?

Glycine (Gly, G)   Alanine (Ala, A)   Glycylalanine (Gly–Ala, GA)

A peptide bond between the ionized structures of glycine and alanine forms the dipeptide glycylalanine.

SAMPLE PROBLEM 18.7 Drawing a Peptide

TRY IT FIRST

Draw the structure and give the name for the tripeptide Gly–Ser–Met.

SOLUTION

ANALYZE THE PROBLEM	Given	Need	Connect
	Gly–Ser–Met	tripeptide structure, name	peptide bonds, amino acid names

STEP 1 Draw the structure for each amino acid in the peptide, starting from the N-terminus.

STEP 2 Remove the O atom from the carboxylate group of the N-terminal amino acid and two H atoms from the ammonium group in the adjacent amino acid. Use peptide bonds to connect the amino acids. Repeat this process until the C-terminal amino acid is reached.

The tripeptide is named by replacing the last syllable of each amino acid name with *yl*, starting at the N-terminus. The amino acid at the C-terminus retains its complete name.

N-terminus   glycine is named glycyl

                serine is named seryl

C-terminus   methionine keeps its complete name

The tripeptide is named glycylserylmethionine.

SELF TEST 18.7

a. Draw the structure and give the name for Phe–Thr, a section in glucagon, which is a peptide hormone that increases blood glucose levels.
b. Draw the structure and give the name for Gln–Asn, a section in oxytocin, a nonapeptide used to induce labor.

ANSWER

a. Phenylalanylthreonine

$$H_3\overset{+}{N}-\overset{CH_2}{\underset{H}{C}}-\overset{O}{\underset{}{C}}-N-\overset{CH_3\ OH}{\underset{H}{\overset{|}{C}H}}-\overset{O}{\underset{H}{C}}-O^-$$

b. Glutaminylasparagine

$$H_3\overset{+}{N}-\overset{CH_2}{\underset{H}{C}}-\overset{O}{\underset{}{C}}-N-\overset{CH_2}{\underset{H}{C}}-\overset{O}{\underset{H}{C}}-O^-$$

PRACTICE PROBLEMS
Try Practice Problems 18.39 and 18.40, 18.43 and 18.44

PRACTICE PROBLEMS

18.4 Amino Acids and Proteins

18.33 Draw the condensed structural formula for each of the following amino acids at physiological pH:
 a. glycine b. threonine c. phenylalanine

18.34 Draw the condensed structural formula for each of the following amino acids at physiological pH:
 a. tyrosine b. leucine c. methionine

18.35 Classify each of the amino acids in problem 18.33 as polar or nonpolar. If polar, indicate if the R group is neutral, acidic, or basic. Indicate if each would be hydrophobic or hydrophilic.

18.36 Classify each of the amino acids in problem 18.34 as polar or nonpolar. If polar, indicate if the R group is neutral, acidic, or basic. Indicate if each would be hydrophobic or hydrophilic.

18.37 Give the name of the amino acid represented by each of the following abbreviations:
 a. Ala b. V c. Lys d. C

18.38 Give the name of the amino acid represented by each of the following abbreviations:
 a. Trp b. M c. Pro d. G

18.39 Draw the condensed structural formula for each of the following peptides, and give its three-letter and one-letter abbreviations:
 a. alanylcysteine b. serylphenylalanine
 c. glycylalanylvaline

18.40 Draw the condensed structural formula for each of the following peptides, and give its three-letter and one-letter abbreviations:
 a. prolylaspartate b. threonylleucine
 c. methionylglutaminyllysine

Applications

18.41 Explain why each of the following pairs are complementary proteins:
 a. corn and peas b. rice and soy

18.42 Explain why each of the following pairs are complementary proteins:
 a. beans and oats b. almonds and peanuts

18.43 Peptides isolated from rapeseed that may lower blood pressure have the following sequence of amino acids. Draw the structure for each peptide and write the one-letter abbreviations.
 a. Arg–Ile–Tyr b. Val–Trp–Ile–Ser

18.44 Peptides from sweet potato with antioxidant properties have the following sequence of amino acids. Draw the structure for each peptide and write the one-letter abbreviations.
 a. Asp–Cys–Gly–Tyr b. Asn–Tyr–Asp–Glu–Tyr

18.5 Protein Structure

LEARNING GOAL Identify the levels of structure of a protein.

A **protein** is a polypeptide of 50 or more amino acids that has biological activity. Each protein in our cells has a unique sequence of amino acids that determines its three-dimensional structure and biological function.

Primary Structure

The **primary structure** of a protein is the particular sequence of amino acids held together by peptide bonds. The first protein to have its primary structure determined was insulin, which was accomplished by Frederick Sanger in 1953. Since that time, scientists have determined the amino acid sequences of thousands of proteins. Insulin is a hormone that regulates the glucose level in the blood. In the primary structure of human insulin, there are two polypeptide chains. In chain A, there are 21 amino acids, and in chain B there are 30 amino acids. The polypeptide chains are held together by *disulfide bonds* formed by the thiol groups of the cysteine amino acids in each of the chains (see **FIGURE 18.9**). Today, human insulin, for the treatment of diabetes, with this exact structure is produced in large quantities through genetic engineering.

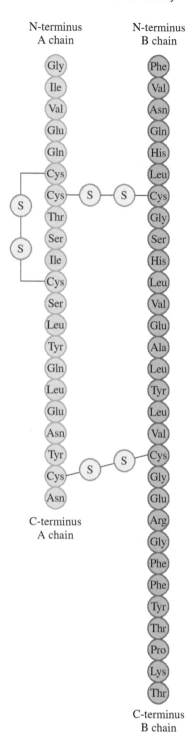

FIGURE 18.9 The sequence of amino acids in human insulin is the primary structure.

PRACTICE PROBLEMS
Try Practice Problems 18.45 to 18.48

- ● Carbon
- ● Oxygen
- ● Nitrogen
- ● R group
- ○ Hydrogen

FIGURE 18.10 The α (alpha) helix acquires a coiled shape from hydrogen bonds between the N—H of the peptide bond in one turn of the polypeptide and the C=O of the peptide bond in the next turn.

Collagen fibers are triple helices of polypeptide chains held together by hydrogen bonds.

Secondary Structure

The **secondary structure** of a protein describes the structure that forms when amino acids form hydrogen bonds between the atoms in the backbone within a single polypeptide chain or between polypeptide chains. The most common types of secondary structures are the *alpha helix* and the *beta-pleated sheet*.

In an alpha helix (α helix), hydrogen bonds form between each N—H group and the oxygen of a C=O group in the next turn of the α helix (see **FIGURE 18.10**). Because there are many hydrogen bonds along the peptide, it has a helical or coiled shape.

In the secondary structure known as the beta-pleated sheet (β-pleated sheet), hydrogen bonds hold polypeptide chains together side by side. The hydrogen bonds holding the sheets tightly in place account for the strength and durability of proteins such as silk (see **FIGURE 18.11**).

Collagen

Collagen, the most abundant protein in the body, makes up as much as one-third of all the protein in vertebrates. It is found in connective tissue, blood vessels, skin, tendons, ligaments, the cornea of the eye, and cartilage. The strong structure of collagen is a result of three polypeptides woven together by hydrogen bonds to form a *triple helix*. When several triple helices wrap together, they form the fibrils that make up connective tissues and tendons. In a young person, collagen is elastic. However, as a person ages, additional bonds form between the fibrils, which make collagen less elastic. Cartilage and tendons become more brittle, and wrinkles are seen in the skin.

Tertiary Structure

The **tertiary structure** of a protein involves attractions and repulsions between the R groups of the amino acids in the polypeptide chain. It is the unique three-dimensional shape of the *tertiary structure* that determines the biological function of the molecule. **TABLE 18.8** lists the interactions that stabilize the tertiary structures of proteins.

Alpha helix

Beta-pleated sheet

Protein with alpha helices
and beta-pleated sheets

The ribbon model of a protein shows regions
of alpha helices and beta-pleated sheets.

Hydrogen bonds
between peptide
backbones

● Carbon
● Oxygen
● Nitrogen
● R group
○ Hydrogen

FIGURE 18.11 In a β (beta)-pleated sheet secondary
structure, hydrogen bonds form between side-by-side
sections of the peptide chains.

TABLE 18.8 Some Interactions That Stabilize Tertiary Structures

Interaction	Nature of Bonding
Hydrophobic	Interactions between nonpolar groups
Hydrophilic	Attractions between polar groups and water
Salt bridges (ionic bonds)	Ionic interactions between acidic and basic amino acids
Hydrogen bonds	Attractions between H and O or N
Disulfide bonds	Strong covalent links between sulfur atoms of two cysteine amino acids

Hydrophilic
interaction
with water

Hydrogen bond

Hydrophobic
interaction

α Helix

Salt
bridge

Disulfide
bonds

β-Pleated sheet

Hydrogen bonds

Hydrogen bond

Interactions between amino acid R groups fold a polypeptide into a
specific three-dimensional shape called its tertiary structure.

ENGAGE 18.13

Why is the interaction between
arginine and aspartate in a tertiary
structure called a salt bridge?

CORE CHEMISTRY SKILL
Identifying the Primary, Secondary, Tertiary, and Quaternary Structures of Proteins

SAMPLE PROBLEM 18.8 Interactions That Stabilize Tertiary Structures

TRY IT FIRST

What type of interaction would you expect between the R groups of the following amino acids?

a. cysteine and cysteine   **b.** glutamate and lysine   **c.** tyrosine and serine

SOLUTION

	Type of R Groups	Type of Interaction	Connect
ANALYZE THE PROBLEM	identify R groups on amino acids	determine type of interaction between R groups	polarity of R groups

a. Two cysteines, each with an R group containing —SH, will form a disulfide bond.
b. The interaction of the —COO⁻ in the R group of glutamate and the —NH₃⁺ in the R group of lysine will form a salt bridge.
c. The R groups in tyrosine and serine both have —OH groups that can form hydrogen bonds.

SELF TEST 18.8

a. What type of interaction would you expect between the R groups of valine and leucine?
b. What type of interaction would you expect between the R groups of threonine and serine?

ANSWER

a. Both are nonpolar and would have hydrophobic interactions.
b. Both contain an —OH group and would form hydrogen bonds.

PRACTICE PROBLEMS
Try Practice Problems 18.49 and 18.50

FIGURE 18.12 In the ribbon model of hemoglobin, the quaternary structure is made up of four polypeptide subunits.

ENGAGE 18.14
What is the difference between a tertiary structure and a quaternary structure?

Quaternary Structure

When a biologically active protein consists of two or more polypeptide subunits, the structural level is referred to as a **quaternary structure**. Hemoglobin, a protein that transports oxygen in blood, consists of four polypeptide chains or subunits (see **FIGURE 18.12**). The subunits are held together in the quaternary structure by the same kinds of interactions that stabilize the tertiary structures. The hemoglobin molecule can bind and transport four molecules of oxygen. **TABLE 18.9** and **FIGURE 18.13** summarize the structural levels of proteins.

TABLE 18.9 Summary of Structural Levels in Proteins

Structural Level	Characteristics
Primary	Peptide bonds join amino acids in a specific sequence in a polypeptide.
Secondary	Hydrogen bonds along or between peptide chains form α helix, or a β-pleated sheet.
Tertiary	A protein folds into a compact, three-dimensional shape stabilized by interactions between R groups of amino acids.
Quaternary	Two or more protein subunits combine to form a biologically active protein stabilized by the same interactions as in the tertiary structure.

SAMPLE PROBLEM 18.9 Identifying Protein Structure

TRY IT FIRST

Indicate whether the following interactions are responsible for the primary, secondary, tertiary, or quaternary structures of proteins:

a. disulfide bonds between portions of a protein chain
b. peptide bonds that form a chain of amino acids
c. hydrogen bonds between the O of a peptide bond and the H of a peptide bond four amino acids away

SOLUTION

ANALYZE THE PROBLEM	Given	Need	Connect
	interactions between amino acids	structural level	identify characteristics

a. Disulfide bonds help to stabilize the tertiary and quaternary levels of protein structure.
b. Peptide bonds form between amino acids in the primary structure of a protein.
c. Hydrogen bonding between peptide bonds in the protein backbone forms the secondary level of protein structure.

INTERACTIVE VIDEO

Different Levels of Protein Structure

SELF TEST 18.9

Identify the structural level represented by each of the following:

a. the grouping of two polypeptides to make an active enzyme
b. the folding of a protein into a compact shape

ANSWER

a. quaternary   b. tertiary

PRACTICE PROBLEMS

Try Practice Problems 18.51 and 18.52

Primary structure

Secondary structure

Tertiary structure

Quaternary structure

FIGURE 18.13 Structural levels for proteins

PRACTICE PROBLEMS

18.5 Protein Structure

18.45 Two peptides each contain one molecule of valine and two molecules of serine. Write the three-letter and one-letter abbreviations for their possible primary structures.

18.46 Two peptides each contain one molecule of alanine, one molecule of glycine, and one molecule of isoleucine. Write the three-letter and one-letter abbreviations for their possible primary structures.

18.47 What is the difference in bonding between an α helix and a β-pleated sheet?

18.48 How is the structure of a β-pleated sheet different from that of a triple helix?

18.49 What type of interaction would you expect between the R groups of the following amino acids in a tertiary structure?
a. cysteine and cysteine   b. glutamate and arginine
c. serine and aspartate   d. leucine and leucine

18.50 What type of interaction would you expect between the R groups of the following amino acids in a tertiary structure?
a. phenylalanine and isoleucine
b. aspartate and histidine
c. asparagine and tyrosine
d. alanine and proline

18.51 Indicate whether each of the following statements describes the primary, secondary, tertiary, or quaternary protein structure:
  a. R groups interact to form disulfide bonds or salt bridges.
  b. Peptide bonds join the amino acids in a polypeptide chain.
  c. Several polypeptides in a β-pleated sheet are held together by hydrogen bonds between adjacent chains.

18.52 Indicate whether each of the following statements describes the primary, secondary, tertiary, or quaternary protein structure:
  a. Hydrogen bonding between amino acid R groups in the same polypeptide gives a coiled shape to the protein.
  b. Hydrophilic amino acids move to the polar aqueous environment outside the protein.
  c. An active protein contains four subunits.

18.6 Proteins as Enzymes

LEARNING GOAL Describe the role of an enzyme in an enzyme-catalyzed reaction.

Every second, thousands of chemical reactions occur in our cells. For example, reactions occur to digest the food we eat, convert the digestion products to chemical energy, and synthesize proteins and other molecules in our cells. Although these reactions can occur outside the body, they occur at rates that are too slow to meet our physiological and metabolic needs. In our bodies, catalysts known as **enzymes** increase the rates at which most biological reactions occur.

An enzyme has a unique three-dimensional shape that recognizes and binds a small group of reacting molecules called *substrates*. In a catalyzed reaction, an enzyme must first bind to a substrate in a way that favors catalysis.

A typical enzyme is much larger than its substrate. However, within its tertiary structure is a region called the **active site** that binds a substrate or substrates and catalyzes the reaction. This active site is often a small pocket within the larger tertiary structure that closely fits the substrate (see **FIGURE 18.14**).

FIGURE 18.14 On the surface of an enzyme, a small region called the active site binds a substrate and catalyzes a reaction of that substrate.

Enzyme-Catalyzed Reaction

The combination of an enzyme (E) and a substrate (S) within the active site forms an *enzyme–substrate* (ES) *complex* that provides an alternative pathway for the reaction with lower activation energy. Within the active site, the amino acid R groups catalyze the reaction to give an *enzyme–product* (EP) *complex*. Then the products are released, and the enzyme is available to bind to another substrate molecule.

$$E + S \rightleftharpoons ES \text{ complex} \longrightarrow EP \text{ complex} \longrightarrow E + P$$

Enzyme and Substrate   Enzyme–Substrate Complex   Enzyme–Product Complex   Enzyme and Product

Models of Enzyme Action

An early theory of enzyme action, called the *lock-and-key model*, described the active site as having a rigid, nonflexible shape. According to the lock-and-key model, the shape of the active site was analogous to a lock, and its substrate was the key that specifically fit that lock. However, this model was a static one that did not allow for the flexibility of the tertiary shape of an enzyme and the way we now know that the active site can adjust to the shape of a substrate.

In the dynamic model of enzyme action, called the **induced-fit model**, the flexibility of the active site allows it to adapt to the shape of the substrate. At the same time, the shape of the substrate may be modified to better fit the geometry of the active site. As a result, the fit of both the active site and the substrate provides the best alignment for the catalysis of the reaction of the substrate. In the induced-fit model, substrate and enzyme work together to acquire a geometrical arrangement that lowers the activation energy.

In the hydrolysis of the disaccharide lactose by the enzyme lactase, a molecule of lactose binds to the active site of lactase. As the lactose binds to the enzyme, both the active site and the substrate lactose change shape. In this ES complex, the glycosidic bond of lactose is in a position that is favorable for *hydrolysis*, which is the splitting by water of a large molecule into smaller parts. The R groups on the amino acids in the active site then catalyze the hydrolysis of lactose which produces the monosaccharides glucose and galactose. Because the structures of the products are no longer attracted to the active site, they are released, which allows lactase to react with another lactose molecule (see **FIGURE 18.15**).

FIGURE 18.15 The induced-fit model for enzyme action

PRACTICE PROBLEMS

Try Practice Problems 18.53 to 18.56

PRACTICE PROBLEMS

18.6 Proteins as Enzymes

18.53 Match the terms, (1) enzyme, (2) enzyme–substrate complex, and (3) substrate, with each of the following:
 a. has a tertiary structure that recognizes the substrate
 b. has a structure that fits the active site of an enzyme
 c. the combination of an enzyme with the substrate

18.54 Match the terms, (1) active site, (2) lock-and-key model, and (3) induced-fit model, with each of the following:
 a. the portion of an enzyme where catalytic activity occurs
 b. an active site that adapts to the shape of a substrate
 c. an active site that has a rigid shape

18.55 a. Write an equation that represents an enzyme-catalyzed reaction.
 b. How is the active site different from the whole enzyme structure?

18.56 a. How does an enzyme speed up the reaction of a substrate?
 b. After the products have formed, what happens to the enzyme?

FIGURE 18.16 The general structure of a nucleotide includes a nitrogen-containing base, a sugar, and a phosphate group.

18.7 Nucleic Acids

LEARNING GOAL Describe the structure of the nucleic acids in DNA and RNA.

Nucleic acids are large molecules, found in the nuclei of cells, that store information and direct activities for cellular growth and reproduction. There are two closely related types of nucleic acids: *deoxyribonucleic acid* (**DNA**) and *ribonucleic acid* (**RNA**). Deoxyribonucleic acid, the genetic material in the nucleus of a cell, contains all the information needed for the development of a complete living organism. Ribonucleic acid interprets the genetic information in DNA for the synthesis of protein.

Both DNA and RNA are unbranched chains of repeating monomer units known as *nucleotides*. Each nucleotide has three components: a base that contains nitrogen, a five-carbon sugar, and a phosphate group (see **FIGURE 18.16**). A DNA molecule may contain several million nucleotides; smaller RNA molecules may contain up to several thousand.

Bases

The nitrogen-containing *bases* in nucleic acids are derivatives of *pyrimidine* or *purine*. A pyrimidine has a single ring with two nitrogen atoms, and a purine has two rings each with two nitrogen atoms. In DNA, the pyrimidine bases with single rings are cytosine (C) and thymine (T), and the purine bases with double rings are adenine (A) and guanine (G). RNA contains the same bases, except thymine (5-methyluracil) is replaced by uracil (U) (see **FIGURE 18.17**).

Pyrimidine Bases in Nucleic Acids

Pyrimidine | Cytosine (C) (DNA and RNA) | Uracil (U) (RNA only) | Thymine (T) (DNA only)

PRACTICE PROBLEMS
Try Practice Problems 18.57 and 18.58

Pentose Sugars in RNA and DNA

Purine Bases in Nucleic Acids

Purine | Adenine (A) (DNA and RNA) | Guanine (G) (DNA and RNA)

FIGURE 18.17 DNA contains the bases A, G, C, and T; RNA contains A, G, C, and U.

Pentose Sugars

In RNA, the five-carbon sugar is *ribose*, which gives the letter R in the abbreviation RNA. The atoms in the pentose sugars are numbered with primes (1′, 2′, 3′, 4′, and 5′) to differentiate them from the atoms in the bases. In DNA, the five-carbon sugar is *deoxyribose*, which is similar to ribose except that there is no hydroxyl group (—OH) on C2′. The *deoxy* prefix means "without oxygen" and provides the D in DNA (see **FIGURE 18.18**).

Nucleosides and Nucleotides

A **nucleoside** is produced when a pyrimidine or purine forms a glycosidic bond to C1′ of a sugar, either ribose or deoxyribose. For example, adenine, a purine, and ribose form a nucleoside called adenosine.

FIGURE 18.18 The five-carbon pentose sugar found in RNA is ribose, and in DNA, deoxyribose.

Sugar   +   Base $\longrightarrow$ Nucleoside   + H_2O

A base forms an *N*-glycosidic bond with ribose or deoxyribose to form a nucleoside.

Nucleotides are produced when the C5′ —OH group of ribose or deoxyribose in a nucleoside forms a phosphate ester. All the nucleotides in RNA and DNA are shown in **FIGURE 18.19**.

Adenosine monophosphate (AMP)
Deoxyadenosine monophosphate (dAMP)

Guanosine monophosphate (GMP)
Deoxyguanosine monophosphate (dGMP)

Cytidine monophosphate (CMP)
Deoxycytidine monophosphate (dCMP)

Uridine monophosphate (UMP)

Deoxythymidine monophosphate (dTMP)

FIGURE 18.19 The nucleotides of RNA (shown in black) are similar to those of DNA (shown in magenta), except in DNA the sugar is deoxyribose and deoxythymidine replaces uridine.

The name of a nucleoside that contains a purine ends with *osine*, whereas a nucleoside that contains a pyrimidine ends with *idine*. The names of nucleosides of DNA add *deoxy* to the beginning of their names. The corresponding nucleotides in RNA and DNA are named by adding *monophosphate* to the end of the nucleoside name. Although the letters A, G, C, U, and T represent the bases, they are often used in the abbreviations of the respective nucleotides.

ENGAGE 18.15

What are two differences in the nucleotides of RNA and DNA?

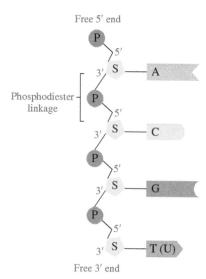

Free 5' end

Phosphodiester linkage

Free 3' end

In the primary structure of nucleic acids, each sugar in a sugar–phosphate backbone is attached to a base.

Structure of Nucleic Acids

The **nucleic acids** are unbranched chains of many nucleotides in which the 3' hydroxyl group of the sugar in one nucleotide bonds to the phosphate group on the 5' carbon atom in the sugar of the next nucleotide. This connection between the sugars in adjacent nucleotides is referred to as a *phosphodiester linkage*. As more nucleotides are added, a backbone forms that consists of alternating sugar and phosphate groups. The bases, which are attached to each sugar, extend out from the sugar–phosphate backbone.

In any nucleic acid, the sugar at one end has a free 5' phosphate group, and the sugar at the other end has a free 3' hydroxyl group. A nucleic acid sequence is read from the free 5' phosphate end to the free 3' hydroxyl end using only the letters of the bases. For example, the nucleotide sequence in the section of RNA shown in **FIGURE 18.20** is —A C G U—.

FIGURE 18.20 In the primary structure of RNA, the nucleotides are connected by phosphodiester linkages.

DNA Double Helix: A Secondary Structure

In 1953, James Watson and Francis Crick proposed that DNA was a **double helix** that consists of two polynucleotide strands winding about each other like a spiral staircase. The sugar–phosphate backbones are analogous to the outside railings, with the bases arranged like steps along the inside.

Complementary Base Pairs

Each of the bases along one polynucleotide strand forms hydrogen bonds to a specific base on the opposite DNA strand. Adenine only bonds to thymine, and guanine only bonds to cytosine (see **FIGURE 18.21**). The pairs AT and GC are called **complementary base pairs**. The specific pairing of the bases occurs because adenine and thymine form two hydrogen bonds, while cytosine and guanine form three hydrogen bonds. No other stable base pairs occur.

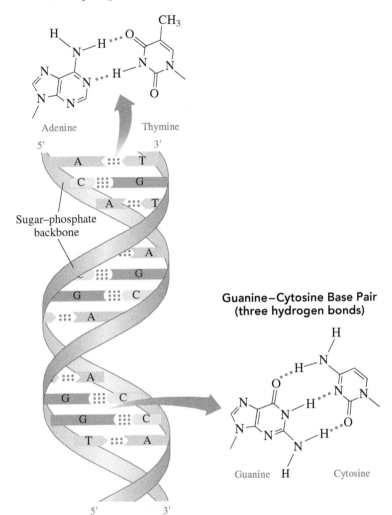

FIGURE 18.21 In the model shown, the sugar–phosphate backbone is represented by a ribbon with hydrogen bonds between complementary base pairs.

SAMPLE PROBLEM 18.10 Complementary Base Pairs

TRY IT FIRST

Write the complementary base sequence for the following segment of a strand of DNA:

A C G A T C T

SOLUTION

In the complementary strand of DNA, the complementary base pairs are AT and CG.

Original segment of DNA: A C G A T C T

Complementary segment: T G C T A G A

SELF TEST 18.10

Write the complementary base sequence for each of the following DNA segments:
a. G G T T A A C C
b. G A T T A C A C G

ANSWER
a. C C A A T T G G
b. C T A A T G T G C

DNA Replication

In DNA **replication**, the strands in the original or *parent* DNA separate, which allows the synthesis of complementary strands of DNA. The process begins when an enzyme catalyzes the unwinding of a portion of the double helix by breaking the hydrogen bonds between the complementary bases. These resulting strands or parent DNA now act as templates for the synthesis of new complementary strands of DNA (see **FIGURE 18.22**).

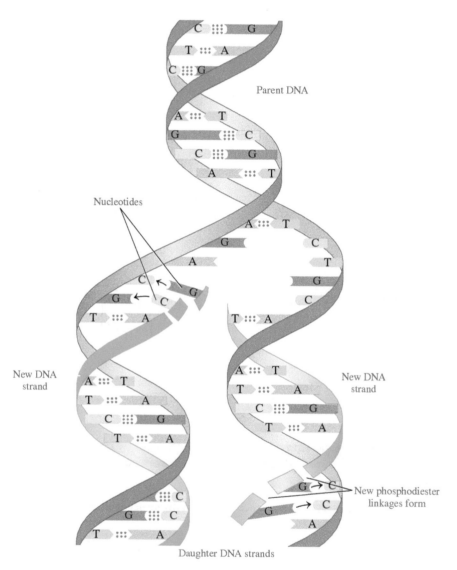

FIGURE 18.22 In DNA replication, the separate strands of the parent DNA are the templates for the synthesis of complementary strands, which produces two exact copies of DNA.

Eventually, the entire double helix of the parent DNA is copied. In each new DNA molecule, one strand of the double helix is from the original DNA and one is a newly synthesized strand. This process produces two new DNAs called *daughter DNA* that are identical to each other and exact copies of the original parent DNA. In DNA replication, complementary base pairing ensures the correct placement of bases in the daughter DNA strands.

PRACTICE PROBLEMS

Try Practice Problems 18.59 to 18.64

PRACTICE PROBLEMS

18.7 Nucleic Acids

18.57 Identify each of the following bases as a component of RNA, DNA, or both:
   a. thymine
   b.   NH$_2$

18.58 Identify each of the following bases as a component of RNA, DNA, or both:
   a. guanine
   b.   NH$_2$

18.59 How are the two strands of nucleic acid in DNA held together?

18.60 What is meant by complementary base pairing?

18.61 Write the base sequence in a complementary DNA segment if each original segment has the following base sequence:
   a. A A A A A A   **b.** G G G G G G
   c. A G T C C A G G T   **d.** C T G T A T A C G T T

18.62 Write the base sequence in a complementary DNA segment if each original segment has the following base sequence:
   a. T T T T T T   **b.** C C C C C C C C C
   c. A T G G C A   **d.** A T A T G C G C T A

18.63 What process ensures that the replication of DNA produces identical copies?

18.64 What is daughter DNA?

18.8 Protein Synthesis

LEARNING GOAL Describe the synthesis of protein from mRNA.

Ribonucleic acid, RNA, which makes up most of the nucleic acid found in the cell, is involved with transmitting the genetic information needed to operate the cell. Similar to DNA, RNA molecules are polymers of nucleotides. However, RNA differs from DNA in several important ways:

1. The sugar in RNA is ribose rather than the deoxyribose found in DNA.
2. In RNA, the base uracil replaces thymine.
3. RNA molecules are single stranded, not double stranded.
4. RNA molecules are much smaller than DNA molecules.

Types of RNA

There are three major types of RNA in the cells: *messenger RNA*, *ribosomal RNA*, and *transfer RNA*, which are classified according to their function, as shown in TABLE 18.10.

TABLE 18.10 Types of RNA Molecules in Humans

Type	Abbreviation	Function in the Cell	Percentage of Total RNA
Messenger RNA	mRNA	Carries information for protein synthesis from the DNA to the ribosomes	5
Ribosomal RNA	rRNA	Major component of the ribosomes; site of protein synthesis	80
Transfer RNA	tRNA	Brings specific amino acids to the site of protein synthesis	15

In replication, the genetic information in DNA is reproduced by making identical copies of DNA. In **transcription**, the information contained in DNA is transferred to mRNA molecules. In **translation**, the genetic information now present in the mRNA is used to build the sequence of amino acids of the desired protein (see FIGURE 18.23).

ENGAGE 18.16

What is the difference between transcription and translation?

FIGURE 18.23 The genetic information in DNA is replicated in cell division and used to produce messenger RNAs that code for the amino acids needed for protein synthesis.

Transcription begins when the section of DNA to be copied unwinds. One strand of DNA acts as a template for the synthesis of mRNA as bonds are formed to each complementary base: C is paired with G, T pairs with A, and A pairs with U (not T).

SAMPLE PROBLEM 18.11   RNA Synthesis

TRY IT FIRST

The sequence of bases in a segment of the DNA template strand for mRNA is C G A T C A. What corresponding mRNA is produced?

SOLUTION

To form the mRNA, each base in the DNA template strand is paired with its complementary base: G with C, C with G, T with A, and A with U.

DNA template strand:	C G A T C A
Transcription	↓ ↓ ↓ ↓ ↓ ↓
Complementary base sequence in mRNA:	G C U A G U

SELF TEST 18.11

Write the DNA base sequence that codes for each of the following mRNA segments:

a. G G G U U U A A A   **b.** G A U U A C A C G

ANSWER

a. C C C A A A T T T   **b.** C T A A T G T G C

The Genetic Code

The **genetic code** consists of a series of three nucleotides (triplets) in mRNA, called **codons** that specify the amino acids and their sequence in a protein. Early work on protein synthesis showed that repeating triplets of uracil (UUU) produced a polypeptide that contained only phenylalanine. Therefore, a sequence of UUU UUU UUU codes for three phenylalanines.

Codons in mRNA:	UUU UUU UUU
	↓ ↓ ↓
Amino acid sequence:	Phe — Phe — Phe

Codons have now been determined for all 20 amino acids. A total of 64 codons is possible from the triplet combinations of A, G, C, and U. Three of these, UGA, UAA, and UAG, are stop signals that code for the termination of protein synthesis. All the other codons shown in TABLE 18.11 specify amino acids. Thus, one amino acid can have several codons. For example, glycine has four codons: GGU, GGC, GGA, and GGG. The triplet AUG has two roles in protein synthesis. At the beginning of an mRNA, the codon AUG signals the start of protein synthesis. In the middle of a series of codons, AUG codes for the amino acid methionine.

TABLE 18.11 mRNA Codons: The Genetic Code for Amino Acids

First Letter	Second Letter				Third Letter
	U	C	A	G	
U	UUU } Phe (F) UUC UUA } Leu (L) UUG	UCU } UCC UCA } Ser (S) UCG	UAU } Tyr (Y) UAC UAA STOP[b] UAG STOP[b]	UGU } Cys (C) UGC UGA STOP[b] UGG Trp (W)	U C A G
C	CUU } CUC CUA } Leu (L) CUG	CCU } CCC CCA } Pro (P) CCG	CAU } His (H) CAC CAA } Gln (Q) CAG	CGU } CGC CGA } Arg (R) CGG	U C A G
A	AUU } AUC } Ile (I) AUA } AUG START[a]/ Met (M)	ACU } ACC ACA } Thr (T) ACG	AAU } Asn (N) AAC AAA } Lys (K) AAG	AGU } Ser (S) AGC AGA } Arg (R) AGG	U C A G
G	GUU } GUC GUA } Val (V) GUG	GCU } GCC GCA } Ala (A) GCG	GAU } Asp (D) GAC GAA } Glu (E) GAG	GGU } GGC GGA } Gly (G) GGG	U C A G

START[a] codon signals the initiation of a peptide chain.
STOP[b] codons signal the end of a peptide chain.

Protein Synthesis: Translation

Once the mRNA is synthesized, it migrates out of the nucleus into the cytoplasm to the ribosomes. At the ribosomes, the *translation* process converts the codons on mRNA into amino acids to make a protein.

Protein synthesis begins when the mRNA combines with a ribosome. There, tRNA molecules, which carry amino acids, align with mRNA, and a peptide bond forms between the amino acids. After the first tRNA detaches from the ribosome, the ribosome shifts to the next codon on the mRNA. Each time the ribosome shifts and the next tRNA aligns with the mRNA, a peptide bond joins the new amino acid to the growing polypeptide chain. After all the amino acids for a particular protein have been linked together by peptide bonds, the ribosome encounters a stop codon. Because there are no tRNAs to complement the termination codon, protein synthesis ends, and the completed polypeptide chain is released from the ribosome. Then interactions between the amino acids in the chain form the protein into the three-dimensional structure that makes the polypeptide into a biologically active protein (see FIGURE 18.24).

CORE CHEMISTRY SKILL

Writing the Amino Acid for an mRNA Codon

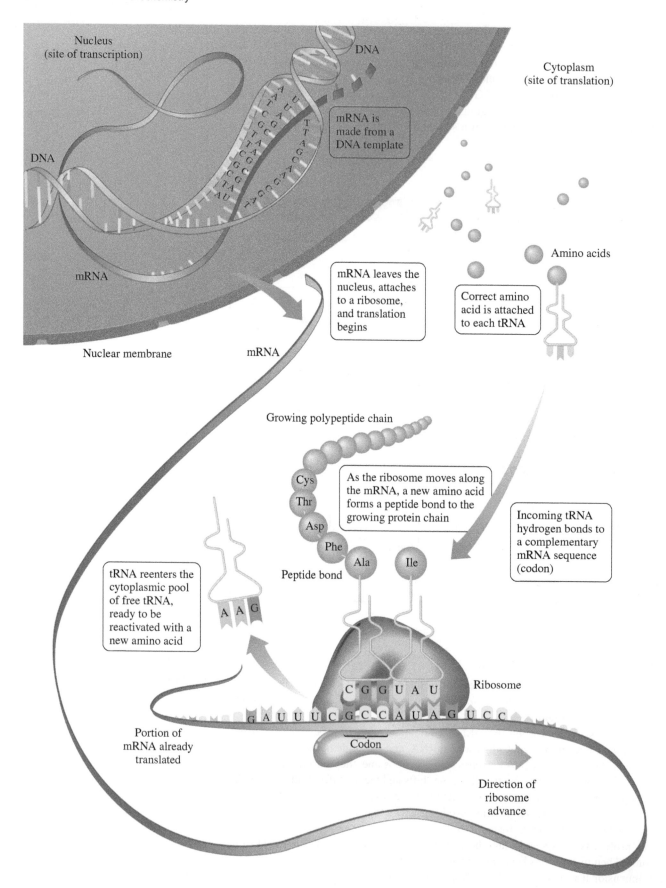

FIGURE 18.24 In the translation process, the mRNA synthesized by transcription attaches to a ribosome, and tRNAs pick up their amino acids, bind to the appropriate codon, and place them in a growing peptide chain.

SAMPLE PROBLEM 18.12 Protein Synthesis: Translation

TRY IT FIRST

Use three-letter and one-letter abbreviations to write the amino acid sequence for the peptide from the mRNA sequence of UCA AAA GCC CUU.

SOLUTION

Each of the codons specifies a particular amino acid. Using Table 18.11, we write the codons and the amino acids in the peptide.

ANALYZE THE PROBLEM	Given	Need	Connect
	mRNA UCA AAA GCC CUU	amino acid sequence	genetic code

mRNA codons: UCA AAA GCC CUU

Amino acid sequence: Ser—Lys—Ala—Leu, SKAL

SELF TEST 18.12

Write three-letter and one-letter abbreviations of the amino acid sequence for the peptide from each of the following mRNA sequences:

a. GGG AGC AGU GAG GUU   **b.** UAU GAU UCC ACG CCG

ANSWER

a. Gly–Ser–Ser–Glu–Val, GSSEV   **b.** Tyr–Asp–Ser–Thr–Pro, YDSTP

PRACTICE PROBLEMS
Try Practice Problems 18.67 to 18.70

PRACTICE PROBLEMS

18.8 Protein Synthesis

18.65 Write the corresponding section of mRNA produced from the following section of DNA template:

 C C G A A G G T T C A C

18.66 Write the corresponding section of mRNA produced from the following section of DNA template:

 T A C G G A A G C T A

18.67 What amino acid is coded for by each of the following RNA codons?
 a. CUG **b.** CCU
 c. GGU **d.** AGG

18.68 What amino acid is coded for by each of the following RNA codons?
 a. AAG **b.** GUC
 c. CGG **d.** GCA

Applications

18.69 The following is a segment of the DNA template that codes for human insulin:

 TTT GTG AAC CAA CAC CTG

 a. Write the corresponding mRNA segment.
 b. Write the three-letter and one-letter abbreviations for the corresponding peptide segment.

18.70 The following is a segment of the DNA template that codes for human insulin:

 TGC GGC TCA CAC CTG GTG

 a. Write the corresponding mRNA segment.
 b. Write the three-letter and one-letter abbreviations for the corresponding peptide segment.

UPDATE   Kate's Program for Type 2 Diabetes

At Kate's next appointment, Paula shows Kate how to use a glucose meter. She instructs Kate to measure her blood glucose level before and after breakfast and dinner each day. Paula explains to Kate that her pre-meal blood glucose level should be 110 mg/dL or less, and if it increases by more than 50 mg/dL after the meal, she needs to lower the amount of carbohydrates she consumes.

Kate and Paula plan several meals. They combine fruits and vegetables that have high and low levels of carbohydrates in the same meal to stay within the recommended range of about 45 to 60 g of carbohydrate per meal. Kate and Paula also discuss the fact that complex carbohydrates in the body take longer to break down into glucose and, therefore, raise the blood sugar level more gradually.

After her appointment, Kate increases her exercise to walking 30 minutes twice a day. She begins to change her diet by eating six small meals a day consisting mostly of small amounts of fruits, vegetables without starch such as green beans and broccoli, small servings of whole grains, and chicken or fish. Kate also decreases the amounts of breads and pasta she eats because she knows carbohydrates will raise her blood sugar.

After three months, Kate reports that she lost 10 lb and that her blood glucose dropped to 146 mg/dL. Her blurry vision improved, and her need to urinate decreased.

Applications

18.71 Kate's blood volume is 3.9 L. Before treatment, if her blood glucose was 178 mg/dL, how many grams of glucose were in her blood?

18.72 Kate's blood volume is 3.9 L. After three months of diet and exercise, if her blood glucose is 146 mg/dL, how many grams of glucose are in her blood?

18.73 For breakfast, Kate had 1 cup of orange juice (23 g carbohydrate), 2 slices of wheat toast (24 g carbohydrate), 2 tablespoons of grape jam (26 g carbohydrate), and coffee with sugar substitute (0 g carbohydrate).

a. Has Kate remained within the limit of 45 to 60 g of carbohydrate?

b. Using the energy value of 4 kcal/g for carbohydrate, calculate the total kilocalories from carbohydrates in Kate's breakfast, rounded to the tens place.

18.74 The next day, Kate had 1 cup of cereal (15 g carbohydrate) with skim milk (7 g carbohydrate), 1 banana (17 g carbohydrate), and 1/2 cup of orange juice (12 g carbohydrate) for breakfast.

a. Has Kate remained within the limit of 45 to 60 g of carbohydrate?

b. Using the energy value of 4 kcal/g for carbohydrate, calculate the total kilocalories from carbohydrates in Kate's breakfast, rounded to the tens place.

CONCEPT MAP

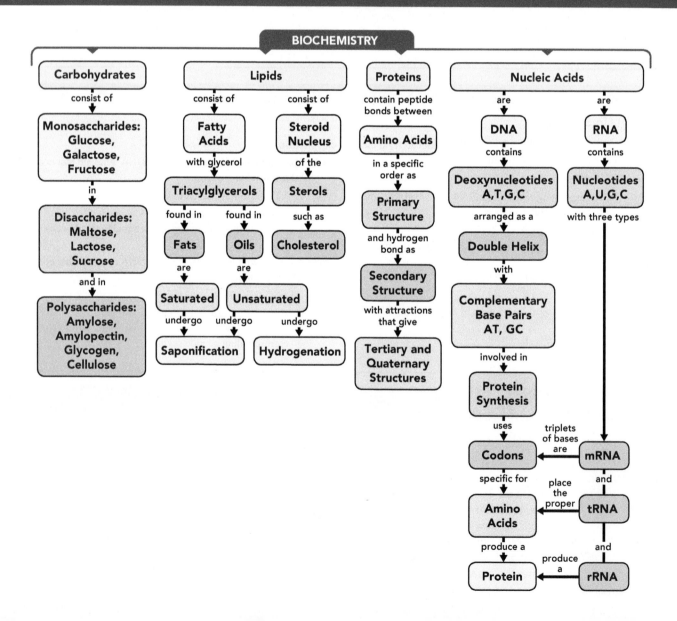

CHAPTER REVIEW

18.1 Carbohydrates

LEARNING GOAL Classify a carbohydrate as an aldose or a ketose; draw the open-chain and Haworth structures for monosaccharides.

- Carbohydrates are composed of carbon, hydrogen, and oxygen.
- Monosaccharides are polyhydroxy aldehydes (aldoses) or ketones (ketoses).
- Monosaccharides are also classified by their number of carbon atoms: triose, tetrose, pentose, or hexose. Important monosaccharides are glucose, galactose, and fructose.
- The predominant form of monosaccharides is the cyclic form of five or six atoms.
- The Haworth structure forms when an —OH group (usually the one on carbon 5 in hexoses) reacts with the carbonyl group of the same molecule.

18.2 Disaccharides and Polysaccharides

LEARNING GOAL Describe the monosaccharide units and linkages in disaccharides and polysaccharides.

- Disaccharides are two monosaccharide units joined together by a glycosidic bond.
- In the most common disaccharides maltose, lactose, and sucrose, there is at least one glucose unit.
- Polysaccharides are polymers of monosaccharide units. Amylose is an unbranched polymer of glucose, and amylopectin is a branched polymer of glucose. Glycogen, the storage form of glucose in animals, is similar to amylopectin with more branching. Cellulose is also a polymer of glucose, but in cellulose the glycosidic bonds are β bonds rather than α bonds.

18.3 Lipids

LEARNING GOAL Draw the condensed structural or line-angle formula for a fatty acid, a triacylglycerol, and the products of hydrogenation or saponification. Identify the steroid nucleus.

- Lipids are nonpolar compounds that are not soluble in water.
- Classes of lipids include waxes, fats, oils, and steroids.
- Fatty acids are long-chain carboxylic acids that may be saturated or unsaturated.
- Triacylglycerols are esters of glycerol with three fatty acids.
- Fats contain more saturated fatty acids and have higher melting points than most vegetable oils.
- The hydrogenation of unsaturated fatty acids converts double bonds to single bonds.
- In saponification, a fat heated with a strong base produces glycerol and the salts of the fatty acids (soap).
- Steroids are lipids containing the steroid nucleus, which is a fused structure of four rings.

18.4 Amino Acids and Proteins

LEARNING GOAL Describe protein functions and draw structures for amino acids and peptides.

- Some proteins are enzymes or hormones, whereas others are important in structure, transport, protection, storage, and muscle contraction.

- A group of 20 amino acids provides the molecular building blocks of proteins.
- Attached to the central (alpha) carbon of each amino acid are an ammonium group, a carboxylate group, a hydrogen atom, and a unique R group.

- Peptides form when an amide bond links the carboxylate group of one amino acid and the ammonium group of a second amino acid.

18.5 Protein Structure

LEARNING GOAL Identify the levels of structure of a protein.

- Long chains of amino acids that are biologically active are called proteins.
- The primary structure of a protein is its sequence of amino acids joined by peptide bonds.
- In the secondary structure, hydrogen bonds between atoms in the peptide bonds produce a characteristic shape such as an α helix or a β-pleated sheet.
- The most abundant protein in the body is collagen, which is composed of fibrils of triple helices that are hydrogen-bonded.
- A tertiary structure is stabilized by interactions between R groups of amino acids in one region of the polypeptide chain with R groups in different regions of the protein.
- In a quaternary structure, two or more subunits combine for biological activity.

18.6 Proteins as Enzymes

LEARNING GOAL Describe the role of an enzyme in an enzyme-catalyzed reaction.

Enzyme–Substrate Complex (ES)

- Enzymes are proteins that act as biological catalysts by accelerating the rate of cellular reactions.
- Within the tertiary structure of an enzyme, a small pocket called the active site binds the substrate.
- In the lock-and-key model, an early theory of enzyme action, a substrate precisely fits the shape of the active site.
- In the induced-fit model, both the active site and the substrate undergo changes in their shapes to give the best fit for efficient catalysis. In the enzyme–substrate complex, catalysis takes place when the amino acid R groups in the active site of an enzyme react with a substrate.
- When the products of catalysis are released, the enzyme can bind another substrate molecule.

18.7 Nucleic Acids

LEARNING GOAL Describe the structure of the nucleic acids in DNA and RNA.

Sugar–phosphate backbone

- Nucleic acids, such as deoxyribonucleic acid (DNA) and ribonucleic acid (RNA), are polymers of nucleotides.
- A nucleotide is composed of three parts: a base, a pentose sugar, and a phosphate group.
- In DNA, the sugar is deoxyribose, and the base can be adenine, thymine,

guanine, or cytosine. In RNA, the sugar is ribose and uracil replaces thymine.

- Each nucleic acid has its own unique sequence of bases.
- A DNA molecule consists of two strands of nucleotides that are wound around each other like a spiral staircase.
- The two strands are held together by hydrogen bonds between complementary base pairs AT and GC.

18.8 Protein Synthesis

LEARNING GOAL Describe the synthesis of protein from mRNA.

UCA   AAA   GCC   CUU

Ser — Lys — Ala — Leu

- The bases in the mRNA are complementary to the bases in DNA, except A in DNA is paired with U in RNA.
- The genetic code consists of a series of codons, which are sequences of three bases that specify the amino acids in a protein.
- Proteins are synthesized at the ribosomes.
- During translation, tRNAs bring the appropriate amino acids to the mRNA at the ribosome and peptide bonds form.
- When the polypeptide is released, it takes on its secondary and tertiary structures and becomes a functional protein in the cell.

KEY TERMS

active site A pocket in the tertiary enzyme structure that binds substrate and catalyzes a reaction.

amino acid The building block of proteins, consisting of a hydrogen atom, an ammonium group, a carboxylate group, and a unique R group attached to the alpha carbon.

amylopectin A branched-chain polymer of starch composed of glucose units joined by $\alpha(1\rightarrow4)$- and $\alpha(1\rightarrow6)$-glycosidic bonds.

amylose An unbranched polymer of starch composed of glucose units joined by $\alpha(1\rightarrow4)$-glycosidic bonds.

carbohydrate A simple or complex sugar composed of carbon, hydrogen, and oxygen.

cellulose An unbranched polysaccharide composed of glucose units linked by $\beta(1\rightarrow4)$-glycosidic bonds that cannot be hydrolyzed by the human digestive system.

codon A sequence of three bases in mRNA that specifies a certain amino acid to be placed in a protein. A few codons signal the start or stop of protein synthesis.

complementary base pairs In DNA, adenine is always paired with thymine (A and T or T and A), and guanine is always paired with cytosine (G and C or C and G). In forming RNA, adenine is paired with uracil (A and U).

disaccharide A carbohydrate composed of two monosaccharides joined by a glycosidic bond.

DNA Deoxyribonucleic acid; the genetic material of all cells containing nucleotides with deoxyribose, phosphate, and the four bases: adenine, thymine, guanine, and cytosine.

double helix The helical shape of the double chain of DNA that is like a spiral staircase with a sugar–phosphate backbone on the outside and base pairs like stair steps on the inside.

enzyme A protein that catalyzes a biological reaction.

fat A triacylglycerol that is solid at room temperature and usually comes from animal sources.

fatty acid A long-chain carboxylic acid found in many lipids.

fructose A ketohexose which is combined with glucose in sucrose.

galactose An aldohexose that is combined with glucose in lactose.

genetic code The sequence of codons in mRNA that specifies the amino acid order for the synthesis of protein.

glucose An aldohexose found in fruits, vegetables, corn syrup, and honey that is also known as blood sugar and dextrose. The most prevalent monosaccharide in the diet. Most polysaccharides are polymers of glucose.

glycogen A polysaccharide formed in the liver and muscles for the storage of glucose as an energy reserve. It is composed of glucose in a highly branched polymer joined by $\alpha(1\rightarrow4)$- and $\alpha(1\rightarrow6)$-glycosidic bonds.

Haworth structure The ring structure of a monosaccharide.

induced-fit model A model of enzyme action in which the shape of a substrate and the active site of the enzyme adjust to give an optimal fit.

lactose A disaccharide consisting of glucose and galactose found in milk and milk products.

lipids A family of compounds that is nonpolar in nature and not soluble in water; includes fats, oils, waxes, and steroids.

maltose A disaccharide consisting of two glucose units; it is obtained from the hydrolysis of starch.

monosaccharide A polyhydroxy compound that contains an aldehyde or a ketone group.

nucleic acid A large molecule composed of nucleotides; found as a double helix in DNA and as the single strands of RNA.

nucleoside The combination of a pentose sugar and a base.

nucleotide Building block of a nucleic acid consisting of a base, a pentose sugar (ribose or deoxyribose), and a phosphate group.

oil A triacylglycerol that is usually a liquid at room temperature and is obtained from a plant source.

polysaccharide A polymer of many monosaccharide units, usually glucose. Polysaccharides differ in the types of glycosidic bonds and the amount of branching in the polymer.

primary structure The specific sequence of the amino acids in a protein.

protein A term used for biologically active polypeptides that have many amino acids linked together by peptide bonds.

quaternary structure A protein structure in which two or more protein subunits form an active protein.

replication The process of duplicating DNA by pairing the bases on each parent strand with their complementary bases.

RNA Ribonucleic acid; a type of nucleic acid that is a single strand of nucleotides containing ribose, phosphate, and the four bases: adenine, cytosine, guanine, and uracil.

secondary structure The formation of an α helix or β-pleated sheet by hydrogen bonds.

steroids Types of lipid composed of a multicyclic ring system.

sucrose A disaccharide composed of glucose and fructose; commonly called table sugar or "sugar."

tertiary structure The folding of the secondary structure of a protein into a compact structure that is stabilized by the interactions of R groups such as salt bridges and disulfide bonds.

transcription The transfer of genetic information from DNA by the formation of mRNA.

translation The interpretation of the codons in mRNA as amino acids in a peptide.

triacylglycerols A family of lipids composed of three fatty acids bonded through ester bonds to glycerol, a trihydroxy alcohol.

CORE CHEMISTRY SKILLS

The chapter section containing each Core Chemistry Skill is shown in parentheses at the end of each heading.

Drawing Haworth Structures (18.1)

- The Haworth structure shows the ring structure of a monosaccharide.
- Groups on the right side of the open-chain structure of the monosaccharide are below the plane of the ring, those on the left are above the plane.
- The new —OH group is the α form if it is below the plane of the ring, or the β form if it is above the plane.

Example: Draw the Haworth structure for β-D-idose.

D-Idose

Answer: In the β-form, the new —OH group is above the plane of the ring.

Identifying Fatty Acids (18.3)

- Fatty acids are unbranched carboxylic acids that typically contain an even number (12 to 20) of carbon atoms.
- Fatty acids may be saturated, monounsaturated with one double bond, or polyunsaturated with two or more double bonds.

Example: State the number of carbon atoms, saturated or unsaturated, and name of the following:

Answer: 16 carbon atoms, saturated, palmitic acid

Drawing Structures for Triacylglycerols (18.3)

- Triacylglycerols are esters of glycerol with three long-chain fatty acids.

Example: Draw the line-angle formula of and name the triacylglycerol formed from glycerol and palmitic acid (16:0).

Answer:

Glyceryl tripalmitate (tripalmitin)

Drawing the Structure for an Amino Acid at Physiological pH (18.4)

- The central α carbon of each amino acid is bonded to an ammonium group ($-NH_3^+$), a carboxylate group ($-COO^-$), a hydrogen atom, and a unique R group.
- The R group gives an amino acid the property of being nonpolar, polar, acidic, or basic.

Example: Draw the condensed structural formula for cysteine at physiological pH.

Answer:

Identifying the Primary, Secondary, Tertiary, and Quaternary Structures of Proteins (18.5)

- The primary structure of a protein is the sequence of amino acids joined by peptide bonds.
- In the secondary structures of proteins, hydrogen bonds between atoms in the peptide bonds produce an alpha helix or a beta-pleated sheet.
- The tertiary structure of a protein is stabilized by R groups that have hydrophobic interactions, attractions between groups and water, hydrogen bonds, disulfide bonds, and salt bridges.
- In a quaternary structure, two or more tertiary subunits are combined for biological activity, held by the same interactions found in tertiary structures.

Example: Identify the following as characteristic of the primary, secondary, tertiary, or quaternary structure of a protein:

a. The R groups of two amino acids interact to form a salt bridge.
b. Eight amino acids form peptide bonds.
c. A polypeptide forms an alpha helix.
d. Two amino acids with hydrophobic R groups interact.
e. A protein with biological activity contains four polypeptide subunits.

Answer:

a. tertiary, quaternary b. primary
c. secondary d. tertiary, quaternary
e. quaternary

Writing the Complementary DNA Strand (18.7)

- During DNA replication, new DNA strands are made along each of the original DNA strands.
- The new strand of DNA is made by forming hydrogen bonds with the bases in the template strand: A with T and T with A; C with G and G with C.

Example: Write the complementary base sequence for the following DNA segment:

ATTCGGTAC

Answer: TAAGCCATG

Writing the mRNA Segment for a DNA Template (18.8)

- Transcription is the process that produces mRNA from one strand of DNA.
- The bases in the mRNA are complementary to the DNA except that A in DNA is paired with U in RNA.

Example: Write the corresponding section of mRNA produced from the following section of DNA template:

CGCATGTCA

Answer: GCGUACAGU

Writing the Amino Acid for an mRNA Codon (18.8)

- The genetic code consists of a sequence of three bases (codons) that specifies the order for the amino acids in a protein.
- The codon AUG signals the start of transcription and codons UAG, UGA, and UAA signal it to stop.

Example: Use three-letter and one-letter abbreviations to write the amino acid sequence for the peptide from the mRNA sequence CCG UAU GGG.

Answer: Pro–Tyr–Gly, PYG

UNDERSTANDING THE CONCEPTS

The chapter sections to review are shown in parentheses at the end of each problem.

18.75 Melezitose, a carbohydrate secreted by insects, has the following Haworth structure: (18.1, 18.2)

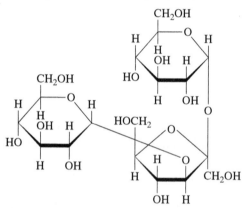

Melezitose

a. Is melezitose a mono-, di-, tri-, or polysaccharide?
b. What ketohexose and aldohexose are present in melezitose?

18.76 What are the disaccharides and polysaccharides present in each of the following? (18.1, 18.2)

(a)

(b)

(c)

(d)

18.77 One of the triacylglycerols in palm oil is glyceryl tripalmitate. Draw the condensed structural formula for glyceryl tripalmitate. (18.3)

The fruit from palm trees are a source of palm oil.

18.78 Jojoba wax in candles consists of a 20-carbon saturated fatty acid and a 20-carbon saturated alcohol. Draw the condensed structural formula for jojoba wax. (18.3)

Candles contain jojoba wax.

18.79 Identify each of the following as a saturated, monounsaturated, or polyunsaturated fatty acid: (18.3)
a. $CH_3-(CH_2)_4-(CH=CH-CH_2)_2-(CH_2)_6-COOH$
b. lauric acid

Salmon is a good source of unsaturated fatty acids.

18.80 Identify each of the following as a saturated, monounsaturated, or polyunsaturated fatty acid: (18.3)
a. $CH_3-(CH_2)_{14}-COOH$
b. $CH_3-(CH_2)_7-CH=CH-(CH_2)_7-COOH$

18.81 Seeds and vegetables are often deficient in one or more essential amino acids. The table below shows which essential amino acids are present in each food. (18.4)

Source	Lysine	Tryptophan	Methionine
Oatmeal	No	Yes	Yes
Rice	No	Yes	Yes
Garbanzo beans	Yes	No	Yes
Lima beans	Yes	No	No
Cornmeal	No	No	Yes

Use the table to decide if each food combination provides the essential amino acids lysine, tryptophan, and methionine.
a. rice and garbanzo beans
b. lima beans and cornmeal
c. garbanzo beans and lima beans

18.82 Seeds and vegetables are often deficient in one or more essential amino acids. Using the table in problem 18.81, state whether each combination provides the essential amino acids lysine, tryptophan, and methionine. (18.4)

Oatmeal is deficient in the essential amino acid lysine.

a. rice and lima beans
b. rice and oatmeal
c. oatmeal and lima beans

18.83 Identify the amino acids and type of interaction that occurs between the following R groups in tertiary protein structures: (18.4, 18.5)

a. $-CH_2-\overset{\overset{\displaystyle O}{\|}}{C}-NH_2$ and $HO-CH_2-$

b. $-CH_2-\overset{\overset{\displaystyle O}{\|}}{C}-O^-$ and $H_3\overset{+}{N}-(CH_2)_4-$

18.84 Identify the amino acids and type of interaction that occurs between the following R groups in tertiary protein structures: (18.4, 18.5)

a. $-CH_2-SH$ and $HS-CH_2-$

b. $-CH_2-\overset{\overset{\displaystyle CH_3}{|}}{CH}-CH_3$ and CH_3-

18.85 Answer the following questions for the given section of DNA: (18.7, 18.8)
a. Complete the bases in the parent and new strands.

b. Using the new strand as a template, write the mRNA sequence.

c. Write the 3-letter abbreviations for the amino acids that would go into the peptide from the mRNA you wrote in part **b**.

18.86 Answer the following questions for the given section of DNA: (18.7, 18.8)
a. Complete the bases in the parent and new strands.

b. Using the new strand as a template, write the mRNA sequence.

c. Write the 3-letter abbreviations for the amino acids that would go into the peptide from the mRNA you wrote in part **b**.

ADDITIONAL PRACTICE PROBLEMS

18.87 What are the structural differences in D-glucose and D-galactose? (18.1)

18.88 What are the structural differences in D-glucose and D-fructose? (18.1)

18.89 Draw the Haworth structure for α-D- and β-D-gulose, whose open-chain structure is shown at right. (18.1)

D-Gulose

18.90 From the compounds shown, select those that match the following: (18.1)

a. an aldohexose   **b.** an aldopentose   **c.** a ketohexose

        1   **2**   **3**

18.91 Gentiobiose, a carbohydrate found in saffron, contains two glucose molecules linked by a β(1→6)-glycosidic bond. Draw the Haworth structure for α-gentiobiose. (18.1, 18.2)

18.92 β-Cellobiose is a disaccharide obtained from the hydrolysis of cellulose. It contains two glucose molecules linked by a β(1→4)-glycosidic bond. Draw the Haworth structure for β-cellobiose. (18.1, 18.2)

18.93 Draw the condensed structural formula for Ser–Lys–Asp at physiological pH. (18.4)

18.94 Draw the condensed structural formula for Val–Ala–Leu at physiological pH. (18.4)

18.95 Identify the base and sugar in each of the following nucleosides: (18.7)

a. deoxythymidine   **b.** adenosine
c. cytidine   **d.** deoxyguanosine

18.96 Identify the base and sugar in each of the following nucleotides: (18.7)

a. CMP   **b.** dAMP   **c.** dGMP   **d.** UMP

18.97 Write the complementary base sequence for each of the following parent DNA segments: (18.7)

a. G A C T T A G G C
b. T G C A A A C T A G C T
c. A T C G A T C G A T C G

18.98 Write the complementary base sequence for each of the following parent DNA segments: (18.7)

a. T T A C G G A C C G C
b. A T A G C C C T T A C T G G
c. G G C C T A C C T T A A C G

18.99 Match the following statements with rRNA, mRNA, or tRNA: (18.8)

a. carries genetic information from the nucleus to the ribosomes
b. combines with proteins to form ribosomes

18.100 Match the following statements with rRNA, mRNA, or tRNA: (18.8)

a. acts as a template for protein synthesis
b. brings amino acids to the ribosomes for protein synthesis

CHALLENGE PROBLEMS

The following problems are related to the topics in this chapter. However, they do not all follow the chapter order, and they require you to combine concepts and skills from several sections. These problems will help you increase your critical thinking skills and prepare for your next exam.

18.101 Raffinose is a trisaccharide found in green vegetables such as cabbage, asparagus, and broccoli. It is composed of three different monosaccharides. Identify the monosaccharides in raffinose. (18.1, 18.2)

Raffinose

18.102 The disaccharide trehalose found in mushrooms is composed of two α-D-glucose molecules joined by an α(1→1)-glycosidic bond. Draw the Haworth structure for trehalose. (18.1, 18.2)

18.103 Sunflower seed oil can be used to make margarine. A triacylglycerol in sunflower seed oil contains two linoleic acids and one oleic acid. (18.3)

Sunflower oil is obtained from the seeds of the sunflower.

a. Draw the condensed structural formulas for two isomers of the triacylglycerol in sunflower oil.
b. Using one of the isomers, write the reaction that takes place when sunflower seed oil is used to make solid margarine.

18.104 What type of interaction would you expect between the following amino acids in a tertiary structure? (18.4, 18.5)

a. threonine and glutamine
b. valine and alanine
c. arginine and glutamate

18.105 What are some differences between each of the following pairs? (18.4, 18.5)

a. secondary and tertiary protein structures
b. essential and nonessential amino acids
c. polar and nonpolar amino acids

18.106 What are some differences between each of the following pairs? (18.4, 18.5)

a. a salt bridge and a disulfide bond
b. an α helix and a β-pleated sheet
c. tertiary and quaternary structures of proteins

Applications

18.107 Endorphins are polypeptides that reduce pain. Use three-letter abbreviations to write the amino acid sequence for the endorphin leucine enkephalin (leu-enkephalin), which has the following mRNA: (18.4, 18.5, 18.8)

AUG UAC GGU GGA UUU CUA UAA

18.108 Endorphins are polypeptides that reduce pain. Use three-letter abbreviations to write the amino acid sequence for the endorphin methionine enkephalin (met-enkephalin), which has the following mRNA: (18.4, 18.5, 18.8)

AUG UAC GGU GGA UUU AUG UAA

18.109 Aspartame, which is used in artificial sweeteners, contains the following dipeptide: (18.4, 18.5)

Some artificial sweeteners contain aspartame, which is an ester of a dipeptide.

a. What are the amino acids in the dipeptide?
b. What is the name of the dipeptide in aspartame?
c. Give the three-letter and one-letter abbreviations for the dipeptide in aspartame.

18.110 The tripeptide shown has a strong attraction for Cu^{2+} and is present in human blood and saliva. (18.4, 18.5)

a. What are the amino acids in the tripeptide?
b. What is the name of this tripeptide?
c. Give the three-letter and one-letter abbreviations for the tripeptide.

18.111 The following sequence is a portion of a DNA template: (18.4, 18.5, 18.8)

GCT TTT CAA AAA

a. Write the corresponding mRNA segment.
b. Write the three-letter and one-letter abbreviations for the corresponding peptide segment.

18.112 The following sequence is a portion of a DNA template: (18.4, 18.5, 18.8)

TGT GGG GTT ATT

a. Write the corresponding mRNA segment.
b. Write the three-letter and one-letter abbreviations for the corresponding peptide segment.

ANSWERS TO ENGAGE QUESTIONS

18.1 The reactants of respiration are glucose and oxygen; the products are carbon dioxide, water, and energy.

18.2 Ribulose has five carbons and contains a carbonyl group bonded to two other carbons. Ribulose is a ketopentose. Glucose has six carbons and contains a carbonyl group bonded to one carbon and one hydrogen. Glucose is an aldohexose.

18.3 The open-chain structure of D-galactose has the hydroxyl group on carbon 4 on the left. The open-chain structure of D-glucose has the hydroxyl group on carbon 4 on the right.

18.4 In α-maltose, the —OH group on carbon 1 is below the plane of the ring. In β-maltose, the —OH group on carbon 1 is above the plane of the ring.

18.5 In lactose, the galactose and glucose are linked by a $\beta(1\rightarrow4)$ glycosidic bond.

18.6 Sucrose is composed of glucose and fructose.

18.7 In amylose and amylopectin, the glucose molecules are linked by $\alpha(1\rightarrow4)$ glycosidic bonds. In amylopectin, there are also branches of glucose molecules linked by $\alpha(1\rightarrow6)$ glycosidic bonds.

18.8 Humans do not have the enzymes needed to break the $\beta(1\rightarrow4)$ glycosidic bonds in cellulose.

18.9 Stearic acid has only carbon–carbon single bonds; it is saturated. Oleic acid has one carbon–carbon double bond; it is monounsaturated.

18.10 Butter contains over 60% saturated fatty acids and is a solid at room temperature. Canola oil contains only about 5% saturated fatty acids and is a liquid at room temperature.

18.11 Valine has a hydrocarbon R group: it is a nonpolar amino acid. Histidine has a charged R group: it is a polar amino acid.

18.12 Valine is at the N-terminus because it is written first. Threonine is at the C-terminus because it is written last.

18.13 Arginine is a basic amino acid, while aspartate is an acidic amino acid. The ionic interaction between acidic and basic amino acids is called a salt bridge.

18.14 The tertiary structure is the compact shape of a protein. The quaternary structure is the combination of protein subunits.

18.15 The nucleotides in DNA contain the sugar deoxyribose and the bases adenine, guanine, cytosine, and thymine. The nucleotides in RNA contain the sugar ribose and the bases adenine, guanine, cytosine, and uracil.

18.16 In transcription, genetic information is transferred to mRNA. In translation, the genetic information is used to form a protein.

ANSWERS TO SELECTED PROBLEMS

18.1 Hydroxyl groups are found in all monosaccharides along with a carbonyl on the first or second carbon.

18.3 A ketopentose contains hydroxyl and ketone functional groups and has five carbon atoms.

18.5 In the ring portion of the Haworth structure of glucose, there are five carbon atoms and an oxygen atom.

18.7 a. α form   **b.** α form

18.9 a. ketohexose   **b.** aldopentose

18.11 In galactose, the —OH group on carbon 4 extends to the left. In glucose, this —OH group extends to the right.

18.13 a. one molecule of glucose and one molecule of galactose; $\beta(1\rightarrow4)$-glycosidic bond; β-lactose
   b. two molecules of glucose; $\alpha(1\rightarrow4)$-glycosidic bond; α-maltose

18.15 a. disaccharide   **b.** two molecules of glucose
   c. $\alpha(1\rightarrow6)$-glycosidic bond   **d.** α

18.17 a. sucrose   **b.** lactose
   c. maltose   **d.** lactose

18.19 a. cellulose   **b.** amylose, amylopectin
   c. amylose   **d.** glycogen

18.21 a. saturated
   b. polyunsaturated
   c. saturated

18.23

$$CH_2-O-\overset{\overset{\displaystyle O}{\|}}{C}-(CH_2)_6-CH_3$$
$$CH-O-\overset{\overset{\displaystyle O}{\|}}{C}-(CH_2)_6-CH_3$$
$$CH_2-O-\overset{\overset{\displaystyle O}{\|}}{C}-(CH_2)_6-CH_3$$

18.25 Safflower oil contains polyunsaturated fatty acids; olive oil contains a large amount of monounsaturated fatty acids. Molecules with more unsaturated fatty acids have lower melting points.

18.27

$$CH_2-O-\overset{\overset{\displaystyle O}{\|}}{C}-(CH_2)_7-CH=CH-(CH_2)_5-CH_3$$
$$CH-O-\overset{\overset{\displaystyle O}{\|}}{C}-(CH_2)_7-CH=CH-(CH_2)_5-CH_3 + 3H_2 \xrightarrow{\text{Ni}}$$
$$CH_2-O-\overset{\overset{\displaystyle O}{\|}}{C}-(CH_2)_7-CH=CH-(CH_2)_5-CH_3$$

$$CH_2-O-\overset{\overset{\displaystyle O}{\|}}{C}-(CH_2)_{14}-CH_3$$
$$CH-O-\overset{\overset{\displaystyle O}{\|}}{C}-(CH_2)_{14}-CH_3$$
$$CH_2-O-\overset{\overset{\displaystyle O}{\|}}{C}-(CH_2)_{14}-CH_3$$

18.29

$$CH_2-O-\overset{\overset{\displaystyle O}{\|}}{C}-(CH_2)_{12}-CH_3$$
$$CH-O-\overset{\overset{\displaystyle O}{\|}}{C}-(CH_2)_{12}-CH_3 + 3NaOH \xrightarrow{\text{Heat}}$$
$$CH_2-O-\overset{\overset{\displaystyle O}{\|}}{C}-(CH_2)_{12}-CH_3$$

$$CH_2-OH$$
$$CH-OH + 3Na^+ \ ^-O-\overset{\overset{\displaystyle O}{\|}}{C}-(CH_2)_{12}-CH_3$$
$$CH_2-OH$$

18.31

18.33 a.

$$H_3\overset{+}{N}-\overset{\overset{\displaystyle H}{|}}{\underset{\underset{\displaystyle H}{|}}{C}}-\overset{\overset{\displaystyle O}{\|}}{C}-O^-$$

b.

$$H_3\overset{+}{N}-\overset{\overset{\displaystyle CH_3 \ OH}{\overset{|}{CH}}}{\underset{\underset{\displaystyle H}{|}}{C}}-\overset{\overset{\displaystyle O}{\|}}{C}-O^-$$

c.

$$H_3\overset{+}{N}-\overset{\overset{\displaystyle CH_2}{|}}{\underset{\underset{\displaystyle H}{|}}{C}}-\overset{\overset{\displaystyle O}{\|}}{C}-O^-$$

18.35 a. nonpolar, hydrophobic
   b. polar, neutral, hydrophilic
   c. nonpolar, hydrophobic

18.37 a. alanine   **b.** valine
   c. lysine   **d.** cysteine

18.39 a.

$$\overset{+}{H_3N}-\underset{\underset{H}{|}}{\overset{\overset{CH_3}{|}}{C}}-\overset{\overset{O}{||}}{C}-\underset{\underset{H}{|}}{N}-\underset{\underset{H}{|}}{\overset{\overset{SH}{|}}{\underset{CH_2}{|}}{C}}-\overset{\overset{O}{||}}{C}-O^-$$

Ala–Cys, AC

b.

$$\overset{+}{H_3N}-\underset{\underset{H}{|}}{\overset{\overset{OH}{|}}{\underset{CH_2}{|}}{C}}-\overset{\overset{O}{||}}{C}-\underset{\underset{H}{|}}{N}-\underset{\underset{H}{|}}{\overset{\overset{\bigcirc}{|}}{\underset{CH_2}{|}}{C}}-\overset{\overset{O}{||}}{C}-O^-$$

Ser–Phe, SF

c.

$$\overset{+}{H_3N}-\underset{\underset{H}{|}}{\overset{\overset{H}{|}}{C}}-\overset{\overset{O}{||}}{C}-\underset{\underset{H}{|}}{N}-\underset{\underset{H}{|}}{\overset{\overset{CH_3}{|}}{C}}-\overset{\overset{O}{||}}{C}-\underset{\underset{H}{|}}{N}-\underset{\underset{H}{|}}{\overset{\overset{CH_3\ CH_3}{|}}{\underset{CH}{|}}{C}}-\overset{\overset{O}{||}}{C}-O^-$$

Gly–Ala–Val, GAV

18.41 a. Corn contains methionine which is lacking in peas, while peas contain lysine and tryptophan, lacking in corn.
   b. Rice contains methionine which is lacking in soy, while soy contain lysine, lacking in rice.

18.43 a.

RIY

b.

VWIS

18.45 Val–Ser–Ser, VSS; Ser–Val–Ser, SVS; or Ser–Ser–Val, SSV

18.47 In the α helix, hydrogen bonds form between the oxygen atom in the C=O group and hydrogen in the N—H group in the next turn of the chain. In the β-pleated sheet, side-by-side hydrogen bonds occur between parallel peptides or across sections of a long polypeptide chain.

18.49 a. disulfide bond   **b.** salt bridge
   c. hydrogen bond   **d.** hydrophobic interaction
18.51 a. tertiary and quaternary   **b.** primary
   c. secondary
18.53 a. (1) enzyme   **b.** (3) substrate
   c. (2) enzyme–substrate complex
18.55 a. E + S $\rightleftharpoons$ ES $\longrightarrow$ EP $\longrightarrow$ E + P
   b. The active site is a region or pocket within the tertiary structure of an enzyme that accepts the substrate, aligns the substrate for reaction, and catalyzes the reaction.
18.57 a. DNA   **b.** both DNA and RNA
18.59 The two DNA strands are held together by hydrogen bonds between the complementary bases in each strand.
18.61 a. T T T T T T   **b.** C C C C C C
   c. T C A G G T C C A   **d.** G A C A T A T G C A A
18.63 The DNA strands separate to allow each of the bases to pair with its complementary base, which produces two exact copies of the original DNA.
18.65 G G C U U C C A A G U G
18.67 a. leucine (Leu)   **b.** proline (Pro)
   c. glycine (Gly)   **d.** arginine (Arg)
18.69 a. AAA CAC UUG GUU GUG GAC
   b. Lys–His–Leu–Val–Val–Asp, KHLVVD
18.71 6.9 g of glucose
18.73 a. Kate's breakfast had 73 g of carbohydrate. She still needs to cut down the amount of carbohydrate.
   b. 290 kcal
18.75 a. Melezitose is a trisaccharide.
   b. Melezitose contains two glucose molecules and a fructose molecule.
18.77

Glyceryl tripalmitate

18.79 a. polyunsaturated   **b.** saturated
18.81 a. yes   **b.** no   **c.** no
18.83 a. asparagine and serine; hydrogen bond
   b. aspartate and lysine; salt bridge
18.85 a.

Parent strand:	G	A	T	C	C	G	T	G	G
New strand:	C	T	A	G	G	C	A	C	C

b.

c. Asp–Pro–Trp
18.87 They differ only at carbon 4 where the —OH group in glucose is on the right side and in galactose it is on the left side.

18.89

α-D-Gulose   β-D-Gulose

18.91

18.93

18.95 **a.** thymine and deoxyribose   **b.** adenine and ribose
c. cytosine and ribose   **d.** guanine and deoxyribose

18.97 **a.** C T G A A T G G C   **b.** A C G T T T G A T G A
c. T A G C T A G C T A G C

18.99 **a.** mRNA   **b.** rRNA

18.101 galactose, glucose, and fructose

18.103 a.

b.

18.105 a. The secondary structure of a protein depends on hydrogen bonds to form an a α helix or a β-pleated sheet. The tertiary structure is determined by the interactions of R groups such as disulfide bonds, hydrogen bonds, and salt bridges.
b. Nonessential amino acids are synthesized by the body, but essential amino acids must be supplied by the diet.
c. Polar amino acids have hydrophilic R groups; nonpolar amino acids have hydrophobic R groups.

18.107 START−Tyr−Gly−Gly−Phe−Leu−STOP

18.109 a. aspartate and phenylalanine
b. aspartylphenylalanine
c. Asp–Phe, DF

18.111 a. CGA AAA GUU UUU
b. Arg–Lys–Val–Phe, RKVF

CI.41 A compound called butylated hydroxytoluene, or BHT, with a molecular formula $C_{15}H_{24}O$, is added to preserve foods such as cereal. As an antioxidant, BHT reacts with oxygen in the cereal container, which protects the food from spoilage. (2.5, 2.6, 7.1, 7.2, 15.1, 17.4)

BHT is an antioxidant added to preserve foods such as cereal.

a. What functional group is present in BHT?
b. Why is BHT referred to as an "antioxidant"?
c. The U.S. Food and Drug Administration (FDA) allows a maximum of 50. ppm of BHT added to cereal. How many milligrams of BHT could be added to a box of cereal that contains 15 oz of dry cereal?

CI.42 Olive oil contains a high percentage of glyceryl trioleate (triolein). (7.2, 11.7, 12.6, 18.3)

One of the triacylglycerols in olive oil is glyceryl trioleate (triolein).

a. Draw the condensed structural formula for glyceryl trioleate (triolein).
b. How many liters of H_2 gas at STP are needed to completely react with the double bonds in 100. g of triolein?
c. How many milliliters of a 6.00 M NaOH solution are needed to completely saponify 100. g of triolein?

CI.43 A sink drain can become clogged with solid fat such as glyceryl tristearate (tristearin). (8.2, 12.6, 18.3)

A sink drain can become clogged with saturated fats.

a. How would adding lye (NaOH) to the sink drain remove the blockage?
b. Use condensed structural formulas to write a balanced equation for the reaction that occurs.
c. How many milliliters of a 0.500 M NaOH solution are needed to completely saponify 10.0 g of tristearin?

CI.44 In response to signals from the nervous system, the hypothalamus secretes a polypeptide hormone known as gonadotropin-releasing factor (GnRF), which stimulates the pituitary gland to release other hormones into the bloodstream.

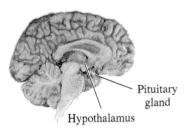

Pituitary gland

Hypothalamus

Gonadotropin-releasing factor (GnRF) is secreted by the hypothalamus.

Two of the hormones are known as gonadotropins, which are the luteinizing hormone (LH) in males, and the follicle-stimulating hormone (FSH) in females. GnRF is a decapeptide with the following primary structure: (18.4, 18.5)

Glu–His–Trp–Ser–Tyr–Gly–Leu–Arg–Pro–Gly

a. What is the N-terminal amino acid in GnRF?
b. What is the C-terminal amino acid in GnRF?
c. Which amino acids in GnRF are nonpolar?
d. Draw the condensed structural formulas at physiological pH for the acidic and basic amino acids in GnRF.
e. Draw the condensed structural formulas at physiological pH for the primary structure of the first three amino acids starting from the N-terminus of GnRF.

CI.45 The plastic known as PETE (**p**oly**e**thylene**te**rephthalate) is a polymer of terephthalic acid and ethylene glycol. PETE is used to make plastic soft drink bottles and containers for salad dressings, shampoos, and dishwashing liquids. Today, PETE is the most widely recycled of all the plastics; in a single year, 1.80×10^9 lb of PETE are recycled. After PETE is separated from other plastics, it can be used in polyester fabric, fill for sleeping bags, door mats, and tennis ball containers. The density of PETE is 1.38 g/mL. (2.5, 2.6, 17.6)

Terephthalic acid   Ethylene glycol

Plastic bottles made of PETE are ready to be recycled.

a. Draw the line-angle formula for the ester formed from one molecule of terephthalic acid and one molecule of ethylene glycol.

b. Draw the line-angle formula for the product formed when a second molecule of ethylene glycol reacts with the ester you drew for the answer in part **a**.

c. How many kilograms of PETE are recycled in one year?

d. What volume, in liters, of PETE is recycled in one year?

e. Suppose a landfill with an area of a football field and a depth of 5.0 m holds 2.7×10^7 L of recycled PETE. If all of the PETE that is recycled in a year were placed in landfills, how many would it fill?

CI.46 Thalassemia is an inherited genetic mutation that limits the production of hemoglobin. If less hemoglobin is produced, there is a shortage of red blood cells (anemia). As a result, the body does not have sufficient amounts of oxygen. In one form of thalassemia, thymine (T) is deleted from section 91 (bold) in the following segment of normal DNA: (18.4, 18.5, 18.8)

89   90   91   92   93   94

AGT CAG **CTG** CAC TGT GAC A....

a. Write the complementary (template) strand for this normal DNA segment.

b. Write the mRNA sequence using the (template) strand in part **a**.

c. What amino acids are placed in the beta chain using the portion of mRNA in part **b**?

d. What is the order of nucleotides after T is deleted?

e. Write the template strand for the mutated DNA segment.

f. Write the mRNA sequence from the mutated DNA segment using the template strand in part **e**.

g. What amino acids are placed in the beta chain by the mutated DNA segment?

h. How might the properties of this segment of the beta chain be different from the properties of the normal protein?

i. How might the level of structure in hemoglobin be affected if beta chains are not produced?

ANSWERS

CI.41 a. The —OH group in BHT is bonded to a carbon atom in an aromatic ring, which means BHT has a phenol functional group.

b. BHT is referred to as an "antioxidant" because it reacts with oxygen in the food container, rather than the food, thus preventing or retarding spoilage of the food.

c. 21 mg of BHT

CI.43 a. Adding NaOH will saponify the glyceryl tristearate (fat), breaking it up into fatty acid salts and glycerol, which are soluble and will wash down the drain.

b.

$$CH_2-O-\overset{\overset{O}{\|}}{C}-(CH_2)_{16}-CH_3$$
$$CH-O-\overset{\overset{O}{\|}}{C}-(CH_2)_{16}-CH_3 + 3NaOH \longrightarrow$$
$$CH_2-O-\overset{\overset{O}{\|}}{C}-(CH_2)_{16}-CH_3$$

$$CH_2-OH$$
$$CH-OH + 3Na^+ \ ^-O-\overset{\overset{O}{\|}}{C}-(CH_2)_{16}-CH_3$$
$$CH_2-OH$$

c. 67.3 mL of a 0.500 M NaOH solution

CI.45 a.

b.

c. 8.18×10^8 kg of PETE

d. 5.93×10^8 L of PETE

e. 22 landfills

Credits

Chapter 11

p. 311 Adam Gault/Science Photo Library/Alamy
p. 313 NASA
p. 315 *top:* GybasDigiPhoto/Shutterstock
p. 315 *bottom:* Kenneth William Caleno/Shutterstock
p. 316 LeventeGyori/Shutterstock
p. 317 Levent Konuk/Shutterstock
p. 320 Steve Bower/Shutterstock
p. 323 *top, right:* Prasit Rodphan/Shutterstock
p. 325 Andrey Nekrasov/Alamy Stock Photo
p. 328 Tomasz Wojnarowicz/Fotolia
p. 330 yuri10b/Getty Images
p. 331 *top:* Jaral Lertjamekorn/Shutterstock
p. 331 *right:* LoloStock/Shutterstock
p. 335 Eric Schrader/Fundamental Photographs, NYC
p. 340 Mic Smith/Alamy Stock Photo
p. 341 Adam Gault/Science Photo Library/Alamy
p. 346 Library of Congress

Chapter 12

p. 350 AJPhoto/Science Source
p. 353 *top:* Thinkstock Images/Getty Images
p. 356 Comstock Images/Getty Images
p. 358 *left:* Dr. Marazzi/Science Source
p. 358 *right:* Remik44992/Shutterstock
p. 359 Mega Pixel/Shutterstock
p. 360 *bottom:* CNRI/Science Source
p. 379 Ecologix Environmental Systems, LLC
p. 380 *top:* Photowind/Shutterstock
p. 380 *bottom:* jane/Getty Images
p. 382 Florida Images/Alamy Stock Photo
p. 385 Thinkstock Images/Getty Images
p. 387 Picsfive/Shutterstock
p. 388 AJPhoto/Science Source
p. 392 *top:* Professor25/Getty Images
p. 392 *bottom:* Brand X Pictures/Getty Images

Chapter 13

p. 398 Dante Fenolio/Science Source
p. 406 Gudellaphoto/Fotolia
p. 417 Incamerastock/Alamy Stock Photo
p. 419 *top:* Sirtravelalot/Shutterstock
p. 419 *bottom:* Photobac/Shutterstock
p. 421 *top:* Scimat/Science Source
p. 421 *bottom:* Taras Verkhovynets/Shutterstock
p. 423 A.P. Fitzpatrick Art Materials/Dorling Kindersley
p. 424 Brian Lasenby/Fotolia

Chapter 14

p. 431 Anna Ivanova/123rf
p. 432 Magicinfoto/Shutterstock
p. 433 *top:* Kul Bhatia/Science Source
p. 433 *bottom:* Lukas Gojda/Shutterstock
p. 438 Magone/123RF
p. 439 *bottom:* Richard Megna/Fundamental Photographs, NYC.
p. 440 Lana Langlois/Shutterstock
p. 448 *vinegar:* George Tsartsianidis/123RF
p. 448 *can:* Fotofermer/Shutterstock
p. 448 *detergent:* PeterG/Shutterstock
p. 448 *bleach:* Luisa Leal/123RF
p. 449 *top:* Alexander Gospodinov/Fotolia
p. 449 *bottom:* Wayne Hutchinson/Alamy Stock Photo
p. 450 Eric Schrader/Pearson Education, Inc.
p. 451 Ana Bokan/Shutterstock
p. 453 Yenyu Shih/Shutterstock
p. 454 Richard Megna/Fundamental Photographs, NYC.
p. 455 Eric Schrader/Pearson Education, Inc.
p. 456 Gerald Bernard/Shutterstock
p. 463 Anna Ivanova/123rf
p. 468 geotrac/Getty Images
p. 470 Universal Images Group/SuperStock
p. 473 *top, left:* Oleksandr Kalinichenko/Shutterstock
p. 473 *center, left:* Patsy Michaud/Shutterstock
p. 473 *bottom:* a-wrangler/Getty Images
p. 473 *center:* isifa Image Service s.r.o./Alamy Stock Photo
p. 473 *center, right:* Rostislav Sedlacek/Fotolia
p. 474 *top:* Charles D. Winters/Science Source
p. 474 *bottom:* Picsfive/Shutterstock

Chapter 15

p. 476 Craig Holmes Premium/Alamy Stock Photo
p. 477 Martyn F. Chillmaid/Science Source
p. 482 Andrew Lambert Photography/Science Source
p. 485 Richard Megna/Fundamental Photographs
p. 488 Charles D. Winters/Science Source
p. 493 *top:* Terekhov igor/Shutterstock
p. 493 *bottom:* Paul Orr/Shutterstock
p. 494 *top:* Yenes/Shutterstock
p. 494 *bottom:* alexlmx/123RF
p. 495 Olivier DIGOIT/Alamy Stock Photo
p. 498 *left:* Jim Barber/Shutterstock
p. 498 *right:* karandaev/123RF
p. 499 *top:* Craig Holmes Premium/Alamy Stock Photo
p. 499 *bottom:* Artproem/Shutterstock
p. 500 *left:* Andrew Lambert Photography/Science Source
p. 500 *top, right:* Richard Megna/Fundamental Photographs
p. 500 *bottom, right:* Charles D. Winters/Science Source
p. 502 *left:* Michal Baranski/123RF
p. 502 *right:* Andrew Lambert/Science Source

Chapter 16

p. 508 Tyler Olson/Shutterstock
p. 510 Celig/Shutterstock
p. 511 Josh Sher/Science Source
p. 513 Uberphotos/Getty Images
p. 514 Photo courtesy of Spruce Environmental Technologies, Inc.

Chapter 17

Chapter 18

Glossary/Index

L

Lactose A disaccharide consisting of glucose and galactose found in milk and milk products, 598

Law of Conservation of Mass In a chemical reaction, the total mass of the reactants is equal to the total mass of the products; matter is neither lost nor gained, 240

Lead storage battery 494–495

Le Châtelier's principle When a stress is placed on a system at equilibrium, the equilibrium shifts to relieve that stress, 414–420

    effect of catalyst on equilibrium, 416

    effect of change in temperature on equilibrium, 418–419

    effect of concentration changes on equilibrium, 415–416

    effect of volume change on equilibrium, 416–417

    effects of condition changes on equilibrium, 418*t*

Length, 29, 40

Lewis structure A structure drawn in which the valence electrons of all the atoms are arranged to give octets except two electrons for hydrogen, 272

Lewis structures for molecules and polyatomic ions, 270–276

    for atoms, 270–271, 270*t*

    for ionic compounds, 271

    for molecular compounds, 271–272, 272*t*

    sharing electrons between atoms of different elements, 272, 272*t*

Lewis symbol The representation of an atom that shows the valence electrons as dots placed around the symbol of an element, 270

Limiting reactant The reactant used up during a chemical reaction, which limits the amount of product that can form, 247–252

    calculating mass of product from limiting reactant, 250–252

    calculating moles of product from limiting reactant, 248–249

Line-angle formula A simplified structure that shows a zigzag line in which carbon atoms are represented as the ends of each line and as corners, 544–545

Linear The shape of a molecule that has two bonded atoms and no lone pairs, 279

Lipids A family of compounds that is nonpolar in nature and not soluble in water; includes fats, oils, waxes, and steroids, 605–612

Liquid A state of matter that takes the shape of its container but has a definite volume, 72

Liter (L) The metric unit for volume that is slightly larger than a quart, 28

Lithium-ion batteries, 494–495

Lock-and-key model of enzyme action, 623

M

Magnetic resonance imaging (MRI), 527

Maltose A disaccharide consisting of two glucose units; it is obtained from the hydrolysis of starch, 598

Mass A measure of the quantity of material in an object, 29

Mass calculation for chemical reactions, 245–247

Mass number The total number of protons and neutrons in the nucleus of an atom, 111

Mass of product calculation from limiting reactant, 250–252

Mass percent (m/m) The grams of solute in 100 g of solution, 364

Mass percent composition The percent by mass of the elements in a formula, 194–197

Mass/volume percent (m/v) The grams of solute in 100. mL of solution, 366

Materials engineer, 125

Math skills, 9–16

Matter The material that makes up a substance and has mass and occupies space, 1, 69–71

    chemical properties and changes, 73–74, 74*t*

    classifications, 69–71

    physical properties and changes, 73, 74*t*

    states and properties, 72–75, 72*t*

Measured number A number obtained when a quantity is determined by using a measuring device, 31

Measurements, 27–67

Medical applications using radioactivity, 526–529, 526*t*

    computed tomography (CT), 527

    magnetic resonance imaging (MRI), 527

    positron emission tomography (PET), 527

    scans with radioisotopes, 527

Medical laboratory technologist, 431

Melting The change of state from a solid to a liquid, 292

Melting point (mp) The temperature at which a solid becomes a liquid (melts). It is the same temperature as the freezing point, 292

Metal An element that is shiny, malleable, ductile, and a good conductor of heat and electricity. The metals are located to the left of the heavy zigzag line on the periodic table, 104

Metallic character A measure of how easily an element loses a valence electron, 147

Metalloid An element with properties of both metals and nonmetals located along the heavy zigzag line on the periodic table, 105

Meter (m) The metric unit for length that is slightly longer than a yard. The meter is the SI standard unit of length, 29

Metric conversion factors, 43

Metric system A system of measurement used by scientists and in most countries of the world, 28, 28*t*

Metric–U.S. system conversion factors, 43

Milliliter (mL) A metric unit of volume equal to one-thousandth of a liter (0.001 L), 28

Mixture The physical combination of two or more substances that are not chemically combined, 69–70

    heterogeneous mixture, 70

    homogeneous mixture, 70

    solution, 70

Molality (m) The number of moles of solute particles in exactly 1 kg of solvent, 380

Molality and freezing point lowering/boiling point elevation, 378–84

Molarity (M) The number of moles of solute in exactly 1 L of solution, 368

Glossary/Index **I-11**

pentose sugars, 624

replication, 628–629

RNA, 624

structure, 626

Nucleoside The combination of a pentose sugar and a base, 624–625

Nucleotide Building block of a nucleic acid consisting of a base, a pentose sugar (ribose or deoxyribose), and a phosphate group, 624–625

Nucleus The compact, extremely dense center of an atom, containing the protons and neutrons of the atom, 109

Number of electrons in sublevels, 134

Nutrition, 87–90, 88*t*

O

Observation Information determined by noting and recording a natural phenomenon, 3–4

Octet rule Elements in Groups 1A to 7A (1, 2, 13 to 17) react with other elements by forming ionic or covalent bonds to produce a stable electron configuration, 157

Oil A triacylglycerol that is usually a liquid at room temperature and is obtained from a plant source, 608

Open-chain structures, 594

Orbital The region around the nucleus of an atom where electrons of certain energy are most likely to be found: *s* orbitals are spherical; *p* orbitals have two lobes, 132

Orbital capacity and electron spin, 134

Orbital diagram A diagram that shows the distribution of electrons in the orbitals of the energy levels, 135

Orbital diagrams and electron configurations, 135–139

Period 1: hydrogen and helium, 136

Period 2: lithium to neon, 136–137

Period 3: sodium to argon, 137–138

Organic chemistry, 541

Organic compounds, 541

Organic compound classes, 552*t*

Osmosis The flow of a solvent, usually water, through a semipermeable membrane into a solution of higher solute concentration, 385–388

dialysis, 387

hypotonic and hypertonic solutions, 386–387

isotonic solutions, 385–386

osmotic pressure, 385

Osmotic pressure The pressure that prevents the flow of water into the more concentrated solution, 385

Oxidation The loss of electrons by a substance. Biological oxidation may involve the addition of oxygen or the loss of hydrogen, 228, 477

Oxidation and reduction, 476–507

Oxidation number A number equal to zero in an element or the charge of a monatomic ion; in molecular compounds and polyatomic ions, oxidation numbers are assigned using a set of rules, 478

Oxidation–reduction equation balancing, 483–488

in acidic solution, 485–486

in basic solution, 486–488

using half-reactions, 483–484

Oxidation–reduction reaction A reaction in which the oxidation of one reactant is always accompanied by the reduction of another reactant, 228–231, 477

in biological systems, 229–230

characteristics, 230*t*

Oxidation–reduction reactions requiring electricity, 497–499

electrolysis, 497

electrolysis of sodium chloride, 497

electrolytic cell, 497

electroplating, 498

Oxidizing agent The reactant that gains electrons and is reduced, 481

Oxidizing and reducing agents, 481–482

Oxygen–hemoglobin equilibrium and hypoxia, 417

P

Partial pressure The pressure exerted by a single gas in a gas mixture, 335–340

air, as mixture of gas, 336–337, 336*t*

Dalton's law, 336

gases collected over water, 338–340

Particles in solution, 380

***p* block** The elements in Groups 3A (13) to 8A (18) in which electrons fill the *p* orbitals, 140

Pentose sugars, 624

Peptides, 615–617

Percentage calculation, 12–13

Percent, ppm, and ppb conversion factors, 44

Percent yield The ratio of the actual yield for a reaction to the theoretical yield possible for the reaction, 252–254

Period A horizontal row of elements in the periodic table, 103

Period 1: hydrogen and helium, 136

Period 2: lithium to neon, 136–137

Period 3: sodium to argon, 137–138

Period 4 and above, 141

Periodic table An arrangement of elements by increasing atomic number such that elements having similar chemical behavior are grouped in vertical columns, 103–107

groups and periods, 103–104

names of groups, 104

pH A measure of the $[H_3O^+]$ in a solution; $pH = -\log[H_3O^+]$, 449

buffer calculation, 460–462, 462*t*

calculation of solutions, 449–450

scale, 447–454

Pharmacist, 156

Photon A packet of energy that has both particle and wave characteristics and travels at the speed of light, 129

pH scale, 447–454

calculation of hydronium from pH, 452–453

calculation of pH of solutions, 449–450

Radon in homes, 514

Rate of reaction The speed at which reactants form products, 399–403

Reactants The initial substances that undergo change in a chemical reaction, 215

Reaction types summary, 226t

Recycling code Arrows in a triangle found on the label or on the bottom of a container, 555

Reducing agent The reactant that loses electrons and is oxidized, 481

Reduction The gain of electrons by a substance. Biological reduction may involve the loss of oxygen or the gain of hydrogen, 228, 477

Registered nurse, 27

Rem (radiation equivalent in humans) A measure of the biological damage caused by the various kinds of radiation (rad × radiation biological factor), 519

Replication The process of duplicating DNA by pairing the bases on each parent strand with their complementary bases, 628–629

Representative element An element in the first two columns on the left of the periodic table and the last six columns on the right that has a group number of 1A through 8A or 1, 2, and 13 through 18, 103

Resonance structures Two or more Lewis structures that can be drawn for a molecule or polyatomic ion by placing a multiple bond between different atoms, 276–279

Respiratory therapist, 311

Reversible reaction A reaction in which a forward reaction occurs from reactants to products and a reverse reaction occurs from products back to reactants, 403

RNA Ribonucleic acid; a type of nucleic acid that is a single strand of nucleotides containing ribose, phosphate, and the four bases: adenine, cytosine, guanine, and uracil, 624

RNA types, 629t

Rounding off, 35–36, 35t

S

Salt An ionic compound that contains a metal ion or NH_4^+ and a nonmetal or polyatomic ion other than OH^-, 454

Saturated solution A solution containing the maximum amount of solute that can dissolve at a given temperature. Any additional solute will remain undissolved in the container, 357

s **block** The elements in Groups 1A (1) and 2A (2) in which electrons fill the *s* orbitals, 140

Scientific method The process of making observations, proposing a hypothesis, and testing the hypothesis; after repeated experiments validate the hypothesis, a conclusion may be drawn as to its validity, 3–5

Scientific notation A form of writing large and small numbers using a coefficient that is at least 1 but less than 10, followed by a power of 10, 17–20, 18t

calculators and, 19

conversion to a standard number, 19–20

powers of 10, 18t

Second (s) A unit of time used in both SI and metric systems, 29

Secondary structure The formation of an α helix or β-pleated sheet by hydrogen bonds, 618

Shapes of molecules and polyatomic atoms, 279–283

bent, 280

central atom with four electron groups (tetrahedral), 280–283

central atoms with three electron groups (trigonal), 279–280

central atoms with two electron groups (linear), 279

trigonal pyramidal, 280–281, 281t

Shapes of orbitals, 132–133

Sharing electrons between atoms of different elements, 272, 272t

Sievert (Sv) A unit of biological damage equal to 100 rem, 520

Significant figures (SFs) The numbers recorded in a measurement, 31, 31t, 34–38

Significant zeroes and scientific notation, 32

Single replacement reaction A reaction in which an element replaces a different element in a compound, 224

Solid A state of matter that has its own shape and volume, 72

Soluble and insoluble ionic compounds, 359–363

Solubility The maximum amount of solute that can dissolve in 100. g of solvent, usually water, at a given temperature, 357–363

effects of temperature, 359

saturated solution, 357

soluble and insoluble ionic compounds, 359–363

solubility rules, 360t

unsaturated solution, 357

Solubility product constant, K_{sp} The product of the concentrations of the ions in a saturated solution of a slightly soluble ionic compound, with each concentration raised to a power equal to its coefficient in the balanced equilibrium equation, 421

Solubility product expression The product of the ion concentrations, with each concentration raised to an exponent equal to the coefficient in the balanced chemical equation, 421

Solubility rules A set of guidelines that states whether an ionic compound is soluble or insoluble in water, 359

Solute The component in a solution that is present in the lesser amount, 351

Solution A homogeneous mixture in which the solute is made up of small particles (ions or molecules) that can pass through filters and semipermeable membranes, 351

Solvent The substance in which the solute dissolves; usually the component present in greater amount, 351

Solving equations, 13–14

Specific gravity (sp gr) A relationship between the density of a substance and the density of water, 57

Specific heat (SH) A quantity of heat that changes the temperature of exactly 1 g of a substance by exactly 1 °C, 82–87

heat equation, 83–85

Metric and SI Units and Some Useful Conversion Factors

Length SI Unit Meter (m)	Volume SI Unit Cubic Meter (m³)	Mass SI Unit Kilogram (kg)
1 meter (m) = 100 centimeters (cm)	1 liter (L) = 1000 milliliters (mL)	1 kilogram (kg) = 1000 grams (g) 1 mg = 1000 mcg
1 m = 1000 millimeters (mm)	1 mL = 1 cm³ 1 dl = 100 mL	1 g = 1000 milligrams (mg) 1 lb = 16 oz
1 cm = 10 mm	1 L = 1.057 quart (qt) 1 mL = 1cc	1 kg = 2.205 lb
1 kilometer (km) = 0.6214 mile (mi)	1 qt = 946.4 mL	1 lb = 453.6 g
1 inch (in.) = 2.54 cm (exact)	1 pint (pt) = 473.2 mL	1 mol = 6.022×10^{23} particles
	1 gal = 3.785 L	**Water**
	1 tsp = 5 mL	density = 1.00 g/mL (at 4 °C)
	1 tbsp (T) = 15 mL	

Temperature SI Unit Kelvin (K)	Pressure SI Unit Pascal (Pa)	Energy SI Unit Joule (J)
$T_F = 1.8(T_C) + 32$	1 atm = 760 mmHg	1 calorie (cal) = 4.184 J (exact)
$T_C = \dfrac{T_F - 32}{1.8}$	1 atm = 101.325 kPa	1 kcal = 1000 cal
$T_K = T_C + 273$	1 atm = 760 Torr	
	1 mol of gas = 22.4 L (STP)	**Water**
	$R = 0.0821$ L · atm/mol · K	Heat of fusion = 334 J/g
	$R = 62.4$ mmHg · atm/mol · K	Heat of vaporization = 2260 J/g
		Specific heat (SH) = 4.184 J/g °C; 1.00 cal/g °C

Metric and SI Prefixes

Prefix	Symbol	Power of Ten
Prefixes That Increase the Size of the Unit		
peta	P	10^{15}
tera	T	10^{12}
giga	G	10^{9}
mega	M	10^{6}
kilo	k	10^{3}
Prefixes That Decrease the Size of the Unit		
deci	d	10^{-1}
centi	c	10^{-2}
milli	m	10^{-3}
micro	μ (mc)	10^{-6}
nano	n	10^{-9}
pico	p	10^{-12}
femto	f	10^{-15}

Formulas and Molar Masses of Some Typical Compounds

Name	Formula	Molar Mass (g/mol)	Name	Formula	Molar Mass (g/mol)
Ammonia	NH_3	17.03	Hydrogen chloride	HCl	36.46
Ammonium chloride	NH_4Cl	53.49	Iron(III) oxide	Fe_2O_3	159.70
Ammonium sulfate	$(NH_4)_2SO_4$	132.15	Magnesium oxide	MgO	40.31
Bromine	Br_2	159.80	Methane	CH_4	16.04
Butane	C_4H_{10}	58.12	Nitrogen	N_2	28.02
Calcium carbonate	$CaCO_3$	100.09	Oxygen	O_2	32.00
Calcium chloride	$CaCl_2$	110.98	Potassium carbonate	K_2CO_3	138.21
Calcium hydroxide	$Ca(OH)_2$	74.10	Potassium nitrate	KNO_3	101.11
Calcium oxide	CaO	56.08	Propane	C_3H_8	44.09
Carbon dioxide	CO_2	44.01	Sodium chloride	NaCl	58.44
Chlorine	Cl_2	70.90	Sodium hydroxide	NaOH	40.00
Copper(II) sulfide	CuS	95.62	Sulfur trioxide	SO_3	80.07
Hydrogen	H_2	2.016	Water	H_2O	18.02

Formulas and Charges of Some Common Cations

Cations (Fixed Charge)

1+		2+		3+	
Li^+	Lithium	Mg^{2+}	Magnesium	Al^{3+}	Aluminum
Na^+	Sodium	Ca^{2+}	Calcium		
K^+	Potassium	Sr^{2+}	Strontium		
NH_4^+	Ammonium	Ba^{2+}	Barium		
H_3O^+	Hydronium	Zn^{2+}	Zinc		
Ag^+	Silver	Cd^{2+}	Cadmium		

Cations (Variable Charge)

1+ or 2+				1+ or 3+			
Cu^+	Copper(I)	Cu^{2+}	Copper(II)	Au^+	Gold(I)	Au^{3+}	Gold(III)
Hg_2^{2+}	Mercury(I)	Hg^{2+}	Mercury(II)				
2+ or 3+				**2+ or 4+**			
Fe^{2+}	Iron(II)	Fe^{3+}	Iron(III)	Sn^{2+}	Tin(II)	Sn^{4+}	Tin(IV)
Co^{2+}	Cobalt(II)	Co^{3+}	Cobalt(III)	Pb^{2+}	Lead(II)	Pb^{4+}	Lead(IV)
Cr^{2+}	Chromium(II)	Cr^{3+}	Chromium(III)				
Mn^{2+}	Manganese(II)	Mn^{3+}	Manganese(III)				
Ni^{2+}	Nickel(II)	Ni^{3+}	Nickel(III)				
3+ or 5+							
Bi^{3+}	Bismuth(III)	Bi^{5+}	Bismuth(V)				

Formulas and Charges of Some Common Anions

Monatomic Ions

F^-	Fluoride	Br^-	Bromide	O^{2-}	Oxide	N^{3-}	Nitride
Cl^-	Chloride	I^-	Iodide	S^{2-}	Sulfide	P^{3-}	Phosphide

Polyatomic Ions

HCO_3^-	Hydrogen carbonate (bicarbonate)	CO_3^{2-}	Carbonate		
$C_2H_3O_2^-$	Acetate	CN^-	Cyanide		
NO_3^-	Nitrate	NO_2^-	Nitrite		
$H_2PO_4^-$	Dihydrogen phosphate	HPO_4^{2-}	Hydrogen phosphate	PO_4^{3-}	Phosphate
$H_2PO_3^-$	Dihydrogen phosphite	HPO_3^{2-}	Hydrogen phosphite	PO_3^{3-}	Phosphite
HSO_4^-	Hydrogen sulfate (bisulfate)	SO_4^{2-}	Sulfate		
HSO_3^-	Hydrogen sulfite (bisulfite)	SO_3^{2-}	Sulfite		
ClO_4^-	Perchlorate	ClO_3^-	Chlorate		
ClO_2^-	Chlorite	ClO^-	Hypochlorite		
OH^-	Hydroxide	CrO_4^{2-}	Chromate		
MnO_4^-	Permanganate	$Cr_2O_7^{2-}$	Dichromate		
SCN^-	Thiocyanate				

Functional Groups in Organic Compounds

Type	Functional Group	Type	Functional Group
Haloalkane	$-F, -Cl, -Br,$ or $-I$	Carboxylic acid	$-\overset{\overset{\displaystyle O}{\|\|}}{C}-OH$
Alkene	$-CH=CH-$		
Alkyne	$-C\equiv C-$		
Aromatic	Benzene ring	Ester	$-\overset{\overset{\displaystyle O}{\|\|}}{C}-O-$
Alcohol	$-OH$		
Ether	$-O-$	Amine	$-NH_2$
Aldehyde	$-\overset{\overset{\displaystyle O}{\|\|}}{C}-H$	Amide	$-\overset{\overset{\displaystyle O}{\|\|}}{C}-NH_2$
Ketone	$-\overset{\overset{\displaystyle O}{\|\|}}{C}-$		